Molecular Breeding for Rice Abiotic Stress Tolerance and Nutritional Quality

Molecular Breeding for Rice Abiotic Stress Tolerance and Nutritional Quality

Edited by

Mohammad Anwar Hossain
Department of Genetics and Plant Breeding
Bangladesh Agricultural University, Mymensingh
Bangladesh

Lutful Hassan
Department of Genetics and Plant Breeding
Bangladesh Agricultural University, Mymensingh
Bangladesh

Khandakar Md. Iftekharuddaula
Plant Breeding Division
Bangladesh Rice Research Institute
Gazipur
Bangladesh

Arvind Kumar
International Rice Research Institute, South Asia Regional Centre
Varanasi
India

Robert Henry
Queensland Alliance for Agriculture and Food Innovation
The University of Queensland
Australia

WILEY Blackwell

Registered Offices
John Wiley & Sons, Inc., 111 River Street, Hoboken, NJ 07030, USA
John Wiley & Sons Ltd, The Atrium, Southern Gate, Chichester, West Sussex, PO19 8SQ, UK

Editorial Office
The Atrium, Southern Gate, Chichester, West Sussex, PO19 8SQ, UK

For details of our global editorial offices, customer services, and more information about Wiley products visit us at www.wiley.com.

Wiley also publishes its books in a variety of electronic formats and by print-on-demand. Some content that appears in standard print versions of this book may not be available in other formats.

Library of Congress Cataloging-in-Publication Data

Names: Hossain, Mohammad Anwar, editor.
Title: Molecular breeding for rice abiotic stress tolerance and nutritional quality / edited by Mohammad Anwar Hossain, Department of Genetics and Plant Breeding, Bangladesh Agricultural University, Mymensingh, Bangladesh, Lutful Hassan, Department of Genetics and Plant Breeding, Bangladesh Agricultural University, Mymensingh, Bangladesh, Khandakar Md. Iftekharuddaula, Plant Breeding Division, Bangladesh Rice Research Institute, Gazipur, Bangladesh, Arvind Kumar, International Rice Research Institute, South Asia Regional Centre, Varanasi, India, Robert Henry, Queensland Alliance for Agriculture and Food Innovation, The University of Queensland, Australia.
Description: Hoboken, NJ : Wiley-Blackwell, 2021. | Includes bibliographical references and index.
Identifiers: LCCN 2020032219 (print) | LCCN 2020032220 (ebook) | ISBN 9781119633112 (cloth) | ISBN 9781119633150 (adobe pdf) | ISBN 9781119633167 (epub)
Subjects: LCSH: Rice–Effect of stress on–Genetic aspects. | Rice–Climatic factors.
Classification: LCC SB191.R5 M577 2021 (print) | LCC SB191.R5 (ebook) | DDC 633.1/8–dc23
LC record available at https://lccn.loc.gov/2020032219
LC ebook record available at https://lccn.loc.gov/2020032220

Cover Design: Wiley
Cover Images: © Bule Sky Studio/Shutterstock,
© zhaojiankang/Getty Images, © marekuliasz/Getty Images,
© StockImageFactory.com/Shutterstock

Set in 9.5/12.5pt STIXTwoText by SPi Global, Chennai, India

C9781119633112_090321

Contents

Editor Biographies *xv*
Preface *xix*
List of Contributors *xxi*

1 Rice Adaptation to Climate Change: Opportunities and Priorities in Molecular Breeding *1*
Vikram Jeet Singh, Kunnummal Kurungara Vinod, Subbaiyan Gopala Krishnan and Ashok K. Singh
Introduction *1*
Rice Production Scenario and Climate Change *4*
Prospects of Molecular Breeding for Climate-Resilient Rice *7*
Molecular Breeding in Rice *9*
Breeding for Abiotic Stress Tolerance *9*
Drought *9*
Salinity *10*
Submergence *11*
High Temperature *12*
Low Temperature *13*
Breeding for Biotic Stress Tolerance *13*
Diseases *14*
Pests *15*
Weed Competitiveness *16*
Conclusion *17*
References *18*

2 Molecular Breeding for Improving Salinity Tolerance in Rice: Recent Progress and Future Prospects *26*
Sandeep Chapagain, Lovepreet Singh, Richard Garcia, Rajat Pruthi, Jonathan Concepcion, Sapphire Coronejo and Prasanta K. Subudhi
Introduction *26*
Salt Tolerance Mechanisms *27*
Root-Level Ion Exclusion *27*
Plant-Level Compartmentation *27*
Cell-Level Compartmentation *27*
Natural Variability for Salt Tolerance in Rice Germplasm *29*

Conventional Tools to Improve Salinity Tolerance *29*
Conventional Breeding *29*
Mutation Breeding *31*
Tissue Culture *32*
Molecular Tools to Improve Salinity Tolerance *32*
QTL Mapping *32*
Marker-Assisted Selection *34*
Candidate Genes and Allele Mining *35*
Genetic Engineering to Improve Salt Tolerance *36*
Novel Tools to Aid in Molecular Breeding *37*
Multi-Parental Advanced Generation Intercross Strategy *37*
Genomic Selection *38*
Phenomics *38*
Next-Generation Sequencing *39*
Genome Editing *40*
Conclusion and Future prospects *41*
Acknowledgments *42*
References *42*

3 Molecular Breeding for Improving Drought Tolerance in Rice: Recent Progress and Future Perspectives *53*
Ratna R. Majumder, Sandeep Sakhale, Shailesh Yadav, Nitika Sandhu, Lutful Hassan, Md. Amir Hossain and Arvind Kumar
Introduction *53*
The Drought Challenges and Its Distribution *53*
Climate Change and Drought *53*
Drought Impacts on the Rice Plant *54*
Drought Phenotyping Platforms *54*
Drought Screening Protocols *54*
Field-Based Phenotyping *54*
Design and Management of Field Drought Experiments *54*
Trait-Based Selection for Drought Resistance *55*
Direct Selection for Yield Under Stress *55*
Molecular Markers and Genotyping Strategies *56*
Genome-Wide Scans with Microsatellite Markers (Simple Sequence Repeats, SSRs) *56*
SNP (Single Nucleotide Polymorphism) Genotyping *56*
Trait-Specific SNPs Used as Diagnostic Markers *57*
Genetic Basis of Drought Tolerance in Rice *57*
Mapping and Deployment of Drought QTLs *57*
Fine Mapping of Identified Genetic Regions *57*
Development of NILs and PLs Carrying Drought QTLs in Background of High Yielding Rice Varieties *57*
MAS Introgression of Multiple Traits *60*
Understanding the Molecular Basis of Complexities of Epistatic Interactions *61*
Modern Breeding Strategies and Programs for Accelerated Genetic Gain *64*
Rapid Breeding Cycles *64*

Novel Breeding Strategies Using GS 65
Deployment of Genome Editing, Especially the CRISPR Technologies 66
Conclusion 67
References 67

4 **Molecular Breeding for Improving Flooding Tolerance in Rice: Recent Progress and Future Perspectives** 75
Ramani K. Sarkar, Jangi N. Reddy and Satya R. Das
Introduction 75
Rice Tolerant to Germination Stage Oxygen Deficiency and Submergence and Stagnant Flooding? 77
Convergence of Conventional and Molecular Breeding 82
Conclusion and Future Perspective 86
Acknowledgment 86
References 86

5 **Molecular Breeding for Improving Heat Stress Tolerance in Rice: Recent Progress and Future Perspectives** 92
Bui Chi Buu, Cho Young Chan and Nguyen Thi Lang
Introduction 92
Climate Change and Heat Tolerance Genotype Needs 93
Recent Progress on Rice Breeding 94
Plant Physiology on Heat-Stress Response 94
Net Photosynthesis Rate 94
Grain Filling 94
Response in Germination 95
Response in Anthesis 96
Amylose Content Response to Heat 97
Heat-Induced Quality Decline of Rice (White-Back Kernel) 97
Source Sink Dynamics 97
Genetic Background 98
Breeding Materials (Donors) 98
Genetic Diversity Analysis 99
Phenotyping 100
Putative and Candidate Genes/QTLs 101
Marker-Assisted Selection (Microsatellites and SNP Arrays) 105
Proteomics 107
Future Perspectives 109
Conclusions 110
References 111

6 **Molecular Breeding for Improving Cold Tolerance in Rice: Recent Progress and Future Perspectives** 120
Ning Xiao and Jian-Min Chen
Introduction 120
Preliminary Mapping of Cold-Tolerant QTLs and Fine Mapping of Major Loci 121
Map-Based Cloning and Molecular Mechanism of Cold-Resistant QTLs 124

Molecular Regulatory Networks for Cold Tolerance in *Oryza sativa* *125*
CBF Regulatory Pathway *125*
MYB Regulatory Pathway *126*
Outlook *126*
References *127*

7 Molecular Breeding for Lower Cadmium Accumulation in Rice Grain: Progress and Perspectives *131*
Dongping Li, Xiaohua Hao and Liangbi Chen
Introduction *131*
Cd Toxicity to Human Health *131*
The Origin of Cd Contamination in Soils *132*
Cd Accumulation in Rice *133*
Cd Uptake and Distribution Patterns During the Whole Growth Cycle of Rice *133*
The Diversity of Cd Accumulation Trait in Different Varieties of Rice *133*
Genetic Basis for Cd Accumulation in Rice *134*
Influence of Environmental Factors on Cd Accumulation in Rice *134*
Influence of Soil Water on Rice Cd Accumulation *134*
Influence of pH Values in Soil and Irrigation Water on Rice Cd Accumulation *139*
Effect of the Application of Other Ions or Chemical Compounds on Rice Cd Accumulation *139*
Influence of Rhizosphere Microorganisms on Rice Cd Accumulation *141*
Influence of Genetic Factors on Cd Accumulation in Rice *141*
Transporters for Cd Uptake and Translocation *142*
Genes Involved in Cd Chelation *146*
Regulating Genes to Mediate Grain Cd Accumulation in Rice *147*
Molecular Breeding for Low-Cd Accumulation in Rice Grains *148*
Screening and Identification of Low-Cd Varieties *148*
Molecular Marker-Assisted Low-Cd Gene Aggregation Breeding *148*
Gene Editing-Based Molecular Breeding *148*
Conclusions and Perspectives *150*
Acknowledgments *150*
References *150*

8 Molecular Breeding for Improving Arsenic Stress Tolerance in Rice: Recent Progress and Future Perspectives *163*
Nourollah Ahmadi and Julien Frouin
Introduction *163*
Tolerance to Arsenic in Rice *164*
The Phytotoxic Effects of Arsenic and the Rice Defense Response *164*
Genetic Diversity for Arsenic Tolerance *165*
Genetic Bases of Arsenic Tolerance *166*
Molecular Breeding Options for Rice Tolerance to Arsenic *169*
Marker-Assisted Recurrent Selection *169*
Genomic Selection *170*
Genomic Selection for Arsenic Tolerance in Rice: A Case Study *170*

Methods *170*
Genomic Prediction Results *171*
Cross-Validation Within the Reference Population *171*
Genomic Prediction Across Populations *171*
Implication for Breeding for Arsenic Tolerance *172*
Future Prospects *173*
References *174*

9 Molecular Breeding for Improving Ozone Tolerance in Rice: Recent Progress and Future Perspectives *180*
Md. Ashrafuzzaman, Robert Henry and Michael Frei
Introduction *180*
Tropospheric Ozone: A Major Threat to Rice Production in Asia *182*
Mechanisms of Ozone Damage *184*
Ozone Tolerance Mechanisms *186*
Ozone-Impact Assessments *187*
Molecular Breeding Strategies for Improved Ozone Tolerance *190*
Breeding Progress and Future Perspectives of Ozone Tolerance in Rice *192*
Conclusion *192*
References *193*

10 Molecular Breeding Strategies for Enhancing Rice Yields Under Low Light Intensity *201*
Mayank Rai, Suvendhu S. Dutta and Wricha Tyagi
Introduction *201*
Effect of Low Light in Rice Physiology, Development and Yield *202*
Candidate Genes and Pathways Imparting Tolerance *202*
Low Light Perception *203*
Molecular Targets for Erect Lamina *205*
Molecular Basis of Delayed Senescence *205*
Cross Talk Between Various Stresses *207*
"Omics" and Mapping Studies *208*
Future Prospects and Molecular Breeding Strategy *211*
References *211*

11 Harnessing Tolerance to Low Phosphorus in Rice: Recent Progress and Future Perspectives *215*
Wricha Tyagi, Erneica N. Nongbri and Mayank Rai
Introduction *215*
Phosphorus Uptake and Assimilation in Plants *215*
Role of Low P in Rice Development and Growth *215*
Uptake and Transport of P in Rice *215*
Signaling in Response to Low Pi *218*
Assimilation of P in Rice *220*
Vegetative Stage *220*
Reproductive Stage *220*
Impact of P Deficiency on Rice *221*
Resources That Have Helped Enhance the Understanding of Low P Tolerance in Rice *221*

QTLs *221*
Functionally Validated Low P Tolerance Potential Candidates *222*
Molecular Targets Available in Rice for Breeding *224*
PUP1 History and Deployment of PUP1 Based Markers for MAS *224*
Other Haplotypes Reported for PUP1 and Implications for Breeding Program *224*
Beyond PSTOL1 *226*
Future Prospects *227*
Acknowledgments *228*
References *228*

12 Molecular Breeding for Improving Nitrogen Use Efficiency in Rice: Progress and Perspectives *234*
Chirravuri N. Neeraja, Sitapati R. Voleti, Subrahmanyam Desiraju, Surekha Kuchi, Sonali Bej, Krishnakanth Talapanti and Raghuveer R. Puskur
Introduction *234*
Evaluation of Genetic Variation for Nitrogen Use Efficiency *235*
Identification of QTL and Genomic Regions through Biparental and Association Mapping *237*
QTL *237*
Association Mapping *239*
Expression Analyses of Candidate Genes Associated with Nitrogen Metabolism *240*
Ways to Go for the Development of Rice Varieties *243*
Conclusions *243*
References *243*

13 Dissecting the Molecular Basis of Drought-Induced Oxidative Stress Tolerance in Rice *249*
Amit K. Pradhan, Sabnoor Y. Jyoti, Zina M. Shandilya, Mehzabin Rehman, Debanjali Saikia, Junu Poudel, Jyotirmay Kalita, Kongkona Borborah, Uma K. Chowra, Jnandabhiram Chutia, Lakshminarayana R. Vemireddy and Bhaben Tanti
Introduction *249*
Effect of Drought Stress in Rice *252*
Sources of ROS Production during Drought Stress in Rice Plants *254*
Antioxidative Defenses and Redox Homeostatic Mechanisms in Rice in Response to Drought *255*
ROS Signaling under Drought Stress *256*
High-Throughput Technologies to Identify Drought-Induced Oxidative Stress-Tolerance Mechanisms *257*
Genome Editing Using CRISPR-Cas9 Technology *258*
Expressed Sequence Tags (ESTs) *258*
Gene Networks *258*
RNA-Seq Technology *259*
Micro-RNA (miRNA) Technology *259*
Virus-Induced Gene-Silencing Machinery *260*
TILLING (Targeting Induced Local Lesions in Genomes) *261*

Identification of Superior Alleles or QTLs Associated with Drought Tolerance *262*
Molecular Mechanism in Tolerant Genotype in Comparison to Sensitive Genotype under Drought *262*
Conclusion and Future Perspectives *265*
References *265*

14 Manipulation of Photosynthesis to Increase Rice Yield Potential *274*
Prabuddha Dehigaspitiya and Saman Seneweera
Introduction *274*
The Relationship Between the Rate of Photosynthesis and Yield in Rice *276*
Understanding the Mode of Photosynthesis and Crop Yield *277*
Genetic Engineering of the C_4 Pathway into C_3 Crops *279*
Mutation Breeding for Improvement of Rice Yield Potential *280*
Marker-Assisted Breeding to Improve Photosynthesis and Yield Potential *281*
Increase in Atmospheric CO_2 Concentration and Future Photosynthesis *282*
Conclusion *283*
References *283*

15 Molecular Breeding for Improved β-carotene Synthesis in Golden Rice: Recent Progress and Future Perspectives *287*
Swapan K. Datta, Shuvobrata Majumder and Karabi Datta
Introduction *287*
The β-carotene Synthesis Pathway in Plant *287*
Metabolic Engineering of β-carotene Pathway in Rice *289*
Endosperm-Specific Expression of β-carotene *289*
Source of β-carotene Synthesis Genes *290*
Rice Genetic Transformation System *290*
Development of Golden Rice (GR) *292*
Homozygous Dihaploid GR Development *294*
GR in Introgression Breeding Programs *294*
Phenotypic and Agronomic Performance of Golden Rice *295*
Recent Progress of GR Research *296*
Better Long-Term Storage of GR (Silencing of *lox*) *296*
Substantial Equivalence Analysis of GR *296*
Metabolic Analysis of GR *297*
Food Safety Analysis of GR *297*
Future Perspective of GR *297*
Confined Field Trial and Regulation of GR *298*
Application of CRISPR/Cas for GR Development *298*
GR Economics *299*
Conclusion *300*
Acknowledgments *300*
References *300*

16 Increasing Grain Zinc Concentration in Rice *304*
Naoya Miyazaki, Miki Ogasawara and Ryo Ishikawa
Role of Zinc in Human Nutrition *304*

Current Progress in Increasing Grain Zinc Concentration in Rice 305
Genes Controlling Grain Zn Content in Rice 305
Transgenic Approaches Increasing Grain Zn Content in Rice 306
Natural Variations Involved in Grain Zn Content in Rice 306
Use of Wild Relatives to Increase Grain Zn Concentration in Rice 307
Control of Grain Zn Concentration with More Yield 308
Acknowledgements 311
References 311

17 Molecular Breeding for Iron Bio-Fortification in Rice Grain: Recent Progress and Future Perspectives *315*
Elssa Pandit, Swapnil Pawar, Priyadarshini Sanghamitra and Sharat K. Pradhan
Introduction 315
Iron : From Soil to Rice Grain 316
Genes, Transporters, and Mechanisms of Fe Transport 317
QTLs Involved for Enhancement of Fe in Rice Grain 319
Approaches for Fe Biofortification in Rice 320
Iron Biofortification in Rice Through Agronomic Practices 320
Limitations of Agronomic Biofortification 323
Iron Biofortification in Rice Through Transgenics/Genetic Engineering 323
Limitations of Fe Biofortification through Genetic Engineering 323
Iron Biofortification in Rice through Conventional Breeding 324
Limitations of Fe Biofortification through Conventional Breeding 324
Iron Biofortification in Rice through Molecular Breeding 324
Strategies Involved in Molecular Breeding 324
Recent Molecular Breeding Approaches to Enhance Grain Fe Content 325
Global Success Towards Development of Fe Fortified Rice Lines 330
Conclusion and Future Perspective 330
Acknowledgments 331
References 331

18 Aromatic Rices: Evolution, Genetics and Improvement through Conventional Breeding and Biotechnological Methods *341*
Lakshminarayana R. Vemireddy, Bhaben Tanti, Lipika Lahkar and Zina M. Shandilya
Aromatic Rices: Background 341
Genetics, Origin, and Evolution of Aromatic Rices 344
Genetics of Aroma 344
Natural Variability in the Aroma (fgr) Gene 344
Additional QTLs for Aroma 345
Origin and Evolution of Aromatic Varieties 346
Biochemical Basis of Aroma 349
Improvement of Aromatic Rice Varieties 350
Conventional Breeding 350
Molecular Breeding of Aromatic Rice 351
Genetic Engineering 352

Trade and Geographical Indication (GI) *353*
Future Prospects of Aromatic Rices *353*
References *354*

19 Genetic Engineering for Increasing Antioxidant Content in Rice: Recent Progress and Future Perspectives *358*
Qinlong Zhu, Jiantao Tan, Bin Wang and Yao-Guang Liu
Introduction *358*
Antioxidant Compounds of Rice *360*
Flavonoids, Anthocyanins, and Proanthocyanidins Composition of Rice *360*
Carotenoids Composition of Rice *361*
Tocopherols and Tocotrienols Composition of Rice *365*
Folate Composition of Rice *366*
Melatonin Composition of Rice *368*
Genetic Engineering for Biofortification with Antioxidants *369*
Flavonoids, Anthocyanins, Proanthocyanidins *369*
Carotenoids *370*
Tocopherols, Tocotrienols *371*
Folates *372*
Melatonin *373*
Conclusions and Perspectives *374*
Acknowledgments *375*
References *375*

20 Molecular Breeding Approaches for Improvement and Development of Water Saving Aerobic Rice *382*
Rahul K. Meena, Kuldeep Kumar, Saurabh K. Dubey, Ashish K. Singh, Adarsh Kumar, Deepanshu Jayaswal, Badal Singh, Rajinder Jain and Sunita Jain
Introduction *382*
Rice Facts and Figure *383*
Concept of Aerobic Rice *385*
Effect of Aerobic Rice on Water Saving and Water Productivity *385*
Greenhouse Gas Emissions under Aerobic Conditions *387*
Diversity in Rice Germplasm *387*
Root Morphology for Aerobic Traits *387*
Mapping of Aerobic Traits *388*
Linking Mapping and Marker-Assisted Selection *388*
Problems in Aerobic Rice *390*
Weed Problems in Aerobic Rice *390*
Nematode Problems in Aerobic Rice *391*
Disease and Insect Pests Constraints in Aerobic Rice Cultivation *392*
Other Problems in Aerobic Rice *393*
Success Achieved *394*
Conclusion and Ways Forward *394*
References *394*

21 Targeting the Ascorbate-Glutathione Pathway and the Glyoxalase Pathway for Genetic Engineering of Abiotic Stress-Tolerance in Rice *398*
Mohammad A. Hossain, Tahsina S. Hoque, Abbu Zaid, Shabir H. Wani, Mohammad G. Mostofa and Robert Henry
Introduction *398*
Accumulation of ROS and MG in Plants in Response to Abiotic Stress *399*
Accumulation of ROS in Plants in Response to Abiotic Stresses *399*
Accumulation of MG in Plants in Response to Abiotic Stresses *401*
Importance of Glyoxalase and AsA-GSH Pathways in Modulating Abiotic Stress Tolerance *403*
Importance of Glyoxalase Pathway in Modulating Abiotic Stress Tolerance *403*
Importance of AsA-GSH Pathway in Modulating Abiotic Stress Tolerance *404*
Interconnection of the Antioxidant Pathway and the Glyoxalase Pathway in Modulating Abiotic Stress Tolerance *406*
Manipulation of Glyoxalase Pathway Genes for Abiotic Stress Tolerance in Rice *410*
Transgenic Plants Over-expressing *gly I* Gene and Abiotic Stress Tolerance in Rice *410*
Transgenic Plants Over-expressing *glyII* Gene and Abiotic Stress Tolerance in Rice *411*
Transgenic Plants Over-expressing *glyI* and *glyII* Gene and Abiotic Stress Tolerance in Rice *412*
Transgenic Plants Over-expressing *glyIII* Gene and Abiotic Stress Tolerance in Rice *412*
Genetic Engineering of the AsA-GSH Pathway Genes for Abiotic Stress Tolerance in Rice *413*
Transgenic Plants Over-expressing APX Gene and Abiotic Stress Tolerance in Rice *413*
Transgenic Plants Over-expressing the MDHAR Gene and Abiotic Stress Tolerance in Rice *414*
Transgenic Plants Over-expressing DHAR Gene and Abiotic Stress Tolerance in Rice *414*
Transgenic Plants Over-expressing the GR Gene and Abiotic Stress Tolerance in Rice *415*
Transgenic Plants Over-expressing GSH Biosynthetic Genes and Abiotic Stress Tolerance in Rice *415*
Transgenic Plants Over-expressing AsA Biosynthetic Genes and Abiotic Stress Tolerance in Rice *416*
Conclusion and Future Perspectives *417*
References *417*

Index *428*

Editor Biographies

Dr. Mohammad Anwar Hossain is a Professor in the Department of Genetics and Plant Breeding, Bangladesh Agricultural University (BAU), Mymensingh, Bangladesh. He received his BSc in Agriculture and MS in Genetics and Plant Breeding from BAU, Bangladesh. He also received an MSc in Agriculture from Kagawa University, Japan, in 2008 and a PhD in Abiotic Stress Physiology and Molecular Biology from Ehime University, Japan, in 2011 through a Monbukagakusho scholarship. As a JSPS postdoctoral researcher he has worked on isolating low phosphorus stress tolerant genes from rice at the University of Tokyo, Japan, during the period of 2015–2017. His current research program focuses on understanding physiological, biochemical, and molecular mechanisms underlying abiotic stresses in plants and the generation of stress-tolerant and nutrient-efficient plants through breeding and biotechnology. He has over 60 peer-reviewed publications and has edited 10 books, including this one, published by CRC press, Springer, Elsevier, and Wiley.

Dr. Lutful Hassan is a Professor in the Department of Genetics and Plant Breeding, BAU, Mymensingh, Bangladesh. Currently he is the Vice-Chancellor of BAU, Bangladesh. He obtained his BSc Agriculture and MSc in Genetics and Plant Breeding from BAU and his PhD in Plant Breeding from the University of Wales, UK. He has conducted post-doctoral research in different countries across the world through Japan Society for the Promotion of Science, Alexander von Humboldt, Norman E. Borlaug, and Royal Society Fellowship. He has also worked at the International Rice Research Institute (IRRI) as a consultant in "Stress to tolerant rice for Africa and South Asia (STRASA)" and "Sustainable Rice Seed Production and Delivery System (SRSPDS)" projects. His current research includes the development of high-yielding stress-tolerant rice and mustard varieties through conventional and molecular breeding approaches. He is a recipient of the John Dillon Memorial Fellow Award for agricultural research management, agricultural policy and/or extension technologies. He has over 150 peer-reviewed publications, and has edited 3 books including this one. He has given over 50 invited presentations nationally and internationally.

Dr. Khandakar Md. Iftekharuddaula is a Chief Scientific Officer and Head in the Division of Plant Breeding, Bangladesh Rice Research Institute, Gazipur, Bangladesh. He received his BSc in Agriculture and MS in Genetics and Plant Breeding from Bangladesh Agricultural University, Bangladesh. He has completed PhD research from the International Rice Research Institute and obtained his degree in Genetics and Plant Breeding from Bangladesh Agricultural University, Bangladesh. He has acted as Collaborative Research Fellow in the Transforming Rice Breeding Project implemented at the Bangladesh Rice Research Institute funded by the Bill and Melinda Gates Foundation for the duration of four years. He has special research experience in the development of submergence and stagnant flooding tolerant, deep-water tolerant, water-saving, irrigated, and premium quality rice varieties utilizing breeding and biotechnological tools. He has published 60 full-length scientific papers, five books/book chapters and six bulletins/proceeding papers.

Dr. Arvind Kumar is the Director of the International Rice Research Institute, South Asia Regional Centre, Varanasi, India. Dr. Kumar has twenty-six years of experience in rice research in South Asia and Southeast Asia. From breeding lines developed by him, 65 rice varieties have been released in 10 different countries. He successfully introgressed drought grain yield QTLs in popular high-yielding varieties following marker-assisted breeding (MAB) and has developed drought-tolerant improved versions of several popular varieties as well as varieties tolerant to multiple abiotic and biotic stresses. Dr. Kumar has identified 14 QTLs for grain yield under drought. Identified QTLs are being used on a large scale in marker-assisted breeding programs all over the world to develop rice varieties with improved yield under drought. Dr. Kumar also has identified seven genes for resistance against rice gall midge, which are used in breeding programs across the world, as well as QTLs for traits enhancing rice adaptability to dry direct-seeded situations. He has implemented 28 research projects, supervised more than 30 scholars, and published more than 30 chapters and 143 research manuscripts. For his varietal development work, he was awarded with the highest award for contribution to Indian agriculture, the Rafi Ahmed Kiwi Award by the Indian Council of Agricultural Research (ICAR), Government of India in 2014. The Nepal Council of Agricultural Research (NARC), Government of Nepal recognized him with honor in 2016 for his contribution to agriculture in Nepal for development of drought-tolerant varieties.

Professor Robert Henry conducts research on the development of new products from plants. His research targets improved understanding of the molecular basis of the quality of products produced from plants and genome analysis to capture novel genetic resources for diversification of food and energy crops. He is the Professor of Innovation in Agriculture and Foundation Director of the Queensland Alliance for Agriculture and Food Innovation, an Institute of the University of Queensland in partnership with the Queensland Government. He was previously director of the Centre for Plant Conservation Genetics at Southern Cross University and Research Program Leader in the Queensland Agricultural Biotechnology Centre. He has been involved in establishing several Cooperative Research Centres in Australia and has contributed to the management of research funding by Rural Research and Development Corporations in Australia. He is a graduate of the University of Queensland (BSc (Hons)), Macquarie University (MSc (Hons)) and La Trobe University (PhD). He was awarded a higher doctorate (DSc) by the University of Queensland for his work on variation in plants, is a Fellow of the Royal Australian Chemical Institute, recipient of the Guthrie Medal for his contributions to cereal chemistry and a Fellow of the Australian Academy of Technological Sciences and Engineering.

Preface

Global population is projected to increase over 9 billion by 2050, and food and feed production will need to increase by 70%. Additionally, to alleviate the world's greatest health and poverty issues (micronutrient malnutrition) the development of nutrient-dense crops is urgently needed. Rice is one of the most important cereal food crops, and it provides food to more than half of the world's population, particularly in many developing countries in Asia, Africa, and Latin America. Globally, rice is grown on approximately 163 million hectares of land of which an estimated 60% or more is affected by various abiotic stresses (salinity, heat, drought, cold, submergence, radiation, heavy metals, etc.) causing significant yield losses. The situation becomes even worse due to climate change, which may multiply the frequency and severity of such abiotic stresses. Importantly, production of rice must continue to increase at the rate of 1 percent a year to maintain food security. Sustainable rice production delivering yields to meet ever-increasing demands and the development of biofortified rice is a major challenge for the scientific community, and will require the combined expertise of agronomists, farmers, breeders, and molecular biologists. Recently, molecular rice breeding in response to global climate change, the increasing fragility of our natural resources and threats to food grain security across the globe, have attracted considerable interest by the scientific community. Since then countless studies in various scientific disciplines dealing with different rice species, in different environments have focused on abiotic stress tolerance, grain and quality improvements, and rice biofortification. Although significant progress has been made over the last few years, there is still a need to bridge the large gap between yields in the most favorable environments and those under stress conditions. Strategies involving bridging the yield gap and increasing yield stability and adaptability to variable environmental conditions are of importance in assuring food security and sustainability in the future. Hence, there is an urgent need to improve our understanding of complex mechanisms regulating abiotic stress tolerance for developing modern rice varieties that are more resilient to abiotic stresses as well as to increase the bioavailable concentrations of essential micronutrients. The discovery of novel genes/QTLs, the analysis of their expression patterns in response to abiotic stress, and the determination of their potential functions in stress adaptation will provide the basis of effective engineering strategies to enhance rice yield under stress and non-stress conditions, to develop biofortified rice as well as sustainable utilization of natural resources.

Over the last decade, tremendous progress has been made in rice genome analysis. The progress has provided powerful tools—DNA markers—for plant genetics and breeding.

Now DNA-based markers have been widely used in the genetic analysis of agronomically important traits regulating abiotic stress tolerance, yield, and quality of rice. Tightly linked DNA markers and causal genes are used in marker-assisted selection in rice-breeding programs and are able to shorten the time of variety development. Another use of DNA-based markers is overcoming the barrier of "linkage drag" which refers to the presence of undesirable genes in the chromosomal region of the target gene, thereby making it difficult to avoid such traits when using conventional breeding. Also, economic analysis has shown the potential impacts of utilizing marker-assisted breeding by overcoming drawbacks of conventional breeding in rice that ultimately reduce the cost of production and promote economic growth.

In this book, *Molecular Breeding for Rice Abiotic Stress Tolerance and Nutritional Quality*, we present a collection of 21 chapters written by leading experts engaged with rice molecular breeding. The chapters of this book aim to contribute the latest understandings of molecular and genetic bases of abiotic stress tolerance, yield, and quality improvement of rice to develop strategies for abiotic stress tolerance and biofortification, which leads to enhanced rice productivity under abiotic stress conditions as well better utilization of natural resources to ensure food security through modern breeding as well as to curb the scourge of micronutrient malnutrition. Finally, this book will be a valuable resource for future environmental stress-related research, and can be considered as a textbook for graduate students and as a reference book for front-line rice researchers around the globe.

Mohammad Anwar Hossain
Lutful Hassan
Khandakar Md. Iftekharuddaula
Arvind Kumar
Robert Henry

List of Contributors

Nourollah Ahmadi
CIRAD, UMR AGAP
Montpellier
France

AGAP, Univ. Montpellier, CIRAD, INRA
Montpellier SupAgro
Montpellier
France

Md. Ashrafuzzaman
Department of Genetic Engineering &
Biotechnology
Shahjalal University of Science &
Technology
Sylhet, Bangladesh

Sonali Bej
ICAR-Indian Institute of Rice Research
Rajendranagar, Hyderabad
India

Kongkona Borborah
Department of Botany
Gauhati University
Guwahati
Assam
India

Bui Chi Buu
High Agricultural Technology Research
Institute for Mekong Delta (HATRI)
Can Tho
Vietnam

Uma K. Chowra
Department of Botany
Gauhati University
Guwahati
Assam
India

Cho Young Chan
National Institute of Crop Science (NICS)
RDA, Wanju
Republic of Korea

Sandeep Chapagain
School of Plant
Environment and Soil Sciences
Louisiana State University Agricultural
Center
Baton Rouge, LA
USA

Jian-Min Chen
Yangzhou University
China

Liangbi Chen
Hunan Province Key Laboratory of Crop
Sterile Germplasm Resource Innovation
and Application College of Life Science
Hunan Normal University
Changsha
P.R. China

Jnandabhiram Chutia
Department of Botany
Gauhati University
Guwahati
Assam
India

Jonathan Concepcion
School of Plant
Environment and Soil Sciences
Louisiana State University Agricultural
Center
Baton Rouge, LA
USA

Sapphire Coronejo
School of Plant
Environment and Soil Sciences
Louisiana State University Agricultural
Center
Baton Rouge, LA
USA

Satya R. Das
Department of Plant Breeding
Orissa University of Agriculture &
Technology
Bhubaneswar
India

Karabi Datta
University of Calcutta
Laboratory of Translational Research on
Transgenic Crops
Kolkata, India

Swapan K. Datta
University of Calcutta
Laboratory of Translational Research on
Transgenic Crops
Kolkata
India

Prabuddha Dehigaspitiya
University of Southern Queensland
Toowoomba, QLD
Australia

Saurabh K. Dubey
Research Scholar
Division of Plant Pathology
Indian Agricultural Research Institute
New Delhi
India

Suvendhu Shekhar Dutta
School of Crop Improvement
College of PG studies in Agricultural
Sciences
Central Agricultural University (Imphal)
Umiam, Meghalaya
India

Michael Frei
Institute of Crop Science and Resource
Conservation (INRES)
University of Bonn
Bonn
Germany

Julien Frouin
CIRAD, UMR AGAP
Montpellier
France

AGAP, Univ. Montpellier, CIRAD, INRA
Montpellier SupAgro
Montpellier
France

Richard Garcia
School of Plant
Environment and Soil Sciences
Louisiana State University Agricultural
Center
Baton Rouge, LA
USA

Xiaohua Hao
Hunan Province Key Laboratory of Crop Sterile Germplasm Resource Innovation and Application College of Life Science
Hunan Normal University
Changsha
P.R. China.

College of Life and Environmental Science
Hunan University of Arts and Science
Changde, China

Lutful Hassan
Bangladesh Agricultural University
Mymensingh
Bangladesh

Robert Henry
Queensland Alliance for Agriculture and Food Innovation
University of Queensland, Brisbane
Australia

Tahsina S. Hoque
Department of Soil Science
Bangladesh Agricultural University
Mymensingh
Bangladesh

Md. Amir Hossain
Bangladesh Agricultural University
Mymensingh
Bangladesh

Mohammad A. Hossain
Department of Genetics and Plant Breeding
Bangladesh Agricultural University
Mymensingh
Bangladesh

Ryo Ishikawa
Laboratory of Plant Breeding, Graduate School of Agricultural Science
Kobe University
Rokkodai-cho, Nada-ku
Kobe
Japan

Rajinder Jain
Department of Molecular Biology and Biotechnology
CCSHAU, Haryana
India

Sunita Jain
Department of Molecular Biology and Biotechnology
CCSHAU, Haryana
India

Deepanshu Jayaswal
Research Scholar
National Institute for Plant Biotechnology
New Delhi
India

Sabnoor Y. Jyoti
Department of Botany
Gauhati University
Guwahati
Assam, India

Jyotirmoy Kalita
Department of Botany
Gauhati University
Guwahati
Assam
India

Krishnakanth Talapanti
ICAR-Indian Institute of Rice Research
Rajendranagar, Hyderabad
India

Subbaiyan Gopala Krishnan
Division of Genetics
ICAR-Indian Agricultural Research Institute
New Delhi
India

Adarsh Kumar
Scientist
ICAR-NBAIM
Uttar Pradesh
India

Arvind Kumar
Rice Breeding Platform
International Rice Research Institute
Manila
Philippines

Kuldeep Kumar
Research Scholar
National Institute for Plant Biotechnology
New Delhi
India

Lipika Lahkar
Department of Botany
Gauhati University
Guwahati
Assam
India

Nguyen Thi Lang
High Agricultural Technology Research Institute for Mekong Delta (HATRI)
Can Tho
Vietnam

Dongping Li
Hunan Province Key Laboratory of Crop Sterile Germplasm Resource Innovation and Application College of Life Science
Hunan Normal University
Changsha
P.R. China

Yao-Guang Liu
College of Life Sciences
South China Agricultural University
Guangzhou
China

Erneica L. Nongbri
School of Crop Improvement
College of PG Studies in Agricultural Sciences
Central Agricultural University (Imphal)
Umiam, Meghalaya
India

Ratna R. Majumder
Rice Breeding Platform
International Rice Research Institute
Manila
Philippines

Bangladesh Agricultural University
Mymensingh
Bangladesh

Shuvobrata Majumder
University of Calcutta
Laboratory of Translational Research on Transgenic Crops
Kolkata
India

Rahul K. Meena
Department of Molecular Biology
Biotechnology and Bioinformatics
CCS Haryana Agricultural University
Hisar, Haryana
India

Naoya Miyazaki
Laboratory of Plant Breeding, Graduate School of Agricultural Science
Kobe University
Rokkodai-cho, Nada-ku
Kobe
Japan

Mohammad G. Mostofa
Department of Biochemistry and Molecular Biology
Bangabandhu Shiekh Mujibur Rahman Agricultural University
Gazipur, Bangladesh

Chirravuri N. Neeraja
ICAR-Indian Institute of Rice Research
Rajendranagar, Hyderabad
India

Miki Ogasawara
Laboratory of Plant Breeding, Graduate School of Agricultural Science
Kobe University
Rokkodai-cho, Nada-ku
Kobe
Japan

Elssa Pandit
Crop Improvement Division
ICAR-National Rice Research Institute
Cuttack, Odisha
India

Department of Biosciences and Biotechnology
Fakir Mohan University
Balasore, Odisha, India

Swapnil Pawar
Crop Improvement Division
ICAR-National Rice Research Institute
Cuttack, Odisha, India

Department of Plant Biotechnology
K. K. Wagh College of Agricultural Biotechnology
Nashik, Maharastra, India

Junu Poudel
Department of Botany
Gauhati University
Guwahati
Assam
India

Amit K. Pradhan
Department of Botany
Gauhati University
Guwahati
Assam
India

Sharat K. Pradhan
Crop Improvement Division
ICAR-National Rice Research Institute
Cuttack, Odisha
India

Rajat Pruthi
School of Plant
Environment and Soil Sciences
Louisiana State University Agricultural Center
Baton Rouge, LA
USA

Mayank Rai
School of Crop Improvement
College of PG studies in Agricultural Sciences
Central Agricultural University (Imphal)
Umiam, Meghalaya
India

Raghuveer R. Puskur
ICAR-Indian Institute of Rice Research
Rajendranagar, Hyderabad
India

Jangi N. Reddy
Division of Crop Improvement
ICAR-National Rice Research Institute
Cuttack
India

Mehzabin Rehman
Department of Botany
Gauhati University
Guwahati
Assam, India

Debanjali Saikia
Department of Botany
Gauhati University
Guwahati
Assam
India

Sandeep Sakhale
Rice Breeding Platform
International Rice Research Institute
Manila
Philippines

Department of Crop Sciences
University of Illinois
Urbana-Champaign, Urbana, Illinois
USA

Nitika Sandhu
Punjab Agricultural University
Ludhiana, Punjab
India

Priyadarshini Sanghamitra
Crop Improvement Division
ICAR-National Rice Research Institute
Cuttack, Odisha, India

Ramani K. Sarkar
Division of Crop Physiology & Biochemistry
ICAR-National Rice Research Institute
Cuttack
India

Saman Seneweera
University of Southern Queensland
Toowoomba, QLD
Australia

National Institute of Fundamental Studies
Kandy
Sri Lanka

Zina M. Shandilya
Department of Botany
Gauhati University
Guwahati
Assam
India

Ashish K. Singh
Scientist
ICAR-VPKAS
Uttarakhand
India

Ashok K. Singh
Division of Genetics
ICAR-Indian Agricultural Research Institute
New Delhi
India

Badal Singh
Scientist
Regional Station Shimla ICAR-NBPGR
Himanchal Pradesh
India

Lovepreet Singh
School of Plant
Environment and Soil Sciences
Louisiana State University Agricultural Center
Baton Rouge, LA
USA

Vikram Jeet Singh
Division of Genetics
ICAR-Indian Agricultural Research Institute
New Delhi
India

Subrahmanyam Desiraju
ICAR-Indian Institute of Rice Research
Rajendranagar, Hyderabad
India

Prasanta Subudhi
School of Plant
Environment and Soil Sciences
Louisiana State University Agricultural Center
Baton Rouge, LA
USA

Surekha Kuchi
ICAR-Indian Institute of Rice Research
Rajendranagar, Hyderabad
India

Jiantao Tan
College of Life Sciences
South China Agricultural University
Guangzhou
China

Bhaben Tanti
Department of Botany
Gauhati University
Guwahati
Assam
India

Wricha Tyagi
School of Crop Improvement
College of PG studies in Agricultural Sciences
Central Agricultural University (Imphal)
Umiam, Meghalaya
India

Lakshminarayana R. Vemireddy
Department of Genetics and Plant Breeding
S.V. Agricultural College
Acharya NG Ranga Agricultural University
Tirupati, Andhra Pradesh
India

Kunnummal K. Vinod
Division of Genetics
ICAR-Indian Agricultural Research Institute
New Delhi
India

Sitapati R. Voleti
ICAR-Indian Institute of Rice Research
Rajendranagar, Hyderabad
India

Bin Wang
College of Life Sciences
South China Agricultural University
Guangzhou, China

Shabir H. Wani
Mountain Research Centre for Field Crops
Sher-e-Kashmir University of Agricultural Sciences and Technology of Kashmir
J&K
India

Ning Xiao
Institute of Agricultural Sciences for Lixiahe Region in Jiangsu
China

Shailesh Yadav
Rice Breeding Platform
International Rice Research Institute
Manila
Philippines

Abbu Zaid
Plant Physiology and Biochemistry Section
Department of Botany
Aligarh Muslim University
Aligarh
India

Qinlong Zhu
College of Life Sciences
South China Agricultural University
Guangzhou
China

1

Rice Adaptation to Climate Change: Opportunities and Priorities in Molecular Breeding

Vikram Jeet Singh, Kunnummal Kurungara Vinod, Subbaiyan Gopala Krishnan and Ashok K. Singh

Division of Genetics, ICAR-Indian Agricultural Research Institute, New Delhi 110012, India

Introduction

Climate change is an unquestionable reality today, as seen from the warming of the global environment. Global warming is principally implicated to human activities, which has led to the description of our era as the Anthropocene. There is an elevation of the CO_2 content in the atmosphere today, surpassing by more than 100% the CO_2 concentration a million years ago (World Bank 2010). In contemporary agriculture, climate change translates into cognizable challenges to sustainable food production. Many people think of climate change as a metaphor by itself, often associating it to localized weather events. Climate is a general expectation of events including weather at a particular place at a particular time. It is both spatial and temporal and is affected by several factors originating from celestial, geologic, geographic, edaphic, hydrologic, biologic, and anthropogenic features. Climate change refers to a shift in the climatic pattern for a significant period, extending to millennia, impacting the natural ecology. As a basic agglomeration of above factors, the impact of climate change is primarily perceived in agriculture, especially in the fragile environments such as the arid and semi-arid tropics, where the annual climatic pattern is the most variable. The quantification of climate change is made through the measurement of shifts in average temperature, rainfall patterns, and sea-level rise that occur on a continual scale for several years. For agriculture, this means a shift in cropping pattern over a longer-term, incited by hydrological and thermal fluctuations. Realized currently as a marked increase in abiotic stress events such as drought, heat, salinity, and waterlogging in shorter time intervals, climate change can potentially reduce crop productivity and threaten production and sustainability. In highly populated Asian countries, climate change can lead to food insecurity, driving inflation of food prices due to reduced availability. Projections show that the demand-supply gap can contribute to a 31% price rise in rice by 2050 (Nelson et al. 2010). Therefore, countries like India are more vulnerable to climatic adversities given their huge agriculture-dependent population. Elevating crop resilience to climate change is critical to ensure food and nutritional security in the most vulnerable areas, and this could be achieved through various breeding strategies.

Molecular Breeding for Rice Abiotic Stress Tolerance and Nutritional Quality, First Edition.
Edited by Mohammad Anwar Hossain, Lutful Hassan, Khandakar Md. Iftekharuddaula, Arvind Kumar and Robert Henry.

As an important dietary source, rice accounts for 23% of global per capita energy and 16% of protein, making it a major staple cereal for around 3.5 billion people globally. During the current 50-year window (2000-2050), it is projected that rice production needs to go up by 40% by 2025, to meet the food demand of the growing world population (Fahad et al. 2018). Cultivated rice belongs to the genus *Oryza* under the grass family *Poaceae*. With 22 co-species, among which *Oryza sativa* and *Oryza glaberrima* are the only species cultivated (Lu 1999), rice plant is a semi-aquatic annual grass, that requires a good quantum of water, sunshine, healthy soil, and warm temperature for its best growth and grain production. Because of these specific ecological needs, rice faces a serious threat from climate change. The estimates now show that the vulnerability of the rice crop could lead to a yield reduction of 11% globally by 2020 (Ramachandran et al. 2017; Halder et al. 2020). Particularly in India, the yield fall may range between 4.5% and 9% by 2039 (Guiteras 2009). In crops like rice, where a profound impact of climate change on productivity is anticipated, technological interventions can ensure sustained food production. The major threat to rice production might come from the hydrological front, manifested in the form of water scarcity, leading to drought. Depletion of water tables around global rice ecosystems is a current reality today, as a severe consequence of global warming leading to aridization and desertification (IPCC 2019). In contrast, floods and inundation are also occurring elsewhere destroying the standing crop abruptly. Another serious threat is the soil salinization, a result of unscrupulous exploitation of groundwater. Additionally, there are threats such as depleting soil fertility, heavy metal toxicity, eutrophication, and environmental degradation that are adversely affecting rice crop worldwide. Concurrent with abiotic stresses, the threat of biotic stresses also affect the biodynamics of flora and fauna in the rice ecosystems. Rice hosts a multitude of organisms in its ecosystem: microbes, insects, birds, and animals, several of which are pests and pathogens. The dynamic equilibrium within which the biota co-exist in the rice ecosystem is disrupted by climate change, triggering adaptational dynamics of the biotic components. This leads to changes in biotic stress levels on the host as well, as is seen with the transformation of several minor pests and diseases such as bakanae (Bashyal et al. 2020) into major threats to rice cultivation.

Rice as a domesticated crop, is under human cultivation for more than 9000 years. Spread all over the tropical world since then (Purugganan 2010), rice is an ancient crop that displays adaptation to highly heterogeneous environments. Although the natural genetic diversity of rice is remarkable, several of these adaptative variations remain unused in the cultivated pool, because humans always strived to provide ideal growing conditions in the cultivation environment to reap greater harvests. Until recently, the yield improvement remained pivotal to rice improvement programs intending to increase food production. Due to the imminent threat of climatic change, the need for diversification to stress adaptation and quality improvement is strongly felt nowadays to sustain rice production in the future. Known as climate-smart agriculture, the approach to combat the negative impact of climate variation on crop production requires cultivars that possess a broad spectrum of stress tolerances such as drought, cold, heat, salinity, flood, submergence, weed competition, pests, diseases, and so on. Furthermore, climatic threats require rapid response, that calls for a realignment of crop breeding to develop stable varieties with a wide range of adaptation features within a short period. The traditional breeding practices in rice are time consuming, and experience shows that using this approach to improve complex traits – such as tolerance to drought, heat, submergence, and salt stresses – has so far met with limited success. The latest advancements in the crop genomics, for which rice played a pivotal role as the model crop, have contributed to the development of several molecular tools that can aid in effective and accelerated breeding, collectively known as molecular breeding. One of the most popular tools in molecular breeding is the use of DNA-based markers that can directly target desirable alleles of beneficial genes for mobilization into elite varieties. Although there are many other gene-transformation systems, several of them remain practically unused for various reasons such as the negative perception of the genetically modified organisms (GMOs).

Marker-assisted breeding (MAB), on the other hand, uses the same traditional breeding route but aids in rapid, precise, and easy identification and selection of desirable plants with the aid of DNA based markers. This ultimately results in a significant saving in the varietal turnover time. MAB has the major advantage of targeting both qualitative and quantitative traits, without any serious interference of dominance, epistatic, and genotype-by-environment interactions enabling the selection of desirable progenies in very early crop growth stages. This can further leverage the handling of a large breeding population within a short period of time, a critical requirement in successful breeding. Besides, MAB can also help in reducing selection cycles (Singh and Singh 2015). Therefore, molecular breeding is one of the major options available in modern day plant breeder's toolkit for addressing climate change, by aiding the transfer of beneficial genes and alleles into desired target varieties to improve stress responsiveness and resilience. Furthermore, recent advancements in phenotyping such as phenomics platforms and in genotyping such as high-throughput sequencing systems have metamorphosed molecular breeding into genomics-assisted breeding (GAB). Experiences from molecular breeding for the last couple of decades shows that conspicuous advancements have been achieved in imparting stress adaptation in several crops. In rice, several genes/alleles/quantitative trait loci (QTLs) governing resistance/tolerance to biotic/abiotic stresses have been introgressed into cultivars to make then more widely adapted to stressful environments. These varieties primarily include existing popular cultivars and mega-varieties that show distinct stress-sensitiveness. Through marker-assisted pyramiding, several improved cultivars have been developed with tolerance to drought, submergence, heat, and salinity. Likewise, molecular breeding approaches have also been successful in improving cultivars with biotic stress tolerance, such as bacterial blight (BB) and blast (Singh et al. 2019). Currently, the arsenal of GAB has been filled with techniques such as marker-assisted selection (MAS), marker-assisted backcross breeding (MABB), genomic selection (GS) and genome editing. Similarly, by genome-deciphering techniques such as linkage mapping, genome wide association studies (GWAS), whole genome sequencing (WGS), and next-generation sequencing (NGS), numerous genes/ QTLs governing tolerance to several abiotic and biotic stresses have been identified in rice. In contemporary rice improvement programs, these resources are being deployed to augment the development of cultivars with resilience and tolerance to adversities that are linked to climate change.

Rice Production Scenario and Climate Change

Around the globe, rice is cultivated in more than 114 countries (Maclean et al. 2002), spread across all six populated continents: Asia, Africa, Australia, Europe, North America, and South America. With a worldwide acreage of 167 million ha accounting for about 12% of the total arable lands, more than 80% of the global rice production is confined to seven Southeast Asian majors such as China, India, Indonesia, Bangladesh, Vietnam, Myanmar, and Thailand. The east, south and southeast Asia jointly share 88.4% of the world's rice-growing area, producing 89.3% of the global rice crop (Yang et al. 2018). The rest of the Asian countries, except Mongolia and central Asian countries, produce the remaining 10% of Asia's share of global production. It is predicted that close to 52% of the rice areas in the Asian countries are climate-sensitive. Beyond Asia, rice is cultivated in Africa, Latin America, and the Caribbean. There are at least 39 counties in Africa (Oteng and Sant'Anna 1999), and 26 countries in Latin America and the Caribbean (Coffman et al. 1992) that grow rice. Unlike other parts of the world, the majority of the rice produced is consumed internally in Asian countries.

Rice is grown in distinct ecologies: upland, lowland, and coastal. Depending upon the water availability, the cultivation environment could be irrigated or rainfed (Figure 1.1). Being a semi-aquatic crop, more than 90% of the rice cultivated worldwide is grown in lowlands, which accounts for more than 90% of the world's rice supply. Altogether, irrigated rice occupies more than 75% of the total rice area that is characterized by a sufficient supply of water to support at least two cropping seasons a year. All the remaining cultivation environments support only a single crop per annum. Each ecosystem has specific characteristics, which may act as hindering factors in other ecosystems, and therefore, genetic adaptation to each ecosystem is under the control of system-specific genes. Therefore, cultivars adapted are specific to each ecosystem and are often non-portable because of their unique adaptation features. In tightly defined ecosystems, any environmental fluctuation can inflict imbalance in the character specificity, that is posed as a new stress to the adapted cultivars. To address the new challenges of stresses would require the mobilization of genes to reinforce adaptive response in the varieties.

China, the most populated country in the world, possesses 7% of the globe's arable land. During the twentieth century, climate change has brought perceptible deviations in average annual surface air temperature in China raising it by 0.5 to 0.8 °C within a span of 100 years. The most perturbing fact is that warming is accelerating, as indicated by an average temperature rise of 1.1 °C in the past 50 years (Ding et al. 2007). Projections for China suggest that the temperature rise over the 30-year average between 1961 and 1990, may go up to 1.5 to 2.1 °C by 2020, 2.3 to 3.3 °C by 2050, and 3.9 to 6.0 °C by 2100. A corresponding increase in the annual precipitation would be 10% to 12% by the year 2100. However, the increase in moisture level is predicted to occur predominantly in the northeast and northwest of China, while central China may move towards drier weather. It is concluded that the forecasted temperature rise may do notable harm in rain-fed lands, but may turn beneficial to the irrigated systems (Wang et al. 2009). Besides the increased water levels, the beneficial effect on crops could also be due to the presence of elevated levels of CO_2 in the atmosphere. Known as carbon fertilization or the CO_2 fertilization effect, this positive effect occurs when the net photosynthesis is improved in the presence of other environmental conditions such as temperature (Long 1991), water and nutrients (Allen,

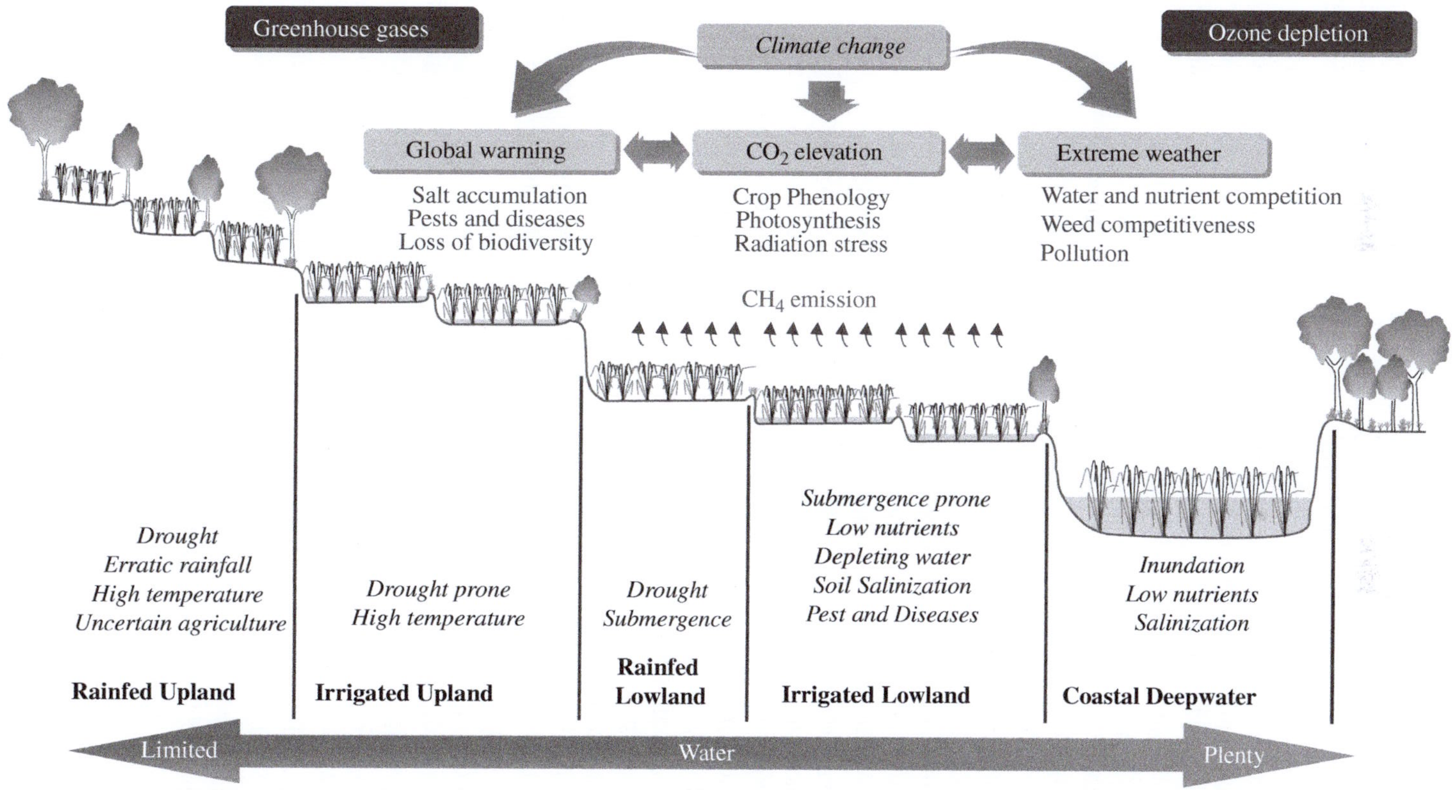

Figure 1.1 Rice ecosystems and its vulnerability to climate change. *Note:* Different rice ecology may face different stresses with rain-fed areas experiencing most of the abiotic stresses. However, biotic stresses may be more under irrigated ecologies. (*See color plate section for color representation of this figure*).

Baker, and Boot 1996). However, the benefit of CO_2 fertilization may not necessarily be reflected in improved growth or yield in all the crop plants. Because of their direct interaction with CO_2 for photosynthesis, studies show that C_3 plants such as wheat and rice could benefit from higher levels of CO_2 concentrations, as against C_4 plants such as maize and sugarcane. Photosynthesis in C_4 system is CO_2 concentration independent (Degener 2015) and therefore may not show apparent benefit of carbon fertilization. In China, projections show that without carbon fertilization, rice yield can decline under rain-fed ecology by 13% by 2020, and up to 14% by 2050 and 29% by 2080s. The analogous decline under the irrigated system is estimated to be 9%, 12%, and 17% for the same period (Xiong et al. 2009; Wang, Huang, and Rozelle 2010). These projections are valid for the A2 scenario of climate change under the standardized reference emission scenarios (SRES). The A2 scenario assumes a heterogeneous world with steady population growth, regionalized economies with self-reliant governance, and slow fragmented technological growth (Nakićenović 2000). Globally, Wang et al. (2009) estimate that without the carbon fertilization effect, rice production would decline by 5%, losing about 4900 thousand tons by the year 2030. Under the same scenario, China's projections for rice production with carbon fertilization would be an increase of 4% and 8% by 2080, correspondingly under rainfed and irrigated systems.

India, as the second-largest rice producer, faces a production challenge to meet the future demand in view of climate change. The temperature rise in India is projected to be between 1.2 and 2.0 °C by 2030, which will increase to 3.3–4.8 °C by 2080 (Chaturvedi et al. 2012). The corresponding increase in average precipitation rate would be 4–5% by 2030 and 6–14% by the 2080s. India may face varying degrees of climate change impacts, owing to its diversified agro-climatic zones, unique production systems, and inter-seasonal variability. With more than 95% of the crop under irrigated ecology that demands about 1200–2500 mm of water to complete a production cycle, rice would be the major crop seriously affected by climate change in India (Palanisami et al. 2019). Further, different regions in India may face different degrees of climate change impacts on local agriculture. According to Wassmann et al. (2009), Punjab and Haryana would experience high temperatures resulting in high grain sterility, together with groundwater depletion, soil organic matter degradation, soil salinization and increased pests and disease incidence. Rain during the maturity phase would be a problem in Uttar Pradesh and Bihar, while Jharkhand and West Bengal may face rains, storms, and flooding, together with drought in certain pockets. Another additional problem would be a shift in pest and disease dynamics, which will attain endemic proportions in certain areas. With the anticipated shortage of irrigation water in key production zones, India's rice production is likely to brace for an intermittent deficit of inputs, price rises, and power scarcity. It is estimated that, in India, the climate change-related production deficit in rice would be about 41% for every rise in temperature by 4 °C (Geethalakshmi et al. 2011). A sector-wise analysis shows a declining trend of rice production in the Indo-Gangetic plains (Aggarwal 2008), and across 16 major rice producing states (Gupta, Sen, and Srinivasan 2012).

In Indonesia, climate change related weather events were identified as the causal factors for the delayed rainfall and reduced area under rice planting during the main rice growing seasons. A risk assessment on the potential impact of El Niño events by 2050 by Naylor et al. (2007), has shown that likelihood of a delayed onset of rain for about 30 days would increase by 9–18% from that of the base year, 2007. With an increase in rains coinciding late cropping season, accompanied by a substantial decrease in rains during the ensuing dry season, climate change would jeopardize cropping pattern of the region in the long run, with flash floods, submergence and drought acting as major adversaries. From the rest of the world, the climate change impact in Indonesia assumes a newer dimension with the threat of sea level rise. If the current level of climate shift continues, it is forecasted that a significant part of the island nation would face the threat of submergence by 2050, particularly around Northern Jakarta. It is reported that some parts of Jakarta are sinking at an alarming rate of 200 mm yr^{-1} (Deltaresw 2013). Rice crops around this region face serious problems of seawater ingression and inundation. The temperature-rise in Indonesia, with climate change, is estimated to be approximately 0.8 °C by 2030 (Oktaviani et al. 2011).

Bangladesh is one of the countries with world's highest per capita daily rice consumption of about 477 g/person in rural areas and about 389 g/person in urban areas (Hossain and Deb 2010). If rice production is impacted significantly, climate change can seriously threaten food security in countries like Bangladesh. Upraising this concern, erratic monsoon has been seen playing havoc in recent times in Bangladesh, seriously affecting rice cultivation due to drought and floods. To sustain the rice production in the future, Bangladesh would require a supply of varieties that are tolerant to drought, submergence, and salinity together with biotic stress tolerance. In countries like Bangladesh, where small rice holdings dominate, countering climatic vagaries can be a major challenge in the future. Small rice holdings of Asia practice subsistence farming, which by itself is constrained with input supplies and suffers greatly under extreme weather, directly threatening the food security of the nation.

Similar weather scenarios are happening in and around neighbouring rice-growing countries too, making the climate change a global challenge. In Japan, extreme summer temperature during 2010 damaged crop quality significantly (Hasegawa et al. 2013). Rice quality is of prime importance to Japanese consumers, as the kernels of *japonica* rice are cooked to preserve its soft and glutinous texture. In the Philippines, estimates show that climate change–associated costs in agriculture may reach more than USD 511 million by 2050 per year (Dikitanan et al. 2017). There was a temperature increase that ranged between 0.10 and 0.18 °C per decade reported from neighbouring Thailand for the past 50 years. But the rise in the average surface air temperature during the past 35 years has been 1 °C. There was an increased incidence of drought recorded in Thailand during this period, which showed little to no deviation in the rainfall pattern.

Prospects of Molecular Breeding for Climate-Resilient Rice

Breeding crop cultivars reinforced with climate resilience is one of the sustainable solutions for future agriculture. Cultivar development is a time-consuming process, both as an art and science, but challenges such as climate change require more precise and rapid scientific responses such as molecular breeding. The pace at which climatic vagaries occur needs immediate attention, and in recent times, the frequency of occurrence of adverse events has increased and become increasingly unpredictable. Therefore, accelerated breeding strategies are to be developed, to bring out a spectrum of resilient cultivars adapted to different scenarios, within a short period. Molecular breeding is an accelerated breeding strategy, that uses DNA-based tools in the breeding process. Since genome-level information is available for most of the primary food crops, it is easier to identify allelic variants at different genetic loci that can define the trait expression. The rice genome, which is 430 Mb long (Eckardt 2000) and harbours about 50,000 genes (Goff, Ricke, and Lan 2002), provides enormous opportunities for molecular breeding. For instance, a single gene controlling the plant height, *sd1*, could spur the green revolution in rice. Likewise, there can be 50,000 opportunities the genes and their allelic variation can offer in providing a multitude of solutions to combat the ill effects of climate change. Although the above prospect is little exaggerated, it illustrates the wider scope of precision breeding in rice. Precision breeding can be done either with one or few genes (mono- or oligo-genic) or a conglomeration of several genes (polygenic) to realize a particular trait expression. This knowledge can be translated into detectable genetic variations (markers) that can be used in the selection process. MAS for improving crop performance, particularly against stress factors, have been successfully utilized in many crops including cereals, millets, oilseeds, pulses, and many others (Collard and Mackill 2008). As a prelude to this process, the understanding of crop diversity at the gene level is critical to identify sources of desirable alleles, and also to design suitable molecular markers to aid in the selection process (Muthamilarasan, Theriappan, and Prasad 2013; Muthamilarasan et al. 2014). The most common approach in molecular breeding is to improve existing elite cultivars by imparting one or more desirable traits, for which the cultivar is deficient. Many such undesirable traits get accumulated during cultivar development, either due to their linkage with desirable target traits or by the lack of selection pressure. The most popular approach for trait introgression in molecular breeding, the MABB, has been successfully demonstrated in rice (Singh et al. 2011, 2019) and several other crops, particularly for the development of cultivars resistant/tolerant to biotic and abiotic stresses. Further, MABB is simple and precise and has shown great promise in the fight against the changing climate.

Molecular breeding is a continuous process and therefore should diversify to encompass several other ancillary techniques to achieve accelerated crop improvement. These include morphological and physiological characterization of stress tolerance, systemic resistance, and induced tolerance. This would help in assembling different sources for physiological variants that may bestow specific adaptation potential under a particular scenario. Genetic mapping of various quantitative and qualitative traits have helped immensely in simultaneous targeting of various genomic locations for targeted breeding using MABB. These QTLs/genes are tagged by a large number of molecular markers such as single nucleotide polymorphism (SNP), simple sequence repeats (SSR), and insertion-deletions (InDel), which would help in quick and robust identification (Subbaiyan et al. 2012). MAS enables rapid identification of selection targets from a large breeding population even before the trait is expressed. Further, the selection process does not go biased by the environmental influence that may alter the trait expression leading to ineffective selection. Additionally, MAS can help in reducing the population size within a short period, saving considerable amounts of money and labour.

To aid quick retrieval of genome wide information of already identified genes/ QTLs and markers there are several public databases made available. These searchable databases are available in the internet environment and are accessible to everyone. These databases aid in quick retrieval of information, as well as provide details for further use in breeding programmes. Most popular rice genome based databases in use are Gramene (www.gramene.org), Oryzabase (shigen.nig.ac.jp/rice/oryzabase), Phytozome (phytozome.jgi.doe.gov/pz/portal.html), Q-TARO (qtaro.abr.affrc.go.jp), Rice genome annotation project (rice.plantbiology.msu.edu), OryGenes (orygenesdb.cirad.fr), OsGDB (www.plantgdb.org/OsGDB), RAP-DB (rapdb.dna.affrc.go.jp), NCBI Genome Data Viewer (www.ncbi.nlm

.nih.gov/genome/gdv/?org=oryza-sativa-japonica-group), SNP-SEEK (snp-seek.irri.org), OMAP (www.omap.org), QTLtoGene (ibi.zju.edu.cn/software/qtltogene), KEGG (www.kegg.jp), ROOTomics (rootomics.dna.affrc.go.jp/en), and funRiceGenes (funricegenes.github.io). Besides, there are several other databases available, wherein rice genome information is incorporated. The assemblage of several techniques, including computational and prediction tools, is currently transforming molecular breeding to genomics-assisted breeding (GAB). GAB includes techniques such as rapid generation turnover, image-based phenotyping, and selection based on genome estimated breeding values (Singh et al. 2019). Further, complicated population improvement methodologies couples with MAS are being practised such as marker-assisted recurrent selection (MARS) and GS. Genomic-assisted prediction of QTLs, markers and mapping using multi-parental populations and identification of genome-wide associations are also now integrated into GAB. In the event of climate change, these techniques promise to protect our future food security.

Molecular Breeding in Rice

Rice is endowed with immense natural genetic diversity, which makes it naturally adapted to environments ranging from hills to deep water, and from temperate to tropical conditions. This offers opportunities to find suitable stress tolerant variants that can tolerate several stresses either singly or in combination, including the abiotic and biotic factors associated with climate change.

Breeding for Abiotic Stress Tolerance

Climate change disturbs the natural equilibrium of rice production systems worldwide. Most of these disturbing factors are abiotic; hence breeding interventions for climate resilience are primarily focused towards imparting tolerance to abiotic stresses such as against drought, high temperature, submergence, salinity, low nutrients, low temperature, and high radiation. To date, there have been several genes/ QTLs mapped for tolerance to each of these stress factors on the rice genome (Table 1.1).

Drought

Under the current climate change scenario, drought and heat co-exist in the cropping systems. Although drought implies a water-deficient situation, it often results from high temperature, causing high evapotranspiration and ultimately water scarcity. Drought is recognized as the most important abiotic stress, limiting rice yields in rainfed ecology particularly under upland conditions (Vinod et al. 2019). Naturally adapted to water rich environments, rice is highly sensitive to drought because of its small fibrous root system, thin cuticle, and swift stomata closure. Therefore, the varieties with improved drought tolerance should fundamentally possess mechanisms to enhance the use of available water. In crops, drought tolerance is a complex character resulting from the crosstalk of several traits spread across different developmental stages (Jakab et al. 2005).

Severe drought can be extremely detrimental to rice and can cause significant damages at different growth stages. While it may cause stunting and poor tillering during the vegetative phase, during the reproductive stage, drought may affect spikelet fertility and

Table 1.1 Number of genes/QTLs mapped for various abiotic stresses affecting rice.

	Chromosomes												
Abiotic stress[a)]	**I**	**II**	**III**	**IV**	**V**	**VI**	**VII**	**VIII**	**IX**	**X**	**XI**	**XII**	**Total**
Drought	134	94	119	76	72	67	62	61	51	36	49	53	874
Salinity	141	98	130	67	78	68	78	58	43	29	49	37	876
Submergence	4	0	2	0	2	0	3	2	3	2	3	2	23
Heat	42	20	29	13	25	11	13	10	13	11	14	7	208
Cold	86	53	69	45	38	40	33	24	28	24	28	23	491
Al toxicity	1	0	0	0	1	0	0	1	1	0	1	0	5
P deficiency	0	0	0	0	0	1	0	0	0	0	0	3	4
Deep water	0	0	0	1	0	1	0	0	0	0	0	0	2
Oxidative stress	50	37	56	37	23	38	43	12	14	15	11	17	353
N deficiency	15	20	13	12	8	10	5	10	5	3	4	8	113

Most of these stresses may get exacerbated under climate change.
a) Al, aluminum; P, phosphorus; N, nitrogen.
Source: Oryzabase (https://shigen.nig.ac.jp/rice/oryzabase).

cause embryo abortion. Depending upon the severity, drought can be either terminal or intermittent. Terminal drought is the most severe, and is often lethal, while intermittent droughts are rather transient and nonlethal (Polania et al. 2017). Drought adaptation in rice occurs through several physiological adjustments, such as osmotic adaptation, cellular elasticity and protoplasmic resistance (Morgan 1984). During osmotic adjustment, several osmolytes such as proline, sucrose, glycine betaine, and other solutes are accumulated to reduce the stress effect (Liang et al. 2013). Therefore, the genes that lead to the metabolism of these solutes can be employed as markers for drought-tolerance screening (Fahramand et al. 2014). There are at least 874 genes/QTLs directly associated with drought tolerance reported in rice and distributed across the whole genome. Besides, there are several other genes implicated indirectly to impart drought tolerance in rice, together with several QTLs mapped. Among these, nine meta QTLs associated with yield under drought, that has been utilized for drought tolerance improvement in rice (Vinod et al. 2019). Genomic resources towards drought tolerance in rice extend opportunities for molecular breeding for climate resilience.

Salinity

Salinity is a serious impediment for rice production worldwide, especially in coastal irrigated lands. Since salt injury causes remarkable yield reduction, the elite salt-tolerant genotypes can help in boosting rice production in salt-affected areas. Rice is sensitive to salt accumulation in both the seedling and the reproductive stages, genetics of which differ significantly (Fageria 1985). Therefore, a comprehensive salt endurance should have integration of both seedling and reproductive stage tolerances. Although there have been several studies to understand the genetics of salt tolerance in rice, attempts to map genes conferring tolerance has begun at the International Rice Research Institute in the

Philippines in the early 1990s. Two tolerant landraces, Pokkali and Nona Bokra, both originating from India, were used in the initial mapping experiments (Vinod et al. 2013). These attempts culminated in the mapping of *Saltol*, a QTL conferring seedling stage salt tolerance. Till today, *Saltol* remains as the first and the most prominent QTL mapped for salt tolerance in rice. Located on the short arm of chromosome 1 (Gregorio 1997), *Saltol* was initially mapped from Pokkali, using a recombinant inbred line population derived from the cross between IR24/Pokkali (Bonilla et al. 2002). Subsequently, two other QTLs have been mapped from Nona Bokra, *qSNC7* for shoot Na^+ and K^+ concentration on chromosome 7 and *qSKC1* for shoot K^+ concentration, mapped on chromosome 1. Subsequently, it is understood that *SKC1* (*OsHKT1;5*) gene is the major player for salt tolerance imparted by *Saltol* (Ren et al. 2005; Thomson et al. 2010).

The fact that *Saltol* governed only seedling-stage salt tolerance in rice has led to the search for reproductive-stage tolerance. Several mapping attempts had been carried out, however, no significant QTLs have been mapped so far. In a bulked RIL analysis derived from the sodicity-tolerant *indica* rice variety CSR11 using a 50k SNP chip identified 34 QTL regions (Tiwari et al. 2016). Similar to drought, several genes/QTLs related to salt tolerance have been reported in rice, providing great opportunities in future crop improvement (Table 1.1). Molecular breeding helps in introgression of salt-tolerance genes/ QTLs through MABB into elite salt sensitive varieties. So far, several attempts have been reported, specifically using the *Saltol* locus originally derived from Pokkali. Most of these studies used a common donor, FL478, an IR29/Pokkali derivative, because FL478 possessed better tolerance than Pokkali. All these methods used a MABB approach and several salt-tolerant lines have been generated (Babu et al. 2017; Singh et al. 2018). Recently, Rana et al. (2019) reported improvement in salt tolerance through efficient SNP MAS coupled with speed breeding, by mobilizing *hst1* gene from Kaijin into high-yielding Yukinko-mai.

Submergence

Although rice is semi-aquatic, most of its cultivars are not adapted to tolerate complete submergence even for a short period. Submergence occurs when a huge volume of water gets accumulated in the rice field, often following heavy rains. Submergence due to erratic weather can either be transient or long lasting. Transient flash floods may cause more mechanical damage than by inundation, however, standing water for a longer period can disturb the physiology of the plant. Submergence significantly affects activities such as water absorption, respiration, and photosynthesis, causing substantial crop loss. The threat of extensive crop loss looms over more than 16 million ha of rice in south and southeast Asia due to irregular distribution of rainfall owing to climate change (Dwiyanti and Yamada 2013). Capricious rainfall can result in periodic incidences of floods causing extensive damage to standing crops endangering rice production. Since most of the cultivated varieties are sensitive to submergence, crop loss can reach near total proportions after a prolonged inundation.

Submergence primarily affects the root system, particularly the gas diffusion process. Rhizosphere aeration through the oxygen diffusion from the aerial part is crucial for the rice plant growth (Colmer and Flowers 2008), in which root aerenchyma cells play a major role. In submerged conditions, oxygen diffusion is affected, thereby root aerenchyma gets enriched with soil borne toxins due to anaerobiosis. This will encourage the accumulation of reactive oxygen species (ROS) leading to disruption of metabolisms and systemic damage in the plants. Therefore, the defence mechanisms include lignification and suberization of peripheral cells that cover the aerenchyma to prevent toxification. Further, other adaptive mechanisms also operate, such as elongation of stems and leaves to facilitate improved aerial oxidation (Bailey-Serres et al. 2010)

Several loci have been mapped for submergence tolerance in rice, most prominent being the QTL, *Sub1* (Xu and Mackill 1996). Mapped from FR13A, a pure line selection from a landrace from India, Dhalputtia, *Sub1* QTL is located on chromosome 9. FR13A can withstand up to two weeks of complete submergence (Vergara and Mazaredo 1975; Xu et al. 2000). *Sub1* QTL contains three complementing ethylene responsive-like transcription factors, that can regulate stress response. In *Sub1* carriers, starch and sucrose metabolism is lowered by suppression of α-amylases and sucrose synthase enzymes to conserve energy (Fukao et al. 2006).

In cultivated rice, the prevalence of *Sub1* alleles are common in *indica* rice but rare in *japonica* types. However, their presence is more common among wild species such as *Oryza nivara* (Niroula et al. 2012) and *O. rufipogon*. For imparting submergence tolerance into popular cultivars, *Sub1* is the only QTL successfully used in molecular breeding so far (Lang, Tao, and Buu 2011; Neeraja et al. 2004). Several improved mega-varieties such as Swarna, Samba Mahsuri, CR1009, IR64, Tadukan, BR1, CO43 etc, have been released through marker assisted introgression of *Sub1* QTL (Kirill et al. 2017). Besides *Sub1*, there are other minor QTLs mapped for submergence tolerance rice, that were found distributed widely, except on chromosomes 2, 4, and 6.

High Temperature

Rice is highly sensitive to high-temperature stress, particularly during the terminal stage. Significant damage can result when drought is experienced at pre- and post-zygotic phases (Cao et al. 2008). During flowering, high temperature can result in poor exsertion, desiccation, and male sterility resulting in severe yield loss. Exposure to elevated heat during anthesis, even for a few hours, can result in several adverse changes such as restricted swelling of lodicules, decreased pollen viability, poor anther dehiscence, desiccation of floral structure, and embryo abortion (Matsui, Omasa, and Horie 2000). Likewise, prolonged heat exposure could lead to tapetal abortion (Endo et al. 2009) leading to reduced numbers for viable florets with a corresponding increase in the number of deformed spikelets (Cao et al. 2008). Heat stress interferes with iron uptake in sensitive genotypes, leading to reduced pollen viability and germination (Jagadish et al. 2010). Besides yield loss, reproductive-stage heat sensitivity in rice can result in severe grain quality deterioration, such as increased chalkiness and grain shrivelling. Corresponding with temperature increase, less prolamin accumulation is observed in grains together with increased chalkiness (Wada et al. 2019) as well as with reduced storage proteins.

Heat stress tolerance in rice is polygenically controlled. When compared to tolerance to drought, salt, and cold, the number of QTLs mapped for heat tolerance has been limited (Zhu et al. 2005). Currently, in Oryzabase, information on 209 genes/QTLs conferring tolerance to high temperature in rice have been assembled. This provides immense opportunities for molecular breeding against heat tolerance. However, unlike other major abiotic stresses, so far, cultivars with heat tolerance through systematic breeding have not been released for commercial cultivation.

Low Temperature

Tropical adaptation of rice, particularly of *indica* subtype, renders it largely sensitive to low temperature. Sensitivity to low temperature can result in stunted growth and crop loss. However, low-temperature adaptation becomes a trait of importance when rice is grown under higher altitude ecology such as hill rice, where a sudden drop in temperature is a routine feature. Additionally, wide temperature fluctuations are predicted to occur under climate change; particularly in elevated regions such as plateaus and montane valleys, where rice is traditionally grown but with low productivity. Hence it is prudent to infuse cold tolerance in rice, to cater regions prone to low temperature as well as to increase rice productivity in such regions by growing improved high yielding cultivars. Cold susceptibility in rice occur mainly due to oxidative stress, response to which can vary with different growth stages. Under low temperature, internal accumulation of reactive oxygen species (ROS) is induced, which drives disruption of cellular machinery. Excess accumulation of oxidative stress-related compounds such as malondialdehyde (MDA) alters membrane lipid composition. This implies that cold tolerance preferably requires mechanisms to scavenge ROS and other reactive species to counter the oxidative stress damages (Sato et al. 2011). Integration of cold tolerance into elite cultivars improves their adaptation potential, thereby helping them to maintain cellular metabolisms leading to elevated plasticity in the structural, catalytic, and enzymatic profile of the plants (Kubien et al. 2003). Similar to heat stress, several genes/QTLs have been mapped against low-temperature stress, out of which some were utilized for varietal improvement through MABB (Zhang et al. 2014; Hasan et al. 2015) and pyramiding (Shinada et al. 2013).

Breeding for Biotic Stress Tolerance

Biotic stresses increase when the dynamics of faunal equilibrium are disturbed. In natural ecosystems, ecosystem elements remain in a state of balance so that every member has an equal chance to survive and persist in the system. These elements maintain a network of interrelations, either symbiotic or parasitic, to maintain the equilibrium. However, when the whole ecosystem encounters stress, such as that due to climate change, different constituents respond differently upsetting the balance, which in turn counterbalance to maintain the equilibrium. During this process, biotic interactions get altered so that pests and pathogens that gain an advantage over their counterparts in invading their hosts, become prominent. The advantage can come from different pest-host interactions, such as pest-parasite, pest-host and pest-pest relations. Therefore, biotic stress related to climate change is often indirect or secondary. However, coupled with abiotic stresses they can impede crop production leading to total crop loss. Several pathogens and pests can be a potential threat in rice, such as BB (*Xanthomonas oryzae* pv. *oryzae*), blast (*Magnaporthe oryzae*), sheath blight (*Rhizoctonia solani*) and pests like brown planthopper (*Nilaparvata lugens*). There are several genes and QTLs governing resistance/tolerance for various biotic stresses reported in rice (Table 1.2), which offers possibilities for molecular breeding for cultivar development. Unlike abiotic stresses, biotic stress tolerance is often governed by the major gene(s), a feature that has been extensively used in varietal improvement. There are newer cultivars developed carrying a single resistance gene or pyramids of genes for one or more diseases.

Table 1.2 Number of genes/QTLs mapped for resistance to biotic stresses in rice.

	Chromosomes												
Biotic stresses	**I**	**II**	**III**	**IV**	**V**	**VI**	**VII**	**VIII**	**IX**	**X**	**XI**	**XII**	**Total**
Blast	10	13	0	6	3	23	1	8	9	1	27	23	124
Bacterial blight	1	0	2	6	1	2	3	2	1	0	27	1	46
Sheath blight	0	0	0	0	1	0	1	0	0	0	0	0	2
Brown leaf spot	0	1	1	0	0	2	0	0	0	0	0	0	4
Rice tungro virus	0	0	1	1	0	0	0	0	0	0	0	0	2
Brown plant hopper	8	9	14	24	14	14	2	4	3	1	7	11	111
White backed plant hopper	1	0	0	0	1	1	1	0	0	0	1	0	5
Green leafhopper	0	0	1	3	2	1	0	1	0	1	2	0	11
Gall midge	0	0	0	4	0	0	0	2	2	0	0	2	10
Stem borer	1	1	2	0	4	2	0	1	0	0	0	1	12
Nematode	5	4	5	3	2	6	1	0	0	1	4	1	32

Biotic stresses may be a secondary effect of abiotic stresses, that may result from different adaptive interactions of species.
Source: Oryzabase (https://shigen.nig.ac.jp/rice/oryzabase).

Diseases

Diseases occur when the environmental conditions are favourable for pathogen development, and when the host is weak in its defence. When an array of favourable factors to the pathogen integrate with factors that are unfavourable to the host, the disease manifests leading to crop loss (Pandey et al. 2017). Warm conditions and humidity are usually the most favourable factors for pathogen proliferation, while poor nutrition and oxidative stress render the host plants defenceless. For instance, the incidence of sheath blight caused by *Rhizoctonia solani*, a minor disease since the 1970s, has been reported to increase under warm humid conditions (Lee and Rush 1983). Similarly, elevated CO_2 in the environment can incite outbreak of leaf blast and sheath blight disease in rice (Kobayashi et al. 2006), which was later practically demonstrated with blast disease (Gória, Ghini, and Bettiol 2013). Rajput et al. (2017) found that sporulation of blast pathogen increased with increasing temperature, evidencing that warming due to climate change can abet blast incidence, especially under cold sub-tropical conditions (Luo et al. 1995; Prabhu, Silva, and Filippi 2008). Further, a temperature range of 18–32 °C with high to moderate humidity can favour blast development (Bevitori and Ghini 2014), although, peak spore liberation may be confined to a temperature below 30 °C with night humidity of about 100%.

Additionally, several environmental factors such as salinity, radiation, submergence, cold, and poor nutrition can cause oxidative stress in plants, making them systematically weak and vulnerable (Guo et al. 2011). Host vulnerability often provides opportunities for otherwise minor pathogens to gain virulence and invade into endemic proportions. For instance, bakanae disease caused by *Fusarium fujikuroi*, an erstwhile insignificant disease has become a major challenge in Asian countries recently (Haq et al. 2011), including India (Bashyal et al. 2016; Gupta et al. 2015). Similarly, widespread incidence of false smut disease caused by *Ustilaginoidea virens* has been reported (Tokpah et al. 2017), particularly from India in recent times. However, a reverse situation also can often be possible, rendering a pathogen weak by changing climate where host gains better defence making a major disease to become minor. Rice tungro disease, one of the dreaded viral disease caused by a complex of rice tungro bacilliform virus (RTBV), and rice tungro spherical virus (RTSV), has become prominently weaker in the recent times, showing lesser incidence worldwide. Mainly attributed to better management practices adopted lately, the reason for this decline could not be due to climate change, but the role of climatic influence cannot be completely ruled out. In a study with rice lines carrying the *Xa7* gene, it was found that the effectiveness of *Xa7* increases with elevated temperature, rendering the rice plants more resistant to bacterial blight pathogen (Webb et al. 2010). Predictions for two diseases, sheath blight and leaf blast, under climate change model, EPIRICE indicate that incidence of these diseases could be lower in South Korea by 2100 (Kim et al. 2015). In case the warming goes beyond the threshold of pathogen fitness, as in the case of *M. oryzae*, the global incidence of rice blast is poised to decline under climate change (Asibi, Chai, and Coulter 2019).

Molecular breeding for biotic stresses, particularly for major diseases such as bacterial blight and rice blast has been very successful (Singh et al. 2011). There are ten improved cultivars released for cultivation, of which six are improved for bacterial blight tolerance and carry genes such as *Xa5*, *xa13*, and *Xa21*, while four are improved with blast resistance, by incorporating genes such as *Pi1*, *Pi2*, *Pi9*, *Pi54*, and *Pita* (Singh et al. 2019). Also, several lines with various combinations of resistant genes are in the pipeline for commercial release in India (Sagar et al. 2020). Other than blight and blast, significant progress has been made in resistance-breeding against diseases such as bakanae and sheath blight. Considering the number of genes identified across the rice genome (Table 1.2), molecular breeding holds greater promise in developing cultivars with biotic-stress resistance in rice.

Pests

Like pathogens, insect pest dynamics of the rice crop are greatly influenced by temperature, humidity, and radiation fluctuations. Insects are highly adaptive and have several mechanisms for resilience such as high fecundity and short life cycles. Abnormal weather conditions exert tremendous pressure on insect populations, triggering adaptive mechanisms like delayed metamorphosis, survival instinct, and increased reproduction rate (Khaliq et al. 2014). In the rice ecosystem, there are several insect pests and beneficial insects live in harmony when conditions are conducive. Climatic stresses affect both the populations, resulting in disruption of population balance, leading to pest outbreaks in epidemic proportions. Among the several pests of rice, the brown planthopper (BPH, *Nilaparvata lugens*) assumes importance for its tremendous adaptive plasticity and for the greater damage it can cause. Recent years have witnessed BPH outbreaks in several rice growing countries such as India, Bangladesh, Malaysia, and China (Ali et al. 2014). In Bangladesh, a ten-year study on the pest dynamics of yellow stem borer (YSB), BPH, green leafhopper (GLH) and their natural enemies indicated that abundance and growth rate of pests increase along with a rise in maximum temperature of about 0.5–1 °C (Ahmed et al. 2013). Such observations lead to conclude that global warming could be a predisposing factor for recent planthopper outbreaks (Ali et al. 2014). Investigations on BPH dynamics in India vis-à-vis elevated

CO_2 conclude that different populations showed varying responses when high CO_2 was coupled with elevated temperature, pointing that some natural populations could be having better adaptability to warming conditions (Pandi et al. 2018; Vailla, Muthusamy, and Konijeti 2019). In the case of YSB, weather parameters such as minimum temperature, humidity, and rainfall are identified as factors influencing population buildup in several studies, implications of which point to influence by climate change (Bhatnagar and Saxena 1999; Sharma, Atsedewoin, and Fanta 2011; Prasannakumar, Chander, and Kumar 2015). During the *Kharif* season of 2019, India witnessed an unprecedented outbreak of gall midge (GM) infestation in rice (Seni and Naik 2019). Naturally, population dynamics of GM is controlled by the density and distribution of natural enemies such as *Paederus fuscipes* as the predator and *Platygaster oryzae* and *Neanastatus grallarius* as parasitoids (Shrivastava et al. 1987). Any imbalance in the dynamics between GM and its natural enemies can break open the equilibrium, resulting in an outbreak. Climate change has been implicated in altering the phenology of host and/or the parasite driving a deranged bioequilibrium (Møller et al. 2013). It would be interesting to study what caused the GM outbreak in 2019, a year in which several climate change related concerns have been reported (Vaughan 2019). Further, there are also several earlier reports indicating the possible influence of climate change on pest dynamics in rice, some of which forewarns possibilities of minor pests such as mealybugs gaining prominence in the long run (Rashid et al. 2017). In yet another example, it is found that the natural population of parasitic nematodes dwindles in rice ecosystems. It is well known that changes in temperature and water regimes can affect reproductive cycles of nematodes, affecting their load on host plants (Colagiero and Ciancio 2011; Somasekhar and Prasad 2012). De Waele and Elsen (2007) note a dramatic increase in the population of root knot nematode (*Meloidogyne graminicola*), the most important parasitic nematode in rice, in response to climate change–related factors in the rice ecosystem. Supporting this, root knot nematode infestation is now seen widespread in all rice ecologies, significantly diversifying from its original confinement, the upland rice (Prasad and Somasekhar 2009; Pankaj, Sharma, and Prasad 2010).

Most of the breeding work to increase pest resistance in rice has gone into developing BPH-resistant cultivars. Beginning in the 1970s, the International Rice Research Institute has developed several cultivars carrying the BPH-resistant genes *Bph1*, *bph2*, *Bph3*, and *bph4* by genetically improving some of the popular varieties such as IR26, IR36, IR50, and IR72, (Jena and Kim 2010). Although the initial works were done through conventional breeding approaches, MABB was used later to transfer *Bph18* into a *japonica* variety, Junambyeo (Suh et al. 2011). Development of several pest resistant hybrids in the background of the popular Chinese hybrid, 9311 carrying several genes/QTLs either alone (Hu, Xiao, and He 2016) or pyramided (Hu et al. 2013) have been reported in China. In a subsequent report, the successful development of an improved CMS restorer KR022 has been described by pyramiding of multiple insect resistance for BPH (*Bph14* and *Bph15*), YSB and leaf folder (*Cry1C*) into a Mianhui 725 derived CMS restorer R022 (Wan et al. 2014).

Weed Competitiveness

Yet another biotic stress that rice crops can experience under changing climate could be from weed invasion. Although all plant species experience the effects of varying environments, the response varies between crop species predisposing their vulnerability to weed invasion. Weed invasion occurs when the level of competition for resources under stress varies significantly between crops and weeds, and weeds get the adaptation advantage. If the crops succumb to weed pressure, significant crop loss can result. Among the various effects of climate change, elevated CO_2 could be of particular importance in driving competitiveness (Korres et al. 2016). CO_2 response varies significantly between C_3 and C_4 plants, where C_3 plants show clear adaptation advantage over C_4 plants under elevated CO_2. In the rice ecosystem, most of the notorious weeds are C_4 plants. Under increased CO_2, being a C_3 plant, rice can exploit the advantage over C_4 weeds to sustain crop production. It is observed that, under high CO_2 conditions, rice produces more tillers but C_4 weeds can produce more biomass and broader leaves, stifling competition (Korres et al. 2017). Notwithstanding, competition between C_3 plants could be the major factor that needs attention under high CO_2 environments. For instance, weedy rice shows a clear advantage over cultivars under increased CO_2 (Ziska, Tomecek, and Gealy 2010), and may smother C_4 weeds to become a major competitor to cultivated rice (Rodenburg and Meinke 2010). Likewise, the weeds adapted with deeper root systems show propensity under receding moisture conditions and can take over shallow-rooted cultivars in uplands, and can invade lowlands as well (Korres et al. 2017).

Although weed management strategies could be enforced to abate crop-weed competition, under stressful environments, this would increase production costs. To ensure sufficient crop production, therefore, it would be prudent to use competitive cultivars in environments where weed menace is a challenge. Unlike other major threats of rice production, crop improvement for weed competitiveness has not received adequate attention so far. Being listed as one of the future challenges of climate change, weed competitiveness needs to be included as a component for the development of climate resilient cultivars. Breeding for weed-competitive rice cultivars may have to look for phenotypes such as high tillering with a deeper root system and possessing slightly broader leaves to ensure adequate competition against the invading non-target species in the ecosystem.

Conclusion

Changing climate all over the world may compel significant changes in the cropping pattern in the future. This is particularly important with rice, because of its extensive cultivation and the huge dependency on it by more than half of the global human population. Climate change scenario, in the long run, may compel shifting rice cultivation to newer locations. In India, receding water levels in the states of Punjab and Haryana are already affecting the rice-wheat cropping system. Although growing short duration rice varieties can be advocated in these regions currently, eventually, a shift to alternate crops may be required. Likewise, cropping systems and cropping pattern also may require revamping to suit changing weather patterns. In rice, water scarcity may compel a shift into crop management changes such as direct seeding, alternate wet and drying, aerobic rice and extension to upland cultivation. Under catastrophic scenarios, it may be required to make cropping shifts within a short period. To thwart such situations, we require a large array of cultivars having the resilience to multiple stresses. With the growing concern of challenges ahead worldwide, accelerated breeding needs to be adopted to sustain food security. The integration of genomics tools with molecular breeding and rapid generation advancement techniques such as speed breeding would help in achieving greater sustainability for rice production for the future world under changing climate.

References

Aggarwal, P.K. (2008). Global climate change and Indian agriculture: impacts, adaptation and mitigation. *Indian J. Agric. Sci.* 78: 911–919.

Ahmed, N., Holmquist, B., Nasrin, S., et al. (2013). Impact of climate change on rice insect pests and their natural enemies. International Conference on Climate Change Impact and Adaptation (I3CIA-2013). Gazipur, Bangladesh: Center for Climate Change and Sustainability Research. 15–17 November 2013. 245–255.

Ali, M.P., Huang, D., Nachman, G. et al. (2014). Will climate change affect outbreak patterns of planthoppers in Bangladesh? *PLoS One* 9: e91678. https://doi.org/10.1371/journal.pone.0091678.

Allen, L.H., Baker, J.T., and Boot, K.J. (1996). The CO_2 fertilization effect: higher carbohydrate production and retention as biomass and seed yield. In: *Global Climate Change and Agricultural Production. Direct and Indirect Effects of Changing Hydrological, Pedological and Plant Physiological Processes* (eds. F. Bazzaz and W. Sombroek), 75–116. Rome, Italy: Food and Agriculture Organization of the United Nations.

Asibi, A.E., Chai, Q., and Coulter, J.A. (2019). Rice blast: a disease with implications for global food security. *Agronomy* 9: 451. https://doi.org/10.3390/agronomy9080451.

Babu, N.N., Krishnan, S.G., Vinod, K.K. et al. (2017). Marker aided incorporation of *Saltol*, a major QTL associated with seedling stage salt tolerance, into *Oryza sativa* 'Pusa Basmati 1121'. *Front. Plant Sci.* 8: 41. https://doi.org/10.3389/fpls.2017.00041.

Bailey-Serres, J., Fukao, T., Ronald, P. et al. (2010). Submergence tolerant rice: SUB1's journey from landraces to modern cultivars. *Rice* 3: 138–147. https://doi.org/10.1007/s12284-010-9048-5.

Bashyal, B.M., Aggarwal, R., Sharma, S. et al. (2016). Occurrence, identification and pathogenicity of *Fusarium* species associated with bakanae disease of basmati rice in India. *Eur. J. Plant Pathol.* 144 (2): 457–466. https://doi.org/10.1007/s10658-015-0783-8.

Bashyal, B.M., Aggarwal, R., Rawat, K. et al. (2020). Genetic diversity and population structure of *Fusarium fujikuroi* causing bakanae, an emerging disease of rice in India. *Ind. J. Exp. Biol.* 58: 45–52.

Bevitori, R. and Ghini, R. (2014). Rice blast disease in climate change times. *J. Rice Res.* 3: 1. https://doi.org/10.4172/2375-4338.1000e111.

Bhatnagar, A. and Saxena, R.R. (1999). Environmental correlates of population buildup of rice insect pests through light trap catches. *Oryza* 36: 241–245.

Bonilla, P., Dvorak, J., Mackill, D.J. et al. (2002). RFLP and SSLP mapping of salinity tolerance genes in chromosome 1 of rice (*Oryza sativa* L.) using recombinant inbred lines. *Philipp. Agric. Sci.* 85: 68–76.

Cao, Y.Y., Duan, H., Yang, L.N. et al. (2008). Effect of heat stress during meiosis on grain yield of rice cultivars differing in heat tolerance and its physiological mechanism. *Acta Agron. Sin.* 34: 2134–2142. https://doi.org/10.1016/S1875-2780(09)60022-5.

Chaturvedi, R.K., Joshi, J., Jayaraman, M. et al. (2012). Multi-model climate change projections for India under representative concentration pathways. *Curr. Sci.* 103: 791–802.

Coffman, W.R., Olufowote, J., Grau, P., and Villareal, R. (1992). A new approach to rice improvement in Latin America. In: *Rice in Latin America: Improvement, Management, and Marketing* (ed. F.E. Cuevas-Pérez), 31–42. Cali, Columbia: International Center for Tropical Agriculture.

Colagiero, M. and Ciancio, A. (2011). Climate changes and nematodes: expected effects and perspectives for plant protection. *J. Zool.* 94: 113–118.

Collard, B.C.Y. and Mackill, D.J. (2008). Marker-assisted selection: an approach for precision plant breeding in the twenty-first century. *Philos. Trans. R. Soc. Lond., B, Biol. Sci.* 363: 557–572. https://doi.org/10.1098/rstb.2007.2170.

Colmer, T.D. and Flowers, T.J. (2008). Flooding tolerance in halophytes. *New Phytol.* 179: 964–974. https://doi.org/10.1111/j.1469-8137.2008.02483.x.

De Waele, D. and Elsen, A. (2007). Challenges in tropical plant nematology. *Annu. Rev. Phytopathol.* 45: 457–485. https://doi.org/10.1146/annurev.phyto.45.062806.094438.

Degener, J.F. (2015). Atmospheric CO_2 fertilization effects on biomass yields of 10 crops in northern Germany. *Front. Environ. Sci.* 3: 48. https://doi.org/10.3389/fenvs.2015.00048.

Deltaresw (2013). *Sinking Cities: An Integrative Approach towards Solution*. Utrecht, The Netherlands: Deltares- Taskforce Subsidence.

Dikitanan, R., Grosjean, G., Nowak, A., and Leyte, J. (2017). *Climate-resilient agriculture in Philippines*. Manila: International Center for Tropical Agriculture (CIAT), Department of Agriculture – Adaptation and Mitigation Initiative in Agriculture. Government of the Philippines.

Ding, Y., Ren, G., Zhao, Z. et al. (2007). Detection, causes and projection of climate change over China: an overview of recent progress. *Adv. Atmos. Sci.* 24: 954. https://doi.org/10.1007/s00376-007-0954-4.

Dwiyanti, M.S. and Yamada, T. (2013). Molecular mapping and breeding for genes/QTLs related to climate change. In: *Genomics and breeding for climate-resilient Crops*, vol. 1 (ed. C. Kole), 179–212. Berlin; Heidelberg: Springer-Verlag https://doi.org/10.1007/978-3-642-37045-8_4.

Eckardt, N.A. (2000). Sequencing the rice genome. *Plant Cell* 12: 2011–2018. https://doi.org/10.1105/tpc.12.11.2011.

Endo, M., Tsuchiya, T., Hamada, K. et al. (2009). High temperatures cause male sterility in rice plants with transcriptional alterations during pollen development. *Plant Cell Physiol.* 50: 1911–1922. https://doi.org/10.1093/pcp/pcp135.

Fageria, N.K. (1985). Salt tolerance of rice cultivars. *Plant and Soil* 88: 237–243. https://doi.org/10.1007/BF02182450.

Fahad, S., Ihsan, M.Z., Khaliq, A. et al. (2018). Consequences of high temperature under changing climate optima for rice pollen characteristics-concepts and perspectives. *Arch. Agron. Soil Sci.* 64: 1473–1488. https://doi.org/10.1080/03650340.2018.1443213.

Fahramand, M., Mahmoody, M., Keykha, A. et al. (2014). Influence of abiotic stress on proline, photosynthetic enzymes and growth. *Int. Res. J. Appl. Basic Sci.* 8: 257–265.

Fukao, T., Xu, K., Ronald, P.C., and Bailey-Serres, J. (2006). A variable cluster of ethylene response factor-like genes regulates metabolic and development acclimation responses to submergence in rice. *Plant Cell* 18: 2021–2034. https://doi.org/10.1105/tpc.106.043000.

Geethalakshmi, V., Lakshmanan, A., Rajalakshmi, D. et al. (2011). Climate change impact assessment and adaptation strategies to sustain rice production in Cauvery Basin of Tamil Nadu. *Curr. Sci.* 101: 342–347.

Goff, S.A., Ricke, D., and Lan, T.H. (2002). A draft sequence of the rice genome (*Oryza sativa* L. ssp. *japonica*). *Science* 296: 92–100. https://doi.org/10.1126/science1068275.

Gória, M.M., Ghini, R., and Bettiol, W. (2013). Elevated atmospheric CO_2 concentration increases rice blast severity. *Trop. Plant Pathol.* 38: 253–257. https://doi.org/10.1590/S1982-56762013005000010.

Gregorio, G.B. (1997). Tagging salinity tolerance genes in rice using amplified fragment length polymorphism (AFLP). Ph.D. thesis. Los Baños: University of the Philippines.

Guiteras, R. (2009). The impact of climate change on Indian agriculture. Working Paper. Department of Economics. University of Maryland.

Guo, M., Chen, Y., Du, Y. et al. (2011). The bZIP transcription factor MoAP1 mediates the oxidative stress response and is critical for pathogenicity of the rice blast fungus *Magnaporthe oryzae*. *PLoS Pathog.* 7: e1001302. https://doi.org/10.1371/journal.ppat.1001302.

Gupta, S., Sen, P., and Srinivasan, S. (2012). *Impact of Climate Change on Indian Economy: Evidence from Food Grain Yields.* Delhi: Centre for Development Economics.

Gupta, A.K., Solanki, I.S., Bashyal, B.M. et al. (2015). Bakanae of rice—an emerging disease in Asia. *J. Anim. Plant Sci.* 25: 1499–1514.

Halder, D., Kheroar, S., Srivastava, R.K., and Panda, R.K. (2020). Assessment of future climate variability and potential adaptation strategies on yield of peanut and Kharif rice in eastern India. *Theor. Appl. Climatol.* https://doi.org/10.1007/s00704-020-03123-5.

Haq, M., Mia, M.A.T., Rabbi, M.F., and Ali, M.A. (2011). Incidence and severity of rice diseases and insect pests in relation to climate change. In: *Climate Change and Food Security in South Asia* (eds. R. Lal, M.V.K. Sivakumar, S.M.A. Faiz, et al.), 445–457. Springer Science+Business Media https://doi.org/10.1007/978-90-481-9516-9_27.

Hasan, M.M., Rafii, M.Y., Ismail, M.R. et al. (2015). Marker-assisted backcrossing: a useful method for rice improvement. *Biotechnol. Biotechnol. Equip.* 29: 237–254. https://doi.org/10.1080/13102818.2014995920.

Hasegawa, T., Sakai, H., Tokida, T. et al. (2013). Rice cultivar responses to elevated CO_2 at two free-air CO_2 enrichment (FACE) sites in Japan. *Funct. Plant Biol.* 40: 148–159. https://doi.org/10.1071/FP12357.

Hossain, M. and Deb, U. (2010). Volatility in rice prices and policy responses in Bangladesh. In: *Rice Crisis: Markets, Policies and Food Security* (ed. D. Dawe), 91–108. London: EarthScan.

Hu, J., Cheng, M.X., Gao, G.J. et al. (2013). Pyramiding and evaluation of three dominant brown planthopper resistance genes in the elite *indica* rice 9311 and its hybrids. *Pest Manag. Sci.* 69: 802–808. https://doi.org/10.1002/ps.3437.

Hu, J., Xiao, C., and He, Y. (2016). Recent progress on the genetics and molecular breeding of brown planthopper resistance in rice. *Rice* 9: 30. https://doi.org/10.1186/s12284-016-0099-0.

IPCC (2019). *Climate Change and Land: An IPCC Special Report on Climate Change, Desertification, Land Degradation, Sustainable Land Management, Food Security, and Greenhouse Gas Fluxes in Terrestrial Ecosystems.* Geneva: Intergovernmental Panel on Climate Change.

Jagadish, S.V.K., Muthurajan, R., Oane, R. et al. (2010). Physiological and proteomic approaches to address heat tolerance during anthesis in rice (*Oryza sativa* L.). *J. Exp. Bot.* 61: 143–156. https://doi.org/10.1093/jxb/erp289.

Jakab, G., Ton, J., Flors, V. et al. (2005). Enhancing *Arabidopsis* salt and drought stress tolerance by chemical priming for its abscisic acid responses. *Plant Physiol.* 139: 267–274. https://doi.org/10.1104/pp.105.065698.

Jena, K.K. and Kim, S. (2010). Current status of brown plant hopper (BPH) resistance and genetics. *Rice* 3: 161–171. https://doi.org/10.1007/s12284-010-9050-y.

Khaliq, A., Javed, M., Sohail, M., and Sagheer, M. (2014). Environmental effects on insects and their population dynamics. *J. Entomol. Zool. Stud.* 2: 1–7.

Kim, K.H., Cho, J., Lee, Y.H., and Lee, W.S. (2015). Predicting potential epidemics of rice leaf blast and sheath blight in South Korea under the RCP 4.5 and RCP 8.5 climate change scenarios using a rice disease epidemiology model, EPIRICE. *Agr. Forest Meteorol.* 203: 191–207. https://doi.org/10.1016/j.agrformet.2015.01.011.

Kirill, V.A., Usatov, A.V., Andrey, V.A., and Kostylev, P. (2017). Validation of SSR markers associated with submergence tolerance in rice. *Am. J. Agric. Biol. Sci.* 11: 142–147. https://doi.org/10.3844/ajabssp.2016.142.147.

Kobayashi, T., Ishiguro, K., Nakajima, T. et al. (2006). Effects of elevated atmospheric CO_2 concentration on the infection of rice blast and sheath blight. *Phytopathology* 96: 425–431. https://doi.org/10.1094/PHYTO-96-0425.

Korres, N.E., Norsworthy, J.K., Tehranchian, P. et al. (2016). Cultivars to face climate change effects on crops and weeds: a review. *Agron. Sustain. Dev.* 36: 12. https://doi.org/10.1007/s13593-016-0350-5.

Korres, N.E., Varanasi, V.K., Gitsopoulos, T.K., and Bararpour, M.T. (2017). Climate change effects on rice, weeds and weed management in Asian-Pacific region. In: *Weed Management in Rice in the Asian-Pacific Region* (eds. A.N. Rao and H. Matsumoto), 42–72. Asian-Pacific Weed Science Society.

Kubien, D.S., von Caemmerer, S., Furbank, R.T., and Sage, R.F. (2003). C_4 photosynthesis at low temperature: a study using transgenic plants with reduced amounts of Rubisco. *Plant Physiol.* 132: 1577–1585. https://doi.org/10.1104/pp.103.021246.

Lang, N.T., Tao, N.V., and Buu, B.C. (2011). Marker-assisted backcrossing (MAB) for rice submegence tolerance in Mekong delta. *Omonrice* 18: 11–21.

Lee, F.N. and Rush, M.C. (1983). Rice sheath blight: a major rice disease. *Plant Dis.* 67: 823–832. https://doi.org/10.1094/PD-67-829.

Liang, X., Zhang, L., Natarajan, S.K., and Becker, D.F. (2013). Proline mechanisms of stress survival. *Antioxid. Redox Signal.* 19: 998–1011. https://doi.org/10.1089/ars.2012.5074.

Long, S.P. (1991). Modification of the response of photosynthetic productivity to rising temperature by atmospheric CO_2 concentrations: has its importance been underestimated? *Plant Cell Environ.* 14: 729–739. https://doi.org/10.1111/j.1365-3040.1991.tb01439.x.

Lu, B.R. (1999). Taxonomy of the genus *Oryza* (Poaceae): historical prospective and current status. *Int. Rice Res. Newslett.* 24: 4–8.

Luo, Y., TeBeest, D.O., Teng, P.S., and Fabellar, N.G. (1995). Simulation studies on risk analysis of rice blast epidemics associated with global climate in several Asian countries. *J. Biogeogr.* 22: 673–678. https://doi.org/10.2307/2845969.

Maclean, J.L., Dawe, D.C., Hardy, B., and Hettel, G.P. (2002). *Rice Almanac: Source Book for the Most Important Economic Activity of Earth*. Wallingford, UK: CABI Publishing.

Matsui, T., Omasa, K., and Horie, T. (2000). High temperature at flowering inhibits swelling of pollen grains, a driving force for thecae dehiscence in rice (*Oryza sativa* L.). *Plant Prod. Sci.* 3: 430–434. https://doi.org/10.1626/pps.3.430.

Møller, A.P., Merino, S., Soler, J.J. et al. (2013). Assessing the effects of climate on host-parasite interactions: a comparative study of European birds and their parasites. *PloS one* 8: e82886. https://doi.org/10.1371/journal.pone.0082886.

Morgan, J.M. (1984). Osmoregulation and water stress in higher plants. *Annu. Rev. Plant. Physiol.* 35: 299–319. https://doi.org/10.1146/annurev.pp.35.060184.001503.

Muthamilarasan, M., Theriappan, P., and Prasad, M. (2013). Recent advances in crop genomics for ensuring food security. *Curr. Sci.* 105: 155–158.

Muthamilarasan, M., Suresh, B.V., Pandey, G. et al. (2014). Development of 5123 intron-length polymorphic markers for large-scale genotyping applications in foxtail millet. *DNA Res.* 21: 41–52. https://doi.org/10.1093/dnares/dst039.

Nakićenović, N. (2000). Greenhouse gas emissions scenarios. *technol. forecast. Soc. Change* 65: 149–166. https://doi.org/10.1016/s0040-1625(00)00094-9.

Naylor, R.L., Battisti, D.S., Vimont, D.J. et al. (2007). Assessing risks of climate variability and climate change for Indonesian rice agriculture. *Proc. Natl. Acad. Sci. U. S. A.* 104: 7752–7757. https://doi.org/10.1073/pnas.0701825104.

Neeraja, C., Maghirang-Rodriguez, R., Pamlpona, A. et al. (2004). A marker assisted backcross approach for developing submergence tolerance rice cultivars. *Theor. Appl. Genet.* 115: 767–776. https://doi.org/10.1007/s00122-007-0607-0.

Nelson, G.C., Rosegrant, M.W., Palazzo, A. et al. (2010). *Food Security, Farming, and Climate Change to 2050: Scenarios, Results, Policy Options*. Washington, DC: International Food Policy Research Institute https://doi.org/10.2499/9780896291867.

Niroula, R.K., Pucciariello, C., Ho, V.T. et al. (2012). SUB 1A-dependent and independent mechanism are involved in the flooding tolerance of wild rice species. *Plant J.* 72: 282–293. https://doi.org/10.1111/j.1365-313X.2012.05078.x.

Oktaviani, R., Amaliah, S., Ringler, C. et al. (2011). *The Impact of Global Climate Change on the Indonesian Economy*. Washington, DC: International Food Policy Research Institute.

Oteng, J.W. and Sant'Anna, R. (1999). Rice production in Africa: current situation and issues. *Int. Rice Comm. Newslett.* 48: 41–51.

Palanisami, K., Kakumanu, K.R., Nagothu, U.S., and Ranganathan, C.R. (2019). Climate change and agriculture in India. In: (eds. K. Palanisami et al.), 1–6. Singapore: Springer https://doi.org/10.1007/978-981-13-8363-2_1.

Pandey, P., Irulappan, V., Bagavathiannan, M.V., and Kumar, M.S. (2017). Impact of combined abiotic and biotic stresses on plant growth and avenues for crop improvement by exploiting physio-morphological traits. *Front. Plant Sci.* 8: 537. https://doi.org/10.3389/fpls.2017.00537.

Pandi, G.P.G., Chander, S., Singh, M.P. et al. (2018). Impact of elevated CO_2 and temperature on brown planthopper population in rice ecosystem. *Proc. Natl. Acad. Sci., India, Sect. B Biol. Sci.* 88: 57–64. https://doi.org/10.1007/s40011-016-0727-x.

Pankaj, Sharma, H.K., and Prasad, J.S. (2010). The rice root-knot nematode, *Meloidogyne graminicola*: an emerging problem in rice-wheat cropping system. *Ind. J. Nematol.* 40: 1–11.

Polania, J., Rao, I.M., Cajiao, C. et al. (2017). Shoot and root traits contribute to drought resistance in recombinant inbred lines of MD 23–24 × SEA 5 of common bean. *Front. Plant Sci.* 8: 296. https://doi.org/10.3389/fpls.2017.00296.

Prabhu, A.S., Silva, C.S., and Filippi, M.C. (2008). Impacto potencial das mudanças climáticas sobre as doenças de arroz no Brasil. In: *Mudanças Climáticas: Impactos Sobre Doenças de Plantas no Brasil* (eds. R. Ghini and E. Hamada), 141–158. Brasília: Embrapa Meio Ambiente: *Embrapa Informação Tecnológica*.

Prasad, J.S., and Somasekhar, N. (2009). Nematode pests of rice: diagnosis and management. Technical Bulletin No. 38. Hyderabad: Directorate of Rice Research.

Prasannakumar, N.R., Chander, S., and Kumar, L.V. (2015). Development of weather based rice yellow stem borer prediction model for the Cauvery command rice areas, Karnataka, India. *Cogent Food Agric.* 1: 995281. https://doi.org/10.1080/23311932.2014.995281.

Purugganan, M.D. (2010). The evolution of rice: molecular vignettes on its origins and spread. *Archaeol. Anthropol. Sci.* 2: 61–68. https://doi.org/10.1007/s12520-010-0026-z.

Rajput, L.S., Sharma, T., Madhusudhan, P., and Sinha, P. (2017). Effect of temperature on growth and sporulation of rice leaf blast pathogen (*Magnoporthe oryzae*). *Int. J. Curr. Microbiol. Appl. Sci.* 6: 6394–6401. https://doi.org/10.20546/ijcmas.2017.603.045.

Ramachandran, A., Praveen, D., Jaganathan, R. et al. (2017). Spatiotemporal analysis of projected impacts of climate change on the major C_3 and C_4 crop yield under representative concentration pathway 4.5: insight from the coasts of Tamil Nadu, South India. *Plos One* 12: e0180706. https://doi.org/10.1371/journal.pone.0180706.

Rana, M.M., Takamatsu, T., Baslam, M. et al. (2019). Salt tolerance improvement in rice through efficient SNP marker-assisted selection coupled with speed-breeding. *Int. J. Mol. Sci.* 20: 2585. https://doi.org/10.3390/ijms20102585.

Rashid, M.M., Jahan, M., Islam, K.S. et al. (2017). Ecological fitness of brown planthopper, *Nilaparvata lugens* (Stål), to rice nutrient management. *Ecol. Process.* 6: 15. https://doi.org/10.1186/s13717-017-0080-x.

Ren, Z.H., Gao, J.P., Li, L.G. et al. (2005). A rice quantitative trait locus for salt tolerance encodes a sodium transporter. *Nat. Genet.* 37: 1141–1146. https://doi.org/10.1038/ng1643.

Rodenburg, J., and Meinke, H. (2010). Adapting weed management in rice to changing climates. In: Proceedings of Second Africa Rice Congress: Innovation and partnerships to realize Africa's rice potential. Bamako, Mali, March 22–26. 5.6.1–5.6.6.

Sagar, V., Dhawan, G., Gopala Krishnan, S. et al. (2020). Marker assisted introgression of genes governing resistance to bacterial blight and blast diseases into an elite Basmati rice variety, "Pusa Basmati 1509.". *Euphytica* 216: 16. https://doi.org/10.1007/s10681-019-2549-4.

Sato, Y., Masuta, Y., Saito, K. et al. (2011). Enhanced chilling tolerance at the booting stage in rice by transgenic overexpression of the ascorbate peroxidase gene OsAPXa. *Plant Cell Rep.* 30: 399–406. https://doi.org/10.1007/s00299-010-0985-7.

Seni, A. and Naik, B.S. (2019). Evaluation of rice germplasm against rice gall midge, *Orseolia oryzae* (Wood-Mason). *J. Entomol. Zool. Stud.* 7: 516–520.

Sharma, M.K., Atsedewoin, A., and Fanta, S. (2011). Forewarning models of the insects of the paddy crop. *Int. J. Biodiver. Conserv.* 3: 367–375.

Shinada, H., Iwata, N., Sato, T., and Fujino, K. (2013). Genetical and morphological characterization of cold tolerance at fertilization stage in rice. *Breed. Sci.* 63: 197–204. https://doi.org/10.1270/jsbbs.63.197.

Shrivastava, S.K., Shukla, B.C., Kittur, S.U., and Agrawal, R.K. (1987). Seasonal incidence of rice gall midge and its natural enemies in Madhya Pradesh, India. *Trop. Pest Manag.* 33: 52–54. https://doi.org/10.1080/09670878709371115.

Singh, B.D. and Singh, A.K. (2015). *Marker Assisted Plant Breeding: Principles and Practices.* New Delhi: Springer.

Singh, A.K., Gopalakrishnan, S., Singh, V.P. et al. (2011). Marker assisted selection: a paradigm shift in Basmati breeding. *Ind. J. Genet. Plant Breed.* 71: 120–128.

Singh, V.K., Singh, B.D., Kumar, A. et al. (2018). Marker-assisted introgression of *Saltol* QTL enhances seedling stage salt tolerance in the rice variety "Pusa Basmati 1.". *Int. J. Genomics* 2018: 8319879. https://doi.org/10.1155/2018/8319879.

Singh, A.K., Gopala Krishnan, S., Vinod, K.K. et al. (2019). Precision breeding with genomic tools: a decade long journey of molecular breeding in rice. *Ind. J. Genet. Plant Breed.* 79 (1): 181–191. https://doi.org/10.31742/IJGPB.79S.1.7.

Somasekhar, N. and Prasad, J.S. (2012). Plant–nematode interactions: consequences of climate change. In: *Crop Stress and Its Management: Perspectives and Strategies* (eds. B. Venkateswarlu, A. Shanker, C. Shanker and M. Maheswari), 547–564. Dordrecht: Springer https://doi.org/10.1007/978-94-007-2220-0_17.

Subbaiyan, G.K., Waters, D.L.E., Katiyar, S.K. et al. (2012). Genome-wide DNA polymorphisms in elite *indica* rice inbreds discovered by whole-genome sequencing. *Plant Biotechnol. J.* 10: 623–634. https://doi.org/10.1111/j.1467-7652.2011.00676.x.

Suh, J.P., Yang, S.J., Jeung, J.U. et al. (2011). Development of elite breeding lines conferring *Bph18* gene-derived resistance to brown planthopper (BPH) by marker-assisted selection and genome-wide background analysis in japonica rice (*Oryza sativa* L.). *Field Crop Res.* 120: 215–222. https://doi.org/10.1016/j.fcr.2010.10.004.

Thomson, M.J., de Ocampo, M., Egdane, J. et al. (2010). Characterizing the *Saltol* quantitative trait locus for salinity tolerance in rice. *Rice* 3: 148–160. https://doi.org/10.1007/s12284-010-9053-8.

Tiwari, S., Krishnamurthy, S.L., Kumar, V. et al. (2016). Mapping QTLs for salt tolerance in rice (*Oryza sativa* L.) by bulked segregant analysis of recombinant inbred lines using 50K SNP chip. *PLoS One* 11: e0153610. https://doi.org/10.1371/journal.pone.0153610.

Tokpah, D.D., Kwoseh, C., Tokpah, E.S., and Kolleh, D. (2017). Rice false smut and its management in major rice growing areas in Ashanti region of Ghana. *Afr. J. Agric. Res.* 12: 3129–3136. https://doi.org/10.5897/AJAR2017.12508.

Vailla, S., Muthusamy, S., and Konijeti, C. (2019). Effects of elevated carbon dioxide and temperature on rice brown planthopper, *Nilaparvata lugens* (Stål) populations in India. *Curr. Sci.* 116: 988–996. https://doi.org/10.18520/cs/v116/i6/988-996.

Vaughan, A. (2019). The year the world woke up to climate change. *New Sci.* 244: 20–21. https://doi.org/10.1016/s0262-4079(19)32417-0.

Vergara, B.S. and Mazaredo, A. (1975). Screening for resistance to submergence under greenhouse conditions. In: *Proceedings International Seminar on Deepwater Rice*, 67–70. Dhaka, Bangladesh: Bangladesh Rice Research Institute.

Vinod, K.K., Krishnan, S.G., Babu, N.N., and Nagarajan, M. (2013). Improving salt tolerance in rice: looking beyond the conventional. In: *Salt Stress in Plants: Signalling Omics and Adaptations* (eds. P. Ahmad, M.M. Azooz and M.N.V. Prasad), 219–260. New York: Springer Science https://doi.org/10.1007/978-1-4614-6108-1_10.

Vinod, K.K., Krishnan, S.G., Thribhuvan, R., and Singh, A.K. (2019). Genetics of drought tolerance, mapping QTLs, candidate genes and their utilization in rice improvement. In: *Genomics Assisted Breeding of Crops for Abiotic Stress Tolerance*, vol. II (eds. V.R. Rajpal, D. Sehgal, A. Kumar and S.N. Raina), 145–186. Cham, Switzerland: Springer Nature https://doi.org/10.1007/978-3-319-99573-1_9.

Wada, H., Hatakeyama, Y., Onda, Y. et al. (2019). Multiple strategies for heat adaptation to prevent chalkiness in the rice endosperm. *J. Exp. Bot.* 70: 1299–1311. https://doi.org/10.1093/jxb/ery427.

Wan, B.L., Zha, Z.P., Li, J.B. et al. (2014). Development of elite rice restorer lines in the genetic background of R022 possessing tolerance to brown planthopper, stem borer, leaf folder and herbicide through marker-assisted breeding. *Euphytica* 195: 129–142. https://doi.org/10.1007/s10681-013-0988-x.

Wang, J., Mendelsohn, R., Dinar, A. et al. (2009). The impact of climate change on China's agriculture. *Agr. Econ.* 40: 323–337. https://doi.org/10.1111/j.1574-0862.2009.00379.

Wang, J., Huang, J., and Rozelle, S. (2010). *Climate Change and China's Agricultural Sector: An Overview of Impacts, Adaptation and Mitigation*. Washington, DC: International Food and Agriculture Trade Policy Council.

Wassmann, R., Jagadish, S.V.K., Heuer, S. et al. (2009). Climate change affecting rice production: the physiological and agronomic basis for possible adaptation strategies. *Adv. Agron.* 101: 59–122. https://doi.org/10.1016/S0065-2113(08)00802-X.

Webb, K.M., Oña, I., Bai, J. et al. (2010). A benefit of high temperature: increased effectiveness of a rice bacterial blight disease resistance gene. *New Phytol.* 185: 568–576. https://doi.org/10.1111/j.1469-8137.2009.03076.x.

World Bank (2010). *Development and Climate Change. World Development Report 2010*. Washington, DC: The International Bank for Reconstruction and Development.

Xiong, W., Conway, D., Lin, E.D., and Holman, I. (2009). Potential impacts of climate change and climate on China's rice yield and production. *Climate Res.* 40: 23–35. https://doi.org/10.3354/cr00802.

Xu, K. and Mackill, D. (1996). A major locus for submergence tolerance mapped on rice chromosome 9. *Mol. Breed.* 2: 219–224. https://doi.org/10.1007/BF00564199.

Xu, K., Xu, X., Ronald, P., and Mackill, D.J. (2000). A high-resolution linkage map of the vicinity of the rice submergence tolerance locus Sub1. *Mol. Gen. Genet.* 263: 681–689. https://doi.org/10.1007/s004380051217.

Yang, M., Xiao, W., Zhao, Y. et al. (2018). Assessment of potential climate change effects on the rice yield and water footprint in the Nanliujiang Catchment, China. *Sustainability* 10: 242. https://doi.org/10.3390/su10020242.

Zhang, S., Zheng, J., Liu, B. et al. (2014). Identification of QTL for cold tolerance at seedling stage in rice (*Oryza sativa* L.) using two distinct methods of cold treatment. *Euphytica* 195: 95–104. https://doi.org/10.1007/s10681-013-0977-0.

Zhu, C.L., Xiao, Y.H., Wang, C.M. et al. (2005). Mapping QTL for heat-tolerance at grain filling stage in rice. *Rice Sci.* 1: 33–38.

Ziska, L.H., Tomecek, M.B., and Gealy, D.R. (2010). Competitive interactions between cultivated and red rice as a function of recent and projected increases in atmospheric carbon dioxide. *Agron. J.* 102: 118–123. https://doi.org/10.2134/agronj2009.0205.

2

Molecular Breeding for Improving Salinity Tolerance in Rice: Recent Progress and Future Prospects

Sandeep Chapagain, Lovepreet Singh, Richard Garcia, Rajat Pruthi, Jonathan Concepcion, Sapphire Coronejo and Prasanta K. Subudhi

School of Plant, Environmental and Soil Sciences, Louisiana State University Agricultural Center, Baton Rouge, LA 70803, USA

Introduction

Salinity is a major abiotic constraint limiting rice production worldwide. It affects 800 million hectares of land throughout the world (Munns 2005). Forty-five million hectares of irrigated land, which constitutes 20% of total irrigated land, have serious issues of soil degradation (FAO 2008). The major drivers for expansion of saline areas are low and erratic rainfall, increased evaporation, depletion of ground water resources, and inappropriate irrigation practices in irrigated land (Munns and Gilliham 2015). Soil salinity along with drought is expected to be exacerbated in the future due to climate change. Salt stress not only has negative ecological and environmental consequences but may also threaten food security in salt-affected regions due to reduced crop productivity.

Rice (*Oryza sativa* L.) is a major food staple for the world's population and is widely grown in Asia. It is considered to be a salt-sensitive crop with a salinity threshold of 3 $dS\ m^{-1}$ (Chinnusamy et al. 2005). It is most vulnerable at both seedling and reproductive stages (Zeng and Shannon 2000). Soil salinity affects plant growth and productivity, resulting in substantial loss in grain yield in rice. In addition to yield loss, rice grain quality is also affected by increased salinity levels (Thitisaksakul et al. 2015). Ionic imbalance and disruption in metabolic activities are the major consequences of salt stress leading to the deleterious effects on growth and productivity (Munns and Tester 2008). Development of salt-tolerant rice varieties is a logical and environmental-friendly option to minimize the impact of salinity stress on global food production. Adoption of salt-tolerant cultivars can help expand rice cultivation to less favorable areas and sustain productivity on salt-affected areas due to the need for less leaching to wash away salt from topsoil (Munns et al. 2006; Ismail et al. 2007). Although our understanding of plant adaptation under salinity stress and the underlying mechanisms has been enhanced, progress in developing salt-tolerant rice varieties has been modest (Shahbaz and Ashraf 2013).

Many different approaches such as quantitative trait loci (QTL) mapping, marker-assisted breeding, and genetic engineering have been employed to develop salt-tolerant varieties

Molecular Breeding for Rice Abiotic Stress Tolerance and Nutritional Quality, First Edition.
Edited by Mohammad Anwar Hossain, Lutful Hassan, Khandakar Md. Iftekharuddaula, Arvind Kumar and Robert Henry.

(Kumar et al. 2013). There is a need to exploit genomic resources, salt-responsive QTLs, and molecular markers to characterize genes to gain deep insights into mechanisms that are useful for developing salt-tolerant varieties. Therefore, increasing rice production in salt-affected areas will require utilization of both traditional and modern genomic tools based on our understanding of the salinity-tolerance mechanisms. This chapter will cover the progress made, methods employed, potential opportunities, and future challenges in expanding rice cultivation in salt-affected areas.

Salt Tolerance Mechanisms

Accumulation of salt in rice fields affects growth and yield of rice plants. This salinization of soil water inhibits plant growth by reducing the plant's capacity to take up water and injuring cells in transpiring leaves via transpiration stream, which are known as osmotic and ion-excess effects, respectively (Munns 2005). Rice plants respond to salt stress through several protective mechanisms starting from root-level ion exclusion to both cellular and plant-level compartmentations (Figure 2.1). These mechanisms operating at different levels can be categorized into ion exclusion, osmotic tolerance, and tissue tolerance (Roy et al. 2014; Reddy et al. 2017).

Root-Level Ion Exclusion

About 98% of salt in the rhizosphere are excluded by roots, allowing only minute amounts to move across the root tissues into the xylem via symplastic and apoplastic pathways. The structural barrier (i.e., Casparian strips and suberin lamellae) in the endodermis plays an important role in restricting salts to move passively into the xylem; hence limiting the amount of Na^+ to freely move from root to shoot (Das et al. 2015).

Plant-Level Compartmentation

In response to salt stress, plant compartmentalizes toxic Na^+ into the older leaves and leaf sheaths, which signals an early leaf senescence at the expense of lower ion concentration in younger leaves (i.e., flag leaves) and reproductive tissues; hence preventing further damage and yield loss. This maintains the rates of leaf initiation at least equal to the rates of leaf death, which is enough for the photosynthesis needed during flowering and grain filling (Munns 2002; Negrão et al. 2011).

Cell-Level Compartmentation

Salt tolerance of rice plants acts at the cellular level, where each cell responds to salt stress by working synergistically via ion exclusion, osmotic tolerance, and tissue tolerance. Ion exclusion entails the Na^+ transport processes that mostly occur in the root system. It prevents accretion of toxic concentrations of Na^+ in the upper shoot system of the plant by minimizing the Na^+ movement into the xylem and transporting back Na^+ from shoots to roots and/or efflux of Na^+ back to the soil (Negrão et al. 2011; Roy et al. 2014; Ismail and

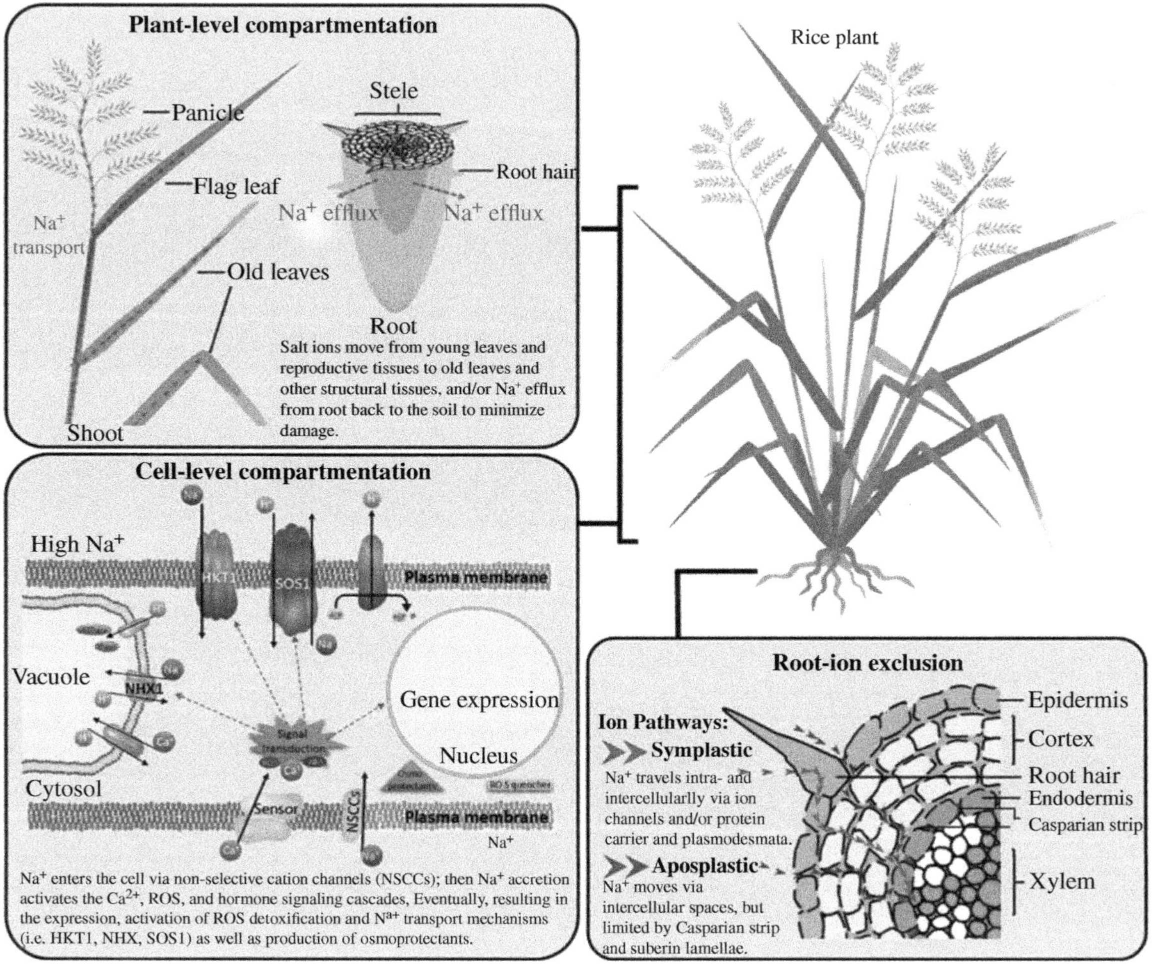

Figure 2.1 Overview of salt tolerance mechanism in rice. (*See color plate section for color representation of this figure*).

Horie 2017). Osmotic tolerance is regulated by sensing molecules (i.e., Ca^{2+} ions, phospholipids, phytohormones), abscisic acid (ABA)-signaling pathways, and transcription factors that are activated before the occurrence of shoot Na^+ accretion (Kumar et al. 2013; Reddy et al. 2017). Tissue tolerance involves mobilization of Na^+ from cytoplasm into the vacuoles of old leaves and other structural tissues, production of compatible solutes such as proline, myo-inositol, trehalose, glycine betaine, and enzymes for detoxification of reactive oxygen species (Ismail and Horie 2017; Reddy et al. 2017). With the understanding of the salt-tolerance mechanisms, improvement of rice growth and yield can be achieved through genetic enhancement of salt exclusion and tolerance mechanisms via modern and/or conventional breeding programs.

Natural Variability for Salt Tolerance in Rice Germplasm

Successful breeding programs depend heavily on the utilization of both existing and novel sources of genetic variation from diverse germplasm (McCouch and Kovach 2008). Numerous studies demonstrated that there is large phenotypic variability among the rice germplasm (Quijano-Guerta and Kirk 2002; Negrão et al. 2011; De Leon et al. 2015; Sakina et al. 2016). Overall, the *indica* varieties are more tolerant to salt stress compared with *japonica* varieties due to the increased efficiency in excluding Na^+, absorbing more K^+, and maintaining a low Na/K ratio (Lee et al. 2003b). "Pokkali" and "Nona Bokra" have been used as donors all over the world due to their high salinity tolerance and adaptation in coastal saline areas. In addition, several landraces and cultivars with moderate to high tolerance to salinity stress have been identified and developed for cultivation in saline-prone areas (Table 2.1). Mishra et al. (2016) evaluated Indian wild rice germplasm and identified 2 highly tolerant, 14 tolerant, and 28 moderately tolerant lines, and the majority of tolerant wild rice lines are from *Oryza rufipogon* and *Oryza nivara*. The salt-tolerant line DJ15 was developed from a cross between a wild rice line of *O. rufipogon* (Dongxiang) and a *japonica* variety (Ninjing 16). Recently, Prusty et al. (2018) identified seven wild rice accessions such as *Oryza alta, Oryza latifolia, Oryza coarctata, Oryza rhizomatis, Oryza eichingeri, Oryza minuta, Oryza grandiglumis* with higher or similar levels of salt tolerance as Pokkali, Nona Bokra, and FL478. These tolerant wild species exhibited high levels of tissue tolerance and chlorophyll retention in young leaves. *O. coarctata,* a salt marsh native from the east coast of India, can tolerate salinity up to 20–40 $dS\,m^{-1}$ or more without any deleterious effects (Sengupta and Majumder 2010). The salt-tolerant landraces and cultivars have successful adaptation mechanisms in saline-prone areas. Further, these saline-tolerant genotypes are potential donors for developing breeding lines and the genes identified from wild rice can be introgressed to improve salt tolerance in cultivated rice.

Conventional Tools to Improve Salinity Tolerance

Conventional Breeding

Conventional plant breeding is a well-established methodology for genetic improvement of any crop species by combining the desirable attributes. Various selection strategies have

Table 2.1 List of rice landraces or released varieties with tolerance to salinity and their country of origin or release.

Salt-tolerant land races or released varieties	Country of release/origin
Nona Bokra, Narendra Usar Dhan 2008, NC 1281, CORH-2, Jagjeevan, ASD 16, Sumati, Karma Mahsuri, CST 7-1, Warangal Sannalu, Nalini, Lunisree, Swarna, Chaitanya, Deepti (MTU-4870), TRY 1 to 3, CSR 1, CSR 2, CSR 3, CSR 4, CSR 10, CSR 13, CSR 20, CSR 22, CSR 23, CSR 27, CSR 30, CSR 36, Basmati CSR 30, Vytilla to 5, Panvel 1 to 3, SR26B, Narendra Usar 2, Narendra Usar 3, Savitri, Ponni, Swarnadhan, Mansatovar, Salivahan, Pavizham, BR 10, Patnai 23, PVR 1, PY 1, Co 43, AD 85002, IET 8113, Getu, Kalarata 1-24, Cheriviruppu, Damodar, Pat, Sollar	India (Singh et al. 2016a; Shahbaz and Ashraf 2013; Kanawapee et al. 2011; Lin et al. 2004; Negrão et al. 2011)
Shaheem Basmati, Jhona 349	Pakistan (Sakina et al. 2016; Akbar et al. 1972)
BR23,BRRI dhan40, BRRI dhan41, BRRI dhan47, BRRI dhan BRRI dhan53, BRRI dhan54, BRRI dhan55, BRRI dhan61, BRRI dhan67, BRRI dhan73, BRRI dhan78, Binadhan-8, Binadhan-10	Bangladesh (Islam and Gregorio 2013; Rashid and Nasrin 2015; Ahmed et al. 2016)
Panthum Thani1, RD10, RD15, RD23, RD25, Leuang Anan, Leuang Tahmo, Nam Sa-Gui19, Sew Mae Jun, SPR 90, Khao Seeta	Thailand (Kanawapee et al. 2011)
Matatag2, FL496, FL473, FL530, FL478, IR45427-2B-2-2B-1-1, IR51500-AC11–1 (PSB Rc50/Bicol), IR51500-AC17 (CSR21), IR51485-AC6534–4 (CSR28), IR63731–1–14-3-2, IR65192-4B-10-3	IRRI (Philippines) (Kanawapee et al. 2011; Negrão et al. 2011)
Tarome-Malaei, Tarommahalli	Iran (Ahmadi and Fotokian 2011; Sabouri and Sabouri 2008)
Jiucaiqing, Tesanai 2	China (Ammar et al. 2009; Masood et al. 2004)
Ketumbar	Indonesia (Negrão et al. 2011)
Moroberekan	West Africa (Kim et al. 2009)
AS996, Soc Nau	Vietnam (Lang et al. 2008; Negrão et al. 2011)
Nipponbare, Jingxil17	Japan (Takehisa et al. 2004; Gong et al. 1999)
Hasawi	Saudi Arabia (Negrão et al. 2011)
Harra	Spain (Negrão et al. 2011)
Agami	Egypt (Negrão et al. 2011)
Daeyabyeo	Korea (Negrão et al. 2011)
Moreoberekan	Guinea (Negrão et al. 2011)
Pokkali	Sri Lanka (Gong et al. 1999)
NSIC 2016 Rc480, Hagonoy, Sipocot, Matnog, Naga, Buguey, Sumilao, Anahawan, Salinas 1 to 25, IR8, IR64, M-20	Philippines (Fan et al. 2016; Fotoukian et al. 2005; Zang et al. 2008; Prasad et al. 2000; Akbar et al. 1972)

been employed for improving salt tolerance in rice. However, improvement of salt-tolerant rice varieties by conventional breeding is an arduous job due to difficulty in establishing the best defined physiological attributes for salt tolerance and involvement of multiple genes (Fita et al. 2015). Several salt-tolerant rice varieties were developed using local and traditional salt-tolerant varieties and were commercialized in several countries in Asia, including Bangladesh, India, and the Philippines (Ismail and Horie 2017). For example, salinity-tolerant varieties, CSR-1, CSR-2, and CSR-3, were developed by pure line selection from local cultivars in India. Similarly, SR-26B, Hamilton, Patnai-23, and Jhona-349 were developed by site-specific selection in different countries in improved backgrounds following recombination breeding (Gregorio et al. 2002). The salt-tolerant varieties of rice, CSR23 and CSR27, were developed by a shuttle-breeding approach at International Rice Research Institute (IRRI)(Mishra 1994). CSR36 that showed tolerance to salinity at the vegetative stage was released in India, and Binadhan 10 that showed tolerance to salinity at the reproductive stage, was released in Bangladesh (Ismail and Horie 2017).

Although several success stories using conventional breeding methods have been reported, there are challenges in breeding high-yielding rice varieties with a higher level of salt tolerance. Most of the salt-tolerant cultivars have poor agronomic traits such as low yield, tall plant stature, and photosensitivity (Reddy et al. 2014). Since many of the physiological attributes that contribute to salt tolerance response are controlled by many genes or QTLs, prioritizing and combining the desirable attributes via classical breeding have been difficult. Therefore, integrating molecular tools in breeding programs should accelerate the development of salt-tolerant rice varieties with high yield and desirable agronomic attributes.

Mutation Breeding

Mutation breeding is an efficient approach to develop salt-tolerant high-yielding rice varieties due to not having any ethical issues compared with the transgenic approach (Das et al. 2014). It is particularly useful to improve the adapted cultivars that are deficient in one trait, such as salt susceptibility. The use of ionizing radiation, such as gamma rays, X-rays, neutrons, and chemical mutagens, is an established technique for mutagenesis. Few studies revealed that mutagenesis has increased genetic variability for salinity tolerance in rice. Salt-tolerant rice varieties like Shua 92 (Mustafa et al. 1997) and Basmati 370 (Saleem et al. 2005) have been produced using gamma irradiation. Takagi et al. (2015) discovered a loss-of-function mutation responsible for the salt tolerance and developed a salt-tolerant variety named "Kaijin". Furthermore, rice seeds irradiated with carbon and neon ions have generated a mutant variety with enhanced salt tolerance (Hayashi et al. 2007). NIAB Rice-1 and PSR 1-84 rice varieties developed through gamma radiation showed higher yield under saline conditions compared with the salt-tolerant check varieties Pokkali and Johna 349 (Anonymous 1986). Similarly, chemical mutagenesis has been used to enhance salinity tolerance in rice. A salt-tolerant rice mutant *rst1* was identified from ethyl-methansulfonate (EMS) mutant library, which showed significantly high chlorophyll content and shoot biomass but lower electrolyte leakage and lipid peroxidation under salinity stress (Deng et al. 2015). Ashraf (1994) reported enhanced survival at the seedling stage in 0.5% NaCl in N-methyl-N-nitrosourea treated Taichung 65 (salt-sensitive rice cultivar).

Tissue Culture

Tissue culture can accelerate development of rice breeding lines that are tolerant of abiotic stresses (Zhang et al. 2014). Double haploid technology shortens the development of homozygous and uniform breeding lines by five to seven years compared with conventional breeding (Germana 2011). Anther culture success depends on donor plant genotype, microspore developmental stage, cold treatment, and induction and differentiation media. The *indica* rice varieties have relatively low callus induction and regeneration efficiency, which serves as a bottleneck in double haploid line breeding (Cha-um et al. 2009). Atabaki et al. (2018) developed salt-tolerant lines from Malaysian rice cultivar MR263 via somatic embryogenesis (seed culture). Calli were generated in culture media containing 100–300 mM NaCl solution to generate salt-tolerant calli, and regenerants. Furthermore, induced somaclonal variation in rice cultivar Wagwag resulted in development and release of moderate-salinity-tolerant Salinas 14 in the Philippines (Ismail and Horie 2017). *In vitro* mutagenesis (combined *in vitro* culture and mutagenesis through physical mutagens) has been successfully utilized in enriching genetic variability and developing abiotic-stress-tolerant crops, including rice (Lee et al. 2003a). Various tissue culture techniques have successfully developed salt-tolerant lines for adaptation in saline-prone rice-growing areas. Improved tissue culture–derived lines have potential to be used as donors for salt tolerance in various breeding programs.

Molecular Tools to Improve Salinity Tolerance

QTL Mapping

Quantitative trait locus mapping has greatly improved our understanding of the genetic architecture of salt tolerance and the underlying mechanisms. Most of the QTL studies have used recombinant inbred lines (RIL), doubled haploid (DH) or introgression lines (IL) mapping populations. Multiple studies have identified several QTLs for traits contributing toward salt tolerance using different donors in the past decades (reviewed by Karan and Subudhi 2011; Choudhary et al. 2019). A major QTL, *Saltol* associated with Na-K ratio and seedling-stage salinity tolerance, was identified on chromosome 1 with 43% of the phenotypic variation (Bonilla et al. 2002). Later, Thomson et al. (2010) detected *Saltol* locus in a RIL population of IR29 and Pokkali using 100 simple sequence repeat (SSR) markers and characterized its role to control Shoot Na^+/K^+ homeostasis. Similarly, another major QTL for shoot K^+ concentration on chromosome 1 (*SKC1)* and shoot Na^+ concentration on chromosome 7 (*qSNC7)* with 40.1% and 48.5% phenotypic variance suggested involvement of different genes to control transport of Na^+ and K^+ (Lin et al. 2004). Furthermore, Ren et al. (2005) validated *qSKC1* by cloning, and it encodes a sodium transporter that maintains K^+ homeostasis under salinity stress. Bimpong et al. (2013) identified seven significant QTLs for root length, plant height, shoot dry weight, and shoot fresh weight using 384 single nucleotide polymorphism (SNP) markers in a RIL population derived from the cross between IR29 and Hasawi.

De Leon et al. (2016) reported 16 large-effect QTLs for seedling-stage salinity tolerance traits in a RIL population developed from the Bengal × Pokkali cross using a genotyping-by-sequencing (GBS) based saturated SNP linkage map. Gimhani et al. (2016) identified 83 QTLs for 11 morpho-physiological traits associated with seedling-stage salinity tolerance with 12.5–46.5% phenotypic variation in a RIL population derived from the cross At354 × Bg352, and the majority of QTLs were clustered on 14 genomic regions. Bizimana et al. (2017) reported 20 new QTLs for different salinity traits on chromosomes 1, 2, 4, 6, 8, 9 using a RIL population of the cross Hasawi x IR29 and Hasawi did not contain the same salinity tolerance allele as Nona Bokra and Pokkali at *SKC1* locus and *Saltol*. The same population was evaluated in different environments leading to identification of common regions for multiple QTLs on chromosomes 1, 4, 6, 8, and 12, suggesting the presence of novel salt-tolerance alleles in Hasawi (Rahman et al. 2017). There is little work on genetics of reproductive-stage salinity tolerance in rice. Mohammadi et al. (2013) identified 35 QTLs for 11 yield and yield-attributing traits using an F_2 population derived from the cross between Sadri and FL478. Twenty-eight of these QTLs explained 6 to 30% of phenotypic variation. Another study by Hossain et al. (2015) reported 16 QTLs with 4 to 47% phenotypic variation on chromosomes 1, 7, 8, and 10 in an F_2 mapping population developed from a cross between a reproductive-stage salt-tolerant donor Cherivirupu and Pusa Basmati 1 (PB1). It was suggested that Na^+ concentration, pollen fertility, and Na^+/K^+ ratio in flag leaf are the most important mechanisms maintaining reproductive-stage salinity tolerance. Application of a bulk segregant strategy to a "CSR11/MI48" derived RIL population led to identification of three salt-tolerant QTLs at the reproductive stage for grain yield (Tiwari et al. 2016).

Chromosome segment substitution lines (CSSLs) or IL populations are useful to identify and fine-map QTLs for discovery of genes present in the QTL regions (Ali et al. 2010; Thuy et al. 2018). IL population of salt-tolerant donors Pokkali and Nona Bokra, developed in several U.S. cultivar back grounds have been used to identify and validate earlier identified QTLs for seedling salinity in rice. De Leon et al. (2017) detected a total of 18 and 32 QTLs using SSR and SNP markers, respectively, from an IL population of Pokkali in the background of "Bengal" and validated at least 14 QTLs identified from the RIL population of the same cross. Puram et al. (2017) identified 33 additive QTLs using an IL population developed from a salt-tolerant donor Nona Bokra in the U.S. cultivar "Jupiter" background. Similarly, 32 QTLs were detected from an IL population of the same donor Nona Bokra in Cheniere background and suggested shoot Na^+/K^+ homeostatsis, Na^+ exclusion, and compartmentation may act as salt-tolerance mechanisms in Nona Bokra (Puram et al. 2018).

Nounjan et al. (2016) evaluated two CSSLs of "Khao Dawk Mali 105" (KDML105) rice carrying segments of drought-tolerant QTL (DT-QTL8) and observed enhanced tolerance for both drought and salinity stresses. Chutimanukul et al. (2018) reported higher salinity tolerance in CSSL16 derived from KDML105 compared to KDML05 plants. Moreover, 10 major genes were detected from co-expression analysis and suggested that candidate locus *PsbS1* gene played a role in salinity tolerance by maintaining photosynthetic activity. Similarly, CSSLs have been utilized to ascertain the salinity-tolerance mechanisms. Thuy et al. (2018) investigated salinity-tolerance mechanisms by using CSSLs carrying segments of tolerance from Nona Bokra in salt-sensitive background Koshihikari and suggested that those identified chromosome regions were related to Na^+ and Cl^- exclusion.

The genome-wide association study (GWAS) is a recently developed tool for genetic characterization of quantitative traits. GWAS has been successfully applied in rice for different abiotic stresses including salinity stress. An association study using markers associated with 14 QTLs and 65 candidate genes for salinity tolerance in 180 European *japonica* accessions concluded that no single accessions carried the favorable alleles and at least nine accessions were needed to accumulate all favorable alleles to accomplish higher levels of salt tolerance using a marker-assisted population improvement strategy (Ahmadi et al. 2011). Kumar et al. (2015) identified a total of 64 SNPs related to salt tress and Na^+/K^+ ratio. Shi et al. (2017) reported 22 significant SNPs related to salinity tolerance on the basis of stress-susceptibility indices (SISs) and mean germination time. Batayeva et al. (2018) detected 26 QTLs for different traits at seedling salinity stress in a GWAS study of temperate *japonica* rice accessions and 11 of these QTLs collocated with known salt-tolerance genes. Similarly, Frouin et al. (2018) evaluated salt tolerance in 235 temperate *japonica* rice accessions genotyped with 30 000 SNPs and reported that many of these significant QTLs identified by GWAS were located near the genes involved in calcium signaling, metabolism, and kinases. Furthermore, Cui et al. (2018) utilized six multi-locus GWASs at the germination stage under salinity stress using 478 rice accessions in conjunction with 162,529 SNPs and identified 371 quantitative trait nucleotides (QTNs) and 66 candidate genes. Yu et al. (2017) reported 93 candidate genes associated with salinity tolerance in a GWAS study involving a panel of 295 rice accessions. Similarly, Naveed et al. (2018) identified 20 QTNs for 11 traits associated with salinity stress in germination and seedling stage using 395,553 SNP markers in 208 rice accessions. Lekklar et al. (2019) evaluated salinity tolerance at the flowering stage in 104 Thai rice accessions and identified 448 SNPs in 200 loci of which 73% of loci overlapped with earlier reported QTLs. An et al. (2019) detected 54 QTLs associated with salt tolerance using 181 cultivars of which 17 loci were found to be significantly associated with seedling dry weight ratio. They also identified a candidate gene encoding a multidrug and toxic compound extrusion (MATE) protein in novel QTL *qST-7.19* on chromosome 7 and validated using a near-isogenic line. The meta-analysis study of QTLs associated with seedling salinity tolerance by Islam et al. (2019) revealed many candidate genes associated with salinity stress, which can be helpful for fine-mapping, pyramiding, marker-assisted selection (MAS), and cloning to understand the salinity-tolerance mechanisms in rice.

Our understanding of the molecular basis of salt-tolerance mechanisms in rice has been greatly enhanced by QTL mapping. The major factor for successful utilization of the salinity-tolerance QTLs in breeding programs depends on the magnitude of the genetic effect of QTL alleles. Efforts should be made to discover large-effect QTLs by modern high-throughput genomics and phenomics, and bioinformatic tools in conjunction with specialized genetic materials developed from diverse genetic materials.

Marker-Assisted Selection

MAS provides a strategy to accelerate conventional breeding efforts by using DNA markers as indirect selection criteria. The allelic variation in the genes of interest or tightly linked markers may be used to develop molecular tools for developing salt-tolerant cultivars. The use of MAS for simple qualitative traits is straightforward, but selection for complex/polygenic traits like salinity stress tolerance through MAS is complicated and difficult (Ashraf et al. 2012). The MAS technique allows breeders to speed up the breeding process by recovering recurrent genomes in only a few generations. Several studies have reported introgression of salt-tolerance-associated QTLs into high-yielding popular rice varieties through MAS. For example, a major salt-tolerance QTL, "*Saltol*" has been introgressed through marker-assisted backcrossing (MABC) into well-adapted popular rice varieties, namely PB1121 and PB6 (Singh et al. 2011), AS996 (Huyen et al. 2012), Bac Thom 7 (Linh et al. 2012; Vu et al. 2012), BRRI dhan49 (Hoque et al. 2015), and several local varieties: "ADT 45, CR 1009, Gayatri, MTU 1010, PR 114, Pusa 44 and Sarjoo 52" in a mega project in India (Singh et al. 2016b). Punyawaew et al. (2016) successfully transferred *Saltol* QTL to a Thai rice variety KDM105 from FL530 through MABC, and the ILs showed higher periods for survival, less Na^+ uptake, and high K^+ uptake suggesting the role of "*Saltol*" in the Na^+/K^+ homeostasis mechanism. The NIL developed by introgression of "*Saltol*" from FL478 to "IR64" (Ho et al. 2016), "Pusa Basmati" (Singh et al. 2018b), and "Improved White Ponni" (Valarmathi et al. 2019) through MABC showed enhanced tolerance to salinity at the seedling stage. Recently, Rana et al. (2019) transferred a salinity-tolerance gene *(hst1)* from a salt-tolerant donor "Kaijin" into a high-yielding cultivar "Yukinko-mai" through MAS coupled with speed-breeding. This salinity-tolerant population "YNU21-2-4," was highly tolerant during both seedling and reproductive-stress salinity tolerance.

The success of MAS in improving salt tolerance largely depends on several factors such as the heritability of a trait, the number and effect of QTLs, and the population size. With the exception of *Saltol*, most QTLs for salt-tolerance traits have small effects and are largely influenced by environment. Other factors hindering the application of MAS in plant-breeding programs are high cost and large QTL confidence intervals, which may pose difficulty in precise introgression of target traits due to linkage drag (Hoang et al. 2016). Therefore, further technical advances are needed to make MAS more reliable and cost-efficient for large-scale adoption in rice breeding programs to design salt-tolerant varieties through introgression and pyramiding of salt-tolerance genes or QTLs.

Candidate Genes and Allele Mining

Identification and validation of the salinity-related alleles and accumulating them in a single genotype is an ideal approach to develop salinity-tolerant rice cultivars. The discovery of SNPs has enhanced the strength of allele mining in a wide range of rice germplasm so that the desirable alleles for a specific trait can be discovered. Candidate gene-based association analysis was conducted in 392 rice accessions for five known salt-tolerant genes involved in salinity tolerance mechanisms through EcoTILLING and 11 significant SNPs linked to salt-stress traits were identified (Negrão et al. 2013). Many orthologous genes from different rice genotypes have been cloned, sequenced, and correlated with phenotypic attributes under salinity stress and favorable alleles have been identified for rice breeding (Ganie et al. 2017). Few salt-tolerant landraces and a wild species (*O. coarctata*) are considered as good sources of salinity-responsive alleles. Platten et al. (2013) identified two landraces showing high tissue-tolerance from Thailand. Furthermore, they found 10 different alleles of *OsHKT1:5*, which is significantly associated with leaf Na^+ concentration. A GWAS study involving genome resequencing of 295 rice accessions identified 79 candidate genes for salt tolerance at the germination stage (Yu et al. 2018). Some of these include K^+ transporter *OsHAK11* (Os04g0613900), protein argonaute 2 (Os04g0615700), auxin efflux carrier protein *OsPIN1d* (Os12g0133800), germin family protein (Os02g0532500), stress-activated protein kinase pathway regulating phosphatase 1 (Os12g0133700), a zinc-induced facilitator-like gene family (Os12g0133100, Os12g0133300). Zhang et al. (2018) highlighted the importance of K^+ transporters (*OsHKT2;2, OsHKT2;3,* and *OsHKT2;4)* present in weedy rice in improving salt tolerance. GWAS analysis of 208 rice accessions from 25 different countries of Asia, Africa, and Latin America led to the identification of 6 and 14 QTLs for salt-tolerance-related traits at the germination and seedling stages, respectively. By using bioinformatics tools and haplotype information, 22 candidate genes were identified including a HKT-type transporter for K^+ homeostasis (*SKC1),* a member of CCCH-type zinc finger gene family (*OsTZF1*) associated with tillering and panicle branching under salt stress (*OsEATB*) (Naveed et al. 2018).

Progress in high-throughput approaches and availability of public repositories such as the Rice SNP-Seek Database (Mansueto et al. 2017) is expected to facilitate discovery of natural allelic variants present in rice germplasm. Several important salinity-tolerance candidate genes in rice with salt-responsive elements and co-localizing with QTLs were summarized in previous reviews (Negrão et al. 2011; Reddy et al. 2017). These candidate genes represent a useful resource to support molecular breeding efforts through functional validation of desirable haplotypes or allelic variants. Since the complexity of salt tolerance due to involvement of multiple genes is well established, a breeding strategy to pyramid multiple beneficial variants at several important loci should be considered.

Genetic Engineering to Improve Salt Tolerance

Genetic engineering has been successfully employed in producing salt-tolerant plants, either by overexpression or introduction of selected genes into elite cultivars. Since the role of antioxidants, osmo-protectants, ion transport, and various transcription factors in imparting salt tolerance is well known, these were primarily utilized to improve salinity tolerance in rice through genetic engineering (Ashraf et al. 2008). There are numerous transgenic studies to enhance salt tolerance in rice using genes from several sources (reviewed by Bajaj and Mohanty 2005; Kathuria et al. 2007; Hoang et al. 2016; Reddy et al. 2017). Most recent studies using transgenic approaches to improve salinity tolerance are listed in Table 2.2. Transgenic pyramiding of salt-responsive protein 3-1 (*SaSRP3-1*) and vacuolar H^+-ATPase subunit c1 (*SaVHAc1*) enhanced salt tolerance in both seedling and reproductive stage (Biradar et al. 2018). Few studies demonstrated the crucial role of RING E3 ligases in salinity tolerance. For example, the transcript level of *OsRMR1* (Lim et al. 2015) and *OsSIRP2* (Chapagain et al. 2018) were highly expressed in salinity stress and showed enhanced salinity tolerance in transgenic *Arabidopsis* but still need to be validated in rice. Due to insufficient information regarding major genes contributing to salinity tolerance, inadequate transgene expression or transgene silencing and consumer resistance to genetic modification approach, the progress in development of salinity-tolerant rice by genetic engineering is still slow. Efforts should be made to target multiple genes involved in different salt-tolerance mechanisms driven by stress-inducible promoters to enhance salt tolerance.

Table 2.2 List of genes overexpressed in *Oryza sativa* to improve salt tolerance.

Gene	Donor	Impact during salt stress	Reference
OsMYB6	*O. sativa*	Increased ROS quencher and proline content	Tang et al. (2019)
OsGS1;1 and *OsGS2*	*O. sativa*	Higher grain yield and photosynthetic rate	James et al. (2018)
TsPIP1;1	*Thellungiella salsuginea*	Enhanced chlorophyll content, and photosynthetic rate, higher K uptake in shoots and lowered osmotic potential in leaves	Li et al. (2018)
LsEm1	*Lactuca sativa*	Increased chlorophyll content, superoxide dismutase activity and reduced yield loss	Xiang et al. (2018)
AmRosea1	*Antirrhinum majus*	Upregulation of K^+ transporter and catalase activity	Dou et al. (2017)
OsSta2	*O. sativa*	Increased chlorophyll content and grain yield	Kumar et al. (2017)
OsHNX1	*O. sativa*	Lowered leaf damage score and higher chlorophyll content; higher spikelet fertility and grain yield	Amin et al. (2016)
ONAC022	*O. sativa*	Increased root and shoot length with decrease in Na^+ concentration in them	Hong et al. (2016)
EcNAC67	*Eleusine coracana L.*	Longer root and shoot length, and less reduction in biomass as compared to non-transgenic plants	Rahman et al. (2016)
PtCYP714A3	*Populus trichocarpa*	Increased expression of *NHX1* and *SOS1* genes	Wang et al. (2016)
OsPP1a	*O. sativa*	Higher germination rate, kernel weight and increased ascorbate peroxidase, superoxide dismutase levels	Liao et al. (2016)
STRK1	*O. sativa*	Activation of CatC belonging to a family of proteins that reduces the level of hydrogen peroxide	Zhou et al. (2018)

Novel Tools to Aid in Molecular Breeding

Multi-Parental Advanced Generation Intercross Strategy

Multi-parental advanced generation intercross (MAGIC) has been recently utilized for QTL mapping of abiotic stress-tolerance traits. The MAGIC approach consists of inter-crossing of different founder lines involving both modern and traditional varieties that possess important agronomic traits such as grain yield, grain quality, and adaptability under salinity stress (Bandillo et al. 2013). Therefore, the MAGIC population consist of superior breeds of lines with agro-morphologically significant traits that are useful for gene discovery and utilization for rice improvement. However, this technique has not yet been adapted for salinity-tolerance QTL mapping. Bandillo et al. (2013) developed a different set of MAGIC rice population such as "*indica*" MAGIC (8 *indica* parents); MAGIC plus (8 *indica* parents with two traditional rounds of 8 way- F_1 inter-crossing); *japonica* MAGIC (8 *japonica* parents) and Global MAGIC (8 *indica* and 8 *japonica* parents) and mapped QTLs for different biotic and abiotic stresses. For salinity stress, a subset of *indica* MAGIC S4 lines showed tolerance to salinity and identified significant markers on chromosome 1 between 9.2 and 12 Mb adjacent to SALT, *qSKC-1*, and *Saltol* QTL. Samaco et al. (2018) investigated seedling-stage salinity tolerance and trait correlation of 970 selected MAGIC *indica* lines through morpho-physiological measurement and SSR analysis and identified 36 moderate and high salt-tolerant lines.

Genomic Selection

Advancement in next-generation sequencing (NGS) has greatly reduced the price of sequencing and therefore has created the possibility of using thousands of markers for adoption of a novel selection approach at genome level to improve quantitative traits (Breseghello and Guedes 2013). Genomic selection uses genome-wide DNA markers for selection of individuals and can potentially improve the efficiency of breeding of quantitative traits like salt tolerance. It involves prediction of the breeding value of each individual based on all available marker data compared to the use of a limited set of markers in a traditional MAS program (Bhandari et al. 2019). Due to availability of high-density SNP markers and rice genome sequence information, it is possible to capture the effects of many QTLs or genes with both small effects and large effects by using a large number of molecular markers distributed throughout the genome (Grenier et al. 2015). In genomic selection, a genomic estimated breeding value (GEBV) generated by a statistical model from the phenotypic and genotypic data in a training population is used to predict the performance of a breeding population (Bhat et al. 2016). The selection of individuals is based on the allelic composition of the markers or loci correlated with the phenotype in the training population (Spindel and Iwata 2018). The genomic selection can be used to make predictions in consecutive generations of the same breeding population. Wang et al. (2017b) predicted the performance of hybrid rice via genomic selection by using the GBLUP statistical model.

Genomic selection is emerging as a useful tool for improving complex traits such as salt tolerance. However, its implementation in plant-breeding programs needs further research to improve the accuracy of genomic selection models, which depend on the size of the training population, marker density, pedigree, and relationship among the individuals (Robertsen et al. 2019). As we develop a better understanding of the genetic architecture of complex traits, there will be an opportunity to improve the genomic selection models to make this tool valuable for plant breeders.

Phenomics

Understanding of phenotypic characteristics of plants related to specific traits such as salt tolerance is equally important as genotyping for genomics studies in any crop. The specific objective of crop breeding for salt tolerance requires detailed and accurate phenotypic information to evaluate breeding lines under salt stress and to study phenotypic variation due to the interaction between genotype and environment for further selection of improved and tolerant lines (Houle et al. 2010). The breeding cycle of crops could be improved and accelerated by collecting large amount of phenotypic data in precise manner via high-throughput phenomic techniques (Kim et al. 2019). Genomics has made tremendous progress over the past decade due to the development of high-throughput NGS technology; however, advances in phenomics are far from satisfactory, and traditional phenotyping methods are still being used (Guo et al. 2018). This gap between high-throughput sequencing and traditional phenotyping methods is a major bottleneck for molecular breeding programs in field crops (Houle et al. 2010). Most of the screening methods are destructive, and therefore it is difficult to measure plant growth in response to salt stress. Screening of salt-tolerant plants via traditional phenotypic approaches results in poor plant performance. Efforts have been made to improve phenomics platforms in the last few years, and the use of non-destructive sensors and imaging technologies has increased in plant science research to generate phenotypic data on different traits. Various non-destructive phenotyping techniques such as magnetic resonance imaging (MRI), thermal imaging, chlorophyll and photosynthetic fluorescence, light detection and ranging, canopy spectral reflectance, LEAF-E, and digital RGB imaging have been used in screening of different crops (Mir et al. 2019). MRI methods can be used to examine root and shoot physiology (Gupta et al. 2012) and measure the water content in different storage tissues via xylem and phloem water transport (Van As et al. 2009). Thermal imaging is used to detect canopy or leaf temperature and genetic variation in the stomata (Li et al. 2014; Munns et al. 2010). Chlorophyll fluorescence phenotyping of rice varieties has shown a positive correlation between salt tolerance and photosynthetic efficiency under salt stress (Tsai et al. 2019). LIDAR for estimating geometrical and structural parameters of plant canopy (Colaco et al. 2018) and digital imaging for shoot structure and biomass are used to increase the precision and speed of phenotyping (Klose et al. 2009). Tamimi et al. (2016) identified previously unidentified loci in rice for transpiration use efficiency for salinity by using a non-invasive phenomics approach. PlantCV is a recently developed new phenotyping technique that uses microcomputers and cameras to acquire pictures during stress and simultaneously analyze the data using software (Gehan et al. 2017). This tool was used to monitor plant growth under salt stress on a daily basis to distinguish the tolerant plants from the susceptible ones (Rajendran et al. 2009). For characterization and in-depth knowledge of salt-stress response in rice, phenomics could serve as a tool to speed up the breeding process and identify new genes in the NGS era (Singh et al. 2018a,b).

Next-Generation Sequencing

NGS is a novel low-cost sequencing technology to generate massive amounts of sequence data at high speed. Whole-genome resequencing of rice varieties helps in the genetic dissection of complex traits through identification of single-nucleotide polymorphisms (SNPs) and insertions/deletions (InDels) in unlimited numbers covering the whole genome (Huang et al. 2013). Using whole-genome sequence (WGS) data of rice genotypes with contrasting response to salt stress, Jain et al. (2014) generated valuable genomic resources to develop functional markers and promising gene targets for use in breeding programs. Chen et al. (2017) identified 42,359 high-impact variants and 3905 differential expressed transcripts via WGS and transcriptome analysis respectively.

NGS has facilitated the development of new tools such as MutMap, QTL-Seq, RNA-Seq, and Genotyping by Sequencing (GBS) to take advantage of massive amounts of sequencing information. In the MutMap approach, a causal mutation can be rapidly identified by WGS of pooled DNA from a segregating population (Abe et al. 2012). This was successfully used to identify a loss-of-function mutation hst1 for salt tolerance in rice (Takagi et al. 2015). On the other hand, the "QTL-seq" approach combines the power of bulked segregant analysis (BSA) and NGS-assisted high-throughput genotyping to identify the QTLs controlling complex traits such as salt tolerance (Takagi et al. 2013). Unlike the conventional QTL mapping, it does not require marker development for genotyping or development of a large mapping population, which is quite laborious. The GBS approach provides a quick way to generate large numbers of DNA polymorphisms for generating high-resolution linkage maps and for QTL mapping (Elshire et al. 2011).

Among many transcriptome-profiling techniques that can be used to identify differentially expressed genes in response to abiotic stresses including salinity (reviewed by Subudhi 2011; Bansal et al. 2013), the NGS-based RNA-Seq has been accepted as a leading global expression profiling tool due to its low cost, precise measurement of transcripts, and high efficiency resulting from advances in sequencing technology (Oono et al. 2011). It provides a strategy for elucidating novel mechanisms involved in abiotic stress response through accurate detection of changes in expression of transcripts over time, different groups or treatments, and analysis of post-transcriptional modifications, alternative spliced transcripts, and mutations (Shelden and Roessner 2013). Comparative transcriptional profiling suggested that the tolerant line FL478 has efficient mechanisms for signal transduction, transport of K^+ and osmotic homeostasis compared to the susceptible line IR29 (Mansuri et al. 2019). Differential expression analysis during salt stress revealed that genes involved in signal transduction and salt-stress response induces early in tolerant *japonica* cultivars compared to susceptible *japonica* ones (Formentin et al. 2018). Shankar et al. (2016) compared the transcriptomic responses between tolerant and susceptible rice cultivars, and the transcripts involved in response to salinity and regulation of gene expression were identified in the tolerant line Pokkali. Their study could be helpful to identify novel salt-tolerance genes by analyzing cultivar-specific stress-responsive transcripts. In another comparative transcriptomic study involving two *indica* genotypes, Wang et al. (2017a) reported that only 286 of 5273 differentially expressed genes were from the tolerant parent.

NGS technology is instrumental in changing the paradigm of QTL mapping, genome-wide gene expression, and gene cloning effort. Accumulation of genomic resources such as SNPs, InDels, and differentially expressed genes will help in elucidation of various salt-tolerance mechanisms and the underlying genes and QTLs to accelerate the development of salt-tolerant rice varieties.

Genome Editing

Compared to the mutation breeding approach, which involves large numbers of random mutations throughout the genome, genome editing makes changes only at the desired genome location. Genome editing methods such as homing endonucleases (HEs) (Cohen-Tannoudji et al. 1998), zinc finger nucleases (ZFNs), transcription activator-like effector nucleases (TALENs), and type II clustered regularly interspaced short palindromic repeat (CRISPR)/CRISPR-associated protein (Cas) (Cong et al. 2013; Mali et al. 2013) relies on site-specific nucleases (SSNs) for inducing double-stranded breaks (DSB) in DNA (Carroll 2014). These DSBs are either repaired by homology-dependent recombination (HDR) or by nonhomologous end-joining (NHEJ) (Wyman and Kanaar 2006). The HDR method of repair is based on "copy and paste" mechanism in which the broken strand pairs with the complementary DNA sequence and the homologous DNA sequence acts as a template for the repair of DSB. In the case of NHEJ, broken DNA does not require a template for the restoration of DSBs resulting in random insertions and deletions. Due to this, frameshift mutation occurs, which ultimately led to disruption of the coding region. Among the other genome editing techniques, CRISPR/Cas9 is simple, robust, and multiplexed in nature. It utilizes 20 nucleotide long guide RNA preceding a 5′-NGG protospacer adjacent motif (PAM), which is recognized by the Cas9 enzyme. After recognizing the PAM sequence, Cas9 enzyme makes a DSB in the target region, which is followed by DNA repair through HDR or NHEJ methods. With the help of guide RNA (gRNA) and Cas9, we can precisely target any genome location.

Zhang et al. (2019) demonstrated that a mutant of the *OsRR22* gene generated through CRISPR/Cas system showed enhanced salt tolerance in the seedling stage. Moreover, the CRISPR/Cas system has proved to be useful in identifying salt-tolerant genes such as *OsNECD3* (Huang et al. 2018), *OsNAC041* (Wang et al. 2019), and the GT-1 element of *OsRAV2* (Duan et al. 2016). Although, genome editing is a recently developed tool, it is favored by many researchers due to its ability to make precise changes in the targeted regions of the genome without introducing any foreign genes. Some major limitations of genome editing are low efficiency of HDR and occurrence of off-target mutations. Genome editing can be particularly useful in manipulating multiple genes at the same time. Therefore, further advances are needed to increase the efficiency of multi-genes editing in order to improve traits of agronomic importance including salinity tolerance.

Conclusion and Future prospects

As a staple food for more than half of the world population, genetic improvement of rice particularly for salinity tolerance has a major role in ensuring global food security. Since salt stress compromises growth and development of rice plants at both seedling and reproductive stages resulting in drastic reduction in yield worldwide, it is imperative to devise strategies to improve stability and sustainability of rice farming in salinity-impacted areas.

Although enhancing salinity tolerance in rice is challenging due to genetic complexity associated with multiple protective mechanisms, there are several possibilities of making progress in this direction due to advances in genomics. MAS will play a major role in targeting tolerance traits to accelerate designing of salt-tolerant rice varieties. Identification, introgression, and pyramiding of QTLs will be the major focus of the molecular breeders. Novel NGS-based tools such as MutMap, QTL-Seq, RNA-Seq, and GBS will facilitate rapid and systematic dissection of the genetic architecture of salt tolerance. In addition, novel mapping populations developed from multiparental crosses, bioinformatics advances, accurate high-throughput phenotyping tools, increased access to sequence information, and efficient tools for functional validation may expedite QTL discovery and cloning (Collins et al. 2008).

There are many salt-tolerant varieties and landraces in the world germplasm. But there is limited knowledge regarding the salt-tolerance mechanisms operating in these accessions. It is well established that none of these have all the salt-tolerance mechanisms (Ahmadi et al. 2011). Therefore, further research is needed to investigate and catalog the mechanisms contributing to salt tolerance in each accession to help accumulate the desirable alleles associated with different mechanisms from different accessions to attain higher levels of salt tolerance. Genomic tools can also expedite the discovery and utilization of superior alleles for salt tolerance from the wild germplasm. Genome editing technology is emerging as a versatile genome manipulation tool that can help develop salt-tolerant varieties by generating novel variants of genes associated with salt-tolerance traits (Ye and Cui 2019). Finally, with improved understanding of the salt-tolerance mechanisms, the discovery of superior alleles from diverse germplasms, and development of novel genotyping and phenotyping methods, molecular tools are expected to be integrated into the breeding programs for development of high-yielding rice varieties with improved salt tolerance.

Acknowledgments

This work was supported by a grant from the USDA-National Institute of Food and Agriculture (Grant Number: 2018-67013-27618).

References

Abe, A., Kosugi, S., Yoshida, K. et al. (2012). Genome sequencing reveals agronomically important loci in rice using MutMap. *Nat. Biotechnol.* 30: 174–178.

Ahmadi, J. and Fotokian, M.H. (2011). Identification and mapping of quantitative trait loci associated with salinity tolerance in rice (*Oryza sativa)* using SSR markers. *Iran. J. Biotechnol.* 9: 21–30.

Ahmadi, N., Negrão, S., Katsantonis, D. et al. (2011). Targeted association analysis identified *japonica* rice varieties achieving Na^+/K^+ homeostasis without the allelic make-up of the salt tolerant *indica* variety Nona Bokra. *Theor. Appl. Genet.* 123: 881–895.

Ahmed, A.U., Hernandez, R., and Naher, F. (2016). Adoption of stress-tolerant rice varieties in Bangladesh. In: *Technological and Institutional Innovations for Marginalized Smallholders in Agricultural Development* (eds. F. Gatzweiler and J. von Braun). Cham, Springer https://doi.org/10.1007/978-3-319-25718-1_15.

Akbar, M., Yabuno, T., and Nakao, S. (1972). Breeding for saline-resistant varieties of rice. *Jpn. J. Breed.* 22: 277–284.

Ali, M.L., Sanchez, P.L., Yu, S. et al. (2010). Chromosome segment substitution lines: a powerful tool for the introgression of valuable genes from *Oryza* wild species into cultivated rice (*O. sativa*). *Rice* 3: 218.

Amin, U.S.M., Biswas, S., Elias, S.M. et al. (2016). Enhanced salt tolerance conferred by the complete 2.3 kb cDNA of the rice vacuolar Na^+/H^+ antiporter gene compared to 1.9 kb coding region with 5'UTR in transgenic lines of rice. *Front. Plant Sci.* 7: 14.

Ammar, M.H.M., Pandit, A., Singh, R.K. et al. (2009). Mapping of QTLs controlling Na^+, K^+ and Cl^- ion concentrations in salt tolerant *indica* rice variety CSR27. *J. Plant. Biochem. Biotechnol.* 18: 139–150.

An, H., Liu, K., Wang, B. et al. (2019). Genome-wide association study identifies QTLs conferring salt tolerance in rice. *Plant Breed.* 139: 73–82. https://doi.org/10.1111/pbr.12750.

Anonymous (1986). *Fifteen Years of NIAB*. Faisalabad, Pakistan: Nuclear Institute for Agriculture and Biology.

Ashraf, M. (1994). Breeding for salinity tolerance in plants. *Crit. Rev. Plant Sci.* 13: 17–42.

Ashraf, M., Athar, H.R., Harris, P.J.C., and Kwon, T.R. (2008). Some prospective strategies for improving crop salt tolerance. *Adv. Agron.* 97: 45–110.

Ashraf, M., Akram, N.A., Rahman, M.U., and Foolad, M.R. (2012). Marker-assisted selection in plant breeding for salinity tolerance. *Methods Mol. Biol.* 913:305–333.

Atabaki, N., Nulit, R., Kalhori, N. et al. (2018). In vitro selection and development of Malaysian salt tolerant rice (*Oryza sativa* L. cv. MR263) under salinity. *Acta Sci. Agric.* 2: 8–17.

Bajaj, S. and Mohanty, A. (2005). Recent advances in rice biotechnology towards genetically superior transgenic rice. *Plant Biotechnol. J.* 3: 275–307.

Bandillo, N., Raghavan, C., Muyco, P.A. et al. (2013). Multi-parent advanced generation inter-cross (MAGIC) populations in rice: progress and potential for genetics research and breeding. *Rice* 6: 1–15.

Bansal, K.C., Lenka, S.K., and Mondal, T.K. (2013). Genomic resources for breeding crops with enhanced abiotic stress tolerance. *Plant Breed.* 133: 1–11.

Batayeva, D., Labaco, B., Ye, C. et al. (2018). Genome-wide association study of seedling stage salinity tolerance in temperate *japonica* rice germplasm. *BMC Genet.* 19: 2.

Bhandari, A., Bartholome, J., Cao-Hamadoun, T.-V. et al. (2019). Selection of trait-specific markers and multi-environment models improve genomic predictive ability in rice. *PLoS ONE* 14: e02008871.

Bhat, J.A., Ali, S., Salgotra, R.K. et al. (2016). Genomic selection in the era of next generation sequencing for complex traits in plant breeding. *Front. Genet.* 7: 221.

Bimpong, I.K., Manneh, B., El-Namaky, R. et al. (2013). Mapping QTLs related to salt tolerance in rice at the young seedling stage using 384-plex single nucleotide polymorphism SNP marker sets. *Mol. Plant Breed.* 5: 47–63.

Biradar, H., Karan, R., and Subudhi, P.K. (2018). Overexpression of a salt responsive protein3-1 (*SaSRP3-1*) as well as pyramiding with *SaVHAc1* from *Spartina alterniflora* L. enhances salt tolerance in rice. *Front. Plant Sci.* 9: 1304.

Bizimana, J.B., Luzi-Kihupi, A., Murori, R.W., and Singh, R.K. (2017). Identification of quantitative trait loci for salinity tolerance in rice (*Oryza sativa* L.) using IR29/Hasawi mapping salinity population. *J. Genet.* 96: 571–582.

Bonilla, P., Dvorak, J., Mackill, D. et al. (2002). RFLP and SSLP mapping of salinity tolerance genes in chromosome 1 of rice (*Oryza sativa* L.) using recombinant inbred lines. *Philipp Agric. Sci.* 85: 68–76.

Breseghello, F. and Guedes, A.S. (2013). Traditional and modern plant breeding methods with examples in rice (*Oryza sativa* L.). *J. Agric. Food Chem.* 35: 8277–8286.

Carroll, D. (2014). Genome engineering with targetable nucleases. *Annu. Rev. Biochem.* 83: 409–439.

Chapagain, S., Park, Y.C., Kim, J.H., and Jang, C.S. (2018). *Oryza sativa* salt induced RING E3 ligase 2 (*OsSIRP2*) acts as a positive regulator of transketolase in plant response to salinity and osmotic stress. *Planta* 247: 925–939.

Cha-um, S., Srianan, B., Pichakum, A., and Kirdmanee, C. (2009). An efficient procedure for embryogenetic callus induction and double haploid plant regeneration through anther culture of Thai aromatic rice (*Oryza sativa* L. subsp. *indica*). *In Vitro Cell. Dev. Biol.* 45: 171–179.

Chen, R., Cheng, Y., Han, S. et al. (2017). Whole genome sequencing and comparative transcriptome analysis of a novel seawater adapted, salt resistant rice cultivar - sea rice 86. *BMC Genomics* 18: 655.

Chinnusamy, V., Jagendorf, A., and Zhu, J.K. (2005). Understanding and improving salt tolerance in plants. *Crop Sci.* 45: 437–445.

Choudhary, M., Wani, S.H., Kumar, P. et al. (2019). QTLian breeding for climate resilience in cereals: progress and prospects. *Funct. Integr. Genomics* 19: 685–701.

Chutimanukul, P., Kositsup, B., Plaimas, K. et al. (2018). Photosynthetic responses and identification of salt tolerance genes in a chromosome segment substitution line of "Khao Dawk Mali 105" rice. *Environ. Exp. Bot.* 155: 497–508.

Cohen-Tannoudji, M., Robine, S., Choulika, A. et al. (1998). I-SceI-induced gene replacement at a natural locus in embryonic stem cells. *Mol. Cell. Biol.* 18: 1444–1448.

Colaco, A.F., Molin, J.S., Rossel-polo, J.R., and Escola, A. (2018). Application of light detection and ranging and ultrasonic sensors to high-throughput phenotyping and precision horticulture: current status and challenges. *Hortic. Res.* 1: 1.

Collins, N.C., Tardieu, F., and Tuberosa, R. (2008). Quantitative trait loci and crop performance under abiotic stress: where do we stand? *Plant Physiol.* 147: 469–486.

Cong, L., Ran, F.A., Cox, D. et al. (2013). Multiplex genome engineering using CRISPR/Cas systems. *Science* 339: 819–823.

Cui, Y.R., Zhang, F., and Zhou, Y.L. (2018). The application of multi-locus GWAS for the detection of salt-tolerance loci in rice. *Front. Plant Sci.* 9: 1464.

Das, P., Mishra, M., Lakra, N. et al. (2014). Mutation breeding: a powerful approach for obtaining abiotic stress tolerant crops and upgrading food security for human nutrition. In: *Mutagenesis: Exploring Novel Genes and Pathways* (eds. N.B. Tomlekova, M.I. Kozgar and M.R. Wani), 15–36. Wageningen: Wageningen Academic Publisher.

Das, P., Nutan, K.K., Singla-Pareek, S.L., and Pareek, A. (2015). Understanding salinity responses and adopting "omics-based" approaches to generate salinity tolerant cultivars of rice. *Front. Plant Sci.* 6: 712.

De Leon, T.B., Linscombe, S., Gregorio, G.B., and Subudhi, P.K. (2015). Genetic variation in Southern USA rice genotypes for seedling salinity tolerance. *Front. Plant Sci.* 6: 374.

De Leon, T.B., Linscombe, S., and Subudhi, P.K. (2016). Molecular dissection of seedling salinity tolerance in rice (*Oryza sativa* L.) using a high-density GBS-based SNP linkage map. *Rice* 9: 52.

De Leon, T.B., Linscombe, S., and Subudhi, P.K. (2017). Identification and validation of QTLs for seedling salinity tolerance in introgression lines of a salt tolerant rice landrace "Pokkali.". *PLoS ONE* 12: e0175361.

Deng, P., Jiang, D., Dong, Y. et al. (2015). Physiological characterization and fine mapping of salt-tolerant mutant in rice (*Oryza sativa*). *Funct. Plant Biol.* 42: 1026–1035.

Dou, M., Fan, S., Yang, S. et al. (2017). Overexpression of *AmRosea1* gene confers drought and salt tolerance in rice. *Int. J. Mol. Sci.* 18: 2.

Duan, Y.B., Li, J., Qin, R.Y. et al. (2016). Identification of a regulatory element responsible for salt induction of rice *OSRAV2* through ex situ and in situ promoter analysis. *Plant Mol. Biol.* 90: 49–62.

Elshire, R.J., Glaubitz, J.C., Poland, J.A. et al. (2011). A robust, simple genotyping-by-sequencing (GBS) approach for high diversity species. *PLoS ONE* 6: 1–10.

Fan, H., Zhang, J., Liu, X. et al. (2016). Effect of different potassium levels on the growth and photosynthesis of sweet sorghum seedlings under salinity stress. In: *Proc. Intl. Conf. on Energy, Environmental, and Material Science (EEMS 2015)*. Guanghzou, China, August 25-26, 2015, pp 169–173.

FAO (2008). FAO Land and Plant Nutrition Management Service. http://www.fao.org/ag/agl/agll/spush

Fita, A., Rodríguez-Burruezo, A., Boscaiu, M. et al. (2015). Breeding and domesticating crops adapted to drought and salinity: a new paradigm for increasing food production. *Front. Plant Sci.* 6: 978.

Formentin, E., Sudiro, C., Perin, G. et al. (2018). Transcriptome and cell physiological analyses in different rice cultivars provide new insights into adaptive and salinity stress response. *Front. Plant. Sci.* 9: 204.

Fotoukian, M.H., Taleei, A.R., Gharahyazi, B. et al. (2005). QTL mapping of genes affecting salt tolerance in rice (*Oryza sativa* L.) using microsatellite markers. *Iran. J. Crop. Sci.* 6: 361–373.

Frouin, J., Languillaume, A., Mas, J. et al. (2018). Tolerance to mild salinity stress in *japonica* rice: a genome-wide association mapping study highlights calcium signaling and metabolism genes. *PLoS ONE* 13: e0190964.

Ganie, S.A., Pani, D.R., and Mondal, T.K. (2017). Genome-wide analysis of DUF221 domain-containing gene family in *Oryza* species and identification of its salinity stress-responsive members in rice. *PLoS ONE* 12: e0182469.

Gehan, M.A., Fahlgren, N., Abbasi, A. et al. (2017). PlantCV v2: image analysis software for high-throghput plant phenotyping. *Peer J.* 5: e4088.

Germana, M.A. (2011). Gametic embryogenesis and haploid technology as valuable support to plant breeding. *Plant Cell Rep.* 30: 839–857.

Gimhani, D.R., Gregorio, G.B., and Kottearachchi, N.S. (2016). SNP-based discovery of salinity-tolerant QTLs in a bi-parental population of rice (*Oryza sativa*). *Mol. Genet. Genomics* 291: 2081–2099.

Gong, J., He, P., Qian, Q. et al. (1999). Identification of salt-tolerance QTL in rice (*Oryza sativa* L.). *Chin. Sci. Bull.* 44: 68–71.

Gregorio, G.B., Senadhira, D., Mendoza, R.D. et al. (2002). Progress in breeding for salinity tolerance and associated abiotic stresses in rice. *Field Crops Res.* 76: 91–101.

Grenier, C., Cao, T.-V., Ospina, Y. et al. (2015). Accuracy of genomic selection in a rice synthetic population developed for recurrent selection breeding. *PLoS ONE* 11: e0154976.

Guo, Q., Wu, F., Pang, S. et al. (2018). Crop 3D—a LiDAR based platform for 3D high-throughput crop phenotyping. *Sci. China-Life Sci.* 61: 328–339.

Gupta, P.K., Balyan, H.S., Gahlaut, V., and Kulwal, P.L. (2012). Phenotyping, genetic dissection and breeding for drought and heat tolerance in common wheat: status and prospects. In: *Plant Breeding Reviews*, (ed. J. Janick), 36:85–168. Hoboken, NJ: Wiley https://doi.org/10.1002/9781118358566.ch2.

Hayashi, Y., Takehisa, H., Kazama, Y. et al. (2007). Effects of ion beam irradiation on mutation induction in rice. In: *Proceedings of the 18th international conference on Cyclotrons and their applications 2007*. Giardini Naxos, Italy.1–5 Oct. 2007 (eds. L.A.C. Piazza and D. Riffuggiato), 237–239.

Ho, V.T., Thomson, M.J., and Ismail, A.M. (2016). Development of salt tolerant IR64 near isogenic lines through marker-assisted breeding. *J. Crop. Sci. Biotech.* 19: 373–381.

Hoang, T.M., Tran, T.N., Nguyen, T.K.T. et al. (2016). Improvement of salinity stress tolerance in rice: challenges and opportunities. *Agronomy* 6: 54.

Hong, Y., Zhang, H., Huang, L. et al. (2016). Overexpression of a stress-responsive NAC transcription factor gene *ONAC022* improves drought and salt tolerance in rice. *Front. Plant Sci.* 7: 4.

Hoque, A.B.M.Z., Haque, M.A., Sarker, M.R.A., and Rahman, M.A. (2015). Marker-assisted introgression of Saltol locus into genetic background of BRRI dhan49. *Int. J. Biosci.* 6: 71–80.

Hossain, H., Rahman, M.A., Alam, M.S., and Singh, R.K. (2015). Mapping of quantitative trait loci associated with reproductive-stage salt tolerance in rice. *J. Agro. Crop. Sci.* 201: 17–31.

Houle, D., Govindaraju, D.R., and Omholt, S. (2010). Phenomics: the next challenge. *Nat. Rev. Genet.* 11: 855–866.

Huang, X., Lu, T., and Han, B. (2013). Resequencing rice genomes: an emerging new era of rice genomics. *Trends Genet.* 29: 225–232.

Huang, Y., Guo, Y., Liu, Y. et al. (2018). 9-*cis*-Epoxycarotenoid dioxygenase 3 regulates plant growth and enhances multi-abiotic stress tolerance in rice. *Front. Plant Sci.* 9: 162.

Huyen, N.T.L., Cuc, M.L., Ismail, A.M., and Ham, H.L. (2012). Introgression the salinity tolerance QTLs *Saltol* into AS996, the elite rice variety of Vietnam. *Am. J. Plant Sci.* 3: 981–987.

Islam, M.R. and Gregorio, G.B. (2013). Progress of salinity tolerant rice variety development in Bangladesh. *SABRAO J. Breed. Genet.* 45: 21–30.

Islam, S., Ontoy, J., and Subudhi, P.K. (2019). Meta-analysis of quantitative trait loci (QTLs) associated with seedling stage salt tolerance in rice (*Oryza sativa* L.). *Plants* 8: 33.

Ismail, A. and Horie, T. (2017). Genomics, physiology, and molecular breeding approaches for improving salt tolerance. *Annu. Rev. Plant Biol.* 68: 405–434.

Ismail, A.M., Heuer, S., Thomson, M.J., and Wissuwa, M. (2007). Genetic and genomic approaches to develop rice germplasm for problem soils. *Plant Mol. Biol.* 65: 547–570.

Jain, M., Moharana, K.C., Shankar, R. et al. (2014). Genome-wide discovery of DNA polymorphisms in rice cultivars with contrasting drought and salinity stress response and their functional relevance. *Plant Biotechnol. J.* 12: 253–264.

James, D., Borphukan, B., Fartyal, D. et al. (2018). Concurrent overexpression of *OsGS1;1* and *OsGS2* genes in transgenic rice (*Oryza sativa* L.): impact on tolerance to abiotic stresses. *Front. Plant Sci.* 9: 786.

Kanawapee, N., Sanitchon, J., Srihaban, P., and Theerakulpisut, P. (2011). Genetic diversity analysis of rice cultivars (*Oryza sativa* L.) differing in salinity tolerance using RAPD and SSR markers. *Electron. J. Biotechnol.* 14: 1–17.

Karan, R. and Subudhi, P.K. (2011). Approaches to increasing salt tolerance in crop plants. In: *Abiotic Stress Responses in Plants: Metabolism to Productivity* (eds. P. Ahmad and M.N.V. Prasad), 63–88. New York: Springer Science+Business Media. Springer.

Kathuria, H., Giri, J., Tyagi, H., and Tyagi, A. (2007). Advances in transgenic rice biotechnology. *Crit. Rev. Plant Sci.* 26: 65–103.

Kim, D.M., Ju, H.G., Kwon, T.R. et al. (2009). Mapping QTLs for salt tolerance in an introgression lines population between *japonica* cultivars in rice. *J. Crop. Sci. Biotech.* 12: 121–128.

Kim, S.L., Chung, Y.S., Silva, R.R. et al. (2019). The opening of phenome-assisted selection era in the early seedling stage. *Sci. Rep.* 1: 1.

Klose, R., Penlington, J., and Ruckelshausen, A. (2009). Usability study of 3D time-of-flight cameras for automatic plant phenotyping. *Bornimer Agrartech.* 69: 93–105.

Kumar, K., Kumar, M., Kim, S.R. et al. (2013). Insights into genomics of salt stress responses in rice. *Rice* 6: 27.

Kumar, V., Singh, A., Mithra, S.V.A. et al. (2015). Genome-wide association mapping of salinity tolerance in rice (*Oryza sativa*). *DNA Res.* 22: 133–145.

Kumar, M., Choi, J., An, G., and Kim, S.R. (2017). Ectopic expression of OsSta2 enhances salt stress tolerance in rice. *Front. Plant Sci.* 8: 316.

Lang, N.T., Buu, B.C., and Ismai, A. (2008). Molecular mapping and marker-assisted selection for salt tolerance in rice (*Oryza sativa* L.). *Omon Rice* 16: 50–56.

Lee, I.S., Kim, D.S., Hyun, D.Y. et al. (2003a). Isolation of gamma-induced rice mutants with increased tolerance to salt by anther culture. *J. Plant Biotech.* 5: 51–57.

Lee, K.S., Choi, W.Y., Ko, J.C. et al. (2003b). Salinity tolerance of *japonica* and *indica* rice (*Oryza sativa* L.) at the seedling stage. *Planta* 216: 1043–1046.

Lekklar, C., Pongpanich, M., Suriya-arunroj, D. et al. (2019). Genome-wide association study for salinity tolerance at the flowering stage in a panel of rice accessions from Thailand. *BMC Genom.* 20: 76.

Li, L., Zhang, Q., and Huang, D. (2014). A review of imaging techniques for plant phenotyping. *Sensors (Basel)* 11: 20078.

Li, W., Qiang, X.J., Han, X.R. et al. (2018). Ectopic expression of a *Thellungiella salsuginea* aquaporin gene, *TsPIP1;1*, increased the salt tolerance of rice. *Int. J. Mol. Sci.* 19: 2229.

Liao, Y.D., Lin, K.H., Chen, C.C., and Chang, C.M. (2016). *Oryza sativa* protein phosphatase 1a (*OsPP1a*) involved in salt stress tolerance in transgenic rice. *Mol. Breed.* 36: 22.

Lim, S.D., Jung, C.G., Park, Y.C. et al. (2015). Molecular dissection of a rice microtubule-associated RING finger protein and its potential role in salt tolerance in *Arabidopsis*. *Mol. Biol.* 89: 365–384.

Lin, H.X., Zhu, M.Z., Yano, M. et al. (2004). QTLs for Na^+ and K^+ uptake of the shoots and roots controlling rice salt tolerance. *Theor. Appl. Genet.* 108: 253–260.

Linh, H.L., Linh, H.T., Xuan, D.T. et al. (2012). Molecular breeding to improve salt tolerance of rice (*Oryza sativa* L.) in the red river delta of Vietnam. *Int. J. Plant Genomics.* 2012: 949038. https://doi.org/10.1155/2012/949038.

Mali, P., Yang, L., Esvelt, K.M. et al. (2013). RNA-guided human genome engineering via Cas9. *Science* 339: 823–826.

Mansueto, L., Fuentes, R., Borja, F. et al. (2017). Rice SNP-seek database update: new SNPs, indels, and queries. *Nucleic Acids Res.* 45: D1075–D1081.

Mansuri, R.M., Shobbar, Z.S., Jelodar, N.B. et al. (2019). Dissecting molecular mechanisms underlying salt tolerance in rice: a comparative transcriptional profiling of the contrasting genotypes. *Rice* 1: 1.

Masood, M.S., Seiji, Y., Shinwari, Z.K., and Anwar, R. (2004). Mapping quantitative trait loci (QTLs) for salt tolerance in rice (*Oryza sativa* L.). using RFLPs. *Pak. J. Bot.* 36: 825–834.

McCouch, S.R. and Kovach, M.J. (2008). Leveraging natural diversity: back through the bottleneck. *Curr. Opin. Plant Biol.* 11: 193–200.

Mir, R.R., Reynolds, M., Pinto, F. et al. (2019). High-throughput phenotyping for crop improvement in the genomics era. *Plant Sci.* 282: 60–72.

Mishra, B. (1994). Breeding for salt tolerance in crops. In: *Salinity Management for Sustainable Agriculture* (eds. D.L.N. Rao, N.T. Singh, R.K. Gupta and N.K. Tyagi), 226–259. Karnal, India: Central Soil Salinity Research Institute.

Mishra, S., Singh, B., Misra, P. et al. (2016). Haplotype distribution and association of candidate genes with salt tolerance in Indian wild rice germplasm. *Plant Cell Rep.* 35: 2295–2308.

Mohammadi, R., Mendioro, M.S., Diaz, G.Q. et al. (2013). Mapping quantitative trait loci associated with yield and yield components under reproductive stage salinity stress in rice (*Oryza sativa* L.). *J. Genet.* 92: 433–443.

Munns, R. (2002). Comparative physiology of salt and water stress. *Plant Cell Environ.* 25: 239–250.

Munns, R. (2005). Genes and salt tolerance: bringing them together. *New Phytol.* 167: 645–663.

Munns, R. and Gilliham, M. (2015). Salinity tolerance of crops—what is the cost? *New Phytol.* 208: 668–673.

Munns, R. and Tester, M. (2008). Mechanisms of salinity tolerance. *Annu. Rev. Plant Biol.* 59: 651–681.

Munns, R., James, R.A., and Läuchli, A. (2006). Approaches to increasing the salt tolerance of wheat and other cereals. *J. Exp. Bot.* 57: 1025–1043.

Munns, R., James, R.A., Furbank, R.T. et al. (2010). New phenotyping methods for screening wheat and barley for beneficial responses to water deficit. *J. Exp. Bot.* 61: 3499–3507.

Mustafa, G., Soomro, A.M., Baloch, A.W., and Siddiqui, K.A. (1997). "Shua-92," a new cultivar of rice (*Oryza sativa* L) developed through fast neutrons irradiation. Mutation Breeding Newsletter, IAEA, Vienna, Austria, 43: 35–36.

Naveed, S.A., Zhang, F., Zhang, J. et al. (2018). Identification of QTN and candidate genes for salinity tolerance at the germination and seedling stages in rice by genome-wide association analyses. *Sci. Rep.* 8: 6505.

Negrão, S., Courtois, B., Ahmadi, N. et al. (2011). Recent updates on salinity stress in rice: from physiological to molecular responses. *Crit. Rev. Plant Sci.* 30: 329–377.

Negrão, S., Almadanim, M.C., Pires, I.S. et al. (2013). New allelic variants found in key rice salt-tolerance genes: an association study. *Plant Biotechnol. J.* 11: 87–100.

Nounjan, N., Siangliw, J.L., Toojinda, T. et al. (2016). Salt-responsive mechanisms in chromosome segment substitution lines of rice (*Oryza sativa* L. cv. KDML105). *Plant Physiol. Biochem.* 103: 96–105.

Oono, Y., Kawahara, Y., Kanamori, H. et al. (2011). mRNA-Seq reveals a comprehensive transcriptome profile of rice under phosphate stress. *Rice* 4: 50–65.

Platten, J.D., Egdane, J.A., and Ismail, A.M. (2013). Salinity tolerance, Na^+ exclusion and allele mining of *HKT1;5* in *Oryza sativa* and *O. glaberrima*: many sources, many genes, one mechanism? *BMC Plant Biol.* 13: 32.

Prasad, S.R., Bagali, P.G., Hittalmani, S., and Shashidhar, H.E. (2000). Molecular mapping of quantitative trait loci associated with seedling tolerance to salt stress in rice (*Oryza sativa* L.). *Curr. Sci.* 78: 162–164.

Prusty, M.R., Kim, S.R., Vinarao, R. et al. (2018). Newly identified wild rice accessions conferring high salt tolerance might use a tissue tolerance mechanism. *Front. Plant Sci.* 9: 417.

Punyawaew, K., Suriya-arunroj, D., Siangli, M. et al. (2016). Thai jasmine rice cultivar KDML105 carrying *Saltol* QTL exhibiting salinity tolerance at seedling stage. *Mol. Breed.* 36: 150.

Puram, V.R.R., Ontoy, J., Linscombe, S., and Subudhi, P.K. (2017). Genetic dissection of seedling stage salinity tolerance in rice using introgression lines of a salt tolerant landrace Nona Bokra. *J. Hered.* 108: 658–670.

Puram, V.R.R., Ontoy, J., and Subudhi, P.K. (2018). Identification of QTLs for salt tolerance traits and prebreeding lines with enhanced salt tolerance in an introgression line population of rice. *Plant Mol. Biol. Rep.* 36: 695–709.

Quijano-Guerta, C. and Kirk, G.J.D. (2002). Tolerance of rice germplasm to salinity and other soil chemical stresses in tidal wetlands. *Field Crops Res.* 76: 111–121.

Rahman, H., Ramanathan, V., Nallathambi, J. et al. (2016). Over-expression of a NAC 67 transcription factor from finger millet (*Eleusine coracana* L.) confers tolerance against salinity and drought stress in rice. *BMC Biotechnol.* 16 (Suppl 1): 35.

Rahman, M.A., Bimpong, I.K., Bizimana, J.B. et al. (2017). Mapping QTLs using a novel source of salinity tolerance from Hasawi and their interaction with environments in rice. *Rice* 10: 47.

Rajendran, K., Tester, M., and Roy, S.J. (2009). Quantifying the three main components of salinity tolerance in cereals. *Plant Cell Environ.* 32: 237–249.

Rana, M.M., Takamatsu, T., Baslam, M. et al. (2019). Salt tolerance improvement in rice through efficient SNP marker-assisted selection coupled with speed-breeding. *Int. J. Mol. Sci.* 20: 2585.

Rashid, M.H. and Nasrin, S. (2015). Productivity and preference of salt tolerant Boro rice varieties in saline non-gher and gher ecosystems. *Bangladesh Rice J.* 18 (1–2): 18–23.

Reddy, A.M., Francies, R.M., Rasool, S.N., and Reddy, V.R.P. (2014). Breeding for tolerance stress triggered by salinity in rice. *Int. J. Appl. Biol. Pharm.* 5: 167–176.

Reddy, I.N.B.L., Kim, B.K., In-Sun Yoon, I.S., and Kim, K.H. (2017). Salt tolerance in rice: focus on mechanisms and approaches. *Rice Sci.* 24: 123–144.

Ren, Z., Gao, J., Li, L. et al. (2005). A rice quantitative trait locus for salt tolerance encodes a sodium transporter. *Nat. Genet.* 37: 1141–1146.

Robertsen, C.D., Rasmus, L., Hjortshøj, R.L., and Janss, L.L. (2019). Genomic selection in cereal breeding. *Agronomy* 9: 95.

Roy, S.J., Negrão, S., and Tester, M. (2014). Salt resistant crop plants. *Curr. Opin. Biotech.* 26: 115–124.

Sabouri, H. and Sabouri, A. (2008). New evidence of QTLs attributed to salinity tolerance in rice. *Afr. J. Biotechnol.* 7: 4376–4383.

Sakina, A., Ahmed, I., Shahzad, A. et al. (2016). Genetic variation for salinity tolerance in Pakistani rice (*Oryza sativa* L.) germplasm. *J. Agron. Crop. Sci.* 202: 25–36.

Saleem, M.Y., Mukhtar, Z., Cheema, A.A., and Atta, B.M. (2005). Induced mutation and in vitro techniques as a method to induce salt tolerance in Basmati rice (*Oryza sativa* L.). *Int. J. Environ. Sci. Technol.* 2: 141–145.

Samaco, A.M., Villa, N.O., and Gregorio, G.B. (2018). Salinity tolerance and traits correlations of selected magic *indica rice* (*Oryza sativa* L.) genotypes at seedling stage. *Philipp Agric. Sci.* 101: 312–325.

Sengupta, S. and Majumder, A.L. (2010). *Porteresia coarctata* (Roxb.) Tateoka, a wild rice: a potential model for studying salt-stress biology in rice. *Plant Cell Environ.* 33: 526–542.

Shahbaz, M. and Ashraf, M. (2013). Improving salinity tolerance in cereals. *Crit. Rev. Plant Sci.* 32: 237–249.

Shankar, R., Bhattacharjee, A., and Jain, M. (2016). Transcriptome analysis in different rice cultivars provides novel insights into desiccation and salinity stress responses. *Sci. Rep.* 6: 23719.

Shelden, M.C. and Roessner, U. (2013). Advances in functional genomics for investigating salinity stress tolerance mechanisms in cereals. *Front. Plant Sci.* 4: 123.

Shi, Y., Gao, L., Wu, Z. et al. (2017). Genome-wide association study of salt tolerance at the seed germination stage in rice. *BMC Plant Biol.* 17: 92.

Singh, R.K., Gopalakrishnan, S., Singh, V.P. et al. (2011). Marker assisted selection: a paradigm shift in basmati breeding. *Indian J. Genet.* 71: 120–128.

Singh, N., Choudhury, D.R., Tiwari, G. et al. (2016a). Genetic diversity trend in Indian rice variety: an analysis using SSR markers. *BMC Genet.* 17: 1–13.

Singh, R., Singh, Y., Xalaxo, Y. et al. (2016b). From QTL to variety-harnessing the benefits of QTLs for drought, flood and salt tolerance in mega rice varieties of India through a multi-institutional network. *Plant Sci.* 242: 278–287.

Singh, B., Mishra, S., Bohra, A. et al. (2018a). Crop phenomics for abiotic stress tolerance in crop plants. In: *Biochemical, Physiological and Molecular Avenues for Combating Abiotic Stress Tolerance in Plants* (ed. S.H. Wani), 277–296. New York: Elsevier.

Singh, V.K., Singh, B.D., Kumar, A. et al. (2018b). Marker-assisted introgression of Saltol QTL enhances seedling stage tolerance in the rice variety "Pusa Basmati 1.". *Int. J. Genomics* 2018: 8319879. https://doi.org/10.1155/2018/8319879.

Spindel, J. and Iwata, H. (2018). Genomic selection in rice breeding. In: *Rice Genomics, Genetics and Breeding* (eds. T. Sasaki and M. Ashikari), 473–496. Singapore: Springer.

Subudhi, P.K. (2011). Omics approaches for abiotic stress tolerance in plants. In: *Omics and Plant Abiotic Stress Tolerance* (eds. N. Tuteja, S.S. Gill and R. Tuteja), 10–38. Bentham Science Publishers Ltd. https://doi.org/10.2174/97816080505811110101.

Takagi, H., Abe, A., Yoshida, K. et al. (2013). QTL-seq: rapid mapping of quantitative trait loci in rice by whole genome resequencing of DNA from two bulked populations. *Plant J.* 74: 174–183.

Takagi, H., Tamiru, M., Abe, A. et al. (2015). MutMap accelerates breeding of a salt-tolerant rice cultivar. *Nat. Biotechnol.* 33: 445–449.

Takehisa, H., Shimodate, T., Fukuta, Y. et al. (2004). Identification of quantitative trait loci for plant growth of rice in paddy field flooded with salt water. *Field Crops Res.* 89: 85–95.

Tamimi, N.A., Brien, C., Oakey, H. et al. (2016). Salinity tolerance loci revealed in rice using high-throughput non-invasive phenotyping. *Nat. Commun.* 7: 13342.

Tang, Y., Bao, X., Zhi, Y. et al. (2019). Overexpression of a MYB family gene, *OsMYB6*, increases drought and salinity stress tolerance in transgenic rice. *Front. Plant Sci.* 10: 168.

Thitisaksakul, M., Tananuwong, K., Shoemaker, C.F. et al. (2015). Effect of timing and severity of salinity stress on rice (*Oryza sativa* L.) yield, grain composition, and starch functionality. *J. Agric. Food Chem.* 63: 2296–2304.

Thomson, M.J., Ocampo, M., and Egdane, J. (2010). Characterizing the *Saltol* quantitative trait locus for salinity tolerance in rice. *Rice* 3: 148–160.

Thuy, N.T., Tokuyasu, M., Mai, N.T., and Hirai, Y. (2018). Identification and characterization of chromosome regions associated with salinity tolerance in rice. *J. Agric. Sci.* 10 (11): 57–68. https://doi.org/10.5539/jas.v10n11p57.

Tiwari, S., Krishnamurthy, S.L., Kumar, V. et al. (2016). Mapping of QTLs for salt tolerance in rice (*Oryza sativa* L.) by bulked segregant analysis of recombinant inbred lines using 50K SNP chip. *PLoS ONE* 11: e0153610.

Tsai, Y.C., Chen, K.C., Cheng, T.S. et al. (2019). Chlorophyll fluorecence analysis in diverse rice varieties reveals the positive correaltion between the seedlings' salt tolerance and photosynthetic efficiency. *BMC Plant Biol.* 19: 403.

Valarmathi, M., Sasikala, R., Rahman, R. et al. (2019). Development of salinity tolerant version of a popular rice variety improved white ponni through marker assisted backcross breeding. *Plant Physiol. Rep.* 24: 262–271.

Van As, H., Scheenen, T., and Vergeldt, F.J. (2009). MRI of intact plants. *Photosynth. Res.* 102: 213.

Vu, H.T.T., Le, D.D., Ismail, A.M., and Le, H.H. (2012). Marker-assisted backcrossing (MABC) for improved salinity tolerance in rice (*Oryza sativa* L.). to cope with climate change in Vietnam. *Aust. J. Crop Sci.* 6: 1649–1654.

Wang, C., Yang, Y., Wang, H. et al. (2016). Ectopic expression of a cytochrome P450 monooxygenase gene PtCYP714A3 from *Populus trichocarpa* reduces shoot growth and improves tolerance to salt stress in transgenic rice. *Plant Biotechnol. J.* 149: 1838–1851.

Wang, J., Zhu, J., Zhang, Y. et al. (2017a). Comparative transcriptome analysis reveals molecular responses to salinity stress of salt tolerant and sensitive genotypes of *indica* rice at seedling stage. *Sci. Rep.* 8: 2045–2322.

Wang, X., Li, L., Yang, Z. et al. (2017b). Predicting rice hybrid performance using univariate and multivariate GBLUP models based on North Carolina mating design II. *Heredity* 118: 302–310.

Wang, B., Zhong, Z., Zhang, H. et al. (2019). Targeted mutagenesis of NAC transcription factor gene. *OsNAC041,* leading to salt sensitivity in rice. *Rice Sci.* 26: 98–108.

Wyman, C. and Kanaar, R. (2006). DNA double-strand break repair: all's well that ends well. *Annu. Rev. Genet.* 40: 363–383.

Xiang, D.J., Man, L.L., Zhang, C.L. et al. (2018). A new Em-like protein from *Lactuca sativa*, *LsEm1*, enhances drought and salt stress tolerance in *Escherichia coli* and rice. *Protoplasma* 255: 1089–1106.

Ye, J. and Cui, X. (2019). Next-generation crop breeding methods. *Mol. Plant.* 12: 470–471.

Yu, J., Zhao, W., He, Q. et al. (2017). Genome-wide association study and gene set analysis for understanding candidate genes involved in salt tolerance at the rice seedling stage. *Mol. Genet. Genomics* 292: 1391–1403.

Yu, J., Zhao, W., Tong, W. et al. (2018). A genome-wide association study reveals candidate genes related to salt tolerance in rice (*Oryza sativa*) at the germination stage. *Int. J. Mol. Sci.* 19: 3145.

Zang, J.P., Sun, Y., Wang, Y. et al. (2008). Dissection of genetic overlap of salt tolerance QTLs at the seedling and tillering stages using backcross introgression lines in rice. *Sci. China C Life Sci.* 51: 583–591.

Zeng, L. and Shannon, M.C. (2000). Salinity effects on seedling growth and yield components of rice. *Crop. Sci.* 40: 996–100.

Zhang, D., Wang, Z., Wang, N. et al. (2014). Tissue culture-induced heritable genomic variations in rice, and their phenotypic implications. *PLoS ONE* 9: 1–10.

Zhang, Y., Fang, J., Wu, X., and Dong, L. (2018). Na^+/K^+ balance and transport regulatory mechanisms in weedy and cultivated rice (*Oryza sativa* L.) under salt stress. *BMC Plant Biol.* 18: 375.

Zhang, A., Liu, Y., Wang, F. et al. (2019). Enhanced rice salinity tolerance via CRISPR/Cas9-targeted mutagenesis of the *OsRR22* gene. *Mol. Breed.* 39: 47.

Zhou, Y., Liu, C., Tang, D. et al. (2018). The receptor-like cytoplasmic kinase STRK1 phosphorylates and activates CatC, thereby regulating H_2O_2 homeostasis and improving salt tolerance in rice. *Plant Cell* 30 (5): 1100–1118. https://doi.org/10.1105/tpc.17.01000.

3

Molecular Breeding for Improving Drought Tolerance in Rice: Recent Progress and Future Perspectives

Ratna R. Majumder[1,4], *Sandeep Sakhale*[1,2,5], *Shailesh Yadav*[1], *Nitika Sandhu*[3], *Lutful Hassan*[4], *Md. Amir Hossain*[4] *and Arvind Kumar*[1,5]

[1] *Rice Breeding Platform, International Rice Research Institute, Los Banos, Laguna 4031, Philippines*
[2] *Department of Crop Sciences, University of Illinois at Urbana-Champaign, Urbana IL 61801, USA*
[3] *Punjab Agricultural University, Ludhiana, Punjab 141004, India*
[4] *Bangladesh Agricultural University, Mymensingh 2202, Bangladesh*
[5] *IRRI South Asia Regional Centre (ISARC), Varanasi, Uttar Pradesh 221106, India*

Introduction

The Drought Challenges and Its Distribution

Rice is the staple food for more than three billion people in Asia. Ninety percent of the world's rice is produced and consumed in Asia (Li and Xu 2007). Rice is grown on 154 million hectares (Mha) worldwide with 45% of the rice area being rainfed, where yield is seriously affected by drought (IRRI 2002). Rice production in Bangladesh is regularly affected by drought and more than 20 drought events have been reported in last 50 years particularly in the northern parts of Bangladesh, where rainfall is inconsistent and comparatively less.

Climate Change and Drought

Global warming of the lower atmosphere has been predicted to strengthen the hydrologic cycle (Coumou and Rahmstorf 2012) causing dry regions to become drier and wet regions to become wetter. Due to this, warming evaporation and evapotranspiration increase surface drying and, thereby, the intensity and duration of droughts increase (Vogt et al. 2018). Aridity has been increasing between 0.50% and 1.74% per decade between the 1950s and the 1970s with the increase of drought area by 8% since 1980 (Cherlet et al. 2018). In Asia, normal crop cultivation is hampered by extreme dry weather scenarios, making rice cultivation unstable and difficult. Development of high-yielding drought-tolerant rice varieties for drought-prone environments would be the key to improve rice production and ensure food security in Asia.

Molecular Breeding for Rice Abiotic Stress Tolerance and Nutritional Quality, First Edition.
Edited by Mohammad Anwar Hossain, Lutful Hassan, Khandakar Md. Iftekharuddaula, Arvind Kumar and Robert Henry.

Drought Impacts on the Rice Plant

Drought severely affects the biochemical, physiological components of rice plants. Due to water stress, biomass production of plants is hampered, significantly affecting the plant's development and yield of rice as it experiences high rates of transpiration (Sahebi et al. 2018). Drought hampers the cell expansion growth and elongation more than cell division inhibiting rice seedling germination, causing decrease of the tiller numbers and plant height (Ashfaq et al. 2014). The maturity of crops is delayed as the turgor pressure of cell is hampered under stress and inhibits stem reserves (Blum 2011). The effect of drought in different growth stages of plants is different as several studies have reported early morphological changes in rice under drought (Saddique et al. 2018).

Drought Phenotyping Platforms

Drought Screening Protocols

Drought causes significant damage if it continues for more than seven days, which may lead up to 10% yield damage per day in upland conditions. There should be a controlled environment to test the drought-managed trials to get the desired intensity and timing of the drought (Lafitte et al. 2004). Vegetative screening of a large number of genotypes is always worthwhile in the dry season (Pantuwan et al. 2004). Long gaps between spells of rain in the wet season can cause drought stress at vegetative, flowering, and maturity stages. Drought stress at flowering is the most critical for the crop. Late planting to synchronize the reproductive stage with monsoon withdrawal followed by proper drainage 21 days before the initiation of the reproductive stage enhances the chance of imposing proper terminal drought. The establishment of trials in highlands is also a good way to conduct drought trials because it takes a few weeks less time to run out of the standing water than in the lowlands (Kamoshita et al. 2008).

Field-Based Phenotyping

Early segregating generations – F_2 and F_3 are managed in a non-stress environment, and selection is emphasized based on plant height, growth duration, disease-insect scores, and plant yield. Subsequently, plants are subjected to a drought rainout shelter, and a greenhouse is suitable for screening all types of drought irrespective of photo insensitivity (Wade et al. 2000). Proper pest control measures should be applied in these facilities to protect the crop from damage.

Design and Management of Field Drought Experiments

To conduct any breeding trials, the first step is the selection of an appropriate location followed by an appropriate experimental design. The experiments have to be set up in a targeted environment (TE) where the outcomes of the experiments across seasons and years remains similar. Three types of sub ecosystems for conducting drought experiments in Bangladesh include (i) **rainfed-shallow-favorable environments**, where drought and submergence last for only a short period and rainfall is enough for crop growth; (ii) **rainfed-shallow-drought-prone**, where rainfall is very short or long and there are bimodal rainy seasons; (iii) **rainfed-shallow-drought and submergence-prone environments**, where drought and submergence can occur in the same or different seasons. Data from expert farmers and researchers can be taken into consideration for selecting the site, and then data of known varieties in multi-environmental trials for several years can be analyzed to get clusters of the environments based on correlation value (Fischer et al. 2012).

In early generations, augmented Randomized Complete Block (RCB) design can be followed as the seeds are less and experimental error can be minimized due to unreplicated test entries. Checks of the trials should be replicated to identify the spatial variability of the field (Federer and Raghavarao 1975). For two directions fertility gradient Latin square design is more suitable. The advanced yield trials of the late-generations lines trials the alpha lattice design is better than the RCB design.

Trait-Based Selection for Drought Resistance

To screen the genotypes under stress, several secondary traits have been proposed. Fukai and Cooper (2001) focused on the severity and predictability of drought in the targeted areas influencing grain yield. The primary drought-related traits – namely, rooting depth, root thickness, branching angle, and root distribution pattern (Kato et al. 2006) – when subjected to drought stress have an influence on available soil water (Lilley and Fukai 1994). These primary traits then influence other traits such as maintaining water in the plant, canopy temperature, leaf rolling, and leaf death, which affect the crop and its yield. But tolerance to drought stress is always genotypes-dependent due to their internal water status. To be effective, there must be variation among lines for this trait, and those characters must be easily measurable (Fischer et al. 2012).

Direct Selection for Yield Under Stress

Heritable-trait grain yield under drought conditions has been suggested to be the best option for selecting drought-tolerant genotypes (Kumar et al. 2007). The first approach to mapping quantitative traits loci (QTL) for secondary drought-related traits was unsuccessful as the characteristics were not related to yield increases under drought conditions (Kamoshita et al. 2002). Kumar et al. (2007) found higher broad-sense heritability of grain yield under severe terminal drought stress in the CT9993/IR62266 population and lower heritability for the secondary traits under field conditions. Venuprasad et al. (2007) found that the direct selection for yield had 25–34% greater response under reproductive-stage drought stress. Guan et al. (2010) also emphasized selection for yield on 46 pyramided lines (PLs) of IR64 background under stress and non-stress with a focus similar to the target environments conditions.

Molecular Markers and Genotyping Strategies

Genome-Wide Scans with Microsatellite Markers (Simple Sequence Repeats, SSRs)

The SSRs are simple to use and don't require much sophistication; that's the reason they were used for mapping and introgression of QTLs of grain yield under drought in many studies (Bernier et al. 2007; Dixit et al. 2012a; Mishra et al. 2013; Venuprasad et al. 2012; Vikram et al. 2011). Also, due to the co-dominant nature and adequate distribution in the genome, SSR markers were commonly used in genotyping of various mapping populations including whole-genome scan, selective genotyping or bulk segregant (BSA) analysis. However, this approach is time consuming and cost-ineffective due to laborious gel-based genotyping, making it not feasible in the present era of available cheap DNA sequencing technologies. To overcome the limitations of SSR, next-generation markers such as single nucleotide polymorphism (SNPs) are now available to use.

SNP (Single Nucleotide Polymorphism) Genotyping

SNP markers are the marker of choice because of flexibility, speed, and cost-effectiveness when compared to the previous marker system. High-throughput SNP genotyping offers an abundance of markers, rapid processing of large populations, availability of a variety of genotyping platforms, and straightforward allele calling and database storage due to the bi-allelic nature of SNP markers. Large-scale marker deployment has become a reality due to high-throughput SNP genotyping techniques, which enable the scientist to handle large breeding populations (Thomson 2014). SNP genotyping technologies are evolving at a rapid pace, from low-throughput gel-based approaches to high-throughput and ultra-high throughput technologies. Various SNP genotyping technologies include the low-throughput gel-based approach, the cleaved amplified polymorphic sequence (CAPS) marker approach, the polymerase chain reaction (PCR)-based fluorescently-labeled high-throughput methods, high-resolution melting (HRM) curve analysis, TaqMan®, and Kompetitive allele specific PCR (KASP™) assay, fixed-array systems such as Illumina Infinium, Affymetrix Axiom, and next-generation sequencing (NGS)-enabled approaches such as restriction-enzyme-based genotyping by sequencing (GBS). SNPs are used for diverse plant genetic and genomic studies such as linkage mapping, QTL analysis, genome-wide association studies (GWAS), marker-assisted selection (MAS) techniques, and genomic selection (GS) in various crops including rice (Scheben et al. 2017). GBS skim sequencing where low-coverage "skim sequencing" spreads out the sequence reads across the genome and makes it very difficult for SNP calling approach, was used in rice for mapping populations (0.02×–0.13×) and diverse germplasm (1×) (Huang et al. 2010; Xu et al. 2010). Using the GBS approach 14 additive QTLs for drought-related traits were identified in a recombinant inbred line (RIL) population from a cross between Cocodrie and N22 (Bhattarai and Subudhi 2018) and four stable, novel QTLs ($qDTY_{2.4}$, $qDTY_{3.3}$, $qDTY_{6.3}$, *and* $qDTY_{11.2}$) were identified for higher grain yield under drought from two backcross mapping populations of Swarna*2/Dular and IR11N121*2/Aus196.

Trait-Specific SNPs Used as Diagnostic Markers

Use and deployment of trait-specific SNPs on a routine basis requires a low-cost genotyping system coupled with the capacity to process large-sized breeding populations for a small set of SNP markers (Thomson 2014). Platforms such as Fluidigm's Dynamic Arrays™, Douglas Scientific's Array Tape™, and LGC's automated system for running KASP™ markers, are well suited for screening large breeding populations for fixed set of SNPs. With the advances in sequencing technology, a 10-SNPs panel set tightly linked with three drought QTLs ($qDTY_{1.1}$, $qDTY_{3.1}$, $qDTY_{12.1}$) including some other key trait of rice had been developed at the International Rice Research Institute (IRRI) to utilize for a markers-assisted forward breeding program.

Genetic Basis of Drought Tolerance in Rice

Mapping and Deployment of Drought QTLs

Advanced molecular markers and genomics technologies have played a major role in the selection of drought-tolerance traits. Associated chromosomal regions of QTLs/genes have been identified in tolerant genotypes, and significant improvement in grain yield under water-limited conditions has been successfully demonstrated. Identifying QTLs linked to drought tolerance provides opportunities to breed high-yielding rice varieties suitable for drought-prone areas. Several large-effect grain-yield QTLs under reproductive-stage drought were identified (Table 3.1) such as $qDTY_{1.1}$ (Vikram et al. 2011; Ghimire et al. 2012), $qDTY_{2.1}$ (Venuprasad et al. 2009), $qDTY_{2.2}$ (Venuprasad et al. 2007; Swamy et al. 2013), $qDTY_{3.1}$ (Venuprasad et al. 2009), $qDTY_{4.1}$ (Swamy et al. 2013), $qDTY_{6.1}$ (Venuprasad et al. 2012), $qDTY_{10.1}$ (Swamy et al. 2013), and $qDTY_{12.1}$ (Bernier et al. 2007). Each of these identified QTLs has shown a yield advantage of 300–500 kg ha^{-1} under reproductive-stage (RS) drought stress depending upon the severity and timing of the drought occurrence (Kumar et al. 2014, 2018).

Fine Mapping of Identified Genetic Regions

For successful deployment of QTLs through MAS, fine mapping is crucial for the targeted QTLs to be free from any undesirable linkage. The large size of the regions-encompassing QTLs and the likely presence of undesirable linked genes make it essential to fine map such regions to facilitate their precise introgression and to identify candidate genes within these QTLs. Using high-density genetic maps through GBS, some novel candidate genomic regions that contained four stable novel QTLs ($qDTY_{2.4}$, $qDTY_{3.3}$, $qDTY_{6.3}$, and $qDTY_{11.2}$) for grain yield under drought in multiple environments had been identified recently. These fine-mapped stable QTLs were free from undesirable linkages of tall plant height/early maturity (Yadav et al. 2019).

Development of NILs and PLs Carrying Drought QTLs in Background of High Yielding Rice Varieties

Efforts were made to make tolerant versions of different popular local rice varieties by introgressing major-effect grain-yield QTLs under drought. Table 3.2 illustrates various

Table 3.1 QTLs identified for grain yield under drought for different ecosystems.

QTLs	Donor	Recipient	Ecosystem	Chr	Marker interval	*R* Square	References
$qDTY_{1.1}$	Dhagad deshi	Swarna	Lowland	1	RM431-RM104	32 (P)	Ghimire et al. (2012)
$qDTY_{1.1}$	Dhagad deshi	IR64	Lowland	1	RM104-RM12091	9 (P)	
$qDTY_{1.1}$	N22	Swarna	Lowland	1	RM11943-RM12091	13 (P)	Vikram et al. (2011)
$qDTY_{1.1}$	N22	IR64	Lowland	1	RM11943-RM12091	17 (P)	
$qDTY_{1.1}$	N22	MTU1010	Lowland	1	RM11943-RM12091	13 (P)	
$qDTY_{1.1}$	Apo	IR64	Upland	1	RM486-RM472	58 (G)	Venuprasad et al. (2012)
$qDTY_{2.1}$	Apo	Swarna	Lowland	2	RM327-RM262	16 (G)	Venuprasad et al. (2009)
$qDTY_{2.2}$	Aday Sel	IR64	Lowland	2	RM236-RM279	11 (P)	Swamy et al. (2013)
$qDTY_{2.2}$	Aday Sel	IR64	Lowland	2	RM236-RM555	3 (P)	
$qDTY_{2.2}$	Aday Sel	IR64	Lowland	2	RM236-RM555	9 (P)	
$qDTY_{2.2}$	Kali Aus	MTU1010	Lowland	2	RM211-RM233A	16 (P)	Palanog et al. (2014)
$qDTY_{2.2}$	Kali Aus	MTU1010	Upland	2	RM263-RM573	6 (P)	Sandhu et al. (2014)
$qDTY_{3.1}$	Apo	Swarna	Lowland	3	RM520-RM16030	31 (G)	Venuprasad et al. (2009)
$qDTY_{3.1}$	IR55419-04	TDK1	Lowland	3	RM168-RM468	8 (P)	Dixit et al. (2014)
$qDTY_{3.1}$	IR55419-04	TDK1	Upland	3	RM168-RM468	15 (P)	
$qDTY_{4.1}$	Aday Sel.	IR64	Lowland	4	RM551-RM16368	11 (P)	Swamy et al. (2013)
$qDTY_{6.1}$	IR55419-04	TDK1	Lowland	6	RM586-RM217	9 (P)	Dixit et al. (2014)
$qDTY_{6.1}$	IR55419-04	TDK1	Upland	6	RM586-RM217	36 (P)	
$qDTY_{6.1}$	IR55419-04	TDK1	Lowland	6	RM121-RM541	9 (P)	
$qDTY_{6.1}$	IR55419-04	TDK1	Upland	6	RM121-RM541	20 (P)	
$qDTY_{6.1}$	Vandana	IR72	Upland	6	RM589-RM204	40 (G)	Venuprasad et al. (2012)
$qDTY_{6.1}$	Apo	IR72	Upland	6	RM589-RM204	63 (G)	
$qDTY_{10.1}$	MTU1010	N22	Lowland	10	RM216-RM304	5 (G)	Vikram et al. (2011)
$qDTY_{12.1}$	IR74371-46-1-1	Sabitri	Lowland	12	RM28166-RM28199	24 (P)	Mishra et al. (2013)
$qDTY_{12.1}$	Way Rarem	Vandana	Upland	12	RM28048-RM28166	33 (P), 51 (G)	Bernier et al. (2007)

Source: Modified from Kumar et al. (2014).

Table 3.2 List of QTLs introgressed lines in different background of popular rice at IRRI.

Introgressed lines	QTLs	Background	References
IR 102818-10-227-1-2-1-2	$qDTY_{1.1} + qDTY_{11.1}$	Sambha Mahsuri	Sandhu et al. (2018)
R 102818-10-227-1-2-1-6	$qDTY_{1.1} + qDTY_{11.1}$	Sambha Mahsuri	
IR 102818-10-227-1-2-1-9	$qDTY_{1.1} + qDTY_{11.1}$	Sambha Mahsuri	
IR 102821-19-233-2-1-1-10	$qDTY_{1.1} + qDTY_{2.1} +$		
Sambha Mahsuri	$qDTY_{3.1} + qDTY_{11.1}$		
IR 102818-10-276-1-2-2-9	$qDTY_{1.1} + qDTY_{2.1} +$		
Sambha Mahsuri	$qDTY_{11.1}$		
IR 102818-10-276-1-2-2-11	$qDTY_{1.1} + qDTY_{2.1} +$		
Sambha Mahsuri	$qDTY_{11.1}$		
IR 102818-10-266-3-2-2-6	$qDTY_{1.1} + qDTY_{2.1}$	Sambha Mahsuri	
IR 99734:1-33-69-1-39-6	$qDTY_{2.2} + qDTY_{4.1}$	Sambha Mahsuri	
IR 99734:1-33-69-1-12-8	$qDTY_{2.2} + qDTY_{4.1}$	Sambha Mahsuri	
IR 99734:1-33-69-1-12-9	$qDTY_{2.2} + qDTY_{4.1}$	Sambha Mahsuri	
IR 99734:1-33-304-1-5-8	$qDTY_{2.2} + qDTY_{4.1}$	Sambha Mahsuri	
IR 99734:1-33-304-1-5-10	$qDTY_{2.2} + qDTY_{4.1}$	Sambha Mahsuri	
Introgressed lines	$qDTY_{1.1}$	Swarna-Sub1	Singh et al. (2016)
Introgressed lines	$qDTY_{2.1}$	Swarna-Sub1	
Introgressed lines	$qDTY_{3.1}$	Swarna-Sub1	
Introgressed lines	$qDTY_{3.2}$	Swarna-Sub1	
Introgressed lines	$qDTY_{12.1}$	Swarna-Sub1	
Introgressed lines	$qDTY_{1.1}$	Sambha Mahsuri-Sub1	Sandhu et al. (2018)
Introgressed lines	$qDTY_{2.1}$	Sambha Mahsuri-Sub1	
Introgressed lines	$qDTY_{2.2}$	Sambha Mahsuri-Sub1	
Introgressed lines	$qDTY_{3.1}$	Sambha Mahsuri-Sub1	
Introgressed lines	$qDTY_{3.2}$	Sambha Mahsuri-Sub1	
Introgressed lines	$qDTY_{1.1}$	IR64-Sub1	Kumar et al. (2018)
Introgressed lines	$qDTY_{2.2}$	IR64-Sub1	
Introgressed lines	$qDTY_{3.1}$	IR64-Sub1	
Introgressed lines	$qDTY_{9.1}$	IR64-Sub1	
IR 102783:2-70-1-2-1-1	$qDTY_{12.1} + Sub1$	IR64-Sub1	

(Continued)

Table 3.2 (Continued)

Introgressed lines	QTLs	Background	References
IR 102784:2-89-632-2-1-2	$qDTY_{12.1}$ + *Sub1*	IR64-Sub1	
IR 102783:2-70-21-1-1-4	$qDTY_{12.1}$ + *Sub1*	IR64-Sub1	
IR 99784-255-78-2-3-1	$qDTY_{2.2}$ + $qDTY_{3.1}$+ $qDTY_{12.1}$	MR219	Shamsudin et al. (2016)
IR 99784-255-7-1-4-1	$qDTY_{2.2}$ + $qDTY_{3.1}$+ $qDTY_{12.1}$	MR219	
IR 99784-188-202-1-1-1	$qDTY_{2.2}$ + $qDTY_{3.1}$+ $qDTY_{12.1}$	MR219	
IR 99784-156-87-2-4-1	$qDTY_{3.1}$ + $qDTY_{12.1}$	MR219	
IR 99784-255-7-2-4-1	$qDTY_{3.1}$ + $qDTY_{12.1}$	MR219	
IR 99784-255-7-2-6-1	$qDTY_{3.1}$ + $qDTY_{12.1}$	MR219	

QTLs introgressed in different popular but drought-susceptible lines at IRRI. Drought grain yield QTLs $qDTY_{1.1}$, $qDTY_{2.1}$, $qDTY_{2.2}$, $qDTY_{3.1}$, and $qDTY_{4.1}$ were pyramided in Sambha Mashuri background (Sandhu et al. 2018). Swamy et al. (2013) reported that introgressed lines with 2–3 *qDTYs* in an IR64 background gave a yield advantage of 1200–2000 kg ha^{-1} under reproductive-stage drought condition and to that of IR64 under non-stress conditions. The popular Malaysian variety, MR219 was pyramided with $qDTY_{2.2}$, $qDTY_{3.1}$, and $qDTY_{12.1}$, and the effects of single/multiple QTLs were also reported (Shamsudin et al. 2016). Two large-effect QTLs for grain yield under drought, $qDTY_{3.2}$ and $qDTY_{12.1}$ were introgressed in a popular Nepalese variety Sabitri for making a drought-tolerant version of the variety (Dixit et al. 2017).

MAS Introgression of Multiple Traits

Production of rain-fed rice ecosystems is highly vulnerable due to increasing incidences of abiotic/biotic stresses under ongoing climate change. Predictions show that there may be a further decrease of 3.8% in rice production in South and Southeast Asia under the climates of this century, because of water scarcity, flooding, and increased temperatures (Murdiyarso 2000). For example, drought and flood stresses have been reported to affect approximately 40 Mha of total rice area at different crop stages, producing negative impacts on plant growth, development, and yield (Neeraja et al. 2007). Occurrence of both stresses, flood and drought, may occur in the same season at different crop growth stages, causing huge losses to rice production. Recently, high-yielding drought- and submergence-tolerant rice varieties using marker-assisted introgression of $qDTY_{1.1}$, $qDTY_{2.1}$, $qDTY_{3.1}$, and *Sub1* were developed to release in flood- and drought-prone areas of India and Nepal (Sandhu et al. 2019).

There are several other examples of introgression of multiple QTLs/genes for both biotic and abiotic stresses other than drought: these examples include gall midge (Das and Rao 2015), blast (Miah et al. 2016), brown plant hopper (Jairin et al. 2009), and submergence (Septiningsih et al. 2009) in the background of popular high-yielding varieties. Multiple genes stacking were also approached for biotic stresses like bacterial blight (*xa5* + *xa13* + *Xa21*) by Kottapalli et al. (2010) and Singh et al. (2001), (*Xa21* + *xa13*) by Singh et al. (2011) etc. controlled by major genes. Kumar et al. (2018) approached MAS for grain yield under drought in combination with biotic and abiotic traits to combat emerging biotic and abiotic stresses related to climate change. Recently at IRRI, MAS lines had been developed carrying up to 10 QTLs/genes for various biotic and abiotic stresses (Yadav et al. 2019; under communication). Table 3.3 illustrates the near isogenic lines (NILs) developed at IRRI released in different countries, and Table 3.4 shows all the drought-tolerant varieties released in South Asia. These varieties are currently engaged to combat drought tolerance in the South Asia.

Understanding the Molecular Basis of Complexities of Epistatic Interactions

Pyramiding of QTLs using MAS is a feasible breeding strategy to maximize the phenotypic level of a complex trait like grain yield (Ashikari and Matsuoka 2006). Due to various interactions (Sandhu et al. 2018), however, the phenotyping expression of the improvement in yield under drought in the PLs may be lower or higher than the expected level. In this context, genomic interactions such as QTL-QTL and QTL-genetic backgrounds play an important role in judging the performance of MAS lines introgresed with various *qDTY* QTLs. The effect of epistatic interactions of QTLs with background markers has been identified recently for various complex traits in a number of crop species including rice (Bocianowski 2012; Upadhyaya et al. 2011). A number of reports highlighted the capture of such interactions and identified positive QTL-QTL interactions with yield increase under drought stress and without any yield penalty under non-stress conditions (Dixit et al. 2012b; Sandhu et al. 2018). However, negative interactions between introgressed drought QTLs and background markers loci, QTL-QTL interaction and among interacting background markers loci on various chromosomes were also detected in our recent studies performed on drought NILs in multiple genetic backgrounds (Yadav et al. 2019). These common negative interactions prevailing in multiple genetic backgrounds suggested the possibility of devising a suitable strategy to avoid these negative interactions in pyramiding the above-mentioned QTLs in any of the rice genetic backgrounds. To understand the pyramiding effect of cold-tolerance QTLs and to identify the efficient combinations of QTLs for the improvement of cold tolerance in rice, Shinada et al. (2014) compared the haplotype patterns of chromosomal regions around the QTLs among the different rice cultivars.

Negative interactions prevailing in multiple genetic backgrounds suggested a careful selection of suitable strategy to avoid these negative interactions in pyramiding in any of the rice genetic backgrounds. Understanding the genetic composition of identified loci and their effect on grain yield helps in allowing the use of loci to achieve maximum grain yield advantage under RS (Yadav et al. 2019). This process of selection coupled with genotyping needs careful selection of closely linked markers based on the size of the introgressed QTLs to avoid undesirable linkages causing low yield potential, such as, for example,

Table 3.3 NILs in popular genetic backgrounds developed through MAS approach released as MAS variety in various countries for tolerance to drought and flood.

NILs designation	QTL combination	Released as variety	Trait(s)	Targeted country	Year	Background
IR 87707-445-B-B-B	$qDTY_{2.2} + qDTY_{4.1}$	*DRR dhan-42	Drought	India	2014	IR64
IR 96322-34-223-B-1-1-1	$qDTY_{1.1} + qDTY_{2.1} + qDTY_{3.1} + Sub\ 1$	*CR dhan 801	Drought + flood	India	2017	Swarna-Sub1
IR 94391-131-358-19-B-1-1-1	$qDTY_{3.1} + Sub\ 1$	Bahuguni dhan-2	Drought + flood	Nepal	2017	Swarna-Sub1
IR 96321-1447-651-B-1-1-2	$qDTY_{1.1} + qDTY_{3.1} + Sub\ 1$	Bahuguni dhan-1	Drought + flood	Nepal	2017	Swarna-Sub1

Source: Modified from Kumar et al. (2014). * CR- Rice from National Rice Research Institute (NRRI), Cuttack; DRR- Directorate of Rice Research, rice from Indian Institute of Rice research (IIRR).

Table 3.4 Drought tolerant rice varieties released in different countries of South Asia.

Name of the variety	Designation	Country	Year released	Stress
BRRI dhan56	IR74371-70-1-1	Bangladesh	2011	Drought
BRRI dhan66	IR 82635-B-B-75-2	Bangladesh	2014	Drought
BRRI dhan71	IR82589-B-B-84-3	Bangladesh	2014	Drought
CR Dhan 40	CRR 383-22	India	2008	Drought
Sahbhagi Dhan	IR 74371-70-1-1	India	2010	Drought
CR Dhan 201	IR 83380-B-B-124-1	India	2014	Drought
CR Dhan 202	IR 84899-B-154	India	2014	Drought
CR Dhan 204	IR 83927-B-B-279	India	2014	Drought
CR Dhan 205	IR 86931-B-578	India	2014	Drought
DRR Dhan 42	IR 87707-445-B-B-B	India	2014	Drought
DRR Dhan 43	IR 83876-B-R	India	2014	Drought
DRR Dhan 44	IR 93376-B-B-130	India	2014	Drought
DRR Dhan 46	IR 83383-B-B	India	2016	Drought
Swarna Shreya	IR 84899-B-179-16-1-1-1	India	2016	Drought
Sabour Ardhajal	—	India	2013	Drought
Sukha Dhan 1	IR74371-46-1-1	Nepal	2012	Drought
Sukha Dhan 2	IR74371-54-1-1	Nepal	2012	Drought
Sukha Dhan 3	IR 74371-70-1-1	Nepal	2012	Drought
Sukha Dhan 4	IR 87707-446-B-B-B	Nepal	2014	Drought
Sukha Dhan 5	IR 83388-B-B-108-3	Nepal	2014	Drought
Sukha Dhan 6	IR 83383-B-B-129-4	Nepal	2014	Drought
Hardinath 3	IR 93376-B-B-130	Nepal	2017	Drought
Sahod Ulan 1	IRRI 148	Philippines	2009	Drought
Katihan 1	IRRI 155	Philippines	2011	Drought
Sahod Ulan 3	IRRI 161	Philippines	2011	Drought
Sahod Ulan 5	IRRI 162	Philippines	2011	Drought
Sahod Ulan 6	IRRI 163	Philippines	2011	Drought
Sahod Ulan 8	IRRI 164	Philippines	2011	Drought
Sahod Ulan 12	IRRI 175	Philippines	2013	Drought
Katihan 2	IR08L152	Philippines	2014	Drought
Katihan 3	IR10L414	Philippines	2014	Drought
Sahod Ulan 15	IRRI 188	Philippines	2015	Drought
Sahod Ulan 19	IRRI 199	Philippines	2015	Drought
Sahod Ulan 23	IRRI 199	Philippines	2016	Drought
Sahod Ulan 25	IRRI 200	Philippines	2016	Drought
Sahod Ulan 1	IRRI 148	Philippines	2009	Drought

(Continued)

Table 3.4 (Continued)

Name of the variety	Designation	Country	Year released	Stress
Katihan 1	IRRI 155	Philippines	2011	Drought
Sahod Ulan 3	IRRI 161	Philippines	2011	Drought
Sahod Ulan 5	IRRI 162	Philippines	2011	Drought
Sahod Ulan 6	IRRI 163	Philippines	2011	Drought
Sahod Ulan 8	IRRI 164	Philippines	2011	Drought
Sahod Ulan 12	IRRI 175	Philippines	2013	Drought
Katihan 2	IR08L152	Philippines	2014	Drought
Katihan 3	IR10L414	Philippines	2014	Drought
Sahod Ulan 15	IRRI 188	Philippines	2015	Drought
Sahod Ulan 19	IRRI 199	Philippines	2015	Drought
Sahod Ulan 23	IRRI 199	Philippines	2016	Drought
Sahod Ulan 25	IRRI 200	Philippines	2016	Drought

Source: Modified from Dixit et al. (2014); Sandhu and Kumar (2017); Kumar et al. (2014).

very short growth duration and tall plant height as much as possible (Kumar et al. 2018; Vikram et al. 2015).

As breeding trials are laborious and costly, and required large space and population sizes, causing unreliable and imprecise phenotyping screening for a good trial, an efficient economical, fast, accurate breeding selection system is required to increase grain yield potential, and productivity (Sandhu and Kumar 2017; Yadav et al. 2019). Genotyping as well as phenotyping cost savings of 25–68% as compared with the traditional molecular marker breeding approach using the class analysis approach in the early generation followed by advancing only higher-performing classes can reduce negative interactions among of QTLs/genes with backgrounds (Kumar et al. 2018; Vikram et al. 2015). This procedure will save time, labor, resources, and space and will allow breeders to focus only on germplasm with higher value, allowing reduction of the population size for phenotypic and genotypic selection in advanced generations compared with earlier MAS (Begum et al. 2015; McNally et al. 2009; Sandhu et al. 2014).

Modern Breeding Strategies and Programs for Accelerated Genetic Gain

Rapid Breeding Cycles

The most promising innovation for increasing genetic gain is reducing the breeding cycle time requiring the implementation of parent selection on non-inbred progeny, rapid generation advance (RGA), and GS. To implement this careful management of the parental selection process, product development and advancement processes are needed relative to its cost of deployment followed by a well-designed data management system for the selection process. There are several other data-rich, evidence-based, and team-oriented processes for transforming plant breeding programs to increase genetic gain: these processes include advances in statistics (Drovandi et al. 2017), quantitative and population genetics (Hill 2010; Walsh 2014), molecular biology (Moose and Mumm 2008), genomics (Bevan and Uauy 2013; Dekkers 2012), phenomics (Cobb et al. 2013), other -omics (Carnielli et al. 2015; VanEmon 2016). Doubled haploid (Asif 2013; Maluszynski et al. 2003) or single seed descent (Collard et al. 2017) methodologies are effective, to permit the rapid evaluation and promotion of high-performing material in self-pollinated crops like rice. RGA is the most recent iteration mentioned as speed breeding by Watson et al. (2018) and Li et al. (2018) where the system shortens the generation cycle following the early seed harvest in F_2–F_6 generations in a modified environment (Collard et al. 2017). Introduction of new favorable alleles through RGAs may be a possible scenario to boost the rate of gain and achieve the 2050 challenge of food security. Genetic gain in a crop breeding program is determined by the selection intensity, heritability, additive genetic variance, and the length of breeding cycle interval or generation (Li et al. 2018). Germplasm stored in gene banks is a source of genetic variation to potentially raise genetic gain. GS is also known for rapid selection of superior genotypes and accelerates the breeding cycle through heterosis breeding (Crossa et al. 2017; Jiang et al. 2017). Boosting of additive genetic variation using a combination of speed breeding and GS could fast-track gene bank mining (Li et al. 2018). Several approaches were applied depending on the crop biology, suc as harvesting of immature seed in soybeans (Carandang et al. 2006) and pigeon peas (Saxena et al. 2017); light manipulation in chickpeas (Gaur et al. 2007), sorghum (Rizal et al. 2014), and rice (Tanaka et al. 2016). Liu et al. (2016) used stress to induce flowering followed by embryo rescue and seven generations of oats and triticale had been developed following RGA in one year. Six generations per year in wheat, barley, and chickpeas were achieved by RGA (Watson et al. 2018).

Novel Breeding Strategies Using GS

Under drought, a crop has to undergo various levels of drought stress across its developmental life. A genomics technology helps as a path to understand the genetic and environmentally complex nature of drought stress. Langridge and Reynolds (2015) focused on the level of drought stress and the careful selection of the crop and phenotyping techniques while GS. GS utilizes the phenotypes and high-density marker scores to predict the genomic breeding values of lines in a population (Heffner et al. 2009), which is the ideal selection criteria for the breeding populations (Jannink et al. 2010).

MAS is effective for large-effect QTLs, but in case of small-effect QTLs that can easily segregate, reliable effect cannot be achieved (Moreau et al. 2004). Small-effect genes have many complex polygenic traits that are crucial for the success of new crop varieties (Crosbie et al. 2006) but marker-assisted selection works best for major-effects genes (Dekkers and Hospital 2002). Because the bi-parental mapping populations mostly used in QTL studies do not readily translate to breeding applications and statistical methods used to identify target loci are inadequate for improving polygenic traits controlled by many small-effect QTLs. The level of accuracy in GS is good enough to consider selecting for agronomic performance using marker information alone thereby accelerating the breeding cycle, enhancing gains per unit time (Heffner et al. 2009). GS is based on many markers distributed throughout the genome; index selection assumptions are met, providing an opportunity to use index selection theory to predict response to GS whereas in MAS selection is based on only few large-effect loci (Dekkers 2007; Lande and Thompson 1990). More information can be generated by using GS. In this way the value of a genotype can be statistically proved by modeling of the genome sequence data under any environment (Crossa et al. 2017; Jannink et al. 2010). The transcriptome analysis conducted by Wang et al. (2016) showed that overexpression of *OsSAP16* that controls the photosynthesis through regulation of stomatal conductance via a set of stress associated proteins (SAPs); zinc-finger transcription factors regulated the response of rice to drought stress. Vij and Tyagi (2006) identified A20/AN1, a zinc-finger containing stress-associated protein 1 gene (*SAP1*) in rice and proved its function in abiotic stress tolerance. Shikha et al. (2017) found seven GS models namely, ridge regression, Least Absolute Shrinkage and Selection Operator (LASSO), elastic net, random forest, reproducing kernel Hilbert space, Bayes A, and Bayes B to show the highest level of prediction accuracy with the highest Pearson correlation coefficient across environments while studying 240 maize subtropical lines using 29,619 cured SNPs in drought-prone environments. Bhandari et al. (2019) showed that a trait-specific marker selection strategy achieved predictive ability of genomic prediction up to 22% higher under reproducing kernel Hilbert space regressions (RKHS) based models than markers selected on the basis of neutral linkage disequilibrium for drought-tolerance studies in rice.

Deployment of Genome Editing, Especially the CRISPR Technologies

Genome editing technologies were started in the late 1970s and 1980s in mice (Rothstein 1983; Scherer and Davis 1979; Smithies et al. 1985; Thomas et al. 1986). This technology enables scientists to make changes to DNA, leading to changes in physical traits. Genome editing provides information to create new alleles along with correcting defective alleles, pyramids allele to create desired phenotype by removing the genetic drug. Unlike before where gene targeting depended on homologous recombination, due to lack of culturable embryonic cells for reselection and characterization in mammals, gene targeting was not successful in all other species. Current genome-editing technologies resolved this issue, making directed genetic manipulations possible in essentially all types of cells and organisms (Carroll 2014; Gaj et al. 2013).

In the process of successful genome editing, the precise and targeted DNA double-strand break (DSB) in the chromosomal sequence needs to be developed. These targeted intentional DSBs would stimulate targeted mutagenesis and homologous recombination depending on the DNA repair pathway (Carroll 2017; Youds and Boulton 2011), and the ionizing radiation that was induced for DSB creation (Burma et al. 2006) leads to crossover of the sister chromatids (Latt 1981). Highly specific nucleases showed programmable genome editing by localized homologous repair by introduction of desired sequence changes (Carroll 2014, 2017). There are three possible ways to conduct genome editing at any specific target position using site-specific nucleases (SSNs) like zinc-finger nucleases (ZFNs) with limited success (Urnov et al. 2010), transcription activator-like effector nucleases (TALENs) (Sun and Zhao 2013) with a number of successful reports (Sprink et al. 2015), and RNA-mediated SSN CRISPR/Cas9 system, based on the bacterial CRISPR (clustered regularly interspaced short palindromic repeats) system and CRISPR-associated (Cas) nucleases (Endo et al. 2018). When targeted mutagenesis is induced by stably integrated DNA encoding Cas9 and single guide RNA (sgRNA), the process for obtaining targeted mutagenized plants is almost the same as standard rice transformation and regeneration. When CRISPR/Cas9 components are inserted in the genome of a rice cell and begin to function, one or both copies of the target genes may be cleaved and mutated (Zhang et al. 2014). The CRISPR/Cas9 systems for targeted mutation and heritability in mutant rice lines were evaluated for several genes including *OsDERF1, OsPMS3, OsEPSPS, OsMSH1, OsMYB5* for drought tolerance (Jaganathan et al. 2018; Zhang et al. 2014). Lou et al. (2017) used the CRISPR/Cas9 system while studying the loss of function of mutants of osmotic stress/ABA – activated protein kinase 2 (*SAPK2*), which is the key regulator of hyper-osmotic stress signaling and abscisic acid (ABA)-dependent development in rice. The *SAPK2* improved drought tolerance by enhancing the expression of stress-responsive and antioxidant enzyme genes in rice as *SAPK2* acted as an important positive regulator of ABA-dependent abiotic stress responses in rice.

Conclusion

To achieve the challenge of food security in this era of climate change, a combination approach using high-throughput precise phenotyping, appropriate breeding design, statistical analysis and modern genotyping platforms looks feasible for achieving increased drought tolerance, and targeted genetic gain in various crops. New advances in marker development, sequencing, and genomic analysis have permitted the opportunity to create populations to combat drought. The expensive phenotyping step has challenged the analysis of drought tolerance and other phenotyping traits. Therefore, characterization of drought tolerance mechanisms with a cost-effective measure will help to improve the progress on drought breeding.

References

Ashfaq, M., Haider, M.S., Khan, A.S., and Allah, S.U. (2014). Breeding potential of the basmati rice germplasm under water stress condition. *Afr. J. Biotechnol.* 11: 6647–6657.

Ashikari, M. and Matsuoka, M. (2006). Identification, isolation and pyramiding of quantitative trait loci for rice breeding. *Trends Plant Sci.* 11: 344–350. https://doi.org/10.1016/j.tplants.2006.05.008.

Asif, M. (2013). *Progress and Opportunities of Doubled Haploid Production*, Springer Briefs in Plant Science, vol. 6. Heidelberg: Springer https://doi.org/10.1007/978-3-319-00732-8.

Begum, H., Spindel, J.E., Lalusin, A. et al. (2015). Genome-wide association mapping for yield and other agronomic traits in an elite breeding population of tropical rice (*Oryza sativa*). *PLoS One* 10 (3): e0019873. https://doi.org/10.1371/journal.pone.0119873.

Bernier, J., Kumar, A., Venuprasad, R. et al. (2007). A large-effect QTL for grain yield under reproductive-stage drought stress in upland rice. *Crop Sci.* 47 (2): 507. https://doi.org/10.2135/cropsci2006.07.0495.

Bevan, M.W. and Uauy, C. (2013). Genomics reveals new landscapes for crop improvement. *Genome Biol.* 14: 206. https://doi.org/10.1186/gb-2013-14-6-206.

Bhandari, A., Bartholomé, J., Cao-Hamadoun, T.V.I. et al. (2019). Selection of trait-specific markers and multi-environment models improve genomic predictive ability in rice. *PLoS One* (5): 14, e0208871. https://doi.org/10.1371/journal.pone.0208871.

Bhattarai, U. and Subudhi, P.K. (2018). Identification of drought responsive QTLs during vegetative growth stage of rice using a saturated GBS-based SNP linkage map. *Euphytica* 214: 38. https://doi.org/10.1007/s10681-018-2117-3.

Blum, A. (2011). Drought resistance – is it really a complex trait? *Funct. Plant Biol.* 38 (10): 753. https://doi.org/10.1071/FP11101h.

Bocianowski, J. (2012). Using weighted multiple linear regressions in estimation of QTL-by-QTL epistasis effects. *Genet. Mol. Biol.* 35: 802–809.

Burma, S., Chen, B.P.C., and Chen, D.J. (2006). Role of non-homologous end joining (NHEJ) in maintaining genomic integrity. *DNA Repair (Amst)* 5 (9–10): 1042–1048. https://doi.org/10.1016/J.DNAREP.2006.05.026.

Carandang, F.M., Shanmugasundaram, S., and Carpena, A.L. (2006). Rapid generation advancement in soybeans using immature seeds. *Philipp. J. Crop Sci.* 31: 53–59.

Carnielli, C.M., Winck, F.V., and Leme, A.F.P. (2015). Functional annotation and biological interpretation of proteomics data. *Biochim. Biophys. Acta Protein Proteomics* 1854: 46–54. https://doi.org/10.1016/j.bbapa p. 2014.10.019.

Carroll, D. (2014). Genome engineering with targetable nucleases. *Annu. Rev. Biochem.* 83 (1): 409–439. https://doi.org/10.1146/annurev-biochem-060713-035418.

Carroll, D. (2017). Genome editing: past, present, and future. *Yale J. Biol. Med.*. 90 (4): 653–659.

Cherlet, M., Hutchinson, C., Reynolds, J. et al. (eds.) (2018). *World Atlas of Desertification*. Luxembourg: Publication Office of the European Union.

Cobb, J.N., DeClerck, G., Greenberg, A. et al. (2013). Next-generation phenotyping: requirements and strategies for enhancing our understanding of genotype–phenotype relationships and its relevance to crop improvement. *Theor. Appl. Genet.* 126: 867–887. https://doi.org/10.1007/s0012 2-013-2066-0..

Collard, B.C.Y., Beredo, J.C., Lenaerts, B. et al. (2017). Revisiting rice breeding methods: evaluating the use of rapid generation advance (RGA) for routine rice breeding. *Plant Prod. Sci.* 20: 337–352. https://doi.org/10.1080/13439 43X.2017.13917 05.

Coumou, D. and Rahmstorf, S. (2012). A decade of weather extremes. *Nat. Clim. Chang.* 2 (7): 491–496. https://doi.org/101038/nclimate1452.

Crosbi,e, T.M., Eathington, S.R., Johnson, G.R. Sr., et al. (2006). Plant breeding: past, present, and future. In: *Plant Breeding*: The Hallauer AR, International Symposium, 3–50. Ames, IA, USA: Blackwell Publishing.

Crossa, J., Pérez-Rodríguez, P., Cuevas, J. et al. (2017). Genomic selection in plant breeding: methods, models, and perspectives. *Trends Plant Sci.* 22 (11): 961–975. https://doi.org/10.1016/j.tplants.2017.08.011.

Das, G. and Rao, G.J.N. (2015). Molecular marker assisted gene stacking for biotic and abiotic stress resistance genes in an elite rice cultivar. *Front. Plant Sci.* 6: 698.

Dekkers, J.C.M. (2007). Prediction of response to marker-assisted and genomic selection using selection index theory. *J. Anim. Breed. Genet.* 124 (6): 331–341. https://doi.org/10.1111/j.1439-0388.2007.00701.x.

Dekkers, J.C.M. (2012). Application of genomics tools to animal breeding. *Curr. Genom.* 13: 207–212. https://doi.org/10.2174/138920212800543057.

Dekkers, J.C.M. and Hospital, F. (2002). The use of molecular genetics in the improvement of agricultural populations. *Nat. Rev. Genet.* 3 (1): 22–32. https://doi.org/10.1038/nrg701.

Dixit, S., Swamy, B.P.M., Vikram, P. et al. (2012a). Fine mapping of QTLs for rice grain yield under drought reveals sub-QTLs conferring a response to variable drought severities. *Theor. Appl. Genet.* 125: 155–169. https://doi.org/10.1007/s00122-012-1823-9.

Dixit, S., Swamy, B.P.M., Vikram, P. et al. (2012b). Increased drought tolerance and wider adaptability of $qDTY_{121}$ conferred by its interaction with $qDTY_{2.3}$ and $qDTY_{3.2}$. *Mol. Breed.* 30 (4): 1767–1779. https://doi.org/10.1007/s11032-012-9760-5.

Dixit, S., Singh, A., Sta Cruz, M. et al. (2014). Multiple major QTL lead to stable yield performance of rice cultivars across varying drought intensities. *BMC Genet.* 15 (1): 16. https://doi.org/10.1186/1471-2156-15-16.

Dixit, S., Yadaw, R.B., Mishra, K.K., and Kumar, A. (2017). Marker-assisted breeding to develop the drought-tolerant version of Sabitri, a popular variety from Nepal. *Euphytica* 213 (8): 1–16. https://doi.org/10.1007/s10681-017-1976-3.

Drovandi, C.C., Holmes, C., McGree, J.M. et al. (2017). Principles of experimental design for big data analysis. *Stat. Sci.* 3: 385–404. https://projecteuclid.org/download/pdfview_1/euclid.ss/1504253123.

Endo, M., Nishizawa-Yokoi, A., and Toki, S. (2018). Rice genome editing. In: *Rice Genomics, Genetics and Breeding* (eds. T. Sasaki and M. Ashikari), 523–539. Singapore: Springer https://doi.org/10.1007/978-981-10-7461-5_27.

Federer, W.T. and Raghavarao, D. (1975). On augmented designs. *Biometrics* 31: 29–35.

Fischer, K.S., Fukai, S., Kumar, A. et al. (2012). Field phenotyping strategies and breeding for adaptation of rice to drought. *Front. Physiol.* 3: 282. https://doi.org/10.3389/fphys.2012.00282.

Fukai, S. and Cooper, M. (2001). Development of drought-resistant cultivars for rainfed lowland rice: experience from Northeast Thailand and surrounding areas. In: *The Impact of Agricultural Research for Development in Southeast Asia: Proceedings of a Conference Held at the Cambodian Agricultural Research and Development Institute* (eds. P. Cox and R. Chhay), 185–194. Phnom Penh: Cambodian Agriculture Research and Development Institute.

Gaj, T., Gersbach, C.A., and Barbas, C.F. (2013). 3rd ZFN, TALEN, and CRISPR/Cas-based methods for genome engineering. *Trends Biotechnol.* 31 (7): 397–405.

Gaur, P.M., Srinivasan, S., and Gowda, C.L.L. (2007). Rapid generation advancement in Chickpea. *SAT J.* 3 (1): 3.

Ghimire, K.H., Lenie, A.Q., Vikram, P. et al. (2012). Identification and mapping of a QTL (qDTY1.1) with a consistent effect on grain yield under drought. *Field Crop Res.* 131: 88–96.

Guan, Y.S., Serraj, R., Liu, S.H. et al. (2010). Simultaneously improving yield under drought stress and non-stress conditions: a case study of rice (*Oryza sativa* L.). *J. Exp. Bot.* 61 (15): 4145–4156. https://doi.org/10.1093/jxb/erq212.

Heffner, E.L., Sorrells, M.E., and Jannink, J.L. (2009). Genomic selection for crop improvement. *Crop Sci.* 49: 1–12. https://doi.org/10.2135/crops ci200 8.08.0512..

Hill, W.G. (2010). Understanding and using quantitative genetic variation. *Philos. Trans. R. Soc. B Biol. Sci.* 365 (1537): 73–85. https://doi.org/10.1098/rstb.2009.0203.

Huang, X., Wei, X., Sang, T. et al. (2010). Genome-wide association studies of 14 agronomic traits in rice landraces. *Nat. Genet.* 42 (11): 961–967. https://doi.org/10.1038/ng.695.

IRRI (International Rice Research Institute) (2002). *Rice Almanac*. Los Banos: IRRI-WARDA-CIAT-FAO.

Jaganathan, D., Ramasamy, K., Sellamuthu, G. et al. (2018). CRISPR for crop improvement: an update review. *Front. Plant Sci.* 9: 985. https://doi.org/10.3389/fpls.2018.00985.

Jairin, J., Teangdeerith, S., Leelagud, P. et al. (2009). Development of rice introgression lines with brown planthopper resistance and KDML105 grain quality characteristics through marker-assisted selection. *Field Crop Res.* 110 (3): 263–271.

Jannink, J.L., Lorenz, A.J., and Iwata, H. (2010). Genomic selection in plant breeding: from theory to practice. *Brief. Funct. Genom.* 9 (2): 166–177. https://doi.org/10.1093/bfgp/elq001.

Jiang, Y., Schmidt, R.H., Zhao, Y., and Reif, J.C. (2017). Quantitative genetic framework highlights the role of epistatic effects for grain-yield heterosis in bread wheat. *Nat. Genet.* 49 (12): 1741–1746. https://doi.org/10.1038/ng.3974.

Kamoshita, A., Zhang, J., Siopongco, J. et al. (2002). Effects of phenotyping environment on identification of quantitative trait loci for rice root morphology under anaerobic conditions. *Crop Sci.* 42 (1): 255. https://doi.org/10.2135/cropsci2002.2550.

Kamoshita, A., Babu, R.C., Boopathi, N.M., and Fukai, S. (2008). Phenotypic and genotypic analysis of drought-resistance traits for development of rice cultivars adapted to rainfed environments. *Field Crop Res.* 109: 1–23. https://doi.org/10.1016/j.fcr.2008.06.010.

Kato, Y., Abe, J., Kamoshita, A., and Yamagishi, J. (2006). Genotypic variation in root growth angle in rice (*Oryza sativa* L.) and its association with deep root development in upland fields with different water regimes. *Plant Soil* 287 (1–2): 117–129. https://doi.org/10.1007/s11104-006-9008-4.

Kottapalli, K.R., Narasu, M.L., and Jena, K.K. (2010). Effective strategy for pyramiding three bacterial blight resistance genes into fine grain rice cultivar Samba Mahsuri using sequence tagged site markers. *Biotechnol. Lett.*. 32: 989–996. https://doi.org/10.1007/s10529-010-0249-1.

Kumar, R., Venuprasad, R., and Atlin, G.N. (2007). Genetic analysis of rainfed lowland rice drought tolerance under naturally occurring stress in eastern India: heritability and QTL effects. *Field Crop Res.* 103 (1): 42–52. http://dx.doi.org/10.1016/j.fcr.2007.04.013.

Kumar, A., Dixit, S., Ram, T. et al. (2014). Breeding high-yielding drought-tolerant rice: genetic variations and conventional and molecular approaches. *J. Exp. Bot.* 65 (21): 6265–6278. https://doi.org/10.1093/jxb/eru363.

Kumar, A., Sandhu, N., Dixit, S. et al. (2018). Marker-assisted selection strategy to pyramid two or more QTLs for quantitative trait-grain yield under drought. *Rice* 11 (1): 35. https://doi.org/10.1186/s12284-018-0227-0.

Lafitte, H.R., Price, A.H., and Courtois, B. (2004). Yield response to water deficit in an upland rice mapping population: associations among traits and genetic markers. *Theor. Appl. Genet.* 109: 1237–1246.

Lande, R. and Thompson, R. (1990). Efficiency of marker assisted selection in the improvement of quantitative traits. *Genetics* 124 (3): 743–756.

Langridge, P. and Reynolds, M.P. (2015). Genomic tools to assist breeding for drought tolerance. *Curr. Opin. Biotechnol.* 32: 130–135. https://doi.org/10.1016/j.copbio.2014.11.027.

Latt, S.A. (1981). Sister chromatid exchange formation. *Annu. Rev. Genet.* 15 (1): 11–53. https://doi.org/10.1146/annurev.ge.15.120181.000303.

Li, Z.K. and Xu, J.L. (2007). Breeding for drought and salt tolerant rice (*Oryza sativa* L.): progress and perspectives. In: *Advances in Molecular Breeding toward Drought and Salt Tolerant Crops* (eds. M.A. Jenks et al.), 531–564. Dordrecht, USA: Springer https://doi.org/10.1007/978-1-4020-5578-2_21.

Li, H., Rasheed, A., Hickey, L.T., and He, Z. (2018). Fast forwarding genetic gain. *Trends Plant Sci.* 23 (3): 184–186. https://doi.org/10.1016/j.tplants.2018.01.007.

Lilley, J.M. and Fukai, S. (1994). Effect of timing and severity of water deficit on four diverse rice cultivars III Phenological development, crop growth and grain yield. *Field Crop Res.* 37 (3): 225–234. https://doi.org/10.1016/0378-4290(94)90101-5.

Liu, H., Pamela, Z., Haibo, W. et al. (2016). A fast generation cycling system for oat and triticale breeding. *Plant Breed.* 135 (5): 574–579. https://doi.org/10.1111/pbr.12408.

Lou, D., Wang, H., Liang, G., and Yu, D. (2017). OsSAPK2 confers abscisic acid sensitivity and tolerance to drought stress in rice. *Front. Plant Sci.* 8: 993. https://doi.org/10.3389/fpls.2017.00993.

Maluszynski, M., Kasha, K., Forster, B.P., and Szarejko, I. (2003). *Doubled haploid production in crop plants: a manual.* Dordrecht: Springer https://doi.org/10.1007/978-94-017-1293-4.

McNally, K.L., Childs, K.L., Bohnert, R. et al. (2009). Genome wide SNP variation reveals relationships among landraces and modern varieties of rice. *Proc. Natl. Acad. Sci.* 106 (30): 12273–12278. https://doi.org/10.1073/PNAS.0900992106.

Miah, G., Rafii, M.Y., Ismail, M.R. et al. (2016). Marker-assisted introgression of broad-spectrum blast resistance genes into the cultivated MR219 rice variety. *J. Sci. Food Agric.* 97 (9): 2810–2818.

Mishra, K.K., Vikram, P., Yadaw, R.B. et al. (2013). $qDTY_{12.1}$: a locus with a consistent effect on grain yield under drought in rice. *BMC Genet.* 14 (1): 12. https://doi.org/10.1186/1471-2156-14-12.

Moose, S.P. and Mumm, R.T. (2008). Molecular plant breeding as the foundation for 21st century crop improvement. *Plant Physiol.* 147: 969–977. https://doi.org/10.1104/pp.108.11823 2..

Moreau, L., Charcosset, A., and Gallais, A. (2004). Experimental evaluation of several cycles of marker assisted selection in maize. *Euphytica* 137 (1): 111–118. https://doi.org/10.1023/B:EUPH.0000040508.01402.21.

Murdiyarso, D. (2000). Adaptation to climatic variability and change: Asian perspectives on agriculture and food security. *Environ. Monit. Assess.* 61 (1): 123–131. https://doi.org/10.1023/A:1006326404156.

Neeraja, C.N., Maghirang-Rodriguez, R., Pamplona, A. et al. (2007). A marker-assisted backcross approach for developing submergence-tolerant rice cultivars. *Theor. Appl. Genet.* 115 (6): 767–776. https://doi.org/10.1007/s00122-007-0607-0.

Palanog, A.D., Swamy, B.P.M., Shamsudin, N.A.A. et al. (2014). Grain yield QTLs with consistent-effect under reproductive-stage drought stress in rice. *Field Crop Res.* 161: 46–54. https://doi.org/10.1016/j.fcr.2014.01.004.

Pantuwan, G., Fukai, S., Cooper, M. et al. (2004). Yield response of rice (*Oryza sativa* L.) genotypes to drought under rainfed lowlands: 4 vegetative stage screening in the dry season. *Field Crop Res.* 89 (2–3): 281–297. https://doi.org/10.1016/j.fcr.2004.02.007.

Rizal, G., Karki, S.N., and Alcasid, M. (2014). Shortening the breeding cycle of sorghum, a model crop for research. *Crop Sci.* 54: 520–529.

Rothstein, R.J. (1983). One step gene disruption in yeast. *Methods Enzymol.* 101: 202–211.

Saddique, M.A.B., Ali, Z., Khan, A.S. et al. (2018). Inoculation with the endophyte *Piriformospora indica* significantly affects mechanisms involved in osmotic stress in rice. *Rice* 11 (1): 34. 101186/s12284-018-0226-1.

Sahebi, M., Hanafi, M.M., Rafii, M.Y. et al. (2018). Improvement of drought tolerance in rice (*Oryza sativa* L.): genetics, genomic tools, and the WRKY gene family. *Biomed. Res. Int.* 2018: 1–20. 101155/2018/3158474.

Sandhu, N. and Kumar, A. (2017). Bridging the rice yield gaps under drought: QTLs, genes, and their use in breeding programs. *Agronomy* 7 (2): 27. https://doi.org/10.3390/agronomy7020027.

Sandhu, N., Singh, A., Dixit, S. et al. (2014). Identification and mapping of stable QTL with main and epistasis effect on rice grain yield under upland drought stress. *BMC Genet.* 15: 63. 101186/1471-2156-15-63.

Sandhu, N., Dixit, S., Swamy, B.P.M. et al. (2018). Positive interactions of major-effect QTLs with genetic background that enhances rice yield under drought. *Sci. Rep.* 8: 1626. doi: 10.1038/s41598-018-20116-7.

Sandhu, N., Dixit, S., Swamy, B.P.M. et al. (2019). Marker assisted breeding to develop multiple stress tolerant varieties for flood and drought prone areas. *Rice* 12 (1): 8. https://doi.org/10.1186/s12284-019-0269-y.

Saxena, K., Saxena, R.K., and Varshney, R.K. (2017). Use of immature seed germination and single seed descent for rapid genetic gains in pigeonpea. *Plant Breed.* 136: 954–957. https://doi.org/10.1111/pbr.12538.

Scheben, A., Batley, J., and Edwards, D. (2017). Genotyping by sequencing approaches to characterize crop genomes: choosing the right tool for the right application. *Plant Biotechnol. J.* 15 (2): 149–161. https://doi.org/10.1111/pbi.12645.

Scherer, S. and Davis, R.W. (1979). Replacement of chromosome segments with altered DNA sequences constructed in vitro. *Proc. Natl. Acad. Sci. U.S.A.* 76: 4951–4955.

Septiningsih, E.M., Pamplona, A.M., Sanchez, D.L. et al. (2009). Development of submergence tolerant rice cultivars: the *Sub1* locus and beyond. *Ann. Bot.* 103 (2): 151–160. https://doi.org/10.1093/aob/mcn206.

Shamsudin, N.A.A., Swamy, B.P.M., Ratnam, W. et al. (2016). Marker assisted pyramiding of drought yield QTLs into a popular Malaysian rice cultivar, MR219. *BMC Genet.* 17 (1): 30. https://doi.org/10.1186/s12863-016-0334-0.

Shikha, M., Kanika, A., Rao, A.R. et al. (2017). Genomic selection for drought tolerance using genome wide SNPs in maize. Front. *Plant Sci.* 8: 550. https://doi.org/10.3389/fpls.2017.00550.

Shinada, H., Iwata, N., Sato, T., and Fujino, K. (2014). QTL pyramiding for improving of cold tolerance at fertilization stage in rice. *Breed. Sci.* 63 (5): 483–488. https://doi.org/10.1270/jsbbs.63.483.

Singh, S., Sidhu, J.S., Huang, N. et al. (2001). Pyramiding three bacterial blight resistance genes (*xa5 xa13* and *Xa21*) using marker-assisted selection into indica rice cultivar PR106. *Theor. Appl. Genet.* 102: 1011–1015.

Singh, A.K., Gopalakrishnan, S., and Singh, V.P. (2011). Marker assisted selection: a paradigm shift in basmati breeding. *Indian J. Genet. Plant Breed.* 71 (2): 1–9.

Singh, R., Singh, Y., Xalaxo, S. et al. (2016). From QTL to variety-harnessing the benefits of QTLs for drought, flood and salt tolerance in mega rice varieties of India through a multi-institutional network. *Plant Sci.* 242: 278–287. https://doi.org/10.1016/j.plantsci.2015.08.008.

Smithies, O., Gregg, R.G., Boggs, S.S. et al. (1985). Insertion of DNA sequences into the human chromosomal beta-globin locus by homologous recombination. *Nature* 317 (6034): 230–234.

Sprink, T., Metje, J., and Hartung, F. (2015). Plant genome editing by novel tools: TALEN and other sequence specific nucleases. *Curr. Opin. Biotechnol.* 32: 47–53 https://doi.org/10.1016/j.copbio.2014.11.010.

Sun, N. and Zhao, H. (2013). Transcription activator-like effector nucleases (TALENs): a highly efficient and versatile tool for genome editing. *Biotechnol. Bioeng.* 110 (7): 1811–1821. https://doi.org/10.1002/bit.24890.

Swamy, B.P.M., Ahmed, H.U., Henry, A. et al. (2013). Genetic, physiological, and gene expression analyses reveal multiple QTL enhance the yield of rice mega-variety IR64 under drought. *PLoS One* 8 (5): e62795. https://doi.org/10.1371/journal.pone.0062795.

Tanaka, J., Hayashi, T., and Iwata, H. (2016). A practical, rapid generation-advancement system for rice breeding using simplified biotron breeding system. *Breed. Sci.* 66: 542–551. https://doi.org/10.1270/jsbbs.15038.

Thomas, K.R., Folger, K.R., and Capecchi, M.R. (1986). High frequency targeting of genes to specific sites in the mammalian genome. *Cell* 44: 419–428.

Thomson, M.J. (2014). High throughput SNP genotyping to accelerate crop improvement. *Plant Breed. Biotechnol.* 2: 195–212. https://doi.org/10.9787/pbb.2014.2.3.195.

Upadhyaya, H.D., Sharma, S., Singh, S., and Singh, M. (2011). Inheritance of drought resistance related traits in two crosses of groundnut (*Arachis hypogaea* L.). *Euphytica* 177: 55–66.

Urnov, F.D., Rebar, E.J., Holmes, M.C. et al. (2010). Genome editing with engineered zinc finger nucleases. *Nat. Rev. Genet.* 11 (9): 636–646. https://doi.org/10.1038/nrg2842.

VanEmon, J.M. (2016). The omics revolution in agricultural research. *J. Agric. Food Chem.* 64: 36–44. https://doi.org/10.1021/acs. jafc.5b045 15.

Venuprasad, R., Lafitte, H.R., and Atlin, G.N. (2007). Response to direct selection for grain yield under drought stress in rice. *Crop Sci.* 47 (1): 285. https://doi.org/10.2135/cropsci2006.03.0181.

Venuprasad, R., Dalid, C.O., Del Valle, M. et al. (2009). Identification and characterization of large-effect quantitative trait loci for grain yield under lowland drought stress in rice using bulk-segregant analysis. *Theor. Appl. Genet.* 120 (1): 177–190. https://doi.org/10.1007/s00122-009-1168-1.

Venuprasad, R., Bool, M.E., Quiatchon, L. et al. (2012). A large-effect QTL for rice grain yield under upland drought stress on chromosome 1. *Mol. Breed.* 30: 535–547. https://doi.org/10.1007/s00122-009-1168-1.

Vij, S. and Tyagi, A.K. (2006). Genome-wide analysis of the stress associated protein (SAP) gene family containing A20/AN1 zinc-finger(s) in rice and their phylogenetic relationship with *Arabidopsis*. *Mol. Gen. Genomics.* 276 (6): 565–575. https://doi.org/10.1007/s00438-006-0165-1.

Vikram, P., Swamy, B.P.M., Dixit, S. et al. (2011). $qDTY_{1.1}$, a major QTL for rice grain yield under reproductive-stage drought stress with a consistent effect in multiple elite genetic backgrounds. *BMC Genet.* 12: 89. https://doi.org/10.1186/1471-2156-12-89.

Vikram, P., Swamy, B.P.M., Dixit, S. et al. (2015). Drought susceptibility of modern rice varieties: an effect of linkage of drought tolerance with undesirable traits. *Sci. Rep.* 5: 14799. https://doi.org/10.1063/1.4791353.

Vogt, J.V., Naumann, G., Masante, D. et al. (2018). *Drought Risk Assessment and Management – a Conceptual Framework. EUR 29464 EN*. Luxembourg: Publications Office of the European Union https://doi.org/10.2760/057223, JRC113937.

Wade, L.J., Kamoshita, A., Yamauchi, A., and Azhiri-Sigari, T. (2000). Genotypic variation in response of rainfed lowland rice to drought and rewatering. III. Water extraction during the drought period. *Plant Prod. Sci.* 3 (2): 189–196. https://doi.org/10.1626/pps.3.189.

Walsh, B. (2014). Special issues on advances in quantitative genetics: introduction. *Heredity* 112: 1–3. https://doi.org/10.1038/hdy.2013.115.

Wang, F., Coe, R.A., Karki, S. et al. (2016). Overexpression of *OsSAP16* regulates photosynthesis and the expression of a broad range of stress response genes in rice (*Oryza sativa* L.). *PLoS One* 11 (6): e0157244. https://doi.org/10.1371/journal.pone.0157244.

Watson, A., Ghosh, S., Williams, M.J. et al. (2018). Speed breeding is a powerful tool to accelerate crop research and breeding. *Nat. Plants* 4: 23–29. https://doi.org/10.1038/s41477-017-0083-8.

Xu, J., Zhao, Q., Du, P. et al. (2010). Developing high throughput genotyped chromosome segment substitution lines based on population whole-genome re-sequencing in rice (*Oryza sativa* L.). *BMC Genomics* 11: 656. https://doi.org/10.1186/1471-2164-11-656.

Yadav, S., Sandhu, N., Majumder, R.R. et al. (2019). Epistatic interactions of major effect drought QTLs with genetic background loci determine grain yield of rice under drought stress. *Sci. Rep.* 9 (1): 2616. https://doi.org/10.1038/s41598-019-50880-z.

Yadav, S., Sandhu, N., Dixit, S., Singh, V.K., Catolos, M., Majumder, R.R., et al. (2019). Genomics-assisted breeding enables successful development of multiple stress tolerant climate smart rice for South and South East Asia. Plant Genome, (Accepted)

Youds, J.L. and Boulton, S.J. (2011). The choice in meiosis – defining the factors that influence crossover or non-crossover formation. *J. Cell Sci.* 124 (4): 501–513. https://doi.org/10.1242/jcs.074427.

Zhang, H., Zhang, J., Wei, P. et al. (2014). The CRISPR/Cas9 system produces specific and homozygous targeted gene editing in rice in one generation. *Plant Biotechnol. J.* 12 (6): 797–807. https://doi.org/10.1111/pbi.12200.

4

Molecular Breeding for Improving Flooding Tolerance in Rice: Recent Progress and Future Perspectives

Ramani K. Sarkar[1], Jangi N. Reddy[2] and Satya R. Das[3]

[1] *Division of Crop Physiology & Biochemistry, ICAR-National Rice Research Institute, Cuttack, 753 006, Odisha, India*
[2] *Division of Crop Improvement, ICAR-National Rice Research Institute, Cuttack, 753 006, Odisha, India*
[3] *Department of Plant Breeding and Genetics, Orissa University of Agriculture & Technology, Bhubaneswar, 751 003, Odisha, India*

Introduction

Flooding is a serious threat to rice. It may occur at any stage of crop growth leading to mortality and loss of productivity. Occurrence of flooding is an increasing trend due to climate change (Rao et al. 2016). Globally, rice cultivation is at stake. Sequentially, plants experience multiple stresses due to flooding (Septiningsih and Mackill 2018). Soil-flooding during the germination stage restricts oxygen supply to the germinating seeds, which hampers germination and crop establishment (Ray et al. 2016). Complete inundation of the plant reduces light supply to the photosynthetic tissue, which restricts photosynthesis and induces mortality (Panda et al. 2006, 2008a; Sarkar et al. 2014). Due to lack of sufficient oxygen, the plant goes for energy-inefficient anaerobic respiration. Once the submergence event is withdrawn, the plant goes from a low light and low oxygen state to a high light and greater oxygen state. This micro-environment change manifests a large numbers of stresses like oxidative, high light, and dehydration, which triggers senescence and finally mortality (Sarkar et al. 2014; Sasidharan et al. 2018). In contrast to complete inundation, long-term stagnant flooding of one to four months duration happens due to water stagnation (Kuanar et al. 2017). Water stagnation at the root zone or within the 15 cm depth of shoot is not injurious to rice. However, with increased depth of water stagnation rice productivity declines (Fukao et al. 2019). The flash flood for two weeks or more impacts differently as compared to long-term stagnant flooding, and therefore the survival strategies differ in these two conditions. Complete inundation of plants causes greater damage to plants than partial inundation. Most rice genotypes can't tolerate complete submergence of one to two weeks except for a few tolerant ones (Sarkar and Bhattacharjee 2011).

Direct seeding is the main mode of rice cultivation in rain-fed lowland ecology and in the areas where farmers are shifting from transplanting to the direct seeding mode of rice cultivation (Sarkar et al. 1999; Ismail et al. 2009). Direct seeding has a great advantage in reducing cultivation costs (Sarkar and Das 2003; Liu et al. 2015). However, flooding at

Molecular Breeding for Rice Abiotic Stress Tolerance and Nutritional Quality, First Edition.
Edited by Mohammad Anwar Hossain, Lutful Hassan, Khandakar Md. Iftekharuddaula, Arvind Kumar and Robert Henry.

the seeding stage causes severe shortages of oxygen, and only the genotypes tolerant to germination-stage oxygen deficiency (GSOD) can overcome the situation by protruding coleoptiles above the water surface and growth of the root even in the absence of oxygen (Vijayan et al. 2018). Due to submergence at the seedling stage plant suffers from low light, silt, and low-to-super-saturation oxygen concentrations in floodwater (Ramakrishnayya et al. 1999; Das et al. 2009). Plants tolerant to submergence show lesser elongation of shoot, maintain greater concentration of non-structural carbohydrate in it, and continue underwater photosynthesis by preserving chloroplast structural and functional ability (Sarkar 1998; Panda et al. 2008b; Ikmal et al. 2019). The mechanism of tolerance under partial submergence or stagnant flooding is somewhat different as compared to the mechanism of tolerance of submergence. Under long-term stagnant flooding certain parts of the plant body always remain above the water surface. Unlike submergence stress at the germination stage and the vegetative stage, oxygen deprivation is not so acute under stagnant flooding. Rice has unique characteristics that supply oxygen to different parts of body under soil flooding through formation of aerenchymatous tissues and lateral roots and optimization of radial oxygen loss (Colmer et al. 1998). Rice survives well under partial submergence. However, water stagnation has a great effect on productivity. Grain yield declines with increasing depth of water stagnation. The greater the depth of water, the greater the reduction of yield (Sarkar 2016). In contrast to a quiescence strategy, here rice applies an expansion strategy. Elongation of stem, leaf, and leaf sheath occurs due to partial submergence. This helps plants to push certain parts of their body above the water surface. Higher elongation growth is required under floating rice (water depth > 1.5 m) ecology (Hattori et al. 2009), whereas moderate, semi-high, and high elongation growth is useful to push the plant parts above the water surface respectively in medium depth (>0.25 m to =0.50 m), deep water (>0.5 m to = 1.0 m), and very deep (>1.0 m to = 1.5 m) ecologies. Floating rice can grow at rates of 25 cm d^{-1} and within a few days can attain a height of 7 m under partial submergence and be able to tolerate easily up to 4 m water depth (Kende et al. 1998). Elongation rates of 2.0 cm d^{-1} at the vegetative stage is sufficient under medium depth whereas elongation rates of >2.0 to <5.0 cm are optimum for deep water conditions for better productivity (Kato et al. 2014; Vergara et al. 2014; Kuanar et al. 2017). Depending on the flooding pattern and amount of rainfall, the depth of water stagnation varies in different years in the same rice field. It is not uncommon that due to shortage of rainfall water stagnation can be reduced to below 25 cm depth in a field, whereas due to greater precipitation in the same field the stagnation of water exceeds more than 50 cm; otherwise in a normal year water stagnation in the same field remains between 26 and 50 cm (Sarkar 2016). The heterogeneous conditions of rain-fed lowland rice fields and difference in flooding patterns, make it difficult to develop a super rice variety suitable for all types of environment of rain-fed lowland. Our understanding of stagnation patterns of water in rice fields shows that the areas where in normal year water stagnation is more than 1 m in general, even in a drought year the level does not reduce to below 50 cm but inter-conversion from medium depth on either side occurs depending on the intensity of rainfall. Inter-conversion from 50 cm on either side of 25 cm is more common that exceeds to 75 cm or below 25 cm (Sarkar 2016). In South and Southeast Asian countries around 20%, i.e., 9.8 million hectares of land, fall into the medium-depth category (Sarkar 2016). If submergence is taken into consideration along with stagnation of water around 50%, or 48.7 million hectares, of

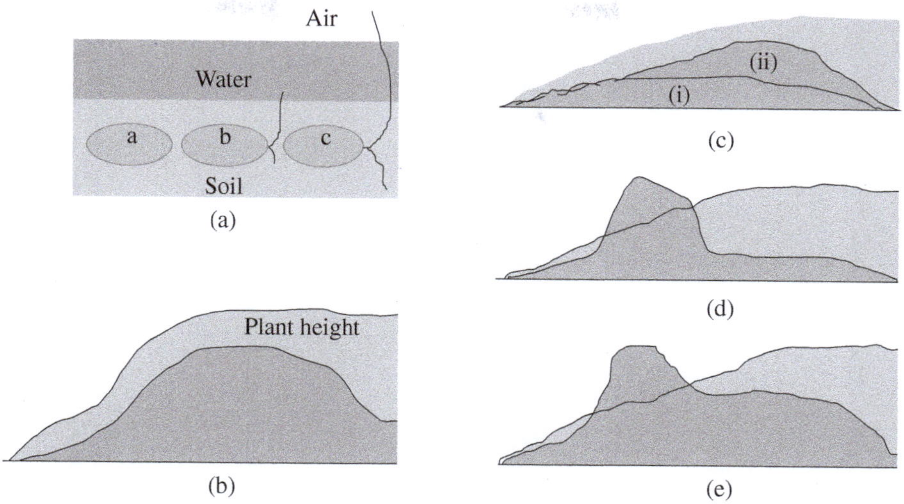

Figure 4.1 Flooding and rice growth. *Note:* Green color, symbol of plant height; Blue color, symbol of water; (a) germination-stage oxygen-deficiency stress; (a) complete cessation of germination; (b) germination happened but coleoptiles growth was not sufficient for establishment; (c) good sign of anaerobic germination potential (AGP), indication of plant establishment; (b) stagnant flooding under deep to very deep floating-rice conditions; water stagnation may persist more than one month to four months or up to maturity; (c) stagnant flooding under shallow (c[i]) and medium-depth (c[ii]) conditions; water stagnation may persist one to two months; (d) flash floods for one to two weeks with more than 1 m depth of water. (e) flash flood for one to two weeks followed by stagnant flooding for one to two months with 30–60 cm depth of water.

land in South and Southeast Asian countries is always under threats of flooding due to complete/ partial inundation/combination of both, that is, submergence followed by stagnant flooding (Figure 4.1). Submergence tolerance is needed in rice grown even in irrigated condition due to unprecedented changes in climate. Flash floods are now more common than before due to climate change (Ghosh et al. 2012). So, for improving flooding tolerance in rice, different strategies have to be followed. Taking farmers' perception into account, it is imperative to emphasize that to get high and stable yields with assured return under flooding stress, it is urgent to design rice plants that can germinate and establish themselves well under oxygen deficiency, tolerate submergence, and stagnant flooding stresses.

Rice Tolerant to Germination Stage Oxygen Deficiency and Submergence and Stagnant Flooding?

To breed a good variety for rain-fed lowlands, at least three distinct traits have to be incorporated in a single plant. Rice variety tolerant to GSOD, submergence, and stagnant flooding (SF) can easily overcome the deleterious effects of flooding stresses in such conditions (Figure 4.2). For rainfed shallow lowlands and irrigated conditions where rice is grown in the transplanted method, submergence tolerance is sufficient, whereas under direct seeded conditions tolerance to GSOD and submergence tolerance is essential. In some cases, combining GSOD and submergence tolerance in a single variety is difficult.

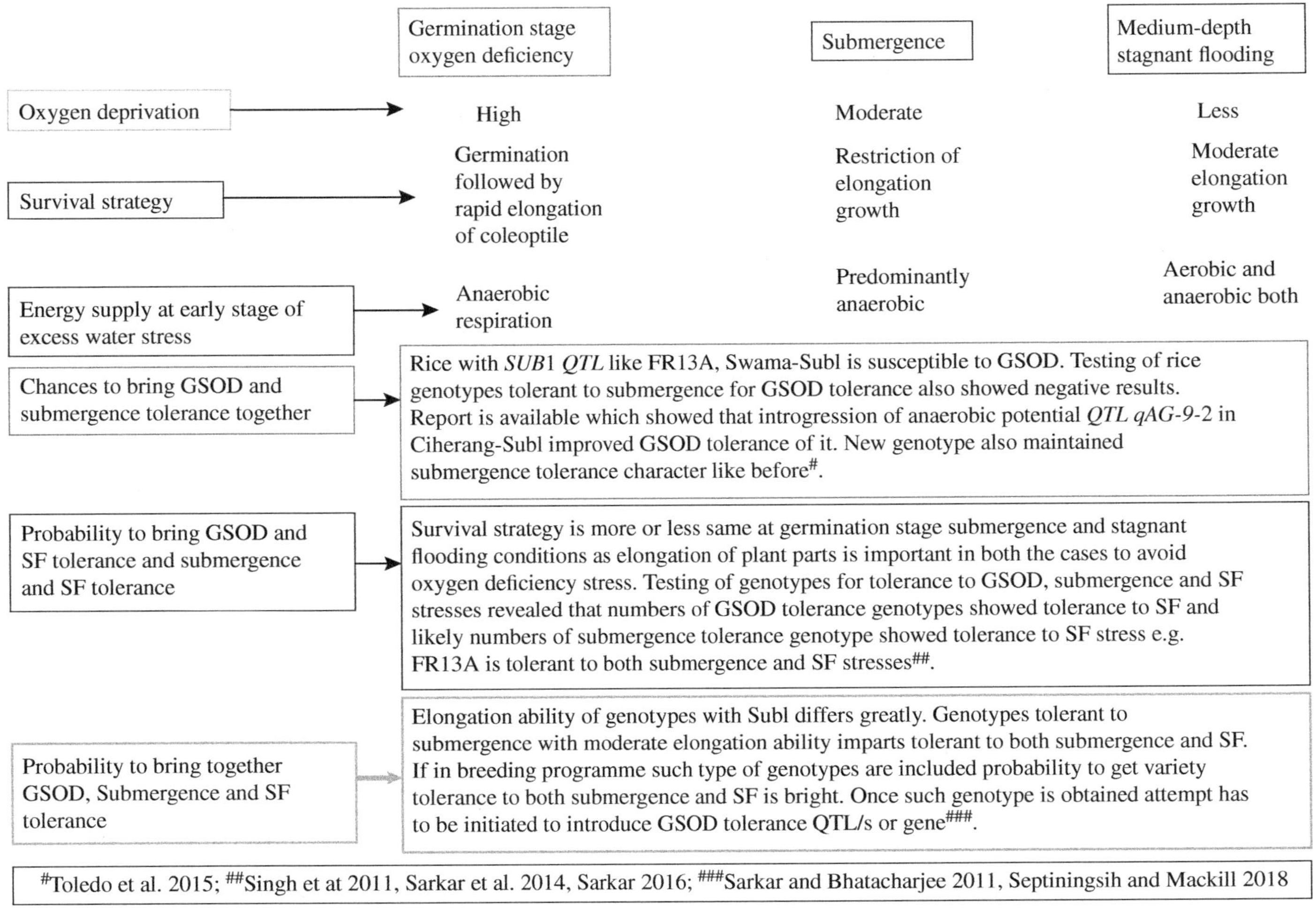

Figure 4.2 Excess water at different growth stages, mechanism of tolerance and probability to get variety with tolerance to GSOD, submergence, and stagnant flooding of medium-depth water stagnation. Source: (Toledo et al. 2015); (Singh et al. 2011); (Sarkar 2016); (Sarkar et al. 2014); (Sarkar and Bhattacharjee 2011); (Septiningsih and Mackill 2018).

The strategy adopted by GSOD tolerance genotype is the elongation of plant parts in search of light and oxygen. GSOD-tolerant genotypes push their coleoptile tip upon germination above the water surface by fast elongation whereas submergence-tolerant plants restrict the elongation growth till the submergence is over. Submergence-tolerant genotypes such as FR13A, Swarna-Sub1, and other genotypes are inherently susceptible to GSOD (Ismail et al. 2009; Sarkar 2012). Recent findings show that introgression of anaerobic germination QTL qAG-9-2 in a SUB1 background variety Ciherang-Sub1 improved GSOD tolerance without any reduction of submergence tolerance strength (Toledo et al. 2015). Though flooding stresses impact differently on different situations their mode of action in most cases are not so different (Figure 4.2). This suggests that it is possible to bring all traits into a single variety by fixing long-term goals. The goal is to be achieved by solving one problem after the other.

History of Flooding Tolerance and Molecular Breeding

Discovery of submergence tolerance *QTL* "*SUB1*" is the golden date of molecular breeding. This particular discovery and the subsequent introduction of this quantitative trait loci (*QTL*) in developing submergence-tolerant rice cultivars led the discovery of other *QTLs* related to many abiotic and biotic stresses. The impact of "*SUB1*" *QTL* was so strong that it made a clear-cut impact in developing submergence-tolerant rice cultivars. The importance of traditional/wild germplasm in developing high-yielding stress-tolerant cultivars was already known (Hawkes 1977); however, later on germplasm evaluation for specific traits, discovery of specific gene/QTL associated with such trait got impetus. Genes/QTLs related to drought, salinity, nutrient deficiency, and other abiotic stresses have been identified from traditional rice genotypes, and numbers of varieties with stress-tolerant characteristics have been developed through molecular breeding for commercial cultivation (Bernier et al. 2007; Angaji et al. 2010; Septiningsih et al. 2013; Gregorio et al. 2013; Sandhu et al. 2019). The first step of such great activity was to transfer submergence-tolerant genes from traditional tall low-yielding photosensitive rice cultivar FR13A (Dhullaputia, a landrace of Orissa, India) to semi-dwarf agronomical superior background through conventional breeding (Mackill et al. 1993). Such agronomical-superior genotypes were the building blocks of the discovery of *SUB1 QTL* and the subsequent development of high-yielding superior rice cultivars tolerant to submergence (Xu and Mackill 1996; Xu et al. 2006). Since 1996 the progress in molecular submergence-tolerant breeding is exemplary. Mechanisms of tolerance were worked out, fine mapping of the gene was completed, new markers associated with submergence-tolerance locus were known, and last of all large numbers of mega-rice varieties susceptible to submergence were converted into submergence-tolerant types (Xu et al. 2004; Fukao and Bailey-Serres 2008; Septiningsih et al. 2009; Bailey-Serres et al. 2010; Fukao et al. 2012; Gonzaga et al. 2017; Azarin et al. 2017; Pucciariello and Perata 2017; Tiwari 2018). Countries of South and Southeast Asia such as Bangladesh (Iftekharuddaula et al. 2011, 2015; Biswas et al. 2015), India (Jena et al. 2015; Rahman et al. 2018; Arya and Shylaraj 2018), Indonesia (Toledo et al. 2015), Nepal (Sah et al. 2013), Malaysia (Ahmed et al. 2016), Philippines (Neeraja et al. 2007; Manzanilla et al. 2014), Sri Lanka (Rathnayake et al. 2012), Thailand (Korinsak et al. 2016), Vietnam (Cuc et al. 2012; Linh et al. 2013), and African countries (AfricaRice 2017) developed and released submergence-tolerant rice for commercial cultivation, and millions of farmers got the benefit of a single gene (Sarkar et al. 2009; Dar et al. 2013, 2017).

Agronomic practices to manipulate germination and establishment of rice seeds under oxygen-deficiency conditions started long ago (Ota and Nakayama 1970). Several attempts had been made during the 1990s to identify genotypes tolerant to GSOD with good anaerobic germination potential (AGP) (Yamauchi et al. 1993; Yamauchi and Chuong 1995; Sarkar et al. 1999). However, the discovery of the *QTL* related to GSOD tolerance is a comparatively recent activity (Jiang et al. 2004, 2006; Angaji et al. 2010). Unlike *SUB1 QTL*, which showed more than 70% phenotypic variation (Xu and Mackill 1996), the *QTL* associated with GSOD tolerance show maximum phenotypic variability of 40%. GSOD-tolerant cultivars assure germination under oxygen-deficiency states. GSOD tolerance *QTLs* identified from different sources are presented in Table 4.1. The study revealed

Table 4.1 *QTLs* linked to GSOD tolerance/anaerobic germination potential (AGP).

QTL/Gene	Cross combination	Origin of tolerant genotype	Chr. No.	LOD score	Phenotypic variation (%)	Reference/s
qAG-1	Kinmaze × DV 85	China/India	1	3.2	12.1	Jiang et al. (2004)
qAG-2			2	4.1	15.6	
qAG-5a			5	2.9	19.6	
qAG-5b			5	2.6	16.6	
qAG-7			7	2.7	10.5	
qAG-5	USSR5 × N22	USSR/India	5	3.8	15.5	Jiang et al. (2006)
qAG-11			11	3.0	11.0	
qAG-1-2	Khao Hlan On × IR64	Myanmar	1	9.8	17.9	Angaji et al. (2010)
qAG-3-1			3	5.7	18.2	
qAG-7-2						
qAG-9-1			7	9.7	19.4	
qAG-9-2			9	16.4	23.6	
			9	20.3	33.5	
qAG7.1	Ma-Zhan Red × IR 42	China	7	14.5	31.7	Septiningsih et al. (2013)
qAG7	Nanhi × IR 64	India	7	13.9	22.3	Baltazar et al. (2014)
qAG1a	Tai Nguyen × Anda	Vietnam/ Korea	1	4.3	5.5	Kim and Reinke (2018)
qAG1b			1	10.1	14.1	
qAG8			8	4.6	5.9	
qAG3	IR 64 × Kharsu 80A	Philippines/ Pakistan	3	5.9	18.4	Baltazar et al. (2019)
qAG7.1			7	5.7	12.5	
qAG7.2			7	7.1	13.5	
qAG7.3			7	7.5	12.6	

that in all the crosses the *QTLs* associated with GSOD tolerance were invariably present on chromosome 7; however, the *QTLs* with greater *LODs* (logarithm of the odds ratio) and phenotypic variations were present on chromosome 9 (Angaji et al. 2010). Fine mapping of *qAG-9-2* (*AG1*) was already completed (Kretzschmar et al. 2015). Introgression of this *QTL* in IR 64 background revealed that newly developed material such as IR 64-AG1 was considerably superior in tolerating oxygen deficiency in the germination stage. The *QTL AG1* was successfully introgressed in the background of Ciherang-Sub1 (Toledo et al. 2015). Kim and Reinke (2018) observed that the combination of three *QTLs* (*qAG1b* + *qAG1a* + *qAG8*) significantly improved under-water establishment in rice as compared to the combination of two *QTLs*. Transcriptomic analysis of diverse rice genotypes with different sensitivity to GSOD identified several genes linked to germination under oxygen deficiency; however, their utilization in molecular breeding has not yet been standardized (Hsu and Tung 2015; Vijayan et al. 2018).

Rice, in general, survives stagnation in water if certain portions of the body parts remain above the water surface (Sarkar 2016). With increasing water level, if the plant elongates to avoid submergence, this certainly helps the plant to survive, but there is a yield penalty. It makes molecular breeding somewhat difficult. If water stagnation on field is roughly divided into five categories such as shallow (≈25 cm), medium-depth (≈26–50 cm), deep water (≈51–100 cm), very deep water (≈101–150 cm), and floating rice (>150 cm), we get different scenarios. The impact of 50 cm water stagnation is different from the impact of 100 cm water stagnation. A plant with 120 cm height has 79%, 58%, and 20% aerial part if water stagnation is 25, 50, and 100 cm, respectively whereas if the water stagnation is more than 120 cm, the plant is underwater. In India another classification, that is, semi-deep has been proposed in between medium-depth and deep-water ecologies where stagnation of water is up to 70 cm. Therefore, when attempting stagnant flooding breeding, an understanding is required for which ecology the variety is to be developed, As such, if 80% of the plant parts remain above the water surface, there is no yield penalty (Sarkar 2016). Crop improvement for shallow rain-fed lowlands does not require any special attention except submergence and GSOD tolerance if direct seeding is taken into consideration, whereas greater attention is required in respect of stagnation of water to breed high-yielding varieties in the rest of the ecologies. Plant height at the initial stage of partial submergence is crucial to realize better yields. If at least 30% of the body parts remain above the water surface at the initial stage of partial submergence, moderate elongation (2.0 cm d^{-1} for initial 30 days of partial submergence) can give better yields under medium-depth stagnant flooding (Kato et al. 2014; Vergara et al. 2014; Kuanar et al. 2017), whereas the elongation may be little bit more under semi-deep (≈70 cm) to deep (≈100 cm) water conditions. Semi-tall to tall varieties such as 125–140 cm, 141–160 cm, and 161–180 cm height are suitable for medium-depth, semi-deep, and deep water ecologies, respectively. Elongation genes such as *Snorkel 1* (*SK1*) and *Snorkel 2* (*SK2*) induce greater elongation, which is suitable for very deep water or floating rice ecology, where water stagnation exceeds more than 150 cm. The impact of three *QTLs* such as *qTIL1*, *qLEI3*, and *qTIL12* is similar to *SK2* (Figure 4.3). *QTL* locates in chromosome 12, that is, *qTIL12* or *SK2* is more effective to induce elongation (Hattori et al. 2008, 2009). The elongation efficiency is greatly decreased in the absence of *qTIL12* or *SK2*. It is likely that two *QTLs,* such as *qTIL1* and *qLEI3,* together impart more elongation compared to *qTIL1* and *qLEI3* alone. Nagai et al. (2014) identified several other *QTLs* along

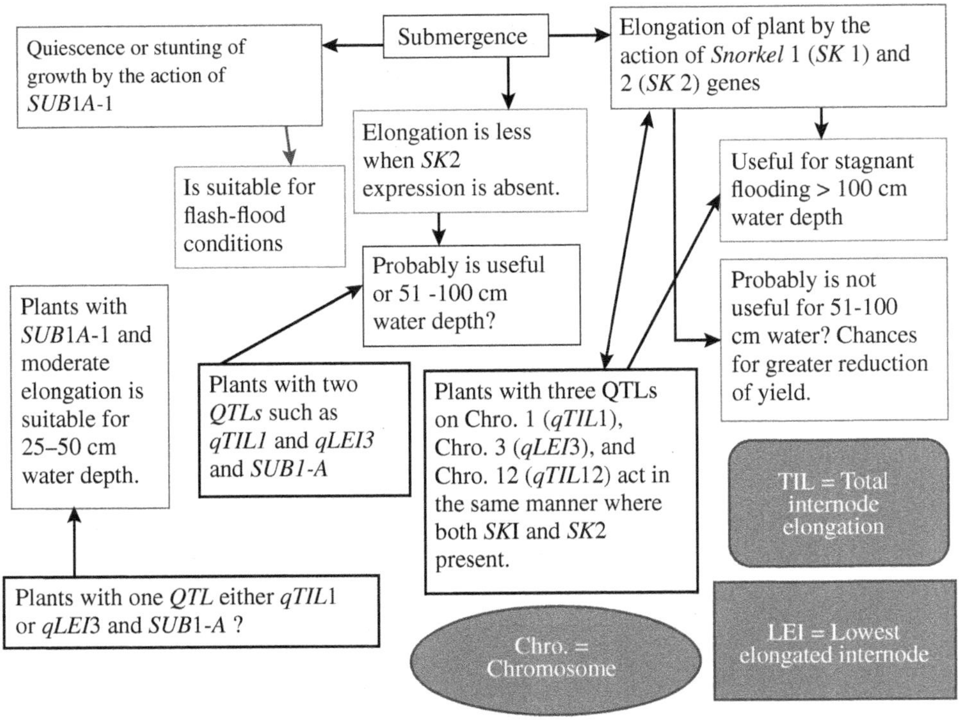

Figure 4.3 QTLs imparting tolerance to submergence and stagnant flooding. *Note:* Based on the elongation ability of different *QTLs* the schematic diagram was made. Everywhere the same elongation ability is not sought for so the probable role of different *QTLs* under different depth of water stagnation was assigned for to get better yield under different depth of water stagnation.

with the *QTLs* reported earlier in chromosome 3 (3 *QTLs*), chromosome 8 (1 *QTL*), chromosome 9 (3 *QTLs*), chromosome 10 (1 *QTL*), and chromosome 12 (1 *QTL*) with varying degrees of intermodal elongation ability. They also observed that *QTLs* of chromosome 3 enhanced the activity of *QTL* of chromosome 1. Elongation ability also differed among the rice genotypes with the presence of *SUB1 QTL* (Sarkar and Bhattacharjee 2011). Submergence tolerance may not be so important under very deep water or floating rice conditions; however, under shallow to deep water conditions (15–100 cm water depth) it is highly relevant (Sarkar et al. 2006). Through the molecular breeding approach, it is now possible to induce different degrees of elongation suitable for diverse types of ecologies (Figure 4.3).

Convergence of Conventional and Molecular Breeding

Significant progress in breeding of submergence-tolerant rice varieties has been achieved during the last decade due to the identification of a major QTL named *SUB1*. This gene was subsequently incorporated with a high level of precision using marker-assisted backcrossing (MABC). As many as eight mega varieties of Asian origin, such as Swarna, Samba Mahsuri, CR 1009, IR64, PSBRc 18, BR 11, Ciherang, and Thadokkham that have been "upgraded" with the *SUB1* and several others are currently being improved (Singh et al.

2015). The varieties with *SUB1* were comprehensively tested in the field, and no negative effects of it were found on other traits including yield under favorable conditions (Sarkar et al. 2009). Furthermore, mega-varieties with *SUB1* have been widely accepted by farmers (Mackill et al. 2012; Collard et al. 2013; Dar et al. 2017). Currently, International Rice Research Institute (IRRI), Philippines and different National Agricultural Research System (NARS) are collaborating to develop submergence-tolerant rice varieties using MABC for Africa and Asian countries such as Bangladesh, Cambodia, India, Nepal, and Pakistan (Collard et al. 2013; Singh et al. 2015).

The submergence-tolerant gene *SUB1* generally confers tolerance to flash floods for up to two weeks, and does not bestow any tolerance to stagnant flooding or immediately after seeding (Ismail et al. 2009; Singh et al. 2011; Sarkar 2012; Kato et al. 2019). The molecular breeding efforts resulted in the introgression of AGP *QTL qAG-9-2* in a *SUB1* background variety Ciherang-Sub1 improved GSOD tolerance without sacrificing submergence tolerance strength (Toledo et al. 2015). As the genetic mechanism controlling tolerance to submergence and stagnant flooding behave in an antagonistic way, combining these two traits requires a precise conventional breeding approach along with molecular breeding. Therefore, if the breeding objective is to combine a set of the above-mentioned traits that make a variety unique or ideal to a given situation, the "convergent breeding" approach is the most appropriate strategy (Siddiq 2013). The ongoing worldwide effort to breed varieties for major rice ecologies requires stepwise accumulation of desirable characteristics that would combine high and stable yield, wide adaptability, and acceptable quality. As marker-assisted selection/backcross breeding is not enough to improve the yield ceiling, there is a need to exploit new germplasm, develop new, precise and efficient molecular markers, and develop efficient phenotyping methods, which would be essential to enhance breeding efficiency, and complement conventional breeding efforts (Collard et al. 2013).

Limited elongation ability along with stem strength/resistance to lodging imparts tolerance to SF (Kuanar et al. 2017; Kato et al. 2019). Besides, new breeding lines must also possess desirable agronomic and quality traits, and possess resistances to important diseases and insects pests. Taking the above-mentioned traits into consideration, a flow chart of a breeding programto integrate submergence tolerance and stagnant flooding has been outlined by Collard et al. (2013) (Figure 4.4).

- To develop new breeding lines, a pedigree method with some alterations can be followed. Testing for submergence tolerance at controlled conditions is done at the F_2 and F_3 stages. After screening of F_3 populations, single plant selection is performed from surviving plants.
- From F_4 to F_6 the materials are tested for resistances to disease and insects. Testing for bacterial blight starts at F_4 while green leaf hopper (GLH), brown plant hopper (BPH), and blast are tested only at the F_6 stage.
- Testing of yield starts under normal and stagnant flooding conditions at the F_6 stage and later on with appropriate check cultivars. Marker-aided selection (MAS) are applied at F_6 stage. At F_8 seeds can be distributed at different testing stations/countries for suitability of the locations. Applying MAS at the early stages is avoided to reduce the cost.
- These approaches take care of submergence and stagnant-flooding tolerance along with diseases and insect problems.

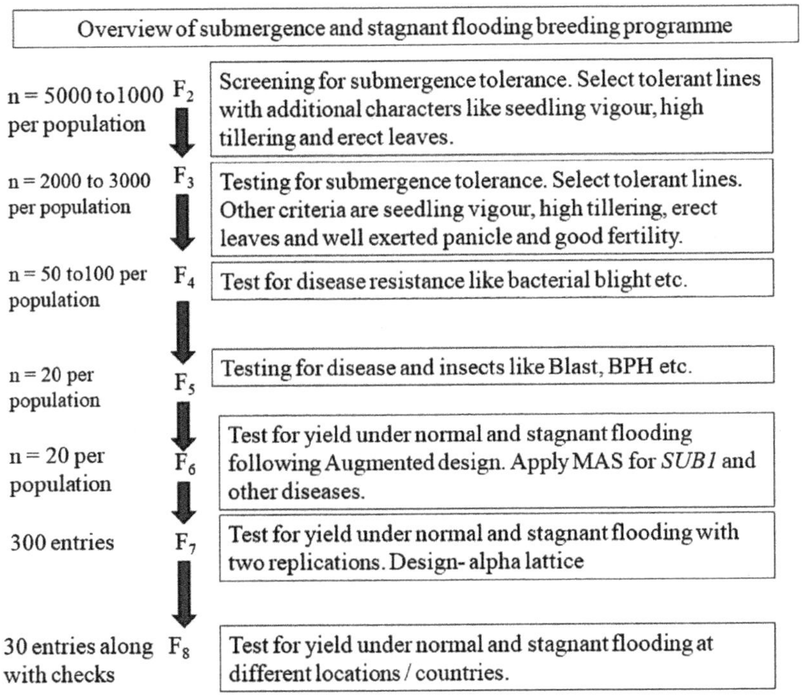

Figure 4.4 Flow chart of breeding methods employed to integrate submergence tolerance with tolerance to stagnant flooding in rice.

One of the major obstacles of breeding long-duration photosensitive varieties for integrating tolerance to submergence and stagnant flooding is the time required for fixing genotypes with the desired character combinations. Several techniques have been developed to reduce the breeding and selection cycles in rice breeding. The single-seed descent method (SSD) coupled with rapid generation advance (RGA) is well suited to breeding long-duration photosensitive varieties under field conditions. This approach has been effectively and widely used by the NARS and IRRI. Ikehasi and Lambers (1977) discussed the details of this method in relation to rice breeding and also proposed an international integrated breeding program for the development of photoperiod-sensitive rice (Figure 4.5). SSD, a modification of the original bulk method, is a breeding procedure in which fixation of traits with minimum bias from potential variability is achieved. To prevent any drift in segregating populations, a single seed should be harvested from each F_2 plant to grow the F_3 population, and that same procedure is repeated until F_3–F_6. RGA is not a breeding method but a way of growing bulk populations. Since the major objective of SSD is to attain homozygosity, it becomes essential for rapid cycling of breeding materials that shortens the growth duration of a given population. A large number of plants can be accommodated in limited space and dense spacing because only one or two seeds from each plant are enough to advance the segregating generations. Thus the SSD method is the most appropriate for and is used as a routine breeding method in order to improve breeding efficiency (Collard et al. 2017). It has been suggested that there

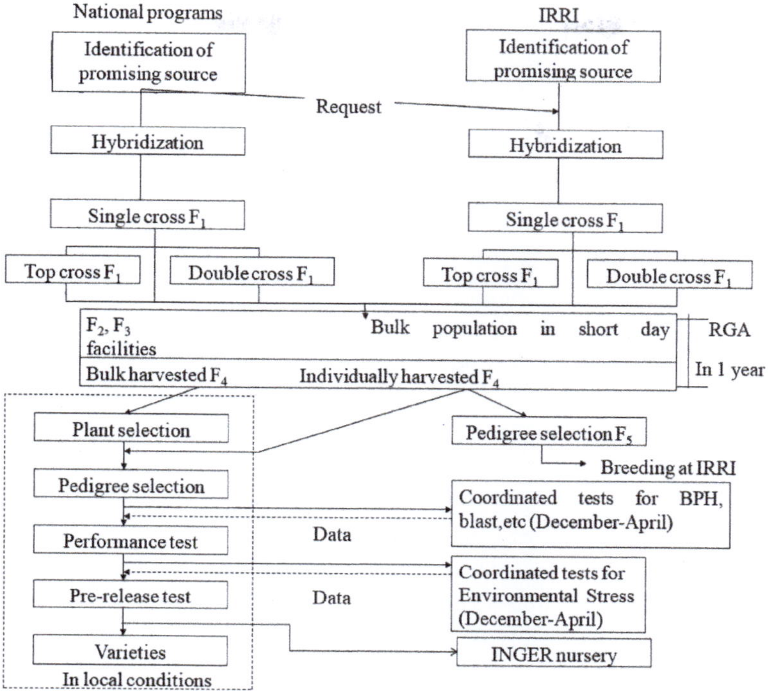

Figure 4.5 Integrated international breeding program for photoperiod-sensitive rice (Ikehasi and Lambers 1977).

would be higher genetic gain by employing the SSD method of breeding coupled with RGA simply by modifying the empirical formula used in selection theory. The formula for genetic gain (ΔG, or response to selection), is given by $\Delta G = (ih^2 \sigma P)/L$, where, i is the selection intensity, h^2 is the trait heritability, σP is the square root of the phenotypic variance, and L is the length of the breeding cycle or generation interval. Genetic gain is increased due to reducing the time of the breeding cycle (Atlin et al. 2017; Collard et al. 2017, 2019).

While concluding the topic we would like to quote and outline the critical considerations made by Collard et al. (2016), who emphasized relooking and reevaluating the breeding methodologies. Single-seed descent or bulk-population methods take less labor and resources. Breeders should look for some traits including yield in early generation materials and primary varietal selection trial data. "The use of new technologies and breeding informatics and use of more efficient experimental designs such as row–column designs or partial replication, or spatial analysis for single site trial analysis could have great potential to improve efficiency along with multi-environmental trials analysis, and proper understanding of target environments. The use of molecular breeding may become more effective when multiple abiotic and biotic stress genes/QTLs segregate within specific populations because multiple target loci can be screened in these populations" (Collard et al. 2016). Collaboration among different partners can help in developing multiple stress-tolerant rice varieties with shortest time period.

Conclusion and Future Perspective

Rice breeders have been very successful in selecting for a wide range of traits using only conventional methods. Breeders do not need to use markers for every trait, even if perfect markers are available. The full advantage of MAS can be realized when markers are used to do something that cannot be done using conventional methods such as MABC or marker-assisted pyramiding. To date, varieties with *SUB1* have been developed using only SSR markers. Utilizing single nucleotide polymorphism (SNP) genotyping can increase selection efficiency of large breeding populations; fastening the molecular breeding, as it is a high through-put technique, takes less time for target gene/s and background selection with lesser cost. Many breeders may not understand topics in molecular breeding in sufficient detail, and molecular geneticists often do not understand breeding processes to identify the most strategic stage at which to apply markers. This also prevents integration between breeding and molecular geneticists including sharing of germplasm such as mapping populations or sharing of information or trait data. Marker-added selection or marker-added backcross breeding does not improved yield potential under normal growth environments. MAS/MAB (marker-assisted breeding) is used to improve the stress-tolerant character of a mega variety. In overall improving the yield potential above the level of the present-day mega-varieties, conventional plant breeding can take a leading role. Here the integration between conventional plant breeders and molecular geneticists is important to narrow down "application gap." Furthermore, the location of breeding and molecular genetics research is often separate, and may occur at different institutions. In order to take advantage of available genomics resources in rice for molecular breeding in the future, breeders will need: planned and well-executed QTL mapping experiments, QTL and marker validation activities, efficient and cheaper genotyping and phenotyping systems, and breeder-friendly analytical tools. Integration between breeders, plant physiologists, and molecular geneticists will be critical to develop new varieties using markers.

Acknowledgment

RKS is grateful to the Indian Council of Agricultural Research, New Delhi, for providing financial help through the Emeritus Scientist project. RKS and JNR are thankful to Director, ICAR-NRRI, India for providing facilities. SRD is grateful to the International Rice Research Institute, Philippines, for financial help, and Vice Chancellor, OUAT, Bhubaneswar, India, for providing facilities for preparation of this manuscript.

References

AfricaRice (2017). Nigeria releases AfricaRice "flood-tolerant" rice strains. http://www.africarice.org/warda/newsrel-flood-tolerant-Aug17.asp.

Ahmed, F., Rafii, M.Y., Ismail, M.R. et al. (2016). The addition of submergence-tolerant *Sub1* gene into high yielding MR219 rice variety and analysis of its BC_2F_3 population in terms of yield and yield contributing characters to select advance lines as a variety. *Biotechnol. Biotechnol. Equip.* 30: 853–863.

Angaji, S., Septiningsih, E.M., Mackill, D.J., and Ismail, A.M. (2010). *QTLs* associated with tolerance of anaerobic conditions during germination in rice (*Oryza sativa* L.). *Euphytica* 172: 159–168.

Arya, K.V. and Shylaraj, K.S. (2018). Introgression of Sub1 QTL (submergence tolerant QTL) into the elite rice variety Jaya by marker assisted backcross breeding. *J. Trop. Agric.* 56: 114–128.

Atlin, G.N., Cairns, J.E., and Das, B. (2017). Rapid breeding and varietal replacement are critical to adaptation of cropping systems in the developing world to climate change. *Glob. Food Sec.* 12: 31–37.

Azarin, K.V., Usatov, A.V., and Kostylev, P.I. (2017). Molecular breeding of submergence-tolerant rice. *Annu. Res. Rev. Biol.* 18: 1–10.

Bailey-Serres, J., Fukao, T., Ronald, P.C. et al. (2010). Submergence tolerant rice: SUB1's journey from landrace to modern cultivar. *Rice* 3: 138–147.

Baltazar, M., Ignacio, J., Thomson, M. et al. (2014). QTL mapping for tolerance of anaerobic germination from IR64 and the Aus landrace Nanhi using SNP genotyping. *Euphytica* 197: 251–260.

Baltazar, M., Ignacio, J.C.I., Thomson, M.J. et al. (2019). QTL mapping for tolerance to anaerobic germination in rice from IR64 and the Aus landrace Kharsu 80A. *Breed. Sci.* 69 (2): 227–233. https://doi.org/10.1270/jsbbs.18159.

Bernier, J., Kumar, A., Ramaiah, V. et al. (2007). A large-effect QTL for grain yield under reproductive-stage drought stress in upland rice. *Crop Sci.* 47: 507–518.

Biswas, A.K.M.A.A., Mamun, M., Rahman, M.S. et al. (2015). Submergence rice cultivation in southern Bangladesh: farmers' opinion and adaptations practices. *Int. J. Environ.* 4: 100–113.

Collard, B.C.Y., Septiningsih, E.M., Das, S.R. et al. (2013). Developing new flood-tolerant varieties at the International Rice Research Institute (IRRI). *SABRAO J. Breed. Genet.* 45: 42–56.

Collard, B.C.Y., Raghavan, C., and Islam, M.R. (2016). Critical considerations for molecular breeding in rice. *Oryza* 53: 346–353.

Collard, B.C.Y., Beredo, J.C., Lenaerts, B. et al. (2017). Revisiting rice breeding methods – evaluating the use of rapid generation advance (RGA) for routine rice breeding. *Plant Prod. Sci.* 20: 337–352.

Collard, B.C.Y., Gregorio, G.B., Thomson, M.J. et al. (2019). Transforming rice breeding: re-designing the irrigated breeding pipeline at the International Rice Research Institute (IRRI). *Crop Breed. Genet. Genom.* 1: e190008.

Colmer, T.D., Gibberd, M.R., Wiengweera, A., and Tinh, T.K. (1998). The barrier to radial oxygen loss from roots of rice (*Oryza sativa* L.) is induced by growth in stagnant solution. *J. Exp. Bot.* 49: 1431–1436.

Cuc, L.M., Huyen, L.T.N., Hien, P.T.M. et al. (2012). Application of marker assisted backcrossing to introgress the submergence tolerance QTL SUB1 into the Vietnam elite rice variety-AS996. *Am. J. Plant Sci.* 3: 528–536.

Dar, M.H., Janvry, A.D., Emerick, K. et al. (2013). Flood-tolerant rice reduces yield variability and raises expected yield, differentially benefitting socially disadvantaged groups. *Sci. Rep.* 3: 1–8.

Dar, M.H., Chakravorty, R., Waza, S.A. et al. (2017). Transforming rice cultivation in flood prone coastal Odisha to ensure food and economic security. *Food Sec.* 9: 711–722.

Das, K.K., Panda, D., Sarkar, R.K. et al. (2009). Submergence tolerance in relation to variable floodwater conditions in rice. *Environ. Exp. Bot.* 66: 425–434.

Fukao, T. and Bailey-Serres, J. (2008). Submergence tolerance conferred by *SUB1A* is mediated by *SLRI* and *SLRLI* restriction of gibberellin responses in rice. *Proc. Natl. Acad. Sci. U.S.A.* 105: 16814–16819.

Fukao, T., Yeung, E., and Bailey-Serres, J. (2012). The submergence tolerance gene SUB1A delays leaf senescence under prolonged darkness through hormonal regulation in rice. *Plant Physiol.* 160: 1795–1807.

Fukao, T., Barrera-Figueroa, B.E., Juntawong, P., and Peña-Castro, J.M. (2019). Submergence and waterlogging stress in plants: a review highlighting research opportunities and understudied aspects. *Front. Plant Sci.* 10: 340.

Ghosh, S., Das, D., Kao, S.C., and Ganguly, A.R. (2012). Lack of uniform trends but increasing spatial variability in observed Indian rainfall extremes. *Nat. Clim. Chang.* 2: 86–91.

Gonzaga, Z.J.C., Carandang, J., Singh, A. et al. (2017). Mapping *QTLs* for submergence tolerance in rice using a population fixed for *SUB1A* tolerant allele. *Mol. Breed.* 37: 47.

Gregorio, G.B., Islam, M.R., Vergara, G.V., and Thirumeni, S. (2013). Recent advances in rice science to design salinity and other abiotic stress tolerant rice varieties. *SABRAO J. Breed. Genet.* 45: 31–41.

Hattori, Y., Nagai, K., Mori, H. et al. (2008). Mapping of three QTLs that regulate internode elongation in deep water rice. *Breed. Sci.* 58: 39–46.

Hattori, Y., Nagai, K., Furukawa, S. et al. (2009). The ethylene response factors SNORKEL1 and SNORKEL2 allow rice to adapt to deep water. *Nature* 460: 1026–1031.

Hawkes, J.G. (1977). The importance of wild germplasm in plant breeding. *Euphytica* 26: 615–621.

Hsu, S.-K. and Tung, C.-W. (2015). Genetic mapping of anaerobic germination-associated *QTLs* controlling coleoptile elongation in rice. *Rice* 8: 38.

Iftekharuddaula, K., Newaz, M., Salam, M. et al. (2011). Rapid and high-precision marker assisted backcrossing to introgress the SUB1 QTL into BR11, the rainfed lowland rice mega variety of Bangladesh. *Euphytica* 178: 83–97.

Iftekharuddaula, K.M., Ghosal, S., Gonzaga, Z.J. et al. (2015). Allelic diversity of newly characterized submergence-tolerant rice (*Oryza sativa* L.) germplasm from Bangladesh. *Genet. Resour. Crop., Evol.* 63: 859–867.

Ikehasi, H., and Lambers, H.R. (1977). . Single seed descent with the use of RGA. Paper Presented at International Rice Research Conference, Philippines. 18–22 April 1997.

Ikmal, A.M., Amira, I., and Shamsuddin, N.A.A. (2019). Morpho-physiological responses of rice towards submergence tolerance. *Int. J. Agric. Biol.* 22: 35–42.

Ismail, A.M., Ella, E.S., Vergara, G.V., and Mackill, D.J. (2009). Mechanisms associated with tolerance for flooding during germination and early seedling growth in rice (*Oryza sativa*). *Ann. Bot.* 103: 197–209.

Jena, P.P., Bharathkumar, S., Reddy, J.N., and Mohapatra, T. (2015). Introgression of Sub1 locus into highly preferred rice cultivars (Pooja and Pratikshya) in eastern region of India for submergence tolerance through marker assisted backcrossing. *Adv. Biores.* 6: 45–53.

Jiang, L., Hou, M., Wang, C., and Wan, J. (2004). Quantitative trait loci and epistatic analysis of seed anoxia germinability in rice (*Oryza sativa*). *Rice Sci.* 11: 238–244.

Jiang, L., Liu, S., Hou, M. et al. (2006). Analysis of QTLs for seed low temperature germinability and anoxia germinability in rice (*Oryza sativa* L.). *Field Crop Res.* 98: 68–75.

Kato, Y., Collard, B.C.Y., Septiningsih, E.M., and Ismail, A.M. (2014). Physiological analyses of traits associated with tolerance of long-term partial submergence in rice. *AoB Plants* 6: 58.

Kato, Y., Collard, B.C.Y., Septiningsih, E.M., and Ismail, A.M. (2019). Increasing flooding tolerance in rice: combining tolerance of submergence and of stagnant flooding. *Ann. Bot.* 124: 1199–1209. https://doi.org/10.1093/aob/mcz118.

Kende, H., van der Knaap, E., and Cho, H.T. (1998). Deepwater rice: a model plant to study stem elongation. *Plant Physiol.* 118: 1105–1110.

Kim, S.-M. and Reinke, R.F. (2018). Identification of QTLs for tolerance to hypoxia during germination in rice. *Euphytica* 214: 160.

Korinsak, S., Siangliw, M., Kotcharerk, J. et al. (2016). Improvement of the submergence tolerance and the brown planthopper resistance of the Thai jasmine rice cultivar KDML105 by pyramiding Sub1 and Qbph12. *Field Crops Res.* 188: 101–112.

Kretzschmar, T., Pelayo, M.A.F., Trijatmiko, K.R. et al. (2015). A trehalose-6-phosphate phosphatase enhances anaerobic germination tolerance in rice. *Nat. Plants* 1: 15124.

Kuanar, S.R., Ray, A., Sethi, S.K. et al. (2017). Physiological basis of stagnant flooding tolerance in rice. *Rice Sci.* 24: 73–84.

Linh, T.-H., Linh, L.-H., Cuc, D.-T.K. et al. (2013). Improving submergence tolerance of Vietnamese rice cultivars by molecular breeding. *J. Plant. Breed. Genet.* 3: 157–168.

Liu, H., Hussain, S., Zheng, M. et al. (2015). Dry direct- seeded rice as an alternative to transplanted-flooded rice in Central China. *Agron. Sustain. Dev.* 35: 285–294.

Mackill, D.J., Amante, M.M., Vergara, B.S., and Sarkarung, S. (1993). Improved semidwarf rice lines with tolerance to submergence of seedlings. *Crop Sci.* 33: 749–753.

Mackill, D.J., Ismail, A.M., Singh, U.S. et al. (2012). Development and rapid adoption of submergence-tolerant (Sub1) rice varieties. *Adv. Agron.* 115: 303–356.

Manzanilla, D.O., Paris, T.R., Vergara, G.V. et al. (2014). Submergence risks and farmers' preferences: implications for breeding Sub1 rice in Southeast Asia. *Agric. Syst.* 104: 335–347.

Nagai, K., Kondo, Y., Kitaoka, T. et al. (2014). QTL analysis of internode elongation in response to gibberellin in deepwater rice. *AoB Plants* 6: plu028.

Neeraja, C.N., Maghirang-Rodriguez, R., Pamplona, A. et al. (2007). A marker-assisted backcross approach for developing submergence tolerant rice cultivars. *Theor. Appl. Genet.* 115: 767–776.

Ota, Y. and Nakayama, M. (1970). Effects of seed coating with calcium peroxide on germination under submerged conditions in rice plant. *Jpn. J. Crop Sci.* 39: 535–536.

Panda, D., Rao, D.N., Sharma, S.G. et al. (2006). Submergence effects on rice genotypes during seedling stage: probing of submergence driven changes of photosystem 2 by chlorophyll a fluorescence induction O-J-I-P transients. *Photosynthetica* 44: 69–75.

Panda, D., Sharma, S.G., and Sarkar, R.K. (2008a). Fast chlorophyll fluorescence transients as selection tools for submergence tolerance in rice (*Oryza sativa*). *Indian J. Agric. Sci.* 78: 933–938.

Panda, D., Sharma, S.G., and Sarkar, R.K. (2008b). Chlorophyll fluorescence parameters, CO_2 photosynthetic rate and regeneration capacity as a result of complete submergence and subsequent re-emergence in rice (*Oryza sativa* L.). *Aquat. Bot.* 88: 127–133.

Pucciariello, C. and Perata, P. (2017). New insights into reactive oxygen species and nitric oxide signalling under low oxygen in plants. *Plant Cell Environ.* 40: 473–482.

Rahman, H., Dakshinamurthi, V., Ramasamy, S. et al. (2018). Introgression of submergence tolerance into CO 43, a popular rice variety of India, through marker-assisted backcross breeding. *Czech J. Genet. Plant Breed.* 54: 101–108.

Ramakrishnayya, G., Setter, T.L., Sarkar, R.K. et al. (1999). Influence of P application to floodwater on oxygen concentrations and survival of rice during complete submergence. *Exp. Agric.* 35: 167–180.

Rao, C.S., Gopinat, K.A., Prasad, J.V.N.S. et al. (2016). Climate resilient villages for sustainable food security in tropical India: concept, process, technologies, institutions, and impacts. *Adv. Agron.* 140: 101–214.

Rathnayake, N.R.M.K.N.D., Bentota, A.P., Dissanayake, D.M.N. et al. (2012). DNA markers RM 464A and RM 219 haplotypes are effective in selecting Sub-1 locus for the introgression of submergence tolerance into new rice varieties. *Ceylon J. Sci. (Bio. Sci.)* 41: 125–136.

Ray, S., Vijayan, J., and Sarkar, R.K. (2016). Germination stage oxygen deficiency (GSOD): an emerging stress in the era of changing trends in climate and rice cultivation practice. *Front. Plant Sci.* 7: 671.

Sah, S.N., Ghimirey, S.K., Ismail, A.B. et al. (2013). Selection of submergence tolerant rice varieties in Nepal through participatory approach. *Nepal Agric. Res. J.* 13: 1–10.

Sandhu, N., Dixit, S., Swamy, B.P.M. et al. (2019). Marker assisted breeding to develop multiple stress tolerant varieties for flood and drought prone areas. *Rice* 12: 8.

Sarkar, R.K. (1998). Saccharide content and growth parameters in relation with flooding tolerance in rice. *Biol. Plant.* 40: 597–603.

Sarkar, R.K. (2012). Seed priming improves agronomic trait performance under flooding and non-flooding conditions in rice with *QTL SUB1*. *Rice Sci.* 19: 286–294.

Sarkar, R.K. (2016). Stagnant flooding tolerance in rice: endeavours and achievements. NRRI Research Bulletin No. 11, ICAR-NRRI. Cuttack, India.

Sarkar, R.K. and Bhattacharjee, B. (2011). Rice genotypes with Sub1 *QTL* differ in submergence tolerance, elongation ability during submergence, and re-generation growth at re-emergence. *Rice* 5: 7.

Sarkar, R. and Das, S. (2003). Yield of rainfed lowland rice with medium water depth under anaerobic direct seeding and transplanting. *Trop. Sci.* 43: 192–198.

Sarkar, R.K., Bera, S.K., and De, R.N. (1999). Rice (*Oryza sativa*) cultivars for anaerobic seeding. *Indian J. Agric. Sci.* 69: 73–76.

Sarkar, R.K., Reddy, J.N., Sharma, S.G., and Ismail, A.M. (2006). Physiological basis of submergence tolerance in rice and implications for crop improvement. *Curr. Sci.* 91: 899–906.

Sarkar, R.K., Panda, D., Reddy, J.N. et al. (2009). Performance of submergence tolerant rice (*Oryza sativa*) genotypes carrying the Sub1 quantitative trait locus under stressed and non-stressed natural field conditions. *Indian J. Agric. Sci.* 79: 876–883.

Sarkar, R.K., Das, K.K., Panda, D., Reddy, J.N., Patnaik, S.S.C., Patra, B.C., and Singh, D.P. (2014). Submergence tolerance in rice: biophysical constraints, physiological basis and identification of donors. CRRI Research Bulletin No. 7, ICAR-CRRI. Cuttack, India

Sasidharan, R., Hartman, S., Liu, Z. et al. (2018). Signal dynamics and interactions during flooding stress. *Plant Physiol.* 176: 1106–1117.

Septiningsih, E.M. and Mackill, D.J. (2018). Genetics and breeding of flooding tolerance in rice. In: *Rice Genomics, Genetics and Breeding* (eds. T. Sasaki and M. Ashikari), Chapter 15. Singapore: Springer Nature Singapore Pvt Ltd.

Septiningsih, E.M., Pamplona, A.M., Sanchez, D.L. et al. (2009). Development of submergence-tolerant rice cultivars: the Sub1 gene and beyond. *Ann. Bot.* 103: 151–160.

Septiningsih, E.M., Ignacio, J.C.I., Sendon, P.M.D. et al. (2013). *QTL* mapping and confirmation for tolerance of anaerobic conditions during germination derived from the rice landrace Ma-Zhan Red. *Theor. Appl. Genet.* 126: 1357–1366.

Siddiq, E.A. (2013). Rice towards food secure India: Potential and Prospects. Dr. K. Ramaiah Memorial Lecture – 2013 on December 24, 2013 at Orissa University of Agriculture & Technology. Bhubaneshwar, Orissa, India.

Singh, S., Mackill, D.J., and Ismail, A.M. (2011). Tolerance of longer-term partial stagnant flooding is independent of *SUB1* locus in rice. *Field Crops Res.* 121: 311–323.

Singh, R., Singh, Y., Xalaxob, S. et al. (2015). From QTL to variety-harnessing the benefits of QTLs for drought, flood and salt tolerance in mega rice varieties of India through a multi-institutional network. *Plant Sci.* 242: 278–287.

Tiwari, D.N. (2018). A critical review of submergence tolerance breeding beyond *Sub 1* gene to mega varieties in the context of climate change. *Int. J. Adv. Sci. Res. Eng.* 4: 140–147.

Toledo, A.M.U., Ignacio, J.C.I., Casal, C. et al. (2015). Development of improved Ciherang-Sub1 having tolerance to anaerobic germination conditions. *Plant Breed. Biotechnol.* 3: 77–87.

Vergara, G.V., Nugraha, Y., Esguerra, M.Q. et al. (2014). Variation in tolerance of rice to long-term stagnant flooding that submerges most of the shoot will aid in breeding tolerant cultivars. *AoB Plants* 6: plu055.

Vijayan, J., Senapati, S., Ray, S. et al. (2018). Transcriptomic and physiological studies identify cues for germination stage oxygen deficiency tolerance in rice. *Environ. Exp. Bot.* 147: 234–248.

Xu, K.N. and Mackill, D.J. (1996). A major locus for submergence tolerance mapped on rice chromosome 9. *Mol. Breed.* 2: 219–224.

Xu, K.N., Deb, R., and Mackill, D.J. (2004). A microsatellite marker and a codominant PCR based marker for marker-assisted selection of submergence tolerance in rice. *Crop Sci.* 44: 248–253.

Xu, K., Xia, X., Fukao, T. et al. (2006). *Sub1A* is an ethylene response factor-like gene that confers submergence tolerance to rice. *Nature* 442: 705–708.

Yamauchi, M. and Chuong, P.V. (1995). Rice seedling establishment as affected by cultivar, seed coating with calcium peroxide, sowing depth, and water level. *Field Crops Res.* 41: 123–134.

Yamauchi, M., Aguilar, A.M., Vaughan, D.A., and Seshu, D.V. (1993). Rice (*Oryza sativa* L.) germplasm suitable for direct sowing under flooded soil surface. *Euphytica* 67: 177–184.

5

Molecular Breeding for Improving Heat Stress Tolerance in Rice: Recent Progress and Future Perspectives

Bui Chi Buu[1], *Cho Young Chan*[2] *and Nguyen Thi Lang*[1]

[1] *High Agricultural Technology Research Institute for Mekong Delta (HATRI), Can Tho, Vietnam*
[2] *National Institute of Crop Science (NICS), RDA, Wanju 55365, Republic of Korea*

Introduction

Rice is one of the most important food crops in the world; it contributes to both food security and income generation, particularly within developing countries (Khush 2005). The high yield of rice plays an important role in food security. However, the high frequency of drought, floods, heat, and biotic stress has brought a huge challenge for rice production. Global warming has become a big concern due to human activities through greenhouse gas emission as CO_2, N_2O, CH_4. Therefore, breeding more rice varieties incorporated with heat tolerance is necessary to cope with future climate change. Rice responses to high temperature differ with the developmental stages. Overall, the reproductive stage, especially the flowering stage, is the most sensitive stage to high temperature (Satake and Yoshida 1978). High temperature over 35 °C at anthesis and lasting for more than one hour causes high spikelet sterility, leading to serious yield losses (Jagadish et al. 2007). It has been proved that the poor anther dehiscence is the main cause of spikelet sterility induced by high temperature, because poor anther dehiscence results in few germinated pollen grains on the stigma (Jagadish et al. 2010b; Matsui and Omasa 2002; Prasad et al. 2006).

The International Rice Research Institute (IRRI) recorded global warming from 0.35 °C rising to 1.13 °C; otherwise, if temperature were to increase 1 °C, rice productivity would decrease by 10% (Peng et al. 2004). The Intergovernmental Panel on Climatic Change (IPCC) observed that every decade had brought a 0.3 °C rise in temperature (Jones et al. 2009). In Japan (2007), the critical temperature at heading stage of >40 °C in Kanto and Tokai had led to 25% unfertilized plants; then, in 2003, 8 million ha in Hubei rice areas, China suffered critical heat at microsporogenesis leading to a high percentage of unfilled grain (Jagadish et al. 2010a).

An IPCC special report on the impacts of global warming of 1.5 °C above pre-industrial levels and related global greenhouse gas emission pathways, examined the context of strengthening the global response to the threat of climate change, sustainable development, and efforts to eradicate poverty (IPCC 2018).

Molecular Breeding for Rice Abiotic Stress Tolerance and Nutritional Quality, First Edition.
Edited by Mohammad Anwar Hossain, Lutful Hassan, Khandakar Md. Iftekharuddaula, Arvind Kumar and Robert Henry.

Climate Change and Heat Tolerance Genotype Needs

In 2015 COP21, also known as the 2015 Paris Climate Conference, aimed to achieve a legally binding and universal agreement on climate, with the aim of keeping global warming below 2 °C. Global warming is projected to have significant impacts on conditions affecting agriculture, including temperature, carbon dioxide, glacial run-off, precipitation, and the interaction of these elements.

To overcome the challenges presented by climate change, it is important to understand how crops respond to high temperatures (Jung et al. 2012). Heat is a serious problem in many areas in the world (Wahid et al. 2007). Because of the global greenhouse effect, heat temperature stress has become increasingly important as a yield-limiting factor, and the increased frequency of short-term but extremely high temperatures is particularly damaging.

Temperature is a highly significant environmental factor in the growth and development of plants. Climate change will lead to drought damage, high temperature, and a spectrum of heat and disease damage in world rice production, especially in tropical areas. The heat stress especially during the flowering stage in rice growth area results in a low seed-setting rate and a loss of grain yield (Morita et al. 2005; Peng et al. 2004). So research on the genetic mechanisms of heat tolerance is becoming more and more important to the utilization of heat-tolerant genes (HT) and the development of new rice varieties with heat tolerance. The most sensitive growth stage of rice to heat stress is the flowering time (Mackill et al. 1982; Kuang et al. 2002). In Vietnam, heat stress at the flowering stage has become a big problem in recent years. A high percentage of unfilled spikelets and a low grain-filling rate (GFR) came about due to heat stress at the flowering stage (Nguyen et al. 2015). A temperature of 35 °C, beyond the threshold level, at flowering stage is considered a critical level that harms floret fertility and grain yield (Matsushima et al. 1982; Li 2003).

The susceptibility of photosynthesis to inhibition by moderate and high temperatures has been attributed to two major causes, depending on the severity of the stress. At temperatures >40 °C, the properties of the lipid bilayer of the thylakoid membrane change, causing a disruption in electron transport and ion transport reactions necessary for photosynthesis, particularly damage to photosystem 5-3 (PS5-3) (Sharkey 2000; Yamori et al. 2014).

The soluble enzyme RUBISCO activase (RCA) impairs photosynthesis by reducing the ability of RCA to maintain ribulose-1,5-bisphosphate carboxylase/oxygenase (RUBISCO) in an active state (Crafts-Brandner and Law 2000; Salvucci et al. 2001; Salvucci and Crafts-Brandner 2004; Sage et al. 2008).

For this reason, research on the genetic mechanisms of heat tolerance is becoming more and more important for the utilization of heat-tolerant genes and the development of new rice varieties with heat tolerance. A high percentage of unfilled spikelets and low GFR came about due to heat stress at flowering stage (Nguyen et al. 2015).

A high temperature, 35 °C beyond threshold level, at flowering can induce floret sterility and consequently high losses in yields (Osada et al. 1973; Satake and Yoshida 1978; Matsushima et al. 1982), with occasionally up to 80% loss (Li 2003). It has been reported that heat stress has significantly reduced anther dehiscence and pollen fertility rates in heat-sensitive varieties, whereas its effects were much smaller in heat-tolerant cultivars (Cao et al. 2008).

This is the main reason why research on the genetic mechanisms of heat tolerance is critical for efficient utilization of the genes and for developing new rice varieties with such traits (Xiao et al. 2011).

Recent Progress on Rice Breeding

Plant Physiology on Heat-Stress Response

Net Photosynthesis Rate

Heat stress results in significant losses in crop production worldwide (Battisti and Naylor 2009; Esmaeili et al. 2019), and it is predicted to become even more serious in the future (Ainsworth and Ort 2010). It has been shown that an increase in air temperature above a plant's optimum growth temperature decreases net photosynthetic rates in plants (Vile et al. 2012) and affects membrane fluidity, which disrupts ion hemostasis in plant cells (Rivero et al. 2014). Increasing heat tolerance in crops, however, may allow plants to maintain metabolic function and growth under heat-stress conditions. Efforts have been made in engineering plants for heat-stress tolerance by overexpressing heat-shock protein genes or heat-shock factor genes in transgenic plants over the last 20 years, yet only limited success has been achieved, and so far, none has been applied in crops. Recent discoveries may change the landscape of transgenic research on heat-tolerant plants. Overexpression of *OsSIZ1*, a *SUMO E3 ligase* gene from rice, was found to confer increased heat and drought tolerance in transgenic creeping bent grass (Li et al. 2013). Overexpression of the Arabidopsis vacuolar *H^+-pyrophosphatase* gene AVP1 increases salt and water-deficit stress tolerance, and overexpression of the rice *SUMO E3 ligase* gene *OsSIZ1* improves heat and water-deficit stress tolerance in transgenic plants (Esmaeili et al. 2019).

Developing rice cultivars with higher (CO_2) responsiveness incorporated with increased tolerance to high temperatures during flowering and grain filling, using donors such as NL-44, will minimize the negative impact of heat stress and increase global food productivity, benefiting from CO_2 rich environments (Chaturvedi et al. 2017).

Grain Filling

High temperatures impair rice grain weight by inhibiting the filling of the caryopses during the milky stage (Liao et al. 2012). Using paired, genetically similar heat-tolerant and heat-sensitive rice lines as plant materials, cDNA-AFLP (amplified fragment length polymorphism) analysis has revealed a total of 54 transcript-derived fragments (TDFs), mainly from the heat-tolerant lines. High temperature stress may trigger a wide range of changes in gene expression in rice caryopses, in turn affecting functions ranging from signal transduction to cellular metabolism. Forty-five of the 54 TDFs were mapped to rice chromosomes (Liao et al. 2012).

Rice grains accumulate starch as their major storage reserve whose biosynthesis is sensitive to heat (Huang et al. 2019). ADP-glucose pyrophosphorylase (AGPase) is among the starch biosynthetic enzymes severely affected by heat stress during seed maturation. To increase the heat tolerance of the rice enzyme, we engineered two dominant AGPase subunits expressed in developing endosperm, the large (L2) and small (S2b) subunits of the cytosol-specific AGPase. Bacterial expression of the rice S2b with the rice L2, potato tuber LS (pLS), or with the mosaic rice-potato large subunits, L2-pLS and pLS-L2, produced heat-sensitive recombinant enzymes, which retained less than 10% of their enzyme activities after five minutes incubation at 55 °C. However, assembly of the rice L2 with the potato tuber SS (pSS) showed significantly increased heat stability comparable to the heat-stable potato pLS/pSS. The S2b assembled with the mosaic L2-pLS subunit showed threefold higher sensitivity to 3-PGA (phosphoglyceric acid) than L2/S2b, whereas the counterpart mosaic pLS-L2/S2b showed 225-fold lower sensitivity. Introduction of a QTC motif into S2b created an N-terminal disulfide linkage that was cleaved by dithiothreitol reduction. The QTC enzyme showed moderate heat stability but was not as stable as the potato AGPase. While the QTC AGPase exhibited an approximately fourfold increase in 3-PGA sensitivity, its substrate affinities were largely unchanged. Random mutagenesis of S2bQTC produced six mutant lines with elevated production of glycogen in bacteria. All six lines contained a L379F substitution, which conferred enhanced glycogen production in bacteria and increased heat stability. The modeled structure of this mutant enzyme revealed that this highly conserved leucine residue is located in the enzyme's regulatory pocket, which provides interaction sites for activators and inhibitors. Our molecular dynamic simulation analysis suggests that introduction of the QTC motif and the L379F mutation improves enzyme heat stability by stabilizing their backbone structures possibly due to the increased number of H-bonds between the small subunits and increased intermolecular interactions between the two SSs and two LSs at elevated temperatures (Huang et al. 2019).

Rice is susceptible to both drought and heat stress, in particular during flowering and grain filling, when both grain yield and quality may be severely compromised (Lawas et al. 2019). Under field conditions, these two stresses rarely occur separately. Drought was applied either during flowering or during early grain filling and resulted in simultaneous heat stress, leading to reduced grain yield and quality. There were distinct metabolic profiles for the three investigated organs (flag leaves, flowering spikelets, developing seeds). Gas chromatography–mass spectrometry resolved metabolic responses to combined drought and heat stress in different organs of field-grown rice. The metabolite profiles can be used to identify potential marker metabolites for yield stability and grain quality that are expected to improve breeding efforts towards developing rice cultivars that are resilient to climate change (Lawas et al. 2019).

Response in Germination

Dry-heat treatment also breaks the seed dormancy of freshly harvested cereal seeds in the breeding program (Lee et al. 2017). Rice (*Oryza sativa* L.) depends on the varietal differences in morphological and physiological characteristics (Morishima and Oka 1981). Among agronomic characteristics, seed dormancy, longevity, and dry-heat tolerance also showed varietal difference, with most *japonica* rice varieties having a lower seed dormancy, longevity, and dry-heat tolerance than *indica* rice varieties (Seshu and Sorells 1986; Lee et al. 2002; Guo et al. 2004). Recently, significant correlation was seen between seed dormancy and dry-heat tolerance of rice seed (Lee et al. 2006). *Indica* varieties were highly tolerant of dry-heat treatment, whereas *japonica* varieties were sensitive (Lee et al. 2002). Dry-heat treatment (65 °C, seven days) resulted in considerably lower fungal incidence, and did not have any adverse effect on seed germination and seedling vigor before or after storage (Dadlani and Seshu 1990). This treatment was better than HNO_3 or hulled treatments in breaking rice seed dormancy (Zhang 1990). In dry-heat treatment, seed viability and seedling vigor are largely reduced with high temperature, but seed germination and seedling emergence are promoted as breaking seed dormancy under the optimum range of dry-heat temperature.

Using 164 recombinant inbred lines (RILs) derived from a cross between Milyang 23 (*indica/japonica*) and Gihobyeo (*japonica*) in rice, dry-heat tolerance was evaluated for the seeds of parents and RILs, whose dormancy was naturally broken in six months after harvesting (Lee et al. 2017). Mapping quantitative trait loci (QTLs) associated with dry-heat tolerance was carried out through interval mapping using QGene 3.0. Seed germination after dry-heat treatments (90 °C for 24 hours) showed a significant difference between the two parents, when evaluated for percentage germination and mean germination time. Milyang 23 was highly tolerant to the dry-heat treatment, while Gihobyeo was sensitive. Three QTLs (*qDHT 1, qDHT 5,* and *qDHT 7*) conferring the dry heat tolerance were mapped to chromosomes 1, 5, and 7, respectively. *qDHT 1* on chromosome 1 was tightly linked at 4 cM from ME1-1. The phenotypic variation explained by the three QTLs was 27.18% of the total variance in the 164 RIL populations, and the parental additive effects of three QTLs affected the Milyang 23 allele increased dry heat tolerance. The detection of new QTLs associated with dry-heat tolerance will provide important information for disease and insect control, using dry-heat treatment in organic or low-input sustainable agriculture (Lee et al. 2017).

Response in Anthesis

The reproductive stage, especially the flowering stage, is the most sensitive stage to high temperature (Zhao et al. 2016; Satake and Yoshida 1978). High temperature over 35 °C at anthesis and lasting for more than one hour causes high spikelet sterility, leading to serious yield losses (Jagadish et al. 2007). It has been proven that poor anther dehiscence is the main cause of spikelet sterility induced by high temperature, because poor anther dehiscence results in few germinated pollen grains on the stigma (Jagadish et al. 2010b; Matsui and Omasa 2002; Prasad et al. 2006). In addition, daily flowering time is also shown to be connected with heat tolerance. The varieties that carry the early-morning flowering (EMF) trait can avoid the high temperature at anthesis thereby mitigating high-temperature-induced spikelet sterility (Ishimaru et al. 2010). The QTL *qEMF3*, identified recently from wild rice *Oryza officinalis*, showed the potential to shift flower opening time of cultivars to earlier in the morning (Hirabayashi et al. 2015). In general, *indica* rice cultivars show much better heat tolerance than *japonica* cultivars (Zhao et al. 2016). Three traits, that is, spikelet fertility, daily flowering time, and pollen shedding level were often introduced for identification of heat tolerance.

Episodes of high temperature at anthesis, which in rice is the most sensitive stage to temperature, are expected to occur more frequently in future climates (Jagadish et al. 2010a). The morphology of the reproductive organs and pollen number, and changes in anther protein expression, were studied in response to high temperature at anthesis in three rice (*O. sativa* L.) genotypes. Plants were exposed to six hours of high (38 °C) and control (29 °C) temperature at anthesis and spikelets collected for morphological and proteomic analysis. Moroberekan was the most heat-sensitive genotype (18% spikelet fertility at 38 °C), while IR64 (48%) and N22 (71%) were moderately and highly heat tolerant, respectively. There were significant differences among the genotypes in anther length and width, apical and basal pore lengths, apical pore area, and stigma and pistil length (Jagadish et al. 2010a).

Among physiological processes occurring at anthesis, anther dehiscence is perceived to be the most critical stage affected by high temperature (Matsui and Omasa 2002; Matsui and Kagata 2003).

Amylose Content Response to Heat

Amylose content mainly controlled by the *Wx* gene is a key physicochemical property for eating and cooking quality in rice. During the grain filling stage, high temperature can harm rice grain quality by significantly reducing the amylose content in many rice varieties (Zhang et al. 2014). Four QTLs, namely *qHAC4, qHAC8a, qHAC8b*, and *qHAC10*, which can reduce the deleterious effects of amylose content at high temperature, were identified and mapped to chromosome 4, 8, 8, and 10, respectively. The major QTL *qHAC8a*, with the highest logarithm of odds (LOD) score of 6.196, was physically mapped to a small chromosome segment (~300 kb). Thus, increasing pre-mRNA processing efficiency of the *Wx* gene could be an important regulation mechanism for maintaining stable amylose content in rice seeds at high temperature (Zhang et al. 2012).

Heat-Induced Quality Decline of Rice (White-Back Kernel)

Decline in the apparent quality of rice (*O. sativa* L.) grain due to high temperatures during ripening recently became a major concern in many areas in Japan (Kobayashi et al. 2013). The occurrence of white-back kernels (WBK) is one of the main problems of heat-induced quality decline. Four QTLs (Hana-echizen × N5-3gata-wase) were identified on chromosomes 3, 4, 6, and 9 (*qWB3, qWB4, qWB6*, and *qWB9*). The combination of *qWB6* and *qWB9* in an F_2 population derived from a cross between a near-isogenic line (NIL) and NW showed that the NW allele at *qWB9* significantly decreased WBK if the *qWB6* allele was homing endonucleases (HE) (Kobayashi et al. 2013).

Source Sink Dynamics

High night temperatures (HNTs) can reduce significantly global rice (*O. sativa*) yield and quality (Lawas et al. 2019; Shi et al. 2013). A systematic analysis of HNT response at the physiological and molecular levels was performed under field conditions. Contrasting rice accessions, N22 (highly tolerant) and Gharib (susceptible), were evaluated at 22 °C (control) and 28 °C (HNT). Nitrogen (N) and nonstructural carbohydrate (NSC) translocation from different plant tissues into grains at key developmental stages, and their contribution to yield, grain-filling dynamics and quality aspects, were evaluated. Proteomic profiling of flag leaf and spikelets at 100% flowering and 12 days after flowering was conducted, and their reprogramming patterns were explored. Grain yield reduction in susceptible Gharib was traced back to the significant reduction in N and NSC translocation after flowering, resulting in reduced maximum and mean grain-filling rate, grain weight, and grain quality. A combined increase in heat shock proteins (HSPs), Ca-signaling proteins and efficient protein modification and repair mechanisms (particularly at the early grain-filling stage) enhanced N22 tolerance for HNT (Lawas et al. 2019). The increased rate of grain filling and efficient proteomic protection, fueled by better assimilate translocation, overcome HNT tolerance in rice. Temporal and spatial proteome programming alters dynamically between key developmental stages and guides future transgenic and molecular analysis targeted towards crop improvement. Gas chromatography–mass spectrometry–based metabolomics were used to elucidate the metabolic responses of flag leaves, flowering spikelets, and developing seeds from three rice cultivars differing in their drought and heat tolerance to rewatering after stress in the field. Then, the identified potential markers can be useful in efforts to breed stress-tolerant rice germplasm (Lawas et al. 2019).

Genetic Background

Breeding Materials (Donors)

Among >20 species in the *Oryza* genus, only two species are cultivated, *O. sativa* and *Oryza glaberrima* (Sweeney and McCouch 2007), and the rest are considered "wild" (Ge et al. 1999; Vaughan et al. 2008) (Table 5.1). Many of the wild species are adapted to adverse environments, as the genus is found across all tropical regions of the earth in a vast array of climates (Atwell et al. 2014) and wild relatives harbor >50% of the genetic diversity found in the genus (Zhu and Ge 2005; Xu et al. 2012).

Scafaro et al. (2018) previously discovered that wild relatives of domesticated rice (*O. sativa*) are more heat tolerant and determined some of the physiological and biochemical factors that confer this tolerance (Scafaro et al. 2010, 2012, 2016). Scafaro et al. (2010) identified RCA as a prime candidate for further investigation because heat enhanced its expression in an Australian wild species of rice (*Oryza meridionalis*) (Table 5.1). Moreover, there is evidence that the RCA protein, its post-translational regulation or aspects of its catalytic behavior are intolerant to temperatures much above 30 °C (Salvucci and Crafts-Brandner 2004; Carmo-Silva and Salvucci 2011; Busch and Sage 2017).

Aus-type rice is highly valuable for breeding applications as a source of novel tolerance traits but also for gene discovery research (Casartelli et al. 2018) with the superior drought tolerance of N22 (IRGC19379) and Dular (IRGC32561) as well as heat tolerance (Bui et al. 2014).

Aus rice ecotype as N22 (Nagina 22) is a deep-rooted upland rice variety, tolerant to various biotic and abiotic stresses such as heat, drought, low-P tolerance, and herbicide resistance were identified (Poli et al. 2013; Mohapatra et al. 2014; Talla et al. 2016; Shoba et al. 2017; Yugandhar et al. 2013, 2018). N22 has been successfully exploited as a parent in a Vietnamese rice breeding program for heat tolerance. Otherwise, Dular has failed due to its poor compatibility instead that good segregation in progenies (Bui et al. 2014). The *aus* rice as Dular has been also used as a donor in some breeding works. Restriction fragment length polymorphism (RFLP) markers were employed to test whether the individuals possess wide compatibility genes (WCGs) in the progenies of Dular, which is one of the typical wide compatibility varieties in ecotype *aus* rice (Yi et al. 2001). It has been concluded that (i) the effect of WCG at S-5 locus is better than the others, (ii) the combination of different wide compatibility loci is obviously effective in increasing spikelet fertility, (iii) intra-allelic interaction can cause *japonica* macrospore abortive, while interallelic interaction has some effects in fertility expression (Wang et al. 1998; Yi et al. 2001). Based on interval analysis with Mapmaker/QTL, 32.3% of the phenotypic variance associated with spikelet fertility was explained by the S5n (LOD = 9.03), and the other two chromosomal segments were responsible for 10.5% (LOD = 2.61) and 10.9% (LOD = 2.14) phenotypic variance,

Table 5.1 Rice breeding materials for heat tolerance analysis.

Designation	Rice ecotype	References
Oryza glaberrima (tolerant)	African cultivable rice (AA)	Sweeney and McCouch (2007)
Oryza officinalis (tolerant)	Asian wild species (CC)	Hirabayashi et al. (2015) (QTL *qEMF3*)
Oryza meridionalis (tolerant)	Australian wild species (AA) closely related to cultivars	Scafaro et al. (2010, 2012, 2016, 2018)
Oryza nivara (tolerant)	(AA) wild rice species	Prasanth et al. (2017) (derivatives: Swarna × *O. nivara*; KMR3 × *O. rufipogon*)
Oryza rufipogon (tolerant)		
Oryza australiensis	(EE) wild rice species	Scafaro et al. (2018) (tolerant due to Rubisco activase)
N22 (tolerant)	*Aus*-type (*O. sativa*)	Casartelli et al. (2018), Poli et al. (2013), Mohapatra et al. (2014), Talla et al. (2016), Shoba et al. (2017), Yugandhar et al. (2013, 2018), Yi et al. (2001), Jagadish et al. (2010a), and Bui et al. (2012, 2013, 2014)
Dular (tolerant and wide-compatible)	*Aus*-type (*O. sativa*)	
Moroberekan (the most sensitive)	*O. sativa* sub. tropical *japonica*	Jagadish et al. (2010a), Bui et al. (2014), Xiao et al. (2011), Zhang et al. (2009), Shanmugavadivel et al. (2017), and Yugandhar et al. (2013, 2018)
N22 (tolerant)	(*Aus*-type) *O. sativa*	
Kasalate (tolerant)	*O. sativa* sub. *indica*	Zhu and Ge (2005)
Gihobyeo (tolerant)	*O. sativa* sub. *japonica*	Lee et al. (2017)
Gayabyeo (sensitive)		
Habataki (tolerant)	*O. sativa* sub. *indica*	Zhao et al. (2016)
Giza178 (tolerant)	*O. sativa* sub. *indica*	Ye et al. (2015)

(AA), same genome AA as well as *Oryza sativa*; (CC, EE) differed from AA genome, needed to have embryo rescue while crossing.

respectively. The results demonstrated that the wide compatibility variety Dular contained three WCGs (Yi et al. 2001). Two germplasm N22 and Dular were considered to be genotypes that are highly tolerant to heat stress at flowering stage (Jagadish et al. 2010a; Bui et al. 2012, 2013). In addition, Moroberekan, the most heat-sensitive genotype, showed 18% spikelet fertility at 38 °C, while IR64, a moderately heat tolerant and N22, a highly tolerant genotype showed 48% and 71% spikelet fertility, respectively (Jagadish et al. 2010a).

Genetic Diversity Analysis

A set of breeding lines and landraces representing 240 germplasm genotypes were studied (Pradhan et al. 2016). Based on spikelet fertility percentage under high temperature, tolerant genotypes were broadly classified into four classes. The analysis of molecular variance revealed 25% variation between population, 61% among individuals, and 14% within individuals in the set. The strongly associated marker RM547 tagged with spikelet fertility under stress and markers such as RM228, RM205, RM247, RM242, INDEL3, and RM314 indirectly controlling high temperature stress tolerance were detected through both mixed linear model and general linear model TASSEL analysis. These markers can be deployed as a resource for marker-assisted breeding programs of high temperature stress tolerance (Pradhan et al. 2016).

Phenotyping

Hakata et al. (2017) reported tolerant cultivars that include three *indica* cultivars,namely IR36, IR24, and IR64, two *japonica* cultivars as Akitakomachi and Koshihikari, and an *aus* cultivar as N22. These rice genotypes were selected using three methods conducted under variable conditions, including temperature gradient chambers (TGC) under natural light, phytotrons, and greenhouses. The traits under given conditions need to be identified as followed:

- Panicles are the rice plant's heat-sensitive organ (Nishiyama and Satake 1981).
- Spikelet sterility caused by high temperatures is also expected to become a serious problem in rice production globally, including Japan, as simulated by Horie et al. (1996). The instability of rice production will be enhanced owing to the increased frequency of heat-induced spikelet sterility (Cruz et al. 2017).
- Heat exposure of panicles during booting and flowering was carried out under a daily maximum temperature of 34–36 °C (Osada et al. 1973; Matsui et al. 2001; Weerakoon et al. 2008). In addition, high humidity promotes spikelet sterility (Weerakoon et al. 2008; Matsui et al. 1997). In addition, low humidity, wind, and high CO_2 concentration also influence sterility (Osada et al. 1973; Matsui et al. 1997).
- The lack of full development of pollen (Matsui et al. 2000) and faulty dehiscence of anthers by high temperatures were reported to reduce the number of healthy pollen grains at the stigma (Satake and Yoshida 1978; Matsui and Omasa 2002).
- Further study of global warming using a crop simulation model suggested that increasing high temperature tolerance by approximately 1.5 °C during the flowering period would greatly reduce the incidence of spikelet sterility (Horie et al. 1996; Nakagawa et al. 2003).
- The panicle top position in the artificial paddies that simulate the environmental conditions of the rice paddy (referred to as "an artificial rice paddy") have been made to be adjustable, so that the position of all cultivars/lines can be uniform (Hakata et al. 2017). Accordingly, rice seeds that were sterilized, then incubated in water at 15 °C for seven days, seeded at 32 °C for two days, and grown in a phytotron (Growth Chamber TGE-3CS) under metal halide lamps (M150FCLSP2-W/BUD, Iwasaki Electric Co., Ltd., Tokyo, Japan) with an intensity of 850 μmol photons m^{-2} s^{-1} PAR at the canopy position.
- They were cultivated in a cycle of day/night air temperatures of 26 °C (13 hours, 5:50–18:50)/22 °C (11 hours, 18:50–5:50) at 60% relative humidity (RH).
- For the light conditions, half of the lamps were turned off for 1 hour at the beginning and end of 13 hours of daytime. The water temperature in the artificial rice paddy (see below) was 24.5 ± 1.2 °C (mean ± SD, $n = 144$), ranging from 22.6 to 26.0 °C. Each plant was grown in a plastic pot (7 cm diameter, 30 cm in height) filled with 1120 ml of rice nursery culture soil containing, 0.20 g of nitrogen and 0.12 g each of phosphate and potassium, and each plant was restricted to the main culm by the removal of the tillers (Hakata et al. 2017).

- Analysis of spikelet fertility and panicle temperature data was conducted using a Tukey's Honestly Significant Difference (HSD) test with JMP software.
- Grain filling under heat stress was considered more important than faulty dehiscence of anthers (Bui et al. 2014). Heading date was recorded for each plant when a panicle emerged. The data were averaged over all plants of each line in each replication. At maturity, five middle plants of each line were harvested in bulk and sun-dried. Five yield traits were measured, including number of panicles per plant, number of grains per panicle, 1000-grain weight (g), spikelet fertility rate (%), and grain yield per plant (g). For measuring 1000-grain weight, fully filled grains, grain filling after two day interval maintaining a constant condition among different samples was used as described previously (Chen et al. 2018; Zhang et al. 2017; Bui et al. 2014)

Putative and Candidate Genes/QTLs

In spite of considerable research in the last several years, the genetic basis of tolerance to heat stress in rice is still largely unknown (Xiao et al. 2011). Molecular mapping efforts on heat tolerance QTLs have been conducted at booting, flowering, and grain filling to ripening stages in rice. A QTL identified to explain the largest phenotype variance for spikelet fertility/seed setting is contributed by IR64 (Cao et al. 2002). The QTL located in the C1100–R1783 region of chromosome 4 derived from Kasalath reduced 3.31% of the grain weight loss under heat stress. Two more QTLs, one located between R1613–C970 on chromosome 1 and the other between C1226–R1440 on chromosome 7, have 2.38% and 2.92% additive effects, respectively (Zhu and Ge 2005). Heat tolerance during the flowering stage in rice was controlled by multiple genes. Previously, it was reported that simple sequence repeats (SSR) markers RM3735 on chromosome 4 and RM3586 on chromosome 3 showed significant association with heat tolerance, which respectively accounted for 17% and 3% of the total variation (Zhang et al. 2009). Likewise, two major QTLs affecting seed set percentage in the field and growth chamber conditions were detected in the interval between RM5687 and RM471 on chromosome 4, and between RM6132 and RM6100 on chromosome 10 (Xiao et al. 2011). The QTL located on chromosome 4 explained 21.3% and 25.8% of the total phenotypic variation in seed storage protein (SSP), respectively in both conditions. The second QTL located on chromosome 10 explained 11.5% in field and 11.6% in growth chamber of the total phenotypic variation in SSP (Xiao et al. 2011). Additional QTLs associated with heat tolerance during flowering have been recently identified from RIL populations developed from *indica*/*japonica* or *indica*/*indica* crosses (Zhang et al. 2009; Chen et al. 2008). In addition, a major dominant locus *OsHTAS* on chromosome 9 was found to confer tolerance to 48 °C at seedling stage (Wei et al. 2013). This locus was mapped on rice chromosome 9 within an interval of 420 kb between markers of InDel5 and RM7364 (Wei et al. 2013).

The QTL *qEMF3*, identified recently from wild rice *O. officinalis*, showed the potential to shift flower opening time of cultivars to earlier in the morning (Hirabayashi et al. 2015). To uncover the genetic basis of heat tolerance at the flowering stage in rice, several independent studies have been conducted, and many QTLs have been identified (Ye et al. 2012). However, there are few reports about the confirmation and fine mapping of the identified QTL for heat tolerance. It is known that mapping populations and accurate phenotyping technology are essential for QTL mappings (Ashikari and Matsuoka 2006; Tuberosa 2012).

Although many kinds of mapping populations, including DH, RILs, F_2, BC_1F_1, and BIL population have been widely used to detect the QTLs associated with heat tolerance (Ye et al. 2012), chromosomal sequence substitution lines (CSSLs) and NILs are rarely used. As a permanent mapping population, these two kinds of populations are ideally suited for quantitative traits QTL detection (Ashikari and Matsuoka 2006). To explore QTLs controlling heat-tolerance at anthesis, a set of CSSLs derived from the backcross between *indica* donor cultivar "Habataki" and *japonica* recipient cultivar "Sasanishiki" were employed for studying the traits across three environments. As for the evaluation index for heat-tolerance identification, spikelet fertility, or some related secondary traits have been usually used. Little attention has been paid to seeking other traits associated with high temperature stress (Ashikari and Matsuoka 2006). In this study, three traits, that is, spikelet fertility, daily flowering time, and pollen shedding level were introduced for identification of heat tolerance. Some main effect QTLs were detected, and the overlapping QTLs from the present study were compared with other research. The results would provide a clue to understand the genetics of rice heat tolerance at the flowering stage.

In the Philippines, QTL linked to HT and EMF were analyzed by selective genotyping of phenotypic extremes in 235 BC_1F_1 plants derived from the cross PSB-Rc82/PSB-Rc82//WAB56-125 under controlled high temperature and 684 BC_1F_2 plants under glasshouse conditions using single marker analysis (Baliuag et al. 2015). Three markers on chromosomes 2, 3, and 8 are possibly linked to pollen fertility, while two markers on chromosomes 7 and 8 are possibly linked to spikelet fertility under high temperature (Baliuag et al. 2015).

Development of rice varieties with tolerance to high temperature stress at the flowering stage is therefore essential for maintaining rice production as the climate continues to normal condition. Four QTLs, *qHAC4, qHAC8a, qHAC8b*, and *qHAC10*, which can reduce the deleterious effects of amylose content at high temperature, were identified and mapped to chromosome 4, 8, 8, and 10, respectively. The major QTL *qHAC8a*, with the highest LOD score of 6.196, was physically mapped to a small chromosome segment (~300 kb) (Zhang et al. 2014). Development in genomics has provided new tools for discovering and tagging novel alleles and genes.

In Vietnam, breeding for tolerance to high temperature (>35 °C) during the reproductive stage in rice (*O. sativa* L.) is one adaptation strategy for mitigating the negative effects of global warming. The objective of this study was to identify potential QTLs associated with heat tolerance and related traits in the OM5930/N22 BC_2F_2 population and to identify markers that could potentially be applied in molecular marker-aided breeding for heat tolerance in rice under conditions in southern Vietnam. A total of 310 lines (BC_2F_2) derived from the cross of OM5930*3/N22 has been evaluated for heat stress at flowering. A genetic map has been set up with 264 polymorphic SSR markers to detect linkage to the target traits. The map covers 2741.63 cM with an average interval of 10.55 cM between marker loci. Markers associated with heat tolerance have been located mostly on chromosomes 3, 4, 6, 8, 10, and 11. The proportions of phenotypic variation explained by each QTL of markers RM3586 and RM160 on chromosome 3 and RM3735 on chromosome 4 were 36.2, 17.1, and 32.6%, respectively. Four QTLs for filled grains per panicle have been detected at the interval of RM468-RM7076 and RM241–RM26212 on chromosome 4, explaining 13.1% and 31.0% of the total phenotypic variation, respectively. Two QTLs controlling unfilled grain percentage have been also detected at loci RM554, RM3686 on chromosome 3 explaining only 25.0% and 11.2% of the total phenotypic variation. One QTL has been detected for 1000-grain weight located at the locus RM103 on chromosome 6, explaining 30.6% of the total phenotypic variation. A single QTL at the locus RM5749 on chromosome 4 was identified explaining 10.8% of the total phenotypic variance of grain yield. Attention has been paid to the interval of RM3586–RM160 on chromosome 3 at the range of 8.1 cM for heat tolerance score (Bui et al. 2014). The subsequent backcross populations derived from the cross of OM5930/N22 have been continued to select via conventional procedures and MAS based on the selected markers. Heat-tolerance rice breeding has been continued through the collaborative research project between National Institute of Crop Science (NICS), RDA (Korea), and CLRRI, IAS (Vietnam); we conducted the breeding program assisted by molecular markers, which were previously identified in QTL mapping and analysis (Bui et al. 2013, 2014). The study aims to select promising new lines adaptable to heat stress in the central coast of southern Vietnam from the crosses with heat tolerance germplasm N22 and Dular based on marker-assisted selection (MSA) using SSR markers. A molecular linkage map (graphic genotyping) on 12 chromosomes in BC_2F_2 population of OM5930/N22 is displayed in Figure 5.1 to address the target sites on chromosome 3 (Bui et al. 2014).

El-Esawi and Alayaf (2019) studied the overexpression of *OsRab7,* then cloned and transformed into rice plants. The expression of four genes encoding reactive oxygen species (ROS)-scavenging enzymes (*OsCATA, OsCATB, OsAPX2, OsSOD-Cu/Zn*) and eight genes conferring abiotic stress tolerance (*OsLEA3, OsRD29A, OsSNAC1, OsSNAC2, OsDREB2A, OsDREB2B, OsRAB16A, OsRAB16C*) was significantly up-regulated in the transformed rice lines as compared to their expression in wild-type. The genes improved heat and drought tolerance (El-Esawi and Alayaf 2019).

"High Resolution Mapping" of QTLs for heat tolerance in rice using a 5 K single nucleotide polymorphism (SNP) array was carried out by Shanmugavadivel et al. (2017). They used 272 F_8 RILs derived from a cross between Nagina 22, a well-known heat-tolerant *Aus* cultivar and IR64, a heat-sensitive popular *indica* rice variety to map the QTLs for heat tolerance. Using the Illumina infinium 5 K SNP array for genotyping the parents and the RILs, and stress susceptibility and stress tolerance indices (SSI and STI) of percent spikelet sterility and yield per plant (g), they identified five QTLs on chromosomes 3, 5, 9, and 12. The identified QTLs explained phenotypic variation in the range of 6.27 to 21.29%. Of these five QTLs, two high-effect QTLs, one novel (*qSTIPSS9.1*), and one known (*qSTIY5.1/qSSIY5.2*), were mapped in less than 400 kb genomic regions, comprising of 65 and 54 genes, respectively. This is the first report of a major QTL for heat tolerance on chromosome 9 of rice. Further, a known QTL for heat tolerance on chromosome 5 was narrowed down from 23 Mb to 331 kb in this study (Shanmugavadivel et al. 2017).

Genes such as HSP binding DnaJ, Hsp70, and temperature-induced lipocalin-2 OsTIL-2 close to these markers are candidates for expression studies and evaluation for use in MSA for heat tolerance (Prasanth et al. 2017). Nine markers, namely, RM243, RM517, RM225, RM518, RM525, RM195, RM282, RM489, and RM570 on chromosomes 1, 2, 3, 4, 6, and 8 showed association with six traits (flag leaf spad, flag leaf thickness, vegetative leaf temperature, plant height, panicle number, and tiller number) under heat stress conditions in both wet and dry seasons (Prasanth et al. 2017).

NILs that were segregated for *qHd1*, were developed from a cross between *indica* rice varieties Zhenshan 97 (ZS97) and Milyang 46 (MY46) (Chen et al. 2018). The thermal

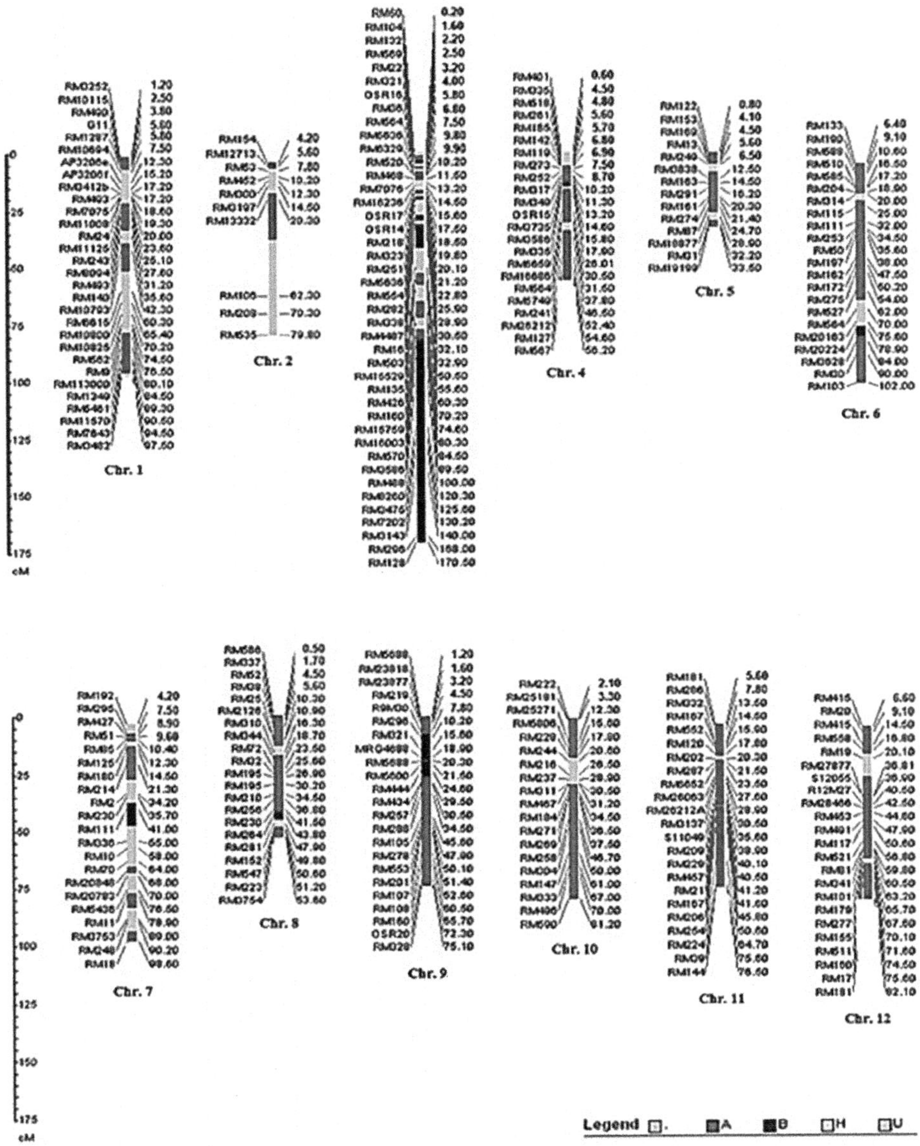

Figure 5.1 Molecular linkage map (graphic genotyping) on 12 chromosomes in BC_2F_2 population of OM5930/N22. *Source:* From Bui (2014). Licensed under CC BY-NC 3.0. *Note:* Blue: homozygous alleles from N22 (tolerance), Red: homozygous alleles from OM5930 (susceptible), Pale green: heterozygotes, Brown: unidentified (Bui et al. 2014). (*See color plate section for color representation of this figure*).

response was confirmed in a temperature-gradient experiment conducted in the phytotron. The *ZS97* allele to heading acceleration at high temperature was associated with higher grain weight that resulted in higher grain yield. By sequence comparison, *OsMADS51* for *qHd1* displayed a 9.5 kb insertion in the first intron of the *ZS97* allele compared to the

MY46 allele to correspond the needs for high-temperature tolerance during the heading and grain-filling stages (Chen et al. 2018).

Zhao et al. (2016) identified the stability of *qPSLht4.1* on chromosome 4 to improve the pollen shedding under heat stress.

Zhu et al. (2017) addressed one of the major-effect QTLs (*qHTB3-3*) detected on the long arm of chromosome 3. It was confirmed using overlapping substituted lines. *qHTB3-3* was finally mapped between the two markers RM3525 and 3-M95, approximately 2.8 Mb apart.

Ye et al. (2015) confirmed the identified QTLs via a three-way-cross population derived from IR64//Milyang23/Giza178, which was genotyped using 6 K SNP chips. The tolerance alleles were from the tolerant parent Giza178. In particular, the QTL on chromosome 4 (*qHTSF4.1*) could be attained through pyramiding validated heat tolerance QTLs (Ye et al. 2015).

Seed germination after dry-heat treatments (90 °C for 24 hours) showed a significant difference between the two parents, when evaluated for percentage germination and mean germination time (Lee et al. 2017). The *qDHT 1* on chromosome 1 was tightly linked at 4 cM from ME1-1. The phenotypic variation explained by the three QTLs was 27.18% of the total variance in the 164 RIL populations from Milyang 23 × Gihobyeo (Lee et al. 2017). Table 5.2 highlights the candidate QTLs or markers conferring to heat tolerance.

Marker-Assisted Selection (Microsatellites and SNP Arrays)

MAS is a process whereby a molecular marker is used for indirect selection of a genetic determinant or determinants of a trait of interest (e.g., abiotic stress tolerance). These tools can enhance the efficiency of breeding programs through their use in MAS (Xu and Couch 2008). The use of the genomic tools is more effectively identifying, quantifying, and characterizing genetic variation from all available germplasm resources (Tanksley et al. 1989; Tanksley and McCouch 1997; Gur and Zamir 2004). The rate of success is likely to increase in gene-base marker development, with more efficient QTL mapping procedures and lower-cost genotyping systems. They need available computational tools, which have tailored molecular breeding programs (Xu and Couch 2008). Many reviews have compared the molecular genetic issues related to different types of markers assay, which can significantly affect the success of MAS (Avise et al. 2009). Marker-assisted introgression has been shown to be an effective approach for precise transfer of genes from wild species with minimum linkage drag (Young and Tanksley 1989), and also it could help in identifying genotypes containing the target gene in early generations even if it is suppressed in a particular genetic background (Hurni et al. 2014).

A total of six markers RM3586 and RM160 on chromosome 3 and RM3735, RM3471, RM3687, and RM3536 on chromosome 4 were used to select promising lines in backcrossing populations for heat tolerance at the flowering stage in rice (Nguyen et al. 2015). A scheme for the development of backcross breeding lines for heat-tolerant QTL using marker-assisted foreground and background selection is shown in Figure 5.2. Fifty lines selected in BC_3F_2, BC_4F_1, and BC_4F_2 and parents were planted in the 2013 and 2014 dry seasons at the CLRRI field under natural heat stress and greenhouse conditions to evaluate heat tolerance at the reproductive period (Figure 5.2). Heat tolerance scoring under field conditions was based on percentage of unfilled grains. All selected lines exhibited their homozygous alleles with two heat-tolerance germplasms N22 or Dular in QTL loci. Twelve

Table 5.2 Identified QTLs, markers conferring heat tolerance.

Chromosome	QTL or Marker	References
4	C1100–R1783	Zhu and Ge (2005) and Zhu et al. (2006)
1	R1613–C970	
7	C1226–R1440	
4	RM3735	Zhang et al. (2008, 2009), Chen et al. (2008), and Bui et al. (2013, 2014)
3	RM3586	
4	RM5687–RM471	Xiao et al. (2011)
10	RM6132–RM6100	
9	InDel5–RM7364 [*OsHTAS*]	Wei et al. (2013) (seedling at 48 °C)
4	qHAC4	Zhang et al. (2014)
8	qHAC8a, qHAC8b	
10	qHAC10	
3	RM3586–RM160	Bui et al. (2013, 2014); Nguyen et al. (2015, 2017)
4	RM468–RM7076	
4	RM241–RM26212	
4	RM5749	
6	RM103	
4	qPSLht4.1	Zhao et al. (2016) (improve the pollen shedding under heat stress)
5	qSTIY5.1/qSSIY5.2 (331 kb)	Shanmugavadivel et al. (2017) (chr.9: the most important with 65 genes) (chr. 5 with 54 genes) from N22
9	qSTIPSS9.1 (400 kb)	
1	qDHT 1	Lee et al. (2017)
5	qDHT 5	
7	qDHT 7	
3	RM3525 to 3-M95 [*qHTB3-3*] Ø2.8 Mb	Zhu et al. (2017)
4	qHTSF4.1	Ye et al. (2015) (response to heat stress at flowering)
1	qDHT 1	Lee et al. (2017) (response to heat stress at seed germination)

lines harboring homozygous alleles to QTL loci RM3586 on chromosome 3 and RM3735 on chromosome 4, respectively were selected and evaluated to agronomic traits and yield potential. Four lines BC4-1-10-1 from OM5930/N22, BC4-5-8 from OM5930/Dular, BC4-5-9-4 from AS996/N22//4*AS996, and BC4-6-3 from AS996/Dular//4*AS996, respectively were finally selected for regional-adaptable tests on the central coast of Vietnam under heat stress conditiona to release to rice farmers (Nguyen et al. 2015, 2017). One rice genotype was released as OM8108. Four elite lines were selected at locus RM3735 on chromosome 4 as HTL1, HTL2, HTL3, and HTL4 (Figure 5.3).

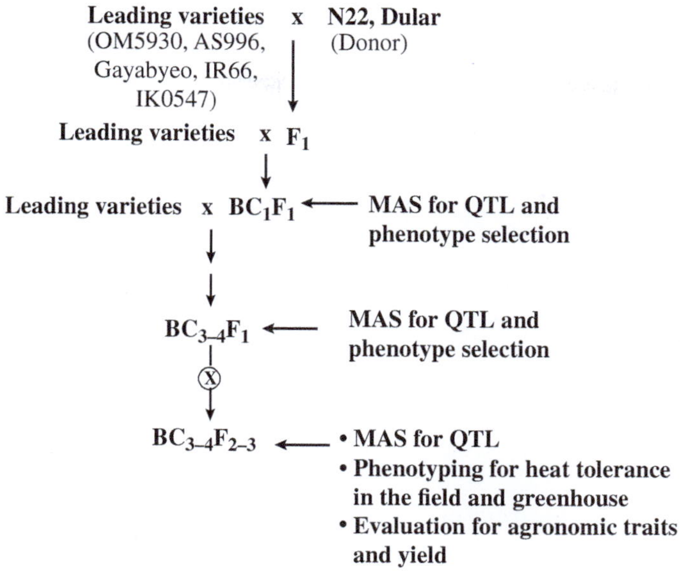

Figure 5.2 Scheme for the development of backcross breeding lines for heat-tolerant QTL using marker-assisted foreground and background selection in Vietnam. *Source:* Nguyen et al. (2015). Licensed under CC BY-NC 3.0.

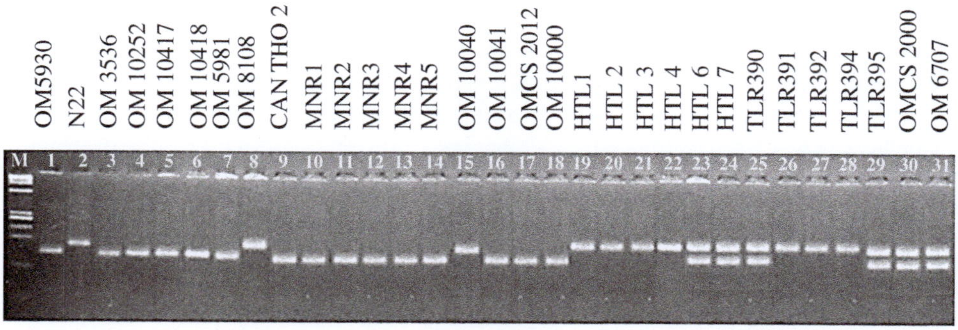

Figure 5.3 PCR products for identifying homozygous at locus RM3735 linked to QTL for heat tolerance from N22 on chromosome 4. *Source:* Nguyen et al. (2015).

Proteomics

Recent proteomics studies in rice have provided evidence for better understanding the mechanisms of thermal adaptation. Heat-stress response in rice is complicated, involving up- or down-regulation of numerous proteins related to different metabolic pathways (Zou et al. 2011). The proteins mainly include protection proteins, proteins involved in protein biosynthesis, protein degradation, energy and carbohydrate metabolism, and redox homeostasis. The complicated heat-stress response of rice involves not only protein up-regulation but also protein down-regulation, indicating that heat tolerance of rice may be attributed to the increase of some proteins abundance and the decrease of others (Zou et al. 2011; Liao et al. 2014). The analysis involved 25 of the 27 differentially displayed proteins that were homologous to known functional proteins to elucidate further the molecular mechanisms underlying the adaptation of rice to high temperature stress (Liao et al. 2014).

To identify transcripts induced by heat stress, 20-day-old rice seedlings of different rice cultivars suffering from heat stress were treated at different times, and differential gene expression analyses in leaves were performed by cDNA-AFLP and further verified by real-time RT-PCR (Cao et al. 2013). Exogenous abscisic acid (ABA) can induce some levels of freezing tolerance in chilling sensitive rice cells and seedlings, probably by eliciting mechanisms different from low-temperature-induced cold acclimation (Shinkawa et al. 2013).

The rice genome contains 32 Hsp70 genes (Sarkar et al. 2013). Promoter and transcript expression analysis divulges that Hsp70 superfamily genes play an important role in heat stress. Three mitochondrial rice Hsp70 sequences (i.e., mtHsp70-1, mtHsp70-2, and mtHsp70-3) complemented the Ssc1 mutation of yeast to differential extents. The information presented in this study provides a detailed understanding of the Hsp70 protein family of rice, the crop species that is the major food for the world population (Sarkar et al. 2013; Sung et al. 2001).

Bui et al. (2016) identified HSP90 expression in tolerant genotypes OM8108 and TLR192 under phytotron heat stress conditions in southern Vietnam through SDS-PAGE and Western blot analysis.

To explore the mechanism of heat stress (HS) tolerance in rice, Mu et al. (2017) conducted a comparative proteomics analysis on the anthers between HS-resistant and HS-sensitive cultivars under different levels of high temperature. Proteomic data showed that HS leads to the degradation of ribosomal proteins in the sensitive cultivar but not in the resistant one, which might result in the injury of protein biosynthetic mechanisms. In contrast, HS induced the increase of sHSP, β-expansins, and lipid transfer proteins in the resistant cultivar, which might contribute to its ability to tolerate heat stress (Mu et al. 2017).

Gonzalez et al. (2016) conducted a genome-wide transcriptome analysis in both tolerant and sensitive varieties. In addition, Kim et al. (2015) conducted a gel-based proteomics study in three rice varieties, including HS high-sensitive, middle-tolerant, and high-tolerant ones. Different techniques have been applied to study the rice HS response, including microarray (Endo et al. 2009), transcriptomics (Zhang et al. 2012), and proteomics (Jagadish et al. 2008; Ye et al. 2015). miRNA in responding to HS were also profiled (Li et al. 2015).

A proteomic analysis of heat-sensitive rice genotype IET 21405 was conducted (Kumar et al. 2017). Two-dimensional electrophoresis (2-DE) and MALDI-TOF/MSbased proteomics approaches revealed a total of 73 protein spots in rice leaf. The protein profiles clearly indicated variations in protein expression between the control and heat-treated rice genotypes. Functional assessment of 73 expressed proteins revealed several mechanisms thought to be involved in high temperature including their putative role in metabolism, energy, protein synthesis, protein transport/storage, and so on. The proteins identified in the present study provide a strong basis to elucidate gene function of these proteins and to explain further the molecular mechanisms underlying the adaptation of rice to high-temperature stress (Kumar et al. 2017).

Polyamines have been reported to be involved in grain filling, contributing to the construction of heat resistance of some cereals. Fu et al. (2019) studied the hybrid rice "YLY 689" to explore the possible effects of exogenous spermidine (Spd) on seed quality under high temperature during the filling stage. The rice plants were transferred to 40 °C for five-day heat treatment. Spermidine significantly increased the peroxidase (POD) activity and decreased the malondialdehyde (MDA) content in seeds. The expression of genes, namely, *spermidine synthase (SPDSYN)* and *spermine synthase* (*SPMS1* and *SPMS2*), significantly increased. The expression of their related synthase genes, *soluble starch synthase 5-3-3* (*SS 5-3-3*) and *granules bound starch synthase I* (*GBSSI*), also increased to some extent (Fu et al. 2019).

Proteomics study provides potential opportunities to elucidate gene functions of HSPs for their functional analysis in rice. Further studies should be targeted towards characterization of these proteins at the molecular level, which may provide a comprehensive picture of these proteins at the cellular level under elevated temperature stress conditions (Kumar et al. 2017).

Future Perspectives

Next Generation Sequencing (NGS): Recent advances in NGS and SNP genotyping promise to greatly accelerate crop improvement if properly deployed (Thomson 2014). High-throughput SNP genotyping offers a number of advantages over previous marker systems, including an abundance of markers, rapid processing of large populations, a variety of genotyping systems to meet different needs, and straightforward allele calling and database storage due to the bi-allelic nature of SNP markers. On the other hand, routine deployment of trait-specific SNP markers requires flexible, low-cost systems for genotyping smaller numbers of SNPs across large breeding populations, using platforms such as Fluidigm's Dynamic Arrays™, Douglas Scientific's Array Tape™, and LGC's automated systems for running KASP™ markers. At the same time, genotyping by sequencing (GBS) is rapidly becoming popular for low-cost high-density genome-wide scans through multiplexed sequencing (Thomson 2014). The identification and isolation of genes underlying QTLs associated with agronomic traits in crops have been recently accelerated thanks to NGS-based technologies combined with plant genetics (Nguyen et al. 2019). Examples of QTLs/genes associated with agronomic traits in crops and identified using different strategies based on whole-genome sequencing (WGS)/whole-genome resequencing (WGR) or RNA-seq are presented and discussed (Nguyen et al. 2019). The NGS would be useful to narrow down the QTL regions accordingly, how to isolate more accurately the candidate genes, which are associated with heat tolerance phenotypes.

Epigenetics: MicroRNAs (miRNAs) are known to regulate expression of genes under stress (Mangrauthia et al. 2017). Genome-wide identification of miRNAs in tolerant (Nagina 22) and susceptible (Vandana) rice cultivars was performed in 16 samples representing root and shoot of 13-day-old seedlings. Expression of these miRNAs was observed only in Nagina 22 during high temperature; this would suggest a specific role of these miRNAs in heat-stress tolerance to look for in future breeding (Mangrauthia et al. 2017).

GWAS: A genome-wide association study (GWAS) has recently been used for assessing associations between genetic markers and abiotic stresses in rice. With its wider use and the development of GWAS modeling, GWAS has become a powerful approach for mapping a number of traits of rice, including agronomic traits (Huang et al. 2010; Zhao et al. 2011) and tolerance to abiotic stress (Zhu et al. 2015; Yan et al. 2016; Wang et al. 2016; Kaler et al. 2017). Because GWASs were conducted using three methods based on single-marker regression, haplotype regression, and simultaneous fitting of all markers, respectively, so that analysis of co-localization of the associated loci with QTLs detected in progenies of bi-parental crosses made it possible to narrow-down the position of the QTLs on spikelet fertility under high temperature (Lafarge et al. 2017).

Genome editing: Genome editing has not been conducted yet to demonstrate the useful tool for heat tolerance. However, CRISPR-Cas9 has been considered as one of the efficient tools in rice breeding for abiotic stress tolerance (Char et al. 2019) including heat tolerance.

Proteomics: To understand the early heat shock (HS)-regulated cellular responses that influence the tolerance of rice plants to high environmental temperatures, two-dimensional difference gel electrophoresis (2-D DIGE) was performed to explore the early HS-regulated proteome (Bui et al. 2016; Zhou et al. 2019). Future analyses of the roles of these OsUBP21-interacting proteins (ubiquitin-specific protease) in plant heat stress (HS) responses will help reveal the protein ubiquitination/deubiquitination-regulated cellular responses induced by HS in rice (Zhou et al. 2019). The rice genome is sequenced and studied more than other crops. However, the molecular mechanism of heat tolerance in rice is unclear. So that the proteomics study of protein family HSPs will elucidate gene functions of HSPs in the rice genome (Kumar et al. 2017). The basic essential functions performed by proteomics are that proteomics can study proteins with specific physiological functions on a large scale, and then clarify the information about the whole protein; (5-3) It can obtain the qualitative and quantitative information of key proteins to effectively study the function and interaction of proteins, protein expression, and post-translational modification, and so on (5-3i). The presence of differentially expressed proteins may lead to differences in the accumulation of secondary metabolites (Zeng et al. 2019).

Conclusions

- The predicted climate change involves a total increase in temperature combined with frequent dry spells. This means that screening to find rice capable of surviving increased temperatures must be considered with spells of drought.
- The reproductive stage, especially the flowering stage, is the most sensitive stage to high temperature. If temperature over 35 °C at anthesis and lasting for more than one hour occurs, it will causes high spikelet-sterility, leading to serious yield losses.
- Research on the genetic mechanism of heat tolerance is getting more important to the utilization of heat-tolerant genes (HT) and the development of new HT rice varieties.
- Plant physiology-based heat-stress responses could be cited as (i) net photosynthesis rate with higher CO_2 responsiveness, (ii) grain filling, (iii) seed dormancy breaking and seedling emergence, (iv) anthesis response via anther dehiscence and spikelet sterility, and (v) HNT response via source and sink dynamics.
- Genetic resource to promote heat stress tolerance is a diverse gene pool from wild rice species to *indica* cultivars. The genotypes, namely, N22 and Dular, which belong to the *aus* ecotype, were mostly exploited by rice breeders and geneticists.
- The QTLs located on chromosomes 1, 3, 4, 5, 7, 8, 9, and 10 were mostly studied, in particular on chromosome 3, 4, 5, and 9. Agronomical characteristics were often recognized as spikelet fertility and grain filling traits.

- MAS can enhance the efficiency of breeding programs. The rate of success is likely to increase in gene-base marker development, with more efficient QTL mapping procedures and lower-cost genotyping systems. NGS and SNP genotyping promise to greatly accelerate rice improvement tolerant to heat stress.
- Proteomic studies in rice have provided evidence for better understanding the molecular mechanisms of thermal adaptation with HSPs expression. Further cooperation at international level must be encouraged.

References

Ainsworth, E.A. and Ort, D.R. (2010). How do we improve crop production in a warming world? *Plant Physiol.* 154: 526–530.

Ashikari, M. and Matsuoka, M. (2006). Identification, isolation and pyramiding of quantitative trait loci for rice breeding. *Trends Plant Sci.* 1: 344–350.

Atwell, B.J., Wang, H., and Scafaro, A.P. (2014). Could abiotic stress tolerance in wild relatives of rice be used to improve *Oryza sativa*? *Plant Sci.* 21: 48–58.

Avise, J., Chen, J., Lamb, B. et al. (2009). Attribution of projected changes in summertime US ozone and PM2. *5 concentrations to global changes. Atmos. Chem. Phys.* 9: 1111–1124.

Baliuag, A.N.N., Redoña, E.D., Hernandez, J.E. et al. (2015). Genetic analysis for heat tolerance and early morning flowering traits at flowering stage in rice (*Oryza sativa* L.). *Philipp. J. Crop Sci.* 40 (3): 62–72.

Battisti, D.S. and Naylor, R.L. (2009). Historical warnings of future food insecurity with unprecedented seasonal heat. *Science* 23: 240–244.

Bui, C.B., Nguyen, V.H., Tran, V.T. et al. (2012). Assessment of breeding materials for heat tolerance rice breeding (*Oryza sativa* L.). *J. Agric. Rural Dev.* 12: 38–46.

Bui, C.B., Pham, T.T.H., Bui, P.T. et al. (2013). Study on genetic variation of heat tolerance trait in BC population of rice (*Oryza sativa* L.). *J. Agric. Rural Dev.* 2: 10–15.

Bui, C.B., Pham, T.T.H., Bui, P.T. et al. (2014). Quantitative trait loci associated with heat tolerance in rice (*Oryza sativa* L.). *Plant Breed. Biotechnol.* 2 (1): 14–24.

Bui, C.B., Luong, T.M., Pham, T.T.H., and Nguyen, T.L. (2016). Rice breeding for heat tolerance in South Vietnam by marker-assisted selection. Final Report of National Project from 2011 to 2015. MARD, Vietnam. 176 pp. (Vietnamese).

Busch, F.A. and Sage, R.F. (2017). The sensitivity of photosynthesis to O_2 and CO_2 concentration identifies strong Rubisco control above the thermal optimum. *New Phytol.* 213: 1036–1051.

Cao, L.Y., Zhu, J., Zhao, S.T. et al. (2002). Mapping QTL for heat tolerance in a DH population from indica-japonica cross of rice (*Oryza sativa* L.). *J. Agric. Biotechnol.* 10: 210–214.

Cao, Y.Y., Duan, H., Yang, L.N. et al. (2008). Effect of heat stress during meiosis on grain yield of rice cultivars differing in heat tolerance and its physiological mechanism. *Acta Agron. Sin.* 34: 2134–2142.

Cao, Y., Zhang, Q., Chen, Y. et al. (2013). Identification of differential expression genes in leaves of rice (*Oryza sativa* L.) in response to heat stress by cDNA-AFLP analysis. *Biomed Res. Int.* 2013: 576189, 11 p.

Carmo-Silva, A.E. and Salvucci, M.E. (2011). The activity of Rubisco's molecular chaperone, Rubisco activase, in leaf extracts. *Photosynth. Res.* 108: 143–155.

Casartelli, A., Riewe, D., Hubberten, H.M. et al. (2018). Exploring traditional aus-type rice for metabolites conferring drought tolerance. *Rice* 11: 9.

Char, S.N., Li, R., and Yang, B. (2019). CRISPR/Cas9 for mutagenesis in rice. In: *Transgenic Plants. Methods in Molecular Biology*, vol. 1864 (eds. S. Kumar, P. Barone and M. Smith), 279–293. New York: Humana Press.

Chaturvedi, A.K., Bahuguna, R.N., Shah, D. et al. (2017). High temperature stress during flowering and grain filling offsets beneficial impact of elevated CO_2 on assimilate partitioning and sink-strength in rice. *Sci. Rep.* 7 (1): 8227.

Chen, Q.Q., Yu, S.B., Li, C.H., and Mou, T.M. (2008). Identification of QTLs for heat tolerance at flowering stage in rice. *Sci. Agric. Sin.* 41 (2): 315–321.

Chen, J.Y., Zhang, H.W., Zhang, H.L. et al. (2018). Natural variation at *qHd1* affects heading date acceleration at high temperatures with pleiotropism for yield traits in rice. *BMC Plant Biol.* 18: 112.

Crafts-Brandner, S.J. and Law, R.D. (2000). Effect of heat stress on the inhibition and recovery of the ribulose-1,5-bisphosphate carboxylase/oxygenase activation state. *Planta* 212 (1): 67–74.

Cruz, R.V., Harasawa, H., Lal, M. et al. (2017). Climate change 2007: impacts, adaptation and vulnerability. In: *Working Group II Contribution to the Intergovernmental Panel on Climate Change Fourth Assessment Report of Intergovernmental Panel on Climate Change* (eds. M.L. Parry, O.F. Canziani, J.P. Palutikof, et al.), 469–506. Cambridge: Cambridge University Press.

Dadlani, M. and Seshu, D.V. (1990). Effect of wet and dry heat treatment on rice seed germination and seedling vigor. *Int. Rice Res. Newslett.* 15 (2): 21–22.

El-Esawi, M.A. and Alayaf, A.A. (2019). Overexpression of rice *Rab7* gene improves drought and heat tolerance and increases grain yield in rice (*Oryza sativa* L.). *Gene* 10: 56.

Endo, M., Tsuchiya, T., Hamada, K. et al. (2009). High temperatures cause male sterility in rice plants with transcriptional alterations during pollen development. *Plant Cell Physiol.* 50: 1911–1922.

Esmaeili, N., Yang, X., Cai, Y. et al. (2019). Co-overexpression of *AVP1* and *OsSIZ1* in arabidopsis substantially enhances plant tolerance to drought, salt, and heat stresses. *Sci. Rep.* 9: 7642.

Fu, Y., Gu, Q., Dong, Q. et al. (2019). Spermidine enhances heat tolerance of rice seeds by modulating endogenous starch and polyamine metabolism. *Molecules* 24: 1395.

Ge, S., Sang, T., Lu, B.R., and Hong, D.Y. (1999). Phylogeny of rice genomes with emphasis on origins of allotetraploid species. *Proc. Natl. Acad. Sci. U.S.A.* 96: 14400–14405.

Gonzalez, N., Dreni, L., Lawas, L.M.F. et al. (2016). Genome-wide transcriptome analysis during anthesis reveals new insights into the molecular basis of heat stress response in tolerant and sensitive rice varieties. *Plant Cell Physiol.* 57: 57–68.

Guo, L., Zhu, L., Xu, Y. et al. (2004). QTL analysis of seed dormancy in rice (*Oryza sativa* L.). *Euphytica* 140: 155–162.

Gur, A. and Zamir, D. (2004). Unused natural variation can lift yield barriers in plant breeding. *PLoS Biol.* 2: e245.

Hakata, M., Wada, H., Masumoto-Kubo, C. et al. (2017). Dev elopment of a new heat tolerance assay system for rice spikelet sterility. *Plant Methods* 13: 34.

Hirabayashi, H., Sasaki, K., Kambe, T. et al. (2015). *qEMF3*, a novel QTL for the early-morning flowering trait from wild rice, *Oryza officinalis*, to mitigate heat stress damage at flowering in rice, *O. sativa*. *J. Ex. Bot.* 66: 1227–1236.

Horie, T., Matsui, T., Nakagawa, H., and Omasa, K. (1996). Effects of elevated CO_2 and global climate change on rice yield in Japan. In: *Climate Change and Plants in East Asia* (eds. K. Omasa, K. Kai, H. Taoda, et al.), 39–56. Springer-Verlag: Tokyo.

Huang, X., Wei, X., Sang, T. et al. (2010). Genome-wide association studies of 14 agronomic traits in rice landraces. *Nat. Genet.* 42 (11): 961–967.

Huang, M., Zhang, H., Zhao, C. et al. (2019). Amino acid content in rice grains is affected by high temperature during the early grain-filling period. *Sci. Rep.* 9: 2700.

Hurni, S., Brunner, S., Stirnweis, D. et al. (2014). The powdery mildew resistance gene *Pm8* derived from rye is suppressed by its wheat ortholog *Pm3*. *Plant J.* 79: 904–913.

IPCC (2018). The IPCC Report Will Have Profound Effects on Climate Governance. By Geraud De Lassus Saint-Geniès. Geneva, Switzerland: IPCC.

Ishimaru, T., Hirabayashi, H., Ida, M. et al. (2010). A genetic resource for early-morning flowering trait of wild rice *Oryza officinalis* to mitigate high temperature-induced spikelet sterility at anthesis. *Ann. Bot.* 106: 515–520.

Jagadish, S.V.K., Craufurd, P.Q., and Wheeler, T.R. (2007). High temperature stress and spikelet fertility in rice. *J. Exp. Bot.* 58: 1627–1635.

Jagadish, S.V.K., Craufurd, P.Q., and Wheeler, T.R. (2008). Phenotyping parents of mapping populations of rice for heat tolerance during anthesis. *Crop Sci.* 48: 1140–1146.

Jagadish, S.V.K., Cairns, J., Lafitte, R. et al. (2010a). Genetic analysis of heat tolerance at anthesis in rice (*Oryza sativa* L.). *Crop Sci.* 50: 1633–1641.

Jagadish, S.V.K., Muthurajan, R., Oane, R. et al. (2010b). Physiological and proteomic approaches to address heat tolerance during anthesis in rice (*Oryza sativa* L.). *J. Exp. Bot.* 61: 143–156.

Jones, J.M., Fogt, R.L., Widmann, M. et al. (2009). Historical SAM variability Part I: century-length seasonal reconstructions. *J. Clim.* 22: 5319–5345.

Jung, K.H., Ko, H.J., Nguyen, M.X. et al. (2012). Genome-wide identification and analysis of early heat stress responsive gene in Rice. *J. Plant Biol.* 55: 458–468.

Kaler, A.S., Ray, J.D., Schapaugh, W.T. et al. (2017). Genome-wide association mapping of canopy wilting in diverse soybean genotypes. *Theor. Appl. Genet.* 130: 2203.

Khush, G.S. (2005). What it will take to feed 5.0 billion rice consumers in 2030. *Plant Mol. Biol.* 59: 1–6.

Kim, M., Kim, H., Lee, W. et al. (2015). Quantitative shotgun proteomics analysis of rice anther proteins after exposure to high temperature. *Int. J. Genomics* 2015: 238704.

Kobayashi, A., Sonoda, J., Sugimoto, K. et al. (2013). Detection and verification of QTLs associated with heat-induced quality decline of rice (*Oryza sativa* L.) using recombinant inbred lines and near-isogenic lines. *Breed. Sci.* 63: 339–346.

Kuang, H.C., Wen, S.S., and Liu, G.M. (2002). Studies on the heat tolerance of Luhui 17 and its cross II you 7 at head sprouting. *SW China J. Agric. Sci. (Chinese)* 15: 106–108.

Kumar, N., Chandra Suyal, D., Sharma, I.P. et al. (2017). Elucidating stress proteins in rice (*Oryza sativa* L.) genotype under elevated temperature: a proteomic approach to understand heat stress response. *3 Biotech* 7: 205.

Lafarge, T., Bueno, C., Frouin, J. et al. (2017). Genome-wide association analysis for heat tolerance at flowering detected a large set of genes involved in adaptation to thermal and other stresses. *PLoS One* 12 (2): 1–27. https://doi.org/10.1371/journal.pone.0171254 (1-27).

Lawas, L.M.F., Erban, A., Kopka, J. et al. (2019). Metabolic responses of rice source and sink organs during recovery from combined drought and heat stress in the field. *Gigascience* 8: 1–15.

Lee, S.Y., Lee, J.H., and Kwon, T.O. (2002). Varietal differences in seed germination and seedling vigor of Korean rice varieties following dry heat treatment. *Seed Sci. Technol.* 30: 311–321.

Lee, S.Y., Ahn, J.H., Kim, H.J., and Eun, M.Y. (2006). Relationships among viviporous germination, dry-heat tolerance and seed longevity in Milyang 23/Gihobyeo RILs. *Korean J. Bioenviron. Control* 15: 421–427.

Lee, S.Y., Kim, Y.H., and Lee, G.S. (2017). Mapping QTLs associated to germination stability following dry-heat treatment in rice seed. *3 Biotech* 7 (3): 220.

Li, C.D. (2003). Analysis on a large of empty grains of rice due to high temperature. *Shanxi J. Agric. Sci.* 49: 45–47.

Li, Z., Hu, Q., Zhou, M. et al. (2013). Heterologous expression of *OsSIZ1*, a rice sumo E3 ligase, enhances broad abiotic stress tolerance in transgenic creeping bentgrass. *Plant Biotechnol. J.* 4: 432–445.

Li, J., Wu, L.Q., Zheng, W.Y. et al. (2015). Genome-wide identification of microRNAs responsive to high temperature in rice (*Oryza sativa*) by high-throughput deep sequencing. *J. Agron. Crop Sci.* 201: 379–388.

Liao, J.L., Zhang, H.Y., Liu, J.B. et al. (2012). Identification of candidate genes related to rice grain weight under high-temperature stress. *Plant Sci.* 196: 32–43.

Liao, J.L., Zhou, H.W., Zhang, H.Y., and Zhong, P.A. (2014). Comparative proteomic analysis of differentially expressed proteins in the early milky stage of rice grains during high temperature stress. *J. Exp. Bot.* 65 (2): 655–671.

Mackill, D.J., Coffman, W.R., and Rutger, J.N. (1982). Pollen shedding and combining ability for high temperature tolerance in rice. *Crop Sci.* 22: 730–733.

Mangrauthia, S.K., Bhogireddy, S., Agarwal, S. et al. (2017). Genome-wide changes in microRNA expression during short and prolonged heat stress and recovery in contrasting rice cultivars. *J. Exp. Bot.* 68 (9): 2399–2412.

Matsui, T. and Kagata, H. (2003). Characteristics of floral organs related to reliable self-pollination in rice (*Oryza sativa* L.). *Ann. Bot.* 91: 473–477.

Matsui, T. and Omasa, K. (2002). Rice (*Oryza sativa* L.) cultivars tolerant to high temperature at flowering: anther characteristics. *Ann. Bot.* 89: 683–687.

Matsui, T., Omasa, K., and Horie, T. (1997). High temperature-induced spikelet sterility of japonica rice at flowering in relation to air temperature, humidity and wind velocity conditions. *Jpn. J. Crop Sci.* 66: 449–455.

Matsui, T., Omasa, K., and Horie, T. (2000). High temperature at flowering inhibits swelling of pollen grains, a driving force for thecae dehiscence in rice (*Oryza sativa* L.). *Plant Prod. Sci.* 3: 430–434.

Matsui, T., Omasa, K., and Horie, T. (2001). The difference in sterility due to high temperatures during the flowering period among japonica-rice varieties. *Plant Prod. Sci.* 4: 90–93.

Matsushima, S., Ikewada, H., Maeda, A. et al. (1982). Studies on rice cultivation in the tropics 1: yielding and ripening responses of the rice plant to the extremely hot and dry climate in Sudan. *Japan. J. Trop. Agric.* 26: 19–25.

Mohapatra, T., Robin, S., Sarla, N. et al. (2014). EMS induced mutants of upland rice variety Nagina22: generation and characterization. *Proc. Indian Nat. Sci. Acad.* 80: 163–172.

Morishima, H. and Oka, H.I. (1981). Phylogenetic differentiation of cultivated rice, XXII. Numerical evaluation of the indica–japonica differentiation. *Jpn. J. Breed.* 31: 402–413.

Morita, S., Yonemaru, J.I., and Takanashi, J.I. (2005). Grain growth and endosperm cell size under high night temperatures in rice (*Oryza sativa* L.). *Ann. Bot.* 95: 695–701.

Mu, Q., Zhang, W., Zhang, Y. et al. (2017). *i*TRAQ-based quantitative proteomics analysis on rice anther responding to high temperature. *Int. J. Mol. Sci.* 18: 1811.

Nakagawa, H., Horie, T., and Matsui, T. (2003). Effects of climate change on rice production and adaptive technologies. In: *Rice Science: Innovations and Impact for Livelihood* (eds. T.W. Mew, D.S. Brar, S. Peng, et al.), 635–658. Los Baños, The Philippines: International Rice Research Institute.

Nguyen, T.L., Pham, T.T.H., Pham, C.T. et al. (2015). Breeding for heat tolerance rice based on marker-assisted backcrossing in Vietnam. *Plant Breed. Biotechnol.* 3 (3): 274–281.

Nguyen, T.L., Pham, T.T.H., Pham, C.T. et al. (2017). Rice breeding for heat tolerance via marker-assisted selection. *J. Vietnam Sci. Technol.* 13 (2): 34–35.

Nguyen, K.L., Grondin, A., Courtois, B., and Gantet, P. (2019). Next-generation sequencing accelerates crop gene discovery. *Trends Plant Sci.* 24 (3): 263–274.

Nishiyama, I. and Satake, T. (1981). High temperature damages in rice plants. *Jpn. J. Trop. Agric.* 25: 14–19.

Osada, A., Sasiprapa, V., and Rahong, M. (1973). Abnormal occurrence of empty grains of indica rice-plants in the dry, hot season in Thailand. *Proc. Crop Sci. Jpn.* 42: 103–109.

Peng, S.B., Huang, J.L., Sheehy, J.E. et al. (2004). Rice yields decline with high temperature from global warming. *Proc. Natl. Acad. Sci. U.S.A.* 101: 9971–9975.

Poli, Y., Basava, R.K., Madhusmita, P. et al. (2013). Characterization of a Nagina22 rice mutant for heat tolerance and mapping of yield traits. *Rice* 6: 36.

Pradhan, S.K., Barik, S.R., Sahoo, A. et al. (2016). Population structure, genetic diversity and molecular marker-trait association analysis for high temperature stress tolerance in rice. *PLoS One* 11 (8): e0160027.

Prasad, P.V.V., Boote, K.J., Allen, L.H. Jr., et al. (2006). Species, ecotype and cultivar differences in spikelet fertility and harvest index of rice in response to high temperature stress. *Field Crops Res.* 95: 398–411.

Prasanth, V.V., Suchandranath Babu, M., Basava, R.K. et al. (2017). Trait and marker associations in *Oryza nivara* and *O. rufipogon* derived rice lines under two different heat stress conditions. *Front. Plant Sci.* 8 (1819): 1–12.

Rivero, R.M., Mestre, T.C., Mittler, R. et al. (2014). The combined effect of salinity and heat reveals a specific physiological, biochemical and molecular response in tomato plants. *Plant Cell Environ.* 5: 1059–1073.

Sage, R.F., Way, D.A., and Kubien, D.S. (2008). Rubisco, Rubisco activase, and global climate change. *J. Exp. Bot.* 59 (7): 1581–1595.

Salvucci, M.E. and Crafts-Brandner, S.J. (2004). Relationship between the heat tolerance of photosynthesis and the thermal stability of rubisco activase in plants from contrasting thermal environments. *Plant Physiol.* 134: 1460–1470.

Salvucci, M.E., Osteryoung, K.W., Crafts-Brandner, S.J., and Vierling, E. (2001). Exceptional sensitivity of Rubisco activase to thermal denaturation in vitro and in vivo. *Plant Physiol.* 127: 1053–1064.

Sarkar, N.K., Kundnani, P., and Grover, A. (2013). Functional analysis of Hsp70 superfamily proteins of rice (*Oryza sativa*). *Cell Stress Chaperones* 18: 427–437.

Satake, T. and Yoshida, S. (1978). High temperature induced sterility in indica rice at flowering. *Jpn. J. Crop Sci.* 47: 6–17.

Scafaro, A.P., Haynes, P.A., and Atwell, B.J. (2010). Physiological and molecular changes in *Oryza meridionalis* Ng., a heat-tolerant species of wild rice. *J. Exp. Bot.* 61: 191–202.

Scafaro, A.P., Yamori, W., Carmo-Silva, A.E. et al. (2012). Rubisco activity is associated with photosynthetic thermotolerance in a wild rice (*Oryza meridionalis*). *Physiol. Plant.* 146: 99–109.

Scafaro, A.P., Gallé, A., Van Rie, J. et al. (2016). Heat tolerance in a wild *Oryza* species is attributed to maintenance of Rubisco activation by a thermally stable Rubisco activase ortholog. *New Phytol.* 211: 899–911.

Scafaro, A.P., Atwell, B.J., Muylaert, S. et al. (2018). A thermotolerant variant of rubisco activase from a wild relative improves growth and seed yield in rice under heat stress. *Front. Plant Sci.* 9: 1633.

Seshu, D.V. and Sorells, M.E. (1986). Genetic studies on seed dormancy in rice. In: *Rice Genetics*, 1e, 369–382. Manila: IRRI.

Shanmugavadivel, P.S., Amitha Mithra, S.V., Prakash, C. et al. (2017). High resolution mapping of QTLs for heat tolerance in rice using a 5K SNP array. *Rice* 10: 28.

Sharkey, T.D. (2000). Some like it hot. *Science* 287: 435–437.

Shi, W., Muthurajan, R., Rahman, H. et al. (2013). Source–sink dynamics and proteomic reprogramming under elevated night temperature and their impact on rice yield and grain quality. *New Phytol.* 197: 825–837.

Shinkawa, R., Morishita, A., Amikura, K. et al. (2013). Abscisic acid induced freezing tolerance in chilling-sensitive suspension cultures and seedlings of rice. *BMC Res. Notes* 6 (351): 2–14.

Shoba, D., Raveendran, M., Manonmani, S. et al. (2017). Development and genetic characterization of a novel herbicide (imazethapyr) tolerant mutant in rice (*Oryza sativa* L.). *Rice* 10: 10.

Sung, D.Y., Vierling, E., and Guy, C.L. (2001). Comprehensive expression profile analysis of the Arabidopsis *Hsp70* gene family. *Plant Physiol.* 126: 789–800.

Sweeney, M. and McCouch, S.R. (2007). The complex history of the rice domestication. *Ann. Bot.* 100: 951–957.

Talla, S.K., Panigrahy, M., Kappara, S. et al. (2016). Cytokinin delays dark-induced senescence in rice by maintaining the chlorophyll cycle and photosynthetic complexes. *J. Exp. Bot.* 67 (6): 1839–1851.

Tanksley, S.D. and McCouch, S.R. (1997). Seed banks and molecular maps: unlocking genetic potential from the wild. *Science* 277: 1063–1066.

Tanksley, S.D., Young, N.D., Paterson, A.H., and Bonierbale, M.W. (1989). RFLP mapping in plant breeding: new tools for an old science. *Nat. Biotechnol.* 7: 257–264.

Thomson, M.J. (2014). High-throughput snp genotyping to accelerate crop improvement. *Plant Breed. Biotechnol.* 2 (3): 195–212.

Tuberosa, R. (2012). Phenotyping for drought tolerance of crops in the genomics era. *Front. Physiol.* 3: 347.

Vaughan, D.A., Lu, B.R., and Tomooka, N. (2008). Was asian rice (*Oryza sativa*) domesticated more than once? *Rice* 1 (1): 16–24.

Vile, D., Pervent, M., Belluau, M. et al. (2012). *Arabidopsis* growth under prolonged high temperature and water deficit: independent or interactive effects? *Plant Cell Environ.* 4: 702–718.

Wahid, A., Gelani, G., Ashraf, M., and Foolad, M.R. (2007). Heat tolerance in plants: an overview. *Environ. Exp. Bot.* 61: 199–223.

Wang, J., Liu, K.D., Xu, C.G. et al. (1998). The high level of wide-compatibility of variety 'Dular' has a complex genetic basis. *Theor. Appl. Genet.* 97 (3): 407–412.

Wang, D., Liu, J., Li, C. et al. (2016). Genome-wide association mapping of cold tolerance genes at the seedling stage in rice. *Rice* 9: 61.

Weerakoon, W.M.W., Maruyama, A., and Ohba, K. (2008). Impact of humidity on temperature-induced grain sterility in rice (*Oryza sativa* L.). *J. Agron. Crop Sci.* 194: 135–140.

Wei, H., Liu, J.P., Wang, Y. et al. (2013). A dominant major locus in chromosome 9 of rice (*Oryza sativa* L.) confers tolerance to 48°C high temperature at seedling stage. *J. Hered.* 104 (2): 287–294.

Xiao, Y.H., Pan, Y., Luo, L.H. et al. (2011). Quantitative trait loci associated with seed set under high temperature stress at the flowering stage in rice (*Oryza sativa* L.). *Euphytica* 178: 331–338.

Xu, Y. and Couch, J.H. (2008). Marker-assisted selection in plant breeding: from publication to practice. *Crop Sci.*. 48: 391–407.

Xu, X., Liu, X., Ge, S. et al. (2012). Resequencing 50 accessions of cultivated and wild rice yields markers for identifying agronomically important genes. *Nat. Biotechnol.* 30: 105–111.

Yamori, W., Hikosaka, K., and Way, D.A. (2014). Temperature response of photosynthesis in C3, C4, and CAM plants: temperature acclimation and temperature adaptation. *Photosynth. Res.* 109: 101–117.

Yan, L.V., Guo, Z., Li, X. et al. (2016). New insights into the genetic basis of natural chilling and cold shock tolerance in rice by genome-wide association analysis. *Plant Cell Environ.* 39 (3): 556–570.

Ye, C., Argayoso, M.A., Redoña, E.D. et al. (2012). Mapping QTL for heat tolerance at flowering stage in rice using SNP markers. *Plant Breed.* 13: 33–41.

Ye, C., Tenorio, F.A., Argayoso, M.A. et al. (2015). Identifying and confirming quantitative trait loci associated with heat tolerance at flowering stage in different rice populations. *BMC Genet.* 16: 41.

Yi, C.D., Yan, C.J., Liang, G.H. et al. (2001). RFLP analysis of the effect of wide compatibility genes in Aus variety 'Dular'. Acta Genetic Sinica - English Abstract. *Yi Chuan Xue Bao* 28 (6): 540–549. (Article in Chinese).

Young, N.D. and Tanksley, S.D. (1989). RFLP analysis of the size of chromosomal segments retained around the *Tm-2* locus of tomato during backcross breeding. *Theor. Appl. Genet.* 77: 353–359.

Yugandhar, P., Basava, R.K., Panigrahy, M. et al. (2013). Characterization of a Nagina22 rice mutant for heat tolerance and mapping of yield traits. *Rice* 6 (1): 36.

Yugandhar, P., Sun, Y., Liu, L. et al. (2018). Characterization of the loss-of-function mutant *NH101* for yield under phosphate deficiency from EMS-induced mutants of rice variety Nagina22. *Plant Physiol. Biochem.* 130: 1–13.

Zeng, R., Farooq, M.U., Wang, L. et al. (2019). Study on differential protein expression in natural selenium-enriched and non-selenium-enriched rice based on iTRAQ quantitative proteomics. *Biomolecules* 9: 130.

Zhang, X.G. (1990). Physiochemical treatments to break dormancy in rice. *Int. Rice Res. Newslett.* 15 (1): 22.

Zhang T, L Yang, KF Jang, M Huang, Q Sun, WF Chen, JK Zheng. 2008. QTL mapping for heat tolerance of the tassel period of rice. *Mol Plant Breed* 6(5):867–873.

Zhang, G.L., Chen, L.Y., Xiao, G.Y. et al. (2009). Bulked segregant analysis to detect QTL related to heat tolerance in rice (*Oryza sativa* L.) using SSR markers. *Agric. Sci. China* 8: 482–487.

Zhang, X.W., Li, J.P., Liu, A.L. et al. (2012). Expression profile in rice panicle: insights into heat response mechanism at reproductive stage. *PLoS One* 7: e49652.

Zhang, H., Duan, L., Dai, J.S. et al. (2014). Major QTLs reduce the deleterious effects of high temperature on rice amylose content by increasing splicing efficiency of *Wx* pre-mRNA. *Theor. Appl. Genet.* 127: 273–282.

Zhang H, L Duan, JS Dai, CQ Zhang, J Li, MH Gu, QQ Liu, Y Zhu. 2014. Major QTLs reduce the deleterious effects of high temperature on rice amylose content by increasing splicing efficiency of Wx pre-mRNA. *Theor Appl Genet* 127:273–282.

Zhang, H.W., Fan, Y.Y., Zhu, Y.J. et al. (2017). Dissection of the *qTGW1.1* region into two tightly-linked minor QTLs having stable effects for grain weight in rice. *BMC Genet.* 17: 98.

Zhao, K., Tung, C.W., Eizenga, G.C. et al. (2011). Genome-wide association mapping reveals a rich genetic architecture of complex traits in *Oryza sativa*. *Nat. Commun.* 2: 467.

Zhao, L., Lei, J., Huang, Y. et al. (2016). Mapping quantitative trait loci for heat tolerance at anthesis in rice using chromosomal segment substitution lines. *Breed. Sci.* 66: 358–366.

Zhou, H., Wang, X., Huo, C. et al. (2019). A quantitative proteomics study of early heat-regulated proteins by two-dimensional difference gel electrophoresis identified OsUBP21 as a negative regulator of heat stress responses in rice. *Proteomics* 19 (20): e1900153.

Zhu, Q. and Ge, S. (2005). Phylogenetic relationships among A-genome species of the genus *Oryza* revealed by intron sequences of four nuclear genes. *New Phytol.* 167: 249–267.

Zhu L, YHX, CM Wang, L Jiang, HQ Zhai, JM Wan. 2006. Mapping QTLs for heat tolerance during grain filling in rice. *Chinese Journal of Rice Science* (19):117–121.

Zhu, Y., Chen, K., Mi, X. et al. (2015). Identification and fine mapping of a stably expressed QTL for cold tolerance at the booting stage using an interconnected breeding population in rice. *PLoS One* 10 (12): e0145704.

Zhu, S., Huang, R., Wai, H.P. et al. (2017). Mapping quantitative trait loci for heat tolerance at the booting stage using chromosomal segment substitution lines in rice. *Physiol. Mol. Biol. Plants* 23 (4): 817–825.

Zou, J., Liu, C., and Chen, X. (2011). Proteomics of rice in response to heat stress and advances in genetic engineering for heat tolerance in rice. *Plant Cell Rep.* 30: 2155–2165.

6

Molecular Breeding for Improving Cold Tolerance in Rice: Recent Progress and Future Perspectives

Ning Xiao[1] and Jian-Min Chen[2]

[1] *Institute of Agricultural Sciences for Lixiahe Region in Jiangsu, China*
[2] *Yangzhou University, China*

Introduction

Oryza sativa is one of the most important grain crops in the world. "Chilling injury" of *O. sativa* refers to the phenomenon of metabolic disorder and irreversible organ damage when the ambient temperature is below the critical point of optimal growth (Wang and Guo 2005). Chilling injury occurs at 10–15 °C during germinating stage and vegetative growth stage of *O. sativa*, resulting in common symptoms of delayed germination and decreased germination rate; yellow leaves, decreased tillers and plant apoptosis in vegetative growth stage (Andaya and Mackill 2003a), while chilling injury occurs at 18–20 °C in reproductive growth period (Yoshida 1981) resulted in spikelet sterility, anther deformity, incomplete heading and ear developmental deformity, delayed, and incomplete grain filling (Suh et al. 2010; Kuroki et al. 2007; Zhou et al. 2010).

Most of the chilling injury occurs in areas with of latitude and high altitude. When the chilling injury is the most serious in northern Japan, the yield of *O. sativa* is reduced by more than 80% (Shimono et al. 2007); the yield of *O. sativa* is reduced by about 26% in South Korea every year (Lee 2001); in tropical Australia, about $0.5 \sim 2\,\mathrm{t\,hm^{-2}}$ of *O. sativa* is lost due to chilling injury during the heading period every year (Singh et al. 2005). As far as China is concerned, the O. sativa area in northeast China suffers from chilling injury once in three to five years on average. In 2002, the affected area of Heilongjiang province reached 1 million $\mathrm{hm^2}$, with an average yield reduction of 30% caused by cold stress. In 2014, 10,000 $\mathrm{hm^2}$ of fields in Jiangsu province in eastern China suffered from low temperature in the late stage of maturity, which caused a 60% decrease of rice production. In addition, the cold stress in the late spring and continuous rainy seasons in the middle and lower reaches of the Yangtze River and South China resulted in a large area of rotten seedlings, dead seedlings, and delayed development of early *O. sativa* (Wang and Guo 2005). It can be seen that chilling injury of *O. sativa* has become a worldwide natural disaster that endangers grain safety. Therefore, it is of great significance to excavate cold-tolerant germplasm and establish their genetic diversity by molecular mapping of the QTLs (quantitative trait loci),

Molecular Breeding for Rice Abiotic Stress Tolerance and Nutritional Quality, First Edition.
Edited by Mohammad Anwar Hossain, Lutful Hassan, Khandakar Md. Iftekharuddaula, Arvind Kumar and Robert Henry.

fine mapping, gene cloning for breeding cold-tolerant varieties, and revealing the molecular mechanism behind cold tolerance.

Preliminary Mapping of Cold-Tolerant QTLs and Fine Mapping of Major Loci

During the chilling injury of *O. sativa*, the permeability of the plasmid membrane of the extracellular membrane and organelle increased (Ji et al. 2003), the content of intracellular reactive oxygen species increased, and photosynthetic respiration decreased, thus resulting in the decline of plant metabolism, slow growth, yellow leaves, or dead leaves (Lee et al. 2009). These physiological symptoms provide the phenotypic basis for the mapping of cold-tolerant QTLs. At present, the cold-tolerant phenotypes that have been mapped include germination rate in bud stage, embryo elongation rate, survival rate in vegetative growth stage, and pollen fertility in reproductive growth stage (Suh et al. 2010).

It has been reported that 57 cold-tolerant QTLs have been mapped to 12 chromosomes, among which 19 are major QTLs with contribution rates greater than 15% (Table 6.1). Various researchers had used different cold treatment conditions and mapping population in order to find the genomic region associated with cold tolerance. In the same period, the overlapping degree of QTL interval for the same identification index is also low, which shows that the cold tolerance of *O. sativa* is a complex genetic character controlled by multiple genes, and different growth stages are controlled by different genes. Therefore, how to select identification index and mapping population is the key for cold-tolerant QTL mapping.

Andaya and Mackill (2003a,2003b) mapped a major QTL (*qCTS12a*, $R^2 = 40.6\%$) related to the degree of leaf curling on chromosome 12 using recombinant inbred lines (RILs) constructed from cold-resistant varieties (M-202), and identified a major QTL (*qCTS12b*, $R^2 = 41.7\%$) related to cold tolerance. The identified cold QTLs were finely mapped to the 55-Kb region, and the study identified candidate gene encoding zeta glutathione S-transferase, within *qCTS12′* QTL region. Koseki et al. (2010) mapped a major QTL (*qCtss11*, $R^2 = 40\%$) related to the survival rate of vegetative growth period on chromosome 11 of F_2 population constructed by *Oryza rufipogon* (W1943), which overlapped with *qCTS11-2*.

After further fine mapping, there were two candidate genes (*Os11g0615600* and *Os*11g0615900) in this region, and the protein function was predicted that *Os11g0615900* contained a NB- ARC functional region. Yang et al. (2013) mapped a QTL (*qCTSS-10*) related to the survival rate of seedlings in the 15.48–21.06 MB range of chromosome 10 of Sun Yat Sen by high-throughput sequencing, which overlapped *qCTSS10* related to the survival rate (Liu et al. 2013) and the *qRC10-2* related to the root conductivity of seedlings to each other (Xiao et al. 2014). Among them, *qRC10-2* was finely mapped in the range of 48.5 KB, and there was only one candidate gene (*Os10g0490100*) differentially expressed in it, whose function was predicted to be pectin-folding protease.

Table 6.1 QTLs related to cold tolerance in rice reported to account for at least 15% of the phenotypic variation.

Identify stage	Located population combination	Chromosome position	PVE (%)	Molecular marker	Candidate gene function/Gene ID	References
Germinating stage	Nipponbare/Kasalath	*4(qLTG-4-1)*	*15*	*C946, C513*		Miura et al. (2011)
	Italica Livorno/Hayamasari	3(*qLTG-3-1*)	35.1	*GBR3001, GBR3002*	*Os03g0103300*	Fujino et al. (2004)
		3(*qLTG-3-2*)	17.4	*RM3436, RM2593*		
Vegetative stage	M-202/IR50	4(*qCTS4-1*)	20.8	*RM335, RM261*		Andaya and Mackill (2003b)
		6(*qCTS6-1*)	15.3	*RM253, RM50*		
		12(*qCTS12a*)	40.6	*RM101, RM247*		
		12(*qCTS12b*)	41.7	*RM101*		
		12(*qCTS12*)	40	*RM5746, RM3103*	*OsGSTZ1/OsGSTZ2*	Koseki et al. (2010)
	Lemont/Teqing	11(*qSCT-11*)	29.8	*RM202, RM209*		Zhang et al. (2005)
	AAV002863/Zhenshan 97B	2(*qCTS-2*)	27.4			Lou et al. (2007)
	Asominori/IR24	1(*qCST-1*)	24.5	*RM297, RM319*		Jiang et al. (2008)
	W1943(*Oryza rufipogon*)/ Guang-lu-ai4	11(*qCtss11*)	40	*S0605, S0485*	*Os11g0615900, Os11g0615600*	Koseki et al. (2010)
	Dongxiang wild rice/Nanjing11	10(*qPSR*)	32.1	*RM25570, RM304*	*Os10g0490100**	Xiao et al. (2018)
	Lijiangxintuanheigu	4(CTB4a)			*Os04g0132500**	

	bZIP73				Os09g0474000*	
	HAN1				Os11g0483000*	
Repro-ductive stage	Norin-PL8/Kirara397	4(*Ctb-1*)	/	*RM151, RM259*	*Os4g0619300, Os4g0619300*	Saito et al. (2004)
	Koshihikari/Akihikari	7(*qCT-7*)	22.1	*R2394, R1422*		Takeuchi et al. (2001)
	M-202/IR50	2(*qCTB2*)	16.8	*RM324, RM301*		Andaya and Mackill (2003b)
		3(*qCTB3*)	16.5	*RM156, RM282*		
	Kunmingxiaobaigu/Towada	7(*qRCT7*)	20.6	*RM182, RG351*		Dai et al. (2004)
	Hokkai-PL9/Hokkai287	8(*qCTB8*)	26.6	*RM5647, PLA61*		Kuroki et al. (2007)
	Kunmingxiaobaigu/Towada	10(*qCTB-10-2*)	15	*RM239, G1010*		Xu et al. (2008)
	Lijiangheigu/Reiziq	10(*qLTSPKST10.1*)	20.5	*S10010, S10014*		Ye et al. (2010)
	ZL1929–4/Towada	7(*qCTB7*)	21	*RM2125, RM2887*	*Os07g0576500/ Os07g0577300/ Os07g0577400*	Zhou et al. (2010)
	TR22183/Dasanbyeo	2(*QTL2.1*)	16.7		*Os07g0576500*	
		8(*QTL8.1*)	24.8		*Os07g0577300 /Os07g0577400*	Jiang et al. (2011)
		10(*QTL10.1*)	22.9			
	Ukei840/Hitomebore	3(*qLTB3*)	24.4	*RM3719, RM7000*	*Os03g0790700*	Shirasawa et al. (2012)

Map-Based Cloning and Molecular Mechanism of Cold-Resistant QTLs

At present, QTLs with cold tolerance cloned in the map have *qLTG3-1* in bud stage, *COLD1* in seedling stage, and *ctb1* in mature stage. *qLTG3-1* is the first cold-tolerant QTL cloned in bud stage. Fujino et al. (2004) mapped a major QTL (*qLTG3-1*, $R^2 = 35.0\%$) related to germination rate on chromosome 3 of RILs preliminary population constructed by Italica Livorno/Hayamasari. It was found that *qLTG3-1* encodes an unknown protein, whose expression is not induced by low temperature; after one day of seed soaking, it is expressed in the aleurone layer of seed coat and the ectoblast of the covering coleoptile, regulating the tissue relaxation caused by cell vacuolation, and improving the germination potential of seeds under low temperature (Fujino et al. 2008; Fujino and Matsuda 2009; Fujino and Sekiguchi 2011).

COLD1 (*Os 04g51180*) is the first seedling QTL that Ma et al. (2015) cloned from RILs population map of Nipponbare/93-11. The *Ctb2* related to pollen fertility at heading stage is in the same interval (Andaya and Mackill 2003b). Gene expression products are G protein signal regulators, subcellular localization in the plasma membrane and endoplasmic reticulum, and can interact with the G-protein α subunit to activate the Ca^{2+} channel to transmit an amplified cold signal. The DNA sequence of *COLD1* was originated from *O. rufipogon* in China, the 15th base in the fourth exon was "A" in the wild type Nipponbare and "T" in the cold sensitive variety 93-11, and the corresponding amino acid changed from "methionine" in Nipponbare to "lysine" in 93-11, which resulted in the significant change of Ca^{2+} concentration around the cell membrane of the cold-resistant variety before and after cold treatment. In that case, a low temperature signal and promoting the GTPase activity of G-protein activate the expression of downstream cold-tolerant genes, so that the plant can obtain cold-tolerant characteristics. Xiao et al. (2018) plants carrying SNP2G survived better as seedlings and showed higher grain weight than plants carrying the SNP2A allele. The CT-related loci identified here and the functional verification of LOC_Os10g34840 will provide genetic resources for breeding cold-tolerant varieties and for studying the molecular basis of CT in rice.

Oryza sativa is more sensitive to chilling injury at booting stage than at vegetative growth stage, and the sterility of gametophyte caused by low temperature can directly result in grain yield reduction. Saito et al. (2004) mapped a major QTL (*Ctb1*) related to pollen fertility at booting stage on the long arm of chromosome 4 by using the near isogenic line derived from the cross Norin Pl8/Silewah. After being finely mapped to the range of 17 kb, there were only two candidate genes (F-box protein and serine/threonine protein kinase) in this interval. The transgenic function verified that F-box protein was a candidate gene for *Ctb1*, and was highly expressed in the young ear. The interaction of *Ctb1* and E3 ubiquitin ligase subunit *Skp1* formed ubiquitin proteasome, which participated in the formation of normal pollen at low temperature during booting stage. The pollen fertility and anther length of over-expressed transgenic plants increased significantly (Saito et al. 2010).

Molecular Regulatory Networks for Cold Tolerance in *Oryza sativa*

Currently, studies have shown that *CBF* (Calmodulin binding transcription activator), *MYB* (v-MYB avian myloblastosis virtual oncogene homing), *ICE* (Inductor of *CBF* expression), and other transcription factors are involved in the regulation of cold-tolerant molecular mechanisms (Ito et al. 2006). The whole process involves cold-signal transmission, transcription-factor regulation, and cold-regulated gene expression (*COR*).

CBF Regulatory Pathway

CBF transcription factors involved in the regulation of cold tolerance were first found in Arabidopsis. According to the differences of gene coding proteins, it can be divided into three types: *CBF1, CBF2,* and *CBF3* (Gilmour et al. 1998, 2004), which belong to the same transcription level of regulation (Fowler and Thomashow 2002). Subsequently, CBF regulatory pathways were found to be prevalent in monocots and dicots (Ruelland et al. 2009). Currently, 12 *CBF* genes have been identified in *O. sativa* (Mao and Chen 2012).

Ca^{2+} is an intracellular second messenger that transmits a low-temperature signal (McCormack et al. 2005). Ca^{2+}, Calmodulin (CAM) and calcium messenger complex composed of Ca^{2+}-ATPase, which actively transports calcium in the membrane system (Phean-O-Pas et al. 2005), in which Ca^{2+} and CAM are responsible for sensing, transmitting, and responding to cold-stress signals. When plants experience low temperature, they transmit cold signals through the Ca^{2+} signaling pathway and regulate the activity of kinases and phosphatases. *ICE1* protein is activated by phosphorylation or ubiquitination, and the continuous expression of the downstream *CBF* gene is activated. The product of *CBF* gene expression regulates the expression of the *COR* gene by combining with the CRT/DRE regulatory element (core sequence is A/GCCGAC) of the upstream promoter of the COR gene (Zhang et al. 2009), and finally enables the plant to obtain cold tolerance. The response of *CBF* to the cold signal is fast and transient. In general, after low-temperature stress for 15 minutes, the expression of *CBF* begins to increase. However, the expression of the *COR* gene regulated by *CBF* begins to increase after two hours (Chinnusamy et al. 2007). However, there are different *CBF* genes involved in the response in different cold-stress periods. The high expression of *CBF1* and *CBF3* is found in 1–4 hours, and *CBF2* is found in 5–8 hours (Zhang et al. 2012). Up-regulation of *CBF1* and *CBF3* can inhibit the expression of CBF2 (Mao and Chen 2012). Therefore, although *CBF1*, *CBF2* and *CBF3* genes have similar domains, they have been functionally differentiated. Xiao et al. (2015) finely mapped a *CBF3* transcription factor related to seedling survival on chromosome 2 of *O. sativa*. The response time of the gene to cold stress in cold-tolerant varieties is earlier than that in cold-sensitive varieties. Therefore, CBF can respond to cold signal quickly and activate the expression of downstream *COR* gene, which plays an important role in obtaining cold resistance.

ICE, as the upstream regulatory gene of *CBF*, can specifically bind *CBF* promoter elements to regulate the expression of *CBF* gene (Miura et al. 2011). Among them, *ICE1* positively regulates *CBF3* (Chinnusamy et al. 2003), *ICE2* positively regulates *CBF2* (Fursova et al. 2009). After *ICE1* mutation, the expression of *CBF3* decreases, and the expression of *COR* gene targeted by *CBF3* decreases. However, the decrease of *ICE1* expression does not result in the decrease of *CBF1* and *CBF2* expression (Chinnusamy et al. 2003). In addition, *CBF* expression is also indirectly regulated by *HOS1, HOS15,* and other genes. *HOS1* is a ring-type ubiquitin E3 ligase, which can ubiquitinate or degrade *ICE1* to negatively regulate *CBF* expression; *HOS15* is a WD40 repeat protein, which can modify chromatin as an inhibitor to regulate *CBF* expression (Zhu et al. 2008); in addition, *HOS9* and *HOS10* can directly and negatively regulate *CBF* targeting *COR* gene expression, so as to regulate plant cold resistance (Zhu et al. 2004, 2005).

MYB Regulatory Pathway

MYB (v-myb avian myeloblastosis viral oncogene homolog) transcription factor is one of the largest members of the plant transcription factor family (Borevitz et al. 2000). *MYB* and *CBF* belong to two different cold-tolerance regulation pathways. *CBF* is usually inhibited when the *MYB* transcription factor expresses. *MYB15* and *MYBS3* negatively regulate the expression of the *CBF* gene (Agarwal et al. 2006; Su et al. 2010). Compared with the quick and transient response of *CBF* to cold stress, the response speed of *MYBS3* to cold stress is slightly slower (generally, it begins to express after 12 hours of cold stress). Overexpression of *CBF1*, *CBF2,* and *CBF3* has no effect on the expression of *MYBC1* (Zhai et al. 2010).

Although over-expression of *CBF* and *MYB* transcription factors can significantly improve the cold tolerance of transgenic plants, it will also have a negative effect on the agronomic characteristics of transgenic plants, often resulting in reduction of plant height and seed-setting rate, prolongation of the growth period, increase of invalid tillers, and so on (Zhang et al. 2009, 2012). Therefore, how to improve the cold resistance of plants while maintaining other excellent agronomic characters is a practical problem to be solved in the cultivation of cold-resistant varieties.

Outlook

Cold tolerance in *O. sativa* is a complex trait controlled by multiple QTLs/genes. Although the molecular basis and regulatory pathway of cold tolerance in *O. sativa* have been revealed by map-based cloning and reverse genetics, the molecular mechanism related to cold stress is still unknown. For the mapping of minor QTL, the traditional QTL method has appeared to be to "pull down one's jacket to conceal the raggedness." With the development of genomic-sequencing technology and the completion of high-quality reference genome, it is very easy to obtain high-density molecular markers covering the whole *O. sativa* genome and the expression spectrum through resequencing (Huang et al. 2013). Using a genome-wide association study (GWAS) to mine the major and minor effect loci of complex agronomic traits and breeding varieties at the genome level has become a new research hotspot (Spinde et al. 2015).

Therefore, GWAS was used to dig out the cold-tolerant gene loci of *O. sativa* in the whole growth period by GWAS and obtain high-throughput expression profiles at the same time, which will provide the technical possibility for revealing the complex cold-tolerant molecular regulatory network of *O. sativa*.

References

Agarwal, M., Hao, Y., Kapoor, A. et al. (2006). A R2R3 type MYB transcription factor is involved in the cold regulation of CBF genes and in acquired freezing tolerance. *J. Biol. Chem.* 281 (49): 37636–37645.

Andaya, V.C. and Mackill, D.J. (2003a). Mapping of QTLs associated with cold tolerance during the vegetative stage in rice. *J. Exp. Bot.* 54 (392): 2579–2585.

Andaya, V.C. and Mackill, D.J. (2003b). QTLs conferring cold tolerance at the booting stage of rice using recombinant inbred lines from a japonica×indica cross. *Theor. Appl. Genet.* 106 (6): 1084–1090.

Borevitz, J.O., Xia, Y.J., Blountb, J. et al. (2000). Activation tagging identifies a conserved MYB regulator of phenylpropanoid biosynthesis. *Plant Cell* 12: 2383–2393.

Chinnusamy, V., Ohta, M., Kanrar, S. et al. (2003). ICE1: a regulator of cold-induced transcriptome and freezing tolerance in Arabidopsis. *Genes Dev.* 17 (8): 1043–1054.

Chinnusamy, V., Zhu, J., and Zhu, J.K. (2007). Cold stress regulation of gene expression in plants. *Trends Plant Sci.* 12 (10): 444–451.

Dai, L.Y., Lin, X.H., Ye, C. et al. (2004). Identification of quantitative trait loci controlling cold tolerance at the reproductive stage in Yunnan landrace of rice, Kunmingxiaobaigu. *Breed. Sci.* 54 (3): 253–258.

Fowler, S. and Thomashow, M.F. (2002). Arabidopsis transcriptome profiling indicates that multiple regulatory pathways are activated during cold acclimation in addition to the CBF cold response pathway. *Plant Cell* 14 (8): 1675–1690.

Fujino, K. and Matsuda, Y. (2009). Genome-wide analysis of genes targeted by qLTG3-1 controlling low-temperature germinability in rice. *Plant Mol. Biol.* 72: 137–152.

Fujino, K. and Sekiguchi, H. (2011). Origins of functional nucleotide polymorphisms in a major quantitative trait locus, qLTG3-1, controlling low-temperature germinability in rice. *Plant Mol. Biol.* 75 (1–2): 1–10.

Fujino, K., Sekiguchi, H., Matsuda, Y. et al. (2008). Molecular identification of a major quantitative trait locus, qLTG3–1, controlling low-temperature germinability in rice. *Proc. Natl. Acad. Sci. U.S.A.* 105 (34): 12623–12628.

Fujino, K., Sekiguchi, H., Sato, T. et al. (2004). Mapping of quantitative trait loci controlling low-temperature germinability in rice (*Oryza sativa* L.). *Theor. Appl. Genet.* 108 (5): 794–799.

Fursova, O.V., Pogorelko, G.V., and Tarasov, V.A. (2009). Identification of *ICE2*, a gene involved in cold acclimation which determines freezing tolerance in *Arabidopsis thaliana*. *Gene* 429: 98–103.

Gilmour, S.J., Fowler, S.G., and Thomashow, M.F. (2004). Arabidopsis transcriptional activators CBF1, CBF2, and CBF3 have matching functional activities. *Plant Mol. Biol.* 54 (5): 767–781.

Gilmour, S.J., Zarka, D.G., Stockinger, E.J. et al. (1998). Low temperature regulation of the *Arabidopsis* CBF family of AP2 transcriptional activators as an early step in cold-induced *COR* gene expression. *Plant J.* 16 (4): 433–442.

Huang, X.H., Lu, T.T., and Han, B. (2013). Resequencing rice genomes: an emerging new era of rice genomics. *Trends in Genet.* 29 (4): 225–232.

Ito, Y., Katsura, K., Maruyama, K. et al. (2006). Functional analysis of rice DREB1/CBF-type transcription factors involved in cold-responsive gene expression in transgenic rice. *Plant Cell Physiol.* 47 (1): 141–153.

Ji, B.H., Cao, Y.Y., Xie, H.S. et al. (2003). Changes in violaxanthin deepoxidase activity and unsaturation of thylakoid membrane lipids in *Indica* and *Japonica* rice under chilling condition and strong light. *Acta Bot. Sin.* 45 (9): 1063–1070.

Jiang, L., Xun, M.M., Wang, J.K. et al. (2008). QTL analysis of cold tolerance at seedling stage in rice (*Oryza sativa* L.) using recombination inbred lines. *J. Cereal Sci.* 48 (1): 173–179.

Jiang, W.Z., Jin, Y.M., Lee, J. et al. (2011). Quantitative trait lLoci for cold tolerance of rice recombinant inbred lines in low temperature environments. *Mol. Cells* 32 (6): 579–587.

Koseki, M., Kitazawa, N., Yonebayashi, S. et al. (2010). Identification and fine mapping of a major quantitative trait locus originating from wild rice, controlling cold tolerance at the seedling stage. *Mol. Genet. Genomics* 284 (1): 45–54.

Kuroki, M., Saito, K., Matsuba, S. et al. (2007). A quantitative trait locus for cold tolerance at the booting stage on rice chromosome 8. *Theor. Appl. Genet.* 115 (5): 593–600.

Lee, M.H. (2001). Low temperature tolerance in rice: the Korean experience. In: *Increased lowland rice production in the Mekong Region or proceedings of an International Workshop held in Vientiane Laos* (eds. F.K. Shu and B. Jaya), 109–117. Canberra, Australia: Australian Center for International Agricultural Research.

Lee, D.G., Ahsan, N., Le,e, S.H. et al. (2009). Chilling stress-induced proteomic changes in rice roots. *J. Plant Physiol.* 166: 1–11.

Liu, F.X., Xu, W.Y., Song, Q. et al. (2013). Microarray-assisted fine-mapping of quantitative trait loci for cold tolerance in rice. *Mol. Plant* 6 (3): 757–767.

Lou, Q.J., Chen, L., Sun, Z.X. et al. (2007). A major QTL associated with cold tolerance at seedling stage in rice (*Oryza sativa* L.). *Euphytica* 158 (1–2): 87–94.

Ma, Y., Dai, X.Y., Xu, Y.Y. et al. (2015). *COLD1* confers chilling tolerance in rice. *Cell* 160 (6): 1209–2221.

Mao, D.H. and Chen, C.Y. (2012). Colinearity and similar expression pattern of rice *DREB1s* reveal their functional conservation in the cold-responsive pathway. *PLoS One* 7 (10): e47275.

McCormack, E., Tsai, Y.C., and Braam, J. (2005). Handling calcium signaling: Arabidopsis CaMs and CMLs. *Trends Plant Sci.* 10 (8): 383–389.

Miura, K., Ohta, M., Nakazawa, M. et al. (2011). ICE1 Ser403 is necessary for protein stabilization and regulation of cold signaling and tolerance. *Plant J.* 67 (2): 269–279.

Phean-O-Pas, S., Punteeranurak, P., and Buaboocha, T. (2005). Calcium signaling-mediated and differential induction of calmodulin gene expression by stress in *Oryza sativa* L. *J. Biochem. Mol. Biol.* 38 (4): 432–439.

Ruelland, E., Vaultier, M.N., Zachowski, A. et al. (2009). Cold signalling and cold acclimation in plants. *Adv. Bot. Res.* 49: 35–150.

Saito, K., Hayano-Saito, Y., Maruyama-Funatsuki, W. et al. (2004). Physical mapping and putative candidate gene identification of a quantitative trait locus *Ctb1* for cold tolerance at the booting stage of rice. *Theor. Appl. Genet.* 109 (3): 515–522.

Saito, K., Hayano-Saito, Y., Kuroki, M. et al. (2010). Map-based cloning of the rice cold tolerance gene Ctb1. *Plant Sci.* 179 (1–2): 97–102.

Shimono, H., Okada, M., Kanda, E. et al. (2007). Low temperature-induced sterility in rice: evidence for the effects of temperature before panicle initiation. *Field Crops Res.* 101 (2): 221–231.

Shirasawa, S., Endo, T., Nakagomi, K. et al. (2012). Delimitation of a QTL region controlling cold tolerance at booting stage of a cultivar, "Lijiangxintuanheigu," in rice, *Oryza sativa* L. *Theor. Appl. Genet.* 124 (5): 937–946.

Singh, R.P., Brennan, J.P., Farrell, T. et al. (2005). Economic analysis of breeding for improved cold tolerance in rice in Australia. *Aust. Agribus. Rev.* 13: 1–9.

Spinde, J., Begum, H., Akdemir, D. et al. (2015). Genomic selection and association mapping in rice (*Oryza sativa*): effect of trait genetic architecture, training population composition, marker number and statistical model on accuracy of rice genomic selection in elite, tropical rice breeding lines. *Plos Genet.* 11 (2): e1004982.

Su, C.F., Wang, Y.C., Hsieh, T.H. et al. (2010). A novel MYBS3-dependent pathway confers cold tolerance in rice. *Plant Physiol.* 153 (1): 145–158.

Suh, J.P., Jeung, J.U., Lee, J.I. et al. (2010). Identification and analysis of QTLs controlling cold tolerance at the reproductive stage and validation of effective QTLs in cold-tolerant genotypes of rice (*Oryza sativa* L.). *Theor. Appl. Genet.* 120 (5): 985–995.

Takeuchi, Y., Hayasaka, H., Chiba, B. et al. (2001). Mapping quantitative trait loci controlling cool-temperature tolerance at booting stage in temperate japonica rice. *Breed. Sci.* 51 (3): 191–197.

Wang, G.L. and Guo, Z.F. (2005). Effects of chilling stress on photosynthetic rate and chlorophyll fluorescence parameter in seedlings of two rice cultivars differing cold tolerance. *Rice Sci.* 12 (3): 187–191.

Xiao, N., Huang, W.N., Zhang, X.X. et al. (2014). Fine mapping of qRC10-2, a quantitative trait locus for cold tolerance of rice roots at seedling and mature stages. *PLoS One* 9 (5): e96046.

Xiao, N., Huang, W.N., Li, A.H. et al. (2015). Fine mapping of the *qLOP2* and *qPSR2-1* loci associated with chilling stress tolerance of wild rice seedlings. *Theor. Appl. Genet.* 128 (1): 173–185.

Xiao, N., Gao, Y., Qian, H. et al. (2018). Identification of genes related to cold tolerance and a functional allele that confers cold tolerance. *Plant Physiol.* 177 (3): 1108–1123.

Xu, L.M., Zhou, L., Zeng, Y.W. et al. (2008). Identification and mapping of quantitative trait loci for cold tolerance at the booting stage in a japonica rice near-isogenic line. *Plant Sci.* 174 (3): 340–347.

Yang, Z.M., Huang, D.Q., Tang, W.Q. et al. (2013). Mapping of quantitative trait loci underlying cold tolerance in rice seedlings via high-throughput sequencing of pooled extremes. *PLoS One* 8 (7): e68433.

Ye, C., Fukai, S., Godwin, I.D. et al. (2010). A QTL controlling low temperature induced spikelet sterility at booting stage in rice. *Euphytica* 176 (3): 291–301.

Yoshida, S. (ed.) (1981). *Fundamentals of Rice Crop Science*, 1–63. Los Banos, Laguna, the Phlippines: International Rice Research Institute.

Zhai, H., Bai, X., Zhu, Y.M. et al. (2010). A single-repeat R3-MYB transcription factor MYBC1 negatively regulates freezing tolerance in Arabidopsis. *Biochem. Biophys. Res. Commun.* 394: 1018–1023.

Zhang, Z.H., Su, L., Li, W. et al. (2005). A major QTL conferring cold tolerance at the early seedling stage using recombinant inbred lines of rice (*Oryza sativa* L.). *Plant Sci.* 168 (2): 527–534.

Zhang, Y., Chen, C., Jin, X.F. et al. (2009). Expression of a rice DREB1 gene, *OsDREB1D*, enhances cold and high-salt tolerance in transgenic Arabidopsis. *BMB Rep.* 42 (8): 486–492.

Zhang, T., Zhao, X.Q., Wang, W.S. et al. (2012). Comparative transcriptome profiling of chilling stress responsiveness in two contrasting rice genotypes. *PLoS One* 7 (8): e43274.

Zhou, L., Zeng, Y.W., Zheng, W.W. et al. (2010). Fine mapping a QTL qCTB7 for cold tolerance at the booting stage on rice chromosome 7 using a near-isogenic line. *Theor. Appl. Genet.* 121 (5): 895–905.

Zhu, J.H., Shi, H.Z., Lee, B.H. et al. (2004). An Arabidopsis homeodomain transcription factor gene, *HOS9*, mediates cold tolerance through a CBF-independent pathway. *Proc. Natl. Acad. Sci. U.S.A.* 101 (26): 9873–9878.

Zhu, J.H., Verslues, P.E., Zheng, X.W. et al. (2005). *HOS10* encodes an R2R3-type MYB transcription factor essential for cold acclimation in plants. *Proc. Natl. Acad. Sci. U.S.A.* 102 (28): 9966–9971.

Zhu, J.H., Jeong, J.C., Zhu, Y.M. et al. (2008). Involvement of *Arabidopsis* HOS15 in histone deacetylation and cold tolerance. *Proc. Natl. Acad. Sci. U.S.A.* 105 (12): 4945–4950.

7

Molecular Breeding for Lower Cadmium Accumulation in Rice Grain: Progress and Perspectives

Dongping Li[1], *Xiaohua Hao*[2] *and Liangbi Chen*[1]

[1] *Hunan Province Key Laboratory of Crop Sterile Germplasm Resource Innovation and Application, College of Life Science, Hunan Normal University, Changsha 410081, P. R. China*
[2] *College of Life and Environmental Science, Hunan University of Arts and Science, Changde 415000, China*

Introduction

Cd is a silvery-white transition metal (atomic number, 48; relative atomic mass, 112.41; density 8.64 g cm^{-3}), which was discovered by a German chemist, Friedrich Stromeyer, in 1817. Cd predominantly exists in zinc sulfide-based ores in the form of cadmium sulfide, and is often isolated from zinc and lead ores as a by-product. Among the naturally occurring elements on earth, Cd ranks the 64th in terms of relative abundance. Furthermore, there is diverse Cd content that is spread widely, from the atmosphere to the earth's crust (Table 7.1) (www.cadmium.org).

Cd metal and certain Cd compounds have applications for their specific physical and chemical properties. Cd is used in rechargeable nickel-cadmium batteries due to its rapid ion electrical exchange activity, and high electrical conductivity (Fan 1991). In addition, Cd can be taken as an efficient anti-corrosive coating to protect iron and steel from corrosion or oxidation (Kautek 1988). Moreover, Cd compounds can also be used as pigments for painting, with strong coloring, high opacity, efficient tinting strength, and resistance to high temperatures (Cesaratto et al. 2013). Finally, Cd can be used as an effective neutron absorber to control the chain fission reaction rate in nuclear reactors (Converse et al. 1987; Larsson and Möller 1968).

Cd Toxicity to Human Health

When Cd accumulates in various organs, it is harmful to human health. The main organs affected by Cd accumulation include the kidneys and liver, and this accounts for approximately 60% of the total Cd present in the body (Hansen et al. 1989; Lech and Sadlik 2017; Satarug et al. 2002). A well-documented Cd-intoxication case, "*Itai Itai*" disease, which was a mass Cd poisoning due to ingesting Cd-contaminated rice and drinking water, was reported in Japan in the 1950s (Horiguchi et al. 1994). Cd is not an essential element for organisms. However, Cd can be absorbed and accumulated in the human body from the

Molecular Breeding for Rice Abiotic Stress Tolerance and Nutritional Quality, First Edition.
Edited by Mohammad Anwar Hossain, Lutful Hassan, Khandakar Md. Iftekharuddaula, Arvind Kumar and Robert Henry.

Table 7.1 Cd concentrations in different environments (www.cadmium.org).

Environmental components	Cd level (ppm)
Atmosphere	0.000 000 1 to 0.000 000 5
Earth's crust	0.1 to 0.5
Marine sediments	1.0
Sea water	0.000 001

Source : Based on International Cadmium Association

environment via food, water, and air. A limited concentration of absorbed Cd is excreted mainly in the urine via the kidney and also in feces via the liver. The biological half-life period of Cd in the human kidney cortex is 10–30 years (Hallenbeck 1984; Järup et al. 1983; Suwazono et al. 2009).

Cd has detrimental toxic effects on the human body. Cd pollution enhances oxidative stress, inhibits DNA synthesis and repair, and ultimately leads to DNA damage in cells (Liu et al. 2009). In addition, Cd can cause damage to multiple tissues (bones, lungs, kidneys, and livers) and multiple systems (the immune system, cardiovascular system, and reproductive system) (Jomova and Valko 2011; Thévenod and Lee 2013). Furthermore, Cd toxicity is closely related to carcinogenesis, including breast cancer (Van Maele-Fabry et al. 2016), pancreatic cancer (Buha et al. 2017), renal cancer (Huff et al. 2007; Il'yasova and Schwartz 2005; Song et al. 2015), bladder cancer (Feki-Tounsi and Hamza-Chaffai 2014), prostate cancer (Rapisarda et al. 2018; Vinceti et al. 2007), and testicular cancer (Goyer et al. 2004).

The Origin of Cd Contamination in Soils

Under natural conditions, the content of Cd in soil is very low. However, Cd levels have increased significantly in Cd polluted areas since the industrial revolution, because of human activities such as industrial production and mining. Globally, approximately 0.235 billion hectares (ha) of arable land were contaminated by heavy metals (Bermudez et al. 2012). In China, approximately 0.278 million ha of agricultural soils were polluted with Cd (Liu et al. 2015). The Cd concentration in agricultural soils in China ranged from 0.01 to 1.34 mg kg^{-1}, with an average of 0.12 mg kg^{-1}. The sources of Cd and other heavy metals in agricultural soils are predominantly from sewage irrigation, atmospheric deposition, and large-scale use of organic fertilizers.

Specifically, sewage irrigation contributes to the majority of the Cd pollution in paddy soils. Sewage, such as industrial and mining wastewater, and urban sewage that is not purified or not treated completely has been irrigated into fields directly and/or discharged into rivers. This results in Cd-contaminated water being continually used for irrigation of the paddy soils. Long-term irrigation with Cd-contaminated water, ultimately leads to serious soil Cd pollution (Khan et al. 2013; Sikka et al. 2008).

Atmospheric deposition is another major contributor in Cd accumulation in soils. A large quantity of Cd-containing exhaust gases were produced by factories, such as enterprises in power generation, in the chemical industry, in building materials production, in oil extraction and processing, and were emitted into the air. As the air moves, Cd falls into the soil as dust and/or rain, causing land pollution. Furthermore, fuel-driven vehicles contribute a considerable percentage of the Cd found in the land close to roads (Cao et al. 2019b; Wu et al. 2013).

Cd pollution caused by fertilizers is also detrimental, specifically phosphate fertilizers. The Cd content in phosphate fertilizer varies from 2 to 200 mg kg^{-1}, depending on its origin and processing method. It has been previously reported that long-term use of high-Cd phosphate fertilizers has led to rapid elevation of Cd levels in soils (Jiao et al. 2012; Li et al. 2017a; Nziguheba and Smolders 2008). With the exception of phosphate fertilizer, Cd concentration is often fixed in organic fertilizers (Nookabkaew et al. 2016).

Cd Accumulation in Rice

Cd Uptake and Distribution Patterns During the Whole Growth Cycle of Rice

Rice is a field crop that can easily absorb Cd from paddy soil. Cd in the soil is first absorbed by the roots, and transported transversely via symplasmic or apoplasmic pathways to the xylem, then translocated to the shoot by xylem loading (Lux et al. 2011). The crop nodes play an important role in Cd distribution towards the shoot (Chen et al. 2019b; Yamaguchi et al. 2012; Yamaji and Ma 2017). During reproductive development, Cd within the grain is transported via the phloem from leaf blades or from xylem-to-phloem transfer in the nodes (Kensuke et al. 2007; Shu et al. (2010); Yoneyama et al. 2015). The general Cd distribution pattern in rice is predominantly in the roots, and less abundant in the stems and leaves during the vegetative developmental stage. Therefore, this correlates with the notion that the further the tissues are from the soil (such as internodes or leaves), the lower the Cd content is. Typically, the Cd content in roots is 100 times higher than that in stems and leaves, and up to 1000 times higher than that in brown rice (Hao et al. 2018). Cd in different tissues of rice seed is also inhomogeneous as the Cd content in the aleurone layer is highest, with the lowest in the glume shell. Rodda et al. (2011) analyzed Cd accumulation patterns in the grain at pre-flowering and post-flowering phases, and estimated that 60% of the final grain Cd could be attributed to Cd remobilization from other tissues stored at the vegetative stage. The rest (40%) could be from root uptake during grain maturation, where the newly-absorbed Cd joins the phloem cycle via structures at vascular tissues of leaves and nodes (Rodda et al. 2011). The finding that the majority, if not all, of the grain Cd has originated from the phloem Cd provides a potential mechanism to reduce grain Cd accumulation by blocking Cd loading into the phloem of the leaves or nodes.

The Diversity of Cd Accumulation Trait in Different Varieties of Rice

A plethora of data from both laboratory and field experiments has proven that Cd accumulation may differ significantly among different cultivars, indicating their genotype divergence

(Arao and Ae 2003; Li et al. 2003; Liu et al. 2007, 2003; Yu et al. 2006; Zeng et al. 2008; Zhan et al. 2013). It is generally accepted that the *japonica* variety of rice can uptake and accumulate less Cd than any other variety of rice. The average grain Cd content from *japonica* varieties is significantly lower than that of *indica* on low or moderate Cd-contamination (Chi et al. 2018; Sun et al. 2016; Xie et al. 2017; Yan et al. 2019) However, a few of the temperate *japonica* varieties showed high Cd-accumulation due to natural variation of the Cd transport gene (Yan et al. 2016). This is still a controversial issue whether the *indica* hybrid rice or super rice accumulates more Cd than the *indica* conventional rice. The paradox is that some studies showed the *indica* hybrid rice or super rice accumulates more Cd (Lin et al. 2018; Shi et al. 2009), while some data opposed this (Li et al. 2014a), or showed no significant differences in Cd content between *indica* hybrid rice and *indica* conventional rice (Liu et al. 2005; Sun et al. 2016). Nevertheless, there are large variations in grain Cd accumulation among the population of either hybrid rice or conventional cultivars, indicating the role of the additive effect of Cd genes, which resulted in the polymorphism of genotypes in Cd accumulation (Yao et al. 2015). The physiological mechanisms of grain Cd accumulation in the low-Cd cultivars were shown to have a stronger capacity in restraining Cd transport to soluble or organelle fractions due to the crucial role of the cell wall to bind more Cd than high grain-Cd-accumulating cultivars. Additionally, a higher transportation rate in the stem and higher Cd outflow in the roots resulted in less injury to roots and, ultimately, lower Cd accumulation in grains (Jie et al. 2009; Wang et al. 2015).

Genetic Basis for Cd Accumulation in Rice

Based on stable low-Cd and high-Cd rice resources, multiple populations were constructed, and tens of Cd-linked quantitative trait loci (QTLs) were mapped to specific regions of the 12 chromosomes (Abe et al. 2013; Ishikawa et al. 2005; Kashiwagi et al. 2009; Pinson et al. 2015; Ueno et al. 2009; Xue et al. 2009; Yan et al. 2019, 2013). Table 7.2 listed Cd-linked QTLs with their traits, chromosomes, markers, and so on, where the QTLs were identified to be associated with various Cd traits such as Cd accumulation, Cd translocation, or Cd tolerance. For example, *qCdT7* /*qGCd7*, which was located on the short arm of the chromosome 7, adjacent to Simple Sequence Repeats (SSR) marker RM8006, was identified as a major QTL for Cd shoot and grain accumulation by three independent groups using different mapping populations (Ishikawa et al. 2010; Tezuka et al. 2010; Ueno et al. 2009). Furthermore, the gene *OsHMA3* was isolated and shown to be responsible for this QTL (Ueno et al. 2010). To date, only three Cd-related QTLs have been successfully mapped to the genes (Ueno et al. 2010; Luo et al. 2018; Yan et al. 2019). The identification of more and more major genes underlying the QTLs could be expected in the future; this will facilitate low-Cd cultivar breeding and phytoremediation.

Influence of Environmental Factors on Cd Accumulation in Rice

Influence of Soil Water on Rice Cd Accumulation

Numerous studies have reported that water management impacts Cd bioavailability in soils and subsequent Cd uptake by rice (Bingham et al. 1976; Hu et al. 2013; Kikuchi et al. 2008;

Table 7.2 QTLs for cadmium accumulation in rice.

QTL names	chromosomes (Chr)	Markers	Parental resources	Populations	Traits	References
—	3	S1513-R663	Koshihikari/Kasalath	39 CSSLs	Low grain Cd	Ishikawa et al. (2005)
—	6	R2171-R2549	Koshihikari/Kasalath	39 CSSLs	High grain Cd	Ishikawa et al. (2005)
—	8	C390-C1121	Koshihikari/Kasalath	39 CSSLs	Low grain Cd	Ishikawa et al. (2005)
qSRR3	3	G144-C848	JX17/ZYQ8	127 DH	Shoot/root Cd	Xue et al. (2009)
qCDR6.1; qCDR6.2	6	CT115-CT506;CT380B-G294	JX17/ZYQ8	127 DH	High root Cd	Xue et al. (2009)
qCDS7	7	RG528-RG769	JX17/ZYQ8	127 DH	High shoot Cd	Xue et al. (2009)
qSH5	5	RG556-RG207	JX17/ZYQ8	127 DH	Cd tolerant (SH)	Xue et al. (2009)
qRL1.3	1	C225-RG536	JX17/ZYQ8	127 DH	Cd tolerant(RL)	Xue et al. (2009)
qSDW3.2	3	G249-G164	JX17/ZYQ8	127 DH	Cd tolerant(SDW)	Xue et al. (2009)
qRDW8.2	8	G278-RG885	JX17/ZYQ8	127 DH	Cd tolerant(RDW)	Xue et al. (2009)
qTDW3	3	G249-G164	JX17/ZYQ8	127 DH	Cd tolerant(TDW)	Xue et al. (2009)
qCC10	10	G2155-C16	JX17/ZYQ8	127 DH	Cd tolerant(CC)	Xue et al. (2009)
qcd4–1;qcd4–2	4	R93-R374: C1100-C1016	Nipponbare/Kasalath	98 BILs/NILs	High grain/shoot	Kashiwagi et al. (2009)
qcd11	11	R728-R1506	Nipponbare/Kasalath	98 BILs/NILs	Low shoot Cd	Kashiwagi et al. (2009)
—	11	RM6623	Shwe War/Badari Dhan	184 F2	High shoot Cd	Ueno et al. (2009)

—	2	RM5378	Shwe War/Badari Dhan	184 F2	High shoot Cd	Ueno et al. (2009)
—	5	RM5329	Shwe War/Badari Dhan	184 F2	High shoot Cd	Ueno et al. (2009)
—	7	RM8006-RM7153	Nipponbare/Anjana Dhan	177 F2	High shoot Cd	Ueno et al. (2009)
OsHMA3[a)]	7	OsHMA3–29-RM21265	Nipponbare/Anjana Dhan	965 F2	High shoot Cd	Ueno et al. (2010)
qCdT7	7	RM6776-RM5436	Akita 63/Cho-Ko-Koku	144 F2	High translocation	Tezuka et al. (2010)
—	4	RM5687	Akita 63/Cho-Ko-Koku	144 F2	Low root Cd	Tezuka et al. (2010)
—	1	RM11008	Akita 63/Cho-Ko-Koku	144 F2	High shoot Cd	Tezuka et al. (2010)
—	12	RM7102	Akita 63/Cho-Ko-Koku	144 F2	High shoot Cd	Tezuka et al. (2010)
qGCd7	7	C383–G1068	Sasanishiki/Habataki	85 BILs	High grain Cd	Ishikawa et al. (2010)
qGCd2	2	R2511–C1470	Sasanishiki/Habataki	85 BILs	High grain Cd	Ishikawa et al. (2010)
qSCd2	2	C1221–R2511	Sasanishiki/Habataki	85 BILs	Shoot Cd	Ishikawa et al. (2010)
qSCd7	7	G383-G1068	Sasanishiki/Habataki	85 BILs	Shoot Cd	Ishikawa et al. (2010)
qSCd12	12	R1709–C1069	Sasanishiki/Habataki	85 BILs	Shoot Cd	Ishikawa et al. (2010)
qCd1	1	C1370	Azucena/Bala	79 RILs	High leaf Cd	Norton et al. (2010)

(Continued)

Table 7.2 (Continued)

QTL names	chromosomes (Chr)	Markers	Parental resources	Populations	Traits	References
qCd3	3	RZ474	Azucena/Bala	79 RILs	High leaf Cd	Norton et al. (2010)
qCd6	6	a12376	Azucena/Bala	79 RILs	High leaf Cd	Norton et al. (2010)
qCd1	1	C949	Azucena/Bala	79 RILs	High grain Cd	Norton et al. (2010)
qCd2	2	RM6	Azucena/Bala	79 RILs	High grain Cd	Norton et al. (2010)
qCd3	3	RG191	Azucena/Bala	79 RILs	High grain Cd	Norton et al. (2010)
qCd7	7	G338	Azucena/Bala	79 RILs	High grain Cd	Norton et al. (2010)
qCd9	9	C506	Azucena/Bala	79 RILs	High grain Cd	Norton et al. (2010)
qCd10	10	AB1001	Azucena/Bala	79 RILs	High grain Cd	Norton et al. (2010)
qCdp7	7	RM3859-RM3635	Koshihikari/Jarjan	103 BILs	High grain Cd	Abe et al. (2011)
qLCdG3	3	RM426-RM6329	Fukuhibiki/LAC23	126 RILs	High grain Cd	Sato et al. (2011)
qLCdG11	11	NBLAC61-NBLAC63	Fukuhibiki/LAC23	126 RILs	Low grain Cd	Sato et al. (2011)
qCdc3	3	G249-G164	JX17/ZYQ8	127 DH	Low grain Cd	Zhang et al. (2011)
qCdc4	4	C513-CT500	JX17/ZYQ8	127 DH	High grain Cd	Zhang et al. (2011)
qCdc6	6	G122-G1314B	JX17/ZYQ8	127 DH	Low grain Cd	Zhang et al. (2011)
qlGCd3	3	Hd6-RM16153	Koshihikari/LAC23	46 CSSLs	Low grain Cd	Abe et al. (2013)

gcc3	3	RM545–RM1338	SNU-SG1/Suwon490	91 RILs	Low grain Cd	Yan et al. (2013)
scc10	10	RM239–RM311	SNU-SG1/Suwon490	91 RILs	Low shoot Cd	Yan et al. (2013)
gcc9	9	RM1328–RM219	SNU-SG1/Suwon490	91 RILs	Low grain Cd	Yan et al. (2013)
gcc11	11	RM286–RM4B	SNU-SG1/Suwon490	91 RILs	Low grain Cd	Yan et al. (2013)
sgr5	5	RM593–RM437	SNU-SG1/Suwon490	91 RILs	Low grain/shoot	Yan et al. (2013)
—	1	RG236	Lemont/TeQing	280 RILs,123 TILs	Grain Cd	Zhang et al. (2013)
—	9	RG570	Lemont/TeQing	280 RILs,123 TILs	Grain Cd	Zhang et al. (2013)
qCd3	3	RM85	Inbred accessions	378 cultivars	Grain Cd	Huang et al. (2015)
qCd5.1;qCd5.2	5	RM459;RM163	Inbred accessions	378 cultivars	Grain Cd	Huang et al. (2015)
qLCd2(CAL1)[a)]	2	RM324-RM5472	Tainan/Chunjiang06	119 DH	High leaf Cd	Luo et al. (2018)
qLCd4	4	RM3276-RM22	Tainan/Chunjiang06	119 DH	High leaf Cd	Luo et al. (2018)
qLCd11	11	RM21-RM3428	Tainan/Chunjiang06	119 DH	Low leaf Cd	Luo et al. (2018)
qCCMR2-1	2	RM13530–RM6	D50/YK17	101 DH	Brown rice Cd	Hu et al. (2018)
qCCBR9-1	9	RM6854–RM24537	D50/YK17	101 DH	Brown rice Cd	Hu et al. (2018)
qCd-2	2	RM250–RM207	Xiang 743/Katy	151 RILs	Grain Cd	Liu et al. (2019)
qCd-7	7	RM6872–RM3718	Xiang 743/Katy	151 RILs	Grain Cd	Liu et al. (2019)
qCd-6.1	6	RM225–RM276	Xiang 743/Katy	151 RILs	Grain Cd	Liu et al. (2019)
qGCD1-qGCD17	1–12	—	Inbred accessions	276 cultivars	Grain Cd	Liu et al. (2019)
QTL3(OsCd1)[a)]	3	—	Inbred accessions	276 cultivars	Grain Cd	Yan et al. (2019)

a) Indicates the QTLs that the Cd-linked genes had been cloned by mapping.

Zhang et al. 2019). When a paddy field is flooded, soil microbes respire by utilizing oxidized soil components, including NO_3^-, SO_4^{2-}, Fe^{3+}, Mn (III/VI) species that are present in oxide phases, as well as the dissimilation products of organic matter. These species receive electrons during the reduction reaction, which generates NO_2^-, S^{2-}, Fe^{2+}, Mn^{2+}, and low molecular-weight organic acids, accompanied by a decrease of the redox potential (Eh) (Livera et al. 2011). Consequently, Cd combines with S^{2-} to form insoluble CdS, and adsorbs Fe and/or Mn oxyhydroxides, which lower the active Cd level in the rhizosphere of rice (Bingham et al. 1976; Sun et al. 2007). In contrast, during the drainage period the paddy soil is in an oxidative state with higher Eh. Cd combines with SO_4^{2-} to form water-soluble $CdSO_4$, leading to a higher active Cd concentration (Sebastian and Prasad 2013). Therefore, water management can influence the Eh state, and therefore can be utilized to regulate the bioavailability of Cd.

Influence of pH Values in Soil and Irrigation Water on Rice Cd Accumulation

Among the factors that affect Cd absorption of plants, the pH value of soil and irrigation water is considered to be a crucial factor, as pH can significantly impact solubility and speciation of Cd in soil solution, and greatly influence the Cd mobilization and grain Cd accumulation in rice (Sarwar et al. 2010; Shi et al. 2019; Zeng et al. 2011; Zhang et al. 2017; Zhu et al. 2016). Generally, the pH of soil and irrigation water is negatively correlated with the availability of heavy metals in plants, including rice (Kirkham 2006; Li et al. 2017b; Zeng et al. 2011). The lower the pH of soil and irrigation water, the more bioavailable the Cd is in the paddy field, and therefore, the higher the Cd accumulation is in grains, or vice versa. At acidic soil pH, the solubility of Cd in solid phases, such as carbonates, hydroxides, and phosphates, is increased (Reddy and Patrick 1977). At alkaline soil pH levels, Cd has the potential to form Cd $(OH)_2$ by hydrolysis, which results in soil Cd precipitating as hydroxides or carbonates and adsorbs tightly to soil colloid, and ultimately leads to decreased Cd mobilization and lower Cd accumulation in rice (Sun et al. 2007). Soil pH can be affected by the soil redox status. For example, after flooding, soil pH can increase gradually to a neutral value in acidic fields, which can be explained by a decrease of Eh caused by oxygen consumption by microbial activity (i.e., increase of reducibility). Under reducing conditions, most reduction reactions consume H^+, resulting in a rise in soil pH (Hindersmann and Mansfeldt 2014; Honma et al. 2016; Narteh and Sahrawat 1999; Pan et al. 2016; Sun et al. 2007). Thus, soil acidification must be avoided to prevent the release of bioavailable Cd in soil (Sebastian and Prasad 2013). Lime application in paddy fields has been shown to be an effective approach to increase the pH of soil and irrigation water and therefore, alleviate bioavailable Cd (Chen et al. 2018a; Liu et al. 2017; Shi et al. 2019; Xiao et al. 2017; Zhu et al. 2016).

Effect of the Application of Other Ions or Chemical Compounds on Rice Cd Accumulation

Numerous studies have shown that different element supplies could impact Cd accumulation in rice due to the alteration of gene expression, enzyme activities, subcellular distribution, or speciation of Cd. For example, excessive nitrate (NO_3^-) supply did not promote rice growth significantly, but increased the uptake of Cd and Fe via nitrate-induced

OsIRT1 expression (Yang et al. 2016). However, ammonium (NH_4^+) not only reduces the absorption and root-to-shoot translocation of Cd^{2+} by decreasing the expression of *OsHMA2* and *OsNramp5* in the root, but also mitigates Cd toxicity via putrescine-dependent alterations of cell wall composition (Wang et al. 2019b; Zhu et al. 2018). To date, with the exception of NH_4^+, many other mineral elements, such as Mn, Fe, Zn, S, B, Ca, or even Se and Si, were also found to alleviate Cd accumulation in rice. Cd entry into rice plants has been shown to be via some Ca channels and/or Mn/Fe transporters in the roots (Chen et al. 2018b; Nakanishi et al. 2006; Sasaki et al. 2012; Takahashi et al. 2011b). Ca application to the soil can lower rice Cd content due to increasing environmental pH as described in the above section. Furthermore, the addition of Mn/Fe in paddy soil competes with Cd transport, leading to decreased Cd accumulation in rice (Sebastian and Prasad 2015). Moreover, the enhanced Fe can cause Fe plaque formation, which can sequester more Cd onto the root surfaces (Gao et al. 2018; Hao et al. 2014; Liu et al. 2010). Similarly, S supply can significantly decrease Cd concentrations in rice plants and grains, which is ascribed to the enhanced formation of Fe plaque on the root surfaces and increased Cd chelation (OsPCS) and vacuolar sequestration (OsHMA3) in roots (Cao et al. 2018; Zhang et al. 2013).

Although Si is not an essential element, it is important for rice growth and development, and stress tolerance (Cao et al. 2018; Zhang et al. 2008). Rice is a typical high Si-accumulating crop, and Si application can reduce uptake and translocation of Cd (Farooq et al. 2016). Si has the potential to obstruct the apoplast bypass flow through the roots, and suppress the apoplastic transport of Cd to decrease rice shoot Cd (Shi et al. 2005). Si up-regulated the expression of vacuolar H^+-pyrophosphatase gene *OVP1*, and inhibited the expression of *Natural resistance-associated macrophage protein 1(Nramp1)*, *Nramp5*, and *Heavy metal-associated domain HMA2*, which caused a decrease in uptake and translocation of Cd in rice (Cao et al. 2019c, 2018; Shao et al. 2017).

B(Boron) is mainly involved in cell wall structure and function due to its capacity to form borate complexes with a variety of polyhydroxy compounds (O'Neill et al. 2004), suggesting that B might have the ability to form a complex with Cd or induce Cd to deposit in the cell walls, and therefore lowering Cd toxicity in rice plants. B application dramatically decreased Cd accumulation in rice by improving oxidative stress and suppressing Cd uptake and transport. The regulating effects of B application on aboveground biomass, chlorophyll content, and Cd accumulation were similar to those of Si addition. Interestingly, the two elements of Si and B showed combined effects on reducing Cd toxicity. Mitigation effects of combined application of Si and B on Cd accumulation and toxicity were more evident than single application of Si or B (Chen et al. 2019a).

Se is not an essential plant nutrient; however, it enhances plant growth within an appropriate physiological concentration (Tamaoki and Maruyama-Nakashita 2017). Many studies have showed that Se application had the potential to decrease Cd accumulation in rice (Huang et al. 2017, 2018; Li et al. 2012b; Liao et al. 2016). However, further analysis found that addition of Se did not significantly reduce Cd uptake and translocation in rice (Wan et al. 2016; Wu et al. 2016; Yang et al. 2019). Se has been suggested to alter the chemical form of Cd, and reduce soil Cd bioavailability, which, in turn, could influence rice Cd accumulation (Huang et al. 2017, 2018).

The combination of some elements showed significant synergistic effects. The single application of elements, such as P and Zn, to soil did not alleviate Cd accumulation in rice (Mao et al. 2019; Zhang et al. 2019). However, foliar application or combination application with other ions, such as "P+S," "P+CaO," "Zn+Mn," or "Zn+P," significantly decreased rice Cd levels (Liu et al. 2019; Lv et al. 2019; Wang et al. 2018; Xiao et al. 2017).

Influence of Rhizosphere Microorganisms on Rice Cd Accumulation

Recently, several microorganisms have been identified that limit Cd accumulation in rice grains (Lin et al. 2016; Punjee et al. 2017; Wang et al. 2019c). For example, Cd-resistant rhizobacterium *Bacillus cereus* promotes the growth of rice and reduces Cd accumulation (Wang et al. 2019a). With the exception of *B.cereus*, many Cd-resistant bacterial species, such as *Bacillus megaterium*, *Bacillus subtilis*, *Cupriavidus taiwanensis*, *Delftia tsuruhatensis*, *Neorhizobium huautlense, Pseudomonas aeruginosa*, and *Stenotrophomonas acidaminiphila*, decreased rice Cd uptake by decreasing Cd bioavailability in soil (Li et al. 2017b; Punjee et al. 2017; Treesubsuntorn et al. 2017). Rhizosphere soil Cd can be immobilized by bacteria mentioned above via three different mechanisms. Firstly, the bacteria can cause Cd ions to be absorbed in cells and subsequently detoxified by binding with intracellular metallothioneins or similar functional polypeptides (Behera et al. 2014; Hrynkiewicz et al. 2012; Ramesh et al. 2009). Secondly, the bacteria can cause Cd ions to be adsorbed by functional groups on some bacterial cell wall surfaces (Liu et al. 2018; Sheng et al. 2016). Finally, the bacteria can cause Cd ions to be chelated by extracellular complexation of bacterial metabolites (Chakraborty and Das 2014).

Besides rhizosphere bacteria, mycorrhizal fungi are also effective at mitigating Cd accumulation in rice. The arbuscular mycorrhizal fungi (AMF) form symbiotic associations with the roots of the majority of plant species (Guo et al. 2013; Shahabivand et al. 2012). The use of AMF has the potential to reduce Cd uptake by rice grown in Cd solutions, through altering subcellular distribution and chemical forms of rice Cd (Li et al. 2016; Zhang et al. 2005). Microbial remediation strategies are attractive treatments in Cd toxicity since they are environmentally friendly and cost-effective.

Influence of Genetic Factors on Cd Accumulation in Rice

Root exudates and microorganisms alter the effectiveness and mobility of metals around the roots. Vascular plants can acquire metal ions and other inorganic nutrients from the soil efficiently. Their absorption and transport depend on the function of multiple transporters in the roots, suggesting that different heavy metal cations are co-transported with nutrients through the plasma membrane. Because toxic heavy metals, such as Cd and Pb, have unknown biological functions, specific transporters may not be present. However, these toxic metal species have been shown to specifically enter the cell via some cation transporters. Plants have evolved a variety of transmembrane transporters that function in ion absorption, transport, or cellular ion homeostasis. These transporters include Natural resistance-associated macrophage protein (NRAMP), ATP-binding cassette (ABC) transporter, cation diffusion facilitator (CDF), Zrt-/Irt-like protein (ZIP), and calcium/cation (CaCA) family members.

Transporters for Cd Uptake and Translocation

The first *Nramp* gene was cloned from mice, and is involved in controlling the natural resistance to infection by ingesting intracellular parasites early during infection (Vidal 1993). Nramp1 functions as a divalent cation transporter (Skamene and Schurr 1998). The NRAMPs group is conserved throughout organisms, with a preserved consensus motif and 10–12 transmembrane domains (Cellier et al. 1995). As transmembrane transporters, NRAMPs were characterized as being responsible for multiple ion transport such as divalent metal Fe^{2+}, Mn^{2+}, Zn^{2+}, Cu^{2+}, Cd^{2+} (Peris-Peris et al. 2017), trivalent Al^{3+} (Li et al. 2014b; Xia et al. 2010), As^{3+} (Tiwari et al. 2014), monovalent Cs^{+} (Yamaki et al. 2017), anion NO_3^- (Unkles et al. 1991), and H^+ (Belouchi et al. 1997), which, in turn, is involved in nutrition homeostasis, detoxification, defense, and/or immune responses.

In rice, the *Nramp* gene family contains seven members encoding NRAMP transporters (Belouchi et al. 1997), three of which have been functionally characterized to mediate Cd uptake and translocation. OsNRAMP1 transports not only Fe and Mn, but also toxic Cd and As^{3+} (Curie et al. 2000; Takahashi et al. 2011b; Tiwari et al. 2014). The amino acid sequence of OsNRAMP1 protein did not differ among the cultivars with different Cd accumulation, but the expression level of OsNRAMP1 protein was significantly different among the cultivars with different Cd accumulation. The tonoplast-localized OsNRAMP2 is involved in grain Cd accumulation by mediating vacuole-to-cytosol transport of Cd (Zhao et al. 2018). OsNRAMP5 functions as a major transporter for uptake and translocation of Mn, Cd, and Fe (Ishimaru et al. 2012a,2012b; Sasaki et al. 2012; Takahashi et al. 2014; Tang et al. 2017). OsNRAMP5 is mainly expressed in rice root tissues, including epidermis, cortex, and near xylem of roots. The *Nramp5* mutant reduced the uptake of Cd by roots, thus lowering the accumulation of Cd in stems and grains.

Heavy metal ATPase (HMA) is a protein that can utilize the energy released by ATP hydrolysis to drive the transmembrane transport of heavy metal ions. It plays an important role in the transport of Zn, Pb, Cd, and Co ions, and widely exists in prokaryotic and eukaryotic organisms (Yamaji et al. 2013b). There are nine members of the HMA family in rice (Table 7.3). OsHMA2, OsHMA3, and OsHMA9 have been reported to transport Cd. OsHMA2 contributes to the transportation of Zn and Cd from root to shoot, and plays an important role in the loading of Zn and Cd to xylem. Interestingly, the Cd content in the grains of *OsHMA2* over-expressed transgenic rice, and *osmha2* mutant was much lower than that of wild type (Satoh-Nagasawa et al. 2012). OsHMA3 is located on the vacuole membrane and is responsible for Cd transport into vacuoles for storage, which reduces the Cd toxicity in plants. The absence of OsHMA3 enhanced Cd transport from root to shoot (Yan et al. 2016). Over-expression of *OsHMA3* could significantly enhance the tolerance to cadmium. Meanwhile, in *OsHMA3* over-expressing plants, the Zn concentration in roots was significantly higher than that in wild type plants, while the Zn concentration in stems was similar to that in wild type plants (Sasaki et al. 2014). OsHMA9

Table 7.3 Genes participated in Cd uptake and accumulation in rice

Gene	Gene family	Location	Function	References
OsNRAMP1	Natural resistance-associated macrophage protein (NRAMP)	Plasma-membrane	Cd, Mn, Fe, and As uptake and transport	Takahashi et al. (2011b), Takahashi et al. (2011a), and Tiwari et al. (2014)
OsNRAMP2	Natural resistance-associated macrophage protein (NRAMP)	Tonoplast	Cd accumulation	Zhao et al. (2018)
OsNRAMP5	Natural resistance-associated macrophage protein (NRAMP)	Plasma-membrane	Cd, Mn and Fe uptake and transport	Ishimaru et al. (2012b), Sasaki et al. (2012), and Yang et al. (2014)
OsHMA2	P-type heavy metal ATPase (HMA)	Plasma-membrane	Cd and Zn translocation	Takahashi et al. (2012a) and Yamaji et al. (2013b)
OsHMA3	P-type heavy metal ATPase (HMA)	Tonoplast	Cd sequestration in root	Sasaki et al. (2012), Ueno et al. (2011) and Ueno et al. (2009)
OsHMA9	P-type heavy metal ATPase (HMA)	Plasma-membrane	Cd efflux	Lee et al. (2007)
OsIRT1	Zinc and iron-regulated transporter (ZIP)	Plasma-membrane	Cd and Fe transport	Lee and An (2009), Nakanishi et al. (2006)
OsIRT2	Zinc and iron-regulated transporter (ZIP)	Plasma-membrane	Cd and Fe transport	Nakanishi et al. (2006)
OsZIP1	Zinc and iron-regulated transporter (ZIP)	Plasma-membrane;ER	Cd, Zn, Cu efflux	Chou et al. (2011), Liu et al. (2019), and Ramesh (2003)
OsZIP3	Zinc and iron-regulated transporter (ZIP)	N/A	Cd and Zn uptake in yeast	Ramesh (2003) and Zheng et al. (2018)
OsZIP6	Zinc and iron-regulated transporter (ZIP)	N/A	Cd, Fe and Co uptake in oocytes	Kavitha et al. (2015)

OsZIP7	Zinc and iron-regulated transporter (ZIP)	Plasma-membrane	Cd and Zn transport	Tan et al. (2019)
OsLCD	Low cadmium	Cytoplasm and nucleus	Cd accumulation	Shimo et al. (2011)
OsLCT1	Low affinity cation transporter	Plasma-membrane	Cd and Ca efflux in phloem	Uraguchi et al. (2011)
OsPCR1	Plant cadmium tolerance	Plasma-membrane	Cd and Zn transport	Song et al. (2015) and Wang et al. (2015)
OsCCX2/ OsCDT1	Cation/calcium exchanger	Plasma- membrane	Cd, Ca efflux	Hao et al. (2018) and Kuramata et al. (2009)
OsABCG43	ATP-binding cassette transporter (ABC transporter)	N/A	Cd compartment?	Oda et al. (2011)
OsABCG36	ABC transporter	Plasma-membrane	Cd efflux	Fu et al. (2019)
OsCd1	Major facilitator superfamily (MFS)	Plasma-membrane	Cd uptake and transport	Yan et al. (2019)
CAL1	Defensin-like protein	Cell wall	Cd-chelation, accmulation in leaf	Luo et al. (2018)
OsMTI-1b	Metallothionein-like protein			Ansarypour and Shahpiri (2017)
RCS1	Cysteine synthase	Cytoplasm	Cd tolerance	Harada et al. (2001)
OsPCS5	Phytochelatin synthase	Cytoplasm	Cd tolerance	Park et al. (2019)
OsPCS15	Phytochelatin synthase	Cytoplasm	Cd tolerance	Park et al. (2019)
OsPCS1	Phytochelatin synthase	Cytoplasm	Cd and As tolerance and mitigation	Das et al. (2017), Yamazaki et al. (2018), Uraguchi et al. (2017)
OsPCS2	Phytochelatin synthase	Cytoplasm	Cd and As tolerance and mitigation	Das et al. (2017), Uraguchi et al. (2017), and Yamazaki et al. (2018)

(Continued)

Table 7.3 (Continued)

Gene	Gene family	Location	Function	References
OsMTP1	Metal tolerance protein	Tonoplast	Cd and Zn tolerance and accumulation	Das et al. (2016), Menguer et al. (2013), and Yuan et al. (2012)
OsHIR1	E3 ligase	Plasma-membrane, nucleus	Regulate Cd and As uptake	Lim et al. (2014)
OsCLT1	CRT-like transporter 1	Plastid	Cd tolerance	Yang et al. (2016)
OsAUX1	Auxin transporter	Plasma-membrane	Cd stress response	Yu et al. (2015)
OVP1	H^+-pyrophosphatase	Tonoplast	Cd sequestration in vacuole	Cao et al. (2019a)
OsHB4	HD-Zip family protein	—	—	Ding et al. (2018)
OsTITANIA (LC5)	PHD-finger protein	Nucleus	Regulate Cd, Cu, Mn uptake	Nobuhiro et al. (2018)
miRNA166	microRNA	—	Regulate Cd	Ding et al. (2018)
miRNA390	microRNA	—	Regulate Cd	Ding et al. (2016)

expression was induced by high concentration of Cu, Zn, and Cd ions, and it was mainly expressed in vascular bundles and anthers. OsHMA9 is located on the plasma membrane and plays an important role in the outward transport of Cd, Zn, and Pb from cells (Lee et al. 2007).

Many ZIP family members were identified to mediate Cd uptake and translocation, such as OsZIP1 (Chou et al. 2011; Liu et al. 2019; Ramesh 2003), OsZIP3 (Ramesh 2003), OsZIP6 (Kavitha et al. 2015), and OsZIP7 (Tan et al. 2019). OsZIP1 functions as a metal-detoxified transporter to mediate excess Zn, Cu, and Cd accumulation in rice by activating its efflux transport activity (Chou et al. 2011; Liu et al. 2019; Ramesh 2003). OsZIP7 plays an integral role as an influx transporter in xylem loading in roots and inter-vascular transfer in nodes to preferentially deliver Zn and Cd to developing tissues and rice grains (Tan et al. 2019). The Fe^{2+} transporters OsIRT1 and OsIRT2, which belong to the ZIP family, transport Cd as well. Fe deficiency enhances OsIRT1-/OsIRT2-mediated Cd uptake and translocation (Nakanishi et al. 2006). OsIRT1-overexpressing rice showed increased sensitivity to excess Zn and Cd (Lee and An 2009).

OsLCT1 (low-affinity cation transporter, LCT) is localized on the plasma membrane and has been shown to be involved in the transportation of Cd in the aboveground part of the plant (Uraguchi et al. 2011). OsLCT1 mainly altered the transport of Cd from phloem to grain in the ganglion.

OsMTP1/OZT1 encodes rice metal tolerance protein. The normal expression of OsMPT1 can endow plants with tolerance to Zn and Cd. OsMPT1 plays an important role in the transport of Zn and Cd and the maintenance of their homeostasis (Lan et al. 2013).

OsCd1, belonging to the major facilitator superfamily, is involved in root Cd uptake and contributes to grain accumulation in rice (Yan et al. 2019).

Interestingly, in some rice cultivar accessions, a single amino acid mutation can change a transporter's Cd-transport activity. For example, a natural variation of OsHMA3 with a single amino acid mutation from Ser to Arg at the 380th position resulted in high Cd accumulation in some *japonica* cultivars (Yan et al. 2016). In the *indica* 9311 cultivar, a single nucleotide polymorphism (SNP) mutation with a single amino acid substitution (Pro236Leu) in OsNRAMP5 resulted in a significant decrease of Cd/Mn-uptake from the external environment (Cao et al. 2019c). Another natural variation in OsCd1 with a missense mutation (Val449Asp) is responsible for the divergence of rice grain Cd accumulation between *indica* and *japonica* (Yan et al. 2019).

Genes Involved in Cd Chelation

Phytochelatins (PCs) are a group of thiol rich small peptides with strong Cd-chelating activity, which are synthesized by phytochelatin synthase (PCS). PCs are characterized by their general structure of $(\gamma\text{-Glu-Cys})n\text{-Gly}(n=2\text{-}11)$, and form complexes with Cd ions at the free thiol group and subsequently sequester the bound Cd inside the vacuole (Grill et al. 1987, 1989). The PCS catalyzes the non-ribosomal synthesis of PCs from glutathiones (GSH), which plays an important role in Cd-chelation and detoxification (Rauser 1995; Zenk 1995). It was previously reported that seed-specific down-regulation of both OsPCS1 and OsPCS2 resulted in significant reduction in the content of Cd and As in grains compared with the control, indicating that PC synthesis contributes to Cd and As accumulation in rice grain

(Das et al. 2017; Li et al. 2007; Uraguchi et al. 2017). The precursors for PC synthesis, such as GSH, glutamate, or cysteine, also play crucial roles in Cd tolerance. The expression changes of the synthase genes for these precursors altered Cd tolerance (Harada et al. 2001; Jiang et al. 2019; Yamazaki et al. 2018).

In addition, heterologous expression of rice metallothionein genes in yeast or bacteria enhanced Cd tolerance, suggesting rice metallothionein may be involved in Cd-chelating (Ansarypour and Shahpiri 2017; Pirzadeh and Shahpiri 2016).

Recently, a defensin-like protein CAL1 has been shown to alter the chelation of Cd and secretion of Cd to the extracellular space, hence lowering cytosolic Cd concentration while enhancing long-distance Cd transport via xylem vessels. CAL1 has been suggested to specifically regulate Cd translocation from roots to shoots via the xylem, but not from shoots to grains via the phloem, leading to leaf but not rice grain Cd accumulation. Therefore, it provides a potential target for breeding dual-function rice cultivars that produce Cd-safe grains while remediating paddy soils (Luo et al. 2018).

Regulating Genes to Mediate Grain Cd Accumulation in Rice

Although many key transporters are responsible for Cd uptake and translocation, little is known on how rice plants respond to regulate Cd transport activities. Transcriptional regulation of gene expression is one of the effective methods of controlling Cd transport activities, where transcriptional regulators and microRNAs (miRNAs) play important roles. The transcriptional regulator OsTITANIA, a rice PHD-finger protein, functions through elevating expression of transporter genes for multiple metals. The mutant *LC5* exhibited a low-Cd accumulation trait (Nobuhiro et al. 2018). Another transcriptional regulator, OsHB4, is involved in the regulation of Cd accumulation as over-expression of *OsHB4* increased Cd sensitivity and Cd accumulation in the leaves and grains. However, knockdown of *OsHB4* enhanced Cd tolerance (Ding et al. 2018). The miR166 targets *OsHB4* to degrade *OsHB4* transcripts, and mitigate Cd accumulation (Ding et al. 2018). Another miRNA, miR390, is also involved in Cd tolerance and accumulation in rice. *OsSRK* is the target of miR390, which was up-regulated by Cd exposure. Over-expressing miR390 displayed reduced Cd tolerance and higher Cd accumulation, with increased levels of OsNRAMP5 and OsHMA2, indicating that miR390 was a negative regulator of Cd stress tolerance in rice (Ding et al. 2016).

The ubiquitin-mediated 26S proteasome degradation is an important mechanisms in post-translational regulation. The rice RING E3 ligase OsHIR1 is involved in As and Cd uptake. The RING E3 ubiquitin ligase OsHIR1interacts with and degrades OsTIP4;1 transporter via the ubiquitin 26S proteasome system. Over-expression of OsHIR1 causes tolerance and significantly reduces the accumulation of Cd and As, indicating that the OsHIR1 E3 ligase positively regulates OsTIP4;1, impacting Cd uptake (Lim et al. 2014).

Disruption of *OsLCD* leads to significant decreases in grain Cd accumulation (Shimo et al. 2011). However, the molecular mechanism of this gene in Cd mediation remains unknown. As OsLCD protein has neither transmembrane domains, nor cysteine-rich elements, it has been suggested that it functions as an adaptor to regulate and interact with some Cd-transporters. Future screening and identification of OsLCD-interacted transporter(s) will aid in the understanding of mechanism behind grain Cd accumulation.

Molecular Breeding for Low-Cd Accumulation in Rice Grains

Screening and Identification of Low-Cd Varieties

Efficient screening and identification for genotyping rice varieties with different Cd accumulation traits have been performed (Arao and Ae 2003; Chi et al. 2018; Duan et al. 2017; Li et al. 2014a, 2012a, 2019; Liu et al. 2003, 2005; Shi et al. 2009; Xie et al. 2017; Zeng et al. 2008; Zhao et al. 2018). Many varieties were reported as low-Cd or high-Cd types (Table 7.4), some of which showed Cd hyperaccumulation traits, such as Cho-ko-koku, Akita 110, TJTT8, and therefore, can be applied for phytoremediation of the polluted soils (Abe et al. 2017; Murakami et al. 2009; Takahashi et al. 2016). High-Cd *japonica* cultivars with loss-of-function of OsHMA3 (temperate *japonica*) or OsNRAMP2 (tropical *japonica*) were identified (Yan et al. 2016; Zhao et al. 2018), while some low-Cd *indica* cultivars were closely related to the function of OsNRAMP5, OsNRAMP2, or OsHMA3 (Cao et al. 2019c; Lu et al. 2019; Wang et al. 2019d).

Molecular Marker-Assisted Low-Cd Gene Aggregation Breeding

Recently, many crucial genes responsible for rice Cd uptake and translocation have been mapped and cloned, and the molecular markers for OsHMA3, OsNRAMP5, OsNRAMP1, OsNRAMP2, LCD have been developed. The *OsNRAMP5* dysfunctional alleles (*lcd1*, *lcd-kmt2*, *LCH3*) have been suggested to significantly decrease Cd absorption by roots, while not impacting Mn uptake and yield if the mutant position occurred near 3′ terminus of the gene (Cao et al. 2019c; Sasaki et al. 2012; Tang et al. 2017; Wang et al. 2019d). In addition, the *japonica* OsHMA3 is functional and could sequester Cd into the root vacuole (Lu et al. 2019; Ueno et al. 2010). These findings have shown the potential for low-Cd rice breeding by aggregating multiple low-Cd genes into one variety. By crossing with low-Cd cultivars and successive backcrossing, many main rice varieties have been greatly improved regarding grain Cd accumulation traits. For hybrid rice breeding, the male sterile line and restorer line were first developed as low-Cd parent lines; then the low-Cd hybrid seeds were obtained by crossing (Hao et al. 2019; Li et al. 2016; Tang et al. 2017). With enhanced understanding of the mechanisms behind rice grain Cd accumulation, commercial low-Cd rice breeding has already commenced, using the low-Cd gene aggregation approach. For example, the super rice JinliangyouHuazhan has shown grain low-Cd traits by aggregating functional *japonica* OsHMA3 with dysfunctional OsNRAMP5 and OsNRAMP1.

Gene Editing-Based Molecular Breeding

CRISPR/Cas9 (Clustered Regularly Interspaced Short Palindromic Repeats/CRISPR-associated Cas9) -based genome editing technology could genetically modify specific genes for the advantages of simplicity, efficiency, and versatility. Furthermore, non-transgenic-element-containing lines could be selected from the offspring of the genetically modified plants. The CRISPR/Cas9 system has been successfully used for low-Cd rice genetic improvement (Tang et al. 2017; Wang et al. 2019d).

Table 7.4 Varieties/lines of different Cd accumulation genotype.

Varieties/lines	Description	Genotype	Major gene	References
Cho-ko-koku	*Indica*, Japan	High-Cd	OsHMA3	Murakami et al. (2009)
Anjana Dhan	*Indica*, Japan	High-Cd	OsHMA3	Ueno et al. (2010)
Jarjan	*Indica*, Japan	High-Cd	OsHMA3	Ueno et al. (2011)
Akita 110	*Indica*, Japan	High-Cd	OsHMA3	Takahashi et al. (2016)
TJTT8	*Indica* with *japonica* consanguinity, Japan	High-Cd	OsHMA3	Abe et al. (2017)
YELICANGHUA	Temperate *japonica*, China	High-Cd	OsHMA3	Yan et al. (2016)
TOX 2104-2-1	Tropical *japonica*, Nigeria	High-Cd	OsNRAMP2-H	Zhao et al. (2018)
BAKUNG (H)	Tropical *japonica*, Malaysia	High-Cd	OsNRAMP2-H	Zhao et al. (2018)
Koshihikari Kan No. 1	*Japonica*, Japan	Low-Cd	OsNRAMP5	Ishimaru et al. (2012a)
BOL ZO	*Indica*, Korea	Low-Cd	OsNRAMP2-L	Zhao et al. (2018)
CHERIVIRUPPU	*Indica*, India	Low-Cd	OsNRAMP2-L	Zhao et al. (2018)
YaHui2816	*Indica*, China	Low-Cd	_	Guo et al. (2019), Zhang et al. (2013)
JinliangyouHuazhan	*Indica* super rice, China	Low-Cd	Gene aggregation	Longping HighTech
lcd1 line	Indica (9311), EMS mutagenesis, China	Low-Cd	OsNRAMP5	Cao et al. (2019c)
LCH3 line	*Indica* (Huanghuazhan), gene edition, China	Low-Cd	OsNRAMP5	Wang et al. (2019d)

Conclusions and Perspectives

Cd exposure could lead to a series of detrimental effects on organisms, including humans. Therefore, it is crucial to understand the mechanisms of Cd uptake, translocation, and accumulation in rice in order to reinforce efficient strategies to reduce Cd in rice. In the past two decades, advances have been made in understanding Cd transport and environmental factors impacting Cd accumulation as several genes that mediate Cd transport have been identified in rice. Furthermore, the influence and mechanisms of main environmental factors on rice Cd accumulation have been clarified, and strategies to alleviate grain Cd accumulation have been developed.

Many field experiments have shown that low-Cd rice varieties, combined with irrigation during the whole growth period of rice, and higher pH regulation could significantly decrease grain Cd accumulation. Some rhizospheric microorganisms also play an important role in alleviating rice Cd uptake and translocation. In addition, many other agronomic measures, such as application of various foliar compound fertilizers, biochar, acid-base modifiers, could also be effective in reducing grain Cd accumulation. For low-Cd variety breeding, four possible sites in rice could be considered to block grain Cd accumulation. Firstly, blocking root Cd-uptake by eliminating Cd influx transport or enhancing efflux transport in the root membranes. Secondly, blocking Cd in the root tissues by sequestrating absorbed Cd to vacuoles or other organelles of root cells. Thirdly, blocking shoot Cd content by inhibiting root-to-shoot translocation via the xylem. Finally, blocking grain Cd accumulation by cutting Cd translocation from leaves and culm via phloem sap. By low-Cd gene aggregation breeding, highly efficient low-Cd varieties could be developed to produce Cd-safe rice under Cd-contaminated paddy soils.

Acknowledgments

Authors acknowledge Mr. Zijing Xie (Hunan Normal University) for help in editing the text during the preparation of this chapter.

References

Abe, T., Nonoue, Y., Ono, N. et al. (2013). Detection of QTLs to reduce cadmium content in rice grains using LAC23/Koshihikari chromosome segment substitution lines. *Breed. Sci.* 63: 284–291.

Abe, T., Taguchi-Shiobara, F., Kojima, Y. et al. (2011). Detection of a QTL for accumulating Cd in rice that enables efficient Cd phytoextraction from soil. *Breed. Sci.* 61: 43–51.

Abe, T., Ito, M., Takahashi, R. et al. (2017). Breeding of a practical rice line "TJTT8" for phytoextraction of cadmium contamination in paddy fields. *Soil Sci. Plant Nutr.* 63: 388–395.

Ansarypour, Z. and Shahpiri, A. (2017). Heterologous expression of a rice metallothionein isoform (*OsMTI-1b*) in *Saccharomyces cerevisiae* enhances cadmium, hydrogen peroxide and ethanol tolerance. *Braz. J. Microbiol.* 48: 537–543.

Arao, T. and Ae, N. (2003). Genotypic variations in cadmium levels of rice grain. *Soil Sci. Plant Nutr.* 49: 473–479.

Behera, M., Dandapat, J., and Rath, C.C. (2014). Effect of heavy metals on growth response and antioxidant defense protection in *Bacillus cereus*. *J. Basic Microbiol.* 54: 1201–1209.

Belouchi, A., Kwan, T., and Gros, P. (1997). Cloning and characterization of the *OsNramp* family from *Oryza sativa*, a new family of membrane proteins possibly implicated in the transport of metal ions. *Plant Mol. Biol.* 33: 1085–1092.

Bermudez, G.M.A., Jasan, R., Plá, R. et al. (2012). Heavy metals and trace elements in atmospheric fall-out: their relationship with topsoil and wheat element composition. *J. Hazard. Mater.* 213–214: 447–456.

Bingham, F.T., Page, A.L., Mahler, R.J. et al. (1976). Cadmium availability to rice in sludge-amended soil under flood and non-flood culture. *Soil Sci. Soc. Am. J.* 40: 715–719.

Buha, A., Wallace, D., Matovic, V. et al. (2017). Cadmium exposure as a putative risk factor for the development of pancreatic cancer: three different lines of evidence. *Biomed Res. Int.* 2017: 1981837.

Cao, Z.Z., Qin, M.L., Lin, X.Y. et al. (2018). Sulfur supply reduces cadmium uptake and translocation in rice grains (*Oryza sativa* L.) by enhancing iron plaque formation, cadmium chelation and vacuolar sequestration. *Environ. Pollut.* 238: 76–84.

Cao, F., Dai, H., Hao, P.F. et al. (2019a). Silicon regulates the expression of vacuolar H(+)-pyrophosphatase 1 and decreases cadmium accumulation in rice (*Oryza sativa* L.). *Chemosphere* 240: 124907.

Cao, X., Tan, C., Wu, L. et al. (2019b). Atmospheric deposition of cadmium in an urbanized region and the effect of simulated wet precipitation on the uptake performance of rice. *Sci. Total Environ.* 700: 134513.

Cao, Z.Z., Lin, X.Y., Yang, Y.J. et al. (2019c). Gene identification and transcriptome analysis of low cadmium accumulation rice mutant (*lcd1*) in response to cadmium stress using MutMap and RNA-seq. *BMC Plant Biol.* 19: 250.

Cellier, M., Prive, G., Belouchi, A. et al. (1995). Nramp defines a family of membrane proteins. *Proc. Natl. Acad. Sci. U.S.A.* 92: 10089–10093.

Cesaratto, A., D'Andrea, C., Nevin, A. et al. (2013). Analysis of cadmium-based pigments with time-resolved photoluminescence. *Anal. Methods-UK* 6: 130–138.

Chakraborty, J. and Das, S. (2014). Characterization and cadmium-resistant gene expression of biofilm-forming marine bacterium *Pseudomonas aeruginosa* JP-11. *Environ. Sci. Pollut. Res.* 21: 14188–14201.

Chen, H., Zhang, W., Yang, X. et al. (2018a). Effective methods to reduce cadmium accumulation in rice grain. *Chemosphere* 207: 699–707.

Chen, X., Younan, O., Yicong, F. et al. (2018b). The pathway of transmembrane cadmium influx via calcium-permeable channels and its spatial characteristics along rice root. *J. Exp. Bot.* 69: 5279–5291.

Chen, D., Chen, D., Xue, R. et al. (2019a). Effects of boron, silicon and their interactions on cadmium accumulation and toxicity in rice plants. *J. Hazard. Mater.* 367: 447–455.

Chen, J., Zou, W., Meng, L. et al. (2019b). Advances in the uptake and transport mechanisms and QTLs mapping of cadmium in rice. *Int. J. Mol. Sci.* 20: 3417.

Chi, Y., Li, F., Tam, N.F.-y. et al. (2018). Variations in grain cadmium and arsenic concentrations and screening for stable low-accumulating rice cultivars from multi-environment trials. *Sci. Total Environ.* 643: 1314–1324.

Chou, T.S., Chao, Y.-Y., Huang, W.-D. et al. (2011). Effect of magnesium deficiency on antioxidant status and cadmium toxicity in rice seedlings. *J. Plant Physiol.* 168: 1021–1030.

Converse, W.E., Lloyd, R.C., Clayton, E.D. et al. (1987). Critical experiments using high-enriched uranyl nitrate with cadmium absorber. *Nucl. Technol.* 79: 82–91.

Curie, C., Alonso, J.M., Jean, M.L. et al. (2000). Involvement of NRAMP1 from *Arabidopsis thaliana* in iron transport. *Biochem. J.* 3: 749–755.

Das, N., Bhattacharya, S., Bhattacharyya, S. et al. (2017). Identification of alternatively spliced transcripts of rice phytochelatin synthase 2 gene *OsPCS2* involved in mitigation of cadmium and arsenic stresses. *Plant Mol. Biol.* 94: 167–183.

Das, N., Bhattacharya, S., and Maiti, M.K. (2016). Enhanced cadmium accumulation and tolerance in transgenic tobacco overexpressing rice metal tolerance protein gene *OsMTP1* is promising for phytoremediation. *Plant. Physiol. Bioch.* 105: 297–309.

Ding, Y., Ye, Y., Jiang, Z. et al. (2016). MicroRNA390 is involved in cadmium tolerance and accumulation in rice. *Front. Plant Sci.* 7: 235.

Ding, Y., Shaohua, G., Yi, W. et al. (2018). MicroRNA166 modulates cadmium tolerance and accumulation in rice. *Plant Physiol.* 177: 1691–1703.

Duan, G., Shao, G., Tang, Z. et al. (2017). Genotypic and environmental variations in grain cadmium and arsenic concentrations among a panel of high yielding rice cultivars. *Rice* 10: 9.

Fan, D. (1991). A mathematical model of a sealed nickel-cadmium battery. *J. Electrochem. Soc.* 138: 17–25.

Farooq, M.A., Detterbeck, A., Clemens, S. et al. (2016). Silicon-induced reversibility of cadmium toxicity in rice. *J. Exp. Bot.* 67: 3573–3585.

Feki-Tounsi, M. and Hamza-Chaffai, A. (2014). Cadmium as a possible cause of bladder cancer: a review of accumulated evidence. *Environ. Sci. Pollut. Res.* 21: 10561–10573.

Fu, S., Lu, Y., Zhang, X. et al. (2019). The ABC transporter OsABCG36 is required for Cd tolerance in rice. *J. Exp. Bot.* 70: 5909–5918.

Gao, H., Xie, W., Yang, C. et al. (2018). NRAMP2, a trans-Golgi network-localized manganese transporter, is required for Arabidopsis root growth under manganese deficiency. *New Phytol.* 217: 179–193.

Goyer, R.A., Liu, J., and Waalkes, M.P. (2004). Cadmium and cancer of prostate and testis. *Biometals* 17: 555–558.

Grill, E., Winnacker, E.-L., and Zenk, M.H. (1987). Phytochelatins, a class of heavy-metal-binding peptides from plants, are functionally analogous to metallothioneins. *Proc. Natl. Acad. Sci. U.S.A.* 84: 439–443.

Grill, E., Lffler, S., Winnacker, E.-L. et al. (1989). Phytochelatins, the heavy-metal-binding peptides of plants, are synthesized from glutathione by a specific γ-glutamylcysteine dipeptidyl transpeptidase (phytochelatin synthase). *Proc. Natl. Acad. Sci. U.S.A.* 86: 6838–6842.

Guo, W., Zhao, R., Zhao, W. et al. (2013). Effects of arbuscular mycorrhizal fungi on maize (*Zea mays* L.) and sorghum (*Sorghum bicolor* (L.) Moench) grown in rare earth elements of mine tailings. *Appl. Soil Ecol.* 72: 85–92.

Guo, J., Li, K., Zhang, X. et al. (2019). Genetic properties of cadmium translocation from straw to brown rice in low-grain cadmium rice (*Oryza sativa* L.) line. *Ecotoxicol. Environ. Saf.* 182: 109422.

Hallenbeck, W.H. (1984). Human health effects of exposure to cadmium. *Experientia* 40: 136–142.

Hansen, J.C., Grøn, P., Jespersen, B.A. et al. (1989). Cadmium exposure in Denmark. Based on analyses of liver and kidney tissues. *Dan. Med. Bull.* 36: 499–502.

Hao, C., Wang, M., Ming, H.W. et al. (2014). Does radial oxygen loss and iron plaque formation on roots alter Cd and Pb uptake and distribution in rice plant tissues. *Plant Soil* 375: 137–148.

Hao, X., Zeng, M., Wang, J. et al. (2018). A node-expressed transporter OsCCX2 is involved in grain cadmium accumulation of rice. *Front. Plant Sci.* 9: 476.

Hao, X., Wu, C., Wang, R. et al. (2019). Association between sequence variants in cadmium-related genes and the cadmium accumulation trait in thermo-sensitive genic male sterile rice. *Breed. Sci.* 69: 455–463.

Harada, E., Choi, Y.-E., Tsuchisaka, A. et al. (2001). Transgenic tobacco plants expressing a rice cysteine synthase gene are tolerant to toxic levels of cadmium. *J. Plant Physiol.* 158: 655–661.

Hindersmann, I. and Mansfeldt, T. (2014). Trace element solubility in a multimetal-contaminated soil as affected by redox conditions. *Water Air Soil Pollut.* 225: 2158.

Honma, T., Ohba, H., Kaneko-Kadokura, A. et al. (2016). Optimal soil Eh, pH, and water management for simultaneously minimizing arsenic and cadmium concentrations in rice grains. *Environ. Sci. Techol.* 50: 4178–4185.

Horiguchi, H., Teranishi, H., Niiya, K. et al. (1994). Hypoproduction of erythropoietin contributes to anemia in chronic cadmium intoxication: clinical study on Itai-itai disease in Japan. *Arch. Toxicol.* 68: 632–636.

Hrynkiewicz, K., Dabrowska, G., Baum, C. et al. (2012). Interactive and single effects of Ectomycorrhiza formation and *Bacillus cereuson* Metallothionein MT1 expression and phytoextraction of Cd and Zn by willows. *Water Air Soil Pollut.* 223: 957–968.

Hu, P., Huang, J., Ouyang, Y. et al. (2013). Water management affects arsenic and cadmium accumulation in different rice cultivars. *Environ. Geochem. Health* 35: 767–778.

Hu, D., Sheng, Z., Li, Q. et al. (2018). Identification of QTLs associated with cadmium concentration in rice grains. *J. Integr. Agric.* 17: 1563–1573.

Huang, Y., Sun, C., Min, J. et al. (2015). Association mapping of quantitative trait loci for mineral element contents in whole grain rice (*Oryza sativa* L.). *J. Agric. Food. Chem.* 63: 10885–10892.

Huang, B., Xin, J., Dai, H. et al. (2017). Effects of interaction between cadmium (Cd) and selenium (Se) on grain yield and Cd and Se accumulation in a hybrid rice (*Oryza sativa*) system. *J. Agric. Food. Chem.* 65: 9537–9546.

Huang, Q., Xu, Y., Liu, Y. et al. (2018). Selenium application alters soil cadmium bioavailability and reduces its accumulation in rice grown in Cd-contaminated soil. *Environ. Sci. Pollut. Res. Int.* 25: 31175–31182.

Huff, T.B., Tong, L., Zhao, Y. et al. (2007). Hyperthermic effects of gold nanorods on tumor cells. *Nanomedicine* 2: 125–132.

Il'yasova, D. and Schwartz, G.G. (2005). Cadmium and renal cancer. *Toxicol. Appl. Pharmacol.* 207: 179–186.

Ishikawa, S., Ae, N., and Yano, M. (2005). Chromosomal regions with quantitative trait loci controlling cadmium concentration in brown rice (*Oryza sativa*). *New phytol.* 168: 345–350.

Ishikawa, S., Abe, T., Kuramata, M. et al. (2010). A major quantitative trait locus for increasing cadmium-specific concentration in rice grain is located on the short arm of chromosome 7. *J. Exp. Bot.* 61: 923–934.

Ishimaru, Y., Bashir, K., Nakanishi, H. et al. (2012a). OsNRAMP5, a major player for constitutive iron and manganese uptake in rice. *Plant Signaling Behav.* 7: 763–766.

Ishimaru, Y., Takahashi, R., Bashir, K. et al. (2012b). Characterizing the role of rice NRAMP5 in manganese, iron and cadmium transport. *Sci. Rep.* 2: 286.

Järup, L., Rogenfelt, A., Elinder, C.G. et al. (1983). Biological half-time of cadmium in the blood of workers after cessation of exposure. *Scand. J. Work Environ. Health* 9: 327–331.

Jiang, M., Jiang, J., Li, S. et al. (2019). Glutamate alleviates cadmium toxicity in rice via suppressing cadmium uptake and translocation. *J. Hazard. Mater.* 28: 121319.

Jiao, W., Chen, W., Chang, A.C. et al. (2012). Environmental risks of trace elements associated with long-term phosphate fertilizers applications: a review. *Environ. Pollut.* 168: 44–53.

Jie, Z., Sun, W., Li, Z. et al. (2009). Cadmium fate and tolerance in rice cultivars. *Agron. Sustainable Dev.* 29: 483–490.

Jomova, K. and Valko, M. (2011). Advances in metal-induced oxidative stress and human disease. *Toxicology* 283: 65–87.

Kashiwagi, T., Shindoh, K., Hirotsu, N. et al. (2009). Evidence for separate translocation pathways in determining cadmium accumulation in grain and aerial plant parts in rice. *BMC Plant Biol.* 9: 8.

Kautek, W. (1988). The galvanic corrosion of steel coatings: aluminum in comparison to cadmium and zinc. *Corros. Sci.* 28: 173–199.

Kavitha, P.G., Kuruvilla, S., and Mathew, M.K. (2015). Functional characterization of a transition metal ion transporter, OsZIP6 from rice (*Oryza sativa* L.). *Plant Physiol. Biochem.* 97: 165–174.

Kensuke, T., Shu, F., Toru, F. et al. (2007). Quantitative estimation of the contribution of the phloem in cadmium transport to grains in rice plants (*Oryza sativa* L.). *Soil Sci. Plant Nutr.* 53: 72–77.

Khan, M.U., Malik, R.N., and Muhammad, S. (2013). Human health risk from heavy metal via food crops consumption with wastewater irrigation practices in Pakistan. *Chemosphere* 93: 2230–2238.

Kikuchi, T., Okazaki, M., Kimura, S.D. et al. (2008). Suppressive effects of magnesium oxide materials on cadmium uptake and accumulation into rice grains II: suppression of cadmium uptake and accumulation into rice grains due to application of magnesium oxide materials. *J. Hazard. Mater.* 154: 294–299.

Kirkham, M.B. (2006). Cadmium in plants on polluted soils: effects of soil factors, hyperaccumulation, and amendments. *Geoderma* 137: 19–32.

Kuramata, M., Masuya, S., Takahashi, Y. et al. (2009). Novel cysteine-rich peptides from *Digitaria ciliaris* and *Oryza sativa* enhance tolerance to cadmium by limiting its cellular accumulation. *Plant Cell Physiol.* 50: 106–117.

Lan, H., Wang, Z., Wang, Q. et al. (2013). Characterization of a vacuolar zinc transporter OZT1 in rice (*Oryza sativa* L.). *Mol. Biol. Rep.* 40: 1201–1210.

Larsson, L.G. and Möller, E. (1968). Measurement of the decay of thermal neutrons in aqueous solutions of the non-1/v neutron absorber cadmium. *Nucl. Sci. Eng.* 33: 218–224.

Lech, T. and Sadlik, J.K. (2017). Cadmium concentration in human autopsy tissues. *Biol. Trace Elem. Res.* 179: 172–177.

Lee, S. and An, G. (2009). Over-expression of *OsIRT1* leads to increased iron and zinc accumulations in rice. *Plant Cell Environ.* 32: 408–416.

Lee, S., Kim, Y.-Y., Lee, Y. et al. (2007). Rice P-1B-type heavy-metal ATPase, OsHMA9, is a metal efflux protein. *Plant Physiol.* 145: 831–842.

Li, Z., Zhang, Y., Pan, G. et al. (2003). Grain contents of Cd, Cu and Se by 57 rice cultivars and the risk significance for human dietary uptake. *Huan Jing Ke Xue* 24: 112–115.

Li, J.C., Guo, J.B., Xu, W.Z. et al. (2007). RNA interference-mediated silencing of phytochelatin synthase gene reduce cadmium accumulation in rice seeds. *J. Integr. Plant Biol.* 49: 1032–1037.

Li, B., Wang, X., Qi, X. et al. (2012a). Identification of rice cultivars with low brown rice mixed cadmium and lead contents and their interactions with the micronutrients iron, zinc, nickel and manganese. *J. Environ. Sci.* 2012: 70–78.

Li, L., Zhou, W., Dai, H. et al. (2012b). Selenium reduces cadmium uptake and mitigates cadmium toxicity in rice. *J. Hazard. Mater.* 235–236: 343–351.

Li, B., Du, R., Wen, D. et al. (2014a). Difference of heavy metal contents in rice of main cultivated rice varieties in Guangdong province. *Chin. J. Trop. Agric.* 34: 5–10.

Li, J.Y., Liu, J., Dong, D. et al. (2014b). Natural variation underlies alterations in Nramp aluminum transporter (NRAT1) expression and function that play a key role in rice aluminum tolerance. *Proc. Natl. Acad. Sci. U.S.A.* 111: 6503–6508.

Li, H., Luo, N., Zhang, L.J. et al. (2016). Do arbuscular mycorrhizal fungi affect cadmium uptake kinetics, subcellular distribution and chemical forms in rice. *Sci. Total Environ.* 571: 1183–1190.

Li, T., Chang, Q., Yuan, X. et al. (2017a). Cadmium transfer from contaminated soils to the human body through rice consumption in southern Jiangsu province, China. *Environ. Sci. Process Impacts* 19: 843–850.

Li, Y., Pang, H.-D., He, L.-Y. et al. (2017b). Cd immobilization and reduced tissue Cd accumulation of rice (*Oryza sativa* wuyun-23) in the presence of heavy metal-resistant bacteria. *Ecotoxicol. Environ. Saf.* 138: 56–63.

Li, D., Wang, L., Wang, Y. et al. (2019). Soil properties and cultivars determine heavy metal accumulation in rice grain and cultivars respond differently to Cd stress. *Environ. Sci. Pollut. Res. Int.* 26: 14638–14648.

Liao, G., Xu, Y., Chen, C. et al. (2016). Root application of selenite can simultaneously reduce arsenic and cadmium accumulation and maintain grain yields, but show negative effects on the grain quality of paddy rice. *J. Environ. Manage.* 183 (pt3): 733–741.

Lim, S.D., Hwang, J.G., Han, A.R. et al. (2014). Positive regulation of rice RING E3 ligase OsHIR1 in arsenic and cadmium uptakes. *Plant Mol. Biol.* 85: 365–379.

Lin, X., Mou, R., Cao, Z. et al. (2016). Characterization of cadmium-resistant bacteria and their potential for reducing accumulation of cadmium in rice grains. *Sci. Total Environ.* 569–570: 97–104.

Lin, X., Zhou, L., Wang, H. et al. (2018). Accumulation of heavy metals in different rice varieties. *Huan Jing Ke Xue* 39: 5198–5206.

Liu, J.G., Liang, J.S., Li, K.Q. et al. (2003). Correlations between cadmium and mineral nutrients in absorption and accumulation in various genotypes of rice under cadmium stress. *Chemosphere* 52: 1467–1473.

Liu, J., Zhu, Q., Zhang, Z. et al. (2005). Variations in cadmium accumulation among rice cultivars and types and the selection of cultivars for reducing cadmium in the diet. *J. Sci. Food Agric.* 85: 147–153.

Liu, J., Qian, M., Cai, G. et al. (2007). Uptake and translocation of Cd in different rice cultivars and the relation with Cd accumulation in rice grain. *J. Hazard. Mater.* 143: 443–447.

Liu, J., Qu, W., and Kadiiska, M.B. (2009). Role of oxidative stress in cadmium toxicity and carcinogenesis. *Toxicol. Appl. Pharmacol.* 238: 209–214.

Liu, J., Cao, C., Wong, M. et al. (2010). Variations between rice cultivars in iron and manganese plaque on roots and the relation with plant cadmium uptake. *J. Environ. Sci.* 22: 1067–1072.

Liu, F., Liu, X., Ding, C. et al. (2015). The dynamic simulation of rice growth parameters under cadmium stress with the assimilation of multi-period spectral indices and crop model. *Field Crop Res.* 183: 225–234.

Liu, S., Li, H., Han, C. et al. (2017). Cd inhibition and pH improvement via a nano-submicron mineral-based soil conditioner. *Environ. Sci. Pollut. Res.* 24: 4942–4949.

Liu, Y.L., Tie, B.Q., Li, Y.X. et al. (2018). Inoculation of soil with Cd-resistant bacterium *Delftia* sp. B9 reduces Cd accumulation in rice (*Oryza sativa* L.) grains. *Ecotox. Environ. Safe.* 163: 223–229.

Liu, J., Hou, H., Zhao, L. et al. (2019). Mitigation of Cd accumulation in rice from Cd-contaminated paddy soil by foliar dressing of S and P. *Sci. Total Environ.* 690: 321–328.

Livera, J.D., McLaughlin, M.J., Hettiarachchi, G.M. et al. (2011). Cadmium solubility in paddy soils: effects of soil oxidation, metal sulfides and competitive ions. *Sci. Total Environ.* 409: 1489–1497.

Lu, C., Zhang, L., Tang, Z. et al. (2019). Producing cadmium-free indica rice by overexpressing *OsHMA3*. *Environ. Int.* 126: 619–626.

Luo, J.-S., Huang, J., Zeng, D.-L. et al. (2018). A defensin-like protein drives cadmium efflux and allocation in rice. *Nat. Commun.* 9: 645.

Lux, A., Martinka, M., Vaculík, M. et al. (2011). Root responses to cadmium in the rhizosphere: a review. *J. Exp. Bot.* 62: 21–37.

Lv, G., Wang, H., Xu, C. et al. (2019). Effectiveness of simultaneous foliar application of Zn and Mn or P to reduce Cd concentration in rice grains: a field study. *Environ. Sci. Pollut. Res. Int.* 26: 9305–9313.

Mao, P., Zhuang, P., Li, F. et al. (2019). Phosphate addition diminishes the efficacy of wollastonite in decreasing Cd uptake by rice (*Oryza sativa* L.) in paddy soil. *Sci. Total Environ.* 687: 441–450.

Menguer, P.K., Farthing, E., Peaston, K.A. et al. (2013). Functional analysis of the rice vacuolar zinc transporter OsMTP1. *J. Exp. Bot.* 64: 2871–2883.

Murakami, M., Nakagawa, F., Ae, N. et al. (2009). Phytoextraction by rice capable of accumulating Cd at high levels: reduction of Cd content of rice grain. *Environ. Sci. Technol.* 43: 5878–5883.

Nakanishi, H., Ogawa, I., Ishimaru, Y. et al. (2006). Iron deficiency enhances cadmium uptake and translocation mediated by the Fe^{2+} transporters OsIRT1 and OsIRT2 in rice. *Soil Sci. Plant Nutr.* 52: 464–469.

Narteh, L.T. and Sahrawat, K.L. (1999). Influence of flooding on electrochemical and chemical properties of West African soils. *Geoderma* 87: 179–207.

Nobuhiro, T., Shimpe,i, U., Masataka, K. et al. (2018). A rice PHD-finger protein OsTITANIA, is a growth regulator that functions through elevating expression of transporter genes for multiple metals. *Plant J.* 96: 997–1006.

Nookabkaew, S., Rangkadilok, N., Prachoom, N. et al. (2016). Concentrations of trace elements in organic fertilizers and animal manures and feeds and cadmium contamination in herbal tea (*Gynostemma pentaphyllum* Makino). *J. Agric. Food. Chem.* 64: 3119–3126.

Norton, G.J., Deacon, C.M., Xiong, L. et al. (2010). Genetic mapping of the rice ionome in leaves and grain: identification of QTLs for 17 elements including arsenic, cadmium, iron and selenium. *Plant Soil* 329: 139–153.

Nziguheba, G. and Smolders, E. (2008). Inputs of trace elements in agricultural soils via phosphate fertilizers in European countries. *Sci. Total Environ.* 390: 53–57.

O'Neill, M.A., Ishii, T., Albersheim, P. et al. (2004). Rhamnogalacturonan II: structure and function of a borate cross-linked cell wall pectic polysaccharide. *Annu. Rev. Plant Biol.* 55: 109–139.

Oda, K., Otani, M., Uraguchi, S. et al. (2011). Rice ABCG43 is Cd inducible and confers Cd tolerance on yeast. *Biosci. Biotechnol., Biochem.* 75: 1211–1213.

Pan, Y., Bonten, L.T.C., Koopmans, G.F. et al. (2016). Solubility of trace metals in two contaminated paddy soils exposed to alternating flooding and drainage. *Geoderma* 261: 59–69.

Park, H.C., Hwang, J.E., Jiang, Y. et al. (2019). Functional characterisation of two phytochelatin synthases in rice (*Oryza sativa* cv. Milyang 117) that respond to cadmium stress. *Plant Biol.* 21: 854–861.

Peris-Peris, C., Serra-Cardona, A., Sánchez-Sanuy, F. et al. (2017). Two NRAMP6 isoforms function as iron and manganese transporters and contribute to disease resistance in rice. *Mol. Plant Microbe Interact.* 30: 385–398.

Pinson, S.R.M., Wang, Y., and Tabien, R.E. (2015). Mapping and validation of quantitative trait loci associated with tiller production in rice. *Crop Sci.* 55: 1537–1551.

Pirzadeh, S. and Shahpiri, A. (2016). Functional characterization of a type 2 metallothionein isoform (OsMTI-2b) from rice. *Int. J. Biol. Macromol.* 88: 491–496.

Punjee, P., Siripornadulsil, W., and Siripornadulsil, S. (2017). Reduction of cadmium uptake in rice endophytically colonized with the cadmium-tolerant bacterium *Cupriavidus taiwanensis* KKU2500-3. *Can. J. Microbiol.* 64: 131–145.

Ramesh, A.S. (2003). Differential metal selectivity and gene expression of two zinc transporters from rice. *Plant Physiol.* 133: 126–134.

Ramesh, G., Podila, G.K., Gay, G. et al. (2009). Different patterns of regulation for the copper and cadmium metallothioneins of the Ectomycorrhizal fungus *Hebeloma cylindrosporum*. *Appl. Environ. Microbiol.* 75: 2266–2274.

Rapisarda, V., Miozzi, E., Loreto, C. et al. (2018). Cadmium exposure and prostate cancer: insights, mechanisms and perspectives. *Front. Biosci. (Landmark Ed.)* 23: 1687–1700.

Rauser, W.E. (1995). Phytochelatins and related peptides: structure, biosynthesis, and function. *Plant Physiol.* 109: 1141–1149.

Reddy, C.N. and Patrick, W.H. (1977). Effect of redox potential and pH on the uptake of cadmium and lead by rice plants. *J. Environ. Qual.* 6: 259–262.

Rodda, M.S., Li, G., and Reid, R.J. (2011). The timing of grain Cd accumulation in rice plants: the relative importance of remobilisation within the plant and root Cd uptake post-flowering. *Plant Soil* 347: 105–114.

Sarwar, N., Saifullah, Malhi, S.S. et al. (2010). Role of mineral nutrition in minimizing cadmium accumulation by plants. *J. Sci. Food Agric.* 90: 925–937.

Sasaki, A., Yamaji, N., Yokosho, K. et al. (2012). Nramp5 is a major transporter responsible for manganese and cadmium uptake in rice. *Plant Cell* 24: 2155–2167.

Sasaki, A., Yamaji, N., and Ma, J.F. (2014). Overexpression of *OsHMA3* enhances Cd tolerance and expression of Zn transporter genes in rice. *J. Exp. Bot.* 65: 6013–6021.

Satarug, S., Baker, J.R., Reilly, P.E.B. et al. (2002). Cadmium levels in the lung, liver, kidney cortex, and urine samples from Australians without occupational exposure to metals. *Arch. Environ. Health* 57: 69–77.

Sato, H., Shirasawa, S., Maeda, H. et al. (2011). Analysis of QTL for lowering cadmium concentration in rice grains from "LAC23.". *Breed. Sci.* 61: 196–200.

Satoh-Nagasawa, N., Mori, M., Nakazawa, N. et al. (2012). Mutations in rice (*Oryza sativa*) heavy metal ATPase 2 (OsHMA2) restrict the translocation of zinc and cadmium. *Plant Cell Physiol.* 53: 213–224.

Sebastian, A. and Prasad, M.N.V. (2013). Cadmium minimization in rice: a review. *Agron. Sustainable Dev.* 34: 155–173.

Sebastian, A. and Prasad, M.N.V. (2015). Iron- and manganese-assisted cadmium tolerance in *Oryza sativa* L.: lowering of rhizotoxicity next to functional photosynthesis. *Planta* 241: 1519–1528.

Shahabivand, S., Maivan, H.Z., Goltapeh, E.M. et al. (2012). The effects of root endophyte and arbuscular mycorrhizal fungi on growth and cadmium accumulation in wheat under cadmium toxicity. *Plant Physiol. Biochem.* 60: 53–58.

Shao, F.J., Che, J., Yamaji, N. et al. (2017). Silicon reduces cadmium accumulation by suppressing expression of transporter genes involved in cadmium uptake and translocation in rice. *J. Exp. Bot.* 68: 5641–5651.

Sheng, Y., Wang, Y., Yang, X. et al. (2016). Cd tolerant characteristic of a newly isolated *Lactococcus lactis* subsp. *lactis*. *Environ. Toxicol. Phar.* 48: 183–190.

Shi, X., Zhang, C., Wang, H. et al. (2005). Effect of Si on the distribution of Cd in rice seedlings. *Plant Soil* 272: 53–60.

Shi, J., Li, L., and Pan, G. (2009). Variation of grain Cd and Zn concentrations of 110 hybrid rice cultivars grown in a low-Cd paddy soil. *J. Environ. Sci.* 21: 168–172.

Shi, L., Guo, Z., Liang, F. et al. (2019). Effect of liming with various water regimes on both immobilization of cadmium and improvement of bacterial communities in contaminated paddy: a field experiment. *Int. J. Environ. Res. Public Health* 16: 498.

Shimo, H., Ishimaru, Y., An, G. et al. (2011). *Low cadmium* (*LCD*), a novel gene related to cadmium tolerance and accumulation in rice. *J. Exp. Bot.* 62: 5727–5734.

Shu, F., Suzui, N., Ishioka, N.S. et al. (2010). Tracing cadmium from culture to spikelet: noninvasive imaging and quantitative characterization of absorption, transport, and accumulation of cadmium in an intact rice plant. *Plant Physiol.* 152: 1796–1806.

Sikka, R., Nayyar, V.K., and Sidhu, S.S. (2008). Monitoring of Cd pollution in soils and plants irrigated with untreated sewage water in some industrialized cities of Punjab, India. *Environ. Monit. Assess.* 154: 53–64.

Skamene, E. and Schurr, E. (1998). Infection genomics: Nramp1 as a major determinant of natural resistance to intracellular infections. *Annu. Rev. Med.* 49: 275–287.

Song, J.K., Luo, H., Yin, X.H. et al. (2015). Association between cadmium exposure and renal cancer risk: a meta-analysis of observational studies. *Sci. Rep.* 5: 17976.

Sun, L., Chen, S., Chao, L. et al. (2007). Effects of flooding on changes in Eh, pH and speciation of cadmium and lead in contaminated soil. *Bull. Environ. Contam. Toxicol.* 79: 514–518.

Sun, L., Xu, X., Jiang, Y. et al. (2016). Genetic diversity, rather than cultivar type, determines relative grain Cd accumulation in hybrid rice. *Front. Plant Sci.* 7: 1407.

Suwazono, Y., Kido, T., Nakagawa, H. et al. (2009). Biological half-life of cadmium in the urine of inhabitants after cessation of cadmium exposure. *Biomarkers* 14: 77–81.

Takahashi, R., Ishimaru, Y., Senoura, T. et al. (2011a). The OsNRAMP1 iron transporter is involved in Cd accumulation in rice. *J. Exp. Bot.* 62: 4843–4850.

Takahashi, R., Ishimaru, Y., Nakanishi, H. et al. (2011b). Role of the iron transporter OsNRAMP1 in cadmium uptake and accumulation in rice. *Plant Signaling Behav.* 6: 1813–1816.

Takahashi, R., Ishimaru, Y., Shimo, H. et al. (2012a). The OsHMA2 transporter is involved in root-to-shoot translocation of Zn and Cd in rice. *Plant Cell Environ.* 35: 1948–1957.

Takahashi, R., Ishimaru, Y., Shimo, H. et al. (2014). From laboratory to field: OsNRAMP5-knockdown rice is a promising candidate for Cd phytoremediation in paddy fields. *PLoS One* 9: e98816.

Takahashi, R., Ito, M., Katou, K. et al. (2016). Breeding and characterization of the rice (*Oryza sativa* L.) line "Akita 110" for cadmium phytoremediation. *Soil Sci. Plant Nutr.* 188: 1–6.

Tamaoki, M. and Maruyama-Nakashita, A. (2017). Molecular mechanisms of selenium responses and resistance in plants. In: *Selenium in plants* (eds. E. Pilon-Smits, L. Winkel and Z.Q. Lin), 35–51. Cham: Springer.

Tan, L., Zhu, Y., Fan, T. et al. (2019). OsZIP7 functions in xylem loading in roots and inter-vascular transfer in nodes to deliver Zn/Cd to grain in rice. *Biochem. Biophys. Res. Commun.* 512: 112–118.

Tang, L., Mao, B., Li, Y. et al. (2017). Knockout of OsNramp5 using the CRISPR/Cas9 system produces low Cd-accumulatingindicarice without compromising yield. *Sci. Rep.* 7: 14438.

Tezuka, K., Miyadate, H., Katou, K. et al. (2010). A single recessive gene controls cadmium translocation in the cadmium hyperaccumulating rice cultivar Cho-Ko-Koku. *Theor. Appl. Genet.* 120: 1175–1182.

Thévenod, F. and Lee, W.-K. (2013). Toxicology of cadmium and its damage to mammalian organs. In: *Cadmium: from toxicity to essentiality* (eds. A. Sigel, H. Sigel and R.K.O. Sigel), 415–490. Dordrecht: Springer Netherlands.

Tiwari, M., Deepika, S., Sanjay, D. et al. (2014). Expression in Arabidopsis and cellular localization reveal involvement of rice NRAMP, OsNRAMP1, in arsenic transport and tolerance. *Plant Cell Environ.* 37: 140–152.

Treesubsuntorn, C., Dhurakit, P., Khaksar, G. et al. (2017). Effect of microorganisms on reducing cadmium uptake and toxicity in rice (*Oryza sativa* L.). *Environ. Sci. Pollut. Res.* 25: 1–12.

Ueno, D., Yamaji, N., and Jian, F.M. (2009). Further characterization of ferric–phytosiderophore transporters ZmYS1 and HvYS1 in maize and barley. *J. Exp. Bot.* 60: 3513–3520.

Ueno, D., Yamaji, N., Kono, I. et al. (2010). Gene limiting cadmium accumulation in rice. *Proc. Natl. Acad. Sci. U.S.A.* 107: 16500–16505.

Ueno, D., Koyama, E., Yamaji, N. et al. (2011). Physiological, genetic, and molecular characterization of a high-Cd-accumulating rice cultivar, Jarjan. *J. Exp. Bot.* 62: 2265–2272.

Unkles, S.E., Hawker, K.L., Grieve, C. et al. (1991). crnA encodes a nitrate transporter in *Aspergillus nidulans*. *Proc. Natl. Acad. Sci. U.S.A.* 88: 204–208.

Uraguchi, S., Kamiya, T., Sakamoto, T. et al. (2011). Low-affinity cation transporter (OsLCT1) regulates cadmium transport into rice grains. *Proc. Natl. Acad. Sci. U.S.A.* 108: 20959–20964.

Uraguchi, S., Sone, Y., Ohta, Y. et al. (2017). Identification of C-terminal regions in *Arabidopsis thaliana* phytochelatin synthase 1 specifically involved in activation by arsenite. *Plant Cell Physiol.* 59: 500–509.

Van Maele-Fabry, G., Lombaert, N., and Lison, D. (2016). Dietary exposure to cadmium and risk of breast cancer in postmenopausal women: a systematic review and meta-analysis. *Environ. Int.* 86: 1–13.

Vidal, S.M. (1993). Natural resistance to infection with intracellular parasites: isolation of a candidate for Bcg. *Cell* 73: 469–485.

Vinceti, M., Venturelli, M., Sighinolfi, M.C. et al. (2007). Case-control study of toenail cadmium and prostate cancer risk in Italy. *Sci. Total Environ.* 373: 77–81.

Wan, Y., Yu, Y., Wang, Q. et al. (2016). Cadmium uptake dynamics and translocation in rice seedling: influence of different forms of selenium. *Ecotoxicol. Environ. Saf.* 133: 127–134.

Wang, F., Wang, M., Liu, Z. et al. (2015). Different responses of low grain-Cd-accumulating and high grain-Cd-accumulating rice cultivars to Cd stress. *Plant Physiol. Biochem.* 96: 261–269.

Wang, H., Xu, C., Luo, Z. et al. (2018). Foliar application of Zn can reduce Cd concentrations in rice (*Oryza sativa* L.) under field conditions. *Environ. Sci. Pollut. Res.* 25: 29287–29294.

Wang, M., Chen, S., Zheng, H. et al. (2019a). The responses of cadmium phytotoxicity in rice and the microbial community in contaminated paddy soils for the application of different long-term N fertilizers. *Chemosphere* 238: 124700.

Wang, C., Chen, X., Yao, Q. et al. (2019b). Overexpression of *TtNRAMP6* enhances the accumulation of Cd in Arabidopsis. *Gene* 696: 225–232.

Wang, C., Liu, Z., Huang, Y. et al. (2019c). Cadmium-resistant rhizobacterium *Bacillus cereus* M4 promotes the growth and reduces cadmium accumulation in rice (*Oryza sativa* L.). *Environ. Toxicol. Pharmacol.* 72: 103265.

Wang, T., Li, Y., Fu, Y. et al. (2019d). Mutation at different sites of metal transporter gene *OsNramp5* affects Cd accumulation and related agronomic traits in rice (*Oryza sativa* L.). *Front. Plant Sci.* 10: 1081.

Wu, G., Yang, C., Guo, L. et al. (2013). Cadmium contamination in Tianjin agricultural soils and sediments: relative importance of atmospheric deposition from coal combustion. *Environ. Geochem. Hlth.* 35: 405–416.

Wu, Z., Wang, F., Liu, S. et al. (2016). Comparative responses to silicon and selenium in relation to cadmium uptake, compartmentation in roots, and xylem transport in flowering Chinese cabbage (*Brassica campestris* L. ssp. *chinensis* var. *utilis*) under cadmium stress. *Environ. Exp. Bot.* 131: 173–180.

Xia, J., Yamaji, N., Kasai, T. et al. (2010). Plasma membrane-localized transporter for aluminum in rice. *Proc. Natl. Acad. Sci. U.S.A.* 107: 18381–18385.

Xiao, R., Zehong, H., Xiaonuo, L. et al. (2017). Lime and phosphate amendment can significantly reduce uptake of Cd and Pb by field-grown rice. *Sustainability* 9: 430–435.

Xie, L.H., Tang, S.Q., Wei, X.J. et al. (2017). The cadmium and lead content of the grain produced by leading Chinese rice cultivars. *Food Chem.* 217: 217–224.

Xue, D., Chen, M., and Zhang, G. (2009). Mapping of QTLs associated with cadmium tolerance and accumulation during seedling stage in rice (*Oryza sativa* L.). *Euphytica* 165: 587–596.

Yamaguchi, N., Ishikawa, S., Abe, T. et al. (2012). Role of the node in controlling traffic of cadmium, zinc, and manganese in rice. *J. Exp. Bot.* 63: 2729–2737.

Yamaji, N. and Ma, J.F. (2017). Node-controlled allocation of mineral elements in *Poaceae. Curr. Opin. Plant Biol.* 39: 18–24.

Yamaji, N., Xia, J., Mitani-Ueno, N. et al. (2013b). Preferential delivery of zinc to developing tissues in rice is mediated by P-type heavy metal ATPase OsHMA2. *Plant Physiol.* 162: 927–939.

Yamaki, T., Otani, M., Ono, K. et al. (2017). Isolation and characterization of rice cesium transporter genes from a rice-transporter-enriched yeast expression library. *Physiol. Plant.* 160: 425–436.

Yamazaki, S., Ueda, Y., Mukai, A. et al. (2018). Rice phytochelatin synthases OsPCS1 and OsPCS2 make different contributions to cadmium and arsenic tolerance. *Plant Direct.* 2: e00034.

Yan, Y., Lestari, P., Lee, K. et al. (2013). Identification of quantitative trait loci for cadmium accumulation and distribution in rice (*Oryza sativa*). *Genome* 56: 227–232.

Yan, J., Wang, P., Wang, P. et al. (2016). A loss-of-function allele of *OsHMA3* associated with high cadmium accumulation in shoots and grain of japonica rice cultivars. *Plant Cell Environ.* 39: 1941–1954.

Yan, H., Xu, W., Xie, J. et al. (2019). Variation of a major facilitator superfamily gene contributes to differential cadmium accumulation between rice subspecies. *Nat. Commun.* 10: 2562.

Yang, M., Zhang, Y., Zhang, L. et al. (2014). OsNRAMP5 contributes to manganese translocation and distribution in rice shoots. *J. Exp. Bot.* 65: 4849–4861.

Yang, Y., Xiong, J., Chen, R. et al. (2016). Excessive nitrate enhances cadmium (Cd) uptake by up-regulating the expression of *OsIRT1* in rice (*Oryza sativa*). *Environ. Exp. Bot.* 122: 141–149.

Yang, B., Yang, C., Shao, Z.Y. et al. (2019). Selenium (Se) does not reduce cadmium (Cd) uptake and translocation in rice (*Oryza sativa* L.) in naturally occurred Se-rich paddy fields with a high geological background of Cd. *B. Environ. Contam. Tox.* 103: 127–132.

Yao, W., Sun, L., Zhou, H. et al. (2015). Additive, dominant parental effects control the inheritance of grain cadmium accumulation in hybrid rice. *Mol. Breed.* 35: 39.

Yoneyama, T., Ishikawa, S., and Fujimaki, S. (2015). Route and regulation of zinc, cadmium, and iron transport in rice plants (*Oryza sativa* L.) during vegetative growth and grain filling: metal transporters, metal speciation, grain Cd reduction and zn and Fe biofortification. *Int. J. Mol. Sci.* 16: 19111–19129.

Yu, H., Wang, J., Fang, W. et al. (2006). Cadmium accumulation in different rice cultivars and screening for pollution-safe cultivars of rice. *Sci. Total Environ.* 370: 302–309.

Yu, C., Sun, C., Shen, C. et al. (2015). The auxin transporter, OsAUX1, is involved in primary root and root hair elongation and in Cd stress responses in rice (*Oryza sativa* L.). *Plant J.* 83: 818–830.

Yuan, L., Yang, S., Liu, B. et al. (2012). Molecular characterization of a rice metal tolerance protein, OsMTP1. *Plant Cell Rep.* 31: 67–79.

Zeng, F., Mao, Y., Cheng, W. et al. (2008). Genotypic and environmental variation in chromium, cadmium and lead concentrations in rice. *Environ. Pollut.* 153: 309–314.

Zeng, F., Ali, S., Zhang, H. et al. (2011). The influence of pH and organic matter content in paddy soil on heavy metal availability and their uptake by rice plants. *Environ. Pollut.* 159: 84–91.

Zenk, M.H. (1995). Heavy metal detoxification in higher plants: a review. *Gene (Amsterdam)* 179: 21–30.

Zhan, J., Wei, S., Niu, R. et al. (2013). Identification of rice cultivar with exclusive characteristic to Cd using a field-polluted soil and its foreground application. *Environ. Sci. Pollut. Res.* 20: 2645–2650.

Zhang, X., Zhu, Y., Chen, B. et al. (2005). Arbuscular mycorrhizal fungi contribute to resistance of upland rice to combined metal contamination of soil. *J. Plant Nutr.* 28: 2065–2077.

Zhang, C., Wang, L., Nie, Q. et al. (2008). Long-term effects of exogenous silicon on cadmium translocation and toxicity in rice (*Oryza sativa* L.). *Environ. Exp. Bot.* 62: 300–307.

Zhang, X., Zhang, G., Guo, L. et al. (2011). Identification of quantitative trait loci for Cd and Zn concentrations of brown rice grown in Cd-polluted soils. *Euphytica* 180: 173–179.

Zhang, M., Pinson, S.R., Tarpley, L. et al. (2013). Mapping and validation of quantitative trait loci associated with concentrations of 16 elements in unmilled rice grain. *Theor. Appl. Genet.* 127: 28.

Zhang, B.L., Ouyang, Y.N., Xu, J.Y. et al. (2017). Cadmium remobilization from shoot to grain is related to pH of vascular bundle in rice. *Ecotoxicol. Environ. Saf.* 147: 913.

Zhang, Q., Chen, H., Huang, D. et al. (2019). Water managements limit heavy metal accumulation in rice: dual effects of iron-plaque formation and microbial communities. *Sci. Total Environ.* 687: 790–799.

Zhao, J., Yang, W., Zhang, S. et al. (2018). Genome-wide association study and candidate gene analysis of rice cadmium accumulation in grain in a diverse rice collection. *Rice* 11: 61.

Zheng, X., Chen, L., and Li, X. (2018). Arabidopsis and rice showed a distinct pattern in *ZIPs* genes expression profile in response to Cd stress. *Bot. Stud.* 59: 22.

Zhu, H., Chen, C., Xu, C. et al. (2016). Effects of soil acidification and liming on the phytoavailability of cadmium in paddy soils of central subtropical China. *Environ. Pollut.* 219: 99–106.

Zhu, C., Cao, X., Zhu, L. et al. (2018). Ammonium mitigates Cd toxicity in rice (*Oryza sativa*) via putrescine-dependent alterations of cell wall composition. *Plant Physiol. Biochem.* 132: 189–201.

8

Molecular Breeding for Improving Arsenic Stress Tolerance in Rice: Recent Progress and Future Perspectives

Nourollah Ahmadi[1,2] *and Julien Frouin*[1,2]

[1] *CIRAD, UMR AGAP, F-34398 Montpellier, France*
[2] *AGAP, University of Montpellier, CIRAD, INRAE, Montpellier SupAgro, Montpellier, France*

Introduction

Arsenic is a ubiquitous metalloid found in a variety of physicochemical forms including simple dissolved inorganic species, complex organic species, and forms associated with diverse colloidal particles. The inorganic forms include the trivalent arsenite *As*(III) and the pentavalent arseniate *As*(V). The organic forms include the mono-methylarsenic ($CH_3 AsO_3H_2$) and the di-methylarsenic ($(CH_3)_2 AsO_3H_2$). These forms illustrate the diversity of states of oxidation of *As*: +III, +V, 0, and −III. The harmful effects of *As* on human health are well established. However, toxicity depends on the form (inorganic or organic) and the oxidation state of *As* compounds. The bioavailability and the associated toxicity is rather limited for the organic forms, much higher for the inorganic forms (Sun et al. 2008; Meharg et al. 2009; EFSA 2009; WHO 2011). Among the inorganic forms, the *As*(III) species, more mobile and endowed with higher ability to form complexes with human coenzymes, is 40–60 times more toxic than the *As*(V) species (Hite 2013). Beyond the acute intoxications due to exposure to very high *As* concentrations, long-term exposure to very low doses of inorganic *As* also constitutes a danger for human health (WHO 2001). The maximal acceptable total *As* concentration in drinking water established by the World Health Organization (WHO) is $10\ mg\ l^{-1}$. On the other hand, so far no international norm exists for *As* content in human foods, including rice. The only country with such a norm for rice is China, where the maximum authorized concentration of inorganic *As* in rice grain is $0.15\ mg\ kg^{-1}$ (Hojsak et al. 2015). The Food and Agriculture Organization (FAO) and WHO have established a tolerable daily uptake of inorganic *As*: 3.0 mg per kg of body weight, per day (WHO 2011).

Pollution of paddy fields and irrigation water by *As* has been reported in more than 70 countries in Asia, America, and Europe (Zavala and Duxbury 2008; Brammer and Ravenscroft 2009; Stroud et al. 2011). The problem, which is often of geological origin, affects several 100 million peoples, especially in Asia (Brammer and Ravenscroft 2009; Meharg et al. 2009; Fan et al. 2017). A survey of total *As* in 901 samples of commercial polished (white) rice collected randomly from *As* contaminated or non-contaminated areas in 10 countries showed sevenfold variations in median total *As* content. The lowest median value

Molecular Breeding for Rice Abiotic Stress Tolerance and Nutritional Quality, First Edition.
Edited by Mohammad Anwar Hossain, Lutful Hassan, Khandakar Md. Iftekharuddaula, Arvind Kumar and Robert Henry.

($0.04\,mg\,kg^{-1}$) was measured in Egypt and the highest in the United States and in France, 0.25 and $0.28\,mg\,kg^{-1}$, respectively (Meharg et al. 2009).

Local and regional surveys have revealed a tight correlation between *As* concentration in the soil, or in the irrigation water, and its concentration in the rice plant (Duxbury et al. 2003; Zavala et al. 2008; Bhattacharya et al. 2009). *As* accumulation in the rice plant is the highest in the roots, followed by the straw and the brown (cargo) grain (CG) (Abedin et al. 2002). Pollution of the paddy field by *As* also affects crop growth and development (lower germination rate, reduced shoot and root growth and biomass production, etc.) and, consequently, crop yield (Abedin et al. 2002).

There are at least three complementary options to mitigate the harmful effects of high soil and/or water *As* concentration: cropping practices, rice cooking methods, and the use of *As*-tolerant rice varieties. The agronomic mitigation options include alternate wetting and drying of the paddy field during the cropping season (Li et al. 2009; Mitra et al. 2017) and application of silicon (*Si*) fertilizer (Li et al. 2009; Tripathi et al. 2012). The cooking-method mitigation relies on washing and/or cooking in excess water (Gray et al. 2015; Liu et al. 2018). Rice varieties tolerant to high soil/water *As* concentration reduce *As* uptake and/or its translocation from the vegetative organs to the grains.

Tolerance to Arsenic in Rice

The Phytotoxic Effects of Arsenic and the Rice Defense Response

Mechanisms of the rice plant in relation to response to excess soil *As* have been reported to be similar to those observed for other types of soil chemical toxicity (Verbruggen et al. 2009; Briat 2010). However, the mechanisms related to the phytotoxic effects of *As* and the rice defense response to *As* remain poorly understood. While in aerobic conditions, the predominant form of soil *As* is arsenate, *As*(V), the uptake of which involves phosphate transporters (Zhao et al. 2009), in paddy fields (anaerobic conditions), the predominant form of *As* is its most mobile and toxic species, arsenite *As*(III) (Xu et al. 2007). It enters root cells through aquaporin-type membrane ports (Ma et al. 2008). Transporters involved in the process include silicon transporters Lsi1 (influx) and Lsi2 (efflux) (Ma and Yamaji 2006; Ma et al. 2008; Zhao et al. 2010c) and several silicon-independent pathways (Rai et al. 2015; Chen et al. 2017). The arsenite is then either extruded back into the rhizosphere (Xu et al. 2007; Zhao et al. 2010b), and/or detoxified by complexation with peptides such as glutathione and phytochelatins, and compartmentalized in the roots' vacuoles (Song et al. 2010), or transported to aboveground organs (Ye et al. 2010; Carey et al. 2011).

The *As* concentration in the rice nodes is much higher than in internodes and in the grains, suggesting it serves as a filter, restricting *As* transfer to the leaves and grains (Song et al. 2014; Yamaji and Ma 2014; Chen et al. 2015). At the reproductive stage, OsABCC1 is expressed in vascular tissues like the uppermost node I, and limits *As* transport to the grains by sequestering *As* in the phloem companion cells (Song et al. 2014). Besides, Lsi2 also shows high-level expression in node I, where it enhances *Si* distribution in rice panicles, but unfortunately also helps *As*(III) transport to rice grains (Ma et al. 2008; Yamaji et al. 2015).

As in rice grain is present primarily as inorganic *As*(III) and *As*(V), but the share of organic *As*, (20–50%), mainly as dimethylarsinic acid (DMA^{V}), is not negligible (Williams et al.

2005). Given the fact that rice is unable to methylate inorganic *As in vivo*, the methylated *As* species found in the grain, most likely come from the rhizosphere via microbial methylation (Lomax et al. 2012; Jia et al. 2013). Methylated *As* uptake by the rice plant is less efficient than *As*(III) and *As*(V) (Abedin et al. 2002). However, once taken up, DMA^V is translocated to the rice grain with over 10 times greater efficiency than inorganic species and is more mobile than *As*(III) in both phloem and xylem transport (Carey et al. 2010).

Genetic Diversity for Arsenic Tolerance

Significant genetic diversity for *As* accumulation has been reported in rice under over-exposure to *As* in both hydroponic cultivation and in field experiments (Dasgupta et al. 2004; Rahman et al. 2007; Norton et al. 2009, 2012b; Frouin et al. 2019). Analysis of grain *As* content in 421 rice accessions grown in six sites distributed in Bangladesh, China, and the United States revealed from 3- to 34-fold variation in each site (Norton et al. 2012b). It also revealed significant effects of location, genotype by environment interaction, and membership of different genetic groups of *Oryza sativa*. In the site with the highest effect, brown rice *As* content varied from 0.363 to 1.266 mg kg^{-1}, and the mean *As* content was 0.676 mg kg^{-1}. On average, accessions belonging to the *Aus* and *japonica* groups had the highest and the lowest *As* contents, respectively. The *japonica* accession Azerbaidjanica, Kon Suito, and Norin 20, had low *As* content across the six environments. Frouin et al. (2019) working with 300 *japonica* accessions grown in a paddy field in France, reported similar large variation of flag leaf *As* content (FL-*As*) and brown (cargo) grain *As* content (CG-*As*). FL-*As* ranged from 1.393 to 15.509 mg kg^{-1} of dry weight and averaged 5.881 mg kg^{-1}. CG-*As* ranged from 0.147 to 0.656 mg kg^{-1} and averaged 0.335 mg kg^{-1} (Figure 8.1). They also reported significantly higher *As* contents among the temperate *japonica* accessions compared to tropical *japonica* accessions. The determination coefficient between FL-*As* and CG-*As* was rather low ($R^2 = 0.204$, $p < 0.0001$), suggesting a rather

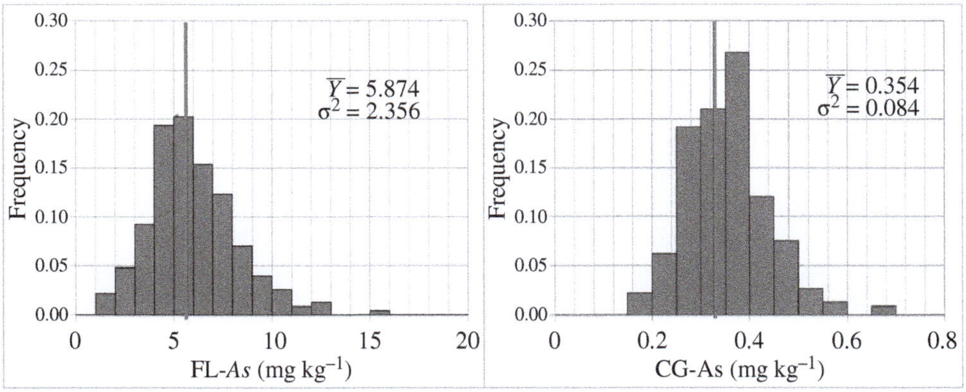

Figure 8.1 Distribution of the flag leaf (FL-*As*), and cargo grain (CG-*As*) arsenic content in a rice diversity panel composed of 300 accessions belonging to *Japonica* group. *Note:* The vertical red line indicates the mean value of *As* content in the population. Data were produced in field experiment conducted in France, where mean soil and water *As* contents were 9.73 mg kg^{-1} and 0.021 mg l^{-1}, respectively.

loose relationship between the two traits. Frouin et al. (2019) reported similar variations of *As* contents in a set of 95 *japonica* advanced inbred lines, evaluated two years later in the same location: variation in CG-*As* ranged from 0.208 to 0.729 mg kg^{-1} and averaged 0.341 mg kg^{-1}. The determination coefficient between the FL-*As* and CG-*As* was, again, low but highly significant ($R^2 = 0.20, p < 0.0001$).

Genetic Bases of Arsenic Tolerance

Approaches mobilized for analysis of genetic bases of *As* tolerance in rice include: linkage mapping, linkage disequilibrium mapping, genome-wide expression analysis, and targeted gene expression analysis.

Linkage mapping approach was implemented using recombinant inbred lines (RIL) of at least three bi-parental crosses. Several quantitative trait loci (QTLs) involved in grain *As* accumulation were mapped (Norton et al. 2010, 2012a; Kuramata et al. 2013; Zhang et al. 2014). The main characteristics of these QTLs are summarized in Table 8.1. Though the heterogeneity of information on the position of the QTLs makes comparison between those studies difficult, there is evidence for co-localization among the total of 17 QTLs reported.

A linkage-disequilibrium approach was implemented by Norton et al. (2014) using a diversity panel of 421 accessions representative of *O. sativa*, and by Frouin et al. (2019) using a diversity panel of 228 accessions representative of the *japonica* group, and a set of 95 *japonica* advanced lines of their breeding program. In each of the three studies genome-wide association analyses (GWAS) detected several significant loci for grain *As* content (Figure 8.2). However, the contributions of individual significant loci to the total variance of the trait (marker R^2) did not exceed 8%. A number of these associations co-localized with QTLs mapped in RIL populations (Figure 8.2), and made it possible to narrow down the position of those QTLs. This was the case, for instance, for the QTLs on chromosomes 2, 3, 9, and 10. Some co-localizations could also be observed between loci significantly associated with grain *As* content in the three independent GWAS studies, using the *O. sativa* or the *japonica* diversity panels, or the set of *japonica* advanced lines. This was the case, for instance, on chromosome 1, 7, 8, and 9 (Figure 8.2). The co-localization rate reached 40% of the significant loci mapped in the *japonica* diversity panel and in the *japonica* set of advanced lines (Frouin et al. 2019).

Genes underlying the significant loci mapped in the two *japonica* populations included a number of candidate genes identified by transcriptome analysis and reverse genetic approaches (see below): ATP binding cassette involved in *As* detoxification (e.g., Os04g0620000), transporters (e.g., phosphate, ammonium, peptide, and efflux transporters multidrug and toxic compound extrusion [MATE]); abiotic stress responsive genes (e.g., several F-box and DUF domain containing proteins, cytochrome P450); and transcription factors (e.g., MBY, zinc finger family protein, ERF) (Frouin et al. 2019). However, none of the significant associations in the three GWAS studies co-localized with genes previously reported to play a central role in rice response to *As* over-exposure such as loci coding for Lsi1 and Lsi2 proteins (Ma et al. 2008; Zhao et al. 2010b).

Analysis of *As*-induced genome-wide modulation of transcriptomes of rice seedling roots revealed up-regulation of several 100 genes, confirming the complexity of the gene network involved in rice response to *As* over-exposure (Chakrabarty et al. 2009; Dubey et al. 2014).

Table 8.1 Main characteristics of *As* tolerance QTLs reported in the literature, using progenies of bi-parental crosses.

Population	Experiment	Mapped trait	QTL	Chrom	Adjacent marker/Nearest marker[a)]		Position (Mb)[b)]		LOD	R^2 (%)	Donor positive allele	References
Bala × Azucena (108 RIL)	Hydroponic	Seedling root length	AsTol	6	RZ516	RG213	2.6	6.3			Bala	Dasgupta et al. (2004)
Padi Perak × Koshihikari (100 F2)	Paddy field	Grain As	qDMAs6.1	6	AE06000003	AE06000040	0.2	2.8	11.1	36	Padi Perak	Kuramata et al. (2013)
			qDMAs6.2	6	AE06001721	AE06003587	4.8	13.1	6.9	25		
			qDMAs8	8	AE08000162	AE08000464	0.5	3.5	3.9	12		
Bala × Azucena (108 RIL)	Field with rainout facility	Leaf As	qAs1	1	R117 − 1		42.7		3.8	16.9	Azucena	Norton et al. (2010)
			qAs3	3	R1618 + 10		30.7		3.9	16	Azucena	
			qAs5	5	C624 + 2		21.4		3.4	12.1	Bala	
			qAs6.1	6	A12361 + 2		6.3		5.4	18	Azucena	
			qAs6.2	6	AB0603 + 0		—		3.6	10.9	Bala	
	Paddy field	Grain As	qAs8	8	R902 − 13		4.7		4.4	22.8	Bala	Norton et al. (2012)
			qAs10	10	A12372 + 0.6		11.5		3.1	8.5	Azucena	
			qAs10	10	C16 − 1.3		20.8		2.9	6.5		
Lemont × TeQing (280 RIL)	Paddy field	Grain As		2	RM341		17.0	25.0				Zhang et al. (2013)
				3	RM16		5.0	17.0				
				5	RM421	RG13	21.0	27.0	3.0	3.7		
				9	RM1189		16.0	22.0				
				11	RM167		0.0	5.0				

a) In some studies, the position of the QTL is given (+) below or above (−) cM of the nearest marker.
b) Physical position of QTLs mapped in Bala × Azucena RILs was established by us, using the Gramene database (www.gramene.org).

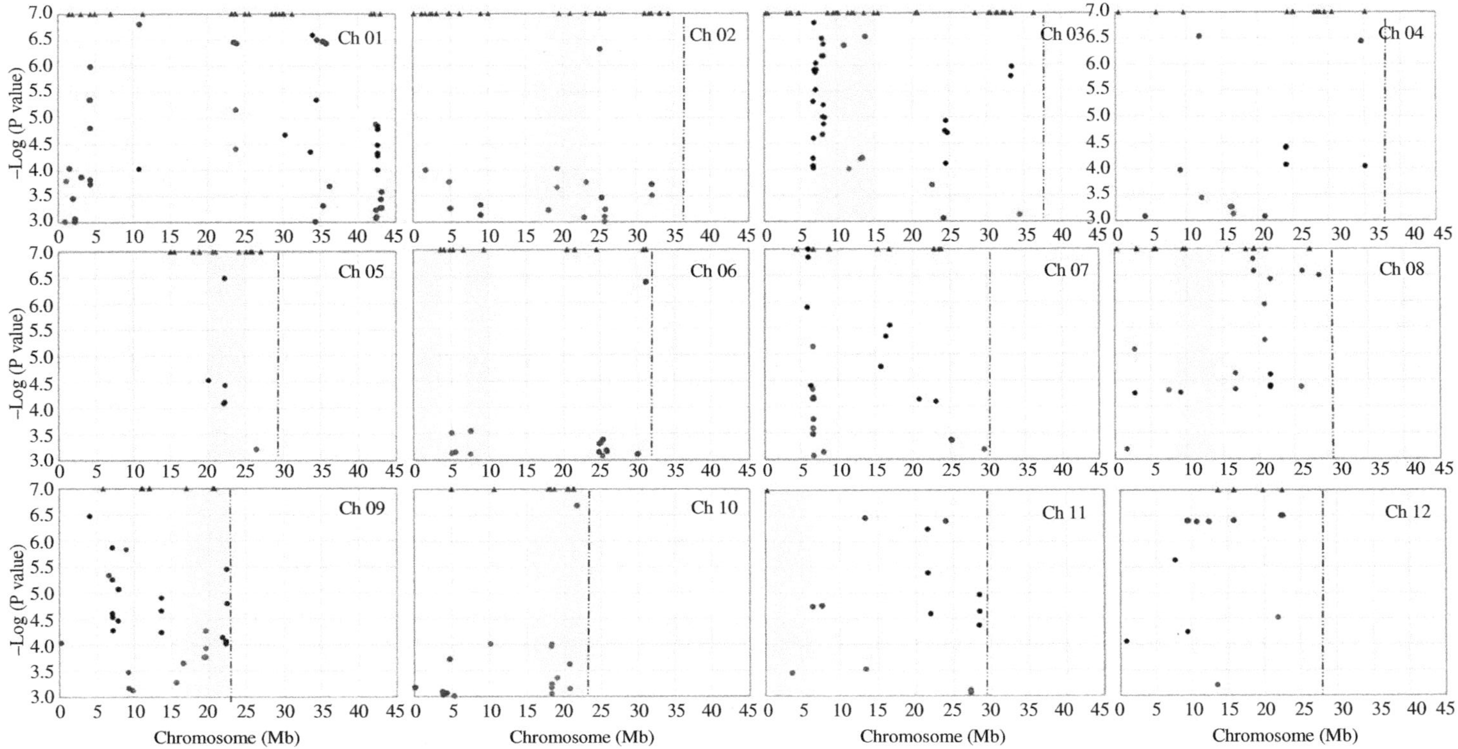

Figure 8.2 Results of association analyses for cargo grain *As* content in a diversity panel of 421 accessions representative of *O. Sativa* (Black Dots), in a diversity panel of 300 *Japonica* Accessions (Red Dots) and 95 *Japonica* advanced line (Blue Dots). *Note:* Data for the *O. sativa* panel were extracted from Norton et al. (2014) when GWAS was implemented with all accessions of the panel. Genomic position of QTLs for grain *As* concentration reported by Zhang et al. (2014) is represented with yellow bars. Genomic position of *As* related candidate genes reports in the literature is represented by brown triangles. The vertical broken line indicates the maximum length of each chromosome. The ordinate axis, −log(*p*-value), indicates the degree of significance of association. The graphic representation draws inspiration from Norton et al. (2014). (*See color plate section for color representation of this figure*)

Gene families with differential gene expression in *As*-tolerant and *As*-susceptible genotypes include glutathione S-transferases, cytochrome P450s, heat shock proteins, metal-binding proteins, and a large number of transporters and transcription factors such as MYBs (Rai et al. 2015). MYB genes may be crucial in *As*(V) stress tolerance as they up-regulate phenylpropanoid and flavonoid biosynthetic pathways (Misra et al. 2010; Pandey et al. 2012). More recently, using a reverse genetics approach, Shi et al. (2016) showed that OsHAC1;1 and OsHAC1;2 (two orthologues of *Arabidopsis thaliana'* HAC1) functioned as *As*(V) reductase and played a role in the control of *As* accumulation in rice. Likewise, Xu et al. (2017) showed that OsHAC4 played a critical role in rice tolerance to arsenate and regulated *As* accumulation in rice. Based on these findings, some authors have recently advocated the use of genome-editing technology to improve rice *As* tolerance (Chen et al. 2017; Mitra et al. 2017).

Molecular Breeding Options for Rice Tolerance to Arsenic

The above-reported review of literature on the genetic bases of rice tolerance to *As* suggests that *As* uptake and accumulation in rice grains are complex processes involving a large number of QTLs and it is unlikely that a single mapping study, whether linkage mapping or GWAS, would make it possible to establish robust and precise genotype–phenotype associations. Indeed, some of the QTLs involved in *As* tolerance are subject to $G \times E$ interactions, others need further validation to ensure their stability across genetic backgrounds. It has also been shown that validation of GWAS results by a second independent GWAS experiment is a complex process with uncertain results (Frouin et al. 2019). Finally, yet importantly, so far all the QTLs mapped for *As* tolerance have relatively small individual effects, dampening their use in breeding for *As* tolerance through the most popular marker-assisted breeding scheme: introgression of the favorable allele of a QTL into the genome of a widely used cultivar through backcrossing. This is probably why, so far, to our knowledge, no such marker-assisted backcrossing for *As* tolerance has been reported in the literature.

Marker-Assisted Recurrent Selection

To overcome the issue of robustness of QTLs detected in experimental bi-parental populations or in diversity panels, one option would be the implementation of a "forward marker-assisted breeding" scheme. In such scheme, QTLs for *As* tolerance would be detected within the segregating progenies of an elite line and one (or a few) donors of *As* tolerance, and then the marker–phenotype information would serve to guide the breeding process. Such breeding schemes involve several successive generations of crossing and therefore constitute what is referred to as marker-assisted recurrent selection (MARS) or genotype construction (Stam 1995; Peleman and van der Voort 2003). Various MARS schemes have been proposed and evaluated using simulation tools (Charmet et al. 1999; Xie and Xu 1999) or direct experiments (Moreau et al. 2004). However, even in such schemes, if the QTLs are numerous but with small individual effects, marker-assisted pyramiding of the favorable alleles would be unpractical.

Genomic Selection

A second option to overcome the issue of robustness of QTLs, and their small individual effect would be the genomic selection (GS) approach. GS has recently emerged as an alternative option to conventional marker-assisted selection targeting mapped QTLs (Meuwissen et al. 2001; Heffner et al. 2009; Lorenz et al. 2011). By shifting the plant breeding paradigm from "breeding by design" to a "genome-wide-approach," GS provides genomic estimates of breeding values (GEBV) based on all available marker data, instead of focusing on a limited set of markers that target putative genes or QTLs. In the last few years, successful proof of the GS concept has been reported in maize (Bernardo and Yu 2007), wheat (Bassi et al. 2005), barley (Sorrells 2015), and oats (Asoro et al. 2011). In rice, moderate to high predictive ability of GEBV has been reported for a variety of quantitative traits in experiments with the progeny of bi-parental crosses and in diverse germplasm collections (for a review, see Ahmadi et al. 2019). In particular, it was shown that rice diversity panels provide accurate genomic predictions for complex traits in the progenies of bi-parental crosses involving members of the panel (Ben Hassen et al. 2017). It was also shown that genomic prediction accounting for genotype by environment interaction offered an effective framework for breeding simultaneously for adaptation to an abiotic stress and for performance under normal cropping conditions in rice (Ben Hassen et al. 2018), and that selection of trait-specific markers improves the predictive ability of genomic selection in rice (Bhandari et al. 2019).

Genomic Selection for Arsenic Tolerance in Rice: A Case Study

We recently evaluated the potential of GS to improve the ability of rice to restrict *As* accumulation in the grains (Frouin et al. 2019). First, genotypic data and field phenotypic data (FL-*As* and CG-*As*) of rice plants from a reference diversity panel of a breeding program, grown in a paddy field with rather high *As* concentrations, were used to train genomic prediction models. Second, using similar phenotypic and genotypic data from a set of advanced lines of the same breeding program, the predictive ability of genomic prediction across the two populations was evaluated. Hereafter we present briefly the methods implemented and the results obtained.

Methods

The plant materiel was composed of 228 accessions of a *japonica* diversity panel, hereafter referred to as reference population (RP), and of 95 *japonica* advanced inbred lines (F5, F6, or F7), hereafter referred to as the validation population (VP). The latter represented the elite lines of the rice breeding program run by the *Centre Français du Riz* (CFR) and Cirad, in the Camargue region of France.

Field trials were conducted at the CFR experimental station, Mas d'Adrien (43°42′ 13.77″N; 4°33′44.71″E; 3 m asl.), under a standard irrigated-rice cropping system. The experimental design was an augmented randomized complete block design, repeated twice for RP, non-repeated for VP. In each field trial, FL-*As* and CG-*As* were measured in each individual plot, using an Inductively Coupled Plasma Mass Spectrum. In each field trial, the soil total *As* content was measured before sowing and after harvest. Likewise, in each field trial, total *As* content of irrigation water was monitored once a month during the rice cropping cycle.

Genotyping by sequencing (GBS) of the two populations, using the restriction enzyme *Apek I* yielded 16,902 single nucleotide polymorphism (SNPs), with minor allele frequency (MAF) above 2.5%, representing an average marker density of one SNP every 17.1 kb. Linkage disequilibrium reached $r^2 = 0.2$ at 1.25 Mb in the RP, at 1.70 Mb in the VP. Both RP and VP populations showed similar MAF patterns, slightly skewed toward low frequencies. The average MAF was 22.2%, and the proportion of loci with MAF < 10% was close to 75%.

Two types of genomic predictions were implemented: cross-validation within the RP, and across-populations prediction using RP as training set and VP as validation set. The cross-validation experiments used 171 (3/4) of the 228 accessions of the RP as the training set and the remaining 57 (1/4) accessions as the validation set. Each cross-validation experiment was repeated 100 times using 100 independent partitioning of the RP into training set and validation set. For each independent partitioning, the correlation between the predicted and the observed phenotype was calculated to obtain 100 correlations for each cross-validation experiment. The predictive ability of each cross-validation experiment was computed as the mean value of the 100 correlations. The predictive ability of genomic prediction across populations was evaluated under three scenarios of composition of the training set. Under the first scenario (S1), the training set included all 228 accessions of RP. Under S2, the training set was composed of 100 accessions of the RP with the lowest average pairwise Euclidian distances with the 95 lines of the VP. Under S3, 100 accessions of the training set were selected among the 228 accessions of RP, using the CDmean method of optimization of the training set (Rincent et al. 2012). Three whole-genome regression models, the genomic best linear unbiased prediction (GBLUP), the Bayesian regression BayesA, and the reproducing kernel Hilbert spaces regressions (RKHS), were used to regress the FL-*As* and CG-*As* phenotype of RP and VP accessions on their genotype at 16 902 SNPs. The three models were implemented using the *BGLR* statistical package (Perez and de Los Campos 2014). The default parameters for prior specification were used, and the number of iterations for the Markov chain Monte Carlo (MCMC) algorithm was set to 25,000 with a burn-in period of 5000.

Genomic Prediction Results

Cross-Validation Within the Reference Population

The six cross-validation experiments involving the three prediction methods and the two phenotypic traits yielded predictive ability ranging from 0.447 to 0.543 (Table 8.2). The average predictive ability was 0.452 for FL-*As* and 0.535 for CG-*As*. Based on ANOVA results, differences in predictive ability between the two traits were highly significant ($P < 0.0001$). Conversely, differences in predictive ability between the three prediction methods were not significant. This was also the case for interactions between prediction methods and phenotypic traits.

Genomic Prediction Across Populations

Under the S1 scenario, using all accessions of the RP as the training set, the predictive ability of GEBV of the 95 lines of VP was on average 0.426 for FL-*As* and 0.476 for CG-*As*

Table 8.2 Predictive ability (*r*) of three methods of genomic prediction for flag leaf (FL-*As*) and cargo grain (CG-*As*) arsenic content in the reference population, based on cross-validation experiments.

Prediction method	Phenotypic traits				
	FL-*As*		CG-*As*		
	r	sd	*r*	sd	Average *r*
BayesA	0.447	0.095	0.530	0.118	**0.435**
GBLUP	0.450	0.097	0.531	0.097	**0.437**
RKHS	0.457	0.094	0.543	0.123	**0.448**
Average	**0.452**	0.095	**0.535**	0.113	**0.440**

sd, standard deviation.

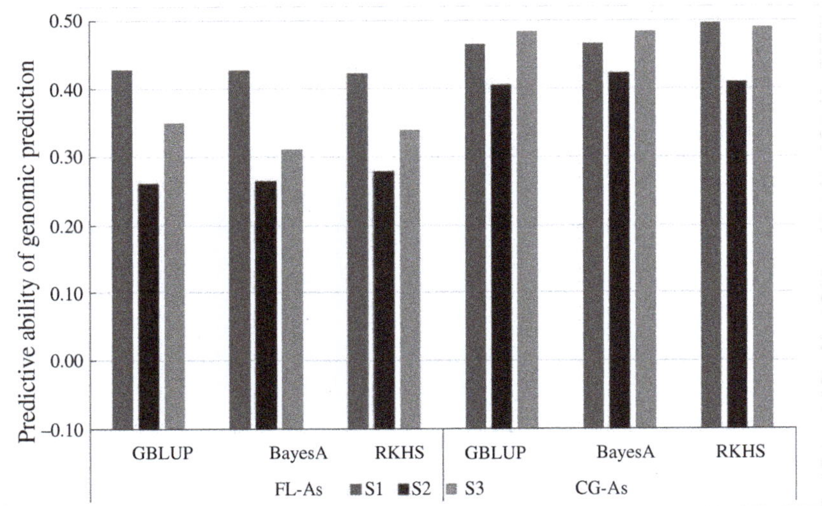

Figure 8.3 Predictive ability of genomic prediction of validation population phenotype for *As* concentration in the flag leaf (FL-*As*) and in the cargo grain (CG-*As*), obtained with three prediction methods (BayesA, GBLUP, and RKHS), under three scenarios of composition of the training set.

(Figure 8.3). The three prediction methods implemented provided similar levels of predictive ability. However, there was some interaction between prediction methods and phenotypic traits. Under the S2 scenario, the predictive ability of GEBV was much lower than under S1 for FL-*As* ($r = 0.266$ in average), comparable to S1 for CG-*As* ($r = 0.411$ in average) (Figure 8.3). Under S3, the average predictive ability was slightly higher than under S1 for CG-*As* (0.491), and much lower than under S1 for FL-*As* (0.341).

Implication for Breeding for Arsenic Tolerance

The GEBV obtained for FL-*As* and CG-*As* were reasonably accurate in both intra-population (cross-validation in the RP) and across-population (RP/VP) prediction experiments. Translation of those predictive abilities into average phenotypic performances of VP lines selected based on their GEBV, by model trained with the RP, is even more encouraging. Indeed, the average FL-*As* and CG-*As* of the 10 best VP lines selected on the basis of phenotypic data were 41% and 65% of the average FL-*As* and CG-*As* of all 95 lines of VP. The average FL-*As* and CG-*As* of the 10 best VP lines selected on the base of GEBV were 55% and 85% of the average FL-*As* and CG-*As* of all the 95 lines of VP. In other words, for a selection rate of approximately 10%, the difference in genetic gain between phenotypic selection and GEBV based selection was approximately 10% for FL-*As* and 5% for CG-*As*, in favor of phenotypic selection. Given these rather small differences in genetic gains, at first sight the choice between phenotypic and GEBV-based selection will depend mainly on the comparative costs of genotyping and phenotyping for *As* content. If the costs are similar, the best option would be GEBV-based selection because genotypic data are a multi-purpose asset that can also be used for genomic prediction of other traits in addition to *As* content. However, integration of GS in a breeding program has multiple implications (Ahmadi et al. 2019; Crossa et al. 2017) that require thorough evaluation.

Future Prospects

Analysis of rice genetic diversity for tolerance to *As* (i.e., low *As* uptake and or restricted *As* accumulation in the grain) has shown the existence of large genetic diversity. Some rice accessions accumulate 20–30 fold less *As* than others (Norton et al. 2012b) suggesting large potential for the development of elite lines combining yield potential and *As* tolerance. Further efforts in enlarging the available genetic diversity for *As* tolerance, should focus on the 3000 rice accessions that were sequenced (Wang et al. 2018) and on wild relatives of *O. sativa*.

Reports on the genetic basis of *As* tolerance, using QTL mapping approaches and transcriptome analysis, suggest it is a complex quantitative trait, which raises questions about the most effective breeding method. Recent GWAS experiments have confirmed the existence of some of the QTLs mapped using the progeny of experimental bi-parental crosses, and have made it possible to narrow down the position of those QTLs. These QTLs constitute reasonably good targets for conventional marker-assisted selection. Conditions for the successful implementation of such a program include the careful choice of the donor parent(s) in terms of membership in different genetic groups of *O. sativa*, and the choice of the target QTLs in terms of robustness and size of the effect. Adoption of the MARS scheme (Moreau et al. 2004) would further ascertain the effectiveness of the target QTLs in the background of the chosen elite line(s).

Reviewing the literature on genes involved in *As* uptake and transport in plants, Chen et al. (2017) proposed to resort to CRISPR/Cas9 (clustered regularly interspaced short palindromic repeats/CRISPR-associated) based gene editing technology (Hsu et al. 2014) to engineer rice with low *As*. They suggested that critical genes responsible for *As* uptake (e.g., OsPht1:8, Lsi1, and Lsi2) might be the early targets for gene editing. Additional targets for genome editing could be OsNRAMP1 and OsABCC1 genes, disruption or modification of which has the potential to develop low-*As* rice. However, Chen et al. (2017) recognized that the manipulation of OsPht1:8, Lsi1, and Lsi2 may also influence phosphorus (*P*) and *Si* uptake, and suggested searching for allelic variations that could selectively transport *P* and *Si* over *As*(V) and *As*(III), thereby reducing *As* uptake. Beside the issue of interference with *P* and *Si* uptake, the implementation of the gene-editing option raises the question of feasibility of dissemination of the low *As* new varieties. Indeed, according to article 3 g of the Cartagena protocol on biosecurity (https://bch.cbd.int/protocol), which was adopted in 2000 and entered into force in 2003), as a "product of modern biotechnology," genome-edited crop varieties are "genetically modified organisms" and as such, their dissemination is subject to biosafety regulations. Thus, in the short-to-medium run, the most profitable use of the genome-editing technology would be the validation of new candidate genes such as the ones identified by Frouin et al. (2019).

Considering the limitations of QTL-based marker-assisted selection and genome editing based varietal development, and considering the rather high predictive ability of genomic prediction we observed, the latter option seems the most promising for breeding for the ability to restrict *As* accumulation in the rice grain. As shown by Ben Hassen et al. (2017), a rice diversity panel could provide accurate genomic predictions for complex traits in the progenies of bi-parental crosses involving members of the panel. In addition, associated with the rapid generation advancement technique (Collard et al. 2017), genomic selection can accelerate the genetic gain of the pedigree breeding scheme, the most common breeding scheme in rice. GS for *As* content can be incorporated in such a breeding program. The main additional cost would be the phenotyping of the diversity/reference panel for *As* content. However, so far, the effectiveness of GS in breeding for *As* tolerance has been tested only in the *japonica* genetic background (Frouin et al. 2019). Additional experiments are needed to confirm the potential of GS to breed rice for low grain *As* content, especially in the *indica* and *Aus* genetic groups, that are the most exposed genetic groups to high soil and water *As* contents in Asia.

References

Abedin, M.J., Cotter, H., and Meharg, A.A. (2002). Arsenic uptake and accumulation in rice (*Oryza sativa* L.) irrigated with contaminated water. *Plant Soil* 240: 311–319.

Ahmadi, N., Bartholomé, J., Cao, T.V. et al. (2019). Genomic selection in rice: empirical results and implications for breeding. In: *Quantitative Genetics, Genomics and Plant Breeding*, 2nd edition (ed. M.J. Kang), 243–258. Wallingford, Oxon: CAB International.

Asoro, F.G., Newell, M.A., Beavis, W.D. et al. (2011). Accuracy and training population design for genomic selection on quantitative traits in elite North American oats. *Plant Genome* 4 (2): 132–144.

Bassi, F.M., Bentley, A.R., Charmet, G. et al. (2005). Breeding schemes for the implementation of genomic selection in wheat (*Triticum* spp.). *Plant Sci.* 242: 23–36. https://doi.org/10.1016/j.plantsci.2015.08.021.

Ben Hassen, M., Cao, T.V., Bartholomé, J. et al. (2017). Rice diversity panel provides accurate genomic predictions for complex traits in the progenies of biparental crosses involving members of the panel. *Theor. Appl. Genet.* 131 (2): 417–435. https://doi.org/10.1007/s00122-017-3011-4.

Ben Hassen, M., Bartholomé, J., Valè, G. et al. (2018). Genomic prediction accounting for genotype by environment interaction offers an effective framework for breeding simultaneously for adaptation to an abiotic stress and performance under normal cropping conditions in rice. *G3* 8 (9): 2319–2332. https://doi.org/10.1534/g3.118.200098.

Bernardo, R. and Yu, J. (2007). Prospects for genomewide selection for quantitative traits in maize. *Crop Sci.* 47: 1082–1090. https://doi.org/10.2135/cropsci2006.11.0690.

Bhandari, A., Bartholomé, B., Cao, T.V. et al. (2019). Selection of trait-specific markers and multi-environment models improve genomic predictive ability in rice. *PLoS One* 14 (5): e0208871. https://doi.org/10.1371/journal.pone.0208871.

Bhattacharya, P., Samal, A.C., Majumdar, J. et al. (2009). Transfer of arsenic from groundwater and paddy soil to rice plant (*Oryza sativa* L.): a micro level study in West Bengal, India. *WJAS* 5: 425–431.

Brammer, H. and Ravenscroft, P. (2009). Arsenic in groundwater: a threat to sustainable agriculture in South and South-East Asia. *Environ. Int.* 35: 647–654.

Briat, J.F. (2010). Arsenic tolerance in plants: "pas de deux" between phytochelatin synthesis and ABCC vacuolar transporters. *Proc. Natl. Acad. Sci. U.S.A.* 107 (49): 20853–20854. https://doi.org/10.1073/pnas.1016286107.

Carey, A.M., Scheckel, K.G., Lombi, E. et al. (2010). Grain unloading of arsenic species in rice. *Plant Physiol.* 152: 309–319. https://doi.org/10.1104/pp.109.146126.

Carey, A.M., Norton, G.J., Deacon, C. et al. (2011). Phloem transport of arsenic species from flag leaf to grain during grain filling. *New Phytol.* 192: 87–98.

Chakrabarty, D., Trivedi, P.K., Misra, P. et al. (2009). Comparative transcriptome analysis of arsenate and arsenite stresses in rice seedlings. *Chemosphere* 74: 688–702.

Charmet, G., Robert, N., Perretant, M.R. et al. (1999). Marker-assisted recurrent selection for cumulating additive and interactive QTLs in recombinant inbred lines. *Theo. Appl. Genet.* 99: 1143–1148.

Chen, Y., Moore, K.L., Miller, A.J. et al. (2015). The role of nodes in arsenic storage and distribution in rice. *J. Exp. Bot.* 66: 3717–3724. https://doi.org/10.1093/jxb/erv164.

Chen, Y., Han, Y.H., Cao, Y. et al. (2017). Arsenic transport in rice and biological solutions to reduce arsenic risk from rice. *Front. Plant Sci.* 8: 268. https://doi.org/10.3389/fpls.2017.00268.

Collard, B., Beredo, J., Lenaerts, B. et al. (2017). Revisiting rice breeding methods – evaluating the use of rapid generation advance (RGA) for routine rice breeding. *Plant Product. Sci.* 20 (4): 337–352. https://doi.org/10.1080/1343943X.2017.1391705.

Crossa, J., Pérez-Rodríguez, P., Cuevas, J. et al. (2017). Genomic selection in plant breeding: methods, models, and perspectives. *Trends Plant Sci.* 22 (11): 961–973. https://doi.org/10.1016/j.tplants.2017.08.011.

Dasgupta, S., Hossain, S.A., Meharg, A.A. et al. (2004). An arsenate tolerance gene on chromosome 6 of rice. *New Phytol.* 163: 45–49.

Dubey, S., Shri, M., Misra, P. et al. (2014). Heavy metals induce oxidative stress and genome-wide modulation in transcriptome of rice root. *Funct. Integr. Genom.* 14: 401–417. https://doi.org/10.1007/s10142-014-0361-8.

Duxbury, J.M., Mayer, A.B., Lauren, J.G. et al. (2003). Food chain aspects of arsenic contamination in Bangladesh: effects on quality and productivity of rice. *J. Environ. Sci. Health. A Tox. Hazard. Subst. Environ. Eng.* 38: 61–69.

European Food Safety Authority (2009). Panel on contaminants in the food chain (CONTAM). Scientific opinion on arsenic in food. *EFSAJ* 7: 1351.

Fan, Y., Zhu, T., Li, M. et al. (2017). Heavy metal contamination in soil and brown rice and human health risk assessment near three mining areas in Central China. *J. Healthc. Eng.* 2017: 4124302. https://doi.org/10.1155/2017/4124302.

Frouin, J., Labeyrie, A., Boisnard, A. et al. (2019). Genomic prediction offers the most effective marker assisted breeding approach for ability to prevent arsenic accumulation in rice grains. *PLoS One* 14 (6): e0217516. https://doi.org/10.1371/journal.pone.0217516.

Gray, P., Conklin, S.D., Todorov, T.I. et al. (2015). Cooking rice in excess water reduces both arsenic and enriched vitamins in the cooked grain. *Food Addit. Contam.* 33 (1): 1–8. https://doi.org/10.1080/19440049.2015.1103906.

Heffner, E.L., Sorrells, M.E., and Jannink, J.L. (2009). Genomic selection for crop improvement. *Crop Sci.* 49 (1): 1–12.

Hite, A.H. (2013). Arsenic and rice: a call for regulation. *Nutrition* 29: 353–354.

Hojsak, I., Braegger, C., Bronsky, J. et al. (2015). Arsenic in Rice – a cause for concern. *J. Pediatr. Gastroenterol. Nutr.* 60: 142–145.

Hsu, P.D., Lander, E.S., and Zhang, F. (2014). Development and applications of CRISPR-Cas9 for genome engineering. *Cell* 157: 1262–1278. https://doi.org/10.1016/j.cell.2014.05.010.

Jia, Y., Huang, H., Zhong, M. et al. (2013). Microbial arsenic methylation in soil and rice rhizosphere. *Environ. Sci. Technol.* 47: 3141–3148. https://doi.org/10.1021/es303649v.

Kuramata, M., Abe, T., Kawasaki, A. et al. (2013). Genetic diversity of arsenic accumulation in rice and QTL analysis of methylated arsenic in rice grains. *Rice* 6 (3): 2–10.

Li, R.Y., Stroud, J.L., Ma, J.F. et al. (2009). Mitigation of arsenic accumulation in rice with water management and silicon fertilization. *Environ. Sci. Technol.* 43: 3778–3783.

Liu, K., Zheng, J., and Chen, F. (2018). Effects of washing, soaking and domestic cooking on cadmium, arsenic and lead bioaccessibilities in rice. *J. Sci. Food Agric.* 98 (10): 3829–3835. https://doi.org/10.1002/jsfa.8897.

Lomax, C., Liu, W.J., Wu, L. et al. (2012). Methylated arsenic species in plants originate from soil microorganisms. *New Phytol.* 193: 665–672. https://doi.org/10.1111/j.1469-8137.2011.03956.x.

Lorenz, A.J., Chao, S., Asoro, F.G. et al. (2011). Genomic selection in plant breeding: knowledge and prospects. *Adv. Agron.* 110: 77–123. https://doi.org/10.1016/B978-0-12-385531-2.00002-5.

Ma, J.F. and Yamaji, N. (2006). Silicon uptake and accumulation in higher plants. *Trends Plant Sci.* 11: 392–397.

Ma, J.F., Yamaji, N., Mitani, N. et al. (2008). Transporters of arsenite in rice and their role in arsenic accumulation in rice grain. *Proc. Natl. Acad. Sci. U.S.A.* 105: 9931–9935.

Meharg, A.A., Williams, P.N., and Adomako, E. (2009). Geographical variation in total and inorganic arsenic content of polished (white) rice. *Environ. Sci. Technol.* 43: 1612–1617.

Meuwissen, T.H., Hayes, B.J., and Goddard, M.E. (2001). Prediction of total genetic value using genome-wide dense marker maps. *Genetics* 157: 1819–1829.

Misra, P., Pandey, A., Tiwari, M. et al. (2010). Modulation of transcriptome and metabolome of tobacco by Arabidopsis transcription factor, AtMYB12, leads to insect resistance. *Plant Physiol.* 152: 2258–2268. https://doi.org/10.1104/pp.109.150979.

Mitra, A., Chatterjee, S., Moogouei, R. et al. (2017). Arsenic accumulation in rice and probable mitigation approaches: a review. *Agronomy* 7 (4): 67. https://doi.org/10.3390/agronomy7040067.

Moreau, L., Charcosset, A., and Gallais, A. (2004). Use of trail clustering to study QTL x environment effects for grain yield and related traits in maize. *Theor. Appl. Genet.* 110: 92–105.

Norton, G.J., Duan, G., Dasgupta, T. et al. (2009). Environmental and genetic control of arsenic accumulation and speciation in rice grain: comparing a range of common cultivars grown in contaminated sites across Bangladesh, China and India. *Environ. Sci. Technol.* 43: 8381–8386.

Norton, G.J., Deacon, C.M., Xiong, L. et al. (2010). Genetic mapping of the rice ionome in leaves and grain: identification of QTLs for 17 elements including arsenic, cadmium, iron and selenium. *Plant Soil* 329: 139–153.

Norton, G.J., Duan, G., Lei, M. et al. (2012a). Identification of quantitative trait loci for rice grain element composition on an arsenic impacted soil: influence of flowering time on genetic loci. *Ann. Appl. Biol.* 161: 46–56.

Norton, G.J., Pinson, S.R.M., Alexander, J. et al. (2012b). Variation in grain arsenic assessed in a diverse panel of rice (*Oryza sativa*) grown in multiple sites. *New Phytol.* 193: 650–664.

Norton, G.T., Douglas, A., Lahner, B. et al. (2014). Genome wide association mapping of grain arsenic, copper, molybdenum and zinc in rice (*Oryza sativa* L.) grown at four international field sites. *PLoS One* 9 (2): 89685.

Pandey, A., Misra, P., Chandrashekar, K. et al. (2012). Development of AtMYB12-expressing transgenic tobacco callus culture for production of rutin with biopesticidal potential. *Plant Cell Rep.* 31: 1867–1876. https://doi.org/10.1007/s00299-012-1300-6.

Peleman, J.D. and van der Voort, J.R. (2003). Breeding by design. *Trends Plant Sci.* 8: 330–334.

Perez, P. and de Los Campos, G. (2014). Genome-wide regression and prediction with the BGLR statistical package. *Genetics* 198 (2): 483–495. https://doi.org/10.1534/genetics.114.164442.

Rahman, M.A., Hasegawa, H., Rahman, M. et al. (2007). Effect of arsenic on photosynthesis, growth and yield of five widely cultivated rice (*Oryza sativa* L.) varieties in Bangladesh. *Chemosphere* 67: 1072–1079. https://doi.org/10.1016/j.chemosphere.2006.11.061.

Rai, A., Bhardwaj, A., Misra, P. et al. (2015). Comparative transcriptional profiling of contrasting rice genotypes shows expression differences during arsenic stress. *Plant Genome* 8 (2): 1–14. https://doi.org/10.3835/plantgenome2014.09.0054.

Rincent, R., Laloë, D., Nicolas, S. et al. (2012). Maximizing the reliability of genomic selection by optimizing the calibration set of reference individuals: comparison of methods in two diverse groups of maize inbreds (*Zea mays* L.). *Genetics* 192 (2): 715–728. https://doi.org/10.1534/genetics.112.141473.

Shi, S., Wang, T., Chen, Z. et al. (2016). OsHAC1;1 and OsHAC1;2 function as arsenate reductases and regulate arsenic accumulation. *Plant Physiol.* 172: -1708, 1719. https://doi.org/10.1104/pp.16.01332.

Song, W.Y., Park, J., Mendoza-Cozatl, D.G. et al. (2010). Arsenic tolerance in Arabidopsis is mediated by two ABCC-type phytochelatin transporters. *Proc. Natl. Acad. Sci. U.S.A.* 107: 21187–21192.

Song, W.Y., Yamaki, T., Yamaji, N. et al. (2014). A rice ABC transporter, OsABCC1, reduces arsenic accumulation in the grain. *Proc. Natl. Acad. Sci. U. S. A.* 111: 15699–15704. https://doi.org/10.1073/pnas.1414968111.

Sorrells, M.E. (2015). Genomic Selection in Plants: Empirical Results and Implications for Wheat and Barley Breeding Programs. Proceedings of the 12th International Wheat Genetics Symposium. doi: https://doi.org/10.1007/978-4-431-55675-6_45

Stam, P. (1995). Marker-assisted breeding. In: *Biometrics In Plant Breeding: Applications of Molecular Markers. Proceedings of the Ninth Meeting of the EUCARPIA Section Biometrics in Plant Breeding* (eds. J.W. Van Ooijen and J. Jansen), 32–44. Wageningen, Netherlands: CPRO-DLO.

Stroud, J.L., Norton, G.J., Islam, M.R. et al. (2011). The dynamics of arsenic in four paddy fields in the Bengal Deltas. *Environ. Pollut.* 159: 947–953.

Sun, G.X., Williams, P.N., and Carey, A.M. (2008). Inorganic arsenic in rice bran and its products are an order of magnitude higher than in bulk grain. *Environ. Sci. Technol.* 42: 7542–7546.

Tripathi, P., Tripathi, R.D., Singh, R.P. et al. (2012). Silicon mediates arsenic tolerance in rice (*Oryza sativa* L.) through lowering of arsenic uptake and improved antioxidant defense system. *Ecol. Eng.* 52: 96–103. https://doi.org/10.1016/j.ecoleng.2012.12.057.

Verbruggen, N., Hermans, C., and Schat, H. (2009). Mechanisms to cope with arsenic or cadmium excess in plants. *Curr. Opin. Plant Biol.* 12: 364–372. https://doi.org/10.1016/j.pbi.2009.05.001.

Wang, W., Mauleon, R., Hu, Z. et al. (2018). Genomic variation in 3,010 diverse accessions of Asian cultivated rice. *Nature* 557 (7703): 43–49. https://doi.org/10.1038/s41586-018-0063-9.

WHO (2001). Arsenic in drinking water. Fact Sheet No. 210. https://apps.who.int/inf-fs/en/fact210.html

WHO (2011). Safety evaluation of certain contaminants in food. WHO Food Additives Series No. 63. Geneva, Switzerland. http://whqlibdoc.who.int/publications/2011/9789241660631_eng.pdf

Williams, P.N., Price, A.H., Raab, A. et al. (2005). Variation in arsenic speciation and concentration in paddy rice related to dietary exposure. *Environ. Sci. Technol.* 39: 5531–5540. https://doi.org/10.1021/es0502324.

Xie, C. and Xu, S. (1999). Mapping quantitative trait loci with dominant and missing markers in four-way crosses. *Theor. Appl. Genet.* 98: 1014–1021.

Xu, X.Y., McGrath, S.P., and Zhao, F.J. (2007). Rapid reduction of arsenate in the medium mediated by plant roots. *New Phytol.* 176: 590–599.

Xu, J., Shi, S., Wang, L. et al. (2017). OsHAC4 is critical for arsenate tolerance and regulates arsenic accumulation in rice. *New Phytol.* 215: 1090–1101. https://doi.org/10.1111/nph.14572.

Yamaji, N. and Ma, J.F. (2014). The node, a hub for mineral nutrient distribution in graminaceous plants. *Trends Plant Sci.* 19: 556–563. https://doi.org/10.1016/j.tplants.2014.05.007.

Yamaji, N., Sakurai, G., Mitani-Ueno, N. et al. (2015). Orchestration of three transporters and distinct vascular structures in node for intervascular transfer of silicon in rice. *Proc. Natl. Acad. Sci. U. S. A.* 112: 11401–11406. https://doi.org/10.1073/pnas.1508987112.

Ye, W.L., Wood, B.A., Stroud, J.L. et al. (2010). Arsenic speciation in phloem and xylem exudates of castor bean. *Plant Physiol.* 154: 1505–1513. https://doi.org/10.1104/pp.110.163261.

Zavala, Y.J. and Duxbury, J.M. (2008). Arsenic in rice: I. Estimating normal levels of total arsenic in rice grain. *Environ. Sci. Technol.* 42: 3856–3860.

Zavala, Y.J., Gerads, R., Gorleyok, H. et al. (2008). Arsenic in rice: II. Arsenic speciation in USA grain and implications for human health. *Environ. Sci. Technol.* 42: 3861–3866.

Zhang, M., Pinson, S.R.M., Tarpley, L. et al. (2014). Mapping and validation of quantitative trait toci associated with concentration of 16 elements in unmilled rice grain. *Theor. Appl. Genet.* 127 (1): 137–165. https://doi.org/10.1007/s0012-013-2207-5.

Zhao, F.J., Ma, J.F., Meharg, A.A. et al. (2009). Arsenic uptake and metabolism in plants. *New Phytol.* 181: 777–794.

Zhao, F.J., McGrath, S.P., and Meharg, A.A. (2010b). Arsenic as a food chain contaminant: mechanisms of plant uptake and metabolism and mitigation strategies. *Annu. Rev. Plant Biol.* 61: 535–559.

Zhao, X.Q., Mitani, N., Yamaji, N. et al. (2010c). Involvement of silicon influx transporter OsNIP2;1 in selenite uptake in rice. *Plant Physiol.* 153: 1871–1877.

9

Molecular Breeding for Improving Ozone Tolerance in Rice: Recent Progress and Future Perspectives

Md. Ashrafuzzaman[1], *Robert Henry*[2] *and Michael Frei*[3]

[1] *Department of Genetic Engineering & Biotechnology, Shahjalal University of Science & Technology, Sylhet 3114, Bangladesh*
[2] *Queensland Alliance for Agriculture and Food Innovation, University of Queensland, Brisbane 4072, Australia*
[3] *Institute for Agronomy and Crop Physiology, Justus Liebig University, 35392, Giessen, Germany*

Introduction

Ozone mainly evolves in two areas of the earth's atmosphere, namely, the stratosphere and the troposphere. Stratospheric ozone (c. 90%) lies between 10 and 17 km above the Earth's surface and extends up to c. 50 km, which is commonly known as the ozone layer (WMO 2014). The remaining ozone in the lower atmospheric region is known as tropospheric or ground-level ozone. The ozone layer in the stratosphere plays an important and beneficial role by absorbing most of the sun's harmful ultraviolet radiation (UV-B) (Donahue 2011). In contrast, elevated tropospheric ozone has been shown to be harmful to human health, vegetation, and crop production (Ashmore et al. 2006; Dentener et al. 2006). While the ozone concentration in the stratosphere is being depleted, tropospheric ozone concentrations are increasing due to increased emissions of its precursor gases as well as global warming (Ashmore 2005; Tai and Val Martin 2017). The formation of ozone is accelerated by hot sunny weather, and therefore tropical regions are at high risk for ozone pollution due to favorable environmental conditions and high emissions of precursor gases (Jain et al. 2005; Tiwari et al. 2008).

Tropospheric ozone is one of the most important environmental pollutants currently adversely affecting vegetation, human health, microbes, and agricultural crop production (Ainsworth et al. 2012; Li et al. 2017; Mills et al. 2018; Ueda et al. 2016). It is also the third-most-potent anthropogenic greenhouse gas, after CO_2 and methane (Myhre et al. 2013). This secondary air pollutant is formed through light-driven reactions between primary air pollutants such as nitrogen oxides (NOx), volatile organic compounds (VOCs), carbon monoxide (CO), and methane (CH_4), which are also termed as ozone precursor gases (Figure 9.1; The Royal Society 2008). Sources of ozone precursor gas include natural and human activities such as energy generation, fossil-fuel combustion, industrialization, transportation, urbanization, deforestation, and rapid population and economic growth

Molecular Breeding for Rice Abiotic Stress Tolerance and Nutritional Quality, First Edition.
Edited by Mohammad Anwar Hossain, Lutful Hassan, Khandakar Md. Iftekharuddaula, Arvind Kumar and Robert Henry.

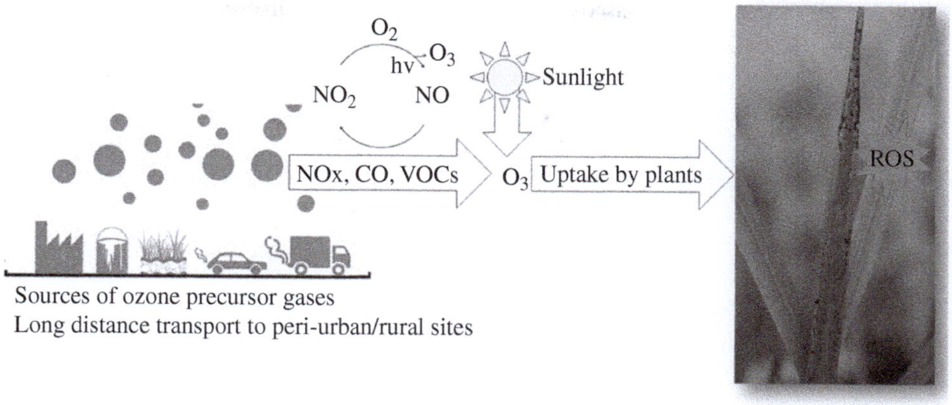

Figure 9.1 A Simplified View of the Major Sources and Formation of Tropospheric Ozone. NOx, nitrogen oxides; CO, carbon monoxide; VOCs, volatile organic compounds; O_3, ozone; ROS, reactive oxygen species. Source: Adapted from the Royal Society (2008). Ground-level ozone in the 21st century: future trends, impacts and policy implications. London: The Royal Society; Climate and Clean Air Coalition, Tropospheric ozone, https://www.ccacoalition.org/ru/slcps/tropospheric-ozone. (*See color plate section for color representation of this figure*)

(Brauer et al. 2016; Cho et al. 2011; IPCC 2014). The phytotoxic effects of elevated tropospheric ozone on different crop and plant species are well documented and significantly affect worldwide crop production. For instance, global modeling studies have demonstrated global yield reduction of around 13% for soybeans, 7% for wheat, 5% for rice, and 6% for maize (Mills et al. 2018), and an annual economic loss totaling USD 14–26 billion (Van Dingenen et al. 2009) due to ozone pollution. Crop damage caused by air pollution was first noticed in the United States in the 1940s (Middleton et al. 1950), and tropospheric ozone pollution impacts were first noted in California during the 1950s as the cause of "weather fleck" in tobacco (Haagen-Smit 1952). The ozone damage symptom of plants was further reproduced in the laboratory by the reaction of organic trace gases or car exhaust with nitrogen oxides (NOx) in the presence of sunlight (Haagen-Smit and Fox 1954; Heggestad and Middleton 1959). Consequently, increasing tropospheric ozone concentrations and adverse effects on plants and vegetation in other parts of North America, Europe, and in Japan were gradually reported throughout the 1970s (The Royal Society 2008).

Ozone pollution is not uniformly distributed over the global land surface, and concentrations tend to be higher during warmer summer months and low during winter and rainy seasons (Ainsworth 2017; Brauer et al. 2016). Mean ozone concentrations range from approximately 20 ppb over parts of Australia and South America to 55–60 ppb in parts of Asia, the Middle East, Europe, and North America (Gillespie et al. 2012). Nevertheless, even relatively low ozone concentrations can have adverse effects on sensitive crop and plant species (Agathokleous et al. 2015; Sugai et al. 2018). Therefore, several ozone phytotoxicity metrics have been proposed and used so far globally. These are principally based on ozone exposure or the accumulated stomatal ozone flux (Musselman et al. 2006; Pleijel et al. 2004). Ozone exposure based metrics are AOT40 (accumulated ozone exposure over a threshold of 40 ppb), SUM06 (sum of hourly ozone concentrations greater than 60 ppb), and W126 (sum of all hourly mixing ratios observed during a particular daily and seasonal time without a threshold value), and the flux-based available metrics are DO_3SE (deposition of ozone and stomatal exchange), AFstY (the cumulative leaf uptake of ozone per total leaf area over time), and POD_Y (phytotoxic ozone dose) (CLRTAP 2017; Emberson et al. 2001; Karlsson et al. 2004; Wang and Mauzerall 2004). Notably, ozone risk assessment based on stomatal uptake or flux models provides a stronger indication of ozone effects in Europe compared to those based on concentrations (Mills et al. 2011; Pleijel et al. 2004; Simpson et al. 2007). In contrast, AOT40 is the most widely adopted critical index recommended by several regulatory agencies, for its simple calculation and suitability for highly polluted regions (Agathokleous et al. 2018a). Thus, many of the productive areas worldwide especially in parts of Asia are endangered by current tropospheric ozone and will be further menaced in the near future by rising ozone trends. Therefore, to protect crops for sustainable production, breeding and biotechnological improvement of crops for ozone tolerance are required, and may be more promising than adaptation to rising ozone concentrations through altered crop management practices (Ainsworth 2008; Frei 2015; Teixeira et al. 2011).

Tropospheric Ozone: A Major Threat to Rice Production in Asia

Several projections have reported that in the past few decades, rising tropospheric ozone concentrations were controlled by adopting various air-quality regulations in North America and Europe (Ainsworth 2017; Chang et al. 2017). In contrast, the same trend is not true in most of the world's populous and developing parts, especially rice-producing countries in Asia, which are currently marked as an ozone hotspot (Mills et al. 2018). The pollution in these parts of the world will continue to grow in the twenty-first century because of rapid urbanization, industrial development, higher emission of precursor gases, and lack of air-quality regulations (Brauer et al. 2016; Maas and Grennfelt 2016). During the summer cropping season, high tropospheric ozone concentrations (c. 100 ppb), far above the damage threshold level (40 ppb) have been reported in many parts of China, India, Pakistan, and Bangladesh (Brauer et al. 2016; Deb Roy et al. 2009; Ran et al. 2009). Brauer et al. (2016) estimated the highest changes in daily maximum ozone concentration in Bangladesh, India, and Pakistan (c. 20%) compared with global changes (c. 9%) between the year of 1990 and 2013 (Figure 9.2). On the other hand, China is the largest emitter of ozone precursor gases (NOx) in Asia (Feng et al. 2015).

Rice (*Oryza sativa* L.) is the staple food crop of half of the world's population and widely grown and consumed in Asia (McCouch et al. 2016). In terms of quantity, rice is the third-highest (c. 750 million tonnes) agricultural commodity in the world after sugarcane and maize (FAOSTAT 2019). The production and consumption (c. 90% of rice globally) is dominated by Asian countries. Among them, China is the leading rice-producing country followed by India (2nd) and Bangladesh (4th), and interestingly these three countries along with Pakistan, Nepal, and Japan produced more than half (c. 63%) of rice globally (Figure 9.3, FAOSTAT 2019). Therefore, rice production in these highly ozone polluted areas is at high risk and will be aggravated in the near future (Frei 2015).

Numerous studies, conducted mostly in controlled environments, have extensively documented the adverse effects of ozone on rice growth, development, yield, and quality (reviewed by Frei 2015; Ainsworth 2008). Rice yield reduction caused by ozone pollution is

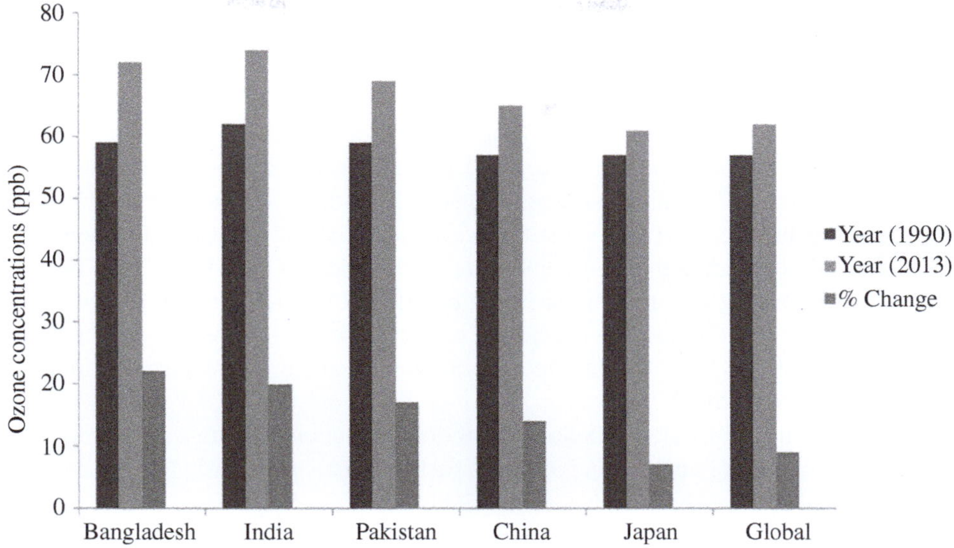

Figure 9.2 Changes of Average One-Hour Daily Maximum Ozone (ppb) Concentrations between the Years 1990 and 2013. Source: Adapted from Brauer, M., Freedman, G., Frostad, J., et al. (2016). Ambient air pollution exposure estimation for the global burden of disease 2013. Environ. Sci. Technol., 50, 79–88.

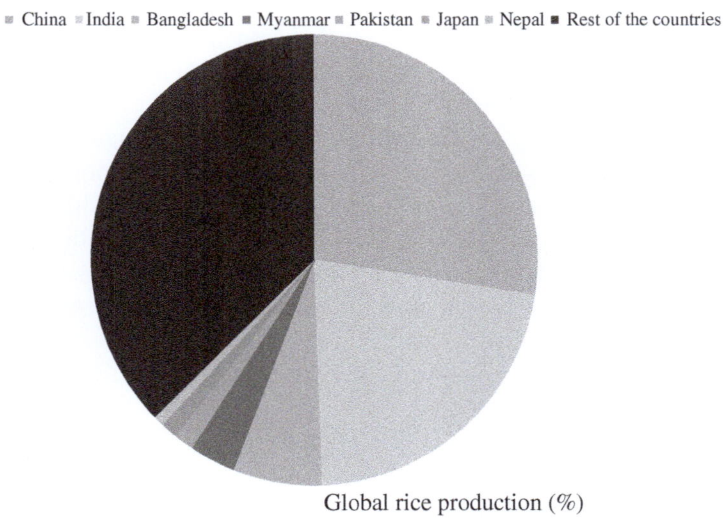

Figure 9.3 Chart Showing % Rice Production by Leading Asian Countries Compared to Global. Source: The chart is derived from global crop production databases of the Food and Agricultural Organization (FAOSTAT 2019).

due to a lower number of spikelets per panicle and individual grain mass (Frei et al. 2012a; Wang et al. 2012), higher spikelet sterility (Yamaguchi et al. 2014), and lower number of tillers (Akhtar et al. 2010; Frei et al. 2008). A meta-analysis has estimated that c. 18% of rice yield reduction may occur with 31–50 ppb of ozone, and a 10% additional loss will be seen at 51–75 ppb ozone concentrations (Feng and Kobayashi 2009). In contrast, more than 10% of regional loss (Van Dingenen et al. 2009), as well as c. 20% of East Asian (Chen et al. 2011) and more than 15% of Indian rice (Debaje 2014; Mills et al. 2018) yield damage is projected to be due to ozone pollution in each year using global crop models. Furthermore, Lal et al. (2017) calculated the total annual rice production loss in India to be 6.7 million tonnes (corresponding to 6.3%) due to ozone pollution, in which the highest reduction (2.6 million tonnes) is estimated in the eastern part, considering the average yield between the year 2011 and 2014. Thus, rising tropospheric ozone is clearly one of the major limiting factors for rice production.

Several projections have clearly indicated that rice production in South Asia especially in the Indian subcontinent's developing countries such as India, Bangladesh, Pakistan, Myanmar, and Nepal are already severely affected by high ozone pollution (Frei 2015; Lal et al. 2017). Ozone levels are expected to increase further until the middle of the twenty-first century due to higher emissions of precursor gases, favorable environmental conditions for ozone formation, and lack of air-quality regulations (Lei et al. 2012; Williams et al. 2014). "Business as usual" scenarios support this trend and indicate that these areas and particularly the Indian subcontinent will be exposed to elevated tropospheric ozone pollution (Pozzer et al. 2012; Van Dingenen et al. 2009). The peak ozone episodes in the Indian subcontinent are seen during the warm and humid summer between the months of March and June (Figure 9.4; Ziemke et al. 2006). Furthermore, rice is produced almost all year round in these regions during two major seasons; the wet season (June to November), in which high ozone levels can overlap with the crucial early vegetative stages including tillering, and the dry season (December to May), in which reproductive growth stages (flowering and grain filling) can coincide with high ozone episodes (Frei 2015). Therefore, both growing seasons are at high risk as different growth phases of rice cultivation overlap with peak ozone episodes, which leads to a critical threat for yield and production (Frei 2015). However, field-based ozone damage evaluation is missing in these highly ozone-affected areas due to lack of infrastructure and ozone research facilities.

Mechanisms of Ozone Damage

Elevated ozone has detrimental effects on plant physiology, and affects numerous metabolic pathways including decreased stomatal conductance and photosynthesis, cell death, plant growth and development, accelerated senescence, and an altered antioxidant system (Frei et al. 2010a; Feng and Kobayashi 2009; Ueda et al. 2015a; Wilkinson et al. 2012). Due to its chemical nature as a molecule composed of three oxygen atoms, ozone is a strong oxidant. Therefore, ozone damage occurs to plants directly through an oxidative burst (Kangasjärvi et al. 2005). It enters plants through the stomata during photosynthetic gas exchange and decomposes into various reactive oxygen species (ROS) such as singlet oxygen ($^1O_2$), hydrogen peroxide (H_2O_2), superoxide (O_2^-), and hydroxyl radicals (·OH) in the humid

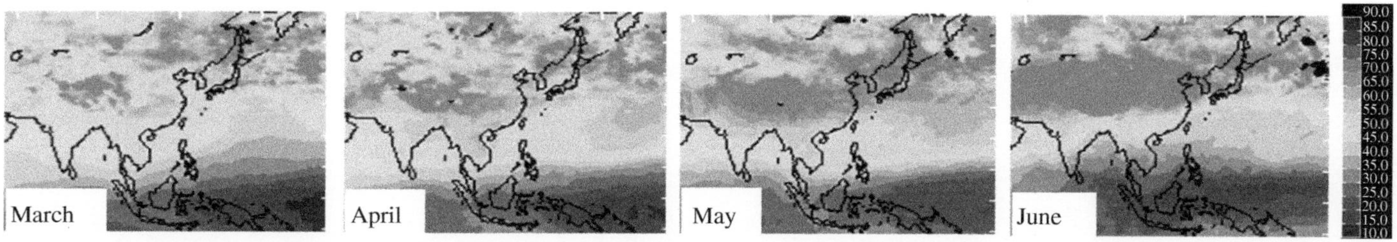

Figure 9.4 Monthly average tropospheric ozone volume mixing ratio (march to june, peak ozone periods in the Indian subcontinents) in Asia in the Year 2016. The color scale indicates the density of ozone divided by the density of all constituents in a unit volume of air (ppbv) in the total tropospheric column. Source: Adapted from Ziemke et al. (2006). (*See color plate section for color representation of this figure*).

apoplastic environment. As plants have not co-evolved with high tropospheric ozone concentrations throughout most of their evolutionary history, specific ozone receptors in the apoplast triggering stress responses are unlikely to exist. Rather, plants respond to ozone with an apoplastic oxidative burst characteristic of a pathogen attack (Kangasjärvi et al. 2005). The signaling cascades triggered by such an oxidative burst induce programmed cell death (PCD), which leads to the formation of characteristic necrotic lesions on plant leaves (Kangasjärvi et al. 2005). In addition, plants close their stomata, sometimes within a few minutes after exposure to high ozone concentrations. The reason for this is contested but may be related to impaired carbon fixation due to oxidative damage of photosynthetic enzymes (Paoletti and Grulke 2005), or the involvement of apoplastic ROS in stomatal apertures (Sierla et al. 2016). Irrespective of the cause of stomatal closure, ozone exposure reduces photosynthetic activity, which leads to reductions in plant growth and biomass (Emberson et al. 2018). In rice, as an indicator of ozone stress, visible brown spots on leaves (commonly termed leaf bronzing), and damage to membrane lipids have been documented in many studies (Frei et al. 2008; Sawada and Kohno 2009).

However, cellular response to ozone varies with the ozone concentrations, and also differs among species (Chen et al. 2009; Grantz and Vu 2012; Yendrek et al. 2015). The duration and extent of ozone-induced cell damage is determined by different plant hormones. Salicylic acid causes PCD; jasmonic acid is required for limiting the lesion spreading, ethylene promotes endogenous ROS formation, and abscisic acid (ABA) is expected to affect stomatal aperture resulting in reduction in leaf lesion formation (Kangasjärvi et al. 2005; McAdam et al. 2017). In contrast, the effects of elevated ozone on crop physiological processes also differ over a growing season. For instance, reductions in photosynthesis due to elevated ozone were higher in the reproductive phases compared to the vegetative phases of development, partly because leaves that developed during late reproductive phases were exposed to elevated ozone at the top of the canopy for an extended period of time (Feng et al. 2008; Morgan et al. 2003). In the reproductive phase, ozone directly affects flowering, pollen sterility, grain injury, and abortion (Wilkinson et al. 2012).

Ozone Tolerance Mechanisms

Tolerance against ozone stress can be achieved via control of ozone entry into the leaf, referred to as stress avoidance. In contrast, stress defense can be associated with the antioxidant system capable of mitigating ROS accumulation thus protecting membranes from lipid degradation (Frei 2015; Krasensky et al. 2017). Stomatal closure can be the first line of defense against ozone stress as the stomatal aperture is the main factor controlling the influx of ozone into plant leaves (Ainsworth 2017; Frei 2015). Vahisalu et al. (2008) described a plasma-membrane slow-anion channel (SLAC1), preferentially expressed in guard cells, which is involved in ozone-induced stomatal closure. Initiation of SLAC1 enhances anion efflux and depolarization of the cell membranes, which subsequently activates potassium-efflux channels. GOLDEN 2-LIKE1 and 2 (GLK1, GLK2) transcription factors act as a positive regulator of potassium influx channels and stomatal movement. Therefore, suppression of GLK1 and GLK2 resulted in decreased stomatal aperture and water loss under ozone stress, resulting in enhanced ozone tolerance (Nagatoshi et al.

2016). On the other hand, the Open Stomata 1 (OST1) gene can play an important role in ABA-dependent stomatal responses by activating several ion channels in guard cells (Vahisalu et al. 2010; Wilkinson and Davies 2010). Brosche et al. (2010) stated that stomatal conductance was the most critical factor determining ozone sensitivity for the model plant *Arabidopsis thaliana* under an acute ozone treatment (350 ppb for seven hours). Consistently, stomatal conductance also showed a high correlation with ozone tolerance in soybeans (Chutteang et al. 2016), and wheat (Biswas et al. 2008), but seemed to be of minor importance in poplar (Ryan et al. 2009) and beans (Guidi et al. 2010). On the other hand, high stomatal conductance is a crucial trait for crop improvement as it is directly related to the entry of CO_2, which ultimately affects productivity.

The second line of defense, enhanced detoxification of ozone or ozone-derived ROS in the apoplast can be initiated after the entry of ozone into the leaf through the stomata (Frei 2015; Kangasjärvi et al. 2005). Among the numerous enzymatic and non-enzymatic antioxidants, ascorbate (AsA) is particularly important for ozone defense, as it is also found in significant amounts in the apoplast (Conklin and Barth 2004; Frei et al. 2012b; Yendrek et al. 2015). Apoplastic AsA detoxifies 30–50% of ozone entering the leaves in different species (Moldau et al. 1997; Turcsanyi et al. 2000). Conklin et al. (1996) performed ozone fumigation as a screening tool to discover AsA biosynthetic genes in *A. thaliana* mutant collections and found a hypersensitive mutant containing only 30% of the wild-type AsA, thereby highlighting that the total AsA pool is an important factor determining ozone tolerance in plants. Based on this finding, the GDP-D-mannose/L-galactose AsA biosynthetic pathway was discovered in plants (Smirnoff et al. 2001; Wheeler et al. 1998). In addition, overexpressing the AsA recycling enzymes dehydroascorbate reductase (DHAR) and monodehydrasorbate reductase enhanced ozone tolerance under acute and chronic exposure (Chen and Gallie 2005; Eltayeb et al. 2007). On the other hand, considerable biomass and yield reduction was observed in a season-long ozone fumigation experiment in a rice mutant lacking expression of an AsA biosynthetic gene (GDP-D-mannose-3′,5′-epimerase) (Frei et al. 2012b). Moreover, improved performance in ozone stress of transgenic rice lines with enhanced foliar AsA was observed when ozone coincided with salinity (Ali et al. 2019). However, some other studies have also stated that apoplastic AsA levels were not related to ozone tolerance (D'Haese et al. 2005), and may play a minor role in ozone detoxification (Sanmartin et al. 2003). Therefore, other antioxidants can play a vital role in ozone tolerance. Although a complete understanding of the metabolome and proteome of the apoplastic compartment is lacking, it is obvious that many additional antioxidant metabolites and enzymes are present in the apoplast, and may be responsible for substantial variation among species (Ainsworth 2017). For example, the enzymes involved in reducing oxidized AsA and glutathione are present in the apoplast in barley (Vanacker et al. 1998), but most of them are missing in the apoplast of peas (Hernandez et al. 2001).

Ozone-Impact Assessments

The primary prerequisite for ozone-tolerance rice breeding is to screen for genetic variation (Frei 2015). Adverse effects of ozone on plants and vegetation are mainly evaluated through sun-lit plant growth chambers, open-top chambers (OTC) and free-air

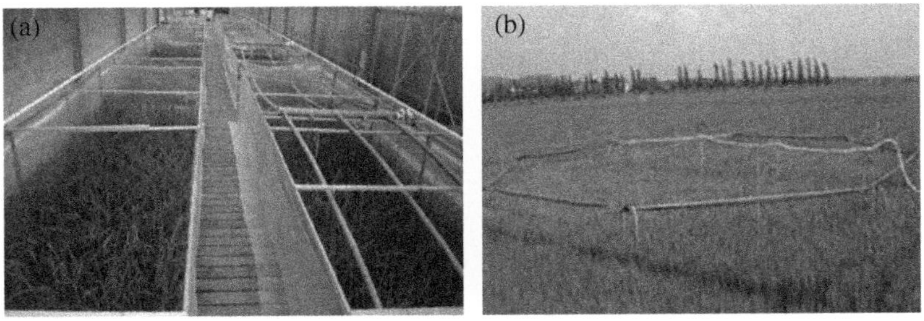

Figure 9.5 Photographs of open top chambers (OTC) and free air concentration enrichment (FACE) experimental system (a) OTC. Source: Dr. Michael Frei. (b) FACE. *Source:* Prof. YANG Lianxin.

concentration-enrichment with ozone (FACE-O_3) experimental systems so far (Kobayashi 2015). Sun-lit chambers are environmentally controlled facilitated chambers, in which ozone concentration can be maintained at a desired level for the purpose of better understanding mechanisms of ozone damage on a smaller scale (Kobayashi 2015). The OTC system, that was first introduced in the early 1970s by Heagle et al. (1973) is the most widely used environmentally controlled system in which ozone concentrations are maintained at desired level with artificially generated ozone blown into the chamber (Figure 9.5a; Frei 2015). FACE is a more realistic chamber-less system for field conditions, in which air enriched with elevated ozone is released into the wind through rings of pipes just above the canopy to achieve a certain ozone level compared with ambient ozone concentrations (Figure 9.5b; Morgan et al. 2004). Nevertheless, both the OTC and FACE-ozone system have a number of limitations. OTC may overestimate ozone impacts by modifying the microclimate (especially, temperature, CO_2 concentration, and humidity) due to "chamber effects," which can differ from actual field conditions (Kobayashi 2015; Piikki et al. 2008). Moreover, the small dimensions are usually not suitable for larger-scale applications (Macháčová 2010). On the other hand, FACE-ozone offers a unique opportunity to screen a large number of genotypes in field conditions but it can only compare between current ambient and elevated ozone concentrations (Ainsworth et al. 2014; Pleijel 2011). Moreover, very few studies have been conducted so far using FACE as only three FACE-ozone system are available globally, mainly in the United States for soybeans (Morgan et al. 2004), in China for rice and wheat (Shi et al. 2009; Tang et al. 2011), and in Italy for tree species (Paoletti et al. 2017). In addition, crop sensitivity towards ozone may differ between FACE and OTC systems (Feng et al. 2018). However, both systems require infrastructure, continuous electricity supply and are expensive, technically challenging, and laborious to maintain. Therefore, they are not easy to use and establish in many areas of the highly ozone-affected rice-growing areas of the developing world (Kobayashi 2015; Oksanen et al. 2013).

As an alternative for ozone biomonitoring, several chemicals have been tested so far (Saitanis et al. 2015). Among them, ethylenediurea (N-[2-(2-oxo-1-imidazolidinyl)ethyl]-N′-phenylurea), abbreviated as (EDU), has been most widely studied and is a long-established chemical compound for evaluation of the differential responses of plants and crop species towards ozone (Figure 9.6; Feng et al. 2010; Manning et al. 2011; Paoletti et al. 2009). EDU was first introduced by Carnahan et al. (1978), and successfully used to protect bean plants against visible ozone foliar injury. After that, numerous studies have demonstrated

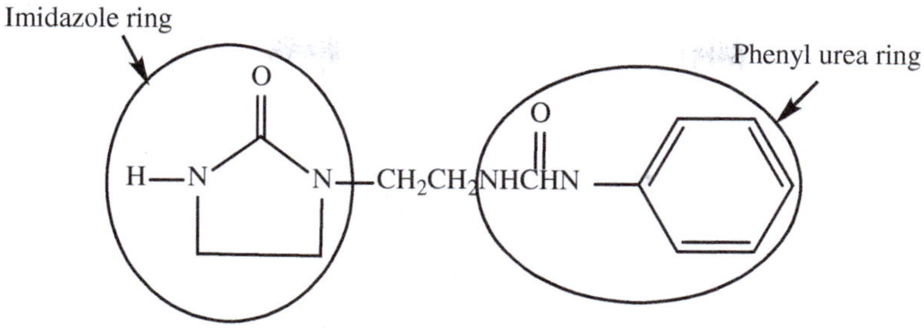

Figure 9.6 The structural formula of ethylenediurea (EDU), chemical formula $C_4H_{10}N_4O_2$. Source: From Singh et al. (2015) © Springer Nature.

and confirmed that EDU has properties that can offset ozone induced phytotoxic effects in several plant species, specifically in the sensitive genotypes. Applications of EDU are mainly carried out through spraying and soil drenching, but stem injections have also been reported in a few cases (Manning et al. 2011; Paoletti et al. 2007). However, the actual mode of action and mechanisms of protection against ozone phytotoxic effects are still unclear (Agathokleous 2017; Tiwari 2017). Two possible hypotheses have been proposed and discussed so far for the EDU-mediated protection: EDU may mediate ozone stress through a direct effect on physiological parameters, or by an antioxidant-mediated defense mechanism (Manning et al. 2011; Pandey et al. 2015). Moreover, several researchers have reported that EDU might act as a foliar fertilizer due to the high percentage (c. 22%) of nitrogen (Godzik and Manning 1998; Manning et al. 2011). Additionally, an important aspect of EDU application is to consider the toxicological aspects, that is, effects on the environment, flora, and fauna. An acute toxicological test was performed with EDU in rats at oral administration LD50 (14000 mg kg^{-1}), and no irritation, injury was reported in skin (Guinea pig) and eye (rabbit), respectively (Manning et al. 2011).

Recently, a series of agronomic and physiological experiments clearly indicated that EDU does not have any constitutive effects on rice, but partly offsets adverse ozone effects in most of the growth, physiological, and yield traits, especially in susceptible varieties. It can thus be used to diagnose ozone damage in field-grown rice and for distinguishing tolerant (less EDU-responsive) and susceptible (more EDU-responsive) genotypes (Ashrafuzzaman et al. 2017). In experiments designed to elucidate the protection mechanisms, transcriptome profiling indicated that several thousand genes responded differentially to ozone treatment, but almost none responded to EDU application (Ashrafuzzaman et al. 2018). On the other hand, plants treated with EDU that were grown under ozone stress had a similar gene expression profile as plants grown without ozone. Therefore, it has been suggested that EDU did not enhance stress-defense pathways in rice plants, but rather acted as a surface protectant against downstream physiological stress reactions (Ashrafuzzaman et al. 2018). Additional experiments demonstrated that EDU might have ozone-degrading properties due to abiotic chemical interactions between ozone and EDU. In addition EDU application did not alleviate the reaction of plants to a number of other abiotic stresses, such as iron toxicity, zinc deficiency, and salinity. Also, no foliar fertilizer effect due to the nitrogen contained in EDU was seen in rice and willow plants with EDU spraying (Agathokleous et al. 2018b; Ashrafuzzaman et al. 2018). Thus, EDU can be considered as a surface protectant that specifically mitigates ozone stress without interfering with the plants' stress-response systems. These properties make it an effective tool for ozone biomonitoring and screening in field-grown rice of highly ozone affected developing countries (Ashrafuzzaman et al. 2018).

Molecular Breeding Strategies for Improved Ozone Tolerance

For sustainable rice production, substantially reducing the anthropogenic emissions that lead to ozone formation should be the primary long-term strategy to reduce ozone-caused damage in crops (Mills et al. 2018). However, in spite of taking different actions, ozone shows an increasing trend globally because of economic development and transboundary movement of the ozone-forming precursor gases among the regions (Yamaji et al. 2012; Yoshitomi et al. 2011). Therefore, as a short-term and rapid solution, the breeding for ozone tolerance in rice is more effective than other management practices (Ainsworth 2017; Frei 2015; Mills et al. 2018).

Identifying intraspecific variation in ozone sensitivity and selecting ozone-tolerant and sensitive genotypes among crop varieties or landraces is the first step towards breeding for ozone tolerance (Frei 2015; Mills et al. 2018). For the selection of ozone tolerance, ideally, crop varieties should be exposed to realistic ozone concentrations in field conditions to assess the ozone-caused yield damage (Mills et al. 2018). Once having the tolerant genotype or gene(s) for breeding, either marker-assisted breeding or transgenic approaches can be followed to develop ozone-tolerant high-yielding rice varieties without compromising the other required agronomic characteristics such as resistance to diseases, grain shattering, and lodging (Mills et al. 2018; Frei 2015). However, given that ozone pollution is heterogeneous and rather unpredictable across space and time, traditional breeding approaches, such as pedigree breeding, seem to be unsuitable for ozone-tolerance breeding as selection requires extensive screening of a large number of plants in multiple locations over an extended period (Frei 2015). Also, transgenic approaches have environmental concerns, and consumers acceptance barriers, so molecular breeding approaches such as marker-assisted selection are the most promising option for developing ozone-tolerant rice because of the possibility of exploitation of sufficient natural variation within the rice species (Frei 2015; Mills et al. 2018). Several attempts have already been reported exploring quantitative traits loci (QTLs) for ozone tolerance followed by candidate gene analysis in rice and are summarized by Frei (2015), though most of the experiments were performed in controlled environments. A number of mapping populations have been developed and screened for ozone tolerance in both acute and chronic ozone treatments (Frei et al. 2008; Kim et al. 2004; Tsukahara et al. 2013, 2015; Ueda et al. 2015a). In total around 20 QTLs for ozone tolerance related to leaf symptoms, leaf greenness, plant height, biomass, tiller number, panicle weight, and grain yield have been reported (Frei 2015). The first attempt at marker-assisted breeding for ozone tolerance in rice was made by Frei et al. (2008, 2010a). These studies identified and confirmed three QTL in bi-parental recombinant inbred lines (RILs) and chromosome segment substitution lines (CSSLs) with the ozone-sensitive *japonica* variety Nipponbare as recurrent parent and the *aus* landrace Kasalath as the tolerant donor. Among the QTLs, two were contrasting for leaf bronzing score located on chromosome 3 (*OzT3*, tolerant allele from Nipponbare), and chromosome 9 (*OzT9*, tolerant allele from

Kasalath) and another QTL correlated to relative biomass on chromosome 8 (*OzT8*, tolerant allele from Kasalath). For the further characterization of the QTL *OzT9* and *OzT8*, they reported that *OzT9* was associated with the ability to avoid oxidative stress by maintaining an enhanced pool of reduced ascorbic acid (Frei et al. 2008, 2010a). A gene underlying this QTL was characterized as *Ozone Responsive Apoplastic Protein1* (*OsORAP1*), It has been suggested that the gene can enhance PCD during ozone stress in rice by involvement in the oxidative cell-death cycle (Ueda et al. 2015b). The sequence of *OsORAP1* had similarities with ascorbate oxidase (AO) proteins, and the protein is localized in the apoplast, but did not possess AO activity. Enhanced tolerance to ozone stress was confirmed in an *OsORAP1* knockout rice line (Ueda et al. 2015b). Further characterization of this gene displayed the enhanced expression of jasmonic acid-responsive genes in the knockout line, which can be responsible for symptom mitigation. Sequence analysis demonstrated that extensive polymorphisms are present in the upstream 5′ promoter region of OsORAP1 between the ozone-tolerant variety Kasalath (Kas-type) and ozone-susceptible cv Nipponbare (NB-type). A diverse set rice of germplasm representing the variable part of tolerant cv Kasalath and susceptible cv Nipponbare also confirmed and validated the effectiveness of the *OsORAP1* gene in differential ozone response phenomenon (author's own data, and not published yet). In contrast, QTL *OzT8* was shown to mitigate the ozone-induced loss of photosynthetic capacity via biochemical acclimation of the photosynthetic apparatus rather than avoidance of ozone via stomatal closure (Chen et al. 2011). Consequently, novel breeding lines were developed, which carried both QTL *OzT8* and *OzT9* from the donor Kasalath in the genetic background of the sensitive parent Nipponbare (Wang et al. 2014). Under a season-long ozone fumigation (100 ppb for seven hours daily), these breeding lines had a clear yield advantage over their parent Nipponbare, lacking the QTL. The lines containing both QTLs performed better in elevated ozone stress in terms of biomass, foliar symptoms, and straw quality, compared to the lines having only one single QTL. Thus, the results demonstrated that the stacking of QTLs can be a viable and vital approach for improving ozone tolerance in rice and ultimately to superior yield (Frei 2015; Frei et al. 2010b, 2011; Wang et al. 2014). Moreover, Genome Wide Association Mapping (GWAS) was conducted to elucidate the mechanisms of ozone tolerance in rice using 328 rice accessions collected from around the world (Ueda et al. 2015a). GWAS demonstrated significant ozone effects and substantial variation in several biomass traits, and reported evidence of functionally relevant mutations in a RING ("really interesting new gene") gene putatively involved in post-translational regulation of ozone stress response (Ueda et al. 2015a). However, the lack of fine mapping of ozone tolerance QTLs in rice, and tests for the stability of ozone-tolerant QTLs in popular mega variety is still the major limitation for ozone tolerance breeding (Frei 2015).

A number of other studies explored transcriptomic responses of plants to ozone through microarrays and RNA-sequencing. In a comparative microarray study using two CSSLs of rice, genes involved in PCD (ethylene or jasmonic acid biosynthesis, mitogen-activated protein (MAP) kinase, and disease resistance), and detoxification of ROS (catalase and peroxidase) were identified as ozone-responsive genes (Frei et al. 2010a). Moreover, a large number of ozone-responsive genes was identified with the help of high-throughput sequencing technique (RNA-Seq): for instance, genes associated with photosynthesis and respiration, oxidative stress, defense response, protein ubiquitination, and organic acid biosynthesis in soybeans (Waldeck et al. 2017), cellular integrity and metabolism (ascorbate-glutathione metabolism, hormone, wax, and cutin biosynthesis) in pak choi

(Zhang et al. 2017), and antioxidant metabolism (ascorbate-glutathione cycle genes) in legume crops (Yendrek et al. 2015). Thus, it is obvious that large numbers of genes are ozone-responsive in different plant species. Although there is a lack of direct evidence of improved ozone tolerance by incorporating tolerant gene(s), transgenic approaches may be an alternative tool in rice breeding, aiming to control the entry of ozone into the leaves or to improve detoxifications of ROS in the apoplast or within the cells or alter the signal transduction pathways.

Breeding Progress and Future Perspectives of Ozone Tolerance in Rice

Adopting air quality regulations aiming to minimize the emission of ozone precursor gases in most of the Asian countries, specifically on the Indian subcontinent, will presumably be very difficult due to rapid population growth and economic expansion and a political climate that is not conducive to the negotiation of international air quality regulations. The previous sections have demonstrated that the breeding of ozone-tolerant rice holds great promise as a short- to mid-term solution for food security in ozone-affected areas. However, the major limiting factor for ozone-tolerance breeding is the translation of controlled experiments into field conditions (Ainsworth 2017). Moreover, ozone-caused yield reductions may be under-estimated by farmers and agronomists due to limited awareness and lack of simple diagnostic tools to detect stress, as compared with other stresses such as salinity, drought, heat, and submergence (Frei 2015). This limitation may be overcome by increased use of EDU as a simple tool for diagnosing ozone stress in field experiments, which will have several important functions: (i) such experiments will help to quantify ozone-induced rice-yield losses; (ii) awareness of the ozone problem will be generated among farmers, agronomists, and policy makers; and (iii) tolerant genotypes can be identified and/or bred.

Another very important consideration for ozone-tolerance breeding is its interaction with other stresses. Mills et al. (2018) suggested a crop ideotype that combines several traits required for ozone tolerance, including short maturity period, high water-use efficiency, low stomatal conductance, high antioxidant capacity, and high root dynamics. These traits may generate synergies or trade-offs with breeding for other types of stress environments. Mills et al. (2018) argued that the above traits would be beneficial in other abiotic stress conditions including drought, heat, and nutrient limitation, while the interactions with biotic stresses are less clear and could be ambivalent (Mills et al. 2018). As the magnitude of ozone-induced yield losses in a similar range as those caused by these other stress factors, breeding for ozone tolerance could in most cases lead to co-benefits for multi-stress tolerance. This idea is also supported by a study by Ali et al. (2019) in which the engineering of rice to contain enhanced levels of the antioxidant ascorbate led to improved tolerance against combined ozone and salinity stress.

Conclusion

Tropospheric ozone concentrations are rapidly increasing in many developing parts of the world and are threatening food security. Therefore, breeding for adapted rice varieties is critical for Asia, especially in the Indian subcontinent, to cope with ozone stress and to secure staple food supply. This chapter provides insights into ozone impacts on rice cultivation and a possible threat to food security in rice-producing countries. National, regional, and global air quality policies that target declines of ozone-precursor emissions are required as sustainable long-term approaches. At the same time ozone-tolerance breeding programs for rice can provide an immediate solution for minimizing the ozone-related yield gap in rice, which will contribute to the food security of many highly populated countries in Asia.

References

Agathokleous, E. (2017). Perspectives for elucidating the ethylenediurea (EDU) mode of action for protection against O_3 phytotoxicity. *Ecotoxicol. Environ. Saf.* 142: 530–537.

Agathokleous, E., Saitanis, C.J., and Koike, T. (2015). Tropospheric O_3, the nightmare of wild plants: a review study. *J. Agric. Meteorol.* 71: 142–152.

Agathokleous, E., Kitao, M., and Kinose, Y. (2018a). A review study on ozone phytotoxicity metrics for setting critical levels in Asia. *Asian J. Atmos. Environ.* 12: 1–16.

Agathokleous, E., Paoletti, E., Manning, W.J. et al. (2018b). High doses of ethylenediurea (EDU) as soil drenches did not increase leaf N content or cause phytotoxicity in willow grown in fertile soil. *Ecotoxicol. Environ. Saf.* 147: 574–584.

Ainsworth, E.A. (2008). Rice production in a changing climate: a meta-analysis of responses to elevated carbon dioxide and elevated ozone concentration. *Glob. Chang. Biol.* 14: 1642–1650.

Ainsworth, E.A. (2017). Understanding and improving global crop response to ozone pollution. *Plant J.* 90: 886–897.

Ainsworth, E., Yendrek, C.R., Sitch, S. et al. (2012). The effects of tropospheric ozone on net primary productivity and implications for climate change. *Annu. Rev. Plant Biol.* 63: 637–661.

Ainsworth, E.A., Serbin, S.P., Skoneczka, J.A. et al. (2014). Using leaf optical properties to detect ozone effects on foliar biochemistry. *Photosynth. Res.* 119: 65–76.

Akhtar, N., Yamaguchi, M., and Inada, H. (2010). Effects of ozone on growth, yield and leaf gas exchange rates of four Bangladeshi cultivars of rice (*Oryza sativa* L.). *Environ. Pollut.* 158: 2970–2975.

Ali, B., Pantha, S., Acharya, R. et al. (2019). Enhanced ascorbate level improves multi-stress tolerance in a widely grown *indica* rice variety without compromising its agronomic characteristics. *J. Plant Physiol.* 240: 152998.

Ashmore, M.R. (2005). Assessing the future global impacts of ozone on vegetation. *Plant Cell Environ.* 28: 949–964.

Ashmore, M., Toet, S., and Emberson, L. (2006). Ozone – A significant threat to future world food production? *New Phytol.* 170: 201–204.

Ashrafuzzaman, M., Lubna, F.A., Holtkamp, F. et al. (2017). Diagnosing ozone stress and differential tolerance in rice (*Oryza sativa* L.) with ethylenediurea (EDU). *Environ. Pollut.* 230: 339–350.

Ashrafuzzaman, M., Haque, Z., Ali, B. et al. (2018). Ethylenediurea (EDU) mitigates the negative effects of ozone in rice: insights into its mode of action. *Plant Cell Environ.* 41 (12): 2882–2898.

Biswas, D.K., Xu, H., Li, Y.G. et al. (2008). Genotypic differences in leaf biochemical, physiological and growth responses to ozone in 20 winter wheat cultivars released over the past 60 years. *Glob. Chang. Biol.* 14: 46–59.

Brauer, M., Freedman, G., Frostad, J. et al. (2016). Ambient air pollution exposure estimation for the global burden of disease 2013. *Environ. Sci. Technol.* 50: 79–88.

Brosche, M., Merilo, E., Mayer, F. et al. (2010). Natural variation in ozone sensitivity among Arabidopsis thaliana accessions and its relation to stomatal conductance. *Plant Cell Environ.* 33: 914–925.

Carnahan, J.E., Jenner, E.L., and Wat, E.K.W. (1978). Prevention of ozone injury to plants by a new protectant chemical. *Phytopathology* 68: 1225–1229.

Chang, K.L., Petropavlovskikh, I., Copper, O.R. et al. (2017). Regional trend analysis of surface ozone observations from monitoring networks in eastern North America, Europe and East Asia. *Elementa.* 5: 50.

Chen, Z. and Gallie, D.R. (2005). Increasing tolerance to ozone by elevating foliar ascorbic acid confers greater protection against ozone than increasing avoidance. *Plant Physiol.* 138: 1673–1689.

Chen, C.P., Frank, T.D., and Long, S.P. (2009). Is a short, sharp shock equivalent to long-term punishment? Contrasting the spatial pattern of acute and chronic ozone damage to soybean leaves via chlorophyll fluorescence imaging. *Plant Cell Environ.* 32: 327–335.

Chen, C.P., Frei, M., and Wissuwa, M. (2011). The *OzT8* locus in rice protects leaf carbon assimilation rate and photosynthetic capacity under ozone stress. *Plant Cell Environ.* 34: 1141–1149.

Cho, K., Tiwari, S., Agrawal, S.B. et al. (2011). Tropospheric ozone and plants: absorption, responses, and consequences. *Rev. Environ. Contam. Toxicol.* 212: 61–111.

Chutteang, C., Booker, F.L., Na-Ngern, P. et al. (2016). Biochemical and physiological processes associated with the differential ozone response in ozone-tolerant and sensitive soybean genotypes. *Plant Biol.* 18: 28–36.

CLRTAP (2017). Mapping critical levels for vegetation. *LRTAP convention modelling and mapping manual.* Chapter 3. Gina Mills, Harry Harmens, Felicity Hayes, Håkan Pleijel, Patrick Büker and Ignacio González-Fernández et al. International Cooperative Programme on Effects of Air Pollution on Natural Vegetation and Crops. Retrieved from https://icpvegetation.ceh.ac.uk/sites/default/files/FinalnewChapter3v4Oct2017_000.pdf.

Conklin, P.L. and Barth, C. (2004). Ascorbic acid, a familiar small molecule intertwined in the response of plants to ozone, pathogens, and the onset of senescence. *Plant Cell Environ.* 27: 959–970.

Conklin, P.L., Williams, E.H., and Last, R.L. (1996). Environmental stress sensitivity of an ascorbic acid-deficient arabidopsis mutant. *Proc. Natl. Acad. Sci. U. S. A.* 93: 9970–9974.

Deb Roy, S., Beig, G., and Ghude, S.D. (2009). Exposure-plant response of ambient ozone over the tropical Indian region. *Atmos. Chem. Phys.* 9: 5253–5260.

Debaje, S.B. (2014). Estimated crop yield losses due to surface ozone exposure and economic damage in India. *Environ. Sci. Pollut. Res.* 21: 7329–7338.

Dentener, F., Stevenson, D., Ellingsen, K. et al. (2006). The global atmospheric environment for the next generation. *Environ. Sci. Technol.* 40: 3586–3594.

D'Haese, D., Vandermeiren, K., Asard, H. et al. (2005). Other factors than apoplastic ascorbate contribute to the differential ozone tolerance of two clones of *Trifolium repens* L. *Plant Cell Environ.* 28: 623–632.

Donahue, N.M. (2011). The reaction that wouldn't quit. *Nat. Chem.* 3: 98–99.

Eltayeb, A.E., Kawano, N., Badawi, G.H. et al. (2007). Overexpression of monodehydroascorbate reductase in transgenic tobacco confers enhanced tolerance to ozone, salt and polyethylene glycol stresses. *Planta* 225: 1255–1264.

Emberson, L.D., Ashmore, M.R., Simpson, D. et al. (2001). Modelling and mapping ozone deposition in Europe. *Water Air Soil Pollut.* 130: 577–582.

Emberson, L.D., Pleijel, H., Ainsworth, E.A. et al. (2018). Ozone effects on crops and consideration in crop models. *Eur. J. Agron.* 100: 1–16.

FAOSTAT. (2019). FAO statistical databases. Available at: http://www.fao.org/faostat/en/#data (Accessed: 15 August 2019).

Feng, Z. and Kobayashi, K. (2009). Assessing the impacts of current and future concentrations of surface ozone on crop yield with meta-analysis. *Atmos. Environ.* 43: 1510–1519.

Feng, Z.Z., Kobayashi, K., and Ainsworth, E.A. (2008). Impact of elevated ozone concentration on growth, physiology, and yield of wheat (*Triticum aestivum* L.): a meta-analysis. *Glob. Chang. Biol.* 14: 2696–2708.

Feng, Z., Wang, S., Szantoi, Z. et al. (2010). Protection of plants from ambient ozone by applications of ethylenediurea (EDU): a meta-analytic review. *Environ. Pollut.* 158: 3236–3242.

Feng, Z., Hu, E., Wang, X. et al. (2015). Ground-level O_3 pollution and its impacts on food crops in China: a review. *Environ. Pollut.* 199: 42–48.

Feng, Z., Uddling, J., Tang, H. et al. (2018). Comparison of crop yield sensitivity to ozone between open-top chamber and free-air experiments. *Glob. Chang. Biol.* 24: 2231–2238.

Frei, M. (2015). Breeding of ozone resistant rice: relevance, approaches and challenges. *Environ. Pollut.* 197: 144–155.

Frei, M., Tanaka, J.P., and Wissuwa, M. (2008). Genotypic variation in tolerance to elevated ozone in rice: dissection of distinct genetic factors linked to tolerance mechanisms. *J. Exp. Bot.* 59: 3741–3752.

Frei, M., Tanaka, J.P., Chen, C.P. et al. (2010a). Mechanisms of ozone tolerance in rice: characterization of two QTLs affecting leaf bronzing by gene expression profiling and biochemical analyses. *J. Exp. Bot.* 61: 1405–1417.

Frei, M., Makkar, H.P.S., Becker, K. et al. (2010b). Ozone exposure during growth affects the feeding value of rice shoots. *Anim. Feed Sci. Technol.* 155: 74–79.

Frei, M., Kohno, Y., Wissuwa, M. et al. (2011). Negative effects of tropospheric ozone on the feed value of rice straw are mitigated by an ozone tolerance QTL. *Glob. Chang. Biol.* 17: 2319–2329.

Frei, M., Kohno, Y., Tietze, S. et al. (2012a). The response of rice grain quality to ozone exposure during growth depends on ozone level and genotype. *Environ. Pollut.* 163: 199–206.

Frei, M., Wissuwa, M., Pariasca-Tanaka, J. et al. (2012b). Leaf ascorbic acid level: is it really important for ozone tolerance in rice? *Plant Physiol. Biochem.* 59: 63–70.

Gillespie, K.M., Xu, F., Richter, K. et al. (2012). Greater antioxidant and respiratory metabolism in field-grown soybean exposed to elevated O_3 and two CO_2 concentrations. *Plant Cell Environ.* 35: 169–184.

Godzik, B. and Manning, W.J. (1998). Relative effectiveness of ethylenediurea, and constitutent amounts of urea and phenylurea in ethylenediurea, in prevention of ozone injury to tobacco. *Environ. Pollut.* 103: 1–6.

Grantz, D.A. and Vu, H.B. (2012). Root and shoot gas exchange respond additively to moderate ozone and methyl jasmonate without induction of ethylene: ethylene is induced at higher O_3 concentrations. *J. Exp. Bot.* 63: 4303–4313.

Guidi, L., Degl'Innocenti, E., Giordano, C. et al. (2010). Ozone tolerance in *Phaseolus vulgaris* depends on more than one mechanism. *Environ. Pollut.* 158: 3164–3171.

Haagen-Smit, A.J. (1952). Chemistry and physiology of Los Angeles smog. *Ind. Eng. Chem.* 44: 1342–1346.

Haagen-Smit, A.J. and Fox, M.M. (1954). Photochemical ozone formation with hydrocarbons and automobile exhaust. *Air Repair* 4: 105–136.

Heagle, A.S., Body, D.E., and Heck, W.W. (1973). An open-top field chamber to assess the impact of air pollution on plants. *J. Environ. Qual.* 2: 365–368.

Heggestad, H.E. and Middleton, J.T. (1959). Ozone in high concentrations as cause of tobacco leaf Injury. *Science* 129: 208–210.

Hernandez, J.A., Ferrer, M.A., Jimenez, A. et al. (2001). Antioxidant systems and O_2^-/H_2O_2 production in the apoplast of pea leaves: its relation with salt-induced necrotic lesions in minor veins. *Plant Physiol.* 127: 817–831.

IPCC. (2014). Intergovernmental Panel on Climate Change 2014: Mitigation of climate change. IPCC 5th assessment report, Bonn, Germany.

Jain, S.L., Arya, B.C., Kumar, A. et al. (2005). Observational study of surface ozone at New Delhi, India. *Int. J. Remote Sens.* 26: 3515–3524.

Kangasjärvi, J., Jaspers, P., and Kollist, H. (2005). Signalling and cell death in ozone-exposed plants. *Plant Cell Environ.* 28: 1021–1036.

Karlsson, P.E., Uddling, J., Braun, S. et al. (2004). New critical levels for ozone effects on young trees based on AOT40 and simulated cumulative leaf uptake of ozone. *Atmos. Environ.* 38: 2283–2294.

Kim, K.M., Kwon, Y.S., Lee, J.J. et al. (2004). QTL mapping and molecular marker analysis for the resistance of rice to ozone. *Mol. Cells* 17: 151–155.

Kobayashi, K. (2015). FACE-ing the challenges of increasing surface ozone concentration in Asia. *J. Agric. Meteorol.* 71: 161–166.

Krasensky, J., Carmody, M., Sierla, M. et al. (2017). *Ozone and Reactive Oxygen Species*, In: eLS, 1–9. Chichester, UK: Wiley.

Lal, S., Venkataramani, S., Naja, M. et al. (2017). Loss of crop yields in India due to surface ozone: an estimation based on a network of observations. *Environ. Sci. Pollut. Res.* 24: 20972–20981.

Lei, H., Wuebbles, D.J., and Liang, X.Z. (2012). Projected risk of high ozone episodes in 2050. *Atmos. Environ.* 59: 567–577.

Li, P., Feng, Z., Catalayud, V. et al. (2017). A meta-analysis on growth, physiological and biochemical responses of woody species to ground-level ozone highlights the role of plant functional types. *Plant Cell Environ.* 40: 2369–2380.

Maas, R.P., and Grennfelt, R.P. (2016). Towards cleaner air. Scientific assessment report 2016. EMEP steering body and working group on effects of the convention on long range transboundary air pollution, UNECE.

Macháčová, K. (2010). Open top chamber and free air CO_2 enrichment: approafches to investigate tree responses to elevated CO_2. *iForest* 3: 102–105.

Manning, W.J., Paoletti, E., Sandermann, H. et al. (2011). Ethylenediurea (EDU): a research tool for assessment and verification of the effects of ground level ozone on plants under natural conditions. *Environ. Pollut.* 159: 3283–3293.

McAdam, E.L., Brodribb, T.J., and McAdam, S.A.M. (2017). Does ozone increase ABA levels by non-enzymatic synthesis causing stomata to close? *Plant Cell Environ.* 40: 741–747.

McCouch, S.R., Wright, M.H., Tung, C.W. et al. (2016). Open access resources for genome-wide association mapping in rice. *Nat. Commun.* 7: 10532.

Middleton, J.T., Kendrick, J.B.J., and Schwalm, H.W. (1950). Injury to herbaceous plants by smog or air pollution. *Plant Dis.* Reporter. 34: 245–252.

Mills, G., Pleijel, H., and Braun, S. (2011). New stomatal flux-based critical levels for ozone effects on vegetation. *Atmos. Environ.* 45: 5064–5068.

Mills, G., Sharps, K., Simpson, D. et al. (2018). Closing the global ozone yield gap: quantification and cobenefits for multistress tolerance. *Glob. Chang. Biol.* 24: 4869–4893.

Moldau, H., Padu, E., and Bichele, I. (1997). Quantification of ozone decay and requirement for ascorbate in *Phaseolus vulgaris* L. mesophyll cell walls. *Phyton* 37: 175–180.

Morgan, P.B., Ainsworth, E.A., and Long, S.P. (2003). How does elevated ozone impact soybean? A meta-analysis of photosynthesis, growth and yield. *Plant Cell Environ.* 26: 1317–1328.

Morgan, P.B., Bernacchi, C.J., Ort, D.R. et al. (2004). An in vivo analysis of the effect of season-long open-air elevation of ozone to anticipated 2050 levels on photosynthesis in soybean. *Plant Physiol.* 135: 2348–2357.

Musselman, R.C., Lefohn, A.S., Massman, W.J. et al. (2006). A critical review and analysis of the use of exposure and flux-based ozone indices for predicting vegetation effects. *Atmos. Environ.* 40: 1869–1888.

Myhre, G., Shindell, D., Breon, F.M. et al. (2013). Anthropogenic and natural radiative forcing. In: *Climate Change 2013: the Physical Science Basis. Contribution of Working Group I to the Fifth Assessment Report of the Intergovernmental Panel on Climate Change* (eds. T.F. Stocker, D. Qin, G.-K. Plattner, et al.), 659–740. Cambridge: Cambridge University Press.

Nagatoshi, Y., Mitsuda, N., Hayashi, M. et al. (2016). GOLDEN 2-LIKE transcription factors for chloroplast development affect ozone tolerance through the regulation of stomatal movement. *Proc. Natl. Acad. Sci. U. S. A.* 113: 4218–4223.

Oksanen, E., Pandey, V., Pandey, A.K. et al. (2013). Impacts of increasing ozone on Indian plants. *Environ. Pollut.* 177: 189–200.

Pandey, A.K., Majumder, B., Keski-Saari, S. et al. (2015). Searching for common responsive parameters for ozone tolerance in 18 rice cultivars in *India*: results from ethylenediurea studies. *Sci. Total Environ.* 532: 230–238.

Paoletti, E. and Grulke, N.E. (2005). Does living in elevated CO_2 ameliorate tree response to ozone? A review on stomatal responses. *Environ. Pollut.* 137: 483–493.

Paoletti, E., Manning, W.J., Spaziani, F. et al. (2007). Gravitational infusion of ethylenediurea (EDU) into trunks protected adult European ash trees (*Fraxinus excelsior* L.) from foliar ozone injury. *Environ. Pollut.* 145: 869–873.

Paoletti, E., Contran, N., Manning, W.J. et al. (2009). Use of the antiozonant ethylenediurea (EDU) in Italy: verification of the effects of ambient ozone on crop plants and trees and investigation of EDU's mode of action. *Environ. Pollut.* 157: 1453–1460.

Paoletti, E., Materassi, A., Fasano, G. et al. (2017). A new-generation 3D ozone FACE (free air controlled exposure). *Sci. Total Environ.* 575: 1407–1414.

Piikki, K., De Temmerman, L., Högy, P. et al. (2008). The open-top chamber impact on vapour pressure deficit and its consequences for stomatal ozone uptake. *Atmos. Environ.* 42: 6513–6522.

Pleijel, H. (2011). Reduced ozone by air filtration consistently improved grain yield in wheat. *Environ. Pollut.* 159: 897–902.

Pleijel, H., Danielsson, H., Ojanperä, K. et al. (2004). Relationships between ozone exposure and yield loss in European wheat and potato: a comparison of concentration and flux-based exposure indices. *Atmos. Environ.* 38: 2259–2269.

Pozzer, A., Zimmermann, P., Doering, U.M. et al. (2012). Effects of business-as-usual anthropogenic emissions on air quality. *Atmos. Chem. Phys.* 12: 6915–6937.

Ran, L., Zhao, C., Geng, F. et al. (2009). Ozone photochemical production in urban Shanghai, China: analysis based on ground level observations. *J. Geophys. Res.* 114: D15301.

Royal Society (2008). *Ground-Level Ozone in the 21st Century: Future Trends, Impacts and Policy Implications*. London: Royal Society.

Ryan, A., Cojocariu, C., Possell, M. et al. (2009). Defining hybrid poplar (*Populus deltoids* X *Populus trichocarpa*) tolerance to ozone: identifying key parameters. *Plant Cell Environ.* 32: 31–45.

Saitanis, C.J., Lekkas, D.V., Agathokleous, E. et al. (2015). Screening agrochemicals as potential protectants of plants against ozone phytotoxicity. *Environ. Pollut.* 197: 247–255.

Sanmartin, M., Drogoudi, P.D., Lyons, T. et al. (2003). Over-expression of ascorbate oxidase in the apoplast of transgenic tobacco results in altered ascorbate and glutathione redox states and increased sensitivity to ozone. *Planta* 216: 918–928.

Sawada, H. and Kohno, Y. (2009). Differential ozone sensitivity of rice cultivars as indicated by visible injury and grain yield. *Plant Biol.* 11: 70–75.

Shi, G., Yang, L., Wang, Y. et al. (2009). Impact of elevated ozone concentration on yield of four Chinese rice cultivars under fully open-air field conditions. *Agric. Ecosyst. Environ.* 131: 178–184.

Sierla, M., Waszczak, C., Vahisalu, T. et al. (2016). Reactive oxygen species in the regulation of stomatal movements. *Plant Physiol.* 171: 1569–1580.

Simpson, D., Ashmore, M.R., Emberson, L. et al. (2007). A comparison of two different approaches for mapping potential ozone damage to vegetation: a model study. *Environ. Pollut.* 146: 715–725.

Singh, A.A., Singh, S., Agrawal, M. et al. (2015). Assessment of ethylene diurea-induced protection in plants against ozone phytotoxicity. *Rev. Environ. Contam. Toxicol.* 233: 129–184.

Smirnoff, N., Conklin, P.L., and Loewus, F.A. (2001). Biosynthesis of ascorbic acid in plants: a renaissance. *Annu. Rev. Plant Physiol. Plant Mol. Biol.* 52: 437–467.

Sugai, T., Kam, D.G., Agathokleous, E. et al. (2018). Growth and photosynthetic response of two larches exposed to O3 mixing ratios ranging from preindustrial to near future. *Photosynthetica* 56: 901–910.

Tai, A.P.K. and Val Martin, M. (2017). Impacts of ozone air pollution and temperature extremes on crop yields: spatial variability, adaptation and implications for future food security. *Atmos. Environ.* 169: 11–21.

Tang, H., Liu, G., Han, Y. et al. (2011). A system for free-air ozone concentration elevation with rice and wheat: control performance and ozone exposure regime. *Atmos. Environ.* 45: 6276–6282.

Teixeira, E., Fischer, G., van Velthuizen, H. et al. (2011). Limited potential of crop management for mitigating surface ozone impacts on global food supply. *Atmos. Environ.* 45: 2569–2576.

Tiwari, S. (2017). Ethylenediurea as a potential tool in evaluating ozone phytotoxicity: a review study on physiological, biochemical and morphological responses of plants. *Environ. Sci. Pollut. Res.* 24: 14019–14039.

Tiwari, S., Rai, R., and Agrawal, M. (2008). Annual and seasonal variations in tropospheric ozone concentrations around Varanasi. *Int. J. Remote Sens.* 29: 4499–4514.

Tsukahara, K., Sawada, H., Matsumura, H. et al. (2013). Quantitative trait locus analyses of ozone-induced grain yield reduction in rice. *Environ. Exp. Bot.* 88: 100–106.

Tsukahara, K., Sawada, H., Kohno, Y. et al. (2015). Ozone-induced rice grain yield loss is triggered via a change in panicle morphology that is controlled by *ABERRANT PANICLE ORGANIZATION1* gene. *PLoS One* 10: e0123308.

Turcsanyi, T., Lyons, T., Plochl, M. et al. (2000). Does ascorbate in the mesophyll cell walls form the first line of defence against ozone? Testing the concept using broad bean (*Vicia faba* L.). *J. Exp. Bot.* 51: 901–910.

Ueda, Y., Frimpong, F., Qi, Y. et al. (2015a). Genetic dissection of ozone tolerance in rice (*Oryza sativa* L.) by a genome-wide association study. *J. Exp. Bot.* 66 (1): 293–306.

Ueda, Y., Siddique, S., and Frei, M. (2015b). A novel gene, ozone-responsive apoplastic protein1, enhances cell death in ozone stress in rice. *Plant Physiol.* 169: 873–889.

Ueda, Y., Frindte, K., Knief, C. et al. (2016). Effects of elevated tropospheric ozone concentration on the bacterial community in the phyllosphere and rhizoplane of rice. *PLoS One* 11 (9): e0163178.

Vahisalu, T., Kollist, H., Wang, Y.F. et al. (2008). SLAC1 is required for plant guard cell S-type anion channel function in stomatal signalling. *Nature* 452: 487–491.

Vahisalu, T., Puzorjova, I., Brosche, M. et al. (2010). Ozone-triggered rapid stomatal response involves the production of reactive oxygen species, and is controlled by SLAC1 and OST1. *Plant J.* 62: 442–453.

Van Dingenen, R., Dentener, F.J., Raes, F. et al. (2009). The global impact of ozone on agricultural crop yields under current and future air quality legislation. *Atmos. Environ.* 43: 604–618.

Vanacker, H., Carver, T.L.W., and Foyer, C.H. (1998). Pathogen-induced changes in the antioxidant status of the apoplast in barley leaves. *Plant Physiol.* 117: 1103–1114.

Waldeck, N., Burkey, K., Carter, T. et al. (2017). RNA-Seq study reveals genetic responses of diverse wild soybean accessions to increased ozone levels. *BMC Genomics* 18: 498.

Wang, X. and Mauzerall, D.L. (2004). Characterizing distributions of surface ozone and its impact on grain production in China, Japan and South Korea: 1990 and 2020. *Atmos. Environ.* 38: 4383–4402.

Wang, X., Zhang, Q., Zheng, F. et al. (2012). Effects of elevated O3 concentration on winter wheat and rice yields in the Yangtze River Delta, China. *Environ. Pollut.* 171: 118–125.

Wang, Y., Yang, L., Höller, M. et al. (2014). Pyramiding of ozone tolerance QTLs *OzT8* and *OzT9* confers improved tolerance to season-long ozone exposure in rice. *Environ. Exp. Bot.* 104: 26–33.

Wheeler, G.L., Jones, M.A., and Smirnoff, N. (1998). The biosynthetic pathway of vitamin C in higher plants. *Nature* 393: 365–369.

Wilkinson, S. and Davies, W.J. (2010). Drought, ozone, ABA and ethylene: new insights from cell to plant to community. *Plant Cell Environ.* 33: 510–525.

Wilkinson, S., Mills, G., Illidge, R.J. et al. (2012). How is ozone pollution reducing our food supply? *J. Exp. Bot.* 63: 527–536.

Williams, J.E., Hodnebrog, Ø., Van Velthoven, P.F.J. et al. (2014). The influence of future non-mitigated road transport emissions on regional ozone exceedences at global scale. *Atmos. Environ.* 89: 633–641.

WMO. (2014). World Meteorological Organization. Assessment for decision-makers: scientific assessment of ozone depletion: Global ozone research and monitoring project—report no. 56, Geneva, Switzerland.

Yamaguchi, M., Hoshino, D., Inada, H. et al. (2014). Evaluation of the effects of ozone on yield of Japanese rice (*Oryza sativa* L.) based on stomatal ozone uptake. *Environ. Pollut.* 184: 472–480.

Yamaji, K., Uno, I., and Irie, H. (2012). Investigating the response of East Asian ozone to Chinese emission changes using a linear approach. *Atmos. Environ.* 55: 475–482.

Yendrek, C.R., Koester, R.P., and Ainsworth, E.A. (2015). A comparative analysis of transcriptomic, biochemical, and physiological responses to elevated ozone identifies species-specific mechanisms of resilience in legume crops. *J. Exp. Bot.* 66: 7101–7112.

Yoshitomi, M., Wild, O., and Akimoto, H. (2011). Contributions of regional and intercontinental transport to surface ozone in the Tokyo area. *Atmos. Chem. Phys.* 11: 7583–7599.

Zhang, L., Xu, B., Wu, T. et al. (2017). Transcriptomic analysis of Pak Choi under acute ozone exposure revealed regulatory mechanism against ozone stress. *BMC Plant Biol.* 17: 236.

Ziemke, J.R., Chandra, S., Duncan, B.N. et al. (2006). Tropospheric ozone determined from Aura OMI and MLS: evaluation of measurements and comparison with the Global Modeling Initiative's Chemical Transport Model. *J. Geophys. Res.* 111: D19303. (accessed 15 August 2019).

10

Molecular Breeding Strategies for Enhancing Rice Yields Under Low Light Intensity

Mayank Rai, Suvendhu S. Dutta and Wricha Tyagi

School of Crop Improvement, College of PG studies in Agricultural Sciences, Central Agricultural University (Imphal), Umiam, Meghalaya, Meghalaya 793103, India

Introduction

Optimum light intensity or photon flux density (PFD) is critical for maintaining the central biochemical and physiological processes in plants. Low light intensity (shading) adversely affects crop growth and yield by inhibiting the photosynthetic processes, sugar metabolism, and translocation, leading to poor vegetative growth, and poor grain filling. In the case of rice, it has been seen globally that regions that receive higher solar radiations and longer day lengths during the rice growing season generally have higher productivity. In India, states like Punjab, Haryana, Telangana, and Tamil Nadu that receive more irradiance (380–460 cal cm^{-2} h^{-1}) have higher rice productivity as compared to eastern and northeastern states that receive irradiance in the range of 250–350 cal cm^{-2} h^{-1} during *kharif* season (Venkateswarlu and Visperas 1987). Boro rice in eastern India returns higher productivity as compared to Sali (monsoon) rice, primarily due to persistent cloud cover leading to low light intensity during the monsoon season. Unlike other stresses like drought and salinity, low light intensity cannot be ameliorated agronomically under open field conditions, and therefore breeding for and selecting low light tolerant/light use efficient (LUE) genotypes is key to enhancing rice productivity, especially in regions that receive excessive cloud cover during the rice growing season. Low light tolerance can be defined as the ability of a plant to maintain yield levels under suboptimal irradiance.

Several studies have focused on understanding the physiological and biochemical effects of low light/shading in several model plants including rice, and we have a fairly good understanding of how plants respond to low light. However, we are just beginning to understand the genetic mechanisms governing tolerance to low light. In the following sections, we discuss various morphological traits, physiological, and biochemical processes that are affected by / respond to low light intensity and whether genotypic variations for these traits can be used for developing low light tolerant varieties.

Molecular Breeding for Rice Abiotic Stress Tolerance and Nutritional Quality, First Edition.
Edited by Mohammad Anwar Hossain, Lutful Hassan, Khandakar Md. Iftekharuddaula, Arvind Kumar and Robert Henry.

Effect of Low Light in Rice Physiology, Development and Yield

Photosynthetic yield under low irradiance depends on the efficiency of light-energy capture by antenna pigments and its delivery to reaction centers. Light-energy capture by photoreceptors depends on PFD and spectral quality. Low PFD is reported to cause damage of PSII, thereby inhibiting the photochemical processes due to closure of reaction centers (Quero et al. 2019). This in turn adversely affects sugar metabolism.

The reproductive stage in rice is critically affected by light intensity. It has been reported that low light from booting to maturity stage causes a yield loss of up to 55%, which is primarily due to reduced seed setting (spikelet fertility) and test weight (grain filling) (Liu et al. 2014). This loss in yield is primarily because of reduced carbohydrate assimilation in source (leaves, culm, and sheath) and its translocation to the sink (seed). Low light intensity during grain filling leads to reduced head rice recovery, amylase content, and gel consistency. It is also reported to cause increased grain chalkiness (Yuan et al. 2005).

For screening and selection of low light intensity tolerant/LUE genotypes, differential light intensity treatments can be achieved by the use of meshes of variable density (Murchie et al. 2002) (Figure 10.1). Our studies have revealed high genotype X environment interaction for plant biomass, effective tillers, spikelet fertility, and grain yield under optimum and low light conditions (Dutta et al. 2018), suggesting the scope for selecting low light intensity-tolerant genotypes. Similar screening and selection for LUE on the basis of the above-mentioned traits have been conducted by several other groups (Horton 2000; Long et al. 2006; Liu and Sun 2013).

Candidate Genes and Pathways Imparting Tolerance

The amount of light or fluence and fluence rate or irradiance determines the occurrence of photoresponse. This section describes the progress made in understanding the molecular and physiological basis of low light intensity perception and response. An ideal phenotype

Figure 10.1 Screening of rice germplasm for low light intensity tolerance under field conditions.

for tolerance would include higher photosynthetic efficiency, erect leaf, delayed senescence, and higher grain yield under low light conditions (Figure 10.2).

Low Light Perception

The primary photoreceptor for red and far-red light in flowering plants is phytochrome (PHY). This receptor can exist in two forms; the inactive Pr (red light-absorbing form) and physiologically active form Pfr (far-red light-absorbing form). Absorbance of far-red light by Pfr converts the active phytochrome to its inactive form Pr (Li et al. 2011). The phytochromes are encoded by three genes, PhyA to PhyC, in rice, and PhyA is responsible mainly for mediating photomorphogenic responses (Takano et al. 2009; Roy et al. 2013). The red : far-red (R : FR) ratio is affected by presence of canopy because chlorophyll only absorbs in R and not in FR light. Responses induced by phytochrome can be (i) rapid biochemical or (ii) slower morphological changes. The perception of blue light by cryptochromes (CRY) promotes flowering. Studies using mutants having defects in various phytochrome and cryptochrome pathway/receptor genes have confirmed that phytochrome interacts with cryptochrome to regulate flowering (Hughes 2013).

Phytochrome signaling leads to either (i) rapid turgor responses due to ion fluxes or (ii) changes in gene expression. The later can be primary or secondary response genes. With the cloning of *nph* (non-phototropic hypocotyl) mutants, it was revealed that phototropin (*PHOT*) is the photoreceptor for phototropism (Takemiya et al. 2005). The two genes, PHOT1 and PHOT2, underlying the mutant *nph* are found in angiosperms. Photosynthetic efficiency and growth under low light conditions are regulated by PHOT1 in rice (Jain et al. 2007; Goh et al. 2009). The physiological observation of chloroplast accumulation at the lower and upper sides of palisade cells in response to low light conditions is also mediated by phototropins via F-actin filaments (Wada 2013). Phototropins also cause activation of H^+ ATPases present on plasma membrane leading to stomatal opening (Takemiya et al. 2013).

Under low R:FR ratio (shade avoidance), accumulation of phytochrome interacting factors (PIFs) occurs, which is in turn dependent on the ratio of Pr to Pfr (de Wit et al. 2016). The PIFs are negatively regulated by DELLA proteins (components of gibberrellic acid [GA] pathway) as well. This leads to increase in stem growth or vegetative phase at the cost of reproductive phase (Casal 2013). In shade avoidance, in order to optimize photosynthesis, light gradients that are perceived by leaves induce the growth of petioles or leaf blades toward higher light (Briggs 2014).

Apart from PIFs, cryptochromes discussed above, photomorphogenesis is also affected by auxin and brassinosteroids (BRs).

It has been shown in *Arabidopsis* and *Brassica* that localized FR enrichment at the lamina tip induces hyponasty (upward leaf movement) (Pantazopoulou et al. 2017). This is caused by PIF-dependent special auxin dynamics and leads to optimizing the leaf movement timing.

With the recent discovery of various light-quality-induced signals that travel between organs and tissue layers and have various effects on cell growth, it is important that future plant photobiology research increase its focus on spatial signaling at the whole plant level

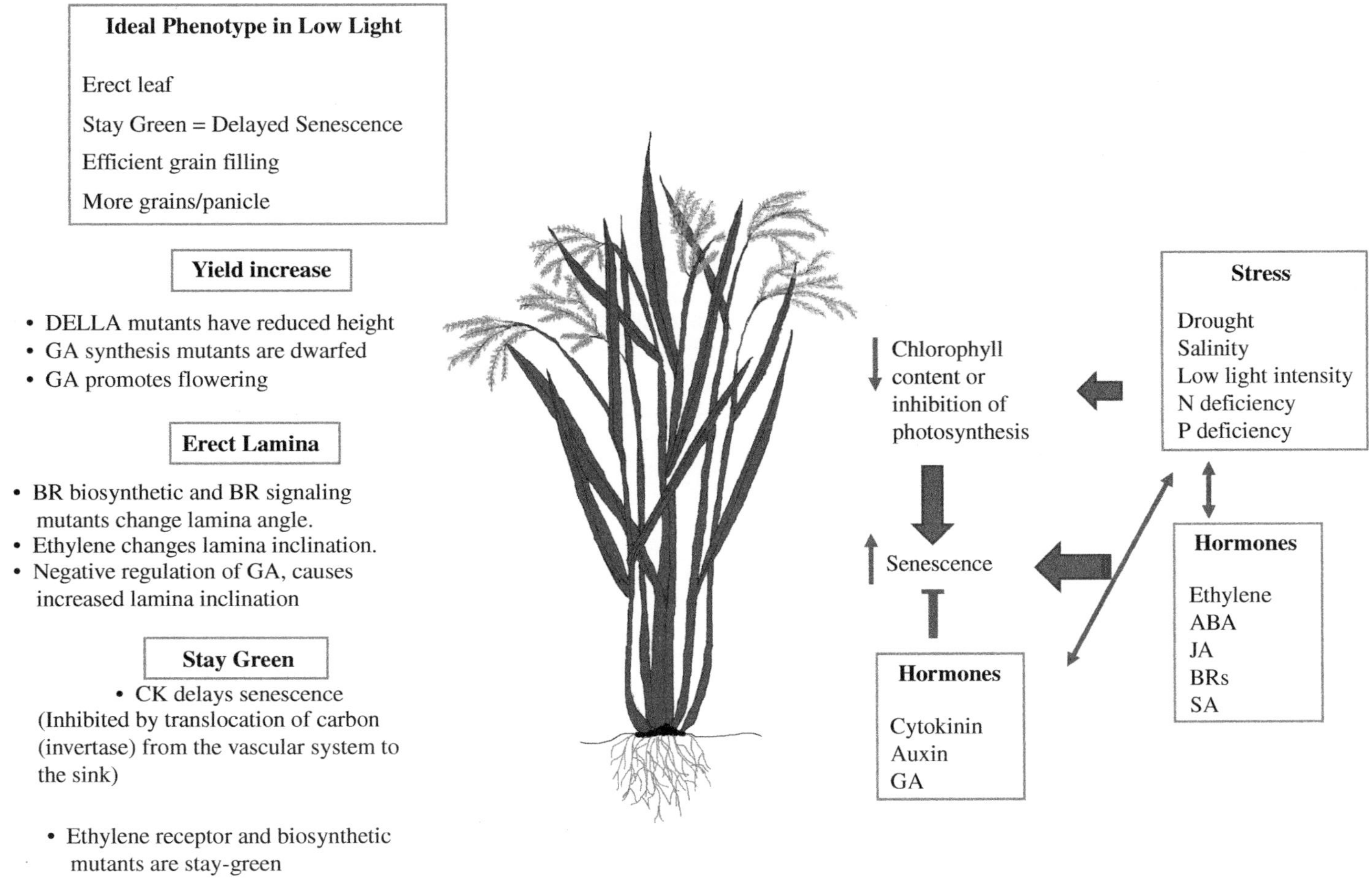

Figure 10.2 Factors affecting rice yield under low light intensity. Bold arrows indicate the response; thin upward and downward arrows indicate up- and down-regulation, respectively. Two-way arrows indicate cross-talk.

(Küpers et al. 2018). It is also known now that kin recognition in the field may lead to increase in grain yield in rice (Yang et al. 2018). This could also be a potential area of interest.

Molecular Targets for Erect Lamina

One of the key phenotypic criteria for a rice plant showing tolerance to low light will be the ability of leaves to maximize harvest of available photons under low light conditions. Apart from PIFs and auxin, genes involved in BRs pathway have been identified with a role in imparting erect lamina in rice (Table 10.1). Most of the mutants identified to date are gain of function but still cannot be used for breeding purposes as the functional "loss of function" allele has not been identified.

Molecular Basis of Delayed Senescence

Leaf senescence can occur prematurely under unfavorable environmental conditions including low light intensity. The chloroplast contains 70% of total leaf protein, most of which is Ribulose-1, 5-bisphosphate carboxylase/oxygenase (RUBISCO) present in stroma and Light harvesting Complex II (LHCII) localized in the thylakoid membrane. One of the earliest signals of leaf senescence involves decrease in rates of photosynthesis and transition of leaf from nitrogen sink to nitrogen source. This involves loss of chlorophyll. As chlorophyll and its primary degradation products are extremely photoactive, two types of autophagic structures, that is, rubisco-containing bodies and senescence-associated vacuoles are involved in the breakdown of proteins. The STAY-GREEN (SGR) protein is a chloroplast protein required for catabolism of chlorophyll and LHCPII recycling. The *sgr* mutants retain their chlorophyll but exhibit decline in photosynthetic efficiency during senescence and therefore are non-functional (Table 10.1). It has now been shown in various crops like wheat (Luo et al. 2013), rice (Ramkumar et al. 2019) among various others that the "functional stay-green" trait requires intact chloroplasts as well as PSII functioning.

Extensive work on leaf senescence has revealed a complex regulatory network integrating various external and internal inputs via gene expression. The role of members of two transcription families, NAC (OsNAC5) and WRKY (WRKY53), in leaf senescence is well documented (Sperotto et al. 2009; Jeong et al. 2013; Song et al. 2011; Takasaki et al. 2010). One of the internal signals to promote senescence is hydrogen peroxide (H_2O_2), a reactive oxygen species (ROS). Under low nitrogen availability high sugar levels can trigger senescence. Hormones are key developmental signals that accelerate (ethylene, abscisic acid (ABA), jasmonate (JA), BRs, and salicylic acid [SA]) or delay (cytokinin, auxin, GAs) senescence. A meta-analysis of key genes involved in shade response in *Arabidopsis* has revealed that a large proportion of shade up-regulated genes are direct targets of PIFs and AUXIN RESPONSE FACTOR6 (ARF6) (Sellaro et al. 2017) suggesting a key role of auxin in low light response.

Yield under stress is a plant breeding target, and genic regions for drought (Kumar et al. 2014), low soil nitrogen (N) levels (Rao et al. 2018), and low soil phosphorus (P) (Wissuwa et al. 2016) have been successfully identified and deployed under field conditions. As yield is a multi-component complex trait, spikelet fertility percentage (SF%), panicle number (PN),

Table 10.1 List of STAY-GREEN and erect leaf mutants reported in rice.

Mutant/gene name	Molecular annotation (Locus ID)	Loss/gain of function	Phenotype	References
esl (*early senescence leaf*)	NA	Loss	Early senescence	Lin et al. (2018)
SGM-3	NA	Gain	Delayed senescence; higher yield; harvest index	Ramkumar et al. (2019)
sgr (*stay green*)	Chr 9	Loss	Reduced photosynthetic activity	Cha et al. (2002), Jiang et al. (2007)
NYC1	Chl b reductase	Loss	Decrease in the chlorophyll a and retention of chlorophyll b and LHC (light harvesting complex) II	Tanaka and Tanaka (2006)
SNU-SG1	Chr 9	Gain	Delayed	Yoo et al. (2007)
OsBUL1	Atypical HLH (Os02g51320)	Loss	Erect leaves; smaller grains	Jang et al. (2017)
SPINDY		Loss	Increased lamina angle	Shimada et al. (2006)
OsLIGULELESS1	SQUAMOSA promoter binding domain protein,	Loss	Erect leaf	Lee et al. (2007)
OsIBH1 (ILI1-BINDING HLH1)		Gain	Erect leaves	Zhang et al. (2009)
IL1	Atypical HLH (Os04g54900)	Gain	Increased lamina angle	Zhang et al. (2009)
PGI	Atypical HLH	Gain	Increased lamina angle	Heang and Sassa (2012)
BUI	Atypical HLH (Os06g12210)	Gain	Increased lamina angle	Tanaka et al. (2009)
LAX PANICLE		Gain	Increased lamina angle	Komatsu et al. (2003)
OsWRKY11	WRKY protein	Loss	Increased lamina angle	Wang et al. (2005)
OsMADS22, *OsMADS55*, and *OsMADS47*	MADS-box protein	Loss	Increased lamina angle	Lee et al. (2008)
LEAF AND TILLER ANGLE INCREASED	CCCH-type zinc-finger protein	Loss	Increased lamina angle	Wang et al. (2008)
RAV6 (RELATED TO ABSCISIC ACID INSENSITIVE3/VIVIPAROUS1 [VP1] 6)	B3 DNA-binding domain-containing protein	Gain	Increased lamina angle; smaller grain size	Zhang et al. (2015)

and 1000 grain weight (GW) are the three key targets for molecular breeding, and in the case of rice some of the genes underlying these traits have been deployed in the field (Wang et al. 2015; Kim et al. 2016). Recently, five more genotypes/genes/alleles have been further characterized such as sucrose synthase (Fan et al. 2019); *osmgd2* (Basnet et al. 2019); Monster rice 1 (Nomura et al. 2019); *d60* (Tonita and Ishimito 2019); *OsSPL14*^{WFP} (Wang et al. 2019), which can be potential targets for higher yields under low light intensity as well.

Cross Talk Between Various Stresses

Data have now emerged that have proved that light is the most important environmental signal that shapes adventitious root architecture in rice during flooding, as it is an indirect detector of oxygen availability (Lin and Sauter 2018). The decision of adventitious roots between upward and downward growth is regulated by light signaling through red and blue light receptors. This study also showed that CPT1, a rice ortholog of *Arabidopsis* NPH3 (Haga et al. 2005), which in turn is required for degradation of PHOT1, has a role in negative phototropism of the primary root of rice seedlings in response to unilateral blue light. Previously, it had been revealed that adventitious root formation on stem nodes of rice is triggered by ethylene and is related to reactive O_2 species (ROS)-mediated epidermal cell death and mechanical signaling (Steffens et al. 2012).

One of the biomolecules linked with regulation of energy homeostasis is trehalose-6-phosphate (reviewed by Lawlor and Paul 2014). The ANAEROBIC GERMINATION1 (AG1) locus of rice, which enables direct seeding of rice into shallow flooded fields, encodes a trehalose-6-P phosphatase (TPP7). The presence of the TPP7 in rice accessions was associated with increased sink strength of elongating coleoptiles, enabling more rapid escape from submergence (Kretzschmar et al. 2015).

It is now well known that flooding-stress signaling works through perturbations in light, oxygen, ethylene, and nitric oxide (Voesenek et al. 2016).

A genome-scale model of rice metabolism under changing light availability has revealed that electron transport between the chloroplast, cytosol, and mitochondrion may play a significant role at low light levels, and a balance between energy demand and availability determine the role of mitochondrial metabolism (Poolman et al. 2013). It has also been suggested that these organelle interactions arise solely as a result of the need for mass and energy balancing without any explicit assumptions concerning kinetic or other regulatory mechanisms.

A profiling of transcripts of rice under four different light treatments (blue, green, red, and white) and dark has led to a genome-scale metabolic model of rice cells, iOS2164, containing 2164 unique genes, 2283 reactions, and 1999 metabolites (Lakshmanan et al. 2015). The study revealed that ABA, ethylene, GA, and JA, are the key biomarkers of light-mediated regulation.

A study on two rice cultivars with different levels of tolerance to chilling has shown that salt and low light stress during vegetative growth increased the susceptibility of rice plants to chilling damage during panicle development (Koumoto et al. 2016). While shade and salt stress significantly reduced stomatal conductance, only low light decreased the specific leaf weight and the leaf sugar and starch contents.

Experimental data on three rice genotypes in three N and three shading treatments have shown that N application could alleviate the detrimental effects of shading on the number of productive tillers and grain yield in rice (Pan et al. 2017). The study suggests that adoption of optimum agronomic measurements such as silicon or organic fertilizer application to cope with low light stress is also beneficial in mitigating low light stress in rice.

"Omics" and Mapping Studies

A genome-wide association mapping approach using 368 rice accessions was conducted to identify alleles and haplotypes for breeding for high photosynthetic efficiency (Zhao et al. 2019). Association studies of six indices for chlorophyll content and stay green phenotype identified 14 pleiotropic regions. Twenty-five known genes and candidate gene *OsSG1* (*Oryza sativa Stay Green 1*; LOC_Os07g27790; involved in glutathione biosynthesis) accounted for natural variation in the two traits. The analysis identified 20 non-synonymous SNPs within 6 known genes and 3 SNPs in the promoter of *OsSG1* as potential SNPs significantly associated with phenotypic differences. Superior haplotypes were also identified based on these potentially functional SNPs, which might be beneficial in breeding for high-photosynthetic efficiency in rice.

A marker trait association study for low light intensity tolerance under field conditions using 110 diverse rice genotypes led to identification of contrasting genotypes (Dutta et al. 2018) with Swarnaprabha, IR 8, Danteswari, Kunti, Megha Rice-1 identified as tolerant to low light intensity. The study also identified a set of 90 potential candidate genes, which were narrowed to 13 genic targets for possible marker–trait association. Six markers [HvSSR02-44 (biological yield), HvSSR02-52 (spikelet fertility), HvSSR02-54 (grain yield), HvSSR06-56 (spikelet fertility), HvSSR06-69 (spikelet fertility; biological yield), HvSSR09-45 (spikelet fertility)] lying on chromosomes 2, 6 and 9 showing significant association ($R^2 > 0.1$) for traits like grain yield/plant, biological yield, and spikelet fertility under low light were also identified. The work also led to identification of eight rice genes [including a member of BBX (B-box) family] known for their role in response to stress or change in plant architecture lying within 10 kb distance of the identified markers.

Microarray studies on panicles of the tolerant rice genotype, Swarnprabha, grown under prolonged shade conditions have revealed up-regulation of most of the ethylene and cytokinin pathway genes (Panigrahy et al. 2019). Significant up-regulation of *ethylene response element binding rotein-2*, *mother of flowering time 1*, and *short panicle1* genes have been suggested to contribute to higher yield in shade.

Low PFD has been reported to damage PSII and lead to decrease in photochemical processes due to closure of reaction centers (Quero et al. 2019). An RNA-seq analysis for comparing leaf transcriptomes at the active vegetative stage of two contrasting varieties Swarnaprabha (tolerant) and IR8 (susceptible) revealed up-regulation of four PSII-associated proteins in Swarnaprabha along with six chlorophyll a-b binding proteins that are part of light harvesting complex (LHC) (Sekhar et al. 2019). Swarnaprabha alleles of these genes can be mined and used in marker assisted breeding programs.

The OGRO (Overview of functionally characterized Genes in Rice Online) tool of the QTARO database (http://qtaro.abr.affrc.go.jp) (Yamamoto et al. 2012) was used to identify

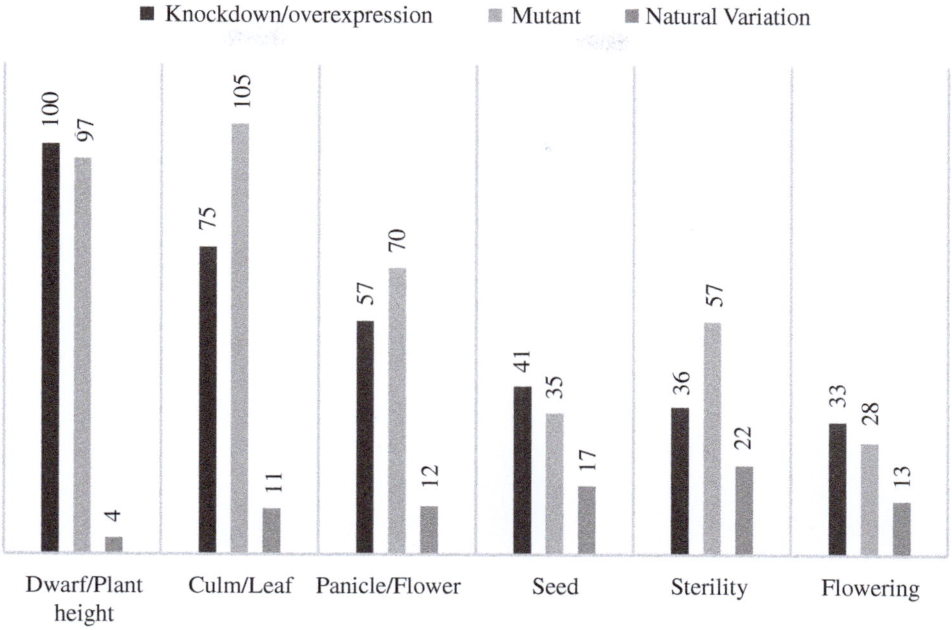

Figure 10.3 Summary of functionally annotated rice genes for six morpho-physiological traits important for low light intensity tolerance in rice. The summary has been compiled based on information available in the Q-TARO rice database.

potential rice genes/regions for low light response based on the morphological and physiological traits affected by low light. A total of 813 functionally annotated rice genes for six morphological/physiological trait classes, namely, seeds, sterility, dwarf/plant height, flowering, panicle flower, and culm leaf were identified (Figure 10.3). Gain or loss of function phenotypes for these genes has been reported based on knockdown, overexpression or mutant analysis. Natural variations have been reported for 79 genes.

Most of these genes have not been analyzed for their role in imparting tolerance under low light conditions, and therefore, their association with higher yield under low light conditions needs to be tested before they can be used for molecular breeding purposes. There are several QTLs reported for these morphological traits. A total of 47 QTLs have been reported for dwarf/plant height, out of which, six are fine mapped and four are physically mapped. Similarly, 66 QTLs were identified for culm/leaf, out of which, three are fine mapped. A total of 111 QTLs are reported for panicle/flower, out of which, 15 have been fine mapped and 1 has been physically mapped. Ninety-three QTLs have been identified for subcategory seed, out of which, three are physically mapped and only one is fine mapped. The physically mapped QTLs can be screened for natural allelic variation across a panel of rice genotypes showing contrasting response to low light. Such allelic variations can be used for designing functional markers for use in molecular breeding.

Based on limited information available of genic regions playing a role in imparting low light tolerance under field conditions, a comprehensive rice chromosome map is presented in Figure 10.4.

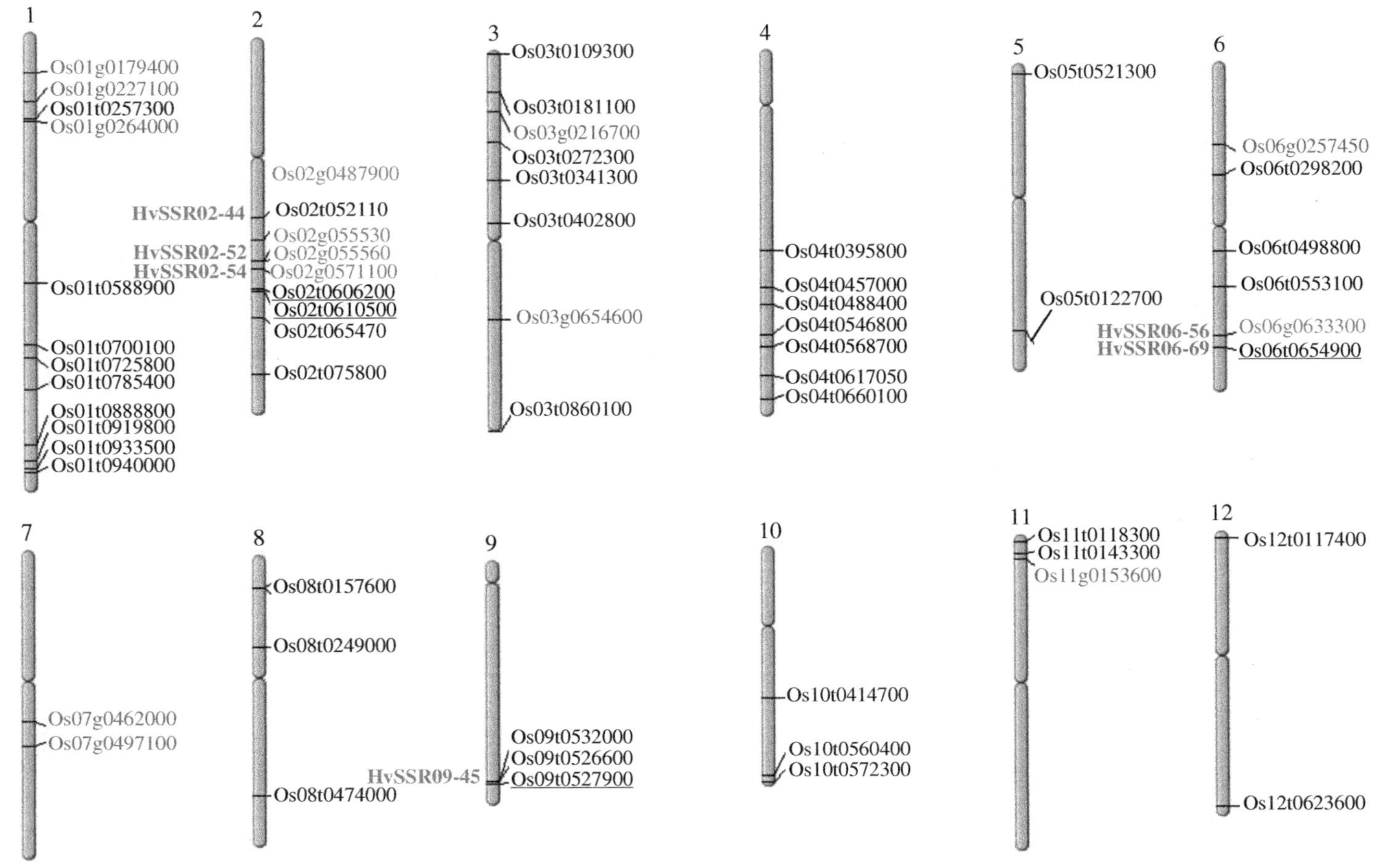

Figure 10.4 Rice chromosomal map showing genic regions and markers identified for low light intensity tolerance in rice through transcriptome or marker trait association studies. Genes (RAP ID) and markers are indicated on the right and left of the chromosomes, respectively. Colors black, red, and blue correspond to the studies by Panigrahy et al. (2019), Zhao et al. (2019), and Dutta et al. (2018), respectively. Locus IDs underlined are common genes identified as potential targets by two studies independently, i.e. Dutta et al. (2018) and Panigrahy et al. (2019). (*See color plate section for color representation of this figure*)

Future Prospects and Molecular Breeding Strategy

Most of the molecular works on low light intensity or light per se have identified genes controlling a desired phenotype such as high chlorophyll content, stay green, erect lamina, and higher grain yield under stress, but natural variation for the genes reported has not been studied. Therefore, detailed information on natural variation and identification of functional alleles for genes underlying the biochemical and morpho-physiological traits identified is required in order to identify or design superior cultivars with highly efficient photosynthetic capacity under low light intensity. The few targets that have been recently reported need to be validated at multiple locations and the underlying identified genomic regions should be gainfully incorporated in breeding programs.

References

Basnet, R., Zhang, J., Hussain, N. et al. (2019). Characterisation and mutational analysis of a monogalactosyldiacylglycerol synthase gene *OsMGD2* in rice. *Front Plant Sci.* 10: 992.

Briggs, W.R. (2014). Phototropism: some history, some puzzles, and a look ahead. *Plant Physiol.* 164: 13–23.

Casal, J.J. (2013). Photoreceptor signalling networks in plants respond to shade. *Annu. Rev. Plant Biol.* 64: 403–427.

Cha, K.W., Lee, Y.J., Koh, H.J. et al. (2002). Isolation, characterization, and mapping of the stay green mutant in rice. *Theor. Appl. Genet.* 104: 526–532.

Dutta, S.S., Tyagi, W., Pale, G. et al. (2018). Marker-trait association for low light intensity tolerance in rice genotypes from Eastern India. *Mol. Genet. Genomics* 293: 1493–1506.

Fan, C., Wang, Y., Zhang, R. et al. (2019). Sucrose synthase enhances hull size and grain weight by regulating cell division and starch accumulation in transgenic rice. *Int. J. Mol. Sci.* 20: 4971.

Goh, C.H., Jang, S., Jung, S. et al. (2009). Rice phot1a mutation reduces plant growth by affecting photosynthetic responses to light during early seedling growth. *Plant Mol. Biol.* 69: 605–619.

Haga, K., Takano, M., Neumann, R. et al. (2005). The rice *COLEOPTILE PHOTOTROPISM1* gene encoding an ortholog of *Arabidopsis* NPH3 is required for phototropism of coleoptiles and lateral translocation of auxin. *Plant Cell* 17: 103–115.

Heang, D. and Sassa, H. (2012). Antagonistic actions of HLH/bHLH proteins are involved in grain length and weight in rice. *PLoS One* 7: e31325.

Horton, P. (2000). Prospects for crop improvement through the genetic manipulation of photosynthesis: morphological and biochemical aspects of light capture. *J. Exp. Bot.* 51: 475–485.

Hughes, J. (2013). Phytochrome cytoplasmic signaling. *Annu. Rev. Plant Biol.* 64: 377–402.

Jain, M., Sharma, P., Tyagi, S.B. et al. (2007). Light regulation and differential tissue-specific expression of phototropin homologues from rice (*Oryza sativa* ssp. *indica*). *Plant Sci.* 172: 164–171.

Jang, S., An, G., and Li, H.-Y. (2017). Rice leaf angle and grain size are affected by the OsBUL1 transcriptional activator complex. *Plant Physiol.* 173: 688–702.

Jeong, J.S., Kim, Y.S., Redillas, M.C. et al. (2013). OsNAC5 overexpression enlarges root diameter in rice plants leading to enhanced drought tolerance and increased grain yield in the field. *Plant Biotechnol. J.* 11: 101–114.

Jiang, H., Li, M., Liang, N. et al. (2007). Molecular cloning and function analysis of the stay green gene in rice. *Plant J.* 52: 197–209.

Kim, S.R., Ramos, J., Ashikari, M. et al. (2016). Development and validation of allele-specific SNP/indel markers for eight yield enhancing genes using whole genome sequencing strategy to increase yield potential of rice, *Oryza sativa L. Rice* 1: 12.

Komatsu, K., Maekawa, M., Ujiie, S. et al. (2003). LAX and SPA: major regulators of shoot branching in rice. *Proc. Natl. Acad. Sci. U.S.A.* 100: 11765–11770.

Koumoto, T., Saito, N., Aoki, N. et al. (2016). Effects of salt and low light intensity during the vegetative stage on susceptibility of rice to male sterility induced by chilling stress during the reproductive stage. *Plant Prod. Sci.* 19: 497–507.

Kretzschmar, T., Pelayo, M.A.F., Trijatmiko, K.R. et al. (2015). A trehalose-6-phosphate phosphatase enhances anaerobic germination tolerance in rice. *Nat. Plants* 1: 15124.

Kumar, A., Dixit, S., Ram, T. et al. (2014). Breeding high-yielding drought-tolerant rice: genetic variations and conventional and molecular approaches. *J. Exp. Bot.* 65: 6265–6278.

Küpers, J.J., van Gelderen, K., and Pierik, R. (2018). Location matters: canopy light responses over spatial scales. *Trends Plant Sci.* 23: 865–873.

Lakshmanan, M., Lim, S.H., Mohanty, B. et al. (2015). Unravelling the light-specific metabolic and regulatory signatures of rice through combined in silico modeling and multiomics analysis. *Plant Physiol.* 169: 3002.

Lawlor, D.W. and Paul, M.J. (2014). Source/sink interactions underpin crop yield: the case for trehalose 6-phosphate/SnRK1 in improvement of wheat. *Front. Plant Sci.* 5: 418.

Lee, J., Park, J.-J., Kim, S.L. et al. (2007). Mutations in the rice liguleless gene result in a complete loss of the auricle, ligule, and laminar joint. *Plant Mol. Biol.* 65: 487–499.

Lee, S., Choi, S.C., and An, G. (2008). Rice SVP-group MADS-box proteins, OsMADS22 and OsMADS55, are negative regulators of brassinosteroid responses. *Plant J.* 54: 93–105.

Li, J., Li, G., Wang, H. et al. (2011). Phytochrome signaling mechanisms. *Arabidopsis Book* 9: e0148.

Lin, C. and Sauter, M. (2018). Control of adventitious root architecture in rice by darkness, light, and gravity. *Plant Physiol.* 176: 1352–1364.

Lin, W., Guo, X., Pan, X. et al. (2018). Chlorophyll composition, chlorophyll fluorescence, and grain yield change in *esl* mutant rice. *Int. J. Mol. Sci.* 19: 2945.

Liu, Z. and Sun, N. (2013). Enhancing photosynthetic CO_2 use efficiency in rice: approaches and challenges. *Acta Physiol. Plant* 35: 1001–1009.

Liu, Q.-H., Xiu, W., Bo-Cong, C. et al. (2014). Effects of low light on agronomic and physiological characteristics of rice including grain yield and quality. *Rice Sci.* 21: 243–251.

Long, S.P., Zhu, X., Naidu, S.L. et al. (2006). Can improvement in photosynthesis increase crop yields? *Plant Cell Environ.* 29: 315–330.

Luo, P.G., Deng, K.J., Hu, X.Y. et al. (2013). Chloroplast ultrastructure regeneration with protection of photosystem II is responsible for the functional “stay-green” trait in wheat. *Plant Cell Environ.* 36: 683–696.

Murchie, E.H., Hubbart, S., Chen, Y. et al. (2002). Acclimation of rice photosynthesis to irradiance under field conditions. *Plant Physiol.* 130: 1999–2010.

Nomura, T., Arakawa, N., Yamamoto, T. et al. (2019). Next-generation long-culm rice with superior lodging resistance and high grain yield Monster Rice 1. *PLoS One* 8: e0221424.

Pan, S., Liu, H., Mo, Z. et al. (2017). Corrigendum: effects of nitrogen and shading on root morphologies, nutrient accumulation, and photosynthetic parameters in different rice genotypes. *Sci. Rep.* 7: 45611.

Panigrahy, M., Ranga, A., Das, J. et al. (2019). Shade tolerance in Swarnaprabha rice is associated with higher rate of panicle emergence and positively regulated by genes of ethylene and cytokinin pathway. *Sci. Rep.* 9: 6817.

Pantazopoulou, C.K., Bongers, F.J., Küpers, J.J. et al. (2017). Neighbor detection at the leaf tip adaptively regulates upward leaf movement through spatial auxin dynamics. *Proc. Natl. Acad. Sci. U.S.A.* 114: 7450–7455.

Poolman, M.G., Kundu, S., Shaw, R. et al. (2013). Responses to light intensity in a genome-scale model of rice metabolism. *Plant Physiol.* 162: 1060–1072.

Quero, G., Bonnecarrere, V., Fernandez, S. et al. (2019). Light – use efficiency and energy partitioning in rice is cultivar dependent. *Photosyn. Res.* 140: 51–63.

Ramkumar, M.K., Kumar, S.S., Gaikwad, K. et al. (2019). A novel stay-green mutant of rice with delayed leaf senescence and better harvest index confers drought tolerance. *Plants* 8: 375.

Rao, R., Neeraja, C.N., Srikanth, B. et al. (2018). Identification of rice landraces with promising yield and the associated genomic regions under low nitrogen. *Sci. Rep.* 8: 9200.

Roy, A., Sahoo, D., and Tripathy, B.C. (2013). Involvement of phytochrome A in suppression of photomorphogenesis in rice seedling grown in red light. *Plant Cell Environ.* 36: 2120–2134.

Sekhar, S., Panda, D., Kumar, J. et al. (2019). Comparative transcriptome profiling of low light tolerant and sensitive rice varieties induced by low light stress at active tillering stage. *Sci. Rep.* 9: 5753.

Sellaro, R., Pacin, M., and Casal, J.J. (2017). Meta-analysis of the transcriptome reveals a core set of shade-avoidance genes in *Arabidopsis*. *Photochem. Photobiol.* 93: 692–702.

Shimada, A., Ueguchi-Tanaka, M., Sakamoto, T. et al. (2006). The rice SPINDLY gene functions as a negative regulator of gibberellin signaling by controlling the suppressive function of the DELLA protein, SLR1, and modulating brassinosteroid synthesis. *Plant J.* 48: 390–402.

Song, S.Y., Chen, Y., Chen, J. et al. (2011). Physiological mechanisms underlying OsNAC5-dependent tolerance of rice plants to abiotic stress. *Planta* 234: 331–345.

Sperotto, R.A., Ricachenevsky, F.K., Duarte, G.L. et al. (2009). Identification of up-regulated genes in flag leaves during rice grain filling and characterization of OsNAC5, a new ABA-dependent transcription factor. *Planta* 230: 985–1002.

Steffens, B., Kovalev, A., Gorb, S.N. et al. (2012). Emerging roots alter epidermal cell fate through mechanical and reactive oxygen species signaling. *Plant Cell* 24: 3296–3306.

Takano, M., Inagaki, N., Xie, X. et al. (2009). Phytochromes are the sole photoreceptors for perceiving red/far-red light in rice. *Proc. Natl. Acad. Sci. U.S.A.* 106: 14705–14710.

Takasaki, H., Maruyama, K., Kidokoro, S. et al. (2010). The abiotic stress-responsive NAC-type transcription factor OsNAC5 regulates stress-inducible genes and stress tolerance in rice. *Mol. Genet. Genomics* 284: 173–183.

Takemiya, A., Inoue, S.-I., Doi, M. et al. (2005). Phototropins promote plant growth in response to blue light in low light environments. *Plant Cell* 17: 1120–1127.

Takemiya, A., Sugiyama, N., Fujimoto, H. et al. (2013). Phosphorylation of BLUS1 kinase by phototropins is a primary step in stomatal opening. *Nature* 4: 1–8.

Tanaka, A. and Tanaka, R. (2006). Chlorophyll metabolism. *Curr. Opin. Plant Biol.* 9: 248–255.

Tanaka, A., Nakagawa, H., Tomita, C. et al. (2009). *BRASSINOSTEROID UPREGULATED1*, encoding a helix-loop-helix protein, is a novel gene involved in brassinosteroid signaling and controls bending of the lamina joint in rice. *Plant Physiol.* 151: 669–680.

Tonita, M. and Ishimito, K. (2019). Rice novel semidwarfing gene *d60* can be as effective as green revolution gene *sd1*. *Plants* 8: E464.

Venkateswarlu, B., and Visperas, R.M. (1987). Solar radiation and rice productivity. *IRRI* Rice Research Paper Series. Vol. 129.

Voesenek, L.A., Sasidharan, R., Visser, E.J.W. et al. (2016). Flooding stress signaling through perturbations in oxygen, ethylene, nitric oxide and light. *New Phytol.* 209: 39–43.

Wada, M. (2013). Chloroplast movement. *Plant Sci.* 210: 177–182.

Wang, D., Zhang, H., Hu, G. et al. (2005). Genetic analysis and identification of a large leaf angles (lla) mutant in rice. *Chin. Sci. Bull.* 50: 492–494.

Wang, L., Xu, Y., Zhang, C. et al. (2008). OsLIC, a novel CCCH-type zinc finger protein with transcription activation, mediates rice architecture via brassinosteroid signaling. *PLoS One* 3: e3521.

Wang, S., Li, S., Liu, Q. et al. (2015). The *OsSPL16-GW7* regulatory module determines grain shape and simultaneously improves grain yield and grain quality. *Nat. Gen.* 47: 949–954.

Wang, P., Gong, R., Yang, Y. et al. (2019). GHd8 controls rice photoperiod sensitivity by forming complex that interacts with Ghd7. *BMC Plant Biol.* 19: 462.

Wissuwa, M., Kretzschmar, T., and Rose, T.J. (2016). From promise to application: root traits for enhanced nutrient capture in rice breeding. *J. Exp. Bot.* 67: 3605–3615.

de Wit, M., Galvão, V.C., and Fankhauser, C. (2016). Light-mediated hormonal regulation of plant growth and development. *Annu. Rev. Plant Biol.* 67: 513–537.

Yamamoto, E., Yonemaru, J., Yamamoto, T. et al. (2012). Rice OGRO: the overview of functionally characterized genes in rice online database. *Rice* 5: 26.

Yang, X.F., Li, L.L., Xu, Y. et al. (2018). Kin recognition in rice (*Oryza sativa*) lines. *New Phytol.* 220: 567–578.

Yoo, S.C., Cho, S.H., Zhang, H. et al. (2007). Quantitative trait loci associated with functional stay-green SNU-SG1 in rice. *Mol. Cells* 24: 83–94.

Yuan, J.C., Ding, Z.Y., Zhao, C. et al. (2005). Effects of sunshine-shading, leaf-cutting and spikelet-removing on yield and quality of rice in the high altitude region. *Acta Agron. Sin.* 31: 1429–1436.

Zhang, L.-Y., Bai, M.-Y., Wu, J. et al. (2009). Antagonistic HLH/bHLH transcription factors mediate brassinosteroid regulation of cell elongation and plant development in rice and Arabidopsis. *Plant Cell* 21: 3767–3780.

Zhang, X., Sun, J., Cao, X. et al. (2015). Epigenetic mutation of RAV6 affects leaf angle and seed size in rice. *Plant Physiol.* 169: 2118–2128.

Zhao, Y., Qiang, C., Wang, X. et al. (2019). New alleles for chlorophyll content and stay-green traits revealed by a genome wide association study in rice (*Oryza sativa*). *Sci. Rep.* 9: 2541.

11

Harnessing Tolerance to Low Phosphorus in Rice: Recent Progress and Future Perspectives

Wricha Tyagi, Erneica N. Nongbri and Mayank Rai

School of Crop Improvement, College of PG studies in Agricultural Sciences, Central Agricultural University (Imphal), Umiam, Meghalaya 793103, India

Introduction

The major challenge agriculture faces in the twenty-first century is how to feed nine billion people. In the scenario of shrinking arable land and exhaustion of non-renewable sources of fertilizers, it is imperative that the focus shift to sustainable agriculture. The breeding/agriculture goal now must be to get more yield per unit of input. This will require understanding of basic mechanisms of uptake and assimilation of major nutrients in plants.

Phosphorus Uptake and Assimilation in Plants

Role of Low P in Rice Development and Growth

One of the key nutrients required for plant growth is phosphorus (P). Because it is highly reactive, P exists in more than 170 mineral or inorganic forms in soil (Holford 1997). Plants uptake P in the form of inorganic phosphate (Pi), that is, orthophosphates from soil ($H_2PO_4^-$ and HPO_4^{2-}), but a large fraction of soil P (50–80%) occurs as organic-P and is not readily available for root uptake (Wang et al. 2010). As P is an essential component of cell membrane and essential for energy currency (ATP), its deficiency leads to major developmental and growth defects, ultimately leading to severe yield loss. In the case of rice, low Pi levels produce recognizable symptoms as early as 30 days after transplanting (DAT). Phosphorus deficiency is widespread in all major rice ecosystems and typical P deficiency symptoms like stunted plants, reduced tillering, dark green, narrow, short, very erect leaves, thin and spindly stems, and reduced number of panicles and grains are commonly observed under field conditions (Das et al. 2017) (Figure 11.1).

Uptake and Transport of P in Rice

As against the typical soil concentration of around 10 μM of soluble Pi, plants maintain a much higher cellular concentration of >10 mM (>3 g kg^{-1}, on a biomass/dry-weight

Molecular Breeding for Rice Abiotic Stress Tolerance and Nutritional Quality, First Edition.
Edited by Mohammad Anwar Hossain, Lutful Hassan, Khandakar Md. Iftekharuddaula, Arvind Kumar and Robert Henry.

Figure 11.1 Phenotypic differences in rice lines in low P (a) and optimum P (b) soil microplots. *Source:* Mayank Rai.

(DW) basis) (Marschner and Rengel 2012). The uptake of P from soil to plants via roots therefore requires movement of P against the concentration gradient requiring P-specific transporters. Plants exploit both high- and low-affinity Pi-transporter (PT) systems: low-affinity PTs are constitutively expressed and operate at Pi concentrations in the millimolar range, high-affinity PTs contribute to Pi uptake at limiting, micromolar concentrations (Chrispeels et al. 1999; Raghothama 2000). The 26 rice PTs characterized so far can be divided into four subfamilies (PHT1, PHT2, PHT3, and PHT4) distributed across the 12 rice chromosomes (Liu et al. 2011). Fifteen *OsPT* genes (*OsPT1* to *OsPT4*, *OsPT8*, *OsPT15* to *OsPT19*, and *OsPT22* to *OsPT26*) showed high or detectable expression in root and other tissues tested and also in response to various hormones. The PHT1 gene family members localize to the plasma membrane, and most of the PHT1 genes display a phosphate starvation-induced (PSI) expression preferentially in roots, and are broadly involved in Pi uptake and translocation (Gu et al. 2016). Most of the PHT1 members have a role in Pi uptake and/or translocation (reviewed extensively in Gu et al. 2016; Table 11.1). The maintenance of cytosolic Pi homeostasis involves transport of Pi to various organelles inside the cell such as mitochondria (PHT3), golgi (PHT4), and plastids (PHT2/4). Proteins harboring SPX (SYG1/PHO81/XPR1) domain and MFS (Major Facilitator Superfamily) at N and C terminals respectively are involved in transport of Pi across vacuoles (Wang et al. 2012). OsSPX-MFS1 and OsSPX-MFS2 facilitate import and export of Pi, respectively, across the tonoplast (Wang et al. 2012, 2015). The precise role of these two proteins is yet to be determined. However, in the case of *Arabidopsis*, the ortholog of OsSPX-MFS3, VPT1 (Vacuolar Phosphate Transporter 1) is reported to be responsible for transporting Pi into vacuoles (Liu et al. 2015) and is vital for Pi accumulation and plant adaptation to fluctuating Pi availability in the environment.

The genes PHO1 (coding a membrane protein) and PHO2 (E2 conjugase) control Pi homeostasis as they ensure proper loading of Pi in either roots or leaves, respectively (Poirier et al. 1991) by controlling the transcription of PHT1;8 and PHT1;9, respectively (Rouached et al. 2010). The EXS domain (corresponding to ERD1 and SYG1 proteins of yeast and XPR1 protein of mammals) present in PHO1 (Wang et al. 2004) play another crucial role as a repressor signal. During phosphate starvation, the inhibition of phosphate starvation response (PSR) via EXS domain of PHO1 is switched off (Wege et al. 2016)

Table 11.1 Mapping studies for P deficiency tolerance in rice.

Traits	Population used for the study	Parents involved in the cross	No. of QTLs (chromosome number)	Percentage phenotypic variation (%)	LOD score	References
DW, PUP	NILs (98)	Nipponbare/Kasalath	4 (2, 6, 10, 12)	45.4 and 54.5	2.30, 2.19	Wissuwa et al. (1998)
RTA, RSDW, RRDW	RILs (285)	IR20/IR55178-3B-9-3	1 (12)	54, 60.8, and 44.2	>3	Ni et al. (1998)
PUP, TN	NILs	Nipponbare/Kasalath	1 (12)			Wissuwa et al. (2002)
RE	F3 (82)	Gimbozu/Kasalath	1 (6)	20	4.5	Shimizu et al. (2004)
RSL	BILs (225)	OM2395/AS996	1 (12)	11.67	>3	Lang and Buu (2006)
REP	CSSLs (54)	Nipponbare/Kasalath	1			Shimizu et al. (2008)
PH, MRL, RN, RV, RFW, RDW, SDW, TDW, RS	ILs (441)	Yuefa/IRAT109	24	Additive and epistatic gene interaction	—	Li et al. (2009)
DW, HI, gPUE, PHI, PT, PTE, PUP, PUEb, PUEg, strPUE	RILs	Zhensan97/Minghui63	36	34, 20.5, 1.4; 42.8, 14.5, 34.1, 13.6, 6.4, 32.6, and 15.1		Wang et al. (2014b)
RRL, RRDW, RsDW, RTDW, RSDW	BC2F4	Shuhui527/Minghui 86	48			Chao et al. (2015)
RSP	Association panel (196)	Diverse rice genotypes	5 (1, 6, 7, 11, 12)	75		Mori et al. (2016)
RN; RDW	CSSLs (76)	XieqingzaoB/Zhonghui9308	7 (3, 5, 6, 8, 10)	10.82 and 18.46	2.0, 3.2	Anis et al. (2018)

Note: DW, plant dry weight; PUP, phosphorus uptake; RTA, relative tillering ability; RSDW, root/shoot dry weight ratio; RRDW, relative root dry weight; TN, tiller number; RE, root elongation; RSL, relative shoot length; REP, root elongation under phosphorus deficiency; CSSLs, chromosomal sequence substitution lines; PH, plant height; MRL, maximum root length; RN, root number; RV, root volume; RFW, root fresh weight; RDW, root dry weight, SDW, shoot dry weight; TDW, total plant dry weight; HI, harvest index; gPUE, P use efficiency for grain yield based on P accumulation in grains; PHI, P harvest index; PT, P translocation; PTE, P translocation efficiency; PUEb, P use efficiency for DW accumulation; PUEg, P use efficiency for grain yield; strPUE, P use efficiency for straw dry weight based on P accumulation in straw; RRL, relative root length; RRDW, RsDW, relative shoot dry weight; RTDW, relative total dry weight; RSP, μg shoot P content per cm^2 root surface area; LOD, logarithm of odds.

suggesting its role in root-to-shoot signaling (Figure 11.2) both under Pi-deficient as well as sufficient conditions.

Organic bound forms of P in the soil majorly include phosphate monoesters (derivatives of inositol hexakisphosphate), phosphate diesters (nucleic acids, phospholipids, cyclic phosphates), and phosphonates (George and Richardson 2008). Consequently, despite having significant amounts of P in the form of organic complexes in soil, plants experience its deficiency (Gepstein et al. 2003; Hurley et al. 2010). Plants have devised several strategies to solubilize these P forms, which involve exudation of organic acids and protons, modification of root system architecture, association with mycorrhiza and secretion of acid phosphatises APases and phytases (Shen et al. 2011). PAPs (purple acid phosphatases) are the largest group of APases and are involved in Pi acquisition and utilization in plants (Kuang et al. 2009). These PAPs catalyze the hydrolysis of several P-containing organic compounds at acidic pH (pH 4–7). Intracellular PAPs remobilize Pi from cellular reserves, whereas secreted ones hydrolyze P compounds in apoplast and rhizosphere. In rice, 26 PAPs have been identified, of which 10 are induced significantly under Pi deficiency (Zhang et al. 2011). Rice PAPs, OsPAP10a, OsPAP10c, and OsPAP21b have been shown to utilize external organic phosphorous when over-expressed in rice (Lu et al. 2016; Tian et al. 2012; Mehra et al. 2017).

Phosphite (Phi) is a non-metabolizable form of Pi, which cannot substitute for Pi in plants and is known to interfere or suppress the expression of many low Pi-inducible genes (Varadarajan et al. 2002). Using this form of Pi, it is possible to identify genes responding to Pi status of plants as these genes in presence of Phi will not respond to local availability of Pi in media.

Signaling in Response to Low Pi

The cellular sensor for Pi are OsPHR1/OsPHR2 (*Oryza sativa* PHosphate starvation Response 1/2), MYB domain containing transcription factors (Puga et al. 2014; Wang et al. 2014a) and the level of Pi in cells is determined by inositol polyphosphate-signaling molecules (InPs) (Wild et al. 2016). The concentrations of InPs vary according to levels of available Pi, and InPs trigger the association of SPX proteins with transcription factors to regulate Pi starvation responses. The three OsPHR genes (*OsPHR1*, *OsPHR2*, and *OsPHR3*) are functionally redundant in Pi signaling and determine root hair morphology (Guo et al. 2015). However, only OsPHR3 over-expressing lines exhibit enhanced growth (based on DW measurements and grain filling) under Pi-deficient hydroponic as well as field conditions (Guo et al. 2015), suggesting that *OsPHR3* could be a molecular breeding target in rice.

Alterations in membrane lipid composition under limited P supply have also been observed. Degradation of phospholipids into Pi and diacylglycerol is mediated by phospholipases C and D, which are essential for lipid turnover in plants adjusting to P deficiency (Li et al. 2006). MicroRNAs like miR399, miR6250 (Secco et al. 2013), miR2111, and miR827 have been reported to be involved in the regulation of gene expression in phosphate starvation.

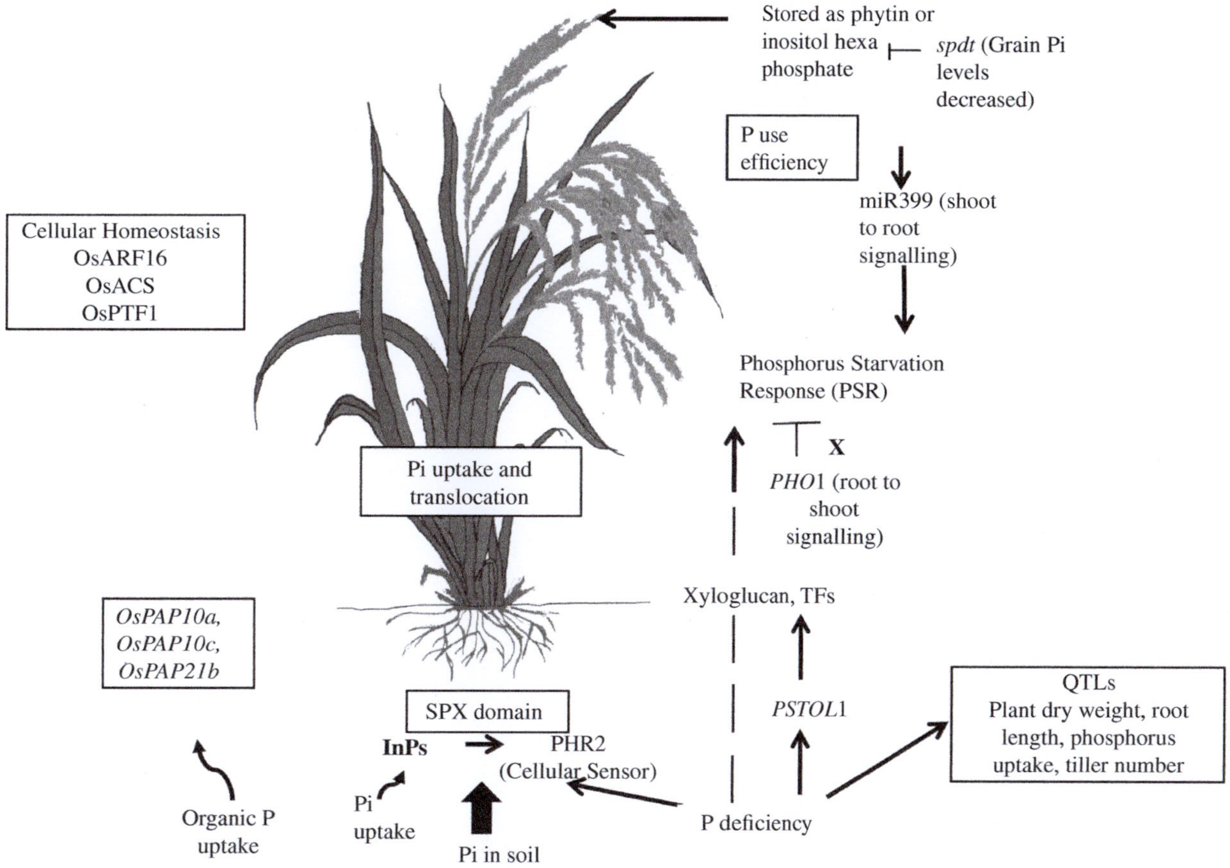

Figure 11.2 Key molecular players in uptake, assimilation, and homeostasis of P in rice. *Note:* The key players include genes involved in making Pi available like purple acid phosphatise (PAP); detecting the availability of Pi within the cell (through SPX domain), i.e. PHR2; Pi translocation through various transporters within the plant; maintenance of cellular homeostasis (via auxin [ARF], ethylene [ACS], and Pi-specific genes [PTF]) and storage of P in grains in the form of phytin. The response of rice to low P levels involves other molecular players also like xyloglucans, starvation response genes, miR399, *PHO1*, *PSTOL1*, etc. among many other yet unidentified molecular targets underlying reported QTLs.

Assimilation of P in Rice

The development of P-deficiency-tolerant crops that grow and yield well in soils deficient in plant-available P is a cost-effective means of improving crop yields in low-input farming systems (Ismail et al. 2007; Wissuwa et al. 2009). Such P-efficient crops would ideally combine high P uptake (phosphorus uptake efficiency, PUP) with very efficient use of P in DW accumulation (phosphorus utilization efficiency, PUE). Plants producing equal DW at lower P concentrations (compared to other plants) or higher DW at equal P concentrations would be considered to have high PUE. The estimates of PUE are confounded if genotypes differ in P acquisition efficiency (Rose et al. 2011). Therefore, if molecular mechanisms involved in PUE are to be truly understood, then the screening protocols should (i) avoid competition for P between individual genotypes to assure equal P supply translated to equal P uptake, (ii) nullify genotypic differences in seed P content or seedling vigor (Rose et al. 2011), and (iii) include multiple entries from different rice genetic groups (Tyagi and Rai 2017; Famoso et al. 2011; Wissuwa et al. 2016).

The sections below summarize the progress made so far in understanding the assimilation of P after uptake in different plant parts at two different stages of rice growth.

Vegetative Stage

The uptake of P is followed by its assimilation to various parts of rice including young and old leaves, flag leaf, and ultimately mature grains. OsSPX1 and OsSPX2 domain proteins play a key a role in P homeostasis during vegetative growth (Secco et al. 2013), while *OsPHO1* and *OsPHO2* play a role in signaling involving P status. Rice SPX-domain-containing proteins are involved in Pi xylem loading (OsPHO1;2), leaf Pi reallocation (OsSPX1, OsSPX-MFS1), and the negative regulation of the expression of downstream PSI genes (OsSPX1).

Reproductive Stage

Phosphorus is rapidly loaded into grains of rice genotypes between 6 and 15 days after flowering (Ogawa et al. 1979; Wang et al. 2016). The mechanism of Pi uptake, assimilation, and partitioning during grain filling of rice is just beginning to be understood (Julia et al. 2016; Jeong et al. 2018; Wang et al. 2016). It was found that 6 DAA (days after anthesis) the flag leaf P concentration was stable, and it was around 15 DAA when the flag leaf started acting as a P source for developing grains (Jeong et al. 2017). Using RNA-seq study, a set of 26 rice genes up-regulated at 15 DAA, including *OsPHO1*, *OsPHO2*, *OsSPX1*, and *OsSPX2* were identified (Jeong et al. 2017) suggesting overlap of genes involved in P assimilation between vegetative and reproductive stage.

With the use of radioactive isotope of P (^{33}P), it has been revealed that the potential contribution of P re-mobilization from vegetative tissues to grain P was around 20% of grain P, and the majority of Pi taken up by roots during grain filling remained in the roots up to maturity (Wang et al. 2016). Plants partitioned the majority of above-ground P to the grains. This suggests that post-flowering P uptake does not move directly to the grains but that grain filling involves indirect P fluxes originating from P previously taken up and located in the vegetative tissues. Therefore, breeding efforts to lower grain P in order to avoid problems associated with high levels of phytic acid in seeds and eutrophication of water bodies, could feasibly focus on reducing the re-partitioning of P from vegetative tissues, rather than limiting post-anthesis P uptake from the soil (Wang et al. 2016). In fact, work by Yamaji et al. (2017) on *OsSPDT* (SULTR-like phosphorus distribution transporter [SULTR – sulfate transporter]) has proved the feasibility of the approach. The knockout of SPDT resulted in a 20% reduction of total P in the grain and increased P in the straw without grain yield penalty.

Impact of P Deficiency on Rice

Rice, the staple food crop for more than half of the world's population, has only ~25% P use efficiency (PUE) (Vinod and Heuer 2012). Therefore, developing rice genotypes for improved PUE with efficient PUP has been recognized as a critical step towards its sustainable production.

Resources That Have Helped Enhance the Understanding of Low P Tolerance in Rice

Numerous quantitative trait loci, mutants as well as genes, have been identified by various groups for low P tolerance.

QTLs

Various studies have identified quantitative traits loci (QTLs) associated with component agronomic traits that are known to contribute towards higher yields under P-deficient conditions in rice. Besides P uptake, translocation and use efficiency, traits related to plant DW, various root characteristics, tillering ability, and so forth have been mapped under P-deficient conditions (Table 11.1). A major QTL for PUP was identified on the long arm of chromosome 12 in rice using near isogenic lines (NILs), population derived from a cross between P-deficiency susceptible *japonica* cultivar, Nipponbare, and P-efficient *indica* landrace Kasalath (Wissuwa et al. 1998, 2002). This QTL was found to contribute 54.5% of the total phenotypic variance with a logarithm of odds (LOD) score of 2.19. Three QTLs for plant dry weight (DW) were also identified in this population on chromosomes 2, 6, and 10 with a cumulative phenotypic variance explained of 45.4%. Using 285 recombinant inbred lines (RILs), generated from a cross derived from P-deficiency-tolerant variety, IR20, and P-deficient-sensitive variety, IR55178-3B-9-3, QTLs with a LOD score >3 for relative tillering ability (RTA), root/shoot dry weight ratio (RSDW), and relative root dry weight (RRDW) were detected on chromosome 12 explaining phenotypic variation of 54%, 60.8%, and 44.2%, respectively (Ni et al. 1998). A variation in root elongation (RE) was mapped on chromosome 6 using 82 F_3 lines from Gimbozu/Kasalath cross (Shimizu et al. 2004). Twenty percent phenotypic variance was explained with a LOD score of 4.5. A QTL for relative shoot length (RSL) was detected on chromosome 12 using 225 BILs lines derived from a cross between OM2395/AS996 explaining a phenotypic variation of 11.67% with a threshold of LOD >3.0 (Lang and Buu 2006). A distinct QTL for root elongation under P deficiency (REP) was detected in the long arm of chromosome 6 using 54 CSSLs (chromosomal sequence substitution lines); in which chromosomal segments of a susceptible genotype, Nipponbare, were partially substituted from Kasalath (Shimizu et al. 2008). A set of additive QTLs for plant height (PH), root traits, and DW under P deficiency was found using 271 introgression lines (ILs), derived from Yuefa/IRAT109 cross, explaining 24% of the total phenotypic variance (Li et al. 2009). A total of 36 QTLs with contribution to phenotypic variance ranging from 1.4 to 34.1% were detected for total plant DW, harvest index (HI), P use efficiency for grain yield based on P accumulation in grains (gPUE), P harvest index (PHI), P translocation (PT), P translocation efficiency (PTE), PUP, PUE for DW accumulation (PUEb), PUE for grain yield (PUEg), and PUE for straw dry weight based on P accumulation in straw (strPUE) for RIL population derived from Zhensan97/Minghui63 cross (Wang et al. 2014b). Across 114 BC_2F_4 ILs, developed from Shuhui 527 and Minghui 86, 48 main QTLs were identified for low P tolerance targeting traits like relative root length (RRL), root dry weight (RDW), shoot dry weight (SDW), and RSDW and their phenotypic variations ranged from 7.7% to 16.6% (Chao et al. 2015). A set of 196 diverse rice varieties were evaluated under P deficiency and five QTLs for RDW were detected on chromosomes 1, 6, 7, 11, and 12, cumulatively explaining 75% phenotype variation (Mori et al. 2016). A total of 7 QTLs were reported on chromosomes 3, 5, 6, 8, and 10 across 76 CSSLs derived from a cross between P stress tolerant variety, Xieqingzao B, and a P stress sensitive variety, Zhonghui 9308. Their contributions to phenotypic variance ranged from 10.82% to 18.46%, with LOD values ranging from 2.0 to 3.32 (Anis et al. 2018).

Functionally Validated Low P Tolerance Potential Candidates

Several intra- and inter-cellular transporter molecules have been implicated to play a role in low P tolerance in rice (Table 11.2; for details see the sections above).

Transcription factors classified as MYB family (PHR1, PHR2, PHR3, MYB2P-1), basic helix-loop-helix (bHLH) (OsPTF1), auxin response factors (ARF16), containing SPX-domain (SPX); phosphatases (PAP21b), serine threonine kinase (PSTOL1), etc have been identified. The role of MYB and SPX family transcription factors is essentially in Pi homeostasis. Genes involved in various hormone pathways, like *ARF16* and *OsACS*, play a role in remodeling root system architecture and cellular P homeostasis. Previously, it was not clear whether low Pi stimulates or inhibits biosynthesis of ethylene (Borch et al. 1999; Drew et al. 1989; Kim et al. 2008). However, recently it has been demonstrated that the key rate-limiting enzyme of the ethylene biosynthetic pathway (1-aminocyclopropane-1-carboxylic acid (ACC) synthase (ACS)) is involved in altering root system architecture in response to Pi deficiency (Lee et al. 2019). Detailed analysis of mutants generated by CRISPR/Cas9 (clustered regularly interspaced short palindromic repeat (CRISPR)/CRISPR-associated protein 9) genome editing revealed that *OsACS1* and *OsACS2* are involved in remodeling of root system architecture, transcript regulation of Pi starvation-induced genes, and cellular phosphorus homeostasis.

Apart from the PT gene family, another phosphate transporter, OsPHF1 (PHosphate transporter traffic Facilitator) is also reported in rice (Chen et al. 2011). This transporter is not regulated by OsPHR2. OsPHF1 regulates the localization of both the low-affinity

Table 11.2 Identified rice genes with role in P uptake, translocation or homeostasis.

Genes	Locus ID	Function	References
PSTOL1	No gene available in the reference genome	Early regulator of crown development and root hair	Gamuyao et al. (2012)
VPT1 (OsSPX-MFS3)		Transporting Pi into vacuole	Liu et al. (2015)
PaP21b	Os011g05400	Pi acquisition and utilization from organophosphates	Mehra et al. (2017)
SPX1	Os06g40120	Phosphate homeostasis in leaves	Liu et al. (2010) and Wang et al. (2014b)
PHR2 (1R type)			
AUX1	Os01g63770	Root hair elongation in response to low P	Giri et al. (2018)
OsPht1;1 (OsPT1)	Os03g05620	Pi uptake and translocation in Pi-replete rice	Sun et al. (2012)
ARF16	Os06g09660	Auxin and phosphate starvation response	Shen et al. (2013)
OsPT11	Os01g46860	Symbiotic phosphate uptake (high affinity Pi transporter)	Paszkowski et al. (2002)
OsPT1;2 (OsPT2)	Os03g05640	Pi uptake and translocation throughout the plant	Ai et al. (2009)
OsPT1;6 (OsPT6)	Os08g45000	Low-affinity Pi transporter functioning in translocation of stored Pi	Ai et al. (2009)
OsMYB2P-1 (R2R3 MYB)		Regulation of phosphate starvation and root architecture	Dai et al. (2012)
OsPHR1	*AK063486*	Regulates Pi signaling and homeostasis	Guo et al. (2015)
MAIF1 (F box protein)	Os02g44990	Reduction of abiotic stress tolerance and root growth promotion	Yan et al. (2011)
OsPht1;8 (OsPT8)	Os10g30790	Pi homeostasis in rice	Jia et al. (2017)
OsPHT1;4	Os04g10750	Yield under P deficiency	Zhang et al. (2015)
PHF1	Os07g09000	Plasma membrane localization of *OsPht1;1*	Chen et al. (2011)
PTF1		Regulator of Pi starvation response	Yi et al. (2005)
OsACS		Ethylene biosynthesis	Lee et al. (2019)

PT, OsPT2, and high-affinity PT, OsPT8, to the plasma membrane and determines Pi uptake and translocation. A bHLH family gene member, OsPTF1 (*Oryza sativa* Phosphate Starvation Induced Transcription Factor 1), has been reported to control root hair length (Yi et al. 2005).

Several studies have focused on mutagenesis as an approach to study the underlying mechanism of phenotypes of interest (Table 11.3). This has resulted in identification of gain-of-function mutants in Nagina 22 background (Panigrahy et al. 2014). Several mutants (*ltn1*, *adw1*, *adw2*, *adw3*, etc.) for one of the key signaling components for low P response, *OsPHO2*, have been identified and characterized (Hu et al. 2011; Takehisa and Sato 2019). Mutants *adw1-1* and *adw1-2* have a Tos17 transposon insertion and *adw1-3* has a 61-bp deletion in the *OsPHO2* exon (Takehisa and Sato 2019). It has been shown that these mutant rice plants have potential to acquire more P from soil. It is also important that the grain Pi levels do not increase beyond a certain threshold. This is to ensure optimum Pi cycling, minimizing eutrophication and avoiding problems associated with high phytic acid levels in seeds. *lpa* mutants are one of the most well studied till date (Frank et al. 2007; Larson et al. 2000; Zhao et al. 2016). One of the mutants, *lpa 1-1*, has reduced phytate and phosphorus concentration in grains and codes for sulphate transporter gene (Zhao et al. 2016). A study of three Tos-17 mutants in rice SULTR-like phosphorus distribution transporter (*OsSPDT*) gene has revealed that distribution of P in different organs was greatly altered (Yamaji et al. 2017). The concentration of phytate was 25–32% lower in the mutants than in the wild-type rice without compromising yield.

Apart from genes and mutants, using an in silico approach along with GWAS (genome wide association mapping), candidate genes for P use efficiency (Wang et al. 2015; Wissuwa et al. 2015) and root efficiency (Mori et al. 2016) have also been identified.

Molecular Targets Available in Rice for Breeding

PUP1 History and Deployment of PUP1 Based Markers for MAS

The only extensively deployed QTL for low P tolerance via molecular markers is PUP1. This QTL was discovered using a bi-parental population derived from Kasalath (low P tolerant) and Nipponbare (low P susceptible) (Wissuwa et al. 1998). This study was followed by fine mapping using a NIL population (Wissuwa et al. 2002) and then identification of markers (Chin et al. 2011) and finally, confirmation of underlying gene *PSTOL1* (Gamuyao et al. 2012). PUP1-based markers are being used for MAS and MABC in several breeding programs (Chin et al. 2011; Chithrameenal et al. 2018). However, field trials in Indonesia have revealed that genetic background is crucial in determining the success of PUP1 in field (Wissuwa et al. 2015).

Other Haplotypes Reported for PUP1 and Implications for Breeding Program

A novel PSTOL1 allele in upland NERICAs (new rice for Africa) inherited from the *Oryza glaberrima* parent CG14 has been reported (Pariasca-Tanaka et al. 2014). This allele showed a 35 base-pair substitution when aligned to the Kasalath allele, but maintained a fully

Table 11.3 Rice mutants identified with a putative role in low P tolerance mechanism.

Mutant	Type gene	Function	References
OsA8	LOC_Os03g01120	Uptake and translocation	Chang et al. (2008)
N4101; NH776, NH710 and NH719	*NA*	Better performance under low P conditions	Panigrahy et al. (2014)
ltn1	*LTN1 (OsPHO2)*	Leaf-tip necrosis predominantly in mature leaves	Hu et al. (2011)
d3	*NA*	Leaf senescence in response to P deficiency	Yamada et al. (2014)
lpa 1-1	*SULTR3:3*	Reduced phytate and phosphorus concentration in grains	Larson et al. (2000) and Zhao et al. (2016)
lpa-xs-110-1	*NA*	Low phytic acid in seeds under field condition	Frank et al. (2007)
lpa-xs-110-2	*NA*		
spdt	*SPDT*	Translocation of Pi to leaves	Yamaji et al. (2017)
Lsi		Defective in Si uptake; decreased P uptake by downregulating the expression of OsPT6	Hu et al. (2018)
adw1	*OsPHO2*	More P acquisition from soil	Takehisa and Sato (2019)

NA, information not available.

conserved kinase domain, and is present in most *O. glaberrima* accessions evaluated. In silico and marker analysis indicated that many other genes of the Kasalath Pup1 locus were missing in the *O. glaberrima* genome, including the dirigent-like gene OsPupK20-2, which was shown to be downstream of PSTOL1. Several allele-specific markers (K46-CG1 and K46-CG2) and a novel marker K46-3, targeting a conserved region and amplifying both PSTOL1 alleles equally for the use for molecular breeding to transfer the PSTOL1 gene from Kasalath to African mega-varieties, including NERICAs are also available (Pariasca-Tanaka et al. 2014). Non-parental introgressions were also reported in several NERICAs. This study also evaluated a set of 76 wild rice accessions using the PSTOL1 allele-specific markers. Within *Oryza nivara* and *Oryza rufipogon*, the ancestors of *O. sativa*, no consistent genotype was detected, and a putative novel allele which was amplified with both, the Kasalath and CG14-specific markers was also detected. The novel allele was also detected in *Oryza longistaminata*, but not in *Oryza barthii*, the *O. glaberrima* ancestor. The presence of *PSTOL1* was restricted to *Oryza* species belonging to the AA genome, since no allele could be detected in accessions with BB ~ HHJJ genomes.

Seventeen different haplotypes were obtained upon sequencing of 975 bp sequence from 67 different *O. rufipogon* accessions (Neelam et al. 2017). Three *O. rufipogon* accessions, namely, IRGC 106506, IRGC 81989, and IRGC 104639 with haplotypes H17, H11, and H3, respectively were identified to be superior, when validated by hydroponics-based low P screening. Whether these three haplotypes outperform Kasalath (carrying the *original PSTOL1*) in the field needs further investigation. Also, as the flanking regions of *PSTOL1* were not sequenced, it is difficult to ascertain whether the functional alleles are just limited to *PSTOL1* or the entire 90 kb PUP1 region is distinct from *O. sativa* in at least these three *O. rufipogon* accessions.

Novel haplotypes in *indica* rice genotypes adapted to acidic soils have also been identified (Yumnam et al. 2017). Sahbhagi Dhan (SD) is a released variety performing well in the acidic soils, carries the PSTOL1 gene but has a different allele across *PupK20* gene (Yumnam et al. 2017), suggesting that though sequence of SD is exactly same as Kasalath for *PSTOL1* gene, it harbors differences for at least one other gene reported for low P tolerance. The reports discussed above suggest the possibility of exploitation of haplotypes superior to Kasalath *PUP1/PSTOL1* locus in future MAS programs. Evidence suggesting balancing selection at this locus also suggests that further phenotypic characterization of allelic variants is needed and may be useful for crop improvement (Vigueira et al. 2016).

Beyond PSTOL1

Our own studies have identified several rice genotypes that lack PSTOL1 but perform at par with Kasalath under low P acidic soils (Tyagi et al. 2012; Yumnam et al. 2017).

When performance of genotypes under poor soil conditions (lacking in major or minor nutrients) is considered, tolerance can be achieved through better acquisition and efficient internal/cellular utilization. In low Pi conditions, too, it has been suggested that tolerance can be achieved through better Pi acquisition and/or efficient cellular Pi utilization (Heuer et al. 2017; Rose et al. 2011; Wang et al. 2010). Higher Pi acquisition efficiency is related to enhanced Pi uptake, root architectural modifications, and solubilization of bound Pi in soil. On the other hand, Pi utilization efficiency requires efficient remobilization of cellular Pi. Breeding under high fertilizer input has led to selection of varieties that are highly "Pi responsive" due to higher PUP, whereas, many traditional landraces, like Chakhao Poreiton (CP), Dular, etc., are "Pi efficient" (Gerloff 1977; Wissuwa and Ae 2001; Yumnam et al. 2017). These genotypes are now slowly being understood at the molecular level, and would serve as a different and distinct molecular resource for deeper understanding of low Pi tolerance mechanisms. Root transcriptome data of one such genotype CP (a specialty rice from a northeast state of India, Manipur) has revealed a distinct mechanism for P deficiency tolerance (Tyagi and Rai 2017). The genotype CP lacks the entire PSTOL1 region (Tyagi et al. 2012) but still showed PUP at par with Kasalath in lowland acidic fields. Root length in response to 15-day low P treatment in case of CP is maintained, whereas the RDW is decreased (Tyagi and Rai 2017). It is hypothesized that suberin-mediated cell wall modification, releasing internally bound Pi and transcriptional modifications using dehydration response-based signals seems to be key to low P tolerance in CP.

Data is also emerging from other labs in India and other countries, which has led to identification of rice genotypes that either carry PSTOL1 but are "less efficient" in low P conditions (Pusa Basmati 1) or carry PSTOL1 (e.g., Dular) but show a mechanism that is distinct from RSA mechanism mediated by serine-threonine kinase, PSTOL1 (Mehra et al. 2016). "Dular" is a traditional landrace carrying PSTOL1 and is reported to be one of the most tolerant rice genotypes for P deficiency in field conditions (Chin et al. 2010; Wissuwa and Ae 2001). "Dular" showed higher expression of lipid remodeling genes and a greater metabolic flexibility under low Pi, which was confirmed by lipid phenotyping (Mehra et al. 2016). Sequencing of Dular and PB1 led to identification of numerous SNPs and InDels in promoter regions of lipid remobilizing genes, OsPT6, SPX domain containing proteins, various organic acid genes and PAPs. This was followed by confirmation of the role of OsPAP21b (PAP) as a useful candidate for improving organic P acquisition and utilization in rice as the OsPAP21b-overexpressing transgenics showed increased root length, DW and lateral roots, APase activity, and P content under low Pi conditions both under hydroponics and soil (Mehra et al. 2017).

Two contrasting rice genotypes, DJ123 (DJ; tolerant) and Nerica4 (N4; sensitive) (Mori et al. 2016; Rose et al. 2016), were used to detect potential genotypic differences (Pujol and Wissuwa 2018) by screening rice seedlings under high and low P conditions. Whole root porosity increase was found to be more pronounced in DJ123 and the formation of aerenchyma in low P conditions was only enhanced in DJ123. Although the study did not provide a direct link between aerenchyma induction and P deficiency tolerance observed in DJ, it is reported that aerenchyma has the potential to improve PUE by reducing cell cost maintenance and recycling nutrients (Lynch 2015; Postma and Lynch 2011). It has been previously shown that DJ possesses high internal PUE in root and shoot tissues (Rose et al. 2016) and carries a rare haplotype associated with increased PUE located on chromosome 11 (Wissuwa et al. 2015). DJ is also known to develop longer root hairs in P-deficient conditions while N4 does not (Nestler and Wissuwa 2016). DJ's inducible aerenchyma could be one component leading to higher PUE and possibly to longer root hairs (Pujol and Wissuwa 2018). Such a trait could therefore be beneficial in breeding programs to develop rice varieties with tolerance to P deficiency.

Future Prospects

In spite of many QTLs, genes, and mutants reported for low P tolerance in rice, marker-assisted breeding efforts in rice are so far restricted to PSTOL1 only. The reason for this is threefold: (i) over-expression of many Pi transporters and SPX domain transcription factors has been reported to lead to Pi toxic effects (reviewed extensively in Gu et al. 2016) or no effect under low P conditions as is the case with overexpression studies of PHR1 and PHR2; (ii) identified mutants do not have a gain of function allele reported; and (iii) QTLs have not been narrowed down to candidate genes or have not been evaluated across multiple genetic backgrounds/environments. The underlying reason for this is lack of complete understanding of the molecular mechanism of this complex quantitative trait. Recently, genes like *OsSPDT*, *OsPHR3*, *VPT1*, etc have been suggested as some of the potential targets for introducing low Pi tolerance and regulating grain Pi levels. As more information emerges, the possibility of a suite of genes and donors with robust markers will become a reality.

Acknowledgments

Authors would like to thank the Indian Council of Agricultural Research (ICAR), Department of Biotechnology (DBT, GOI), and Central Agricultural University (Imphal) (CAU (I)) for providing financial and physical support, respectively for research on low phosphorus tolerance.

References

Ai, P., Sun, S., Zhao, J. et al. (2009). Two rice phosphate transporters, OsPht1; 2 and OsPht1; 6, have different functions and kinetic properties in uptake and translocation. *Plant J.* 57: 798–809.

Anis, G.B., Zhang, Y., Wang, H. et al. (2018). Genomic regions analysis of seedling root traits and their regulation in responses to phosphorus deficiency tolerance in CSSL population of elite super hybrid rice. *Int. J. Mol. Sci.* 19: 1460.

Borch, K., Bouma, J., Lynch, J.P. et al. (1999). A regulator of root architectural responses to soil phosphorus availability. *Plant Cell Environ.* 22: 425–431.

Chang, C., Hu, Y., Sun, S. et al. (2008). Proton pump OsA8 is linked to phosphorus uptake and translocation in rice. *J. Exp. Bot.* 60: 557–565.

Chao, X., Jie, R., Xiu-qin, Z. et al. (2015). Genetic dissection of low phosphorus tolerance related traits using selected introgression lines in rice. *Rice Sci.* 22: 264–274.

Chen, J., Liu, Y., Ni, J. et al. (2011). OsPHF1 regulates the plasma membrane localization of low- and high-affinity inorganic phosphate transporters and determines inorganic phosphate uptake and translocation in rice. *Plant Physiol.* 157: 269–278.

Chin, J.H., Lu, X., Haefele, S.M. et al. (2010). Development and application of gene-based markers for the major rice QTL phosphorus uptake 1. *Theor. Appl. Genet.* 120: 1073–1086.

Chin, J.H., Gamuyao, R., Dalid, C. et al. (2011). Developing rice with high yield under phosphorus deficiency: Pup1 sequence to application. *Plant Physiol.* 156: 1202–1216.

Chithrameenal, K., Alagarasan, G., Raveeendran, M. et al. (2018). Genetic enhancement of phosphorus starvation tolerance through marker assisted introgression of OsPSTOL1 gene in rice genotypes harbouring bacterial blight and blast resistance. *PLoS One* 13: e0204144.

Chrispeels, M.J., Crawford, N.M., and Schroeder, J.I. (1999). Proteins for transport of water and mineral nutrients across the membranes of plant cells. *Plant Cell* 11: 661–676.

Dai, X., Wang, Y., Yang, A. et al. (2012). OsMYB2P-1, an R2R3 MYB transcription factor, is involved in the regulation of phosphate-starvation responses and root architecture in rice. *Plant Physiol.* 159: 169–183.

Das, S., Tyagi, W., Rai, M. et al. (2017). Understanding Fe^{2+} and P deficiency tolerance in rice for enhancing productivity under acidic soils. *Biotechnol. Genet. Eng. Rev.* 33: 97–117.

Drew, M.C., He, C.J., and Morgan, P.W. (1989). Decreased ethylene biosynthesis, and induction of aerenchyma, by nitrogen- or phosphate-starvation in adventitious roots of *Zea mays* L. *Plant Physiol.* 91: 266–271.

Famoso, A.N., Zhao, K., Clark, R.T. et al. (2011). Genetic architecture of aluminum tolerance in rice (*Oryza sativa*) determined through genome-wide association analysis and QTL mapping. *PLoS Genet.* 7: 1–16.

Frank, T., Meuleye, B.S., Miller, A. et al. (2007). Metabolite profiling of two low phytic acid (lpa) rice mutants. *J. Agric. Food Chem.* 55: 11011–11019.

Gamuyao, R., Chin, J.H., Pariasca-Tanaka, J. et al. (2012). The protein kinase PSTOL1 from traditional rice confers tolerance of phosphorus deficiency. *Nature* 488: 535.

George, T.S. and Richardson, A.E. (2008). Potential and limitations to improving crops for enhanced phosphorus utilization. In: *Ecophysiology of plant phosphorus interactions* (eds. P.J. White and J.P. Hammond), 247–270. Dordrecht: Springer.

Gepstein, S., Sabehi, G., Carp, M.J. et al. (2003). Large-scale identification of leaf senescence-associated associated genes. *Plant J.* 5: 629–642.

Gerloff, S. (1977). Plant efficiencies in use of N, P and K. In: *Plant adaptation to mineral stress in problem soils* (ed. M.J. Wright), 61–74. New York: Cornell University Press Ithaca.

Giri, J., Bhosale, R., and Huang, G. (2018). Rice auxin influx carrier OsAUX1 facilitates root hair elongation in response to low external phosphate. *Nat. Commun.* 9: 1408.

Gu, M., Chen, A., Sun, S. et al. (2016). Complex regulation of plant phosphate transporters and the gap between molecular mechanisms and practical application: what is missing? *Mol. Plant.* 9: 396–416.

Guo, M., Ruan, W., Li, C. et al. (2015). Integrative comparison of the role of the PHOSPHATE RESPONSE1 subfamily in phosphate signaling and homeostasis in rice. *Plant Physiol.* 168: 1762–1776.

Heuer, S., Gaxiola, R., Schilling, R. et al. (2017). Improving phosphorus use efficiency: a complex trait with emerging opportunities. *Plant J.* 90: 868–885.

Holford, I.C.R. (1997). Soil phosphorus: its measurement, and its uptake by plants. *Aust. J. Soil Res.* 35: 227–239.

Hu, B., Zhu, C., Li, F. et al. (2011). LEAF TIP NECROSIS1 plays a pivotal role in the regulation of multiple phosphate starvation responses in rice. *Plant Physiol.* 156: 1101–1115.

Hu, A.Y., Che, J., Shao, J.F. et al. (2018). Silicon accumulated in the shoots results in down-regulation of phosphorus transporter gene expression and decrease of phosphorus uptake in rice. *Plant Soil* 423: 317–325.

Hurley, B.A., Tran, H.T., Marty, N.J. et al. (2010). The dual-targeted purple acid phosphatase isozyme atpap26 is essential for efficient acclimation of *Arabidopsis* to nutritional phosphate deprivation. *Plant Physiol.* 153: 1112–1122.

Ismail, A.M., Heuer, S., Thomson, M.J. et al. (2007). Genetic and genomic approaches to develop rice germplasm for problem soils. *Plant Mol. Biol.* 65: 547–570.

Jeong, K., Baten, A., Waters, D.L. et al. (2017). Phosphorus remobilization from rice flag leaves during grain filling: an RNA-seq study. *Plant Biotechnol. J.* 15: 15–26.

Jeong, K., Pantoja, O., Baten, A. et al. (2018). Transcriptional response of rice flag leaves to restricted external phosphorus supply during grain filling in rice cv. IR64. *PLoS One* 13: e0203654.

Jia, H., Zhang, S., Wang, L. et al. (2017). OsPht1;8, a phosphate transporter, is involved in auxin and phosphate starvation response in rice. *J. Exp. Bot.* 68: 5057–5068.

Julia, C., Wissuwa, M., Kretzschmar, T. et al. (2016). Phosphorus uptake, partitioning and redistribution during grain filling in rice. *Ann. Bot.* 118: 1151–1162.

Kim, H.J., Lynch, J.P., and Brown, K.M. (2008). Ethylene insensitivity impedes a subset of responses to phosphorus deficiency in tomato and petunia. *Plant Cell Environ.* 31: 1744–1755.

Kuang, R., Chan, K.H., Yeung, E. et al. (2009). Molecular and biochemical characterization of AtPAP15, a purple acid phosphatase with phytase activity, in Arabidopsis. *Plant Physiol.* 151: 199–209.

Lang, N.T. and Buu, B.C. (2006). Mapping QTLs for phosphorus deficiency tolerance in rice (*Oryza sativa* L.). *Omonrice* 14: 1–9.

Larson, S.R., Rutger, J.N., Young, K.A. et al. (2000). Isolation and genetic mapping of a non-lethal rice (*Oryza sativa* L.) low phytic acid 1 mutation. *Crop Sci.* 40: 1397–1405.

Lee, H.Y., Chen, Z., Zhang, C. et al. (2019). Editing of the OsACS locus alters phosphate deficiency-induced adaptive responses in rice seedlings. *J. Exp. Bot.* 70: 1927–1940.

Li, M., Welti, R., and Wang, X. (2006). Quantitative profiling of Arabidopsis polar glycerolipids in response to phosphorus starvation: roles of phospholipases Dζ1 and Dζ2 in phosphatidylcholine hydrolysis and digalactosyldiacylglycerol accumulation in phosphorus starved plants. *Plant Physiol.* 142: 750–761.

Li, J., Xie, Y., Dai, A. et al. (2009). Root and shoot traits responses to phosphorus deficiency and QTL analysis at seedling stage using introgression lines of rice. *J. Genet. Genomics* 36: 173–183.

Liu, F., Wang, Z., Ren, H. et al. (2010). OsSPX1 suppresses the function of OsPHR2 in the regulation of expression of OsPT2 and phosphate homeostasis in shoots of rice. *Plant J.* 62: 508–517.

Liu, F., Chang, X.-J., Ye, Y. et al. (2011). Comprehensive sequence and whole-life-cycle expression profile analysis of the phosphate transporter gene family in rice. *Mol. Plant* 4: 1105–1122.

Liu, J.H., Yang, L., Luan, M.D. et al. (2015). A vacuolar phosphate transporter essential for phosphate homeostasis in *Arabidopsis*. *Proc. Natl. Acad. Sci. U.S.A.* 112: E6571–E6578.

Lu, L., Qiu, W., Gao, W. et al. (2016). OsPAP10c, a novel secreted acid phosphatase in rice, plays an important role in the utilization of external organic phosphorus. *Plant Cell Environ.* 39: 2247–2259.

Lynch, J.P. (2015). Root phenes that reduce the metabolic costs of soil exploration: opportunities for 21st century agriculture: new roots for agriculture. *Plant Cell Environ.* 38: 1775–1784.

Marschner, P. and Rengel, Z. (2012). Nutrient availability in soils. In: *Marschner's mineral nutrition of higher plants* (ed. P. Marschner), 315–330. Cambridge, MA: Academic Press.

Mehra, P., Pandey, B.K., and Giri, J. (2016). Comparative morphophysiological analyses and molecular profiling reveal Pi-efficient strategies of a traditional rice genotype. *Front. Plant Sci.* 6: 1184.

Mehra, P., Pandey, B.K., and Giri, J. (2017). Improvement in phosphate acquisition and utilization by a secretory purple acid phosphatase (OsPAP21b) in rice. *Plant Biotechnol. J.*: 1–14.

Mori, A., Fukuda, T., Vejchasarn, P. et al. (2016). The role of root size versus root efficiency in phosphorus acquisition in rice. *J. Exp. Bot.* 67: 1179–1189.

Neelam, K., Thakur, S., Neha et al. (2017). Novel alleles of phosphorus-starvation tolerance 1 gene (*PSTOL1*) from *Oryza rufipogon* confers high phosphorus uptake efficiency. *Front. Plant Sci.* 6: 509.

Nestler, J. and Wissuwa, M. (2016). Superior root hair formation confers root efficiency in some, but not all, rice genotypes upon P deficiency. *Front Plant Sci.* 7: 1935.

Ni, J.J., Wu, P., Senadhira, D. et al. (1998). Mapping QTLs for phosphorus deficiency tolerance in rice (*Oryza sativa* L.). *Theor. Appl. Genet.* 97: 1361–1369.

Ogawa, M., Tanaka, K., and Kasai, Z. (1979). Accumulation of phosphorus, magnesium and potassium in developing rice grains: followed by electron microprobe X-ray analysis focusing on the aleurone layer. *Plant Cell Physiol.* 20: 19–27.

Panigrahy, M., Rao, D.N., Yugandhar, P. et al. (2014). Hydroponic experiment for identification of tolerance traits developed by rice Nagina 22 mutants to low-phosphorus in field condition. *Arch. Agron. Soil Sci.* 60: 565–576.

Pariasca-Tanaka, J., Chin, J.H., Drame, K.N. et al. (2014). A novel allele of the P-starvation tolerance gene *OsPSTOL1* from African rice (*Oryza glaberrima* Steud) and its distribution in the genus *Oryza*. *Theor. Appl. Genet.* 127: 1387–1398.

Paszkowski, U., Kroken, S., Roux, C. et al. (2002). Rice phosphate transporters include an evolutionarily divergent gene specifically activated in arbuscular mycorrhizal symbiosis. *Proc. Natl. Acad. Sci. U.S.A.* 99: 13324–13329.

Poirier, Y., Thoma, S., Somerville, C. et al. (1991). A mutant of *Arabidopsis* deficient in xylem loading of phosphate. *Plant Physiol.* 97: 1087–1093.

Postma, J.A. and Lynch, J.P. (2011). Theoretical evidence for the functional benefit of root cortical aerenchyma in soils with low phosphorus availability. *Ann. Bot.* 107: 829–841.

Puga, M.I., Mateos, I., Charukesi, R. et al. (2014). SPX1 is a phosphate-dependent inhibitor of phosphate starvation response 1 in *Arabidopsis*. *Proc. Natl. Acad. Sci. U.S.A.* 111: 14947–14952.

Pujol, V. and Wissuwa, M. (2018). Contrasting development of lysigenous aerenchyma in two rice genotypes under phosphorus deficiency. *BMC Res. Notes* 11: 60.

Raghothama, K.G. (2000). Phosphate transport and signaling. *Curr. Opin. Plant Biol.* 3: 182–187.

Rose, T.J., Rose, M.T., Pariasca-Tanaka, J. et al. (2011). The frustration with utilization: why have improvements in internal phosphorus utilization efficiency in crops remained so elusive? *Front. Plant Sci.* 2: 73.

Rose, T.J., Mori, A., Julia, C.C. et al. (2016). Screening for internal phosphorus utilisation efficiency: comparison of genotypes at equal shoot P content is critical. *Plant Soil* 401: 79–91.

Rouached, H., Arpat, A.B., and Poirier, Y. (2010). Regulation of phosphate starvation responses in plants: signaling players and cross-talks. *Mol. Plant* 3: 288–299.

Secco, D., Jabnoune, M., Walker, H. et al. (2013). Spatio-temporal transcript profiling of rice roots and shoots in response to phosphate starvation and recovery. *Plant Cell* 25: 4285–4304.

Shen, J., Yuan, L., Zhang, J. et al. (2011). Phosphorus dynamics: from soil to plant. *Plant Physiol.* 156: 997–1005.

Shen, C., Wang, S., Zhang, S. et al. (2013). OsARF16, a transcription factor, is required for auxin and phosphate starvation response in rice (*Oryza sativa* L.). *Plant Cell Environ.* 36: 607–620.

Shimizu, A., Yanagihara, S., Kawasaki, S. et al. (2004). Phosphorus deficiency induced root elongation and its QTL in rice (*Oryza sativa* L). *Theor. Appl. Genet.* 109: 1361–1368.

Shimizu, A., Kato, K., Komatsu, A. et al. (2008). Genetic analysis of root elongation induced by phosphorus deficiency in rice (*Oryza sativa* L.): fine QTL mapping and multivariate analysis of related traits. *Theor. Appl. Genet.* 117: 987–996.

Sun, S., Gu, M., Cao, Y. et al. (2012). A constitutive expressed phosphate transporter, OsPht1;1, modulates phosphate uptake and translocation in phosphate-replete rice. *Plant Physiol.* 159: 1571–1581.

Takehisa, H. and Sato, Y. (2019). Transcriptome monitoring visualizes growth stage-dependent nutrient status dynamics in rice under field conditions. *Plant J.* 97: 1048–1060.

Tian, J., Wang, C., Zhang, Q. et al. (2012). Overexpression of *OsPAP10a*, a root-associated acid phosphatase, increased extracellular organic phosphorus utilization in rice. *J. Integr. Plant Biol.* 54: 631–639.

Tyagi, W. and Rai, M. (2017). Root transcriptomes of two acidic soil adapted indica rice genotypes suggest diverse and complex mechanism of low phosphorus tolerance. *Protoplasma* 254: 725–736.

Tyagi, W., Rai, M., and Dohling, A. (2012). Haplotype analysis for locus in rice genotypes of north eastern and eastern India to identify suitable donors tolerant to low phosphorus. *SABRAO J. Breed. Genet.* 44: 398–405.

Varadarajan, D.K., Karthikeyan, A.S., Matilda, P.D. et al. (2002). Phosphite, an analog of phosphate, suppresses the coordinated expression of genes under phosphate starvation. *Plant Physiol.* 129: 1232–1240.

Vigueira, C.C., Small, L.L., and Olsen, K.M. (2016). Long-term balancing selection at the *phosphorus starvation tolerance 1 (PSTOL1)* locus in wild, domesticated and weedy rice (*Oryza*). *BMC Plant Biol.* 16: 101.

Vinod, K.K. and Heuer, S. (2012). Approaches towards nitrogen- and phosphorus-efficient rice. *AoB Plants* 2012: pls028.

Wang, Y., Ribot, C., Rezzonico, E. et al. (2004). Structure and expression profile of the *Arabidopsis* PHO1 gene family indicates a broad role in inorganic phosphate homeostasis. *Plant Physiol.* 135: 400–411.

Wang, X., Shen, J., and Liao, H. (2010). Acquisition or utilization, which is more critical for enhancing phosphorus efficiency in modern crops? *Plant Sci.* 179: 302–306.

Wang, C., Huang, W., Ying, Y.H. et al. (2012). Functional characterization of the rice SPX-MFS family reveals a key role of OsSPX-MFS1 in controlling phosphate homeostasis in leaves. *New Phytol.* 196: 139–148.

Wang, Z., Ruan, W., Shi, J. et al. (2014a). Rice SPX1 and SPX2 inhibit phosphate starvation responses through interacting with PHR2 in a phosphate-dependent manner. *Proc. Natl. Acad. Sci. U.S.A.* 111: 14953–14958.

Wang, K., Cui, K., Liu, G. et al. (2014b). Identification of quantitative trait loci for phosphorus use efficiency traits in rice using a high density SNP map. *BMC Genet.* 15: 155.

Wang, C., Yue, W.H., Ying, Y.H. et al. (2015). Rice SPX major facility superfamily3, a vacuolar phosphate efflux transporter, is involved in maintaining phosphate homeostasis in rice. *Plant Physiol.* 169: 2822–2831.

Wang, F., Rose, T., Jeong, K. et al. (2016). The knowns and unknowns of phosphorus loading into grains, and implications for phosphorus efficiency in cropping systems. *J. Exp. Bot.* 67: 1221–1229.

Wege, S., Khan, G.A., Jung, J.Y. et al. (2016). The EXS domain of PHO1 participates in the response of shoots to phosphate deficiency via a root-to-shoot signal. *Plant Physiol.* 170: 385–400.

Wild, R., Gerasimaite, R., Jung, J.-Y. et al. (2016). Control of eukaryotic phosphate homeostasis by inositol polyphosphate sensor domains. *Science* 352: 986–990.

Wissuwa, M. and Ae, N. (2001). Genotypic variation for tolerance to phosphorus deficiency in rice and the potential for its exploitation in rice improvement. *Plant Breed.* 120: 43–48.

Wissuwa, M., Yano, M., and Ae, N. (1998). Mapping of QTLs for phosphorus-deficiency tolerance in rice (*Oryza sativa* L.). *Theor. Appl. Genet.* 97: 777–783.

Wissuwa, M., Wegner, J., Ae, N. et al. (2002). Substitution mapping of Pup1: a major QTL increasing phosphorus uptake of rice from a phosphorus-deficient soil. *Theor. Appl. Genet.* 105: 890–897.

Wissuwa, M., Mazzola, M., and Picard, C. (2009). Novel approaches in plant breeding for rhizosphere-related traits. *Plant Soil* 321: 409–430.

Wissuwa, M., Kondo, K., Fukuda, T. et al. (2015). Unmasking novel loci for internal phosphorus utilization efficiency in rice germplasm through genome-wide association analysis. *PLoS One* 10: e0124215.

Wissuwa, M., Kretzschmar, T., and Rose, T.J. (2016). From promise to application: root traits for enhanced nutrient capture in rice breeding. *J. Exp. Bot.* 67: 3605–3615.

Yamada, Y., Furusawa, S., Nagasaka, S. et al. (2014). Strigolactone signaling regulates rice leaf senescence in response to a phosphate deficiency. *Planta* 240: 399–408.

Yamaji, N., Takemoto, Y., Miyaji, T. et al. (2017). Reducing phosphorus accumulation in rice grains with an impaired transporter in the node. *Nature* 541: 92.

Yan, Y.S., Chen, X.Y., Yang, K. et al. (2011). Overexpression of an F-box protein gene reduces abiotic stress tolerance and promotes root growth in rice. *Mol. Plant* 4: 190–197.

Yi, K., Wu, Z., Zhou, J. et al. (2005). OsPTF1, a novel transcription factor involved in tolerance to phosphate starvation in rice. *Plant Physiol.* 138: 2087–2096.

Yumnam, J.S., Rai, M., and Tyagi, W. (2017). Allele mining across two low-P tolerant genes *PSTOL1* and *PupK20-2* reveals novel haplotypes in rice genotypes adapted to acidic soils. *Plant Genet. Resour.* 15: 221–229.

Zhang, Q., Wang, C., Tian, J. et al. (2011). Identification of rice purple acid phosphatases related to phosphate starvation signalling. *Plant Biol.* 13: 7–15.

Zhang, F., Sun, Y.F., Pei, W.X. et al. (2015). Involvement of *OsPht1;4* in phosphate acquisition and mobilization facilitates embryo development in rice. *Plant J.* 82: 556–569.

Zhao, H., Frank, T., Tan, Y. et al. (2016). Disruption of *OsSULTR3;3* reduces phytate and phosphorus concentrations and alters the metabolite profile in rice grains. *New Phytol.* 211: 926–939.

12

Molecular Breeding for Improving Nitrogen Use Efficiency in Rice: Progress and Perspectives

Chirravuri N. Neeraja, Sitapati R. Voleti, Subrahmanyam Desiraju, Surekha Kuchi, Sonali Bej, Krishnakanth Talapanti and Raghuveer R. Puskur

ICAR-Indian Institute of Rice Research, Rajendranagar, Hyderabad 500030, India

Introduction

Rice is a major staple food crop for more than half of the world and is cultivated in >160 million ha with production of >750 million tonnes (FAO 2018). N fertilizers are an essential input of rice cultivation and account for ~15% of the world's N consumption of 109 million tonnes (www.fertilizer.org, www.fao.org). Nitrogen is an essential plant nutrient, being a component of amino acids, nucleic acids, nucleotides, chlorophyll, enzymes, and hormones. It promotes plant growth and grain yield through higher tillering, increased leaf area, grain formation, grain filling, and protein synthesis. N fertilization has become more critical for rice production with the development of fertilizer-responsive semi-dwarf high-yield rice varieties since the Green Revolution. The application rate of N for rice cultivation ranges from 100 to 300 kg ha^{-1} across the world. But the nitrogen use efficient (NUE) for rice is observed to be very low ranging from 30% to 50% especially in irrigated ecologies, as a result of ammonia volatization, denitrification, surface runoff, and leaching (De Datta and Buresh 1989). Thus, NUE in rice becomes important because of the N losses to the environment impacting climate change and increase in the cost of rice cultivation. Rice fields contribute to methane and nitrous oxide, which are important greenhouse gases with nitrous oxide 310 times more potent as a greenhouse gas than CO_2 and 21 times more potent than methane on a 100-year time scale, though atmospheric loading of nitrous dioxide is low (IPCC Second Assessment Report 1995). The demand for N is growing every year ~3.2% from 102.8 MT in 2014 to 116.9 MT in 2018 (Surekha et al. 2019). Significant contributions have been made, and several efforts are being made in rice N fertilization management strategies regarding the application methods and various modified N sources in the field of rice crop production (Neeraja et al. 2019). The genetic solution of development of rice varieties with NUE is sustainable and feasible. With the available germplasm and genome resources, breeding for NUE in rice is another promising approach, and in the present review the status of characterization of genotypic variation and genetic information required for molecular breeding towards NUE of rice is summarized in brief.

Molecular Breeding for Rice Abiotic Stress Tolerance and Nutritional Quality, First Edition.
Edited by Mohammad Anwar Hossain, Lutful Hassan, Khandakar Md. Iftekharuddaula, Arvind Kumar and Robert Henry.

Evaluation of Genetic Variation for Nitrogen Use Efficiency

Before the development of fertilizer-responsive high-yielding rice varieties, the rice genotypes popularly known as landraces were grown by the farmer under low inputs and were adapted accordingly. Though cultivated under low inputs, the farmer would make a conscious selection for yield, quality, and other desirable traits like aroma, grain size, and so on. Thus, the landraces become interesting genetic material for dissecting the NUE, identifying promising donors and genes. High-yielding rice varieties have been selected for high response under N-sufficient conditions and for the development of varieties with high NUE, selection should be done with high response under N-deficient conditions with high uptake efficiency and high internal utilization efficiency. NUE comprises effective N uptake, translocation, assimilation, and remobilization with maximum possible yield as the target. Since several processes are involved, NUE is inherently complex and is governed by several genetic and environmental factors (Tegeder and Masclaux-Daubresse 2018). Characterization of NUE in rice genotypes is difficult unlike the stress resistance or abiotic stress tolerances like drought (DT), and salinity, for which scoring systems for evaluation have been developed and implemented. Lack of selection criteria, other than yield, complicates the process of screening for NUE in rice.

The NUE can best be studied in nitrogen-deficient conditions allowing the genotype to express its ability to adapt to a low N situation. The early NUE evaluation studies in rice were conducted by agronomists, soil scientists, and plant physiologists under differential N treatments consisting of zero, 25%, and 50% of the recommended dose of N in field conditions (Mahajan et al. 2010; Singh et al. 1998). Hydroponic studies were also conducted to understand the mechanism of NUE in rice. Attempts were made to decipher the mechanisms of uptake, translocation, assimilation, and remobilization of N. The number of genotypes ranged from 1 to 472 (Inthapanya et al. 2000; Rao et al. 2018; Singh et al. 1998) and consisted of released varieties, landraces, and breeding lines derived from wild species. Significant genotypic differences for NUE were reported in rice, and significant interactions were also reported among seasons, treatments, and genotypes suggesting extensive variation for the genotypic response to nitrogen dosage in wet and dry seasons. Under low N, plants show overall reduction of biomass, with poor tillering, decreased height, lower flag leaf area, reduction of panicles, spikelets, and grains.

At the International Rice Research Institute, by screening 24 rice genotypes under soils with high and low N fertility, a ranking system was developed, and yield was shown to be the most significant single parameter for selection (De Datta and Broadbent 1990). A set of 180 rice genotypes, comprising lowland *indica* improved varieties or elite breeding lines of different growth durations ranging from 60 to 135 days, were evaluated at suboptimal N levels, and it was observed that some genotypes absorbed similar amounts of N but produced different grain yields and/or total dry matter (TDM) and the genotypes with higher NUE had lower percentage straw N at maturity (Tirol-Padre et al. 1996). Use efficiency and responsiveness to N were studied in 20 varieties, belonging to medium- and long-duration groups and showing genotypic variation in N uptake and efficiency of use (Singh et al. 1998). To study the genotype's ability to extract more nutrients or to use absorbed nutrients more efficiently to produce grain yield, 16 contrasting genotypes were evaluated, and a large effect of genotype was reported (Inthapanya et al. 2000). Twenty-one genotypes tested under three different nitrogen levels, namely, 0, 75, and 150 kg of N levels, and the efficient genotypes were identified (Sedeek et al. 2010). Considering the great amount of basal N fertilizer but lower uptake ability at the rice seedling stage, 199 rice varieties with different genetic backgrounds were evaluated, which showed highly significant genotype differences for grain yield confirming the genotypic variation for NUE (Zhao 2012). A total of 18 genotypes were evaluated for two seasons at N0 and N100 plots, and the promising genotypes for Soil Plant Analysis Development (SPAD) chlorophyll content, plant height, number of tillers, effective booting tillers (EBT), filled grain weight and TDM, harvest index (HI), total nitrogen content (straw + grain) were identified under low N (Swamy et al. 2015).

Biochemical and physiological characterization for NUE in 78 aromatic rice genotypes identified genotypes with contrasting responses (Vijayalakshmi et al. 2015). A set of 472 rice genotypes comprising landraces and breeding lines were evaluated for two seasons under field conditions with low and recommended nitrogen, and >100 landraces were identified with relative higher yield under low nitrogen (LNT) (Rao et al. 2018). Several other studies involving <10 rice genotypes have also characterized the genotypic variation for identification of traits/parameters useful for selection of NUE in rice. Using a modified backcross breeding approach, starting from BC_1F_2 derived from high-yielding variety and a donor, through one generation of selection followed by four generations of screening and selection for high grain yield under low N and other nutrients, promising introgression lines (ILs) were identified (Jewel et al. 2019a).

Nitrogen is essential to the rice plant, with about 75% of leaf N associated with chloroplasts, which are physiologically important in dry-matter production through photosynthesis. Increased N enhances leaf length, width, area, and thickness along with chlorophyll content (Anzoua et al. 2010; Mahajan et al. 2012; Ohnishi et al. 1999; Sedeek et al. 2010; Singh and Verma 2013; Zhang et al. 2013). SPAD has shown positive association with yield for the contrasting genotypes under low N in our study (Rao et al. 2018). The leaf area index was also shown to be positively correlated with high NUE of elite rice varieties under low N (Wu et al. 2016). Though the leaf traits of the efficient genotypes appear to be promising, not all the genotypes with promising leaf traits were efficient under low N (Neeraja et al. 2019). The importance of N for realizing the potential height and number of tillers in the rice plant has been underscored in several studies. Increased availability of N increases of leaf area and photo-assimilates results in more biomass (Singh and Verma 2013; Singh et al. 1998; Tirol-Padre et al. 1996). An increased trend in growth dynamics during active tillering, flowering stage, and biomass partitioning was observed across four doses of N treatment for four rice genotypes (Singh et al. 2014). Increased numbers of tillers in experiments might be due to the greater availability of nitrogen, which plays a vital role in cell division. The stimulation of tillering by nitrogen also enhances the number of panicles. An increase of 39.2% was observed for productive tillers for aromatic rice with 40 kg N ha^{-1} in comparison with zero N application (Mahajan et al. 2010). Significant increases of panicles m^{-2} with the increase of N application were also reported (Kanfany et al. 2014; Sedeek, Abdelkhalik, and Metwali 2010; Singh et al. 1998; Sui et al. 2013; Tirol-Padre et al. 1996; Yoshinaga et al. 2013; Zhang et al. 2013). Grain yield, number of grains per panicle, number of tillers, plant height, length of flag leaf, TDM and shoot dry matter, 1000 grain weight, and HI have always been increased by N application in field conditions as well as pot conditions across the studies. Grain yield of rice is the final product of yield components, namely, the number of panicles per unit area, the number of spikelets per panicle, the percentage of filled spikelets, and the grain weight (Yoshida 1983). The grain yield increments under recommended N were possible mainly through increasing the number of spikelets of the panicle and then the number of the tillers (Anzoua et al. 2010). Nitrogen is essential, initially for the development of spikelets and later for the reduction of spikelet degeneration. A strong interrelationship was found between spikelet production and N application in the late spikelet differentiation stage (Kamiji et al. 2011). The filled grain number has been affected by the N dosage as reported in several studies, and the decrease in the amount of metabolites (carbon and N skeletons) under low N was indicated as the reason (Mahajan et al. 2010; Sedeek et al. 2010). Interestingly, for grain-filling percentage, there have been contrasting reports of increase under low N and significant decrease in spikelet fertility % under low N for medium-duration varieties, and no significant response for long-duration varieties was reported by a few studies (Kanfany et al. 2014; Mahajan et al. 2010, 2012; Singh et al. 1998; Sui et al. 2013; Zhang et al. 2013). The nitrogen content in grain and straw and their association to biomass or yield appears to be highly variable across the locations, genotypes, duration of the genotypes, stage of application, and sampling (De Datta and Broadbent 1990; Inthapanya et al. 2000; Samonte et al. 2006; Zhao 2012; Singh et al. 1998; Sui et al. 2013; Yoshinaga et al. 2013).

Though NUE is the ratio of grain yield to N supplied, the reported nomenclature of NUE indicators sounds similar, namely, agronomic NUE, physiological use efficiency, internal N use efficiency, and so forth, but the definitions and parameters used are highly variable (Fageria and Baligar 2003; Ladha et al. 2005; Mahajan et al. 2010; Samonte et al. 2006; Sedeek et al. 2010; Singh et al. 1998; Sui et al. 2013; Tirol-Padre et al. 1996). NUE indicators have also been calculated for low and recommended N individually and also in relation to low and recommended N. The nitrogen-efficient indicators suggest the relationship of the application of nitrogen to the grain yield obtained. Out of all the traits/parameters/indicators, yield continues to be the most reliable and user-friendly selection criterion of NUE in rice.

Identification of QTL and Genomic Regions through Biparental and Association Mapping

QTL

Many of the agronomically important traits are quantitative in nature controlled by multiple genetic loci or quantitative trait loci (QTL). Each individual QTL generally explains about <10% to >70% of phenotypic variation based upon a trait. QTLs are also affected by environment, and also there are interactions among the QTL. Thus characterizing and deploying a QTL in the development of a variety can be quite challenging. However, in rice, several QTL for many complex traits like yield, drought, submergence, and salinity have been mapped. For QTL identification using bi-parental mapping, use of contrasting parents, and development of populations, namely, recombinant inbred line (RIL), doubled haploid (DH), BC_nF_n are prerequisites. Several types of molecular markers like restriction fragment length polymorphism (RFLP), random amplification of polymorphic deoxy ribonucleic acid (DNA)

(RAPD), and simple sequence repeats (SSR) were/are being used to construct molecular linkage genetic maps for studying the association between the markers and the traits. After identifying the number and position of QTL, deploying the rice genome sequence information, the putative candidate genes are being studied in the identified QTL and validated. A characterized major QTL can be introgressed into the popular adapted varieties as has been demonstrated by Swarna Sub1. *Pup1* is another major characterized QTL for phosphorus uptake, which has been introgressed into many popular rice varieties and is being evaluated under field conditions. Even the QTL with minor effects identified for drought are being characterized and pyramided as demonstrated in rice (Yadav et al. 2019).

Since 2001, several QTL/genomic regions associated with various traits under low N were reported in rice.

- QTL for plant height were identified under two N levels (low and high) in DH lines by Fang and Wu (2001). A total of eight QTL were identified at low N on chromosomes 1, 2, 3, 4, 5, and 6.
- A QTL region identified with panicle number and panicle weight on chromosome 2 showed to a regulator gene (*GS1*) for glutamine synthetase (GS) activity and observed to be associated with increased number of tillers, panicles, and total panicle weight under low N (Obara et al. 2001, 2004).
- (Tong et al. 2006) detected 31 QTL for different agronomic and yield characters with 15 QTL under LNT conditions and 16 QTL under recommended nitrogen conditions in chromosomal segment substitution lines.
- QTL for agronomic, yield traits, and nitrogen concentration characters were identified in RILs of Dasanbyeo/TR22183 under low (9 QTL) and normal nitrogen (8 QTL) levels (Cho et al. 2007).
- QTL for different morphological, agronomic, NUE, and yield traits were identified in DH lines of IR64/Azucena under native (6 QTL), optimum (9 QTL), and high (8 QTL) nitrogen levels (Senthilvel et al. 2008).
- Using F_9 RILs and a Bayesian model selection technique, a total of 28 main-effect QTL and 23 pairs of epistatic QTL were detected for traits associated with nitrogen utilization efficiency (Piao et al. 2009).
- Tong et al. (2011) identified QTL for yield traits under low (19 QTL), medium (23 QTL), and high (15 QTL) nitrogen levels in RILs generated from parents Zhenshan97/HR5. Two main-effect QTL (*qGYPP-4b* and *qGNPP-12*) were identified at low N.
- QTL for shoot nitrogen concentration were identified under low- and high-nitrogen conditions in DH population by Hu et al. (2012).
- QTL for plant height, tillers, and NUE were identified under LNT (5 QTL) and standard nitrogen (8 QTL) levels in RILs by Nguyen et al. (2016).
- Ogawa et al. (2016) identified 8 QTL under two nitrogen levels (native nitrogen, farmers practice N level) in chromosomal segment substitution lines.
- A total of 52 QTL for yield traits distributed in 27 regions on 9 chromosomes were detected, with each QTL explaining 4.93–26.73% of the phenotypic variation under low and normal levels of N (Feng et al. 2015). Eleven QTL were simultaneously detected under the two levels, and 30 different QTL were detected under the two N levels, thereby suggesting that the genetic bases controlling rice growth under the low and normal N levels were different.

- Using bulked segregant analysis (BSA) and genome re-sequencing (WGS) technology, a key QTL, designated as qNUE6 was identified on chromosome 6 and in recombinants from F_2 population. Significant differences in the expression levels of LOC_Os06g15370 (peptide transporter PTR2) and LOC_Os06g15420 (asparagine synthetase) were observed between two parents (Y11 and GH998) (Zhou et al. 2017).
- Evaluation of three sets of trait-specific ILs resulted in the identification of nine drought (DT) QTL and seven LNT QTL by a segregation distortion approach and a genome-wide association study, respectively. Using four pairs of Kompetitive allele specific polymerase chain reaction (PCR) (KASP) Single nucleotide polymorphism (SNP) markers, 66 F_2 individuals with different combinations of the target DT- and LNT-QTL favorable alleles were selected and showed expected improvement in DT and/or LNT, which were further validated by the significant improvement in DT and/or LNT of their F_3 progeny testing (Feng et al. 2018).
- A subset of 96 RIL population derived from Rasi (NUE variety) and Vibhava (high yielding variety) was screened under low and recommended N, and the promising RIL were identified. Association of seven traits under low N with eight SSR markers was observed through mapping analysis (Neeraja et al. 2019).
- A total of 230 BC_1F_5 ILs evaluated under six nutrient conditions and a high-density linkage map with 704 high-quality SNP markers showed 49 main-effect QTL (Jewel et al. 2019b).

Association Mapping

In rice, association mapping (AM) is a proven strategy for identification of genomic regions associated traits of interest (Verdeprado et al. 2018). More recently, genome-wide association studies (GWAS) provide opportunities to explore the enormous allelic diversity existing in natural populations like landraces and to position QTL more precisely. GWAS has become more feasible now with low-cost high-throughput SNP genotyping methods and advances in rice genomics. QTL in GWAS (also called "signals," "peaks," or "hits") are represented using "Manhattan plots" that display the association of markers with the trait along the chromosome. The *y*-axis indicates –log10 (*P* value) for the association plotted against the SNPs along each chromosome on the *x*-axis, so the positions of all the SNPs must be known (Verdeprado et al. 2018). As AM is dependent on Linkage disequilibrium (LD) across the genome, knowledge of extent of LD is a prerequisite. An association is defined when there is significant correlation between the covariance of a marker polymorphism and a trait of interest.

Association analysis on NUE-related traits in a panel of 184 varieties with 157 genome-wide SSR has detected eight statistically significant marker loci associated with NUE-related traits. Two candidate genes, namely, *GS1;2* and *AspAt3* were located in the two QTL regions. KASP SNP markers were developed at a novel NUE-QTL region, and a transcriptional map of NUE-related genes in the novel gene was also developed. The NUE-related alleles of the eight loci were validated for their use in marker-assisted selection (MAS) and the breeding values of all the varieties through genomic prediction approach were estimated for rice NUE enhancement (Liu et al. 2016).

Through AM, using a marker set of 50 rice SSR markers and 472 rice genotypes, 12 genomic regions were identified for yield and yield-associated traits under LNT in our studies. Out of four subpopulations identified in the set of 472 genotypes, no common traits across the seasons or no common marker across all the four subpopulations indicating the complexity of NUE trait. Several positive alleles and a few negative alleles were identified for the traits across subpopulations for low and recommended N. Within alleles of the same locus, some alleles showed negative effect, some showed positive effect, and some showed negative effect for some traits, and positive for some traits in the same population such as RM 271. Four associated genomic regions on chromosomes 5, 7, and 10 were fine mapped, and QTL for yield under low N were identified from the marker-delimited regions. Three candidate genes, namely, 2-oxoglutarate/malate translocator (Os05g0208000), alanine aminotransferase (Os07g0617800), and pyridoxal phosphate-dependent transferase (Os10g0189600) from QTL regions showed enhanced expression in the genotypes with promising yield under low N (Rao et al. 2018).

The identification of QTL/genomic regions associated with yield and its components under low N will be useful in MAS for improving the NUE of rice. No major QTL with >30% phenotypic variance were identified in the QTL studies of bi-parental mapping populations to be deployed in marker-assisted introgression/selection. Validation of the reported QTL is required to use the reported QTL in regular breeding programs because of the use of different donors, mapping populations, experimental conditions, and methods.

Expression Analyses of Candidate Genes Associated with Nitrogen Metabolism

The identification of potential candidate genes/proteins in N metabolism could serve as biomarkers for NUE to screen the genotypes for different alleles and their associations with NUE, though the candidate genes involved in N metabolism in rice are well elucidated.

The N has been taken up by transporters, which can be high-affinity transporter system (HATS) and low-affinity transport system (LATS) depending on the availability of N (Vinod and Heuer 2012). More than 10 rice ammonium-transport (AMT) proteins have been characterized (Li et al. 2009). Regulation of temporal and spatial expression of *OsAMT1;1, OsAMT1;2,* and *OsAMT1;3* was reported (Ding et al. 2011). An up-regulated high-affinity urea transporter (*OsDUR3*) under N deficiency has been characterized (Wang et al. 2012). Ammonium is incorporated into organic molecules by GS and Glutamine oxoglutarate aminotransferase (GOGAT).

Ammonium is incorporated into organic molecules by GS and GOGAT. The family of GS genes has been characterized (Obara et al. 2001; Tabuchi et al. 2005). The two kinds of GOGAT in rice, namely, ferredoxin (Fd)-dependent GOGAT and nicotinamide adenine dinucleotide hydrogen (NADH)-dependent Fd-GOGAT. The role of these enzymes in the status of the re-utilization is unclear, and the precise mechanism is still being deciphered (Yamaya et al. 2002). Several enzymes involved in N metabolism have been thoroughly characterized (Kamachi et al. 1991; Ireland and Lee 1999; Lea et al. 1990; Lam et al. 1996; Xu et al. 2012; Masclaux-Daubresse et al. 2010;; Srikanth et al. 2016).

Genome-wide expression analysis is a promising approach to understand complex traits like NUE and to identify N-responsive genes for their deployment in the development of rice varieties with NUE. Transcriptome analyses using microarrays or ribonucleic acid

(RNA)-seq have identified several differentially expressed genes (DEGs) and pathways in response to differential N in rice.

Expression profiles of *indica* variety were studied after low N with normal N as control using a microarray of 11 494 rice expressed sequence tags (ESTs) representing 10 422 genes. No significant differences were found in the leaf tissue, and 471 ESTs were detected in root tissues with 115 ESTs up-regulated and 358 ESTs down-regulated. The up-regulated genes comprised early response genes involved with biotic and abiotic stress and some transcriptions factors and signal transduction. The down-regulated genes included photosynthesis and energy metabolism, stress response, transcription factors, and signal transduction. Under microarray analysis, no differential expression was found in the genes known to be involved in N uptake and assimilation (Lian et al. 2006). Using Affymetrix Genechip rice arrays, the dynamics of rice transcriptome under N starvation situation, 3518 induced/suppressed genes belonging to cellular metabolic pathways including stress response, primary and secondary metabolism, molecular transport, regulatory process, and organ development representing 10.88% genome were identified; 462 or 13.1% transcripts for N starvation expressed similarly in root and shoot (Cai et al. 2012).

Another Affymetrix Rice Genome array analysis to study the overlap between the DEG response to N and phosphorous (P) starvation conditions after 1 hour, 24 hours, and 7 days in root and shoot studies of seedlings showed the DEGs belonged primarily to six biological process categories: molecular transport, molecular metabolism, regulation and modification, organism development, stress stimuli, and electron transport (Cai et al. 2013). To characterize the molecular genetic basis of adaptation to macronutrient deprivation, microarray analysis of rice root at 6 and 24 hours after nitrogen, phosphorus, and potassium deficiency treatments was performed, and the transcriptome response to nitrogen depletion occurred more rapidly than corresponding responses to P and potassium (K) deprivation. Signaling via reactive oxygen species was observed to be a common feature in response to macronutrient deficiency (Takehisa et al. 2013).

The co-expression network analysis and metabolic profiling performed in rice pinpointed the relevance of signal transduction components and regulation of N mobilization in response to limiting N conditions. Among the identified enriched gene ontology (GO) terms, those related to nucleoside/nucleotide, purine, and ATP binding, defense response, sugar/carbohydrate binding, protein kinase activities, cell-death, and cell-wall enzymatic activity are enriched (Coneva et al. 2014). RNA-Seq data identified 8509 multi-exonic genes with 19,628 alternative splicing events after 12 hours of N-starvation. Among the differentially expressed transcripts, 1158 were differentially expressed in the leaf sheaths (548 up-regulated and 610 down-regulated) and 492 in the roots (276 up, 216 down) (Yang et al. 2015b). RNA seq analysis identified 143 and 251 up- and down-regulated DEGs in roots and 119 up-regulated/349 down-regulated DEGs in shoots. According to Kyoto Encyclopedia of Genes and Genomes (KEGG) analysis, carbohydrate and nitrogen metabolisms were reduced (Yang et al. 2015a). Transcriptome analysis focusing on amino acid analysis and identification of —N-regulated genes in rice roots with N starvation (—N, 1 hour), revealed that genes involved in metabolism, plant hormone signal transduction (e.g., abscisic acid, auxin, and jasmonate), transporter activity, and oxidative stress responses were rapidly regulated by —N. Some of the —N-regulated genes encoded transcription factors, protein kinases, and protein phosphatases, which were proposed to be involved in the regulation of early —N responses in rice roots (Hsieh et al. 2018). Using four different transcriptome approaches, new putative long non-coding RNAs (lncRNAs), and microRNAs (miRNAs) and their target messenger RNAs (mRNAs) in response to N starvation were identified. Analysis of rice genes involved in N metabolism and/or transport using strand-specific RNA-Seq identified 2588 novel putative lncRNA encoding loci. Poly A-primed sequencing (2P-Seq) revealed alternatively polyadenylated isoforms of N starvation-responsive lncRNAs and provided precise 3′ end information on the transcript models of these lncRNAs. Analysis of small RNA-Seq data identified N starvation-responsive miRNAs and down-regulation of miR169 family members, causing de-repression of nuclear transcription factor Y subunit alpha (NF-YA). Profiling of the N starvation-responsive down-regulation of root-specific miRNA, osa-miR444a.4-3p, and Degradome sequencing confirmed MADS25 as a novel target gene (Shin et al. 2018). Genome-wide transcriptome analysis of two rice genotypes, IR 64 and Nagina 22 under optimal (N+) and chronic starvation (N−) of N from 15-day-old root and shoot tissuesrevealed nearly double the number of genes to be differentially expressed (DEGs) in roots (1016) compared to shoots (571). Most of these DEGs belonged to starch and chloroplast metabolism, followed by membrane and signaling proteins (Sinha et al. 2018).

So far, transcriptome studies in rice to determine differential expression were reported only in seedlings under hydroponics and artificial medium. No studies have yet delineated the responses at the transcriptome level, when rice is grown under chronic N stress in field conditions during the panicle-development stage. Most of the transcriptome studies only focused a single genotype except for a few studies in which two or more genotypes have been analyzed under low N.

Two transcriptome data sets (GS FLX 454 Roche and IlluminaHiSeq 2000) derived from the booting stage of panicle development of two efficient and two poor genotypes grown under field conditions of low and recommended nitrogen were assembled de novo in our study for identification of DEGs and pathways. Down-regulation was observed in 30% of the metabolic pathways in the efficient genotypes and is being proposed as an acclimation strategy to low N. The remobilization of stored N and the release of ammonium via the biosynthesis of phenylpropanoids were stimulated by N deprivation. It is evident that plants have evolved regulatory systems to adjust metabolism, conserve resources, and activate the acclimatory pathways enabling them to adapt to N-deficient conditions.

Global gene expression and comparative analysis of genotypes contrasting for NUE would aid in narrowing down the candidate genes. So far, these studies have provided valuable insights into N response and its linkage to other biological pathways. Differential expression indicates the potential target genes for nitrogen-use efficiency improvement of rice. With the available data about the large number of reported QTL/genomic regions for NUE in rice and DEGs in contrasting genotypes under differential N, the convergence of these two datasets is possible to identify potential genomic regions or candidate genes for deployment in rice molecular breeding for NUE.

Ways to Go for the Development of Rice Varieties

Though the selection of donor genotypes and identification of genes is being done under low or zero N, since no farmer would go for zero input cultivation, 50% of the recommended dose of N should be the target N level for the development of NUE rice varieties. Promising donors can be selected from the pool of screened germplasm comprising landraces and varieties for the development of the segregating populations. It was observed that some of the popular released varieties, namely, Swarna, IR64, MTU1010, Varadhan, Jaya were found to be efficient under sub-optimal levels of N and can also be used as donors (Surekha et al. 2019). Thus segregating material can be generated using either varietal combinations or variety and landrace combinations, and field selections can be made from F_2 onwards under sub-optimal N levels. Varietal combinations would lead to the faster development of NUE, as they are already with desirable plant type. The breeding line development from landrace combinations might take two or more extra generations as it could involve backcrosses and additional selections owing to the undesirable plant type of landraces. Evaluation systems for the release of NUE rice varieties should be integrated to national varietal release systems of the relevant countries. In India, rice varieties are released through an All Indian Coordinated Rice Improvement Program (AICRIP). However only in 2018 were varietal trials for nutrient-use efficiency initiated in AICRIP. Thus, the NUE rice breeding lines developed in our study were evaluated under a trial entitled "Evaluation of Radiation and Nitrogen use Efficient Promising Rice Genotypes" conducted by Plant Physiology of AICRIP and two promising lines were identified through evaluation for three years.

Conclusions

Improved nitrogen-use efficiency in rice cultivation is needed to address the impact of climate change and cost of cultivation. The promising genotypes identified under low or sub-optimal levels of N should be utilized for the generation of the breeding material for development of the rice varieties with NUE. The repertoire of QTL information generated for NUE should be validated and deployed for marker-assisted introgression of NUE in rice. Several candidate genes have been identified from the transcriptome data and should be confirmed for their use in allele mining and haplotype breeding for NUE in rice. As nitrogen is the building block of biomass, an optimum level of N is required for realizing the yield. So, the strategy should be maximum uptake, maximum utilization, and maximum remobilization of the optimum N inputs to give maximum possible yield.

References

Anzoua, K.G., Junichi, K., Toshihiro, H. et al. (2010). Genetic improvements for high yield and low soil nitrogen tolerance in rice (*Oryza Sativa* L.) under a cold environment. *Field Crops Res.* 116 (1–2): 38–45. https://doi.org/10.1016/j.fcr.2009.11.006.

Cai, J., Chen, L., Qu, H. et al. (2012). Alteration of nutrient allocation and transporter genes expression in rice under N, P, K, and Mg deficiencies. *Acta Physiol. Plant.* 34 (3): 939–946. https://doi.org/10.1007/s11738-011-0890-x.

Cai, H., Xie, W., and Lian, X. (2013). Comparative analysis of differentially expressed genes in rice under nitrogen and phosphorus starvation stress conditions. *Plant Mol. Biol. Report.* 31 (1): 160–173. https://doi.org/10.1007/s11105-012-0485-8.

Cho, Y.I., Jiang, W., Chin, J.H. et al. (2007). Identification of QTLs associated with physiological nitrogen use efficiency in rice. *Mol.Cells* 23 (1): 72–79.

Coneva, V., Simopoulos, C., Casaretto, J.A. et al. (2014). Metabolic and co-expression network-based analyses associated with nitrate response in rice. *BMC Genomics* 15 (1): 1–14. https://doi.org/10.1186/1471-2164-15-1056.

De Datta, S.K. and Broadbent, F.E. (1990). Nitrogen-use efficiency of 24 rice genotypes on an N-deficient soil. *Field Crops Res.* 23 (2): 81–92. https://doi.org/10.1016/0378-4290(90)90104-J.

De Datta, S.K. and Buresh, R.J. (1989). Integrated nitrogen management in irrigated rice. In: *Advances in Soil Science* (ed. B.A. Stewart), 143–169. New York: Springer https://doi.org/10.1007/978-1-4613-8847-0_4.

Ding, Z., Wang, C., Chen, S., and Yu, S. (2011). Diversity and selective sweep in the OsAMT1;1genomic region of rice. *BMC Evol. Biol.* 11 (1): 61. https://doi.org/10.1186/1471-2148-11-61.

Fageria, N.K. and Baligar, V.C. (2003). Methodology for evaluation of lowland rice genotypes for nitrogen use efficiency. *J. Plant Nutr.* 26 (6): 1315–1333. https://doi.org/10.1081/PLN-120020373.

Fang, P. and Wu, P. (2001). QTL x N-level interaction for plant height in rice (*Oriza sativa* L.). *Plant Soil* 236 (2): 237–242. https://doi.org/10.1023/A:1012787510201.

FAO (2018). *Rice Marker Monitor*, vol. 21, 1. Rome: FAO.

Feng, Y., Zhai, R.R., Lin, Z.C. et al. (2015). Quantitative trait locus analysis for rice yield traits under two nitrogen levels. *Rice Sci.* 22 (3): 108–115. https://doi.org/10.1016/j.rsci.2015.05.014.

Feng, B., Chen, K., Cui, Y. et al. (2018). Genetic dissection and simultaneous improvement of drought and low nitrogen tolerances by designed QTL pyramiding in rice. *Front. Plant Sci.* 9 (March): 1–13. https://doi.org/10.3389/fpls.2018.00306.

Hsieh, P.H., Kan, C.C., Wu, H.Y. et al. (2018). Early molecular events associated with nitrogen deficiency in rice seedling roots. *Sci. Rep.* 8 (1): 1–23. https://doi.org/10.1038/s41598-018-30632-1.

Hu, S.K., Zeng, D.L., Su, Y. et al. (2012). QTL analysis of nitrogen content of plant shoot under two nitrogen conditions in rice (*Oryza sativa* L.). *Aust. J. Crop Sci.* 6 (12): 1737–1744.

Inthapanya, P., Sipaseuth, S.P., Sihathep, V. et al. (2000). Genotype differences in nutrient uptake and utilisation for grain yield production of rainfed lowland rice under fertilised and non-fertilised conditions. *Field Crops Res.* 65 (1): 57–68. https://doi.org/10.1016/S0378-4290(99)00070-2.

IPCC Second Assessment Report (1995). *Climate Change 1995: The Science of Climate Change* (eds. J.T. Houghton, L.G. Meira Filho, B.A. Callander, et al.). Cambridge: Cambridge University Press.

Ireland, R.J. and Lea, P. (1999). The enzymes of glutamine, glutamate, asparagine, and aspartate metabolism. In: *Plant amino acids, biochemistry and biotechnology* (ed. B.K. Singh), 49–109. New York: Marcel Dekker.

Jewel, Z.A., Ali, J., Mahender, A. et al. (2019a). Identification of quantitative trait loci associated with nutrient use efficiency traits, using SNP markers in an early backcross population of rice (*Oryza sativa* L.). *Int. J. Mol. Sci.* 20 (4): 900. https://doi.org/10.3390/ijms20040900.

Jewel, Z.A., Ali, J., Pang, Y. et al. (2019b). Developing green super rice varieties with high nutrient use efficiency by phenotypic selection under varied nutrient conditions. *Crop J.* 7 (3): 368–377. https://doi.org/10.1016/j.cj.2019.01.002.

Kamachi, K., Yamaya, T., Mae, T., and Ojima, K. (1991). A role for glutamine synthetase in the remobilization of leaf nitrogen during natural senescence in rice leaves. *Plant Physiol.* 96 (2): 411–417. https://doi.org/10.1104/pp.96.2.411.

Kamiji, Y., Yoshida, H., Palta, J.A. et al. (2011). N applications that increase plant N during panicle development are highly effective in increasing spikelet number in rice. *Field Crops Res.* 122 (3): 242–247. https://doi.org/10.1016/j.fcr.2011.03.016.

Kanfany, G., El-Namaky, R., Ndiaye, K. et al. (2014). Assessment of rice inbred lines and hybrids under low fertilizer levels in Senegal. *Sustainability* 6 (3): 1153–1162. https://doi.org/10.3390/su6031153.

Ladha, J.K., Pathak, H., Krupnik, T.J. et al. (2005). Efficiency of fertilizer nitrogen in cereal production: retrospects and prospects. *Adv. Agron.* 87: 85–156. https://doi.org/10.1016/S0065-2113(05)87003-8.

Lam, H.M., Coschigano, K.T., Oliveira, I.C. et al. (1996). The molecular-genetics of nitrogen assimilation into amino acids in higher plants. *Annu. Rev. Plant Physiol. Mol. Biol.* 47: 569–559.

Lea, P.J., Joy, K.W., and Blackwell, R.D. (1990). Nitrogen metabolism of mutants of barley lacking key enzymes of ammonia assimilation. In: *Inorganic Nitrogen in Plants and Microorganisms* (eds. W.R. Ullrich, C. Rigano, A. Fuggi and P.J. Aparicio), 115–120. Berlin, Heidelberg: Springer https://doi.org/10.1007/978-3-642-75812-6_17.

Li, B., Merrick, M., Li, S. et al. (2009). Molecular basis and regulation of ammonium transporter in rice. *Rice Sci.* 16 (4): 314–322. https://doi.org/10.1016/S1672-6308(08)60096-7.

Lian, X., Wang, S., Zhang, J. et al. (2006). Expression profiles of 10,422 genes at early stage of low nitrogen stress in rice assayed using a cDNA microarray. *Plant Mol. Biol.* 60 (5): 617–631. https://doi.org/10.1007/s11103-005-5441-7.

Liu, Z., Zhu, C., Jiang, Y. et al. (2016). Association mapping and genetic dissection of nitrogen use efficiency-related traits in rice (*Oryza sativa* L.). *Funct. Integr. Genomics* 16 (3): 323–333. https://doi.org/10.1007/s10142-016-0486-z.

Mahajan, G., Sekhon, N.K., Singh, N. et al. (2010). Yield and nitrogen-use efficiency of aromatic rice cultivars in response to nitrogen fertilizer. *J. New Seeds* 11 (4): 356–368. https://doi.org/10.1080/1522886X.2010.520145.

Mahajan, G., Timsina, J., Jhanji, S. et al. (2012). Cultivar response, dry-matter partitioning, and nitrogen-use efficiency in dry direct-seeded rice in Northwest India. *J. Crop Improv.* 26 (6): 767–790. https://doi.org/10.1080/15427528.2012.686473.

Masclaux-Daubresse, C., Daniel-Vedele, F., Dechorgnat, J. et al. (2010). Nitrogen uptake, assimilation and remobilization in plants: challenges for sustainable and productive agriculture. *Ann. Bot.* 105 (7): 1141–1157. https://doi.org/10.1093/aob/mcq028.

Neeraja, C.N., Voleti, S.R., Subrahmanyam, D. et al. (2019). Breeding rice for nitrogen use efficiency. *Indian J. Genet. Plant Breed.* 79 (01S): 208–215. https://doi.org/10.31742/IJGPB.79S.1.11.

Nguyen, H.T.T., Dang, D.T., Van Pham, C., and Bertin, P. (2016). QTL mapping for nitrogen use efficiency and related physiological and agronomical traits during the vegetative phase in rice under hydroponics. *Euphytica* 212 (3): 473–500. https://doi.org/10.1007/s10681-016-1778-z.

Obara, M., Kajiura, M., Fukuta, Y. et al. (2001). Mapping of QTLs associated with cytosolic glutamine synthetase and NADH-glutamate synthase in rice (*Oryza sativa* L.). *J. Exp. Bot.* 52 (359): 1209–1217. https://doi.org/10.1093/jxb/52.359.1209.

Obara, M., Sato, T., Sasaki, S. et al. (2004). Identification and characterization of a QTL on chromosome 2 for cytosolic glutamine synthetase content and panicle number in rice. *Theor. Appl. Genet.* 110 (1): 1–11. https://doi.org/10.1007/s00122-004-1828-0.

Ogawa, S., Valencia, M.O., Lorieux, M. et al. (2016). Identification of QTLs associated with agronomic performance under nitrogen-deficient conditions using chromosome segment substitution lines of a wild rice relative, Oryza rufipogon. *Acta Physiol. Plant.* 38 (4) https://doi.org/10.1007/s11738-016-2119-5.

Ohnishi, M., Horie, T., Homma, K. et al. (1999). Nitrogen management and cultivar effects on rice yield and nitrogen use efficiency in Northeast Thailand. *Field Crops Res.* 64 (1–2): 109–120. https://doi.org/10.1016/S0378-4290(99)00054-4.

Piao, Z., Li, M., Li, P. et al. (2009). Bayesian dissection for genetic architecture of traits associated with nitrogen utilization efficiency in rice. *Afr.J. Biotechnol.* 8 (24): 6834–6839. https://doi.org/10.5897/AJB2009.000-9532.

Rao, I.S., Neeraja, C.N., Srikanth, B. et al. (2018). Identification of rice landraces with promising yield and the associated genomic regions under low nitrogen. *Sci. Rep.* 8 (1): 1–13. https://doi.org/10.1038/s41598-018-27484-0.

Samonte, S.O., Wilson, L.T., Medley, J.C. et al. (2006). Nitrogen utilization efficiency: relationships with grain yield, grain protein, and yield-related traits in rice. *Agron. J.* 98 (1): 168–176. https://doi.org/10.2134/agronj2005.0180.

Sedeek, S.E.M., Abdelkhalik, A.F., and Metwali, E.M.R. (2010). Genetic behaviour of some rice (*Oryza sativa* L.) genotypes under different treatments of nitrogen levels. *Behaviour* 1 (5): 1266–1278.

Senthilvel, S., Vinod, K.K., Malarvizhi, P., and Maheswaran, M. (2008). QTL and QTL × environment effects on agronomic and nitrogen acquisition traits in rice. *J. Integr. Plant Biol.* 50 (9): 1108–1117. https://doi.org/10.1111/j.1744-7909.2008.00713.x.

Shin, S.Y., Jeong, J.S., Lim, J.Y. et al. (2018). Transcriptomic analyses of rice (*Oryza sativa*) genes and non-coding RNAs under nitrogen starvation using multiple omics technologies. *BMC Genomics* 19 (1): 1–20. https://doi.org/10.1186/s12864-018-4897-1.

Singh, H. and Verma, A. (2013). Characterization and screening of high nitrogen efficient rice genotype to curtail environmental pollution. *Int. J. Agric. Environ. Biotechnol.* 6 (1): 23–30.

Singh, U., Ladha, J.K., Castillo, E.G. et al. (1998). Genotypic variation in nitrogen use efficiency in medium- and long-duration rice. *Field Crops Res.* 58 (1): 35–53.

Singh, H., Verma, A., Ansari, M., and Shukla, A. (2014). Physiological response of rice (*Oryza sativa* L.) genotypes to elevated nitrogen applied under field conditions. *Plant Signal. Behav.* 9 (7): e29015. https://doi.org/10.4161/psb.29015.

Sinha, S.K., Amitha Mithra, S.V., Chaudhary, S. et al. (2018). Transcriptome analysis of two rice varieties contrasting for nitrogen use efficiency under chronic N starvation reveals differences in chloroplast and starch metabolism-related genes. *Genes* 9 (4): 1–22. https://doi.org/10.3390/genes9040206.

Srikanth, B., Subhakara Rao, I., Surekha, K. et al. (2016). Enhanced expression of *OsSPL14* gene and its association with yield components in rice (*Oryza sativa*) under low nitrogen conditions. *Gene* 576 (1): 441–450. https://doi.org/10.1016/j.gene.2015.10.062.

Sui, B., Feng, X., Tian, G. et al. (2013). Optimizing nitrogen supply increases rice yield and nitrogen use efficiency by regulating yield formation factors. *Field Crops Res.* 150: 99–107. https://doi.org/10.1016/j.fcr.2013.06.012.

Surekha, K., Kumar, R.M., Subhramanyam, D. et al. (2019). All India co-ordinated rice improvement programme aprroaches in improving nitrogen use efficiency in rice. *Indian J. Fert.* 15 (4): 64–75.

Swamy, K.N., Kondamudi, R., Vijayalakshmi, P. et al. (2015). A comparative study on nitrogen response among upland, IRHTN, DRR and other released rice groups. *Afr. J. Agric. Res.* 10 (48): 4364–4369. https://doi.org/10.5897/ajar2015.10323.

Tabuchi, M., Sugiyama, K., Ishiyama, K. et al. (2005). Severe reduction in growth rate and grain filling of rice mutants lacking OsGS1;1, a cytosolic glutamine synthetase1;1. *Plant J.* 42 (5): 641–651. https://doi.org/10.1111/j.1365-313X.2005.02406.x.

Takehisa, H., Sato, Y., Antonio, B.A., and Nagamur, Y. (2013). Global transcriptome profile of rice root in response to essential macronutrient deficiency. *Plant Signal. Behav.* 8 (6) https://doi.org/10.4161/psb.24409.

Tegeder, M. and Masclaux-Daubresse, C. (2018). Source and sink mechanisms of nitrogen transport and use. *New Phytol.* 217 (1): 35–53. https://doi.org/10.1111/nph.14876.

Tirol-Padre, A., Ladha, J.K., Singh, U. et al. (1996). Grain yield performance of rice genotypes at suboptimal levels of soil N as affected by N uptake and utilization efficiency. *Field Crops Res.* 46 (1–3, 143): 127. https://doi.org/10.1016/0378-4290(95)00095-x.

Tong, H.H., Mei, H.W., Yu, X.Q. et al. (2006). Identification of related QTLs at late developmental stage in rice (*Oryza sativa* L.) under two nitrogen levels. *Acta Genet. Sin.* 33 (5): 458–467. https://doi.org/10.1016/S0379-4172(06)60073-5.

Tong, H.H., Chen, L., Li, W.P. et al. (2011). Identification and characterization of quantitative trait loci for grain yield and its components under different nitrogen fertilization levels in rice (*Oryza sativa* L.). *Mol. Breed.* 28 (4): 495–509. https://doi.org/10.1007/s11032-010-9499-9.

Verdeprado, H., Kretzschmar, T., Begum, H. et al. (2018). Association mapping in rice: basic concepts and perspectives for molecular breeding. *Plant Prod. Sci.* 21 (3): 159–176. https://doi.org/10.1080/1343943X.2018.1483205.

Vijayalakshmi, P., Vishnukiran, T., Ramana Kumari, B. et al. (2015). Biochemical and physiological characterization for nitrogen use efficiency in aromatic rice genotypes. *Field Crops Res.* 179: 132–143. https://doi.org/10.1016/j.fcr.2015.04.012.

Vinod, K.K. and Heuer, S. (2012). Approaches towards nitrogen- and phosphorus-efficient rice. *AoB Plants* 2012 (0): pls028. https://doi.org/10.1093/aobpla/pls028.

Wang, W.-H., Köhler, B., Cao, F.-Q. et al. (2012). Rice DUR3 mediates high-affinity urea transport and plays an effective role in improvement of urea acquisition and utilization when expressed in Arabidopsis. *New Phytol.* 193 (2): 432–444. https://doi.org/10.1111/j.1469-8137.2011.03929.x.

Wu, L., Yuan, S., Huang, L. et al. (2016). Physiological mechanisms underlying the high-grain yield and high-nitrogen use efficiency of elite rice varieties under a low rate of nitrogen application in China. *Front. Plant Sci.* 7 (2016JULY): 1–12. https://doi.org/10.3389/fpls.2016.01024.

Xu, G., Fan, X., and Miller, A.J. (2012). Plant nitrogen assimilation and use efficiency. *Annu. Rev. Plant Biol.* 63 (1): 153–182. https://doi.org/10.1146/annurev-arplant-042811-105532.

Yadav, S., Sandhu, N., Majumder, R.R. et al. (2019). Epistatic interactions of major effect drought QTLs with genetic background loci determine grain yield of rice under drought stress. *Sci. Rep.* 9 (1): 1–13. https://doi.org/10.1038/s41598-019-39084-7.

Yamaya, T., Obara, M., Nakajima, H. et al. (2002). Genetic manipulation and quantitative-trait loci mapping for nitrogen recycling in rice. *J. Exp. Bot.* 53 (370): 917–925. https://doi.org/10.1093/jexbot/53.370.917.

Yang, J., Chen, X., Zhu, C. et al. (2015a). RNA-seq reveals differentially expressed genes of rice (*Oryza sativa*) spikelet in response to temperature interacting with nitrogen at meiosis stage. *BMC Genomics* 16 (1): 1–18. https://doi.org/10.1186/s12864-015-2141-9.

Yang, W., Yoon, J., Choi, H. et al. (2015b). Transcriptome analysis of nitrogen-starvation-responsive genes in rice. *BMC Plant Biol.* 15 (1): 1–12. https://doi.org/10.1186/s12870-015-0425-5.

Yoshida, S. (1983). Growth and yield of field crops rice. In: Potential Productivity of Field Crops under Different Environments., 103–127. Los Baños, the Philippines: International Rice Research Institute http://books.irri.org/9711041146_content.pdf.

Yoshinaga, S., Takai, T., Arai-Sanoh, Y. et al. (2013). Varietal differences in sink production and grain-filling ability in recently developed high-yielding rice (*Oryza sativa* L.) varieties in Japan. *Field Crops Res.* 150: 74–82. https://doi.org/10.1016/j.fcr.2013.06.004.

Zhang, Z., Chu, G., Liu, L. et al. (2013). Mid-season nitrogen application strategies for rice varieties differing in panicle size. *Field Crops Res.* 150: 9–18. https://doi.org/10.1016/j.fcr.2013.06.002.

Zhao, S.-P. (2012). Genotype variation in grain yield response to basal N fertilizer supply among different rice cultivars. *Afr. J. Biotechnol.* 11 (59): 12298–12304. https://doi.org/10.5897/AJB12.787.

Zhou, Y., Tao, Y., Tang, D. et al. (2017). Identification of QTL associated with nitrogen uptake and nitrogen use efficiency using high throughput genotyped cssls in rice (*Oryza sativa* L.). *Front. Plant Sci.* 8: 1166. https://doi.org/10.3389/fpls.2017.01166.

13

Dissecting the Molecular Basis of Drought-Induced Oxidative Stress Tolerance in Rice

Amit K. Pradhan[1], Sabnoor Y. Jyoti[1], Zina M. Shandilya[1], Mehzabin Rehman[1], Debanjali Saikia[1], Junu Poudel[1], Jyotirmay Kalita[1], Kongkona Borborah[1], Uma K. Chowra[1], Jnandabhiram Chutia[1], Lakshminarayana R. Vemireddy[2] and Bhaben Tanti[1]

[1] *Department of Botany, Gauhati University, Guwahati, Assam 781014, India*
[2] *Department of Genetics and Plant Breeding, S.V. Agricultural College, Acharya NG Ranga Agricultural University, Tirupati, Andhra Pradesh 517502, India*

Introduction

Drought is defined as an inadequate availability of water, which includes basically the precipitation and soil moisture storage capacity in quantity and distribution, during the life cycle of a crop (Sarma et al. 2016; Sinha 1986; Farooq et al. 2009; Prakash et al. 2016). Drought stress impacts the production of food, mostly in areas with inadequate agricultural water affecting both the physiological and developmental processes such as photosynthetic ability, disturbance in maintenance of cell potential, enzyme activities, and mostly the production of secondary metabolites (Cruz de Carvalho 2008; Shandilya and Tanti 2019, 2020). Therefore, drought is considered to be one of the worse stressors impacting plant performance and crop yields. Plants suffer drought-associated stress either when transpiration rates are too high or when the supply of water to the roots is limited (Ji et al. 2012). Drought stress highly induces increase in the rate of reactive oxygen species (ROS) such as the superoxide radical ($O_2^{\cdot -}$), singlet oxygen ($^1O_2$), hydrogen peroxide (H_2O_2), and the hydroxyl radical ($OH^{\cdot}$) causing a major effect of oxidative stress to rice plants (Hasanuzzaman et al. 2012; Nahar et al. 2016, 2018a,b).

According to published statistics, the percentage of acquaintance to drought stress has doubled from that of 1970s to the early 2000s (Isendahl and Schmidt 2006). Of all the cereals, rice (*Oryza sativa*), which requires the highest amount of water for growth is the most susceptible cereal to water deficit. About, 50% of world rice production is estimated to be affected by drought (Lafitte and Bennet 2003; Lilley and Fukai 1994; Li et al. 2015).

As per the reports, the response of plants to drought are generally monitored by analyzing basic morphological, physiological, and molecular phenomenon, which vary from species to species and the tolerant cultivars showing clear indications of drought tolerance such as better plant growth, improved leaf characters, stomatal properties, water relations, antioxidant activities, and higher expression of tolerant genes to alleviate drought stress.

Molecular Breeding for Rice Abiotic Stress Tolerance and Nutritional Quality, First Edition.
Edited by Mohammad Anwar Hossain, Lutful Hassan, Khandakar Md. Iftekharuddaula, Arvind Kumar and Robert Henry.

Rice, a water-loving crop is highly drought-susceptible due to its small root system, sensitive stomatal activity, and with very low cuticular wax layer (Joo et al. 2019). Leaf rolling is one of the primary acclimation responses of rice to drought stress and is used as a criterion for scoring drought tolerance. High induction of spikelet sterility leading to reduction in number of filled spikelets per panicle is also well documented in rice (O'Toole and Moya 1981; Selote and Khanna-Chopra 2004). In addition to the physiological damage, drought in concurrence with pollen development or fertilization can completely damage the entire crop yield (O'Toole and Namuco 1983).

Rice though highly susceptible to drought shows various mechanisms of resistance/tolerance, which include mainly the development of drought-avoidance mechanisms and mechanisms that allow maintaining physiological activity even under reduced cell water potential (Blum and Ebercon 1981). Cell membrane serves as the primary target of drought stress in rice and as a mechanism of tolerance (Levitt 1980), as reported by Iljin (1957) alteration/change in the membrane was observed, which included modification of the cellular compartmentalization leading to increase in activity of hydrolytic enzymes (Stocker 1961; Todd 1972; Wang et al. 2019). Increased alteration or modification of the cellular membrane led to the decrease in membrane integrity/increase in cell permeability, causing increased ion efflux leading to a chaotic redox imbalance in the cell (Kalita et al. 2018; Pradhan et al. 2019; Xie et al. 2019). Consequently, tests based on evaluation of membrane resistance could be used to screen rice varieties for drought tolerance (Figure 13.1).

Imbalance in the redox potential of the cell, led to increase in production of ROS such as $O_2^{\cdot -}$, $^1O_2$, H_2O_2, and $OH^{\cdot}$, under drought stress, causing a major oxidative stress in the crop, which highly hampers its production and development. ROS, which induces the generation of antioxidant system at the homeostasis level serves as an important regulator of crop development, as basal antioxidant capacity is sufficient to restore the original balance between ROS production and scavenging (Tsukagoshi et al. 2010; Ramamurthy and Nina 2003). But the increase in production of ROS molecules leads to a major oxidative damage as per its superior ability to react with most of the cellular components, which provokes a destructive protein modification, mutagenic DNA strand breaks, purine oxidation, and insignificant protein-DNA cross-links in the plant, thus leading to plant death (Van Breusegem and Dat 2006).

Drought, which is a significant limiting factor for the cultivation of rice, generates an imbalance in the development of ROS molecules and its scavenging activity, leading to a robust increase in ROS production in rice organs. Drought-induced oxidative stress with enhanced generation of ROS within cells leads to degradation in protein and nucleic acid, lipid peroxidation, and membrane damage (Tambussi et al. 2000). Lipid peroxidation and membrane damage serves as the major effect of drought induced oxidative stress and thereby is considered to be the scaling factor to determine the tolerance and sensitivity of rice against drought stress (Nahar et al. 2018a; Wang et al. 2019). Scavenging of ROS is performed by various enzymatic and non-enzymatic antioxidants, such as superoxide dismutase (SOD), glutathione (GSH), ascorbate (AsA), and different antioxidant enzymes. SOD comprises a group of metallo-enzymes and catalyzes the dismutation of $O_2^{\cdot -}$ to O_2 and H_2O_2. Three isozymes of SOD are being reported in rice, which include copper/zinc SOD (Cu/Zn-SOD), manganese SOD (Mn-SOD), and iron SOD (Fe-SOD) (Racchi et al.

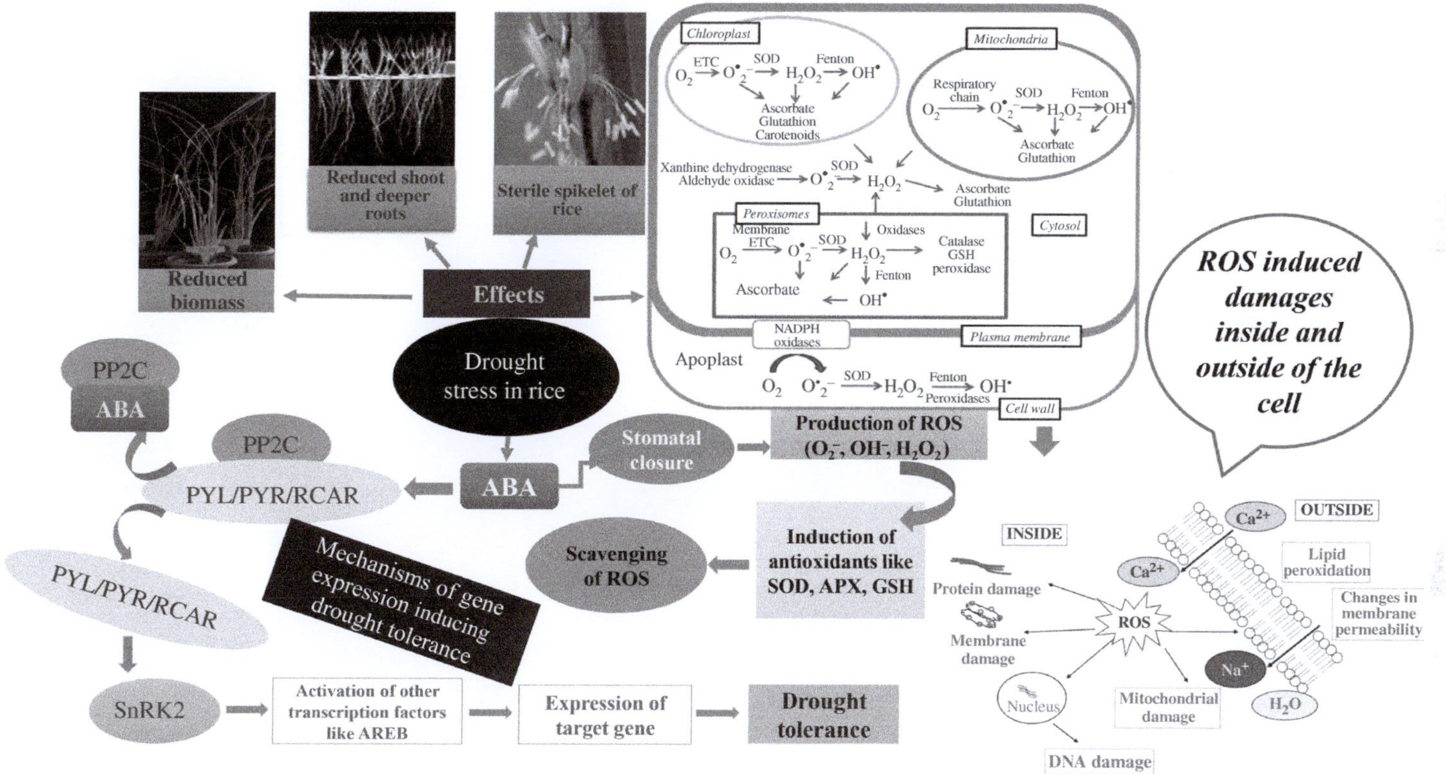

Figure 13.1 ROS generation and defense mechanisms in rice plants in response to drought stress. *Source:* Bhaben Tanti. *Note:* Its consequences for plant physiological processes and mechanisms of drought tolerance, including the regulation of genes associated with tolerance.

2001). When screening for drought-tolerant cultivars, SOD can be used as an indirect selection criterion (Zaefyzadeh et al. 2009), and it is reported that overproduction of SOD results in enhanced oxidative stress tolerance in plants (Gupta et al. 1993).

Several genes that undergo increased expression under drought conditions regulate various important processes and pathways. In recent years, numerous drought-tolerant genes have been identified in rice such as *OsDRAP1*, *OsZFP252* as reported by Huang et al. (2018), with its increased expression showed about 80% chance of survival against drought stress in rice plants. With the aid of available genomic and transcriptomic data, reverse genetic approaches accelerated the investigations of gene-function and gene-expression patterns under different abiotic stresses (Lenka et al. 2011). In various crop improvement techniques, the most widely used approaches involve transgenic approaches, which have been successfully used in many crops. However there are various other techniques that have emerged with time that facilitate producing advantageous alterations in gene sequences that can benefit the drought-affected plants. Some studies conclude that various target-based gene analyses such as virus-induced gene silencing (VIGS), targeting-induced local lesions in genomes (TILLING) and expressed sequence tags (ESTs), clustered regularly interspace short palindromic repeat (CRISPR)-Cas9 system can give prompt results over transgenic techniques for analysis of target gene(s) in collaboration with that of molecular breeding (Sahebi et al. 2018; Zu et al. 2017). With the increase in use of various high-throughput modern techniques in sideways with modern breeding, various drought-tolerant lines have been developed such as *Dular, Nagina 22,* and so on, which produce around 2.0–2.5 tons ha^{-1} yield, are widely cultivated by farmers in drought-prone areas (Biswas et al. 2002). Development of drought-tolerant lines, through the use of modern biotechnological approaches, blending yield and drought-tolerance appropriately is the prime concern for rice cultivation, with water serving as the major requirement for the growth and development of rice. Various approaches to the mechanism of drought-induced oxidative tolerance in the drought-tolerant cultivar as compared to the sensitive rice cultivars using the recent development in techniques and approaches have been discussed below (Figure 13.1).

Effect of Drought Stress in Rice

Drought is one of the major causes of crop-yield reduction in the world (Wang et al. 2019). Rice, a water-needy crop, is the most consumed staple food throughout the globe. But with the increasing problem of climate change and irregular rainfall patterns, water-deficit stress or drought has led to the depletion in the production of this crop (Laxa et al. 2019; Wang et al. 2019).

Drought highly affects the morphological, physiological, and biochemical metabolism of rice plants. Apart from the decreased yield of rice, drought reduces plant height, number of tillers, and overall development of the plant. One of the important effects of drought is stomatal closure due to an increased activity of plant growth factors like abscisic acid (ABA) (Joshi et al. 2016). Leaf rolling, decrease in leaf area, and death of leaves take place in response to drought stress in rice. Reduction of meristematic activity, suberization of the root cells, and formation of deeper roots also occur due to drought. A lower root-to-shoot ratio and reduction in both fresh and dry biomass of rice plants are other consequences of drought stress (Upadhyaya and Panda 2019). Drought also induces sterility in rice by preventing peduncle elongation and spikelet exertion, which areareas on for yield loss (Nahar et al. 2016, 2020). Drought that impacts an increase in ABA levels leading to decrease in relative water content results in a decrease in gibberellic acid content, which causes failure in the panicle exertion of rice (Muthurajan et al. 2011; Nahar et al. 2016). Drought also hampers the water potential and transpiration rate, which imbalances cell turgidity causing chaos in the whole rice metabolism (Upadhyaya and Panda 2019). Due to poor leaf turgor and stomatal closure, CO_2 assimilation in drought-stressed rice plants is reduced, and this result reduces photosynthetic rates (Chaves et al. 2009; Sharma et al. 2012). Photosystem I and II are adversely affected by drought, which is caused in the reduction of Adenosine triphosphate (ATP) synthesis in rice plants (Zlatev 2009). Drought also reduces CO_2 utilization capacity of the mesophyll cells, which also results in reduced activity of chlorophyll molecules. All these damages result in the decreased production of photosynthetic products leading to inhibition or reduction in different vital aspects of the metabolism of rice plants. Drought impairs the balance between the photo-assimilates and photochemical activity in the flag leaf of rice. The oxygen-evolving complex (OEC) and the D1 protein of PSII are also degraded due to drought, which finally inactivates the PSII (Liu et al. 2006; Kawakami et al. 2009). Disruption of photosynthetic electron transport leads to production of ROS, which results in photo-inhibition and oxidative damage (Ashraf 2009; Chutia et al. 2012).

The most important biochemical effect of drought is the production of ROS, which results in oxidative stress. ROS is generated due to electron leakage in different organelles like chloroplasts, mitochondria, peroxisome, or plasma membrane during different metabolic activities. ABA-induced stomatal closure is also a reason for ROS formation. Closed stomata delimit the assimilation of CO_2, which results in the activation of the oxygenase activity of the enzyme RuBisCo in the photorespiration process, where ROS like H_2O_2, $O_2^{\cdot-}$ are generated (Sharma et al. 2012).

The effect of drought stress also includes some responses in the rice plants, which ultimately results in induction of some defense mechanisms against drought and successive oxidative damages. Increase in the level of the enzymatic antioxidants such as SOD, ascorbate peroxidase (APX), catalase (CAT), glutathione peroxidase (GPX), glutathione reductase (GR), dehydroascorbate reductase (DHAR), mono-dehydroascorbate reductase (MDHAR) in rice plants and non-enzymatic antioxidant, such as GSH and AsA (Sharma and Dubey 2005; Hasanuzzaman et al. 2012; Hossain et al. 2015, 2017) are examples of such responses. Antioxidants are meant for the scavenging of ROS to overcome the deleterious effects of oxidative stress. If this defense system fails to delimit the power of ROS, then damage is caused to the plants in the form of lipid peroxidation, protein oxidation, DNA damage, and so forth, and in severe cases death of the plant cells ultimately causing reduction in plant yield (Nahar et al. 2016; Kalita et al. 2018).

There is an increase in the concentration of different osmolytes like proline, phenolic contents, soluble sugar, and total free amino acids, which leads to the scavenging of ROS (Anjum et al. 2017). The scavenging protein of singlet oxygen or ROS mainly includes glycine betaine, which stabilizes the structure of Ribulose-1,5-bisphosphate carboxylase/oxygenase (Rubisco) and thereby defends the photosynthetic machinery (Rezaei et al. 2012).

In rice plants several proteins have been found to be expressed differentially under drought stress. These proteins include RNA binding proteins and the EF-Tu factor involved in protein synthesis, chloroplastic fructose 1,6-biphosphate aldolase, cytosolic, and chloroplastic Cu-Zn SOD, isoflavone reductase like protein, which act as scavengers against various oxidative-stress-inducing abiotic factors. Proteins such as nucleoside diphosphate kinase that maintain the levels of CTP, GTP, and UTP cells and actin depolymerizing factors, involved in remodeling of the cytoskeleton in cells under water deficit are also expressed at higher levels (Salekdeh et al. 2002; Nahar et al. 2016).

Many genes translating into various proteins show increased expression with variable expression patterns in some specific sites and at variable times. For example, genes involved in photosynthesis are generally expressed in leaf tissues, while genes associated with the formation and modification of cell membranes and cell walls are expressed differentially in young panicles and roots. Similarly, with different growth stages like tillering, booting, or panicle exertion different genes have also been found to express differentially in response to drought. On the other hand reciprocal expression of genes like NAC-domain protein, LEA protein, and so forth, are found in different tissues during the same growth stage (Wang et al. 2011).

Drought-induced ABA triggers certain responses, which ultimately leads to the expression of genes responsible for drought tolerance. The signal of ABA induction is received by a receptor complex called PYL/PYR/RCARs combined with protein phosphatase 2C (PP2C). Binding of ABA with PP2C makes PYL/PYR/RCARs release from the receptor complex. As a consequence *SnRK2*, the key regulator of osmotic stress is activated, which further activates the transcription factors like AREB/ABF, which binds to the abscisic acid response element (ABRE) resulting in the expression of genes related to drought tolerance (Joshi et al. 2016; Furlan et al. 2016; Shinozawa et al. 2019; Ye et al. 2011).

Sources of ROS Production during Drought Stress in Rice Plants

Drought, one of the major impacts of climate change, results in decreased production of crops and causes an imbalance in the mechanism of ROS production and its quenching by antioxidants in rice (Wang et al. 2019). Under drought stress the rate of different metabolic activities like photorespiration, beta-oxidation, and so forth, are increased, which results in increased electron transport among different cell organelles involved in these activities. Due to this enhanced transport of electrons, leakage occurs, which is one of the major causes of ROS generation (Gilroy et al. 2016; Kerchev et al. 2016; Kao, 2017). The type of ROS generated depends upon the type of stress. $O_2^{\cdot-}$, $^1O_2$, and H_2O_2 are the type of ROS produced under drought stress (Laloi et al. 2004; Zulfugarov et al. 2014). Generally the balanced production of ROS is normal and is a part of the defense mechanism of plants for enhancing the activities of antioxidants against different biotic and abiotic stresses.

In rice, one of the important sources of ROS is produced in the chloroplast, where both the photosystems of the thylakoids and the antennae molecules of the light-harvesting complex (LHC) are involved in the production of ROS (Zulfugarov et al. 2014; Impa et al. 2012). During the process of phosphorylation in antenna molecules and the photosystems II (PSII) of chloroplasts, the principal ROS produced is singlet oxygen ($^1O_2$), which is known for its strong oxidizability (Impa et al. 2012; Zulfugarov et al. 2014). In recent studies by several workers such as Li et al. (2015), it has been found that the potential of the PSII is decreased under excessive drought stress. PSII is an important regulator for ROS metabolism, which can maintain electron transport even when CO_2 assimilation is declined by the plants under water stress, but with the increase in water deficiency this photosystem undergoes damage, which results in decrease in the antioxidant system, leading to increase in accumulation of ROS molecules.

The other major source of ROS production due to drought stress in rice is the peroxisome that comprises several metabolisms involving redox reactions and various ROS-producing enzymes. H_2O_2 and $O_2^{\cdot -}$ are the ROS mainly produced in peroxisomes (Del Rio et al. 2006). Rice plants suffer due to the insufficiency of CO_2in water-deficit conditions; the process of carboxylation under the calvin cycle by RuBisCo is hindered. As a result an alternative oxygenation process of the RuBP is activated, that is, photorespiration in which H_2O_2 is generated via the enzyme glycolate peroxidase (Impa et al. 2012). The other pathways where ROS can be generated in peroxisomes include β-oxidation of fatty acids, disproportionation of $O_2^{\cdot -}$ radicals, flavin oxidase reactions,and so forth (Del Rio et al. 2006; Palma et al. 2009). $O_2^{\cdot -}$ in this organelle can be produced inside the matrix from xanthin and hypo-xanthin via the enzyme xanthin oxidase. $O_2^{\cdot -}$ can also be generated in the membrane of peroxisomes where flavoprotien Nicotinamide adenine dinucleotide (NADH), ferricyanide reductase and cytochrome b are found to be involved in an electron transport chain (Impa et al. 2012).

In addition to the chloroplasts, mitochondria are also responsible for production of ROS (Davidson and Schiestl 2001; Miller et al. 2010), which is comparatively less (Foyer and Noctor 2005). The three different complexes, namely-complex I, complex II, and complex III of the mitochondrial electron transport chain (METC) involved in respiration—are the principal sites for the production of ROS in mitochondria (Das and Roychoudhury 2014).The intermediate between complex I and complex III, ubi-semiquinone serves as a donor of electron to generate ROS (Impa et al. 2012).

In response to drought, ROS is also generated at the plasma membrane and in the diffusible space surrounding the plasma membrane known as the apoplast. In the plasma membrane the Nicotinamide adenine dinucleotide phosphate (NADPH) oxidase or respiratory burst oxidase homolog (RBOH) are the major producers of ROS. Drought stress considerably enhances the activity of total NADPH oxidase, which in turn increases the production of ROS (Wang et al. 2015). According to a report from Wang et al. (2015) the expression of the *OsRbohA* gene encoding RBOH is increased under drought stress in rice plants. In addition to the above there are several other sources of ROS production inside plant cells such as the endoplasmic reticulum and cell wall. Of the different ROS generated, H_2O_2 is found to be produced in all of these organelles, whereas $O_2^{\cdot -}$ is produced in the plasma membrane and cell wall, and $OH^{\cdot}$ is produced in the endoplasmic reticulum (Mittler 2002; Del Rio et al. 2006; Hossain et al. 2011; Kao 2014; Das and Roychoudhury 2014).

Antioxidative Defenses and Redox Homeostatic Mechanisms in Rice in Response to Drought

Plants have evolved antioxidant defense mechanisms that help them in containing the extent of oxidative injury induced by ROS. A number of enzymatic and non-enzymatic antioxidants play important roles in quenching ROS and protecting plant cells against oxidative stress. Drought increases GPX activity and reduces the GSH/GSSG ratio in drought-sensitive rice varieties, for example, *IR-29* and *Pusa Basmati*. The levels of antioxidants such as flavonoids and phenolics increase in rice plants with higher concentration in tolerant cultivars as compared to that of the sensitive ones (Das and Roychoudhury 2014). Activities of both enzymatic antioxidants such as SOD, CAT, APX, and levels of non-enzymatic antioxidants such as AsA, GSH, and so forth, increase under drought stress in tolerant cultivars, thereby limiting the extent of oxidative damage and promoting better plant growth with decrease in ROS accumulation, lipid peroxidation, and leakage of electrolytes (Guo et al. 2006).

The redox reaction that includes the transfer of electrons during the transduction of energy in the cell organelles is a natural cellular metabolic process, and its homeostatic maintenance is based on the balanced production of antioxidants. Redox disequilibrium is the result of an imbalance between pro-oxidants and antioxidants in rice, and redox regulation is the key to improve growth and development of rice plants under drought stress (Suzuki et al. 2011; Upadhyaya and Panda 2019). ROS and antioxidants respond to the redox state of cells during stress and perform many of their activities through regulation of hormone signaling (Noctor et al. 2014). Electron flux of the photosynthetic and respiratory systems leads to generation of superoxide from which subsequently other forms of ROS are produced (Kapoor et al. 2015). ROS production takes place within the cells even during normal conditions, but increases during stress. During drought, H_2O_2 production increases in peroxisomes of C_3 plants due to the activity of glycolate oxidase in photorespiration as intercellular CO_2 concentration decreases with stomatal closure. Moreover, the redox reactions in the chloroplasts give rise to H_2O_2, $O_2^{\cdot-}$ and $^1O_2$ in chloroplasts. Changes in dark respiration as well as photorespiration induced by drought also lead to ROS production in the mitochondria (Noctor et al. 2014). These ROS are quenched by the cellular antioxidant system through maintaining various cellular components in an active state, which prevents their accumulation (Kapoor et al. 2015). Tocopherols synthesized by plants also serve as important liposoluble antioxidants that scavenge singlet oxygen and other ROS (Foyer and Noctor 2005). Therefore, to maintain rice growth under drought-induced oxidative stress, a highly coordinated system needs to be implemented, that is, maintenance of homeostatis between the ROS-generating organelles, reversible phosphoproteins regulations, and activation of both enzymatic, non-enzymatic activities with increased expression of antioxidant regulatory genes such as *OsDRAP1*, *OsZFP252* and various ROS- antioxidant interactions (Foyer and Noctor 2005).

ROS Signaling under Drought Stress

Drought-induced production of ROS is a by-product of increased aerobic metabolism in various plant species including rice (Das and Roychoudhury 2014; Miller et al. 2010). These ROS cause considerable oxidative damage to cellular bio-polymers and regulate cation channels, which regulate cell swelling, shrinkage, apoptosis, and activation of Ca^+ dependent transcription factors (Demidchik 2015; Lang et al. 2005). Recently, ROS have been shown to play significant roles as signal transduction molecules involved in activating plant responses such as programmed cell death, responses to pathogen attacks, and responses to environmental stresses, helping in the acclimation of plants to stressful conditions (Miller et al. 2010). As reported by Nahar et al. (2016), in rice ROS generation is managed by both enzymatic and non-enzymatic components, indicating an increased amount in tolerant cultivar as compared to sensitive ones, which is also co-related to the differential expression of antioxidants genes activated with generation of ROS signaling cascades. Accumulation of ROS promotes the increased activity of antioxidants such as SOD, CAT, GPX, and so forth (Das and Roychoudhury 2014). ROS as a whole serves as the redox sensors that correspond to the redox-regulatory networks that include redox input elements, redox transmitters,and so forth, and thus develop a co-network with other signaling pathways (Dietz et al. 2016; Duan et al. 2012). ROS as signaling molecules require the presence of redox-sensitive proteins such as respiratory burst oxidase homologues proteins, capable of oxidizing, and reducing either directly or indirectly, based on the different enzymatic and non-enzymatic antioxidant compounds such as GSH/thioredoxins (Laxa et al. 2019; Wang et al. 2019). ROS molecules serving as signaling molecules include the following actions: initially, activating a compartment-based cellular signaling network with a required specificity with the signal molecules as oxidized proteins, fragments of cell wall, hormones, and so forth; secondly developing a specific oscillation wave carrying a decoded signal such as the Ca^{2+} signals. Then, based on the signal frequency, amplitude, or the source of signal generation a specific gene expression gets triggered, leading to acclimation of plants against oxidative stress (Sahebi et al. 2018; Upadhyaya and Panda 2019).

The generation of ROS and its serving as signaling molecules include two phases of ROS generation, that is, the primary rapid phase and the secondary phase. As reported by Mittler et al. (2010), the primary phase of ROS generation undergoes rapidly within minutes of exposure to stress followed by the secondary one after hours/days of exposure. With the rapid burst of ROS and its activating, the secondary ROS generation serves as the major signal regulators for the downstream signaling pathways, thereby leading to tolerance of plants against stress specially drought (Suzuki et al. 2011; Lahkar and Tanti 2017, 2018; Lahkar et al. 2020). The primary phase ROS generally serves in ROS sensing, and it further enhances the secondary phase generation amplifying its downstream signaling activity and hence further transducing response to ameliorate the stress effects, but its activity remains constrained to a specific non-toxic level (Furlan et al. 2016). The development of oxidative stress with increased production of ROS under drought stress may also serve as a phenomenon of variation between the tolerant and sensitive cultivar, based on the capacity of cultivars in the initial sensing of ROS molecules and ROS specificity, dynamicity, ability of interaction with other signaling pathways, and its signaling capacity within and across the different cells (Mittler et al. 2010).

High-Throughput Technologies to Identify Drought-Induced Oxidative Stress-Tolerance Mechanisms

Drought is one of the most devastating abiotic stresses faced by plants all over the world. Over the past few decades, change in global climate has led to increased frequency of drought, leading to reduction in yield. The resistance of rice against such environmental problems, by improving the agronomically important traits to increase their economic value, is a worldwide concern, and there is a need to implement efficient and sustainable solutions. The genetic traits responsible for enhanced tolerance against drought-induced oxidative stress as well as promoting high yield are of foremost interest in the field of crop management (Prakash et al. 2016). The molecular mechanism of drought-induced oxidative tolerance is characterized by a cascade of genetic and cellular pathways. The advancement of technology and the emergence of high-throughput techniques have helped to uncover the underlying mechanisms involved in response to such stress conditions (Ashraf 2009). Therefore, to unravel the exact mechanism and the metabolites generated in relation to that of the candidate genes associated with drought tolerance, various high-throughput molecular approaches have been made to explore their functionality. Some of the approaches developed are discussed below.

Genome Editing Using CRISPR-Cas9 Technology

Genome editing is a promising tool that can be used to make specific changes to the DNA of a cell or organism. The most efficient and cost-effective genome editing technology used so far is the CRISPR-Cas9 system (Arora and Narula 2017). It is an anti-viral defense mechanism that uses a Cas9 endonuclease protein and a homing device (guide RNA) that is responsible for cleaving target DNA both in natural and artificial conditions. It is suitable for monocots with high GC content such as rice. So far various reports have demonstrated the adaptation of CRISPR-Cas9 genome editing system in plant cells (Shi et al. 2017). The successful transformation of ABA-responsive element-binding protein1/ABRE binding factor (AREB1/ABF2), which is a key positive regulator of the drought stress response, revealed that plants such as rice, maize, and tomatoes so generated showed higher chlorophyll content and a faster stomatal aperture under water-deficit condition, and, in addition, a better survival rate under drought stress (Sun et al. 2016; Lou et al. 2017).

Expressed Sequence Tags (ESTs)

ESTs are mRNA fragments that have been generated from single sequencing of cDNA clones (Kanth et al. 2015). Since these are generated from mRNA fragments, they are an important means to detect a gene's functional properties and enable responsive transcriptome analysis of an organism. cDNA libraries have been prepared from the plants that were subjected to such drought conditions. There are millions of ESTs of various crops available at the Expressed Sequence Tags Database of the National Centre of Biotechnology Information (NCBI). Various ESTs from several cloned cDNA libraries have been identified for candidate abiotic stress-tolerance genes associated with ROS networks, signal transduction, and the accumulation of osmolytesin crops subjected to drought-stress conditions (Ergen and Budak 2009). Such studies of genes as functional markers have been undertaken for various crops such as rice, barley, sorghum, and wheat, which provide a marker for identification and better insight towards the mining of candidate genes responsible for drought tolerance.

Gene Networks

Gene microarrays are another useful tool that enables researchers to detect thousands of genes at the same time. Probes are used to detect DNA, messenger RNAs/transcriptome (most commonly cDNAs derived from mRNAs) that may or may not be translated into proteins. There are two different types of microarray technologies available, namely, the cDNA microarray and the oligonucleotide microarray, the most predominant being the "Affymetrix GeneChip." The Affymetrix Gene Chip array ismore precise and has the greatest range in terms of the magnitude of expression levels, which is followed by oligonucleotide arrays. Rice gene expression under drought stress was first analyzed using cDNA microarrays. An Affymetrix study revealed a number of genes that are differentially expressed under drought stress. Most of these genes were tissue- or stage-specific which showed distinct functional categories such as photosynthesis-related genes that are prevalent in leaf, and the genes associated with cell membrane biogenesis, and cell wall modification present in roots and young panicles. In drought conditions, several genes, such as *GA2ox*, *SAP15*, and *Chitinase III*, were regulated in a reciprocal way in two tissues at the same development stage. A number of transcription factor genes were found to be differentially regulated by drought stress. Most of them were also regulated in a tissue- or stage-specific manner again. A cis-element containing "CGCG" box was identified to over-present in the upstream of some common induced genes, and it may be very important for rice plants responding to drought conditions (Sahebi et al. 2018; Zu et al. 2017). Such microarray studies have been a very important tool to detect various numbers of genes and their expression profiling at the same time.

RNA-Seq Technology

RNA-Seq technology is another powerful tool in terms of transcriptome analysis. It determines the quantity and the presence of RNA coding for a specific gene in a particular sample under a specific condition using next-generation sequencing (Dubouzet et al. 2003). The transcriptome studies have helped researchers through the global genomes' depth to at once detect the allele among different rice genotypes responsible for specific traits under that particular condition (Shinizaki and Shinozaki 2007). Through the reports by Iquebal et al. (2019), transcriptomic analysis of drought-tolerant and drought-sensitive cultivars was made for the identification of novel genetic regulatory mechanisms and suggested up-regulation of gene responsible for maintenance of water potential during drought. Similarly Yoo et al. (2017) reported the identification of *OsPhyB*, responsible for induction of novel regulatory mechanisms for drought tolerance in rice, using global RNA-seq transcriptome analysis. The RNA-Seq analysis technique, which helps to determine the exact phenomenon of molecular-induced tolerance mechanisms such as pattern of gene expression and so forth, has been carried out in order to understand the molecular mechanism in certain rice seedlings that were subjected to drought stress (Zombori 2008; Dash et al. 2017; Kumar and Dash 2019).

Micro-RNA (miRNA) Technology

Cereal crops such as rice develop various mechanisms of defense of which micro RNAs-based reprogramming of gene expression serves as an important mechanism of defense in rice. Micro RNA is the small noncoding regulatory RNA that regulates gene expression through mRNA cleavage or translational inhibition and is largely preserved in plant species, so understanding the regulatory mechanism of miRNA-based genetic reprogramming helps to increase the potential for drought tolerance (Ferdous et al. 2015). The mechanism of micro RNA (miRNA) occurs in the cytoplasm wherein dsRNA in the cell is cleaved by RNase DICER into short dsRNA fragments known as miRNAs, which are then loaded with a protein component on one of the strands, known as argonaute forming an RNA-induced silencing complex (RISC), which later then targets mRNA to prevent translation (Zhou et al. 2010). However, during a stress condition, methylated micro RNAs are produced, which follow the same pathway, but the mature miRNAs that are incorporated into the RNA-induced silencing complex (RISC) guide the RNA silencing activity of AGO1 (argonaute) to partially complementary mRNA. The miRNA then targets the abiotic stress-responsive mRNA, and that causes translation repression and mRNA degradation. miRNA expression profiling has been done in many *O. sativa* and *Arabidopsis* under drought stress (Faisal et al. 2017; Balyan et al. 2017). The miR169, miR396, artificial miRNA, miR171, miR319, miR393, miR156, miR158 are known to be drought responsive. However, it has been reported that miR167 was down-regulated by treatment of ABA in rice seeding, in contrast, to its expression, that is, up-regulated by drought stress in *Arabidopsis* (Table 13.1).

Virus-Induced Gene-Silencing Machinery

Virus-induced gene-silencing (VIGS) is an effective anti-viral defense mechanism that has been exploited in plants. It provides a powerful approach for genetic and functional

Table 13.1 Some of the miRNA induced under drought stress in rice and its functions.

Sl. No.	Micro RNA	Target gene	Functions	References
1	MiR169	*CBF/DREBs* TFs	It regulates the CBF/DREBs (dehydration- responsive element) transcriptions factors and plays an important role in drought stress responses in rice	Zhao and Srivastava (2007)
2	MiR396	Growth regulating factor 6 (*OsGRF6*) gene	Alters crop yield by regulating shape and architecture of inflorescence	Zhou et al. (2010)
3	MiR398	Copper binding Cu/ZnSODs, CCS	Reduction is SOD activity resulting high ROS levels leading to enhanced stomata closure in the drought-tolerant rice cultivar	Balyan et al. (2017)
4	A miRNA (artificial miRNA)	DST gene	Loss of DST function increases stomatal closure and reduces stomatal density via regulation of genes involved in ROS homeostasis and consequently enhances drought tolerance in rice over-expressing DST-amiRNA	Faisal et al. (2017)

characterization of genes in plants. Post-transcriptional gene silencing (PTGS), an epigenetic phenomenon responsible for the sequence specific-endogenous mRNAs degradation (Smith et al. 2006), allows the targeted down-regulation of a targeted gene through the degradation of its transcripts. Viruses alone can trigger PTGS as an anti-viral mechanism in certain plant species (Baulcombe 1999; Guo et al. 2010; Wang et al. 2016a). The main molecule that induces the function of PTGS is double-stranded RNA, dsRNA. This dsRNA is cleaved to produce small guide molecules called short interfering RNA (siRNA). The siRNA associates with the RISC that ultimately degrades the target homologous RNA (Marathe et al. 2000; Bartel 2004). In plant species that are difficult to transform stably, VIGS serves as an important approach for functional genomics studies and characterization of genes in various abiotic stresses, including drought stress (Wang et al. 2016b). There are three major steps for a successful VIGS technology application; engineering viral genomes to include fragments of host genes that are targeted to be silenced, infecting the appropriate plant hosts, and as part of the defense mechanism of the plant against virus infection silencing the target genes (Senthilkumar and Mysore 2011). VIGS has several advantages as it is easy and rapid gene silencing, no need of stable plant transformation, less sequence information is needed, can be used for both forward and reverse genetics and so forth. For functional validation of drought-responsive genes that are identified from transcript profiling of plants exposed to drought stress, VIGS is a powerful tool. A gene encoding late embryogenesis-abundant protein (LEA); silencing of lea4, by TRV-VIGS silencing resulted in increased susceptibility of tomato plants to drought stress (Gopalakrishna et al. 2001). A study conducted by Wang et al. (2019) indicated that in wheat a histone H2B family gene, TaH2B-7D, is responsible for drought tolerance in wheat. VIGS technology was used to verify the function of the TaH2B-7D gene in drought tolerance. In a study on a *GLUTAREDOXIN* gene, *SlGRX1*, it has been showed that chlorophyll and relative water content seem to be decreased in vector control plants under drought stress (Guo et al. 2010). In a recent study conducted on monocotyledonous crop species, two drought-stress-responsive genes, *ENHANCED RESPONSE TO ABSCISIC ACID* (*Era1*) and *INOSITOL POLYPHOSPHATE 1-PHOSPHATASE* (*Sal1*), were silenced in wheat using BSMV-VIGS vector (Manmathan et al. 2013). The loss-of-function of *Era1* induces enhanced ABA sensitivity and hence reduces stomatal conductance, which leads to water loss (Wang et al. 2005). Similarly, loss-of-function of *Sal1* genes has shown increased sensitivity of plants to drought stress. In another study, on wheat LEA class of proteins of *Hordeum vulgarae, ABUNDANT PROTEIN* (*HvHVA1*) and *DEHYDRIN 6* (*HvDhn6*) were individually silenced using the BSMV-based VIGS vector (Liang et al. 2012). Both *HVA1*- and *Dhn6*-silenced plants showed lower survival rates under submerged stress than vector control plants. Similarly, BSMV-VIGS-mediated silencing of AUTOPHAGY-RELATED 8 (TdAtg8) of a wild wheat (*Triticum dicoccoides*) resulted in reduced chlorophyll content and an increase in malondialdehyde (MDA) content that damages membrane lipids by the effect of ROS in BSMV-VIGS- silenced plants under drought stress (Kuzuoglu-Ozturk et al. 2012).

TILLING (Targeting Induced Local Lesions in Genomes)

Drought-induced mutations or any molecular changes in rice cause various effects some of which increases the tolerance capacity whereas some causes sensitivity. TILLING is a non-transgenic reverse genetic high-throughput approach that identifies SNPs (single nucleotide polymorphism)s, INDELS (insertions/deletions) in drought-induced genes in mutagenized plant species regardless of its mating, pollinating system, ploidy level, or genome size. In rice, particular research programs on physiological, molecular, and genetic aspects of drought-tolerance response cannot suppress drought susceptibility (Roy-chowdhury and Tah 2013). TILLING is used to locate allelic series induced point mutation in a gene of interest. For this approach, a TILLING population needs to be generated, which is introduced in most of the studies by a chemical mutagen ethyl-methansulfonate (EMS). To study the polymorphism, due to environmental effects a modified approach has been adapted, which is Eco-TILLING. It is more appropriate to study the functional behavior of abiotic stress-related genes (Zhang and Kirkham 1994). In a chemically mutagenized population TILLING is a simple procedure that allows rapid, high-sensitivity, high-efficiency, cost-effective detection, gene discovery, DNA polymorphism detection, and crop improvement programs (Zhou et al. 2005).

Identification of Superior Alleles or QTLs Associated with Drought Tolerance

The identification quantitative trait loci (QTLs) responsible for specific traits of tolerance against drought help in understanding the presence of a specific candidate gene and its performance in tolerant cultivars or sensitive ones (Prakash et al. 2016). QTLs correlate to specific alleles in the chromosome, and understanding the variation between tolerant and sensitive cultivars will help in development-specific marker-based analysis for the exact differentiation of tolerant from sensitive ones. Recent advances in plant systematic genomics, phenomics, and physiology have further enhanced understanding of the complex drought-tolerance network and its relation to different traits, providing increased efficiency in the selection and identification of candidate QTLs and genes responsible to it. In modern research, study of the association mapping using either of the two phenomena, that is, genome-wide association mapping or candidate gene-based mapping has further enhanced modern molecular breeding, leading to identification of the specific traits present in the tolerant rice species, showing improved expression of genes responsible for drought tolerance either due to the SNPs or variable insertion/deletion (*indel*). Identification of the SNPs using various mapping approaches in a genome-wide basis further enhances the specific identification of tolerant cultivars linked to a specific quantitative trait (Hoang et al. 2019). Different studies have reported identification of various genes with specific quantitative trait loci, such as *OsALDH2-2*, aldehyde dehydrogenase, *OsbZIP80* related to drought-induced tolerance in rice (Sahebi et al. 2018). Prakash et al. (2016), in theirwork on two rice cultivars—that is, one as tolerant and other as sensitive—observed increased GR enzyme activity under both the cultivars, but the tolerant one showed more increased activity due to the presence of one extra GR activity enzyme as compared to sensitive, indicating the presence of superior alleles leading to drought-induced oxidative stress tolerance.

Molecular Mechanism in Tolerant Genotype in Comparison to Sensitive Genotype under Drought

Tolerant cultivars show better ROS scavenging mechanisms to counter the oxidative stress caused by drought. Functional proteins such as dehydrins have been identified with protective functions against drought stress. Dehydrins act as molecular chaperones in the protection of proteins, nucleic acids, and membrane integrity under drought. Over-expression of the drought-inducible *OsDhn1* gene, which encodes the drought-responsive dehydrin protein results in better drought tolerance in rice plants as indicated by higher chlorophyll content and fluorescence, increased biomass, increased survivability, and increased ROS-scavenging activities under drought (Kumar et al. 2014). A myo-inositoloxygenase gene, *OsMIOX*, plays a positive role in increasing drought tolerance in over-expressing rice plants. Its expression is induced by drought stress primarily in the roots, and leads to better growth, leaf survivability and increased proline content, increased activities of CAT, and peroxidase (POD) antioxidant enzymes and increase in the expression of ROS-scavenging genes *POD1, POD2, CatB, APX1,*and *APX2* under the stress (Table 13.2). Oxidative damage in terms of MDA content was found to reduce in *OsMIOX* over-expressing rice plants under drought (Faisal et al. 2017). LEA proteins have been found to be involved in plant tolerance against dehydration. The drought-induced LEA protein gene *OsLEA3-1* over-expressing rice has shown increased tolerance to drought as indicated by yield (Xiao et al. 2007). Wang et al. (2007) identified 34 LEA genes in the rice genome that showed diverse patterns of expression, some of which are constitutively expressed, some regulated by ABA or GA, while some (*OsLEA3, OsLEA7, OsLEA21, OsLEA22, OsLEA27,* and *OsLEA28*) are drought inducible.

Regulation of gene expressions plays important roles in protection of plants against oxidative injury induced by drought. NAC genes, characterized in rice and other plants, and involved in developmental and stress-responsive functions are important regulators of gene expression providing drought tolerance. An important transcription factor, *SNAC3*, has been reported in rice, which regulates the expression of several ROS-scavenging genes. Over-expression of the *SNAC3* gene leads to increased expression of several ROS-scavenging genes, leading to enhanced tolerance to drought in terms of lower H_2O_2 concentration, lesser lipid peroxidation, and electrolyte leakage, while its silencing by RNAi has led to increased susceptibility to drought (Fang et al. 2015). Drought also induces expression of the *SNAC1* gene primarily in guard cells. Its over-expression leads to enhanced expression of an NAC transcription factor showing transactivation activity and leading to improved tolerance to drought stress with delayed leaf rolling, higher spikelet fertility, higher seed setting, and decreased transpirational water loss due to stomatal closure (Hu et al. 2006). A mitogen-activated protein kinase (MAPK) gene DSM1 encoding for a Raf-like MAPKKkinase (MAPKKK) protein is involved in tolerance response to drought-induced oxidative stress in rice. Mutants for the gene as well as DSM1-RNA interference lines are hypersensitive to oxidative injury during drought stress. Mutants showed down-regulation of *POX* genes and increased damage due to ROS (Ning et al. 2010). MAPK cascades are involved in transduction of extracellular signals to the nucleus and regulate the antioxidant defense system during stress in plants (Sinha et al. 2011). Dehydration-responsive elements (DRE) play important roles in regulation of gene expression and have been identified in various plants. DRE-binding proteins are encoded by DREB genes, which are induced by drought. *DREB1/DREB2 gene* homologues have been isolated in rice. Expression of *OsDREB2A* has been found to be increased during drought stress (Lata and Prasad 2011). ABA-mediated signaling pathways during abiotic stress are regulated by the bZIP transcription factors family. An *OsbZIP71* gene encoding a bZIP transcription factor has been identified in rice,

Table 13.2 Genes responsible for drought-induced oxidative stress in rice.

Genes	Target species	Functions	References
OsCATB	*O. sativa*	Under drought conditions, ABA prevents H_2O_2 accumulation through induction of CAT *OsCATB* expression and protects cells against ROS oxidative damage	Huang et al. (2019)
RBOHH	*O. sativa*	Knockout of *RBOHH* by CRISPR/Cas9 reduces ROS accumulation and inducible aerenchyma formation in rice roots, which is essential for rice to adapt to flooding and other oxygen-deficient conditions	Huang et al. (2019)
ONAC066	*O. sativa*	Over-expression of *ONAC066* in transgenic rice improved drought and oxidative stress tolerance and increased ABA sensitivity, accompanied with decreased rate of water loss, increased contents of proline and soluble sugars, decreased accumulation of ROS and up-regulated expression of stress-related genes under drought stress conditions	Yuan et al. (2019)
RRTF1	*O. sativa*	Stimulation of ROS production by RRTF1 maybe important for a rapid and transient establishment of local ROS maxima to induce appropriate downstream responses	Huang et al. (2019)
OsSRT1	*O. sativa*	Rice plants over-expressing *OsSRT1*, a SILENT INFORMATION REGULATOR2 (SIR2)-related HDAC gene, have shown an enhanced tolerance to oxidative stress	Huang et al. (2019)
OsDREB1F	*O. sativa*	*OsDREB1F* is highly induced by drought stress and exogenous ABA application and over-expression of *OsDREB1F* is reported to lead to enhanced tolerance	Wang et al. (2007)
OsDhn1	*O. sativa*	Overexpression of drought-inducible *OsDhn1* gene, which encodes drought-responsive dehydrin protein results in better drought tolerance in rice plants as indicated by higher chlorophyll content and fluorescence, increased biomass, increased survivability, and increased ROS-scavenging activities under drought	Kumar et al. (2014)
OsLEA3	*O. sativa*	Over-expressing *OsLEA3* in rice enhanced drought tolerance in the field response to water deficit stress	Xiao et al. (2007)
OsOAT	*O. sativa*	Over-expressing *OsOAT* gene increased the level of proline and improved drought-induced oxidative-stress tolerance	Oladosu et al. (2019)
SNAC1	*O. sativa*	*SNAC1*-over-expressing plants significantly enhance drought resistance by increasing stomatal closure to prevent water loss	Li et al. (2015)

which is highly induced by drought and ABA treatments. Over-expression of *OsbZIP71* has been demonstrated to result in increased tolerance to drought stress in rice (Liu et al. 2013). Therefore, gene expression and regulation is a primary determinant in the tolerance of rice cultivars to drought stress (Table 13.2).

Conclusion and Future Perspectives

Drought stress induced in rice plants due to limitation in water supply leads to imbalance in ROS generation and scavenging activity by antioxidants, both enzymatic and non-enzymatic. This imbalance induces increased oxidative stress in rice plants, causing various abnormalities. To overcome the stress, rice undergoes various phenomena of tolerance such as increased expression of tolerant genes, increased production of antioxidants, early sensing of ROS molecules, and various other morphological and physiological changes. The tolerant cultivars show improved performance as reported in many studies showing increase in various biochemical, molecular phenomena, that is, its ability to withstand the stress, but the sensitive ones remain almost opposite to that of the tolerant. Recent advancements in various high-throughput techniques and development in exact mechanisms to identify the factors responsible for stress tolerance have been positive phenomena for further crop improvement. Understanding the co-relation between the various gene networks to that of the phenotype generated along with the exact trait loci will further enhance in development of tolerant rice genotypes with the help of modern molecular breeding. Recent identification of oxidized molecules serving as secondary messengers inducing various tolerance-signaling pathways in the tolerant cultivar has also enhanced the capacity to ameliorate the causal agents of oxidative stress. Therefore, in connection to the vast findings against drought-induced oxidative stress and its variation among the sensitive and tolerant cultivars, along with the various factors causing increased oxidation, that is, development of co-networks with other factors causing oxidative stress would promote better progress on understanding the exact phenomena of stress in rice plants. Along with the difference in response of drought stress in two alternate cultivars combining the whole phenomenon, that is, the specific candidate gene, its effects and the changes would promote better understanding.

References

Anjum, S.A., Ashraf, U., Tanveer, M. et al. (2017). Drought induced changes in growth, osmolyte accumulation and antioxidant metabolism of three maize hybrids. *Front. Plant Sci.* 8: 69.

Arora, L. and Narula, A. (2017). Gene editing and crop improvement using CRISPR-Cas9 system. *Front. Plant Sci.* 8: 1932.

Ashraf, M. (2009). Biotechnological approach of improving plant salt tolerance using antioxidants as markers. *Biotechnol. Adv.* 27: 84–93.

Balyan, S., Kumar, M., Mutum, R.D. et al. (2017). Identification of miRNA-mediated drought responsive multi-tiered regulatory network in drought tolerant rice, *Nagina 22. Sci. Rep.* 7 (1): 15446.

Bartel, D.P. (2004). Micro RNAs: genomics, biogenesis, mechanism, and function. *Cell* 116: 281–297.

Baulcombe, D.C. (1999). Fast forward genetics based on virus induced gene silencing. *Curr. Opin. Plant Biol.* 2: 109–113.

Biswas, J., Chowdhury, B., Bhattacharya, A., and Mandal, A.B. (2002). In vitro screening for increased drought tolerance in rice. *In Vitro Cell. Dev. Biol. Plant* 38: 525–530.

Blum, A. and Ebercon, A. (1981). Cell membrane stability as a measure of drought and heat tolerance in wheat. *Crop Sci.* 21: 43–47.

Chaves, M.M., Flexas, J., and Pinheiro, C. (2009). Photosynthesis under drought and salt stress: regulation mechanisms from whole plant to cell. *Ann. Bot.* 103: 551–560.

Chutia, J., Borah, S.P., and Tanti, B. (2012). Effect of drought stress on protein and proline metabolism in seven traditional rice (*Oryza sativa* Linn.) genotypes of Assam, India. *J. Res. Biol.* 2 (3): 206–214.

Cruz de Carvalho, M.H. (2008). Drought stress and reactive oxygen species: production, scavenging and signaling. *Plant Signaling Behav.* 3 (3): 156–165.

Das, K. and Roychoudhury, A. (2014). Reactive oxygen species (ROS) and response of antioxidants as ROS scavengers during environmental stress in plants. *Front. Environ. Sci.* 2: 53. https://doi.org/10.3389/fenvs.2014.00053.

Dash, M., Yordanov, Y.S., Georgieva, T. et al. (2017). Poplar *PtabZIP1*-like enhances lateral root formation and biomass growth under drought stress. *Plant J.* 89 (4): 692–705.

Davidson, J.F. and Schiestl, R.H. (2001). Mitochondrial respiratory electron carriers are involved in oxidative stress during heat stress in *Saccharomyces cerevisiae*. *Mol. Cell. Biol.* 21: 8483–8489.

Del Rio, L.A., Sandalio, L.M., Corpas, F.J. et al. (2006). Reactive oxygen species and reactive nitrogen species in peroxisomes, production, scavenging, and role in cell signaling. *Plant Physiol.* 141: 330–335.

Demidchik, V. (2015). Mechanisms of oxidative stress in plants: from classical chemistry to cell biology. *Environ. Exp. Bot.* 109: 212–228.

Dietz, K.J., Turkan, I., and Krieger-Liszkay, A. (2016). Redox-and reactive oxygen species-dependent signaling into and out of the photosynthesizing chloroplast. *Plant Physiol.* 171 (3): 1541–1550.

Duan, J., Zhang, M., Zhang, H. et al. (2012). *OsMIOX*, a myo-inositol oxygenase gene, improves drought tolerance through scavenging of reactive oxygen species in rice (*Oryza sativa* L.). *Plant Sci.* 196: 143–151.

Dubouzet, J., Sakuma, Y., Ito, Y., and Kasuga, M. (2003). *OsDREB* genes in rice, *Oryza sativa* L., encode transcription activators that function in drought-, high-salt- and cold-responsive gene expression. *Plant J.* 33: 751–763.

Ergen, N.Z. and Budak, H. (2009). Sequencing over 13,000 expressed sequence tags from six subtractive cDNA libraries of wild and modern wheats following slow drought stress. *Plant Cell Environ.* 32 (3): 220–236.

Faisal, A.R., Biswas, S., Zerin, T. et al. (2017). Downregulation of the DST transcription factor using artificial microRNA to increase yield, salt and drought tolerance in rice. *Am. J. Plant Sci.* 8 (9): 2219.

Fang, Y., Liao, K., Du, H. et al. (2015). A stress-responsive NAC transcription factor SNAC3 confers heat and drought tolerance through modulation of reactive oxygen species in rice. *J. Exp. Bot.* 66: 6803–6817.

Farooq, M., Wahid, A., Kobayashi, N. et al. (2009). Plant drought stress: effects, mechanisms and management. In: *Sustainable Agriculture* (eds. E. Lichtfouse, M. Navarrete, P. Debaeke, et al.), 153–188. Dordrecht: Springer.

Ferdous, J., Hussain, S.S., and Shi, B.J. (2015). Role of micro RNA s in plant drought tolerance. *Plant Biotechnol. J.* 13 (3): 293–305.

Foyer, C.H. and Noctor, G. (2005). Oxidant and antioxidant signaling in plants: a re-evaluation of the concept of oxidative stress in a physiological context. *Plant Cell Environ.* 29: 1056–1107.

Foyer, C.H. and Noctor, G. (2009). Redox homeostasis and antioxidant signaling: a metabolic interface between stress perception and physiological responses. *Plant Cell* 17: 1866–1875.

Furlan, A.L., Bianucci, E., and Castro, S. (2016). Signaling role of ROS in modulating drought stress tolerance. In: *Drought Stress Tolerance in Plants*, vol. 1 (eds. M.A. Hossain, S.H. Wa, S. Bhattacharjee, et al.), 309–330. Switzerland: Springer International Publishing.

Gilroy, S., Białasek, M., Suzuki, N. et al. (2016). ROS, calcium and electric signals: key mediators of rapid systemic signaling in plants. *Plant Physiol.* 171: 1606–1615.

Gopalakrishna, R., Kumar, G., Krishnaprasad, B.T. et al. (2001). A stress-responsive gene from groundnut, Gdi-15, is homologous to flavonol 3-O glucosyltransferase involved in anthocyanin biosynthesis. *Biochem. Biophys. Res. Commun.* 284: 574–579.

Guo, Z., Ou, W.Z., Lu, S.Y., and Zhong, Q. (2006). Differential responses of antioxidative system to chilling and drought in four rice cultivars differing in sensitivity. *Plant Physiol. Biochem.* 44: 828–836.

Guo, Y., Huang, C., Xie, Y. et al. (2010). A tomato glutaredoxin gene *SlGRX1* regulates plant responses to oxidative, drought and salt stresses. *Planta* 232: 1499–1509.

Gupta, S., Heinen, J.L., Holaday, A.S. et al. (1993). Increased resistance to oxidative stress in transgenic plants that overexpress chloroplastic Cu/Zn superoxide dismutase. *Proc. Natl. Acad. Sci. U.S.A.* 90: 1629–1633.

Hasanuzzaman, M., Hossain, M.A., Teixeira da Silva, J.A., and Fujita, M. (2012). Plant response and tolerance to abiotic oxidative stress: antioxidant defense is a key factor. In: *Crop Stress and its Management: Perspectives and Strategies* (eds. B. Venkateswarlu, A.K. Shanker, C. Shanker and M. Maheswari), 261–315. New York: Springer.

Hoang, G.T., Van Dinh, L., Nguyen, T.T. et al. (2019). Genome-wide association study of a panel of Vietnamese rice landraces reveals new QTLs for tolerance to water deficit during the vegetative phase. *Rice* 12 (1): 4.

Hossain, M.A., Teixeira da Silva, J.A., and Fujita, M. (2011). Glyoxalase system and reactive oxygen species detoxification system in plant abiotic stress response and tolerance: an intimate relationship. In: *Abiotic Stress in Plants-Mechanisms and Adaptations* (eds. A.K. Shanker and B. Venkateswarlu), 235–266. Rijeka, Croatia: INTECH-Open Access Publisher.

Hossain, M.A., Bhattacharjee, S., Armin, S.M. et al. (2015). Hydrogen peroxide-priming modulates abiotic oxidative stress tolerance: insights from ROS detoxification and scavenging. *Front. Plant Sci.* 6: 420.

Hossain, M.A., Li, Z.G., Hoque, T.S. et al. (2017). Heat or cold priming-induced cross-tolerance to abiotic stresses in plants: key regulators and possible mechanisms. *Protoplasma* 255: 399–412.

Hu, H., Dai, M., Yao, J. et al. (2006). Overexpressing a NAM, ATAF, and CUC (NAC) transcription factor enhances drought resistance and salt tolerance in rice. *Proc. Natl. Acad. Sci. U.S.A.* 103: 12987–12992.

Huang, L., Wang, Y., Wang, W. et al. (2018). Characterization of transcription factor gene *OsDRAP1* conferring drought tolerance in rice. *Front. Plant Sci.* 9: 94.

Huang, H., Ullah, F., Zhou, D.X. et al. (2019). Mechanisms of ROS regulation of plant development and stress responses. *Front. Plant Sci.* 10.

Iljin, W.S. (1957). Drought resistance in plants and physiological processes. *Annu. Rev. Plant Physiol.* 8: 257–274.

Impa, S.M., Nadaradjan, S., and Jagadish, S.V.K. (2012). Drought stress induced reactive oxygen species and anti-oxidants in plants. In: *Abiotic Stress Responses in Plants: Metabolism, Productivity and Sustainability* (eds. P. Ahmad and M.N.V. Prasad). New York: Springer https://doi.org/10.1007/978-1-4614-0634-1_7.

Iquebal, M.A., Sharma, P., Jasrotia, R.S. et al. (2019). RNAseq analysis reveals drought-responsive molecular pathways with candidate genes and putative molecular markers in root tissue of wheat. *Sci. Rep.* 9 (1): 1–18.

Isendahl, N. and Schmidt, G. (2006). *Drought in the Mediterranean-WWF policy proposals. World Wide Fund for Nature*. Madrid: Adena.

Ji, K., Wang, Y., Sun, W. et al. (2012). Drought-responsive mechanisms in rice genotypes with contrasting drought tolerance during reproductive stage. *J. Plant Physiol.* 169 (4): 336–344.

Joo, J., Choi, D.H., Lee, Y.H. et al. (2019). The rice SUMO conjugating enzymes *OsSCE1* and *OsSCE3* have opposing effects on drought stress. *J. Plant Physiol.* 11: 152993.

Joshi, R., Wani, S.H., Singh, B. et al. (2016). Transcription factors and plants' response to drought stress: current understanding and future directions. *Front. Plant Sci.* 7: 1029. https://doi.org/10.3389/fpls.2016.01029.

Kalita, J., Pradhan, A.K., Shandilya, Z.M., and Tanti, B. (2018). Arsenic stress responses and tolerance in rice: physiological, cellular and molecular approaches. *Rice Sci.* 25 (5): 235–249.

Kanth, B.K., Kumari, S., Choi, S.H. et al. (2015). Generation and analysis of expressed sequence tags (ESTs) of *Camelinasativa* to mine drought stress-responsive genes. *Biochem. Biophys. Res. Commun.* 467 (1): 83–93.

Kao, C.H. (2014). Role of hydrogen peroxide in rice plants. *Crop Enviorn. Bioinform.* 11: 1–10.

Kao, C.H. (2017). Mechanism of salt tolerance in rice plants: cell wall related genes and expansins. *J. Taiwan Agric. Res.* 66: 87–93.

Kapoor, D., Sharma, R., Handa, N. et al. (2015). Redox homeostasis in plants under abiotic stress: role of electron carriers, energy metabolism mediators and proteinaceousthiols. *Front. Environ. Sci.* 3: 13.

Kawakami, K., Umenab, Y., Kamiyab, N., and Shen, J. (2009). Location of chloride and its possible functions in oxygen-evolving photosystem II revealed by X-ray crystallography. *Proc. Natl. Acad. Sci. U.S.A.* 106: 8567–8572.

Kerchev, P., Waszczak, C., Lewandowska, A. et al. (2016). Lack of GLYCOLATE OXIDASE1, but not GLYCOLATE OXIDASE 2, attenuates the photorespiratory phenotype of CATALASE2- deficient *Arabidopsis*. *Plant Physiol.* 171: 1704–1719.

Kumar, A. and Dash, P.K. (2019). *Transcriptome Analysis for Abiotic Stresses in Rice (Oryza sativa L.)*. Intech Open https://doi.org/10.5772/intechopen.84955.

Kumar, M., Lee, S.C., Kim, J.Y. et al. (2014). Over-expression of dehydrin gene, *OsDhn1*, improves drought and salt stress tolerance through scavenging of reactive oxygen species in rice (*Oryza sativa* L.). *J. Plant Biol* 57: 383–393.

Kuzuoglu-Ozturk, D., Cebeci Yalcinkaya, O., Akpinar, B.A. et al. (2012). Autophagy-related gene, *TdAtg8*, in wild emmer wheat plays a role in drought and osmotic stress response. *Planta* 236: 1081–1092.

Lafitte, H. and Bennet, J. (2003). Requirement for aerobic rice. Physiological and molecular considerations. In: *Water - Wise Rice Production* (eds. B.A.M. Bouman, H. Hengsdijk and B. Hardy). Los Baños, Philippines: IRRI.

Lahkar, L. and Tanti, B. (2017). Study of morphological diversity of traditional aromatic rice landraces (*Oryza sativa* L.) collected from Assam, India. *Ann. Plant Sci.* 6 (12): 1855–1861.

Lahkar, L. and Tanti, B. (2018). Morpho-physicochemical and cooking characteristics of traditional aromatic Joha rice (*Oryza sativa* L.) of Assam, India. *Biocatal. Agric. Biotechnol.* 16: 644–654.

Lahkar, L., Hazarika, G., and Tanti, B. (2020). Proximate composition, physicochemical and antioxidant properties revealed the potentiality of traditional aromatic (Joha) rice as functional food. *Vegetos* 33: 40–51. https://doi.org/10.1007/s42535-019-00076-7.

Laloi, C., Apel, K., and Danon, A. (2004). Reactive oxygen signaling: the latest news. *Curr. Opin. Plant Biol.* 7: 323–328.

Lang, F., Shumilina, E., Ritter, M. et al. (2005). Ion channels and cell volume in regulation of cell proliferation and apoptotic cell death. *Contrib. Nephrol.* 152: 142–160.

Lata, C. and Prasad, M. (2011). Role of DREBs in regulation of abiotic stress responses in plants. *J. Exp. Bot.* 62: 4731–4748.

Laxa, M., Liebthal, M., Telman, W. et al. (2019). The role of the plant antioxidant system in drought tolerance. *Antioxidants* 8 (4): 94.

Lenka, S.K., Katiyar, A., Chinnusamy, V., and Bansal, K.C. (2011). Comparative analysis of drought-responsive transcriptome in Indica rice genotypes with contrasting drought tolerance. *Plant Biotech.* J9 (3): 315–327.

Levitt, J. (1980). Responses of plants to environmental stresses. In: *Water, Radiation, Salt and Other Stresses*, vol. 2, 5502–5506. New York: Academic Press.

Li, J., Cang, Z., Jiao, F. et al. (2015). Influence of drought stress on photosynthetic 5 characteristics and protective enzymes of potato at 6 seedling stage. *J. Saudi Soc. Agric. Sci.* 16 (1): 82–88. https://doi.org/10.1016/j.jssas.2015.03.001.

Liang, J., Deng, G., Long, H. et al. (2012). Virus-induced silencing of genes encoding LEA protein in Tibetan hulless barley (*Hordeum vulgare* ssp. *vulgare*) and their relationship to drought tolerance. *Mol. Breed.* 30: 441–445.

Lilley, J.M. and Fukai, S. (1994). Effect of time and severity of water deficit on four diverse rice cultivars: phenological development, crop growth and grain yield. *Field Crop Res.* 37: 225–234.

Liu, N., Ko, S., Yeh, K.C., and Charng, Y. (2006). Isolation and characterization of tomato Hsa32 encoding a novel heat-shock protein. *Plant Sci.* 170: 976–985.

Liu, M., Li, X., Liu, Y., and Cao, B. (2013). Regulation of flavanone 3-hydroxylase gene involved in the flavonoid biosynthesis pathway in response to UV-B radiation and drought stress in the desert plant, Reaumuriasoongorica. *Plant Physiol. Biochem.* 73: 161–167.

Lou, D., Wang, H., Liang, G., and Yu, D. (2017). *OsSAPK2* confers abscisic acid sensitivity and tolerance to drought stress in rice. *Front. Plant Sci.* 8: 993.

Manmathan, H., Shaner, D., Snelling, J. et al. (2013). Virus-induced gene silencing of *Arabidopsis thaliana* gene homologues in wheat identifies genes conferring improved drought tolerance. *J. Exp. Bot.* 64: 1381–1392.

Marathe, R., Anandalakshmi, R., Smith, T.H. et al. (2000). RNA viruses as inducers, suppressors and targets of post-transcriptional gene silencing. *Plant Mol. Biol.* 43: 295–306.

Miller, G., Suzuki, N., Ciftci-Yilmaz, S., and Mittler, R. (2010). Reactive oxygen species homeostasis and signalling during drought and salinity stresses. *Plant Cell Environ.* 33: 453–467.

Mittler, R. (2002). Oxidative stress, antioxidants and stress tolerance. *Trends Plant Sci.* 7: 405–410.

Mittler, R., Vanderauwera, S., Suzuki, N. et al. (2010). ROS signaling: the new wave? *Trends Plant Sci.* 16 (6): 300–309.

Muthurajan, R., Shobba, Z.S., Jagadish, S.V.K. et al. (2011). Physiological and proteomic responses of rice peduncles to drought stress. *Mol. Biotechnol.* 48: 173–182.

Nahar, S., Kalita, J., Sahoo, L., and Tanti, B. (2016). Morphophysiological and molecular effects of drought stress in rice. *Ann. Plant Sci.* 5: 1409–1416.

Nahar, S., Sahoo, L., and Tanti, B. (2018a). Screening of drought tolerant rice through morpho-physiological and biochemical approaches. *Biocat. Agri. Biotechnol.* 15: 150–159.

Nahar, S., Vemireddy, L.R., Sahoo, L., and Tanti, B. (2018b). Antioxidant protection mechanisms reveal significant response in drought-induced oxidative stress in some traditional rice of Assam, India. *Rice Sci.* 25 (4): 185–196.

Nahar, S., Lahkar, L., Islam, M.A. et al. (2020). Genetic diversity based on osmotic stress tolerance-related morpho-physiological traits and molecular markers in traditional rice cultivars. *Biologia* 75: 669–679. https://doi.org/10.2478/s11756-020-00443-9.

Ning, J., Li, X., Hicks, L.M., and Xiong, L. (2010). A Raf-like MAPKKK gene DSM1 mediates drought resistance through reactive oxygen species scavenging in rice. *Plant Physiol.* 152: 876–890.

Noctor, G., Mhamdi, A., and Foyer, C.H. (2014). The roles of reactive oxygen metabolism in drought: not so cut and dried. *Plant Physiol.* 164: 1636–1648.

Oladosu, Y., Rafii, M.Y., Samuel, C. et al. (2019). Drought resistance in rice from conventional to molecular breeding: a review. *Int. J. Mol. Sci.* 20 (14): 3519.

O'Toole, J.C. and Moya, T.B. (1981). Water deficits and yield in upland rice. *Field Crop Res.* 4: 247–259.

O'Toole, J.C. and Namuco, O.S. (1983). Role of panicle exertion in water stress induced sterility. *Crop Sci.* 23: 1093–1097.

Palma, J.M., Corpas, F.J., and del Río, L.A. (2009). Proteome of plant peroxisomes: new perspectives on the role of these organelles in cell biology. *Proteomics* 9: 2301–2312.

Pradhan, A.K., Shandilya, Z.M., Lahkar, L. et al. (2019). Comparative metabolomics approach towards understanding chemical variation in rice under abiotic stress. In: *Advances in Rice Research for Abiotic Stress Tolerance* (eds. M. Hasanuzzaman, M. Fujita, K. Nahar and J.K. Biswas), 537–550. Cambridge, UK: Woodhead Publishing.

Prakash, C., Mithra, S.A., Singh, P.K. et al. (2016). Unraveling the molecular basis of oxidative stress management in a drought tolerant rice genotype Nagina 22. *BMC Genomics* 17 (1): 774.

Racchi, M.L., Bagnoli, F., Balla, I., and Danti, S. (2001). Differential activity of catalase and superoxide dismutase in seedlings and in vitro micropropagated oak (*Quercus robur* L.). *Plant Cell Rep.* 20: 169–174.

Ramamurthy, M. and Nina, F. (2003). Stress response, cell death and signalling: the many faces of reactive oxygen species. *Plant Physiol.* 119: 56–68.

Rezaei, M.A., Jokar, I., Ghorbanli, M. et al. (2012). Morpho physiological improving effects of exogenous glycine betaine on tomato (*Lycopersicum esculentum* Mill.) cv. PS under drought stress conditions. *Plant Omics* 5: 79–86.

Roychowdhury, R. and Tah, J. (2013). Mutagenesis-a potential approach for crop improvement. In: *Crop Improvement* (eds. K.R. Hakeem, P. Ahmad and M. Ozturk). Berlin: Springer.

Sahebi, M., Hanafi, M.M., Rafii, M.Y. et al. (2018). Improvement of drought tolerance in rice (*Oryza sativa* L.): genetics, genomic tools, and the WRKY gene family. *Biomed. Res. Int.* 7: 3158474.

Salekdeh, G.H., Siopongco, J., Wade, L.J. et al. (2002). Proteomic analysis of rice leaves during drought stress and recovery. *Proteomics* 2: 1131–1145.

Sarma, B., Basumatary, N.R., Nahar, S., and Tanti, B. (2016). Effect of drought stress on morphophysiological traits in some traditional rice cultivars of Kokrajhar district, Assam, India. *Ann. Plant Sci.* 5 (08): 1402–1408.

Selote, D.S. and Khanna-Chopra, R. (2004). Drought-induced spikelet sterility is associated with an inefficient antioxidant defence in rice panicles. *Physiol. Plant.* 121 (3): 462–471.

Senthil Kumar, M. and Mysore, K.S. (2011). New dimensions for VIGS in plant functional genomics. *Trends Plant Sci.* 16: 656–665.

Shandilya, Z.M. and Tanti, B. (2019). Hydroponic screening of traditional rice varieties in Assam, India to estimate their potential resistance to Al toxicity under P deficiency. *Acta Agrobot.* 72 (4): 1793. https://doi.org/10.5586/aa.1793.

Shandilya, Z.M. and Tanti, B. (2020). Comparative biochemical, histochemical and expression analysis of SOD gene in a few traditional rice varieties under aluminium toxicity and phosphorous deficiency in Assam, India. *Vegetos* 33: 145–157. https://doi.org/10.1007/s42535-019-00092-7.

Sharma, P. and Dubey, R.S. (2005). Drought induces oxidative stress and enhances the activities of antioxidant enzymes in growing rice seedlings. *Plant Growth Regul.* 46: 209–221.

Sharma, P., Jha, A.B., Dubey, R.S., and Pessarakli, M. (2012). Reactive oxygen species, oxidative damage and antioxidative defense mechanism in plants under stressful conditions. *J. Bot.* 2012: 217037. https://doi.org/10.1155/2012/217037.

Shi, J., Gao, H., Wang, H. et al. (2017). ARGOS 8 variants generated by CRISPR-Cas9 improve maize grain yield under field drought stress conditions. *Plant Biotechnol. J.* 15 (2): 207–216.

Shinizaki, K. and Shinozaki, K.Y. (2007). Gene networks involved in drought stress response and tolerance. *J. Exp. Bot.* 58: 221–227.

Shinozawa, A., OTake, R., Takezawa, D. et al. (2019). SnRK2 protein kinase represent an ancient system in plants for adaptation to a terrestrial environment. *Commun. Biol.* 2: 30. https://doi.org/10.1038/s42003-019-0281-1.

Sinha, S.K. (1986). *Approaches for Incorporating Drought and Salinity Resistance in Crop Plants* (eds. V.L. Chopra and R.S. Paroda), 58–78. New Delhi: Oxford and IBH Publishing.

Sinha, A.K., Jaggi, M., Raghuram, B., and Tuteja, N. (2011). Mitogen-activated protein kinase signaling in plants under abiotic stress. *Plant Signaling Behav.* 6: 196–203.

Smith, T.M.B., Schiff, M., Liu, Y., and Kuma, S.P.D. (2006). Breakthrough technologies: efficient virus induced gene silencing in Arabidopsis. *Plant Physiol.* 142: 21–27.

Stocker, O. (1961). Contributions to the problem of drought resistance of plants. *Indian J. Plant Physiol.* 4: 87–102.

Sun, Y., Zhang, X., Wu, C. et al. (2016). Engineering herbicide-resistant rice plants through CRISPR/Cas9-mediated homologous recombination of acetolactate synthase. *Mol. Plant* 9 (4): 628–631.

Suzuki, T., Yokoyama, A., Tsuji, T. et al. (2011). Identification and characterization of genes involved in glutathione production in yeast. *J. Biosci. Bioeng.* 112 (2): 107–113.

Tambussi, E.A., Bartoli, C.G., Beltrano, J. et al. (2000). Oxidative damage to thylakoid proteins in water-stressed leaves of wheat (*Triticum aestivum*). *Physiol. Plant.* 08: 398–404.

Todd, G.W. (1972). Water deficits and enzymatic activity. In: *Water Deficits and Plant Growth* (ed. T.T. Kozlowski), 177–216. New York, London: Academic Press.

Tsukagoshi, H., Busch, W., and Benfey, P. (2010). Transcritional regulation of ROS controls transition from proliferation to differentiation in the root. *Cell* 143: 606–616.

Upadhyaya, H.,and Panda, S.K. (2019). Drought stress responses and its management in rice. In: *Advances in Rice Research for Abiotic Stress Tolerance* Ed: Hasanuzzaman, M., Fujita, M., Nahar K., Biswas, J.K. (eds.). Cambridge, UK: Woodhead Publishing, 177–200.

Van Breusegem, F. and Dat, J.F. (2006). Reactive oxygen species in plant cell death. *Plant Physiol.* 141: 384–390.

Wang, Y., Ying, J., Kuzma, M. et al. (2005). Molecular tailoring of farnesylation for plant drought tolerance and yield protection. *Plant J.* 43: 413–424.

Wang, X.S., Zhu, H.B., Jin, G.L. et al. (2007). Genome-scale identification and analysis of LEA genes in rice (*Oryza sativa* L.). *Plant Sci.* 172: 414–420.

Wang, D., Pan, Y., Zhao, X. et al. (2011). Genome-wide temporal-spatial gene expression profiling of drought responsiveness in rice. *BMC Genomics* 12: 149. https://doi.org/10.1186/1471-2164-12-149.

Wang, F., Chen, H. W., Li, Q. T., Wei, W., Li, W., Zhang, W. K., et al. (2015). GmWRKY27 interacts with GmMYB174 to reduce expression of GmNAC29 for stress tolerance in soybean plants. *Plant J.* 83, 224–236. doi: 10.1111/tpj.12879.

Wang, R., Yang, X., Wang, N. et al. (2016a). An efficient virus-induced gene silencing vector for maize functional genomics research. *Plant J.* 86: 102–115.

Wang, X., Cai, X., Xu, C. et al. (2016b). Drought-responsive mechanisms in plant leaves revealed by proteomics. *Int. J. Mol. Sci.* 17 (10): 1706.

Wang, X., Liu, H., Yu, F. et al. (2019). Differential activity of the antioxidant defence system and alterations in the accumulation of osmolyte and reactive oxygen species under drought stress and recovery in rice (*Oryza sativa* L.) tillering. *Sci. Rep.* 9: 8543.

Xiao, B., Huang, Y., Tang, N., and Xiong, L. (2007). Over-expression of a LEA gene in rice improves drought resistance under the field conditions. *Theor. Appl. Genet.* 115: 35–46.

Xie, X., He, Z., Chen, N. et al. (2019). The roles of environmental factors in regulation of oxidative stress in plant. *Biomed. Res. Int.* 2019: 9732325. https://doi.org/10.1155/2019/9732325.

Ye, N., Zhu, G., Liu, Y. et al. (2011). ABA controls H_2O_2 accumulation through the induction of *OsCATB* in rice leaves under water stress. *Plant Cell Physiol.* 52: 689–698. https://doi.org/10.1093/pcp/pcr028.

Yoo, Y.H., Nalini Chandran, A.K., Park, J.C. et al. (2017). *OsPhyB*-mediating novel regulatory pathway for drought tolerance in rice root identified by a global RNA-Seqtranscriptome analysis of rice genes in response to water deficiencies. *Front. Plant Sci.* 8: 580.

Yuan, X., Wang, H., Cai, J. et al. (2019). Rice NAC transcription factor *ONAC066* functions as a positive regulator of drought and oxidative stress response. *BMC Plant Biol.* 19 (1): 278.

Zaefyzadeh, M., Quliyev, R.A., Babayeva, S.M., and Abbasov, M.A. (2009). The effect of the interaction between genotypes and drought stress on the superoxide dismutase and chlorophyll content in durum wheat landraces. *Turk. J. Biol.* 33: 1–7.

Zhang, J. and Kirkham, M.B. (1994). Drought-stress-induced changes in activities of superoxide dismutase, catalase, and peroxidase in wheat species. *Plant Cell Physiol.* 35: 785–791.

Zhao, Y. and Srivastava, D. (2007). A developmental view of microRNA function. *Trends Biochem. Sci.* 32 (4): 189–197.

Zhou, L.M., Wang, L., Palais, R. et al. (2005). High resolution DNA melting analysis for simultaneous mutation scanning and genotyping in solution. *Clin. Chem.* 51: 1770–1777.

Zhou, L., Liu, Y., Liu, Z. et al. (2010). Genome-wide identification and analysis of drought-responsive microRNAs in *Oryza sativa*. *J. Exp. Bot.* 61 (15): 4157–4168.

Zlatev, Z. (2009). Drought-induced changes in chlorophyll fluorescence of young wheat plant. *Biotechnology* 23: 437–441.

Zombori, Z. (2008). Investigation of the effect of drought stress on the rice transcriptome. *Acta Biol. Szeged.* 52 (1): 143–145.

Zu, X., Lu, Y., Wang, Q. et al. (2017). A new method for evaluating the drought tolerance of upland rice cultivars. *Crop J.* 5 (6): 488–498.

Zulfugarov, I.S., Tovuu, A., Eu, Y.J. et al. (2014). Production of superoxide from Photosystem II in a rice (*Oryza sativa* L.) mutant lacking PsbS. *BMC Plant Biol.* 14: 242.

14

Manipulation of Photosynthesis to Increase Rice Yield Potential

Prabuddha Dehigaspitiya[1] *and Saman Seneweera*[1,2]

[1] *Centre for Crop Health, Institute of Life Sciences and the Environment, University of Southern Queensland, Toowoomba, Queensland 4350, Australia*
[2] *Director, National Institute of Fundamental Studies, Kandy 20000, Sri Lanka*

Introduction

Rice (*Oryza sativa* L.) is one of the most important staple cereals in the world, catering to approximately half of the earth's population (Hibberd et al. 2008). Based on production volume, rice is the third most important cereal crop in the world, after maize and wheat, playing an essential role in global food security. During the 2015–2016 growing season, global milled rice production was about 491 million metric tons and is predicted to be a little over 500 million metric tons in 2020. Asian countries such as India, China, Bangladesh, Vietnam, Indonesia, Thailand, Myanmar, Pakistan, Philippines, Japan, and Korea, are accountable for approximately 85–90% of the world's total rice production. Among rice-growing countries, Australia is the most efficient, having an average yield of 11 tons per hectare. Although rice plays a vital role in global food security, the current rate of production is estimated to be insufficient to meet the requirements of future rice demand.

Global food security is threatened by current population growth and the adverse effects of climate change. It is predicted that global population will increase up to 9.5 billion within the next three decades (Xu et al. 2020; Dehigaspitiya et al. 2019). At the end of 2016, Asia accounted for approximately 60% of the global population, and is undergoing a continuous growth (Prosekov and Ivanova 2016). In order to cater to the growing Asian population, 60% improvement in the yield of rice may possibly be needed in the near future. This translates to catering for 43 people from a rice hectare instead of the current 27 (Dehigaspitiya et al. 2019; Xu et al. 2020). A dramatic increase in yield of critical C_3 cereals like wheat (*Triticum aestivum* L.) and rice was achieved during the Green Revolution between 1950s and 1980s mainly through the introduction of dwarf varieties. The Green Revolution helped to satisfy the global rice demand for many years. However, feeding the future global population seems impossible without another huge breakthrough similar to the Green Revolution, as the genetic yield potential of current rice varieties using conventional breeding has almost plateaued. Meeting future rice production goals has become even more challenging because of the predicted changes in climate, such as water scarcity and global warming.

Molecular Breeding for Rice Abiotic Stress Tolerance and Nutritional Quality, First Edition.
Edited by Mohammad Anwar Hossain, Lutful Hassan, Khandakar Md. Iftekharuddaula, Arvind Kumar and Robert Henry.

The Green Revolution increased the harvest index of rice with little change to the total above-ground biomass (Dehigaspitiya et al. 2019; Seneweera et al. 2011; Evans 2013; Rathnathunga et al. 2019). However, a further increase in the harvest index, may result in decreased rice yield as it comes at the expense of rice leaves; the primary source of carbon fixation. As a resolution to this phenomenon, it was suggested that the total rice biomass must be increased to achieve higher yields, which would require an increase in the photosynthetic efficiency, and/or enrichment in the light interception of the rice plant throughout the growing season. In cereals, accumulated photosynthate over the growing season and their subsequent partitioning into grains, determines the yield. Among different strategies to increase rice biomass, increasing photosynthetic efficiency merits investigation, as there appears little possibility of further improvement to light interception.

Photosynthesis is the key determinant of crop yield potential and has the possibility to increase crop yield by increasing photosynthetic CO_2 assimilation per unit leaf area (Faralli and Lawson 2020). It is well known that there is a significant natural variation of the photosynthetic pathways in angiosperms as an evolutionary adaptation to different environmental conditions. However, the intraspecific variation of photosynthesis, which is known to exist in different organs of the same plant, provides an unexploited target for crop improvement (Faralli and Lawson 2020; Rangan et al. 2016). However, determining this intraspecific variation of photosynthesis in crop plants has become challenging due to the large number of factors that influence photosynthetic rates in dynamic field conditions; for example, fluctuating temperatures and light intensities, and complex biochemistry of different plant organs (Faralli and Lawson 2020; Seneweera and Conroy 2005).

Genetic variability of a crop species is also a priority for crop improvement. As rice is a diploid species with a comparatively smaller genome compared to most other cultivated cereals, it has shown the most advances in functional genomics in the past decades (Bernardi et al. 2019). Harnessing the existing genetic diversity of rice to improve yield potential has shown a rapid development with the expansion of next-generation sequencing technologies (Subedi et al. 2019; Imaizumi et al. 1990). However, because of the narrowed genetic variability due to high selection pressure applied in rice breeding since its domestication thousands of years ago, obtaining new rice cultivars with higher yield potential has become a significant challenge for rice breeders (Bernardi et al. 2019; Moin et al. 2017). Therefore, a significant number of recent studies have focused on analyzing and improving the genetic variability of rice, which has shown substantial success (Viana et al. 2019). Of the few different approaches to improving the genetic variability of rice, inducing genetic variability through mutations has become one of the most successful tools. This is further facilitated by the small genome size of rice, which shows successful gene modifications through random and targeted mutations, providing genetically varied rice varieties for future breeding programs (Viana et al. 2019).

Rice has become one of the most vulnerable crop species to the effects of climate change that is being experienced at present and predicted in the future. Along with climate change, many socio-economic changes could also negatively affect the traditional rice system. Although there are some minor benefits of the current puddled transplanted system of rice cultivation, it is a highly water, labor, and energy-intensive system with considerable adverse effects on the environment such as methane emission (Subedi et al. 2019;

Bouwman 1991). Due to these inefficiencies of traditional rice cultivation, a successful transition from puddled to a dry, direct-seeded rice cultivation system seems a must to ensure global food security (Bouman et al. 2002; Subedi et al. 2019). As a tool for identifying significant marker-trait associations for traits enhancing rice adaptability to a direct-seeded rice system, genome-wide association studies can be utilized (Subedi et al. 2019; Perera et al. 2019). These studies can be further extended to identify important marker-traits associations for traits enhancing pest and disease tolerance, and adaptability to future atmospheric CO_2 conditions, which could be much higher than those of the present.

The Relationship Between the Rate of Photosynthesis and Yield in Rice

Increased use of nitrogen fertilizers, genetic improvement, and improved management practices since the 1950s remarkably increased the potential and realized rice yield in the world (Long et al. 2006). Yield potential is defined as the yield of a crop when grown in an adapted environment where water and nutrients are not limited, and with optimum management practices (effective pest and disease control) (Long et al. 2006). When examining the opportunities for further improvements of yield potential of rice, increasing the rate of photosynthesis of rice leaves is the only remaining trait available (Dehigaspitiya et al. 2019; Seneweera et al. 2002).

Photosynthesis is the biological process that produces organic carbon compounds using water, atmospheric carbon dioxide, and solar energy. In this process, atmospheric carbon dioxide after going through a series of reductions, is converted to sugars. With the advent of portable infrared gas analyzers (IRGAs), the rate of photosynthesis can be easily measured, and therefore, selecting crop genotypes based on the leaf photosynthetic rates is possible and has been widely used as a trait in modern plant breeding.

Initially, it was thought that the higher rates of photosynthesis may have a positive correlation with grain yield (Long et al. 2006). However, in most of the modern crop varieties, there has been little change or even declines in the rate of leaf photosynthesis per unit area compared to its progenitors to accompany the genetic increase in grain yield (Long et al. 2006; Lo et al. 2016). This lack of correlation between the photosynthetic rate of leaves and crop yield is due to the genetic difference of modern crop varieties to their progenitors. Although the rate of photosynthesis is crucial for crop yield, this is the efficiency of photosynthesis of the whole plant averaged over time, instead of the rate of photosynthesis under saturated light in a single leaf at a single stage in plant development (Long et al. 2006).

Hence, using photosynthetic data from portable IRGAs for crop breeding programs is very tricky, as the relationship between single-leaf, light-saturated and single-growth-stage measurements can be highly different from the whole crop performance over time under light-limiting conditions. For instance, approximately 50% of the plant's carbon may be assimilated under light-limiting conditions, which varies widely both physiologically and biochemically, to the plants at light-saturation (Long et al. 2006; Wang et al. 2017). Also, increase in leaf area of modern crops may often be achieved through decreased investment per unit leaf area (thin leaves). Hence, the light-saturated photosynthetic rate may be lower in crop species with thinner leaves as the photosynthetic apparatus is spread more thinly (Long et al. 2006).

Understanding the Mode of Photosynthesis and Crop Yield

Different modes of photosynthesis can be seen in higher plants and is considered an evolutionary adaptation to diverse ecosystems (Dehigaspitiya et al. 2019; von Caemmerer et al. 2012). Currently, four major photosynthetic modes have been documented in angiosperms: C_3, crassulacean acid metabolism (CAM), C_4, and C_3-C_4 intermediates (Figure 14.1). Of these the most common mode of photosynthesis is the C_3 pathway where Ribulose Bisphosphate Carboxylase/Oxygenase (Rubisco) fixes atmospheric carbon dioxide into a three-carbon compound; phosphoglycerate (Figures 14.1 and 14.2). In the C_4 pathway, the first stable compound of photosynthesis is the four-carbon molecule oxaloacetate, and C_4 is considered the most efficient photosynthetic pathway in a majority of current atmospheric conditions (Figures 14.1 and 14.2) (Rangan et al. 2016; Dehigaspitiya et al. 2019). Because of the higher photosynthetic efficiency of the C_4 pathway, bioengineering C_4 photosynthesis into photosynthetically inefficient C_3 rice, is intensely studied as a solution for the foreseen food crisis. However, due to a lack of knowledge of specific gene regulation in both photosynthetic pathways, bioengineering a complete C_4 pathway into C_3 has proven elusive (Dehigaspitiya et al. 2019; Wang et al. 2011).

Rubisco, the primary enzyme of atmospheric carbon fixation of C_3 rice, is a poor catalyst of CO_2 at current atmospheric compositions (von Caemmerer et al. 2012). It often confuses its substrate CO_2 (less abundant) with the more abundant O_2 (oxygenation activity of Rubisco). Also, Rubisco is a very slow catalyst of CO_2 and turns over only once or twice per second (von Caemmerer et al. 2012; Dehigaspitiya et al. 2019). Oxygenation activity of Rubisco catalyzes a reaction known as photorespiration, which requires recycling a toxic by-product, phosphoglycolate, resulting in a loss of previously fixed CO_2, which is an energy-costing process. It has been estimated that photorespiration can decrease the rate of photosynthesis to 30% of its potential. To overcome these inefficiencies of Rubisco, many photosynthetic organisms have developed carbon-concentrating mechanisms (CCMs), and the C_4 pathway is the ultimate result of this photosynthetic evolution (Dehigaspitiya et al. 2019; von Caemmerer et al. 2012).

The C_4 pathway has evolved from C_3 photosynthesis simply by the coordination of existing biochemical and anatomical traits that were not effectively utilized in the C_3 pathway (Christin and Osborne 2013). It is believed that the C_4 mechanism may have evolved independently up to 60 times during angiosperm evolution (Sage 2004; Dehigaspitiya et al. 2019; von Caemmerer et al. 2012). To acquire CCMs in the C_4 pathway, Rubisco must be isolated from the intercellular air spaces to avoid photorespiration. In most cases, CCM of C_4 photosynthesis is a result of distinct biochemical activities in two types of cells; bundle sheath cells (BSCs) and mesophyll cells (MCs), specifically arranged around the vascular tissue (Kranz anatomy) (Figure 14.2) (Voznesenskaya et al. 2001). During C_4 photosynthesis, atmospheric CO_2 is primarily fixed in MCs and produces oxaloacetate, a four-carbon molecule, which then diffuses into, and is decarboxylated in, BSCs. This process significantly increases the CO_2 concentration within the BSCs allowing Rubisco to perform close to its catalytic maximum (V_{max}), minimizing photorespiration (von Caemmerer and Furbank 2003; Dehigaspitiya et al. 2019). As a result of this difference between C_3 and C_4 photosynthetic pathways, C_4 crops such as maize and sorghum show higher yields and increased water and nitrogen-use efficiency in comparison to C_3 crops such as rice (von

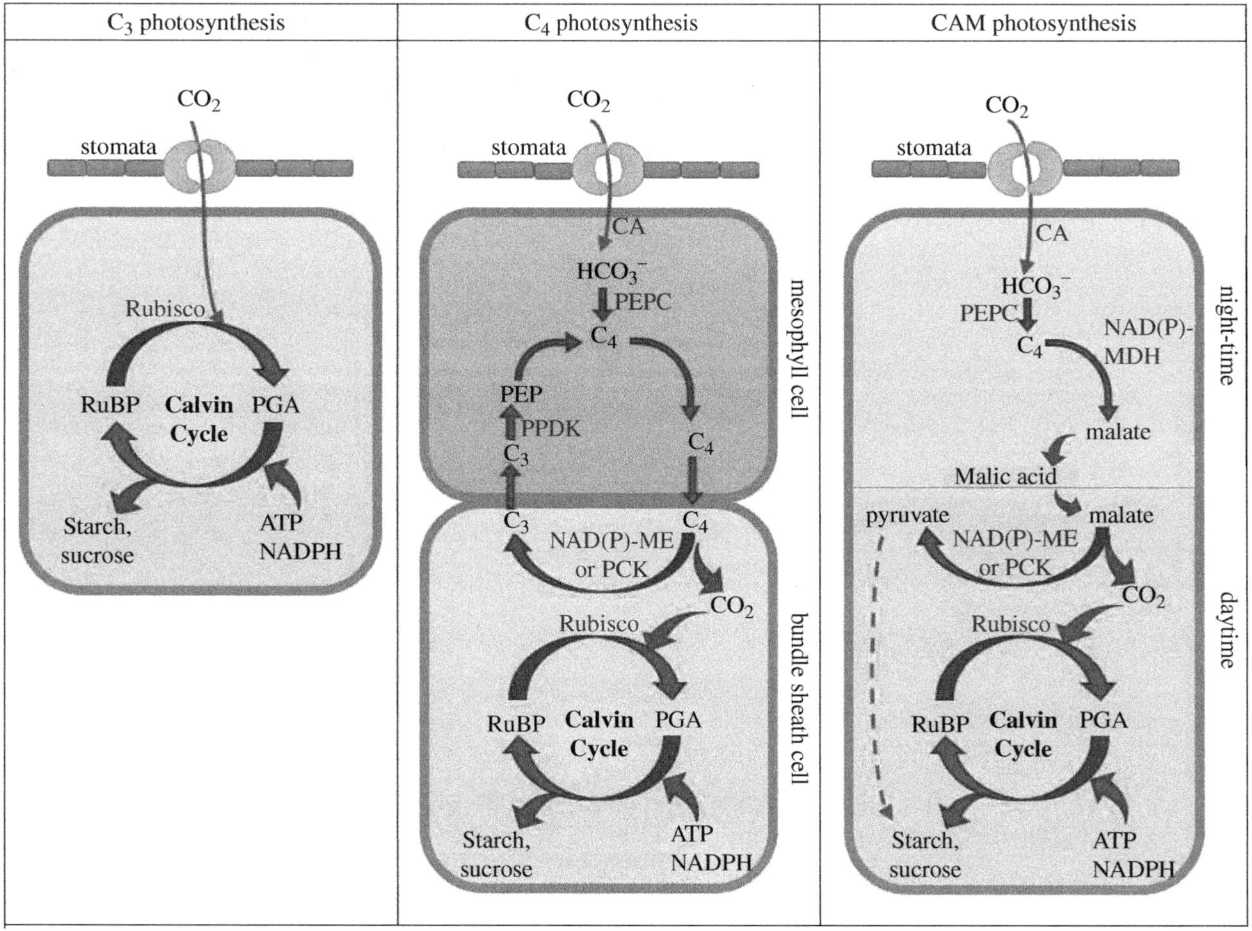

Figure 14.1 Schematic diagram of C_3 CAM and C_4 photosynthesis. Rubisco ribulose-1, 5-bisphosphate carboxylase/oxygenase, RuBP ribulose 1,5-bisphosphate, PGA phosphoglyceric acid, NADPH nicotinamide adenine dinucleotide phosphate hydrogen, HCO_{3-} bicarbonate, CA carbonic anhydrase, NADP nicotinamide adenine dinucleotide phosphate, ATP adenosine tri phosphate, NADP-MDHNADP-malate dehydrogenase, NADPME:NADP-dependent malic enzyme, PCK phosphoenolpyruvate carboxykinase, PEPC phosphorenolpyruvate carboxylase, PEP phosphorenolpyruvate, PPDK pyruvate orthophosphate dikinase.

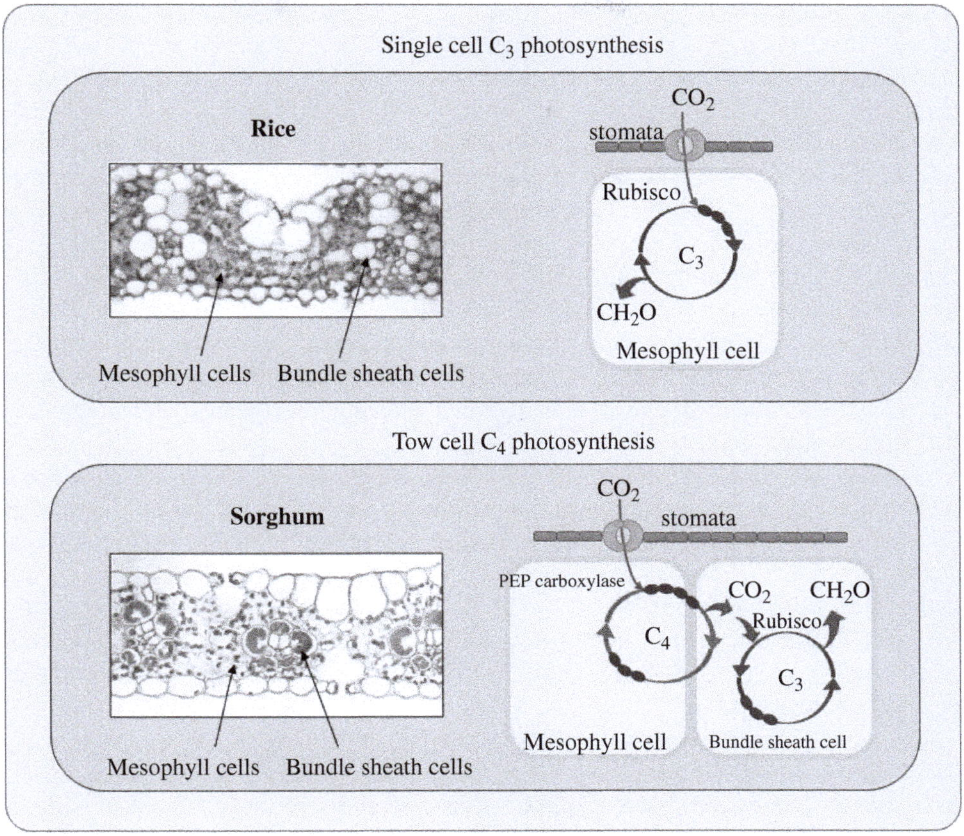

Figure 14.2 In C_3 photosynthesis, atmospheric CO_2 is primarily fixed by the enzyme Rubisco. On the other hand, in the C_4 pathway, the primary CO_2-fixing enzyme is phosphoenolpyruvate carboxylase (PEP Carboxylase). Also, in the C_4 photosynthesis, the entire photosynthetic pathway is spatially separated between mesophyll cells and the bundle sheath cells. This unique cell arrangement is termed as Kranz anatomy.

Caemmerer et al. 2012; Langdale 2011; Hibberd et al. 2008). The importance of C_4 photosynthesis to mitigate the foreseen food crisis is further illustrated by the fact that although only 3% of all plant species show C_4 photosynthesis, they contribute to 25% of global photosynthetic productivity (Dehigaspitiya et al. 2019; Sage et al. 2012).

Genetic Engineering of the C_4 Pathway into C_3 Crops

It is widely accepted that altering the photosynthetic pathway of classical C_3 cereals like rice and wheat into more efficient photosynthetic modes (C_4 photosynthesis) could improve global food production by at least 30% (Wang et al. 2017). However, current knowledge about the specific gene regulation of both C_3 and C_4 pathways is inadequate to achieve this ambitious goal (Schuler et al. 2016).

Current advances in techniques of molecular biology, like Next Generation Sequencing, have provided new insights into bioengineering CCMs of the C_4 pathway into C_3 food crops (Dehigaspitiya et al. 2019). Complete genome sequencing of genetically closely related C_3 and C_4 plants could be an important step in gaining a detailed understanding of both photosynthetic pathways (Zhang et al. 2012). The importance of transitioning from the C_3 pathway to "proto-Kranz" anatomy before bioengineering a fully functioned Kranz system, was suggested by Wang et al. (2017). The distinguishing characteristics of proto-Kranz anatomy are the comparatively large organelles in sheath cells surrounding vascular tissue of leaves (Wang et al. 2017). Showing a significant success in transforming C_3 rice into proto-Kranz, by manipulating the expression of maize *Golden2* like genes in rice, Wang et al. (2017) were able to induce chloroplast and mitochondrial development in rice vascular sheath cells, with higher organelle volumes and accumulation of photosynthetic enzymes. This is considered a crucial intermediate step towards bioengineering a complete C_4 Kranz anatomy into C_3 food crops.

With the determination of releasing a commercial rice variety that performs C_4 photosynthesis, the C_4 rice consortium has commenced many projects in the recent past (von Caemmerer et al. 2012). From the four different subtypes of C_4 photosynthesis, nicotinamide adenine dinucleotide phosphate-malic enzyme (NADP-ME) was selected as the prototypical pathway, as to that date, NADP-ME was the best characterized of the four C_4 subtypes (Kajala et al. 2011). These approaches have contributed in narrowing down the knowledge gap of C_3 and C_4 photosynthesis. However, the possibility of releasing a commercial rice variety that performs CCMs like C_4 Kranz is still uncertain.

As a novel approach of improving photosynthesis in crop plants, Dehigaspitiya et al. (2019) have discussed the importance of site-specific photosynthesis for crop improvement. According to Dehigaspitiya et al. (2019), site-specific photosynthesis is the possibility of expressing different photosynthetic properties at different sites of the same plant. For example, although rice is considered as a classical C_3 crop based on the leaf photosynthesis, Imaizumi et al. (1990) showed the possibility of having C_4-like photosynthesis in rice panicles during grain filling. Similarly, grain-specific C_4 photosynthesis was reported by Rangan et al. (2016) in C_3 wheat. The mounting evidence of site-specific photosynthesis provides a novel way of improving crop photosynthesis and thereby the total yield of critical C_3 food crops including rice and wheat.

Identifying prominent photosynthetic organs that appreciably contribute to rice yield and introduction of photosynthetically efficient traits in those organs via molecular breeding may significantly improve rice yield (Dehigaspitiya et al. 2019). This may also be much easier than altering the entire photosynthetic system of a C_3 crop into C_4 photosynthesis. Further, to utilize the site-specific photosynthesis in crop improvements, knowledge of physiological and biochemical changes related to efficient carbon fixation is increasingly important and merits investigation (Dehigaspitiya et al. 2019).

Mutation Breeding for Improvement of Rice Yield Potential

Among most cultivated cereals, rice has shown the highest advances in functional genomics in recent decades, due to its small genome in comparison to other critical cereals and its diploidy (Moin et al. 2017). Although there was significant genetic variation in pre-historic rice species, modern rice cultivars show narrow genetic variability due to repeated use of the same genotypes over centuries (Viana et al. 2019). Therefore, increasing genetic variability is identified as an essential step for breeding a high-yielding rice variety to address future food security.

Mutations can be considered as the primary source of genetic variability and can also be used as a tool for gene functional studies (Lo et al. 2016; Viana et al. 2019). Analysis of genetic mutations along with phenotypic differences, such as forward and reverse genetics, can be used to understand the function of a particular gene, which is an important step in molecular breeding for crop improvement (Lo et al. 2016). A mutation is a natural phenomenon, which could be seen in almost all living organisms and viruses. However, the rate of spontaneous mutations in higher plants is very low, ranging from 10^{-5} to 10^{-8} (Jiang and Ramachandran 2010). Therefore, mutagenesis or induced mutation is an essential and important strategy to increase genetic variability of a crop. This enables the studies of functional genomic aspects as well.

Mutations can be induced by mutagens and may lead to a gain or loss of gene function. Random mutations can occur due to DNA insertions and genotoxic agents. Direct mutation is another form of mutation, which is used to change the structure of the target DNA. There are three main tools for mutagenesis: chemical agents such as alkylating agents, physical agents such as ionizing radiation, and biological agents such as transposons, retrotransposons, and T-DNA (Jiang and Ramachandran 2010).

The primary importance of rice mutants is the possibility of understanding the function of a particular gene that can be used in breeding programs. Also, mutations increase the genetic variability of rice leading to the development of new cultivars. With the current advances in molecular biology, tools such as the CRISPR/Cas9 system can be applied to improve modern rice varieties further. Also, unveiling the genetic mechanisms of DNA damage and repair systems have facilitated an increase in mutation rates in rice (Viana et al. 2019). However, more research must be conducted in the areas of rice mutagenesis with the emphasis on improving photosynthetic efficiency as the current genetic diversity of rice for the development of a superior rice variety seems inadequate. This may also lead to the understanding of the specific gene regulation of different photosynthetic mechanisms, which will finally be directed towards the ultimate goal of bioengineering the C_4 pathway into C_3 rice.

Marker-Assisted Breeding to Improve Photosynthesis and Yield Potential

Understanding the genetic source of desired phenotypic traits of economically important cereals such as rice has been a crucial focus to improve crop productivity, yield potential, and tolerance to biotic and abiotic stresses and diseases (Brachi et al. 2011). In crops, mapping of quantitative trait loci (QTL) of bi-parental crosses was used to get an initial idea of the genetic source of phenotypic variation. However, the use of QTL mapping in extensive crop improvements has been limited due to the restricted allelic diversity and limited genomic resolution (Borevitz and Nordborg 2003).

To overcome the limitations of traditional QTL mapping, comparatively modern tools like genotyping by sequencing (GBS) have been widely used. They allow plant breeders to detect chromosome segments and marker-trait associations in QTL mapping and perform genome-wide associations (Korte and Farlow 2013). Genome-wide associations are used to eliminate several limitations of traditional gene mapping by providing much higher genomic resolution, mostly at the gene level, and providing the possibility of using samples from previously studied populations in which, the most common genetic variations can be associated with phenotypic diversity (Brachi et al. 2011).

Genome-wide association mapping has also been conveyed as an effective tool to map the complex and economically important traits in critical crops such as photosynthetic efficiency and growth responses to elevated atmospheric CO_2 concentrations, contributing to crop yield potential (Zhao et al. 2011; Gamage et al. 2018; Thilakarathne et al. 2013; Seneweera et al. 2005). Furthermore, genome-wide associations can be used in the studies of statistical marker-traits associations involving genetically varied germplasms, multi-parent breeding populations, landraces, and nested association mapping populations (Rafalski 2010). Improvement of rice yield potential can benefittby the use of association studies by accessing allelic variation and by determining the best alleles to be assembled in developing a rice variety with a much more efficient photosynthesis (Rafalski 2010).

Increase in Atmospheric CO_2 Concentration and Future Photosynthesis

Since the industrial revolution, atmospheric CO_2 concentrations have rapidly increased, rising from 300 ppm to the current 400 ppm and are predicted to continuously rise in the future (Thompson et al. 2017; Shukla et al. 2019). Parallel to that, global warming has increased due to excess emission of greenhouse gases. Increasing CO_2 concentration in the atmosphere and high temperatures significantly influence plant growth and development both positively and negatively. Much higher photosynthetic rates together with reduced stomatal conductance, have been identified as the crucial factors that stimulate plant growth, above ground biomass and yield in high-CO_2 environments (Gamage et al. 2018). However, it has been predicted that by the end of this century, the yield of staple cereals including rice and maize could decline by 20–40% due to global warming (Zhu et al. 2018). As the increasing CO_2 concentrations in the atmosphere and global warming have become global concerns when feeding the future population, breeding critical food crops for the future atmospheric conditions has become imperative.

High atmospheric CO_2 concentrations increase plant photosynthesis, mainly by minimizing the oxygenation activity of Rubisco and therefore photorespiration by increasing the rate of Rubisco carboxylation (Kanno et al. 2017). Most of the C_3 species will be facilitated by the future elevated CO_2 conditions. However, recent studies have shown that plants that perform C_4 photosynthesis show insignificant or no yield improvements under elevated atmospheric CO_2 (Long et al. 2006). Studies by Zhu et al. (2018) and Nakandalage et al. (2016) demonstrate that even though photosynthetic rates and the total yield of rice can increase at elevated CO_2 concentrations, there would be negative effects on the grain protein, micro-nutrient, and vitamin content of rice grain with potential health consequences for the poorest rice-dependent countries in the world. This has raised the importance of considering future atmospheric conditions when utilizing molecular breeding approaches to manipulate photosynthetic performance of rice for yield improvements.

Conclusion

Among various cereal crop species, rice is the third most important cereal in the world, being the primary food source for more than two billion people. To assure global food security, 60% of rice yield improvement is needed within the next few decades. As rice yield improvement by conventional methods has plateaued, feeding the future Asian population seems impossible without a significant leap in science. Of the many different approaches being studied, manipulating photosynthetic performance of rice through molecular breeding is an essential step in addressing the foreseeable risk to global food security. As the C_4 pathway is considered the most efficient photosynthetic mechanism, bioengineering a C_4-like CCM into C_3 species has been intensely studied. However, due to huge knowledge gaps in the specific gene regulation of both C_3 and C_4 photosynthetic pathways, the timeline for releasing a commercial C_4 rice variety is unclear. To close the knowledge gap of both C_3 and C_4 photosynthetic pathways, the concept of site-specific photosynthesis can be used, and it can be further extended into molecular breeding to improve photosynthesis. As the future atmosphere could be significantly different from the present, atmospheric changes such as CO_2 concentration and mean temperatures must be accounted for when planning molecular breeding programs to improve rice photosynthesis for higher yields in cereal crops.

References

Bernardi, A., Härter, C.J., Silva, A.W. et al. (2019). A meta-analysis examining lactic acid bacteria inoculants for maize silage: effects on fermentation, aerobic stability, nutritive value and livestock production. *Grass Forage Sci.* 74 (4): 596–612.

Borevitz, J.O. and Nordborg, M. (2003). The impact of genomics on the study of natural variation in Arabidopsis. *Plant Physiol.* 132 (2): 718–725.

Bouman, B., Wang, H., Yang, X. et al. (2002). Aerobic rice (Han Dao): a new way of growing rice in water-short areas. In: *Proceedings of the 12th International Soil Conservation Organization Conference*, vol. 26, 31. Beijing: Tsinghua University Press.

Bouwman, A. (1991). Agronomic aspects of wetland rice cultivation and associated methane emissions. *Biogeochemistry* 15 (2): 65–88.

Brachi, B., Morris, G.P., and Borevitz, J.O. (2011). Genome-wide association studies in plants: the missing heritability is in the field. *Genome Biol.* 12 (10): 232.

von Caemmerer, S. and Furbank, R.T. (2003). The C_4 pathway: an efficient CO_2 pump. *Photosynth. Res.* 77 (2–3): 191.

von Caemmerer, S., Quick, W.P., and Furbank, R.T. (2012). The development of C_4 rice: current progress and future challenges. *Science* 336: 6089, 1671–1672.

Christin, P.-A. and Osborne, C.P. (2013). The recurrent assembly of C_4 photosynthesis, an evolutionary tale. *Photosynth. Res.* 117 (1–3): 163–175.

Dehigaspitiya, P., Milham, P., Ash, G.J. et al. (2019). Exploring natural variation of photosynthesis in a site-specific manner: evolution, progress, and prospects. *Planta* 250: 1033–1050.

Evans, J.R. (2013). Improving photosynthesis. *Plant Physiol.* 162 (4): 1780–1793.

Faralli, M. and Lawson, T. (2020). Natural genetic variation in photosynthesis: an untapped resource to increase crop yield potential? *Plant J.* 101 (3): 518–528.

Gamage, D., Thompson, M., Sutherland, M. et al. (2018). New insights into the cellular mechanisms of plant growth at elevated atmospheric carbon dioxide concentrations. *Plant Cell Environ.* 41 (6): 1233–1246.

Hibberd, J.M., Sheehy, J.E., and Langdale, J.A. (2008). Using C_4 photosynthesis to increase the yield of rice—rationale and feasibility. *Curr. Opin. Plant Biol.* 11 (2): 228–231.

Imaizumi, N., Usuda, H., Nakamoto, H., and Ishihara, K. (1990). Changes in the rate of photosynthesis during grain filling and the enzymatic activities associated with the photosynthetic carbon metabolism in rice panicles. *Plant Cell Physiol.* 31 (6): 835–844.

Jiang, S.-Y. and Ramachandran, S. (2010). Assigning biological functions to rice genes by genome annotation, expression analysis and mutagenesis. *Biotechnol. Lett.* 32 (12): 1753–1763.

Kajala, K., Covshoff, S., Karki, S. et al. (2011). Strategies for engineering a two-celled C_4 photosynthetic pathway into rice. *J. Exp. Bot.* 62 (9): 3001–3010.

Kanno, K., Suzuki, Y., and Makino, A. (2017). A small decrease in Rubisco content by individual suppression of *RBCS* genes leads to improvement of photosynthesis and greater biomass production in rice under conditions of elevated CO_2. *Plant Cell Physiol.* 58 (3): 635–642.

Korte, A. and Farlow, A. (2013). The advantages and limitations of trait analysis with GWAS: a review. *Plant Methods* 9 (1): 29.

Langdale, J.A. (2011). C_4 cycles: past, present, and future research on C_4 photosynthesis. *Plant Cell* 23 (11): 3879–3892.

Lo, S.F., Fan, M.J., Hsing, Y.I. et al. (2016). Genetic resources offer efficient tools for rice functional genomics research. *Plant Cell Environ.* 39 (5): 998–1013.

Long, S.P., Zhu, X.G., Naidu, S.L., and Ort, D.R. (2006). Can improvement in photosynthesis increase crop yields? *Plant Cell Environ.* 29 (3): 315–330.

Moin, M., Bakshi, A., Saha, A. et al. (2017). Gain-of-function mutagenesis approaches in rice for functional genomics and improvement of crop productivity. *Brief. Funct. Genom.* 16 (4): 238–247.

Nakandalage, N., Nicolas, M., Norton, R.M. et al. (2016). Improving rice zinc biofortification success rates through genetic and crop management approaches in a changing environment. *Front. Plant Sci.* 7: 764.

Perera, I., Fukushima, A., Arai, M. et al. (2019). Identification of low phytic acid and high Zn bioavailable rice (*Oryza sativa* L.) from 69 accessions of the world rice core collection. *J. Cereal Sci.* 85: 206–213.

Prosekov, A.Y. and Ivanova, S. (2016). Providing food security in the existing tendencies of population growth and political and economic instability in the world. *Foods Raw Mater.* 4 (2): 202–211.

Rafalski, J.A. (2010). Association genetics in crop improvement. *Curr. Opin. Plant Biol.* 13 (2): 174–180.

Rangan, P., Furtado, A., and Henry, R.J. (2016). New evidence for grain specific C_4 photosynthesis in wheat. *Sci. Rep.* 6: 31721.

Rathnathunga, E.U.U., Senanayake, G., Seneweera, S., and Geekiyanage, S. (2019). Varietal diversity of Sri Lankan traditional rice based on sensitivity to temperature and photoperiod at vegetative stage. *J. Natl. Sci. Found.* 47 (1): 51–68.

Sage, R.F. (2004). The evolution of C_4 photosynthesis. *New Phytol.* 161 (2): 341–370.

Sage, R.F., Sage, T.L., and Kocacinar, F. (2012). Photorespiration and the evolution of C_4 photosynthesis. *Annu. Rev. Plant Biol.* 63: 19–47.

Schuler, M.L., Mantegazza, O., and Weber, A.P. (2016). Engineering C_4 photosynthesis into C_3 chassis in the synthetic biology age. *Plant J.* 87 (1): 51–65.

Seneweera, S.P. and Conroy, J.P. (2005). Enhanced leaf elongation rates of wheat at elevated CO_2: is it related to carbon and nitrogen dynamics within the growing leaf blade? *Environ. Exp. Bot.* 54 (2): 174–181.

Seneweera, S.P., Conroy, J.P., Ishimaru, K. et al. (2002). Changes in source–sink relations during development influence photosynthetic acclimation of rice to free air CO_2 enrichment (FACE). *Funct. Plant Biol.* 29 (8): 945–953.

Seneweera, S., Makino, A., Mae, T., and Basra, A.S. (2005). Response of rice to p (CO_2) enrichment: the relationship between photosynthesis and nitrogen metabolism. *J. Crop Improv.* 13 (1–2): 31–53.

Seneweera, S., Makino, A., Hirotsu, N. et al. (2011). New insight into photosynthetic acclimation to elevated CO_2: the role of leaf nitrogen and ribulose-1, 5-bisphosphate carboxylase/oxygenase content in rice leaves. *Environ. Exp. Bot.* 71 (2): 128–136.

Shukla, P., Skeg, J., Buendia, E.C., Masson-Delmotte, V., Pörtner, H.-O., Roberts, D., Zhai, P., Slade, R., Connors, S., and van Diemen, S. (2019). Climate Change and Land: An IPCC Special Report on Climate Change, Desertification, Land Degradation, Sustainable Land Management, Food Security, and Greenhouse Gas Fluxes in Terrestrial Ecosystems.

Subedi, S.R., Sandhu, N., Singh, V.K. et al. (2019). Genome-wide association study reveals significant genomic regions for improving yield, adaptability of rice under dry direct seeded cultivation condition. *BMC Genomics* 20 (1): 471.

Thilakarathne, C.L., Tausz-Posch, S., Cane, K. et al. (2013). Intraspecific variation in growth and yield response to elevated CO_2 in wheat depends on the differences of leaf mass per unit area. *Funct. Plant Biol.* 40 (2): 185–194.

Thompson, M., Gamage, D., Hirotsu, N. et al. (2017). Effects of elevated carbon dioxide on photosynthesis and carbon partitioning: a perspective on root sugar sensing and hormonal crosstalk. *Front. Physiol.* 8: 578.

Viana, V.E., Pegoraro, C., Busanello, C., and de Oliveira, A.C. (2019). Mutagenesis in rice: the basis for breeding a new super plant. *Front. Plant Sci.* 10: 1326.

Voznesenskaya, E.V., Franceschi, V.R., Kiirats, O. et al. (2001). Kranz anatomy is not essential for terrestrial C_4 plant photosynthesis. *Nature* 414 (6863): 543–546.

Wang, L., Peterson, R.B., and Brutnell, T.P. (2011). Regulatory mechanisms underlying C_4 photosynthesis. *New Phytol.* 190 (1): 9–20.

Wang, S., Tholen, D., and Zhu, X.G. (2017). C_4 photosynthesis in C_3 rice: a theoretical analysis of biochemical and anatomical factors. *Plant Cell Environ.* 40 (1): 80–94.

Xu, J., Henry, A., and Sreenivasulu, N. (2020). Rice yield formation under high day and night temperatures—a prerequisite to ensure future food security. *Plant Cell Environ* 43 (7): 1595–1608.

Zhang, S.-B., Guan, Z.-J., Sun, M. et al. (2012). Evolutionary association of stomatal traits with leaf vein density in Paphiopedilum, Orchidaceae. *PLoS One* 7 (6): e40080.

Zhao, K., Tung, C.-W., Eizenga, G.C. et al. (2011). Genome-wide association mapping reveals a rich genetic architecture of complex traits in *Oryza sativa*. *Nat. Commun.* 2 (1): 1–10.

Zhu, C., Kobayashi, K., Loladze, I. et al. (2018). Carbon dioxide (CO_2) levels this century will alter the protein, micronutrients, and vitamin content of rice grains with potential health consequences for the poorest rice-dependent countries. *Sci. Adv.* 4 (5): eaaq1012.

15

Molecular Breeding for Improved β-carotene Synthesis in Golden Rice: Recent Progress and Future Perspectives

Swapan K. Datta, Shuvobrata Majumder and Karabi Datta

Laboratory of Translational Research on Transgenic Crops, University of Calcutta, Kolkata 700019, India

Introduction

Golden rice (GR) is one of the "top 10 Biotech" projects in the world, according to the list made by the Project Management Institute, USA, and it is the only plant-based influential Biotech project (https://mip.pmi.org/biotech-top-ten, access on October 10, 2019). According to WHO, half of the world's children suffering from vitamin A deficiency (VAD) live in Southeast Asia and Sub-Saharan Africa regions (Ghosh et al. 2019). Half of the world's population depend on rice; therefore GR could be a sustainable solution to supply adequate amount of carotenoids (provitamin-A) to biofortify their diet to address VAD. Dietary carotenoids have various health benefits. The study shows that a number of carotenoids help in human health, antioxidant properties cell differentiation, synthesis of glycoprotein, growth and development of bones, decreasing the risk of eye disease, cancer and nutritional benefits, like the β-carotene, lycopene, lutein, and zeaxanthin (Ghosh et al. 2019; Johnson 2002). Carotenoids are not present in rice grain, though it is present in other parts of the rice plant. Specifically, β-carotene is beneficial as it can convert to vitamin A, which is also not present in rice grains. GR is an example of successful metabolic engineering of seed-specific carotenoid biosynthesis pathway in rice.

After 20 years of GR development, dismayingly, it is still not available in the market for human consumption because of different issues that are mostly based on regulatory and non-scientific issues including anti-GM protests. Recently some positive movements towards its availability have been observed through international agencies and countries like Bangladesh, Australia, Canada, and the Philippines. Here, we have discussed the current scenario of worldwide GR research since its development, including seed quality, metabolomics, and successful field trials.

The β-carotene Synthesis Pathway in Plant

Plants produce provitamin A carotenoids – α-carotene, β-carotene, γ-carotene, and β-cryptoxanthin. First two of them (α-carotene and β-carotene) accumulate in plants, and

Molecular Breeding for Rice Abiotic Stress Tolerance and Nutritional Quality, First Edition.
Edited by Mohammad Anwar Hossain, Lutful Hassan, Khandakar Md. Iftekharuddaula,
Arvind Kumar and Robert Henry.

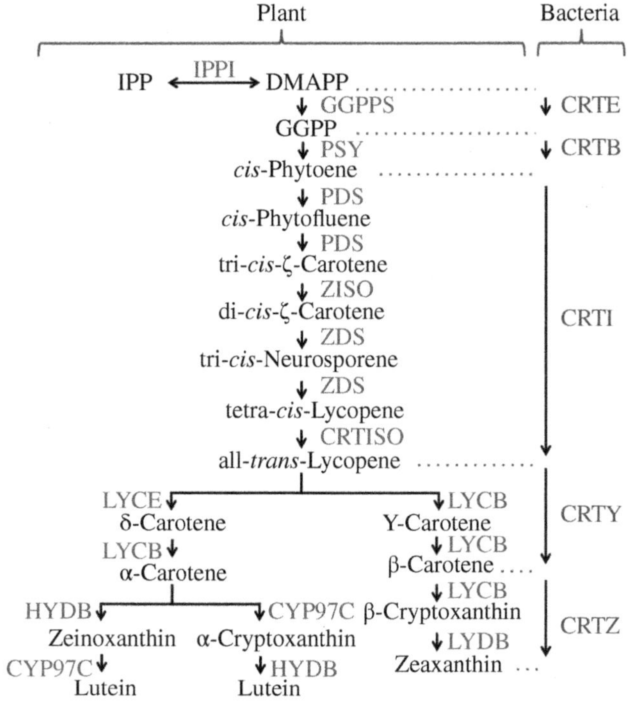

Figure 15.1 Carotenoid biosynthesis pathway in plants with equivalent steps in bacteria. *Source:* Adapted from Bai et al. (2011) and Ghosh et al. (2019). *Note:* Enzymes involved in plant are in red color and in bacteria are marked in blue color. *IPPI:* isopentenyl diphosphate isomerase; *DMAPP*: dimethylallyl diphosphate; *IPP*: isopentenyl diphosphate; *GGPP*: geranylgeranyl diphosphate; *GGPPS:* GGPP synthase; *PSY:* phytoene synthase; *PDS:* phytoene desaturase; *ZISO:* ζ-carotene isomerase; *ZDS:* ζ-carotene desaturase; *CRTISO:* carotenoid isomerase; *LYCB:* lycopene β-cyclase; *LYCE:* lycopene ε-cyclase; *CYP97C:* carotene ε-ring hydroxylase; *HYDB:* β-carotene hydroxylase; *CRTE*: bacterial GGPP synthase; *CRTB*: bacterial phytoene synthase; *CRTI*: bacterial phytoene desaturase/isomerase; *CRTY:* bacterial lycopene cyclase; *CRTZ:* bacterial β-carotene hydroxylase

others (γ-carotene and β-cryptoxanthin) are converted into downstream products (Farre et al. 2010; Zhu et al. 2010). In plants, the carotenoid biosynthesis pathway is a multistep process and is completed through multiple enzymes, but in bacteria, fewer enzymes are required to complete this process. Here we have highlighted these key enzymes of carotenoid biosynthesis pathway in plants and bacteria in Figure 15.1.

In plants, through a condensation reaction catalyzed by the geranylgeranyl diphosphate synthase (GGPPS), three IPP (isopentenyl diphosphate) molecules with one molecule of DMAPP (dimethylallyl diphosphate) produce the geranylgeranyl diphosphate (GGPP). IPP and DMAPP are derived predominantly from the plastidial methylerythritol 4-phosphate (MEP) pathway and partially formed by the cytosolic mevalonic acid pathway (Rodriguez-Concepcion and Boronat 2002). In bacteria, GGPPS is encoded by the *crt*E gene (Figure 15.1). Synthesis of carotenoids begins from the GGPP in plants. In a condensation process, two GGPP molecules are converted into 15-cis-phytoene by the enzyme phytoene synthase (PSY) (Misawa et al. 1994). The equivalent enzyme of plant PSY is CrtB in bacteria (Bai et al. 2011).

In plants, four desaturation reactions are carried out to produce all-trans-lycopene from 15-cis-phytoene by phytoene desaturase (PDS), ζ-carotene isomerase (Z-ISO), ζ-carotene desaturase (ZDS), and carotenoid isomerase (CRTISO) (Li et al. 2007; Chen et al. 2010). In bacteria, a single enzyme encoded by the *crt*I gene accomplishes all the above four reaction steps to produce lycopene (Bai et al. 2011) (Figure 15.1). In the carotenoid pathway, lycopene is an important branch point because it functions as the substrate for two competing enzymes, lycopene β-cyclase (LYCB), and lycopene ε-cyclase (LYCE) (Cunningham et al. 1996). In one pathway lycopene is converted into γ-carotene, which is rapidly converted into β-carotene (has pro-vitamin A activity) by LYCB. The bacterial gene *crt*Y encodes LYCB to produce β-carotene. Alternatively, lycopene generates δ-carotene (which has no pro-vitamin A activity) by LYCE (Bai et al. 2011). In subsequent steps, α-carotene produces lutein, and β-carotene produces zeaxanthin (through intermediate β-cryptoxanthin), which can be converted into more forms of carotenoids, but they have not showed pro-vitamin A activity (Tian et al. 2003). The bacterial gene *crt*Z encodes β-carotene hydroxylase (BCH) (Figure 15.1). Ultimately, zeaxanthin enters into the xanthophyll cycle and gets converted into abscisic acid, one of the important plant hormones (Seo and Koshiba 2002).

Metabolic Engineering of β-carotene Pathway in Rice

Metabolic engineering has played an important role to express β-carotene in rice grain. In rice grain (endosperm) phytoene synthase (*psy*) is unable to express, and carotenoid biosynthesis is blocked at the first enzymatic step (Figure 15.1). Expression of carotenoids in endosperm has been achieved through endosperm specific promoters and the use of bacterial carotenoid biosynthesis genes.

Endosperm-Specific Expression of β-carotene

To express carotenoids or any other proteins in rice endosperm, promoter sequences of seed storage proteins (SSPs), namely, albumins, globulins, prolamins, and glutelins are always a prime choice for genetic engineering. Glutelin is the major SSP in rice. In rice, about 60–80% by weight of total seed protein is glutelin, and about 20–30% are prolamin (Kawakatsu et al. 2008). Expression of genes under rice glutelin promoter has been elaborately studied as compared to prolamin, globulin, and albumin promoters. The order of promoter activity and endosperm specific expression efficiency was in the decreasing order of glutelin, prolamin, globulin, and albumin genes (Wu et al. 1998). Six promoters of rice glutelin genes (*GluA-1*, *GluA-2*, *GluA-3*, *GluB-3*, *GluB-5*, and *GluC*) were isolated and examined in rice (Qu et al. 2008). Study revealed that *GluA-1*, *GluA-2*, and *GluA-3* promoters directed transgene expression in the peripheral region of the endosperm whereas the *GluB-5* and *GluC* promoters were able to express in the whole endosperm. The *GluB-3* promoter expressed solely in aleurone and subaleurone layers of rice grain. These findings could be helpful for future choices of promoters for endosperm-specific gene expression in rice. After successfully rice genome sequencing, many new potential promoter sequences have been identified. A rice genome-wide analysis identified (through microarray and gene ontology analyses) 151 endosperm-specific genes and their promoters; they could be useful in future for expression of carotenoids (Nie et al. 2013). PROLAM26 (LOC_Os07g10580), RAL2 (LOC_Os07g11330), RAL4 (LOC_Os07g11380), and CAPIP (LOC_Os06g33640) are also newly reported promoters of such categories and could be useful in the future (Xu et al. 2017). So far rice *Glu-1* promoter has been extensively used for carotenoid expression in rice (Table 15.1).

Source of β-carotene Synthesis Genes

Plants and bacteria are the source of β-carotene synthesis genes that have been used for metabolic engineering. Daffodils (*Narcissus pseudonarcissus*) phytoene synthase (*psy*) gene has been extensively used since Burkhardt et al. (1997) demonstrated its ability to produce carotenoid-specific intermediate phytoene in rice endosperm. Another extensively used gene is bacterial (*Erwinia uredovora*) PDS (*crt*I), used for carotenoid metabolic engineering. A combination of these two genes including daffodils (*N. pseudonarcissus*) lycopene *β* -cyclase (*lcy*) was tried for the first time to develop β-carotene expression in *japonica* rice (mentioned here as GR1) (Ye et al. 2000). In GR1, Gtu-1 promoter was used for *psy* and *lcy* and a constitutive CaMV35S promoter with a plastid-specific transit peptide (TP) used for the expression of *Crt*I gene (Table 15.1). A combination of daffodils *psy* with bacterial *crt*I genes has been tested multiple times for metabolic engineering of β-carotene pathway in *indica* rice endosperm (Krishnan et al. 2003; Datta et al. 2003, 2006; Hoa et al. 2003; Parkhi et al. 2005; Rai et al. 2007). In a combination of maize phytoene synthase (*Zmpsy*) gene with bacterial *crt*I has been expressed under Gtu-1 promoter in rice to develop an improved golden rice variety (mentioned here as GR2) (Paine et al. 2005).

A different combination of genes was tested by Bai et al. (2016) where they tried two combinations (Table 15.1). In a combination, maize *Zmpsy* and bacterial *Pantoea ananatis crt*I (*Pacrt*I) genes with the *Arabidopsis thaliana* genes *AtDXS* (encoding 1-deoxy-D-xylulose 5-phosphate synthase, which supplies metabolic precursors) was introduced in rice and in another combination *Zmpsy, Pacrt*I with *AtOR* (the ORANGE gene, which promotes the formation of a metabolic sink) was introduced to develop golden rice.

Synthetic gene constructs of two carotenoid biosynthetic genes, *psy* from Capsicum (*Capsicum annuum*)and *crt*I from Pantoea were also reported for golden rice development. Two genes were linked via either the synthetic codon-optimized 2A sequence (from foot-and-mouth disease virus) or the IRES sequence (the internal ribosome entry site) to generate PAC (Psy-2A-CrtI) and PIC (Psy-IRES-CrtI) constructs, respectively (Ha et al. 2010, 2019).

Rice Genetic Transformation System

For development of golden rice *Agrobacterium*-mediated transformation has been mostly used (Ye et al. 2000; Datta et al. 2003; Paine et al. 2005; Rai et al. 2007), although biolistic transformation was also attempted (Bai et al. 2016) for golden rice development. Details of rice transformation methods, plant regeneration techniques, promising transgenic events selection, molecular, and biochemical characterization events were described earlier (Datta et al. 1990a, 2003; Datta and Bouis 2000; Vasconcelos et al. 2003; Parkhi et al. 2005; Rai et al. 2003). The majority of gene constructs used for golden rice development have marker genes. Marker genes—specifically antibiotic-resistant genes such as hygromycin-B

Table 15.1 Selected breakthrough reports indicating genes and promoters related to golden rice development.

Genes involved	Promoter involved	Rice variety	Findings	References
Daffodil (*Narcissus pseudonarcissus*) phytoene synthase (*psy*) gene	Seed-specific glutelin promoter (Gt1 P)	Taipei 309	Formation of carotenoid-specific intermediate phytoene	Burkhardt et al. (1997)
Daffodil *psy* and lycopene *β* –cyclase (*lcy*) and bacterial (*Erwinia uredovora*) phytoene desaturase (*crt*I)	Gt1 P for *psy* and CaMV35S+TP (plastid-specific transit peptide) for *Crt*I	Taipei 309	Total carotenoids-1.6 $\mu g\ g^{-1}$ (GR1)	Ye et al. (2000)
Daffodil *psy* with bacterial *crt*I and *lcy*	Gt1 P for *psy* and CaMV35S+TP for *Crt*I	IR64, BR29, Nang Hong Cho Dao and Mot Bui	Total carotenoids-1.05 $\mu g\ g^{-1}$	Datta et al. (2003)
Daffodil *psy* and bacterial *crt*I	Gt1 P for *psy* and CaMV35S+TP for *Crt*I	Taipei 309, IR64 and MTL 250	Total carotenoids-1.2 $\mu g\ g^{-1}$ (Taipei 309) and 0.8 $\mu g\ g^{-1}$ (IR64)	Hoa et al. (2003)
Maize phytoene synthase (*Zmpsy*) and bacterial *crt*I	Gt1 P for *psy* and *Crt*I (with TP)	Asanohikari	Total carotenoids-37 $\mu g\ g^{-1}$ (GR2)	Paine et al. (2005)
Introgression line with *psy* and *crt*I and *lcy*	Gt1 P for *psy* and CaMV35S+TP for *Crt*I	IR64	Total carotenoids-1.06 $\mu g\ g^{-1}$	Baisakh et al. (2006)
Daffodil *psy* and bacterial *crt*I	Gt1 P for *psy* and CaMV35S+TP for *Crt*I	IR64, BR29	Total carotenoids-9.34 $\mu g\ g^{-1}$ (BR29) and 2.32 $\mu g\ g^{-1}$ (IR64)	Datta et al. (2006)
Daffodil *psy* and bacterial *crt*I	Gt1 P for *psy* and CaMV35S+TP for *Crt*I	BR29	Total carotenoids-6.77 $\mu g\ g^{-1}$	Rai et al. (2007)
Anther culture derived line with Daffodil *psy* and bacterial *crt*I	Gt1 P for *psy* and CaMV35S+TP for *Crt*I	BR29	Total carotenoids-3.188 $\mu g\ g^{-1}$	Datta et al. (2014)
Co-expressed *AtDXS*, *ZmPSY*1 and *PaCRT*I	Gt1 P	EYI105	Total carotenoids-31.78 $\mu g\ g^{-1}$	Bai et al. (2016)
Co-expressed *AtOR*, *ZmPSY*1 and *PaCRT*I	Gt1 P	EYI105	Total carotenoids-25.83 $\mu g\ g^{-1}$	

phosphotransferase (*hpt* gene) against hygromycin-B, aminoglycoside phosphotransferase (*nptII* gene) against kanamycin, neomycin, paromomycin, 3″(9)-O-aminoglycoside adenylyltransferase enzyme (*aad* gene) against spectinomycin, and streptomycin and beta lactamase enzyme (*bla* gene) against ampicillin—are commonly used in rice genetic modification (Majumder et al. 2019). Such marker genes are not desirable in golden rice due to the regulatory process and consumer concerns. Parkhi et al. (2005) and Baisakh et al. (2006) demonstrated how a marker gene can be successfully removed from GR through a segregation and selection process, irrespective of the marker gene used for genetic transformation. This "marker free" process was possible as the gene of interest and the marker gene were present in different loci of a chromosome or on different chromosomes of transgenic plants (Parkhi et al. 2005). Datta et al. (2003) applied a non-antibiotic selection system (Positech) for developing GR.

Development of Golden Rice (GR)

After the development of GR1 at the Swiss Federal Institute of Technology, Zurich, Switzerland in *japonica* rice variety Taipei 309 (Ye et al. 2000), the next year in 2001 the developers assigned their patents to Syngenta AG for further modification of the technology (Figure 15.2). A deal was made with Syngenta AG for assisting the inventors' humanitarian and altruistic objectives of distribution of golden rice technology under the inventors' license. The inventors licensed a network of Asian government-owned research institutea under the condition that there would be no charge for the nutritional technology and it would only be introduced to publicly owned rice varieties (Dubock 2019). At the International Rice Research Institute (IRRI), *indica* golden rice varieties were also developed, namely R64, IR68144 (IRRI), BR29, BR28, (Bangladesh), Mot Bui, Nang Huong Cho Dao (NHCD) (Myanmar), and Swarna (*indica*) rice varieties (Datta et al. 2003; Datta et al. 2007). Later in 2005 improvements were made to the nutritional technology by Syngenta AG scientists as they have the only authority for any modification of nutritional technology (Paine et al. 2005), and total carotenoids were increased to 37 $\mu g\,g^{-1}$ in improved GR2. The major differences between GR1 and GR2 are the gene source of phytoene synthase (*psy*) and rice variety used. In GR2 a maize *psy* gene and Asanohikari (*japonica*) variety were used, whereas in the case of GR1 development daffodils *psy* gene and Taipei 309 rice were used (Table 15.1). Syngenta AG provided six selected events (G1, R1, L1, T1, W1, and E1) of the GR2 series to the Golden Rice Humanitarian Board (HumBo) for use as donor parents in rice-breeding applications (Bollinedi et al. 2017). It is assumed that golden rice research is a not-for-profit project and has been funded by the public sector of different countries (such as India, Bangladesh, China, Indonesia, the Philippines, and Vietnam) and philanthropy funding of Rockefeller and Bill & Melinda Gates Foundations.

Many other attempts have been made to improve the carotenoid content in rice specifically provitamin β-carotene in rice grains (Table 15.1). To date the highest expressed amount of total carotenoids in rice seed is 37 $\mu g\,g^{-1}$ (Paine et al. 2005), but nutritional technology has been improved considerably. Different combinations of genes have been applied to develop GR in different rice cultivars; a selected few are listed in Table 15.1. Technological development for better storage of golden rice, nutritional quality, rapid homozygous line development, details metabolic studies of GR, food safety studies of GR, comparative

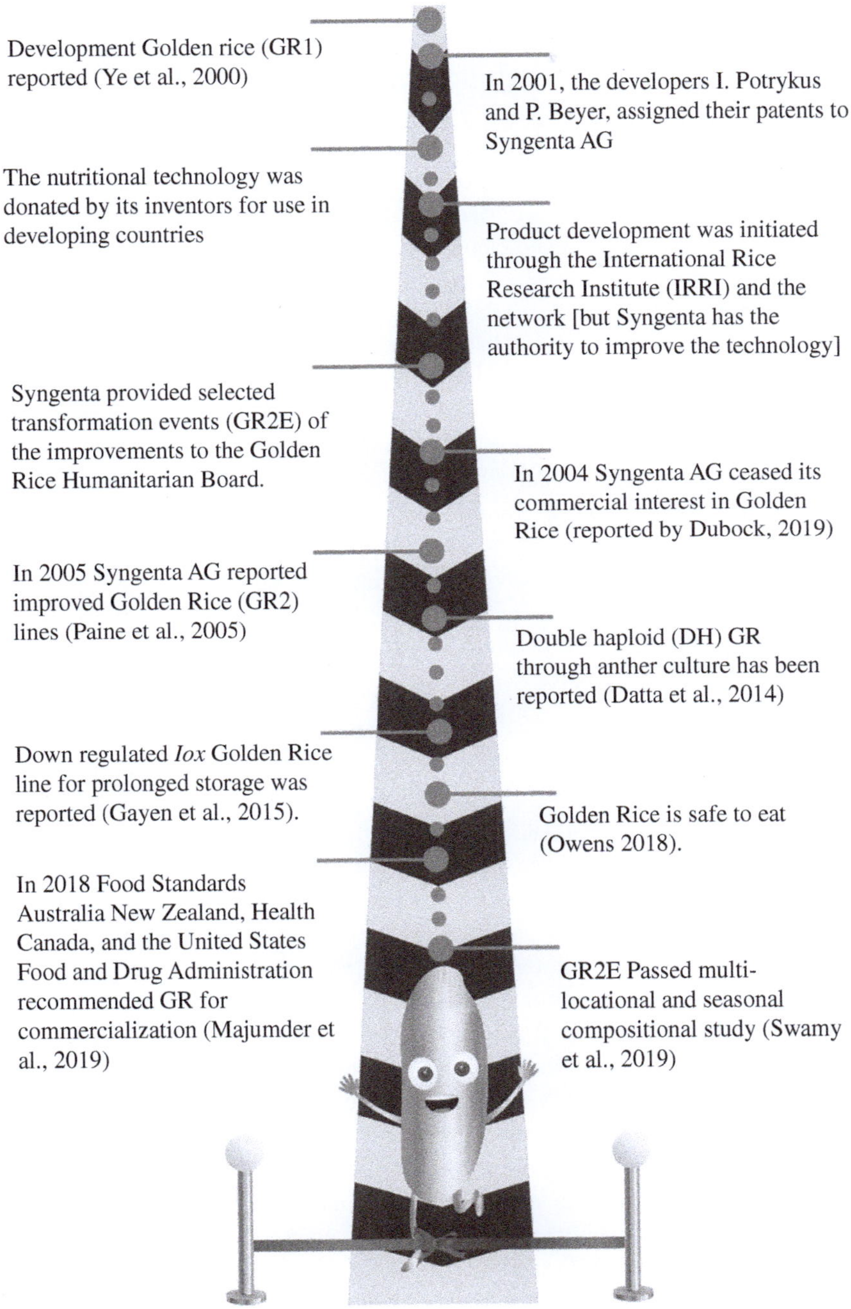

Figure 15.2 After going through a rigorous standard of regulations golden rice is very close to the finish line and almost ready to reach the people who need it most.

analysis of nutritional qualities with other rice, successful application of GR in breeding events are discussed here in the following sections.

Homozygous Dihaploid GR Development

Development of homozygous GR lines requires propagation of multiple generations (at least three generations) depending on the number of transgene inserts (copy number of transgene) and position of the inserted genes in rice chromosome. The dihaploid/double haploid (DH) homozygosity approach is an efficient way of procuring homozygous rice plants. This was achievable by two ways: firstly, GR plants were developed by molecular breeding; then after successful anther or pollen culture from those plants, further multiplication was initiated (Datta et al. 2014). Secondly, anther culture-derived dihaploid (pollen-derived) embryogenic cell lines were used as explant for transformation for GR development (Datta et al. 1990a). Successful rice anther culture depends on several factors, such as genotypic variations, media composition, condition of culture room, and the handling of the cultures, as discussed in Datta and Wenzel (1987), Datta et al. (1990b), and Datta (2005). The authors of this chapter (Swapan K. Datta and Dr. Karabi Datta) have successfully applied this technology to achieve dihaploid GR from the anther cultures of primary GR (BR29) plants (Datta et al. 2014). During development of such GR plants haploid and tetraploid plants were also generated from anther cultures. The High Performance Liquid Chromatography (HPLC) analysis confirmed the events of DH transgenic lines from levels of carotenoids (including β-carotene) in golden rice plants (Figure 15.3).

GR in Introgression Breeding Programs

After public availability of six selected GR2 events through the Golden Rice Humanitarian Board, many breeding programs have been initiated to introgress seed-specific carotenoids synthesis genes in other rice cultivars. In IRRI, Philippines, many attempts were made for the developing BR29, IR64, Swarna GR from the GR2 lines. Few such introgressed GR lines (BR29) have been successfully grown in the field of Bangladesh.

In India, under ICAR-Indian Agricultural Research Institute, New Delhi, similar type of introgression breeding program for Swarna using GR2 lines (event R1) was initiated. Resulting homozygous Swarna GR lines showed poor phenotypes as compared to the recurrent parent Swarna. Introgressed Swarna GR lines were dwarf with pale green leaves, with a drastic decrease in panicle size, grain number, and yield (Bollinedi et al. 2017). Detailed molecular analyses showed that the transgene disrupted a native gene *OsAux1*, resulting in disbalance of plant growth regulators, namely, auxins, gibberellic acid, and abscisic acid. This disruption resulted in manifestation of abnormalities in plant growth and homozygous line development (Bollinedi et al. 2017). Under introgression breeding programs of GR, the development of promising introgressed lines based on plant phenotype and yield, is truly a challenging task. A few breeding programs have been suddenly stopped without disclosing their problems (plant phenotype, carotenoids expression), but quite a few programs have succeeded as well.

Gene stability of the GR events, developed under the introgression breeding program, is an issue due to the different genomic backgrounds. GR1 and GR2 were developed based

Figure 15.3 Development of *Indica* golden rice varieties. *Note:* (a) Team of golden rice researchers (from left to right) Karabi Datta, Swapan K. Datta, Ingo Potrykus and Peter Beyer; Source: Swapan K. Datta (b) A representative photo of DH golden rice (BR29) and its control counterpart (Datta et al. 2007). Source: Datta et al. 2007 © Springer Nature (c) The HPLC estimation shows presence of β-carotene in DH golden rice grains. (Datta et al. 2007). (d) Growing of different *indica* golden rice homozygous lines (for multiple generations) at greenhouse facility of University of Calcutta, Kolkata, India. Source: Swapan K. Datta.

on the *japonica* rice cultivars Taipei 309 and Asanohikari respectively. Later these *japonica* cultivars were also used as parents for developing introgressed *indica* rice cultivars like Swarna, IR64, BR29 under different breeding programs. A study indicated that the intact and/or rearranged copies of the transgenes are clustered in a limited part of the introgressed rice genome, and these changes are very hard to find through random amplification of polymorphic DNA (RAPD) (Rai et al. 2003). However, anther-culture-derived homozygous lines showed better stability of gene and expression of the carotenoids in the progeny and hence were selected for further breeding programs (Datta et al. 2007).

Phenotypic and Agronomic Performance of Golden Rice

Convincing a farmer to cultivate GR only because of its nutritional quality of grain is very difficult. Similar or greater yield performance of GR as compared to popular (control) rice is one of the desirable characters that would assist in its acceptance. Breeders at IRRI, Philippines, Department of Agriculture – Philippine Rice Research Institute (DA-PhilRice); the Bangladesh Rice Research Institute (BRRI); and the Indonesian Center for Rice Research

(ICRR) are developing GR versions from existing popular rice varieties that would retain the same yield along with being pest resistant (https://www.irri.org/golden-rice, access on November 20, 2019).

No adverse or unpleasant observations on plant phenotype or agronomically important parameters (plant height, number of panicles, seeds per panicles, days to flowering, seed set, 100-seed weight, etc.) were reported during GR1 and GR2 development or in case of any other GR events (based on IR64, BR29). In most of the cases, except for a few introgressed GR lines (Bollinedi et al. 2017), GR was found to be at par with its control counterpart based on plant phenotype or agronomically important parameters. However, unlike the detailed scientific report by Bollinedi et al. (2017), reports on field evaluation of introgressed lines in Bangladesh or elsewhere are not yet available.

DH Golden Rice plants (BR29, IR64) grown along with control in the containment greenhouse were found similar and without any phenotypic cost, and showed no segregation in yellow color-seed-containing carotenoids (Datta et al. 2007). Phenotypic and agronomic performance studies on GR by many renowned international institutes including IRRI inferred satisfactory outcome in majority of the cases.

Recent Progress of GR Research

Improvement of GR nutritional technology, its nutritional quality, long-term grain storage, other nutritional analyses are topics of immense interest for the worldwide rice research community. In the following sections selected findings have been reviewed.

Better Long-Term Storage of GR (Silencing of *lox*)

In long-term storage of GR its carotenoids succumb to degradation, and seed quality deteriorates. Carotenoids degradation is mainly caused by the activity of the lipoxygenase (LOX) enzyme. The rice genome contains 14 types of LOX protein-encoding genes; the *r9-LOX1* gene is one of them and is mainly responsible for seed quality deterioration (Gayen et al. 2014). The LOX enzyme in the GR grains catalyzes the insertion of molecular oxygen into polyunsaturated fatty acids (PUFA) and yields conjugated hydroperoxide products (Carrera et al. 2007). These products cause oxidation of carotenoids and deterioration of seed nutritional quality.

RNA interference (RNAi)-mediated down-regulation of *r9-LOX1* gene in GR is a proven approach (Gayen et al. 2014). RNAi-mediated silencing of this gene, under the control of *Oleosin-18* promoter, improved the storage stability and viability of GR seeds (Gayen et al. 2015). This technological improvement of GR could be useful for long-term storage of GR seeds in future.

Substantial Equivalence Analysis of GR

Multiple detailed documentation on substantial equivalence analysis of GR with its control is available. Recently a seed compositional study was reported on GR2E from the Philippines. About 62 compositional parameters for key nutritional components of minerals, vitamins, and antinutrients, proximates, fiber, polysaccharides, fatty acids, and amino acids were analyzed, and no statistically significant differences were found between GR2E and its control (Swamy et al. 2019).

Metabolic Analysis of GR

We have completed a proteome and metabolite based metabolic analysis of our developed GR BR29 event (event name SKBR-244). In SKBR-244 seeds we observed an alteration in carbohydrate metabolism pathways. In 2D gel-electrophoresis-based analysis, up-regulation of carbohydrate metabolism related enzymes like pullulanase, uridine diphosphate (UDP)-glucose pyrophosphorylase, and glucose-1-phosphate adenylyltransferase have been observed (Gayen et al. 2016). In GR, the enzyme- pyruvate phosphate dikinase (PPDK) was also up-regulated (enhances pyruvate biosynthesis), which serves as a precursor of carotenoids pathways. Gas chromatography–mass spectrometry (GC-MS) based metabolic profiling of SKBR-244 shows an increase in the levels of glyceric acid, fructo-furanose, galactose and decrease in galactonic acid and gentiobiose as compared to control BR29 seeds. This study on GR enlightened us with a deeper insight into the adaptation process of golden rice along with balancing the expression of enzymes, and metabolites of other pathways for higher carotenoid biosynthesis in grain (Gayen et al. 2016).

Food Safety Analysis of GR

GR can fulfill a large portion of our recommended daily intake (RDI) of vitamin A, but its food safety must be verified scientifically by trustworthy public/private organizations for assurance and development of trust and confidence about the product. Analyses revealed that, 100 g uncooked GR2E rice could supply about 40% of the RDI of vitamin A for children and about 13% of the RDI for adult women (Swamy et al. 2019). GR has been found safe for consumption according to the judgment of the U.S. Food and Drug Administration (FDA) (Owens 2018). In an experiment reported on consumption of GR in China, a total of 68 children (aged 6–8) were given meals containing golden rice, spinach or β-carotene supplemented oil capsules for lunch every week during the three-week trial (Tang et al. 2012). Results indicated that β-carotene in GR was an effective alternative in fulfilling RDI of vitamin A in children. The nutritional supplement value of GR was as effective as β-carotene present in vitamin A oil (supplements) and better than spinach. Unfortunately, no further analysis has been reported, and the experiment was terminated on principles of ethics and non-compliance of regulatory permissions (the retraction notice at publisher website was prepublished online on July 29, 2015; https://doi.org/10.3945/ajcn.114.093229). Provisions of testing food safety of GR in every aspect should be prioritized to acknowledge the efforts of rice researchers who put in years of research for GR development to help the common masses.

Future Perspective of GR

The future of GR depends on the continuously developing fields of research (for better stable lines of GR) and user friendly GM (genetically modified) crop regulations (Datta et al. 2019).

Currently, the GR regulation framework and process for its release for consumption is far behind specifically, in those Asian countries where it is required the most and especially in the fight against VAD in children and pregnant women. Some progress has been reported from the Philippines and Bangladesh towards certification of some GR events, but they are mostly under field trial.

Confined Field Trial and Regulation of GR

The field trial is an integral part of certification and regulation. Unfortunately, multi-locational and seasonal field trials of GR have been delayed due to lack of proper permission from the regulatory agencies and frequent hyperactivity of anti-GMO (genetically modified organism) activists. Incidents of GR trials vandalized by anti-GMO protesters in the Philippines were reported in BBC News on August 9, 2013 (https://www.bbc.com/news/science-environment-23632042, access on October 5, 2019). This field trial was conducted by IRRI together with Philippine Rice Research Institute (PhilRice).

A fresh application was submitted by PhilRice on February 28, 2017, to the Department of Agriculture-Bureau of Plant Industry (DA-BPI), Philippines, for a biosafety permit to conduct a field trial. The aim of such a field trial is to collect environmental risk-assessment data from GR cultivation (https://www.irri.org/golden-rice, access on November 20, 2019). A joint application for direct use of GR2 (event E) for food and feed or for processing has been fielded by PhilRice and IRRI to the DA-BPI on March 1, 2017 at Philippines (https://www.irri.org/golden-rice, access on November 20, 2019).

According to a report of *BusinessWorld*, published on November 11, 2019, multi-locations and seasons (consecutive two years) GR trials are being conducted in Isabela, Nueva Ecija (https://www.bworldonline.com/golden-rice-undergoing-trials-in-isabela-nueva-ecija; access on November 20, 2019). Two BR29 golden rice lines (developed by Datta's group at IRRI) were transferred from IRRI to Bangladesh for field evaluation (Datta et al. 2006). Further work did not take place due to non-compliance/non-availability of regulatory systems. Recently BRRI, Bangladesh, has been testing dhan 29 (a Syngenta AG-derived IRRI selected GR metalial) at multiple locations and has found no new farming challenges and no significant differences in quality (Stokstad 2019). GR regulatory applications are currently under review in Bangladesh, and they are expected to be approved very soon (Stokstad 2019; Wight 2019).

Application of CRISPR/Cas for GR Development

The CRISPR/Cas9 (clustered regularly interspaced short palindromic repeats – associated endonuclease Cas9) based genome editing tools have tremendous potential for nutrition technology—improvement of grain quality of GR in the near future. Application of CRISPR tools can be extended to inactivate those genes that negatively regulate rice grain quality (collectively mentioned as Q-genes) thus effectively facilitating the future of GR. A list of such Q-genes seen as potential targets for CRISPR tools has been recently reviewed by Fiaz et al. (2019). How CRISPR-mediated silencing of such genes would be possible in rice has been discussed by Bandyopadhyay et al. (2019). A recent study confirms that CRISPR/Cas9 mediated mutation of the rice *ORANGE* (*Osor*) gene leads to accumulation of β-carotene in rice callus (Endo et al. 2019). As a result of the highly accurate and specific nature of CRISPR/Cas9 genome-editing tool, in the near future transgenes can be incorporated in rice genome (via gene knockin approach) in a pretargeted location without disrupting any important gene in the chromosome. For better understanding of carotenoid biosynthesis pathways and their relationship with other pathways in rice, studies via CRISPR/Cas9-mediated mutation (via gene knock-out approach) of single or multiple enzyme encoding genes could be effective in the future.

The regulatory fate of CRISPR/Cas9 genome-edited crops (CRISPR-ed crops) has yet to be decided, and faith in CRISPR-ed GR greatly depends on the regulation process. The USDA (United States Department of Agriculture) has exempted the application of strict GMO regulations in many CRISPR-edited crops. On the other hand, the Court of Justice of the European Union has recently judged that CRISPR-ed crops will also be regulated as per GMO regulation guidelines (Gelinsky and Hilbeck 2018).

GR Economics

Rice is a major source of nutrients for more than half of the world's population. Rice contributes up to 70% of daily calories to its consumers, and golden rice would be one of the sustainable and economical (that medicatial supplements) approaches for people to become healthier as they defeat VAD. Many nutritional and economic ex ante assessment studies have suggested that rice enriched with provitamin A (GR) can reduce the effect of VAD at less cost than any other alternative strategies (Wesseler and Zilberman 2014). Economic benefits of introducing GR in different countries (where it is needed the most) have been studied including India, Bangladesh.

According to a study by Wesseler and Zilberman (2014), India has delayed for over a decade and still has not introduced GR or high iron rice (it was expected to be introduced in 2002) and has caused losses of at least 1,424,680 life years, along with the incurrence of additional perceived costs (other than GR) of at least US$199 million annually. Blind opposition of GM-technology has rendered common people incapable of distinguishing between the technological difference in GR and other GM-crops. Common misconceived narratives against *Bt*-technology (GM cotton) include farmer suicides and dead sheep, environmental damage due to *Bt* brinjal, and health damages linked to antibiotic marker genes used for GM-crop development. Anti-GMO oppositions have successfully developed a small industry around the opposition to GM-crops that survives mainly on donations (Apel 2010). Their activities are used to stir uncertainty among decision makers and meanwhile economic losses continue due to avoidance of GR.

Priority should be given to the adaptation of GR for human benefit where it is required most. Hopefully, with the support from the donors of Intellectual Property Rights (IPR), Syngenta AG, and the funding agencies, the including Melinda and Bill Gates Foundation, the Rockefeller Foundation, and freedom to operate (FTO) licenses for the GR use in India, Bangladesh, and Southeast Asia through the Golden Rice Humanitarian Board could expedite the research, breeding, and its availability for the consumers. In India, the Department of Biotechnology (DBT) and the Indian Council of Agricultural Research

(ICAR) deal with research and policy on agriculture and GM regulations and hope that the regulatory packages soon will be supportive for release of GM products including GR. The ICAR has taken the lead for the biofortification of cereal crops including rice based on earlier research efforts from India and the world (Neeraja et al. 2017).

Conclusion

Golden rice is one of the urgent and top priority projects for our society where malnutrition, VAD, and othersevere health and growth problems in children persist. This single project can facilitate the attainment of 6 out of the 17 most important Sustainable Development Goals 2030. With a proper FTO, modern CRISPR/Cas-based genome editing tools can be applied to improve rice carotenoids level with other seed qualities. GM crops or GR regulatory agencies should look at the glorious and tremendous improvements in the last two decades in this area. Strong data-driven analyses support golden rice for its nutritional quality, viability, and economic importance. Bangladesh is very close to the finishing line of the regulatory procedures followed by the Philippines and very soon these countries are expected to deliver golden rice to those who have been suffering from VAD and hidden hunger.

Acknowledgments

We thank the Department of Biotechnology (DBT), Government of India (GoI) for financial and research support to the University of Calcutta. "Distinguished Biotechnology Research Professor Award" by the DBT to S.K.D. The authors are grateful to Professor Ingo Potrykus and Professor Peter Beyer for providing us the initial support to commence the golden rice project. The authors are also grateful to Dr. Nilanjan Baisakh, Dr. Vilas Parkhi, Dr. Mayank Rai, Dr. Sayeda Rehana, Dr. Md. Alamgir, Dr. Krishnan, Dr. Anindya Bandyopadhyay, Dr. Khalekuzzaman, Dr. Moumita Ganguly, Editha Abrigo, Norman Oliva, Lina Torrizo, Dr. Gayetri Sahoo for their contribution. We are grateful to the Rockefeller Foundation, USA, and particularly to Dr. Gary Toenniessen for encouragement and funding support of our golden rice research at ETH-Zurich and IRRI, Philippines, the DBT and ICAR (GoI) for financial support of GR research at the University of Calcutta, Kolkata, India.

References

Apel, A. (2010). The costly benefits of opposing agricultural biotechnology. *New Biotechnol.* 27: 635–640.

Bai, C., Twyman, R.M., Farre, G. et al. (2011). A golden era-pro-vitamin A enhancement in diverse crops. *In Vitro Cell. Dev. Biol. Plant* 47: 205–221.

Bai, C., Capell, T., Berman, J. et al. (2016). Bottlenecks in carotenoid biosynthesis and accumulation in rice endosperm are influenced by the precursor-product balance. *Plant Biotechnol. J.* 14 (1): 195–205.

Baisakh, N., Rehana, S., Rai, M. et al. (2006). Marker-free transgenic (MFT) near-isogenic introgression lines (NILs) of "golden" *indica* rice (cv IR64) with accumulation of provitamin A in the endosperm tissue. *Plant Biotechnol. J.* 4: 467–475.

Bandyopadhyay, A., Yin, X., Biswal, A. et al. (2019, 1892. Methods Mol. Biol.). CRISPR-Cas9-mediated genome editing of rice towards better grain quality. In: *Rice Grain Quality* (ed. N. Sreenivasulu), 311–336. New York: Humana Press.

Bollinedi, H., Krishnan, G., Prabhu, K.V. et al. (2017). Molecular and functional characterization of GR2-R1 event based backcross derived lines of golden rice in the genetic background of a mega rice variety Swarna. *PLoS One* 12 (1): e0169600.

Burkhardt, P.K., Beyer, P., Wunn, J. et al. (1997). Transgenic rice (*Oryza sativa*) endosperm expressing daffodil (*Narcissus pseudonarcissus*) phytoene synthase accumulates phytoene, a key intermediate of provitamin a biosynthesis. *Plant J.* 11 (5): 1071–1078.

Carrera, A., Echenique, V., Zhang, W. et al. (2007). A deletion at the Lpx-B1 locus is associated with low lipoxygenase activity and improved pasta color in durum wheat (*Triticum turgidum* ssp. durum). *J. Cereal Sci.* 45: 67–77.

Chen, Y., Li, F., and Wurtzel, E.T. (2010). Isolation and characterization of the *Z-ISO* gene encoding a missing component of carotenoid biosynthesis in plants. *Plant Physiol.* 153: 66–79.

Cunningham, F.X. Jr., Pogson, B., Sun, Z. et al. (1996). Functional analysis of the β and ε lycopene cyclase enzymes of Arabidopsis reveals a mechanism for control of cyclic carotenoid formation. *Plant Cell* 8: 1613–1626.

Datta, S.K. (2005). Androgenic haploids: factors controlling development and its application in crop improvement. *Curr. Sci.* 89: 1870–1878.

Datta, S.K. and Bouis, H.E. (2000). Application of biotechnology to improving the nutritional quality rice. *Food Nutr. Bull.* 21: 451–456.

Datta, S.K. and Wenzel, G. (1987). Isolated microspore-derived plant formation via embryogenesis in *Triticum aestivum*. *Plant Sci.* 48: 49–54.

Datta, S.K., Peterhans, A., Datta, K. et al. (1990a). Genetically engineered indica rice recovered from protoplasts. *Nat. Biotechnol.* 8: 736–740.

Datta, S.K., Datta, K., and Potrykus, I. (1990b). Embryogenesis and plant regeneration from microspores of both *Indica* and *Japonica* rice (*Oryza sativa*). *Plant Sci.* 67: 83–88.

Datta, K., Baisakh, N., Oliva, N. et al. (2003). Bioengineered "golden" indica rice cultivars with beta-carotene metabolism in the endosperm with hygromycin and man-nose selection systems. *Plant Biotechnol. J.* 1: 81–90.

Datta, K., Rai, M., Parkhi, V. et al. (2006). Improved "golden" indica rice and post-transgeneration enhancement of metabolic target products of carotenoids (β-carotene) in transgenic elite cultivars (IR64 and BR29). *Curr. Sci.* 91: 935–939.

Datta, S.K., Datta, K., Parkhi, V. et al. (2007). Golden rice: introgression, breeding, and field evaluation. *Euphytica* 154: 271–278.

Datta, K., Sahoo, G., Krishnan, S. et al. (2014). Genetic stability developed for β-carotene synthesis in BR29 rice line using dihaploid homozygosity. *PLoS One* 9: e100212.

Datta, S., Dhillon, B.S., Goutam, P.L. et al. (2019). India needs genetic modification technology in agriculture. *Curr. Sci.* 177 (3): 390–394.

Dubock, A. (2019). Golden rice: to combat vitamin a deficiency for public health. *IntechOpen*: 1–21. https://doi.org/10.5772/intechopen.84445.

Endo, A., Saika, H., Takemura, M. et al. (2019). A novel approach to carotenoid accumulation in rice callus by mimicking the cauliflower *Orange* mutation via genome editing. *Rice* 12: 81.

Farre, G., Sanahuja, G., Naqvi, S. et al. (2010). Travel advice on the road to carotenoids in plants. *Plant Sci.* 179: 28–48.

Fiaz, S., Ahmad, S., Noor, M.A. et al. (2019). Applications of the CRISPR/Cas9 system for rice grain quality improvement: perspectives and opportunities. *Int. J. Mol. Sci.* 20: 888.

Gayen, D., Ali, N., Ganguly, M. et al. (2014). RNAi mediated silencing of *lipoxygenase* gene to maintain rice grain quality and viability during storage. *Plant Cell Tissue Organ. Cult.* 118: 229–243.

Gayen, D., Ali, N., Sarkar, S.N. et al. (2015). Down-regulation of *lipoxygenase* gene reduces degradation of carotenoids of golden rice during storage. *Planta* 242: 353–363.

Gayen, D., Ghosh, S., Paul, S. et al. (2016). Metabolic regulation of carotenoid-enriched golden rice line. *Front. Plant Sci.* 7: 1622.

Gelinsky, E. and Hilbeck, A. (2018). European Court of Justice ruling regarding new genetic engineering methods scientifically justified: a commentary on the biased reporting about the recent ruling. *Environ. Sci. Eur.* 30 (1): 52.

Ghosh, S., Datta, K., and Datta, S.K. (2019). Rice vitamins. In: *Rice Chemistry and Technology*, 4e, vol. 7 (ed. J.(e. Bao), 195–220. Amsterdam: Elsevier Inc.; St. Paul, MN: AACC International.

Ha, S.-H., Liang, Y.S., Jung, H. et al. (2010). Application of two bicistronic systems involving 2A and IRES sequences to the biosynthesis of carotenoids in rice endosperm. *Plant Biotechnol. J.* 8: 928–938.

Ha, S.-H., Kim, J.K., Jeong, Y.S. et al. (2019). Stepwise pathway engineering to the biosynthesis of zeaxanthin, astaxanthin and capsanthin in rice endosperm. *Metab. Eng.* 52: 178–189.

Hoa, T.T.C., Al-Babli, S., Schaub, P. et al. (2003). Golden indica and *japonica* rice lines amenable to deregulation. *Plant Physiol.* 133: 161–169.

Johnson, E.J. (2002). The role of carotenoids in human health. *Nutr. Clin. Care* 5 (2): 47–49.

Kawakatsu, T., Yamamoto, M.P., Hirose, S. et al. (2008). Characterization of a new rice glutelin gene *GluD-1* expressed in the starchy endosperm. *J. Exp. Bot.* 59: 4233–4245.

Krishnan, S., Datta, K., Baisakh, N. et al. (2003). Tissue-specific localization of β-carotene and iron in transgenic indica rice (*Oryza sativa* L.). *Curr. Sci.* 84 (9): 1232–1234.

Li, F., Murillo, C., and Wurtzel, E.T. (2007). Maize *Y9* encodes a product essential for 15-cis-zeta-carotene isomerization. *Plant Physiol.* 144: 1181–1189.

Majumder, S., Datta, K., and Datta, S.K. (2019). Rice biofortification: high iron, zinc, and vitamin-A to fight against "hidden hunger". *Agronomy* 9: 803.

Misawa, N., Truesdale, M.R., Sandmann, G. et al. (1994). Expression of a tomato cDNA coding for phytoene synthase in *Escherichia coli*, phytoene formation *in vivo* and *in vitro*, and functional analysis of the various truncated gene products. *J. Biochem.* 116: 980–985.

Neeraja, C.N., Babu, V.R., Ram, S. et al. (2017). Biofortification in cereals: progress and prospects. *Curr. Sci.* 113 (6): 1050–1057.

Nie, D.-M., Ouyang, Y.-D., Wang, X. et al. (2013). Genome-wide analysis of endosperm-specific genes in rice. *Gene* 530 (2): 236–247.

Owens, B. (2018). Golden Rice is safe to eat, says FDA. *Nat. Biotechnol.* 36: 559–560.

Paine, J.A., Shipton, C.A., Chagger, S. et al. (2005). Improving the nutritional value of Golden rice through increased pro-vitamin A content. *Nat. Biotechnol.* 23: 482–487.

Parkhi, V., Rai, M., Tan, J. et al. (2005). Molecular characterization of marker free transgenic indica rice lines that accumulate carotenoids in seed endosperm. *Mol. Genet. Genomics* 274: 325–336.

Qu, L.Q., Xing, Y.P., Liu, W.X. et al. (2008). Expression pattern and activity of six *glutelin* gene promoters in transgenic rice. *J. Exp. Bot.* 59 (9): 2417–2242.

Rai, M., Datta, K., Baisakh, N. et al. (2003). Agronomic performance of Golden indica rice (cv. IR64). *Rice Genet. Newslett.* 20: 30–33.

Rai, M., Datta, K., Parkhi, V. et al. (2007). Variable T-DNA linkage configuration affects inheritance of carotenogenic transgenes and carotenoid accumulation in transgenic indica rice. *Plant Cell Rep.* 26: 1221–1231.

Rodriguez-Concepcion, M. and Boronat, A. (2002). Elucidation of the methylerythritol phosphate pathway for isoprenoid biosynthesis in bacteria and plastids. A metabolic milestone achieved through genomics. *Plant Physiol.* 130: 1079–1089.

Seo, M. and Koshiba, T. (2002). Complex regulation of ABA biosynthesis in plants. *Trends Plant Sci.* 7: 41–48.

Stokstad, E. (2019). After 20 years, golden rice nears approval. *Science* 366 (6468): 934.

Swamy, B.P.M., Samia, M., Boncodin, R. et al. (2019). Compositional analysis of genetically engineered GR2E "golden rice" in comparison to that of conventional rice. *J. Agric. Food. Chem.* 67 (28): 7986–7994.

Tang, G., Hu, Y., Yin, S.A. et al. (2012). β-carotene in golden rice is as good as β-carotene in oil at providing vitamin A to children. *Am. J. Clin. Nutr.* 96 (3): 658–664.

Tian, L., Magallanes-Lundback, M., Musetti, V., and DellaPenna, D. (2003). Functional analysis of beta- and epsilon-ring carotenoid hydroxylases in Arabidopsis. *Plant Cell* 5: 1320–1332.

Vasconcelos, M., Datta, K., Oliva, N. et al. (2003). Enhanced iron and zinc accumulation in transgenic rice with the *ferritin* gene. *Plant Sci.* 64 (3): 371–378.

Wesseler, J. and Zilberman, D. (2014). The economic power of the golden rice opposition. *Environ. Dev. Econ.* 19: 724–742. https://doi.org/10.1017/S1355770X1300065X.

Wight, A.J. (2019). The precautionary tale of golden rice. *Science* 366 (6462): 192.

Wu, C.-Y., Adachi, T., Hatano, T. et al. (1998). Promoter of rice seed storage protein genes direct endosperm-specific gene expression in transgenic rice. *Plant Cell Physiol.* 39 (8): 855–889.

Xu, R., Li, D., Li, H. et al. (2017). Isolation of four rice seed-specific promoters and evaluation of endosperm activity. *Plant Cell Tissue Organ Cult.* 128: 125–132.

Ye, X., Al-Babili, S., Kloti, A. et al. (2000). Engineering the provitamin A (beta-carotene) biosynthetic pathway into (carotenoid-free) rice endosperm. *Science* 287 (5451): 303–305.

Zhu, C., Bai, C., Sanahuja, G. et al. (2010). The regulation of carotenoid pigmentation in flowers. *Arch. Biochem. Biophys.* 504: 132–141.

16

Increasing Grain Zinc Concentration in Rice

Naoya Miyazaki, Miki Ogasawara and Ryo Ishikawa

Laboratory of Plant Breeding, Graduate School of Agricultural Science, Kobe University, Kobe 657-8501, Japan

Role of Zinc in Human Nutrition

Zinc (Zn) is one of the mineral elements that is important for all living beings. Humans need Zn for activity of more than 1000 transcription factors and 300 enzymes, such as covering oxido-reductases, transferases, hydrolases, lyases, isomerases, and ligases (McCall et al. 2000; Prasad 2013). Zn deficiency leads to reduced Zn-related enzyme activities, affecting most of the protein synthesis pathways. In humans, Zn deficiency is known to cause damage to cells and organs that play a role in protein synthesis. Major health damages due to Zn deficiency in humans are growth failure, loss of appetite, impaired immune function, hair loss, diarrhoea, dermatitis, mental lethargy, and disorders related to skeletal and reproductive systems (Roohani et al. 2013; Sharma et al. 2013). In many children, Zn deficiency prevents them from growing in height because of decreased growth hormone receptors. Concerning hair loss and dermatitis, about 8% of body Zn is concentrated in hair and skin; therefore, Zn deficiency results in critical damage to hair and skin. Deficiency of micronutrients, which causes such disorders is collectively designated as "Hidden Hunger" (Harding et al. 2018). It is one of the serious health problems, worldwide, and about two billion people in the human population are known to suffer from iron (Fe) or Zn deficiencies (Harding et al. 2018). This problem is especially observed in developing countries, affecting many children and pregnant women. People in these countries depend on cereals in their diets, and they do not have many options in food varieties (Shahzad et al. 2014). For people suffering from Zn deficiency, stable intake of Zn in their daily diet is very important to resolve Zn-deficiency-related disorders. Developed countries have adequate access to fortified Zn-containing processed foods, but most such products are not available in developing countries because of ineffective delivery systems (Berti et al. 2014). Hence, biofortification of cereals is useful to improve micronutrient deficiencies (Bouis and Welch 2010). Moreover, biofortification of major cereals with Zn is necessary to efficiently add adequate amounts of Zn in the daily diet of people in developing countries.

Rice (*Oryza sativa*), is one of the major cereals and staple foods, supplying energy to more than half of the world population (Khush 2005). Rice is rich in carbohydrates but contains

Molecular Breeding for Rice Abiotic Stress Tolerance and Nutritional Quality, First Edition.
Edited by Mohammad Anwar Hossain, Lutful Hassan, Khandakar Md. Iftekharuddaula, Arvind Kumar and Robert Henry.

low amounts of essential minerals, such as Zn as compared to other cereals. Edible parts of rice, mainly consisting of endosperms, are known to be poor in Zn or Fe. More than 70% of micronutrients are distributed in the outer layer of endosperms, such as the aleurone layer; thus lower levels of micronutrients are present in the grains after the polishing process (Sharma et al. 2013). Consequently, biofortification of rice with Zn, especially the edible parts, may be an effective solution to Zn deficiency, worldwide, including the developing countries (Swamy et al. 2016).

Current Progress in Increasing Grain Zinc Concentration in Rice

Genes Controlling Grain Zn Content in Rice

To date, several researches have focused on improving grain Zn concentration of rice. Here, we review recent molecular genetics and breeding studies related to the enhancement of grain Zn levels. The genetic identification of major quantitative trait loci (QTLs) or genes followed by the use of maker-assisted selection provides basic breeding strategies for improving grain Zn concentration. Three critical steps are proposed to be important for Zn distribution in rice: transmission of Zn from soil to roots, roots to shoots, and shoots to grains (Swamy et al. 2016). Primarily to increase grain Zn concentration, it is important to enhance the uptake of Zn by the roots from soil. Gene families of Zn-regulated transporters and Iron (Fe)-regulated transporter-like Protein (ZIP), consisting of with 18 members, play important roles in the transportation of metals. The *OsZIP1* gene is expressed in roots and is highly expressed in Zn-deficient soil for uptake and efflux of this metal (Liu et al. 2019b). *OsZIP3*, *OsZIP4*, and *OsZIP8* are also expressed in roots and have important functions in homeostasis related to uptake or distribution of metals (Bashir et al. 2012). *OsZIP7* expressed in parenchyma cells of vascular bundles in roots and nodes plays a role in inter-vascular transfer in nodes to deliver Zn to grains (Tan et al. 2019). Also, *OsZIP6* is related to metal uptake in shoots and roots, but the functions of *OsZIP9* – *OsZIP16* are still unknown. It has been observed that the expression levels of five ZIP genes, *OsZIP4*, *OsZIP5*, *OsZIP8*, *OsZIP9*, and *OsZIP10*, are significantly higher in over-expression line of *OsHMA3*, a gene encoding tonoplast-localized transporter for cadmium (Cd) in the roots of rice (Sasaki et al. 2014). Sperotto et al. (2010) reported about 25 metal-related genes, including *OsYSL6, OsYSL8, OsYSL14, OsNRAMP1, OsNRAMP7, OsNRAMP8, OsNAS1, OsFRO1*, and *OsNAC5* are expressed in flag leaves, and showed correlations with Zn concentrations in the rice grain. These results showed that several genes are involved in the uptake of Zn in roots, which function to regulate Zn homeostasis. Zn-loading is one of the most critical pathways to improve grain Zn concentration. Mutations in the *OsHMA2* gene was found to decrease Zn concentration in the upper organs of plants but resulted in increases in grains (Yamaji et al. 2013); however, over-expression of *OsHMA2* showed no change in grain Zn and other metals concentrations but not for Cd (Takahashi et al. 2012). *OsVIT* (*Vacuolar Iron Transporter*) genes are known to be important Fe and Zn vacuole transporters. Functional disruption of *OsVIT1* and *OsVIT2* in transfer-DNA (T-DNA) insertion lines led to an increase in Zn concentration in grains, but led to a corresponding decrease in flag leaves, demonstrating their roles in Fe/Zn translocation between source and sink organs (Zhang et al. 2012). A recent study using a γ ray-irradiated mutant, designated as Lilizhi (LLZ), showed low Zn concentration in roots, but at least 35% higher Zn concentration in grains as compared to the wild type *japonica* variety, Dongbeixiang (Wang et al. 2017).

Transgenic Approaches Increasing Grain Zn Content in Rice

Several studies have been reported that aimed to increase grain Zn concentration by transgenic approaches using Zn-related genes. Johnson et al. (2011) reported that over-expression of the rice nicotianamine synthase (*OsNAS*) genes has been utilized to increase Fe and Zn concentrations in rice grain. Following this study, Moreno-Moyano et al. (2016) also reported that over-expression of *OsNAS2* resulted in biofortification of rice with Fe and Zn. Similarly, over-expression of *OsIRT1*, which functions as a transporter of Zn and Cd showed high grain Zn concentration, although the plant architecture was altered (Lee and An 2009). Moreover, it was observed that *Plant Cadmium Resistance 1* (*PCR1*) affected both grain Zn concentration and weight, and that the grains of *OsPCR1* knock-down lines exhibited higher Zn concentration and lower Cd concentrations than those of the control plants (Song et al. 2015). These results suggest that transgenic technology, including over-expression of genes and the newly introduced CRISPR (clustered regularly interspaced short palindromic repeats) technology, has a great potential to create new Zn-rich rice varieties. However, cultivation of genetically modified plants still requires deregulation efforts in many countries.

Natural Variations Involved in Grain Zn Content in Rice

The use of natural variations related to grain Zn concentration is one of the possible key solutions for rice biofortification to improve Zn deficiency in humans. Several QTLs pertaining to grain Zn level were identified using various rice resources (Swamy et al. 2016). Norton et al. (2014) reported that a QTL for grain Zn was mapped on chr. 7, between 11.3 and 14.6 Mb, in which a large number of SNPs (single nucleotide polymorphisms) were detected using genome-wide association (GWA) mapping. Another study detected 134 QTLs controlling metal levels including Zn, using recombinant inbred lines (RILs) between "Lemont" and "TeQing" (Zhang et al. 2014). Furthermore, four QTLs on chrs. 3, 6, 7, and 8 were detected in two sets of backcross recombinant-inbred lines (BRILs) between IR75862 (a *japonica* waxy variety) and Ce258 or Zhongguangxiang1 (ZGX1) (Xu et al. 2015). In addition, RILs derived between Madhukar and Swarna were used to map QTLs for Zn and Fe in unpolished rice, six Zn-related QTLs on chrs. 3, 7, and 12 were detected; one QTL was located on chr 3, three on chr. 7, and two on chr. 12, although their phenotypic variances are relatively smaller with 29 to 35% (Anuradha et al. 2012a). These results suggested that many QTLs related to Zn from landrace or cultivated rice were detected; however, their effects on grain Zn concentration were robust. Although the effects of the various QTLs on grain Zn level are relatively small in their individual capacities, their application on improving grain Zn levels can be made possible by marker-assisted pyramiding (Swamy et al. 2016).

Use of Wild Relatives to Increase Grain Zn Concentration in Rice

It is known that Asian cultivated rice, *O. sativa*, was domesticated from one of the wild relatives, *O. rufipogon*. During domestication, several traits essential for the survival of the species of wild rice under severe natural conditions were removed by repeated human interventions (Ishii and Ishikawa 2018). For example, loss of seed shattering and larger grain size were selected to increase yield, and plant architecture was selected for efficient cultivation to maximize yield. These repeated selections of specific agricultural traits led the cultivated rice to lose genetic diversity, while the wild relatives had abundant genetic diversity. Wild rice species are very useful due to their potential to introduce new useful traits, such as disease resistance or environmental stress tolerance, which cultivated rice do not process. Consequently, the wild relatives of rice have a potential for enhanced nutrient accumulation (Ricachenevsky and Sperotto 2016). Several studies have been reported on the comparison of grain Zn concentration in different wild rice species. Brown rice obtained from *O. nivara* and *O. rufipogon* showed a higher grain Zn concentration than that of cultivated rice, *O. sativa* (Anuradha et al. 2012b). Beside the wild rice species processing A genome, *O. officinalis* (C or BC genome) and *O. latifolia* (CD genome) showed higher grain Zn concentration than others (Banerjee et al. 2010). These two studies showed that wild rice species could act as Zn donors to create new high-Zn containing rice. However, the application of these wild rice species to improve grain Zn concentration of cultivated rice is currently limited, because of difficulties in handling samples using genetic methods. In our recent study, we compared the grain Zn concentration of wild rice species, especially those having an A genome that could be crossed with *O. sativa*. Several wild rice species processing the A genome (*O. rufipogon, O. glumaepatula, O. meridionalis*, and *O. barthii*) and two Asian rice cultivars (*O. sativa japonica* "Nipponbare" and *indica* "IR36") were subjected to an evaluation of their grain Zn concentration. Results showed that *O. meridionalis* W1627 showed the highest Zn concentration in brown rice among these species (Ishikawa et al. 2017; Figure 16.1a). To understand the molecular basis underlying the differences in grain Zn concentration, we produced BRILs, randomly having W1627 chromosomal segments in the genetic background of cultivated rice *O. sativa* "Nipponbare." These BRILs showed contentious distribution in grain Zn concentrations. Genome-wide genotype information analyzed by SSR (simple sequence repeat) markers as well as grain Zn concentration were used for QTL analysis. Four QTLs were detected on chrs. 2 (two loci), 9, and 10 (Figure 16.1b). Among these QTLs, a locus on chr. 9, named "*QTL for Grain Zn concentration on chr. 9*" (*qGZn9*), was found to be the most effective QTL in increasing grain Zn concentration (Ishikawa et al. 2017). Further genetic analysis studies revealed that *qGZn9* could be divided into two linked loci, *qGZn9a* and *qGZn9b*. The region of *qGZn9a* was estimated to be of approximately 190 kb, however, no genes involved in metal transport were observed in this region. Interestingly, an introgression line (IL) having the *O. meridionalis* W1627 chromosomal segment covering *qGZn9a* region, IL(*qGZn9a*), showed lower fertility with higher grain Zn level, and the lower fertility of IL(*qGZn9*) is possibly caused by *qGZn9a* (Ishikawa et al. 2017; Figure 16.1c, d). Furthermore, these results suggested that the causal gene of *qGZn9a* might be involved in the process of reproduction, such as pollen or anther functions, which could secondarily affect grain Zn concentration. The *qGZn9b* region was estimated to be of approximately 950 kb, and some genes were shown to be related to the control of grain Zn concentration. The plants carrying a W1627 chromosomal segment covering the *qGZn9b* region showed an increase in grain Zn concentration, about 10 ~ 15 mg kg^{-1} (Figure 16.1d). To overcome the Zn deficiency in the world population, it is essential to up-regulate the grain Zn concentration and high-yield ability of rice. Although, the two above-mentioned QTLs can act to increase grain Zn concentration, the W1627 allele present in *qGZn9a* is not useful to increase grain Zn concentration because of a reduction in mature seeds. In contrast, the W1627 allele at *qGZn9b* has a large potential for the production of new high-Zn rice varieties and is able to cross with high-yield rice. Thus our study demonstrated an example of the use of a wild relative of rice for biofortification to improve grain Zn concentration. It will be quite interesting to search for wild rice species that might have a potential to change mineral elements of rice grains in future studies on breeding. If the difficulties in performing experimental procedures of crossing between wild and cultivated rice species are overcome, the wild alleles conferring favorable effects on grain Zn concentration will be more useful. Since the studies on the use of wild relatives for biofortification of grains are also focused on other crops, such as wheat and barley (Velu et al. 2019; Wiegmann et al. 2019), exploring wild relatives that have high grain Zn level as a natural genetic variation will be an important approach for biofortification of many crops in the future.

Control of Grain Zn Concentration with More Yield

In our recent study, two chromosomal regions at *qGZn9* (*qGZn9a* and *qGZn9b*) that affect grain Zn concentration have been identified in rice (Ishikawa et al. 2017; Figure 16.1). The cultivated rice, *O. sativa* Nipponbare, having a chromosomal segment of wild rice *O. meridionalis* W1627 covering the region of *qGZn9a*, showed higher grain Zn concentration than those of the Nipponbare cultivar (Figure 16.1d). Interestingly, the introgression line showed a lower seed-setting (fertility) rate than the Nipponbare cultivar (Figure 16.1c). This result suggests that an increase in grain Zn concentration is probably caused by a reduction in fertility. Genetic-mapping analysis showed that there are eight candidate genes in the 190-kb region of *qGZn9a* (Ishikawa et al. 2017). However no genes, annotated for the transportation of mineral elements are found in this region. Of these eight genes, we found that *Os09g0384900* showed high gene expression levels in anthers in the Nipponbare cultivar, but this gene is not found in the corresponding region in W1627, implying that the gene might be involved in the function of pollen activities (Ishikawa et al. 2017). Although the involvement of this gene in the control of grain Zn concentration still needs to be tested, reduction of fertility is likely to change the allocation of Zn in grains. One of the possible explanations for the relationship between fertility and grain Zn concentration may be a trade-off between these two parameters. Fertility is one of the most important traits in rice seed production. Fertility mainly depends on the development of pollen and anther dehiscence, which are often affected by several environmental factors, such as a high temperature. Therefore, understanding the relationship between fertility and grain Zn level is important for biofortification of rice grain with high yields.

In our study, we found that the reduction in fertility was associated with increased grain Zn level (Figure 16.1c, d; Figure 16.2). Contrary to a decrease in grain number per panicle

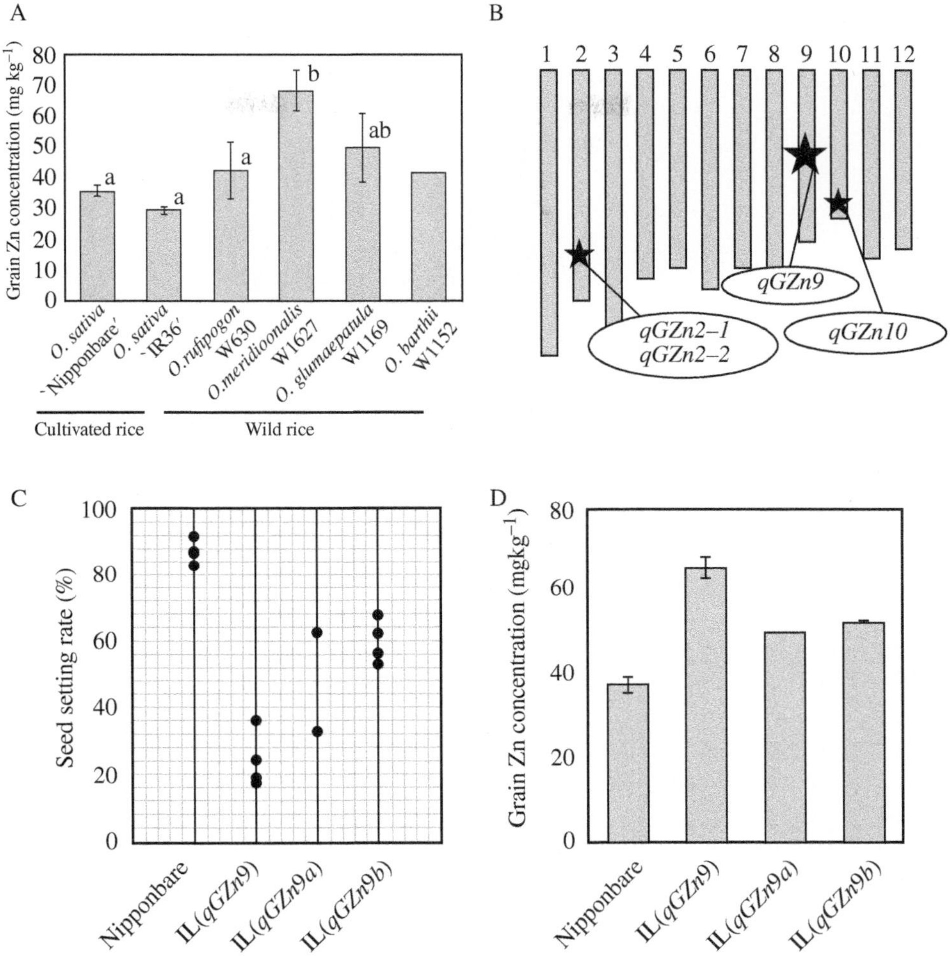

Figure 16.1 Identification of the quantitative trait loci controlling grain Zn concentration between cultivate rice (*Oryza sativa* Nipponbare) and Australian wild rice (*O. meridionalis* W1627). (a) Grain Zn concentrations of two cultivars, *O. sativa* (Nipponbare and IR36) and four wild rice accessions (*O. rufipogon* W630, *O. meridionalis* W1627, *O. glumaepatula* W1169, and *O. barthii* W1152). Data are presented as mean ± S.D. (n = 3). Mean values labeled with different letters are significantly different (Tukey's test, $P < 0.01$). The average grain Zn concentration of two plants is shown for W1152. (b) Positions of the QTL detected for grain Zn concentration are shown on the regions of the chromosomes (1–12). (c, d) Effect of the two linked QTLs, *qGZn9a* and *qGZn9b*, on seed setting ratio (c) and grain Zn concentration (d) using introgression lines carrying W1627 chromosomal segment at the QTLs in Nipponbare genetic background. Data are presented as mean ± S.D. (n = 3).

caused by low fertility rate, if grain number per panicle increases due to larger number of seeds in a panicle, how will Zn concentration of each grain be affected? Therefore, we speculated whether the grain Zn concentration decreases in high-yielding rice varieties (Figure 16.2). Since the world population continues to increase drastically, it is necessary to develop crops with higher yields. Rice is the main crop for about half of world's population, especially in Asia, and to meet the increasing demand, we have to increase the total rice

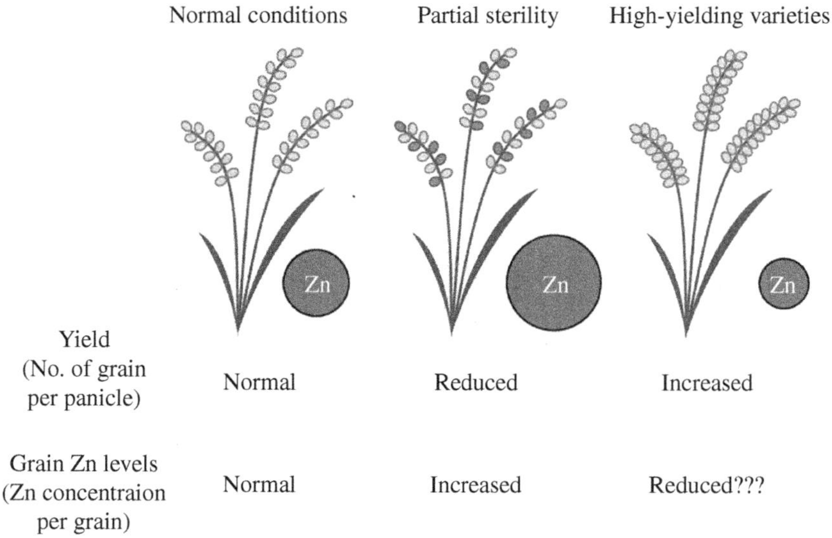

Figure 16.2 Relationship of trade-off between yield and grain Zn concentration in rice. *Source:* Modified from Ishikawa et al. (2017). *Note:* As compared to standard rice cultivars (left), those with reduced number of grains due to low fertility show higher grain Zn concentration (centre) (Ishikawa et al. 2017). In contrast, high-yield cultivars have a larger number of grains, but grain Zn levels may be reduced (right).

production twice as much as that of the present time, up to the year 2030 (Foley et al. 2011). Therefore, immediate breeding of new rice cultivars with high-yielding characteristics is the need of the hour.

The control of yield in rice has been extensively studied so far, and several morphological traits have been observed to be involved in achieving good yield (Xing and Zhang 2010). *GW2* and *qSW5* were previously identified to increase grain size (Song et al. 2007; Shomura et al. 2008). The control of grain numbers per panicle is another important factor to be considered to obtain good yields. *Gn1a* controls the number of grains via change in cytokinin level in inflorescence (Ashikari et al. 2005). *DEP1* also controls grain numbers and the dominant allele acts to enhance meristematic activity (Huang et al. 2009). *WFP / IPA1* and *LAX1* control the number of branches per panicle, resulting in an increased yield (Jiao et al. 2010; Miura et al. 2010; Komatsu et al. 2001). Utility of these natural alleles will be important for breeding new rice varieties with better yields to meet further increase in the world's population. However, the higher yields of rice have not been translated into higher grain Zn level. Consequently, besides the efficient use these genes to increase the yield of rice, their effect on grain Zn level also needs to be investigated.

Balancing yield with other agricultural traits is also a challenge in rice breeding. One of the challenges is to increase both yield and disease resistance of rice. It is known that introduction of disease-resistant genes adversely affects plant architecture and yield (Liu et al. 2019a). This trade-off between growth and defence mechanisms is suggested to be due to a crosstalk between hormones, such as salicilic acid, auxin, jasmonic acid, and gibberellin acid (Robert-Seilaniantz et al. 2011; Pieterse et al. 2012). A recent study revealed a breakthrough to overcome this trade-off by the use of the *Ideal Plant Architecture*

1 (*IPA1*, also known as *WFP /OsSPL14*) encoding a plant SQUAMOSA promoter-binding protein-like transcription factor (Wang et al. 2018). *IPA1/WFP* was identified because it increases grain numbers per panicle and reduces unproductive tillers (Jiao et al. 2010; Miura et al. 2010). Infection by the fungus *Magnaporthe oryzae* induced phosphorylation of IPA1 protein, which subsequently up-regulated the pathogen defence gene, *WRKY45*, by direct promoter binding, resulting in elevated disease resistance (Wang et al. 2018). More recently, a better approach for the control of *IPA1* was reported. Over-expression of *IPA1* showed enhanced resistance to the infection of bacterial blight, *Xanthomonas oryzae pv. oryzae* (*Xoo*), but reduced yield, possibly due to altered gibberellin signaling. Transgenic rice plants, which controlled *IPA1* expression by the inducible promoter of *OsHEN1* in response to a pathogen attack showed both enhanced disease resistance and better yield (Liu et al. 2019a). Uncovering the regulatory role of *IPA1* and its relationship to yield clearly revealed several possibilities to overcome the trade-off between two important agricultural traits. Therefore, understanding the mechanisms of the control of grain Zn distribution will provide possibilities to enhance both the two traits. In addition, these finding will also be useful in the control of other mineral elements in rice grains in the future.

Acknowledgements

The wild rice accessions used in our study were provided by the National Institute of Genetics (Japan) supported by the National BioResource Project (MEXT, Japan). We thank Prof. Takashige Ishii for rice genetic materials for QTL analysis and Prof. Jian Feng Ma for grain Zn measurement. This study is partly supported by the public foundation of Elizabeth Arnold-Fuji and the Joint Usage/Research Center, Institute of Plant Science and Resources, Okayama University.

References

Anuradha, K., Agarwal, S., Rao, Y.V. et al. (2012a). Mapping QTLs and candidate genes for iron and zinc concentrations in unpolished rice of Madhukar×Swarna RILs. *Gene* 508: 233–240.

Anuradha, K., Agarwal, S., Batchu, A.K. et al. (2012b). Evaluating rice germplasm for iron and zinc concentration in brown rice and seed dimensions. *J. Phytol.* 4: 19–25.

Ashikari, M., Sakakibara, H., Lin, S. et al. (2005). Cytokinin oxidase regulates rice grain production. *Science* 309: 741–745.

Banerjee, S., Sharma, D.J., Verulkar, S.B., and Chandel, G. (2010). Use of in silico and semiquantitative RT-PCR approaches to develop nutrient rich rice (*Oryza sativa* L.). *Indian J. Biotechnol.* 9: 203–212.

Bashir, K., Ishimaru, Y., and Nishizawa, N.K. (2012). Molecular mechanisms of zinc uptake and translocation in rice. *Plant Soil* 361: 189–201.

Berti, C., Faber, M., and Smuts, C.M. (2014). Prevention and control of micronutrient deficiencies in developing countries: current perspectives. *Nutr. Diet. Suppl.* 6: 41–57.

Bouis, H.E. and Welch, R.M. (2010). Biofortification—a sustainable agricultural strategy for reducing micronutrient malnutrition in the global south. *Crop Sci.* 50: 20–32.

Foley, J.A., Ramankutty, N., Brauman, K.A. et al. (2011). Solutions for a cultivated planet. *Nature* 478: 337–342.

Harding, K.L., Aguayo, V.M., and Webb, P. (2018). Hidden hunger in South Asia: a review of recent trends and persistent challenges. *Public Health Nutr.* 21: 785–795.

Huang, X., Qian, Q., Liu, Z. et al. (2009). Natural variation at the *DEP1* locus enhances grain yield in rice. *Nat. Genet.* 41: 494–497.

Ishii, T. and Ishikawa, R. (2018). Domestication loci controlling panicle shape, seed shattering, and seed awning. In: *Rice Genomics Genetics and Breeding* (eds. T. Sasaki and M. Ashikari), 207–221. Singapore: Springer.

Ishikawa, R., Iwata, M., Taniko, K. et al. (2017). Detection of quantitative trait loci controlling grain zinc concentration using Australian wild rice, *Oryza meridionalis*, a potential genetic resource for biofortification of rice. *PLoS One* 12: e0187224.

Jiao, Y., Wang, Y., Xue, D. et al. (2010). Regulation of *OsSPL14* by *OsmiR156* defines ideal plant architecture in rice. *Nat. Genet.* 42: 541–544.

Johnson, A.A., Kyriacou, B., Callahan, D.L. et al. (2011). Constitutive overexpression of the *OsNAS* gene family reveals single-gene strategies for effective iron- and zinc-biofortification of rice endosperm. *PLoS One* 6: e24476.

Khush, G.S. (2005). What it will take to feed 50 billion rice consumers in 2030. *Plant Mol. Biol.* 59: 1–6.

Komatsu, M., Maekawa, M., Shimamoto, K., and Kyozuka, J. (2001). The *LAX1* and *FRIZZY PANICLE 2* genes determine the inflorescence architecture of rice by controlling rachis-branch and spikelet development. *Dev. Biol.* 231: 364–373.

Lee, S. and An, G. (2009). Over-expression of *OsIRT1* leads to increased iron and zinc accumulations in rice. *Plant Cell Environ.* 32: 408–416.

Liu, M., Shi, Z., Zhang, X. et al. (2019a). Inducible overexpression of *ideal plant architecture1* improves both yield and disease resistance in rice. *Nat. Plants* 5: 389–400.

Liu, X.S., Feng, S.J., Zhang, B.Q. et al. (2019b). OsZIP1 functions as a metal efflux transporter limiting excess zinc, copper and cadmium accumulation in rice. *BMC Plant Biol.* 19: 283.

McCall, K.A., Huang, C.C., and Fierke, C.A. (2000). Function and mechanism of zinc metalloenzymes. *J. Nutr.* 130: 1437S–1446S.

Miura, K., Ikeda, M., Matsubara, A. et al. (2010). *OsSPL14* promotes panicle branching and higher grain productivity in rice. *Nat. Genet.* 42: 545–549.

Moreno-Moyano, L.T., Bonneau, J.P., Sánchez-Palacios, J.T. et al. (2016). Association of increased grain iron and zinc concentrations with agro-morphological traits of biofortified rice. *Front. Plant Sci.* 7: 1463.

Norton, G.J., Douglas, A., Lahner, B. et al. (2014). Genome wide association mapping of grain arsenic, copper, molybdenum and zinc in rice (*Oryza sativa* L.) grown at four international field sites. *PLoS One* 9: e89685.

Pieterse, C.M., Van der Does, D., Zamioudis, C. et al. (2012). Hormonal modulation of plant immunity. *Annu. Rev. Cell Dev. Biol.* 28: 489–521.

Prasad, A.S. (2013). Discovery of human zinc deficiency: its impact on human health and disease. *Adv. Nutr.* 4: 176–190.

Ricachenevsky, F.K. and Sperotto, R.A. (2016). Into the wild: *Oryza* species as sources for enhanced nutrient accumulation and metal tolerance in rice. *Front. Plant Sci.* 7: 974.

Robert-Seilaniantz, A., Grant, M., and Jones, J.D. (2011). Hormone crosstalk in plant disease and defense: more than just jasmonate-salicylate antagonism. *Annu. Rev. Phytopathol.* 49: 317–343.

Roohani, N., Hurrell, R., Kelishadi, R., and Schulin, R. (2013). Zinc and its importance for human health: an integrative review. *J. Res. Med. Sci.* 18: 144–157.

Sasaki, A., Yamaji, N., and Ma, J.F. (2014). Overexpression of *OsHMA3* enhances Cd tolerance and expression of Zn transporter genes in rice. *J. Exp. Bot.* 65: 6013–6021.

Shahzad, Z., Rouached, H., and Rakha, A. (2014). Combating mineral malnutrition through iron and zinc biofortification of cereals. *Compr. Rev. Food Sci. Food Saf.* 13: 329–346.

Sharma, A., Patni, B., Shankhdhar, D., and Shankhdhar, S.C. (2013). Zinc - an indispensable micronutrient. *Physiol. Mol. Biol. Plants* 19: 11–20.

Shomura, A., Izawa, T., Ebana, K. et al. (2008). Deletion in a gene associated with grain size increased yields during rice domestication. *Nat. Genet.* 40: 1023–1028.

Song, X.J., Huang, W., Shi, M. et al. (2007). A QTL for rice grain width and weight encodes a previously unknown RING-type E3 ubiquitin ligase. *Nat. Genet.* 39: 623–630.

Song, W.Y., Lee, H.S., Jin, S.R. et al. (2015). Rice PCR1 influences grain weight and Zn accumulation in grains. *Plant Cell Environ.* 38: 2327–2339.

Sperotto, R.A., Boff, T., Duarte, G.L. et al. (2010). Identification of putative target genes to manipulate Fe and Zn concentrations in rice grains. *J. Plant Physiol.* 167: 1500–1506.

Swamy, B.P.M., Rahman, M.A., Inabangan-Asilo, M.A. et al. (2016). Advances in breeding for high grain zinc in rice. *Rice* 9: 49.

Takahashi, R., Ishimaru, Y., Shimo, H. et al. (2012). The OsHMA2 transporter is involved in root-to-shoot translocation of Zn and Cd in rice. *Plant Cell Environ.* 35: 1948–1957.

Tan, L., Zhu, Y., Fan, T. et al. (2019). *OsZIP7* functions in xylem loading in roots and inter-vascular transfer in nodes to deliver Zn/Cd to grain in rice. *Biochem. Biophys. Res. Commun.* 512: 112–118.

Velu, G., Crespo Herrera, L., Guzman, C. et al. (2019). Assessing genetic diversity to breed competitive biofortified wheat with enhanced grain Zn and Fe concentrations. *Front. Plant Sci.* 9: 1971.

Wang, Y., Mei, S., Wang, Z. et al. (2017). Metabolite profiling of a zinc-accumulating rice mutant. *J. Agric. Food. Chem.* 65: 3775–3782.

Wang, J., Zhou, L., Shi, H. et al. (2018). A single transcription factor promotes both yield and immunity in rice. *Science* 361: 1026–1028.

Wiegmann, M., Thomas, W.T.B., Bull, H.J. et al. (2019). Wild barley serves as a source for biofortification of barley grains. *Plant Sci.* 283: 83–94.

Xing, Y. and Zhang, Q. (2010). Genetic and molecular bases of rice yield. *Annu. Rev. Plant Biol.* 61: 421–442.

Xu, Q., Zheng, T.Q., Hu, X. et al. (2015). Examining two sets of introgression lines in rice (*Oryza sativa* L.) reveals favorable alleles that improve grain Zn and Fe concentrations. *PLoS One* 10: e0131846.

Yamaji, N., Xia, J., Mitani-Ueno, N. et al. (2013). Preferential delivery of zinc to developing tissues in rice is mediated by P-type heavy metal ATPase OsHMA2. *Plant Physiol.* 162: 927–939.

Zhang, Y., Xu, Y.H., Yi, H.Y., and Gong, J.M. (2012). Vacuolar membrane transporters OsVIT1 and OsVIT2 modulate iron translocation between flag leaves and seeds in rice. *Plant J.* 72: 400–410.

Zhang, M., Pinson, S.R., Tarpley, L. et al. (2014). Mapping and validation of quantitative trait loci associated with concentrations of 16 elements in unmilled rice grain. *Theor. Appl. Genet.* 127: 137–165.

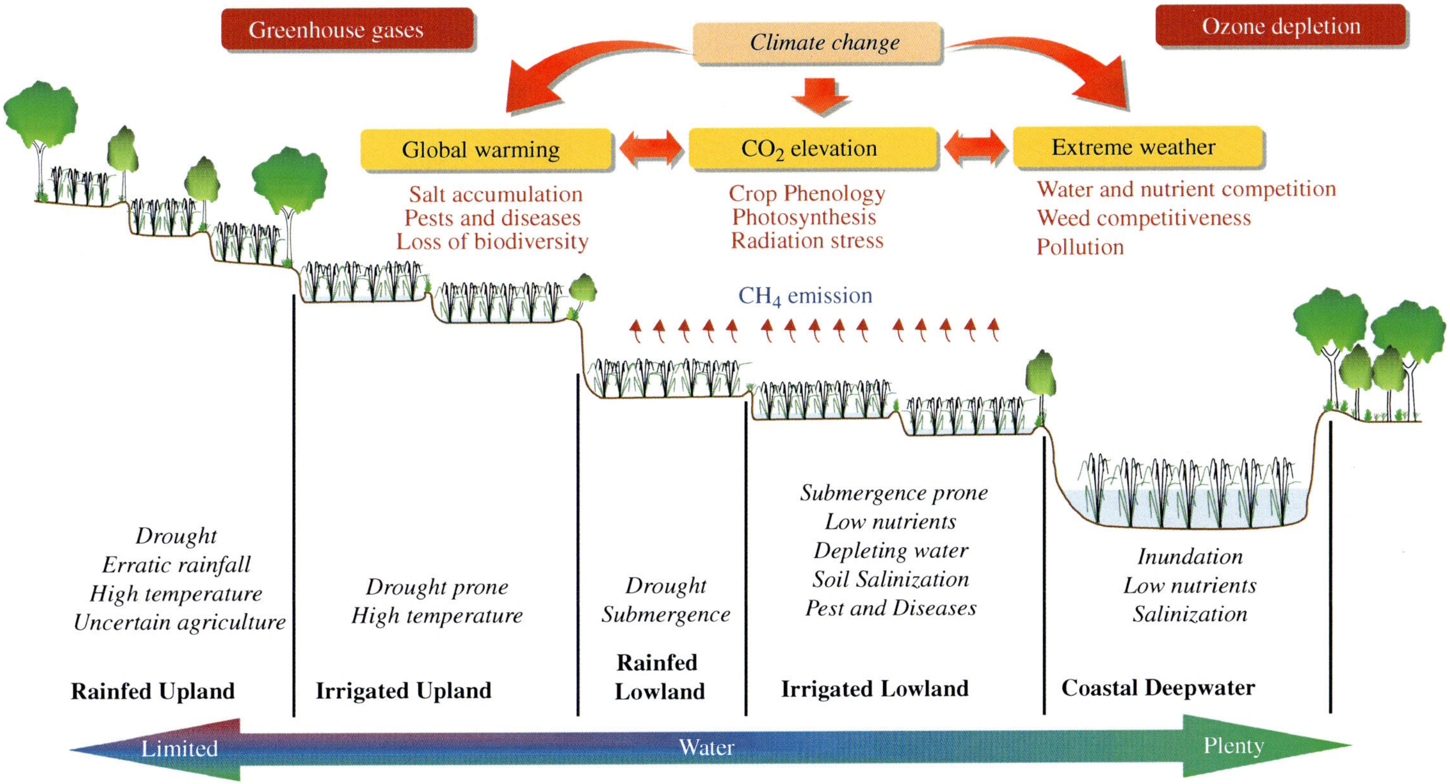

Figure 1.1 Rice ecosystems and its vulnerability to climate change. *Note:* Different rice ecology may face different stresses with rain-fed areas experiencing most of the abiotic stresses. However, biotic stresses may be more under irrigated ecologies.

Molecular Breeding for Rice Abiotic Stress Tolerance and Nutritional Quality, First Edition.
Edited by Mohammad Anwar Hossain, Lutful Hassan, Khandakar Md. Iftekharuddaula, Arvind Kumar and Robert Henry.

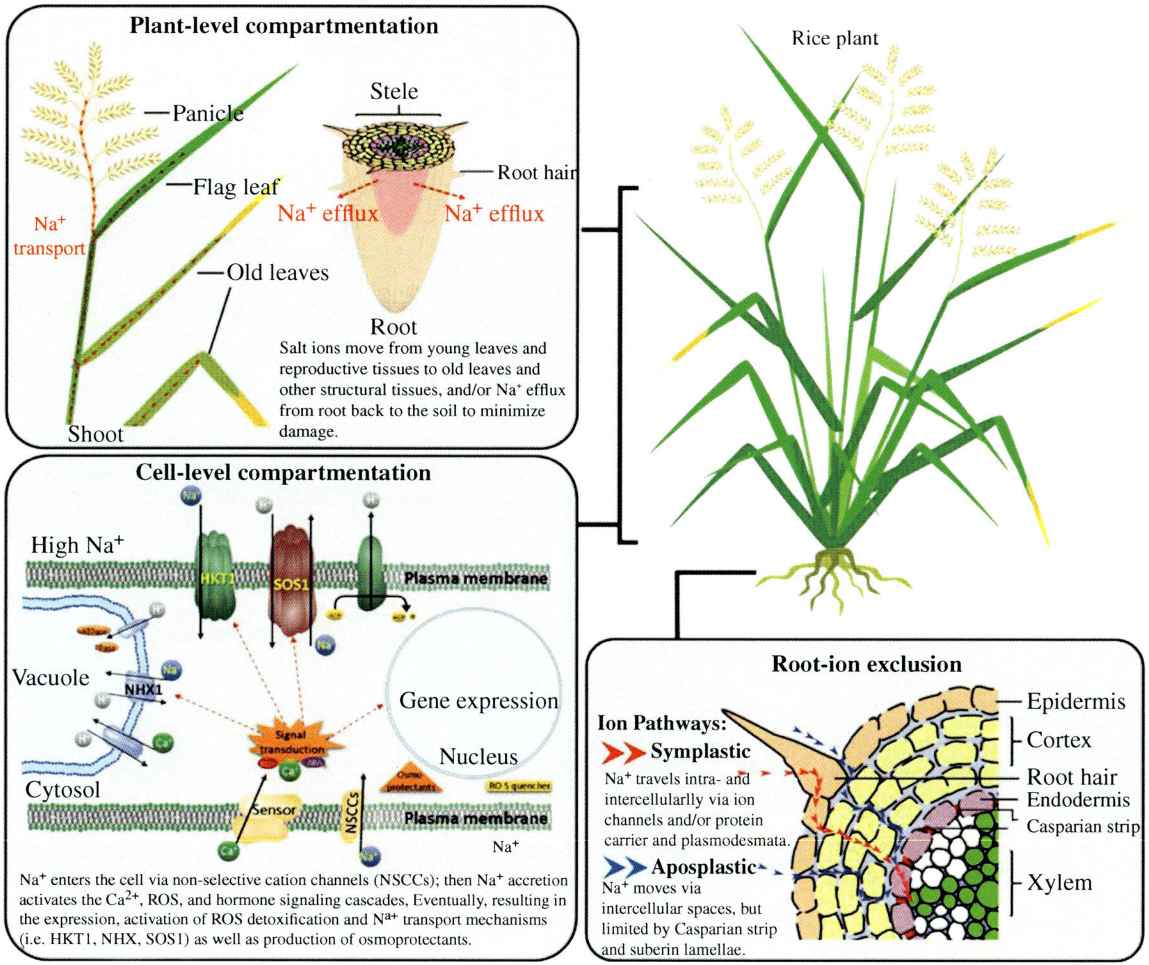

Figure 2.1 Overview of salt tolerance mechanism in rice.

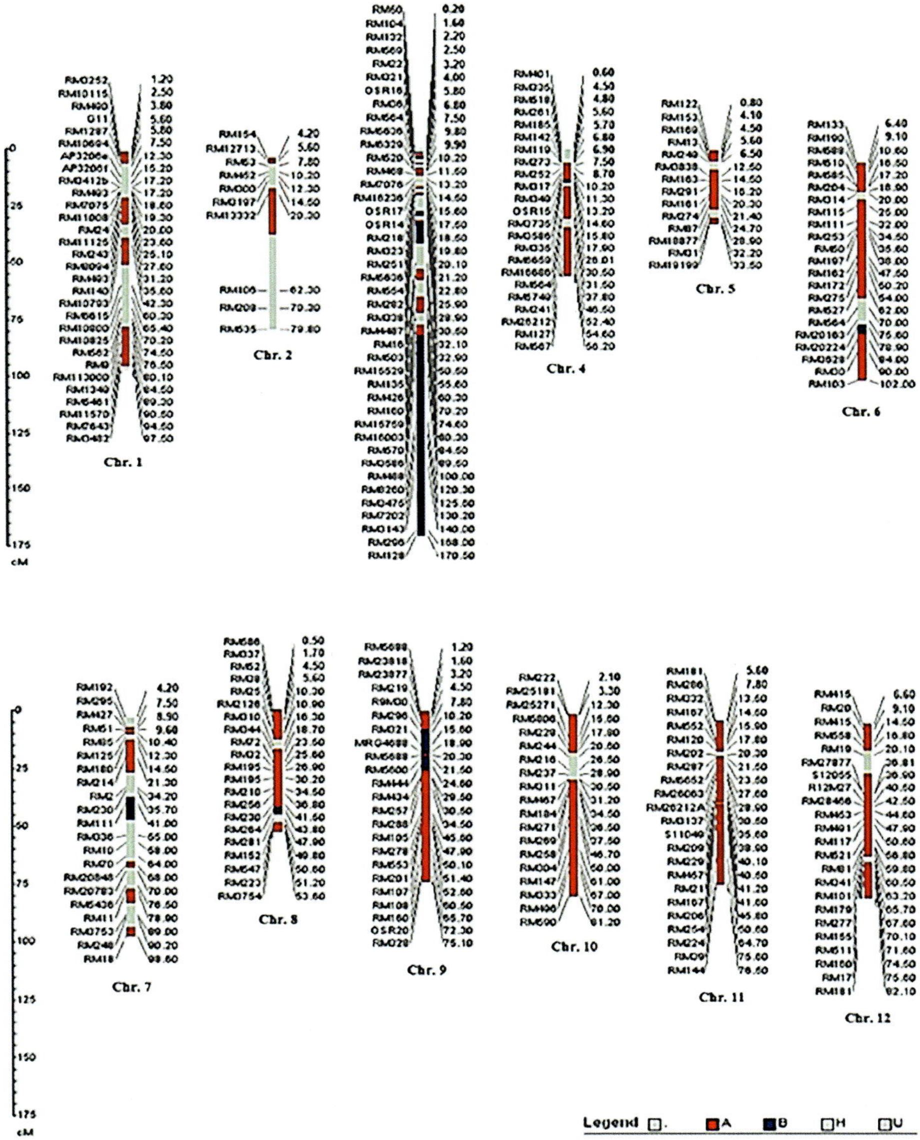

Figure 5.1 Molecular linkage map (graphic genotyping) on 12 chromosomes in BC_2F_2 population of OM5930/N22. *Source:* From Bui (2014). Licensed under CC BY-NC 3.0. *Note:* Blue: homozygous alleles from N22 (tolerance), Red: homozygous alleles from OM5930 (susceptible), Pale green: heterozygotes, Brown: unidentified (Bui et al. 2014).

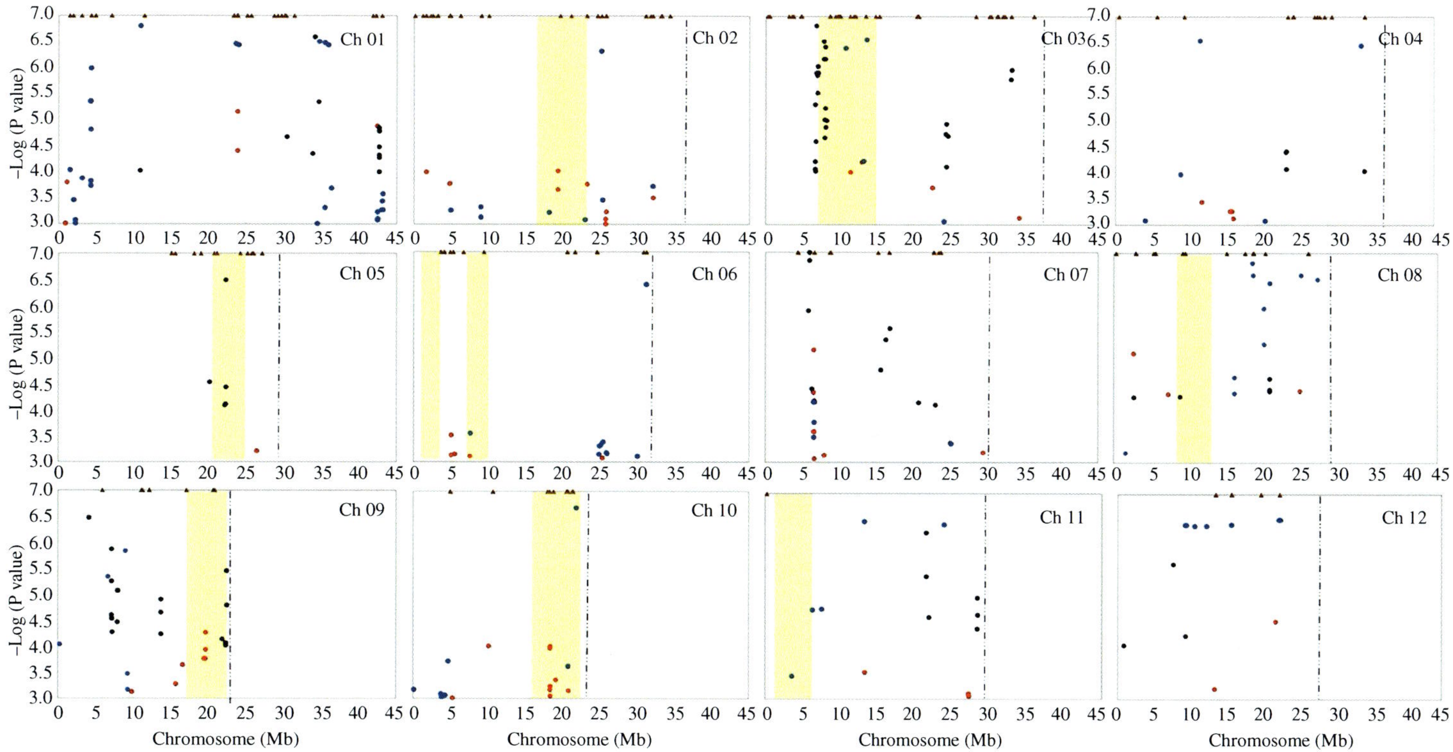

Figure 8.2 Results of association analyses for cargo grain *As* content in a diversity panel of 421 accessions representative of *O. Sativa* (Black Dots), in a diversity panel of 300 *Japonica* Accessions (Red Dots) and 95 *Japonica* advanced line (Blue Dots). *Note:* Data for the *O. sativa* panel were extracted from Norton et al. (2014) when GWAS was implemented with all accessions of the panel. Genomic position of QTLs for grain *As* concentration reported by Zhang et al. (2013) is represented with yellow bars. Genomic position of *As* related candidate genes reports in the literature is represented by brown triangles. The vertical broken line indicates the maximum length of each chromosome. The ordinate axis, –log(*p*-value), indicates the degree of significance of association. The graphic representation draws inspiration from Norton et al. (2014).

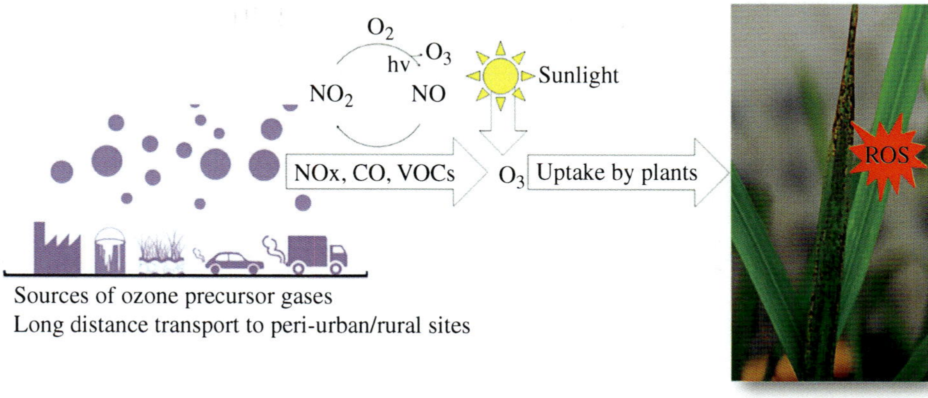

Figure 9.1 A Simplified View of the Major Sources and Formation of Tropospheric Ozone. NOx, nitrogen oxides; CO, carbon monoxide; VOCs, volatile organic compounds; O_3, ozone; ROS, reactive oxygen species. Source: Adapted from the Royal Society (2008). Ground-level ozone in the 21st century: future trends, impacts and policy implications. London: The Royal Society; Climate and Clean Air Coalition, Tropospheric ozone, https://www.ccacoalition.org/ru/slcps/tropospheric-ozone.

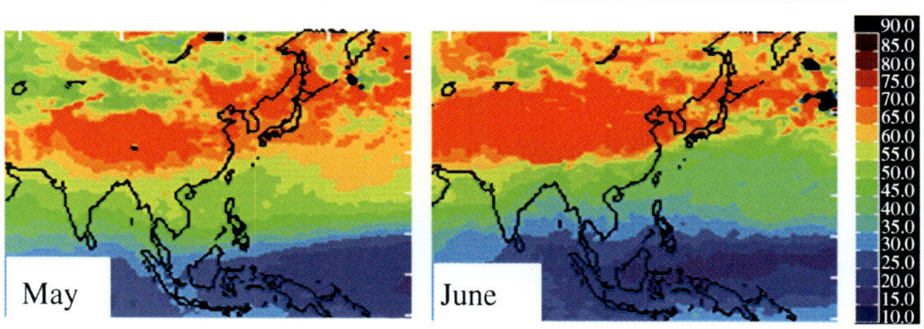

Figure 9.4 Monthly average tropospheric ozone volume mixing ratio (march to june, peak ozone periods in the Indian subcontinents) in Asia in the Year 2016. The color scale indicates the density of ozone divided by the density of all constituents in a unit volume of air (ppbv) in the total tropospheric column. Source: Adapted from Ziemke et al. (2006).

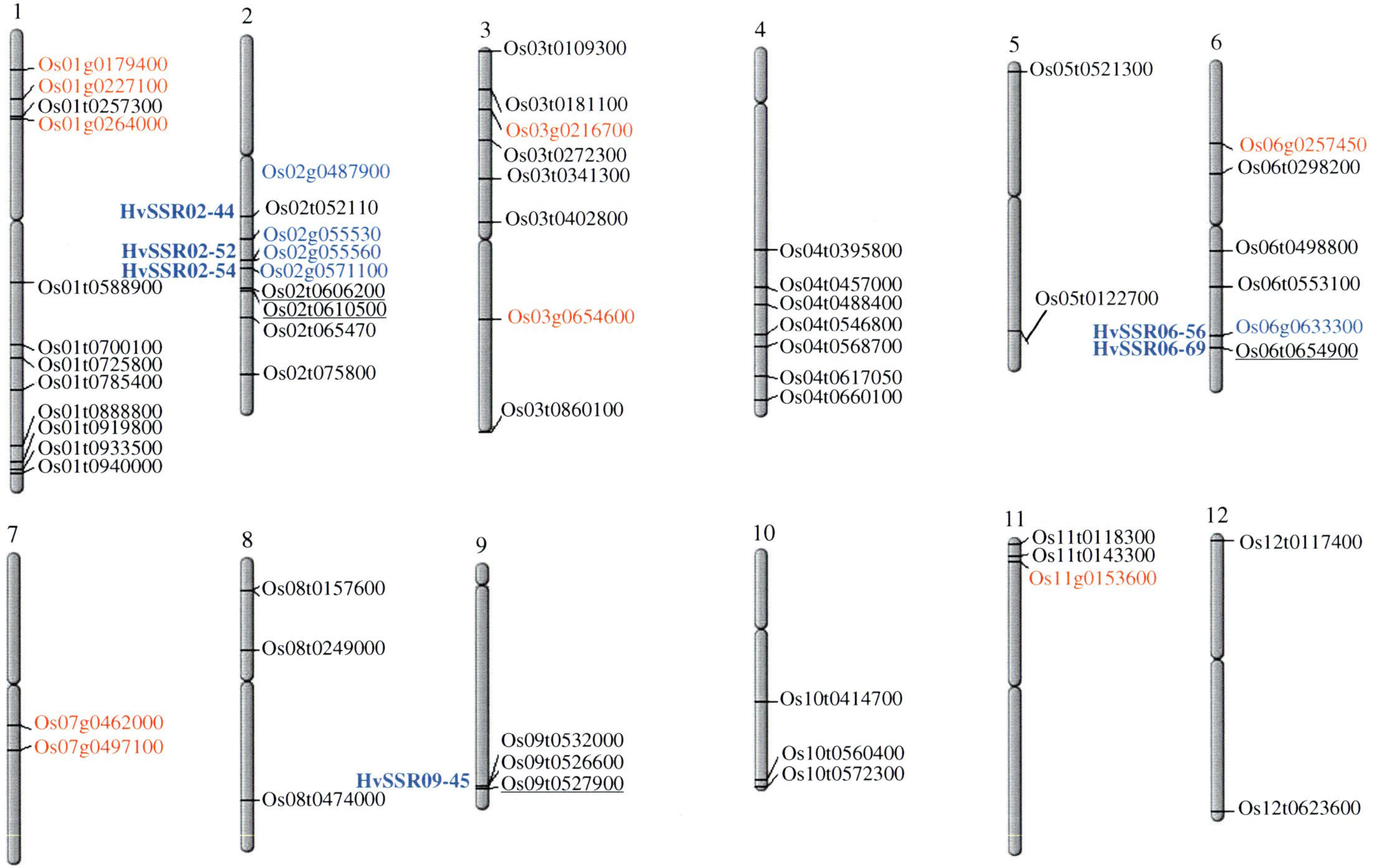

Figure 10.4 Rice chromosomal map showing genic regions and markers identified for low light intensity tolerance in rice through transcriptome or marker trait association studies. Genes (RAP ID) and markers are indicated on the right and left of the chromosomes, respectively. Colors black, red, and blue correspond to the studies by Panigrahy et al. (2019), Zhao et al. (2019), and Dutta et al. (2018), respectively. Locus IDs underlined are common genes identified as potential targets by two studies independently, i.e. Dutta et al. (2018) and Panigrahy et al. (2019).

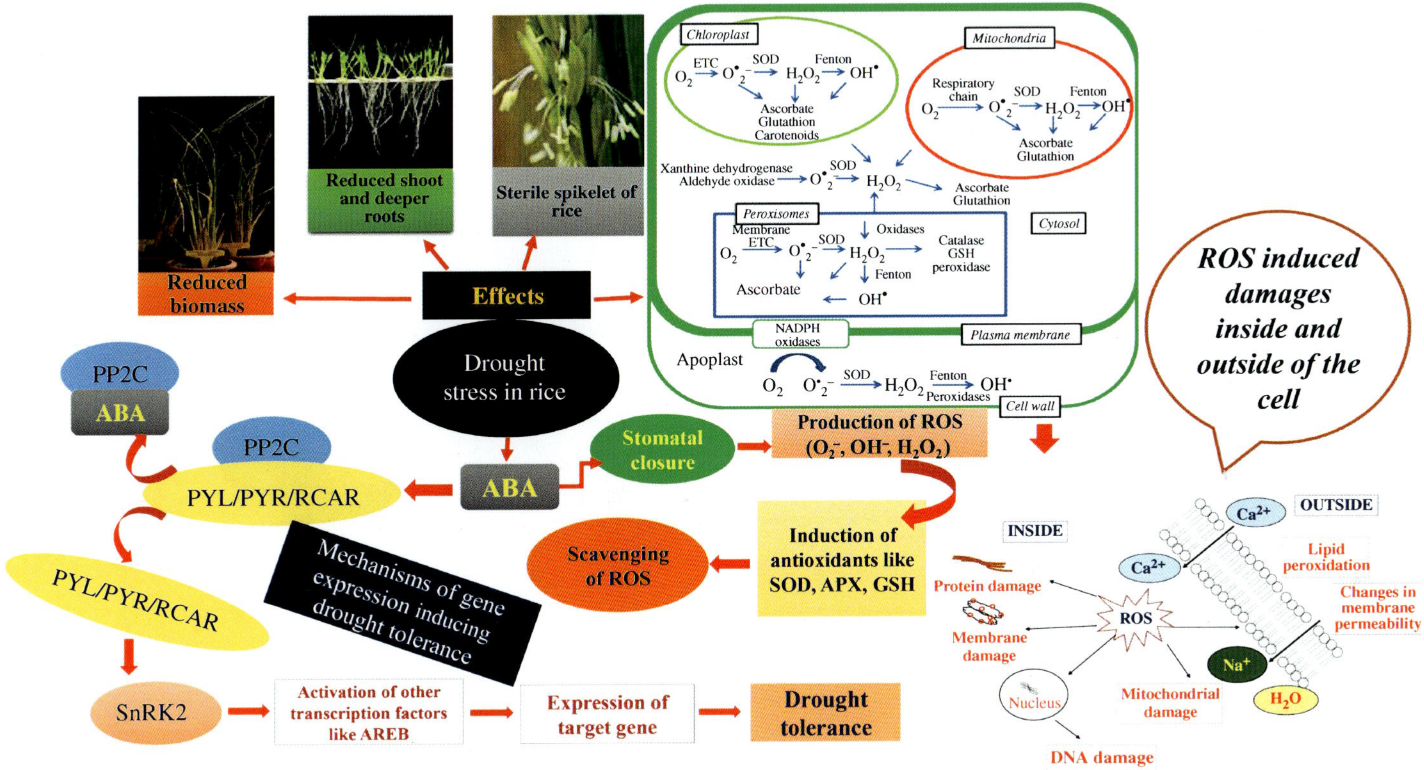

Figure 13.1 ROS generation and defense mechanisms in rice plants in response to drought stress. *Source:* Bhaben Tanti. *Note:* Its consequences for plant physiological processes and mechanisms of drought tolerance, including the regulation of genes associated with tolerance.

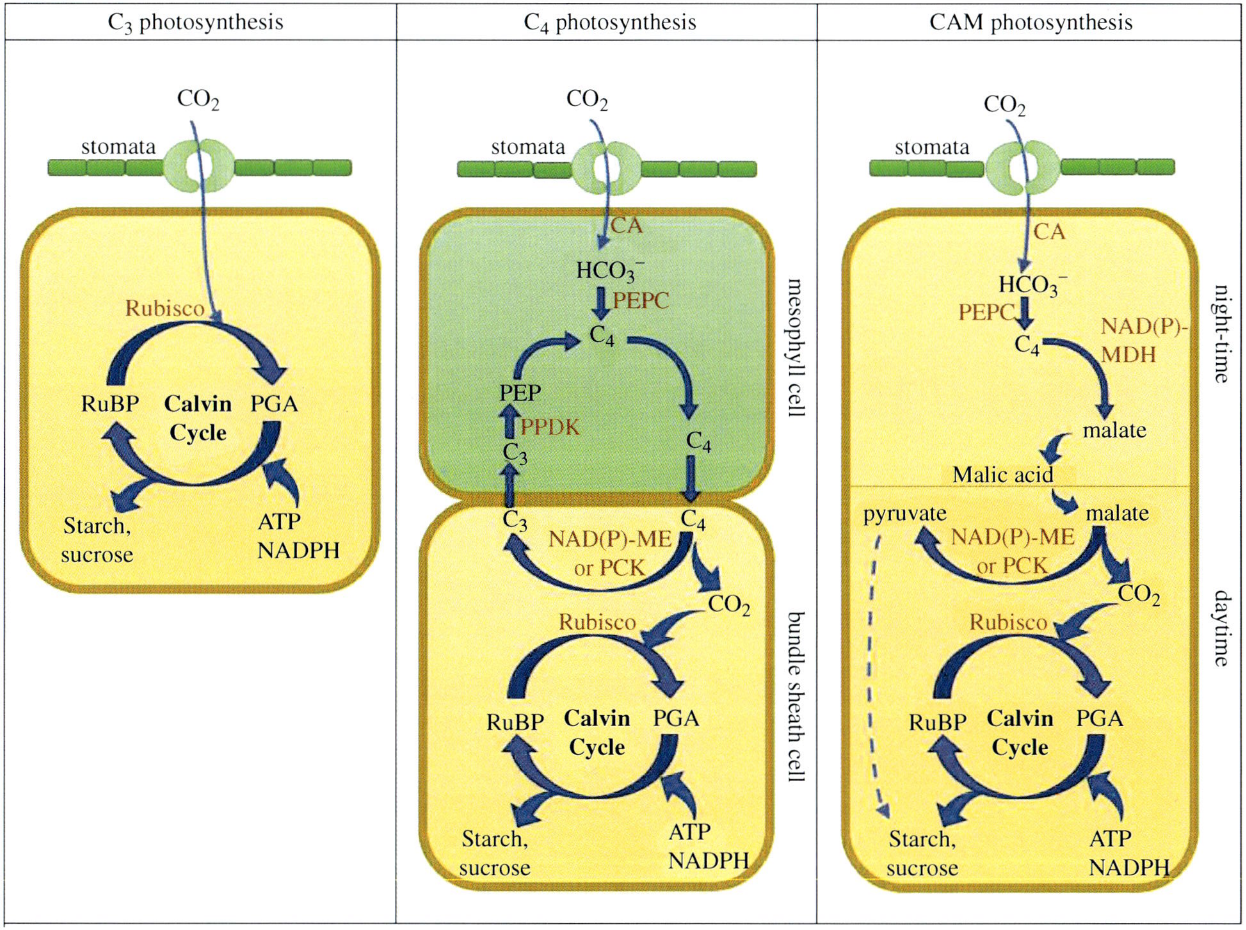

Figure 14.1 Schematic diagram of C_3 CAM and C_4 photosynthesis. Rubisco ribulose-1, 5-bisphosphate carboxylase/oxygenase, RuBP ribulose 1,5-bisphosphate, PGA phosphoglyceric acid, NADPH nicotinamide adenine dinucleotide phosphate hydrogen, $HCO_{3\_}$ bicarbonate,CA carbonic anhydrase, NADP nicotinamide adenine dinucleotide phosphate, ATP adenosine tri phosphate, NADP-MDHNADP-malate dehydrogenase, NADPME:NADP-dependent malic enzyme, PCK phosphoenolpyruvate carboxykinase, PEPC phosphorenolpyruvate carboxylase, PEP phosphorenolpyruvate, PPDK pyruvate orthophosphate dikinase.

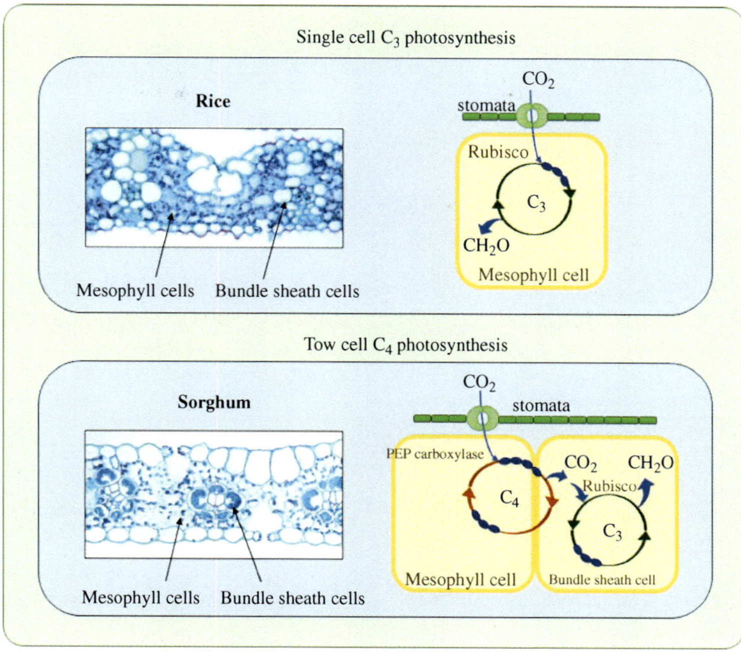

Figure 14.2 In C_3 photosynthesis, atmospheric CO_2 is primarily fixed by the enzyme rubisco. On the other hand, in the C_4 pathway, the primary CO_2-fixing enzyme is phosphoenolpyruvate carboxylase (PEP Carboxylase). Also, in the C_4 photosynthesis, the entire photosynthetic pathway is spatially separated between mesophyll cells and the bundle sheath cells. This unique cell arrangement is termed as Kranz anatomy.

Figure 15.3 Development of *Indica* golden rice varieties. *Note:* (a) Team of golden rice researchers (from left to right) Karabi Datta, Swapan K. Datta, Ingo Potrykus and Peter Beyer; Source: Swapan K. Datta (b) A representative photo of DH golden rice (BR29) and its control counterpart (Datta et al. 2007). Source: Datta et al. 2007 © Springer Nature (c) The HPLC estimation shows presence of β-carotene in DH golden rice grains. (Datta et al. 2007). (d) Growing of different *indica* golden rice homozygous lines (for multiple generations) at greenhouse facility of University of Calcutta, Kolkata, India. Source: Swapan K. Datta.

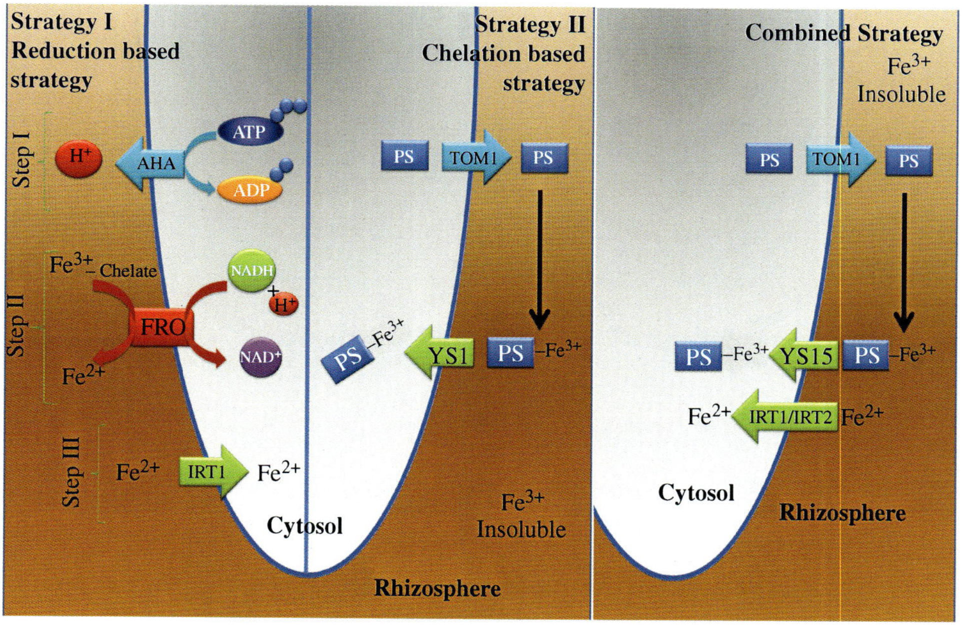

Figure 17.2 Strategies used by rice plants for Fe uptake from rhizosphere to roots.

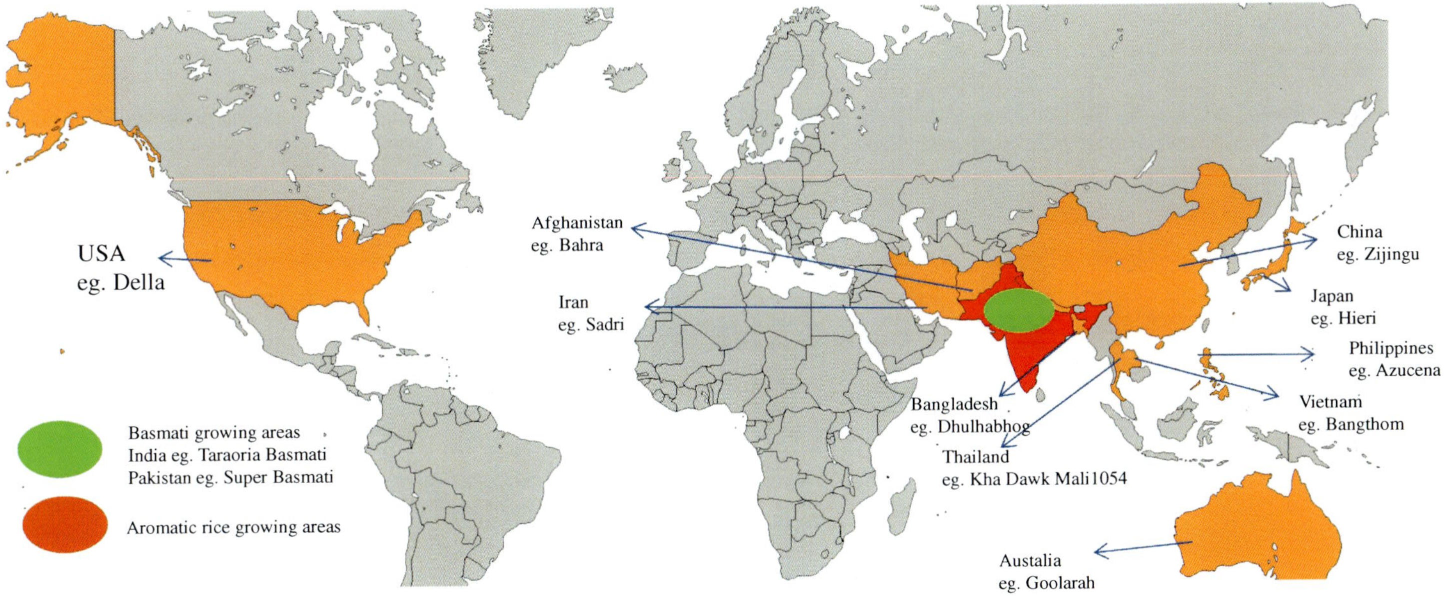

Figure 18.1 Global distribution of aromatic rice varieties.

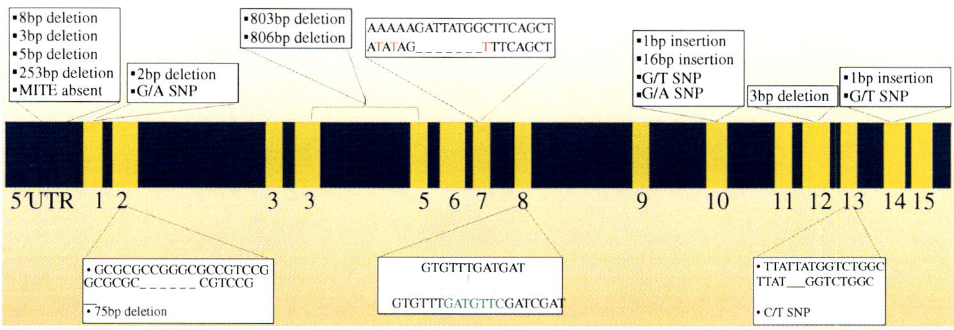

Figure 18.2 Overall distributions of alleles in betain aldehyde dehydragenase (*Badh2*) gene. *Note:* Yellow color lines indicate exons. The dashes (----) indicates deletion, red color nucleotides are SNPs and green color nucleotides are insertion.

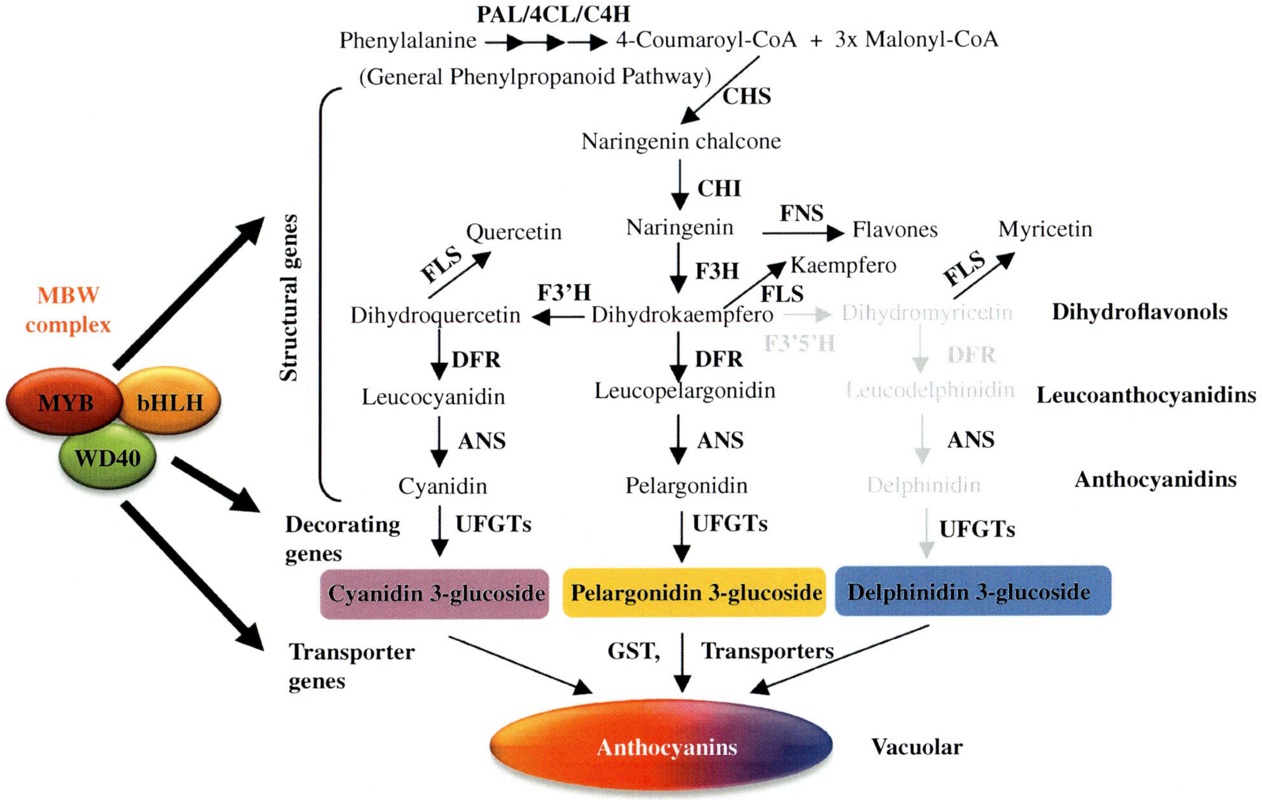

Figure 19.1 Schematic representation of the anthocyanin biosynthetic pathway in plant (rice). *Note:* The general phenylpropanoid pathway includes PAL (phenylalanine ammonia lyase), C4H (cinnamate 4-hydroxylase), and 4CL (4-coumaryol CoA ligase). CHS, chalcone synthase; CHI, chalcone isomerase; F3H, flavanone 3-hydroxylase; F3′H, flavonoid 3′-hydroxylase; F3′5′H, flavonoid 3′,5′-hydroxylase; DFR, dihydroflavonol 4-reductase; ANS, anthocyanidin synthase; UFGT, flavonoid 3-O-glucosyltransferase; FLS, flavonol synthase; FNS, flavone synthase. Gray fonts represent pathways that do not exist in rice due to F3′5′H and DFR no function.

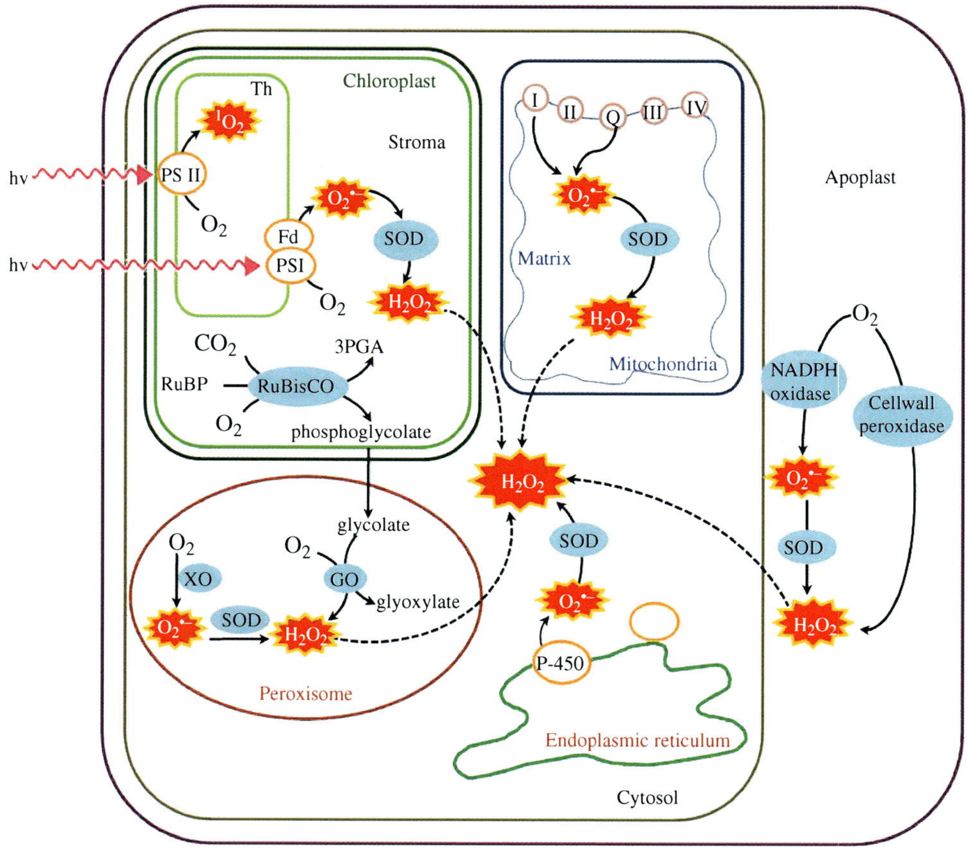

Figure 21.1 Major sites and sources of ROS production in plant cells. *Source:* Adapted from Hossain, M.A., Teixeira da Silva, J.A., and Fujita, M. (2011b).

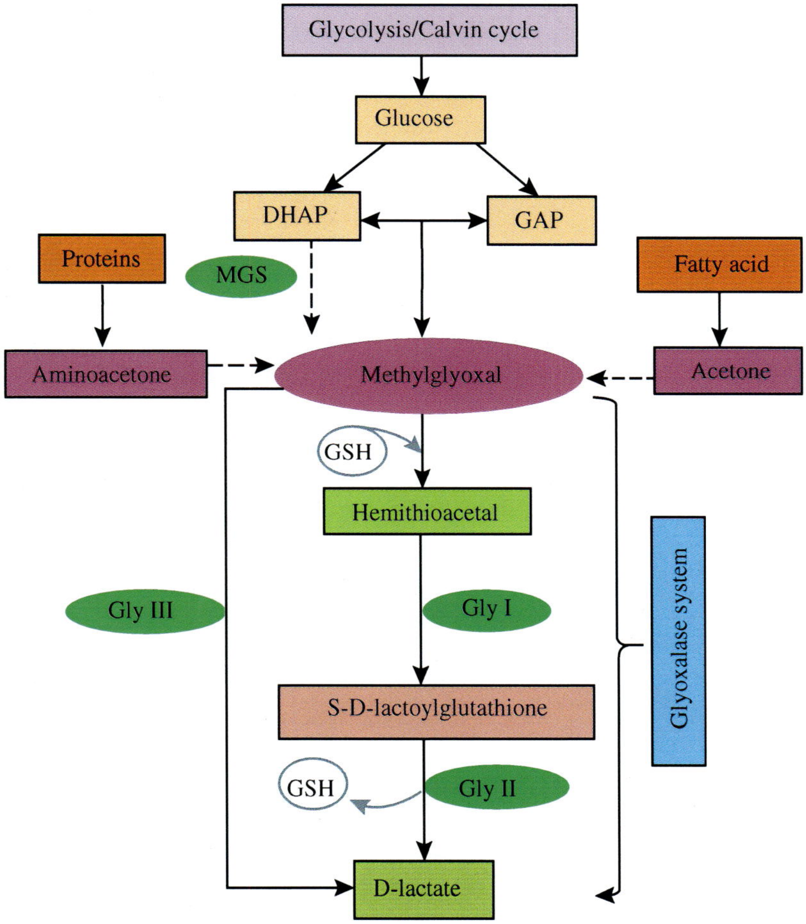

Figure 21.2 Schematic representation of the glyoxalase system in plants. *Source:* Modified from Hossain, M.A., Teixeira da Silva, J.A. and Fujita, M. (2011b) and Bhowal, B., Singla-Pareek, S.L., Sopory, S.K., et al. (2020). *Note:* MG is mainly produced during glycolysis and the Calvin cycle as a by-product of carbohydrate metabolism from triose sugar phosphates namely glyceraldehyde 3-phosphate (GAP) and dihydroxyacetone phosphate (DHAP). A small amount of MG is synthesized from aminoacetone and acetone in protein and lipid metabolism, respectively. Cytotoxic MG is principally detoxified via the glyoxalase system where two enzymes, namely, glyoxalase I (Gly I) and glyoxalase II (Gly II), are associated to transform MG. In this pathway, MG spontaneously reacts with glutathione (GSH) to form hemithioacetal, which is converted to S-D-lactoylglutathione (SLG) by Gly I. Subsequently, this S-D-lactoylglutathione is transformed into D-lactate with regeneration of GSH by the catalytic function of Gly II. Apart from conventional glyoxalase system, glyoxalase III (Gly III) irreversibly converts MG to D-lactate without requiring GSH. The dotted lines indicate the pathways that have not yet been confirmed in the plant system.

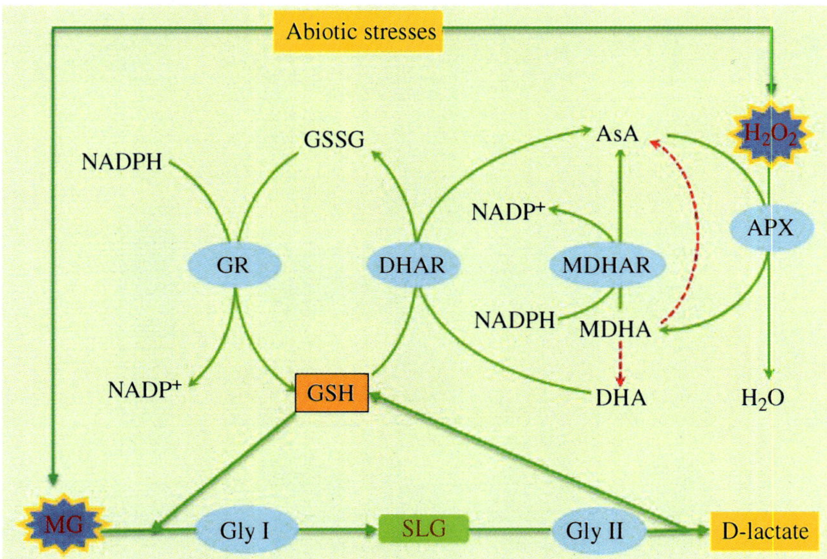

Figure 21.4 Inter-connection of the AsA-GSH pathway and the glyoxalase pathway through GSH.
Note: For further discussion, seeSection 21.4 in the main text.

17

Molecular Breeding for Iron Bio-Fortification in Rice Grain: Recent Progress and Future Perspectives

Elssa Pandit[1,2], *Swapnil Pawar*[1,3], *Priyadarshini Sanghamitra*[1] *and Sharat K. Pradhan*[1]

[1] *Crop Improvement Division, ICAR-National Rice Research Institute, Cuttack 753006, Odisha, India*
[2] *Department of Biosciences and Biotechnology, Fakir Mohan University, Balasore, 756020 Odisha, India*
[3] *Department of Plant Biotechnology, K. K. Wagh College of Agricultural Biotechnology, Nashik, 422003 Maharastra, India*

Introduction

Iron (Fe) is an essential micronutrient for human growth and health. More than two billion people suffer from Fe deficiency annually, which results in mortality of about 0.8 million people in the world (Stoltzfus and Dreyfuss 1998; de Benoist et al. 2008; Wessells and Brown 2012; Arcanjo et al. 2013; Prasad et al. 2017; Simbauranga et al. 2015; Freitas et al. 2016; Trijatmiko et al. 2016; WHO 2016; Santos et al. 2017; Swamy et al. 2018). More than half of the world population, mostly from developing countries, is suffering from bioavailable micronutrient deficiencies (Seshadri 1997; Da Silva 2013; Shahzad et al. 2014; Mahender et al. 2016). Fe-deficiency in the human body has evolved as a constant nutritional deficiency problem in the world, and billions of people are suffering from Fe deficiency anemia, which results in decrease in red blood cells, hemoglobin, and myoglobin (Hallberg 2000). Iron deficiency leads to irregular heartbeats, pregnancy complications, and abnormal growth of infants, and affects other metabolic processes (UNICEF 2017). The symptoms of Fe deficiency include fatigue, pallor, dizziness, weakness, reduced cognition performance, and work capacity (Lynch 2003; Pandav et al. 2013). More than 85% of the pregnant women in South Asia are affected by Fe deficiency, leading to maternal mortality and impaired mental development in children (McMahon 2010). Low hemoglobin and prevalence of anemia is highest among preschool children and women in Southeast Asia, Western Pacific and Africa (Stevens et al. 2013).

Micronutrient supplementation, food fortification, dietary diversification, and biofortification are the strategic ways to triumph over such deficiency-related problems. Management of Fe deficiency disorders through food fortification has not reached the target level in rural areas due to economical inadequacy, unsatisfied marketing, and lack of knowledge. Biofortification increases quantity and quality of nutrients in crops at the genetic level so that it can fulfill demand of nutrients (Bouis et al. 2011). It usually refers to the process of

Molecular Breeding for Rice Abiotic Stress Tolerance and Nutritional Quality, First Edition.
Edited by Mohammad Anwar Hossain, Lutful Hassan, Khandakar Md. Iftekharuddaula, Arvind Kumar and Robert Henry.

enriching food grains for nutritive elements through genetic approaches, which is an easy, effective, and economical way to supplement micronutrient deficiencies. In order to eradicate Fe deficiency–related problems, Fe-biofortification in rice will be a useful strategy for poor people in developing countries where rice is the staple food (Nakandalage et al. 2016; Swamy et al. 2016; Trijatmiko et al. 2016; Garg et al. 2018). More than of half of the world population and around 65% of the population of India depend on rice for their major calorie intake (Prasad et al. 2017). This affirms the credibility of rice as a proper candidate for bio-fortification to combat malnutrition in developing countries.

High-yielding modern rice varieties are poor in grain Fe content (Zimmermann and Hurrell 2002; Brinch-Pedersen et al. 2007; Mahender et al. 2016; Pradhan et al. 2020). However, genetic variations for Fe content in grains exist in rice germplasms (Yang et al. 1998; Jiang et al. 2008; Zeng et al. 2010; Mahender et al. 2016; Trijatmiko et al. 2016). Commonly grown rice varieties contain only 7–8 μg Fe g^{-1}; however, a few germplasm lines contain as high as 147 μg Fe g^{-1} in brown rice (Zeng et al. 2010). The high Fe-containing rice germplasm lines will be useful as potential donors in the biofortification programs of the high-yielding popular rice varieties using various approaches. Also, Fe content in rice grain can be enhanced manyfold using transgenic approaches. Once a stable transgenic donor line is obtained, the increased Fe content can be transferred to high-yielding varieties through molecular-breeding approaches (Masuda et al. 2013). Micronutrient improvement may not be antagonistic to yield enhancement in rice as separate genomic regions control yield and micronutrient traits in major crops (Welch and Graham 2004; Pradhan et al. 2019; Pradhan et al. 2020). The transport mechanism of Fe from soil to rice grain, the genes/QTLs involved in the process, and the different approaches for Fe enhancement with special reference to molecular breeding are described in the following sections of the chapter.

Iron : From Soil to Rice Grain

Understanding of the Fe uptake mechanism is needed for enhancement of Fe content in rice. Iron is present in fairly large amounts in soil, but mostly in ferric form, which is not bio-available to the plants. The plants can absorb only ferrous form of Fe. Hence, plants have developed different mechanisms to absorb Fe from soil. These mechanisms have been reviewed in detail by many authors (Walker and Connolly 2008; Palmer and Guerinot 2009; Morrissey and Guerinot 2009; Bashir et al. 2010). To summarize the process, there are two strategies, namely, the reduction-based strategy and the chelation-based strategy, adopted by plants for Fe uptake. The oxidation state of absorbed Fe differs in both the strategies, that is, in strategy I (reduction based), ferrous (Fe^{2+}) geta absorbed whereas in strategy II (chelation-based) ferric (Fe^{3+}) is used (Römheld and Marschner 1986).

The transport of Fe from soil to seeds is summarized in Figure 17.1. Fe gets diffused from soil to roots by apoplastic movement. Fe enters root epidermis and travels to the stele via the root cortex and endodermis through symplastic movement. Then Fe is exported into the xylem vessel. The casparian strip helps in blocking the apoplastic leakage. After reaching the xylem, Fe gets transported to leaves through the transpiration stream. Then through the sieve element of phloem it reaches the seed by symplastic movement (Kim and Guerinot 2007). There are many transporters and enzymes involved in the entire process of symplastic Fe transport.

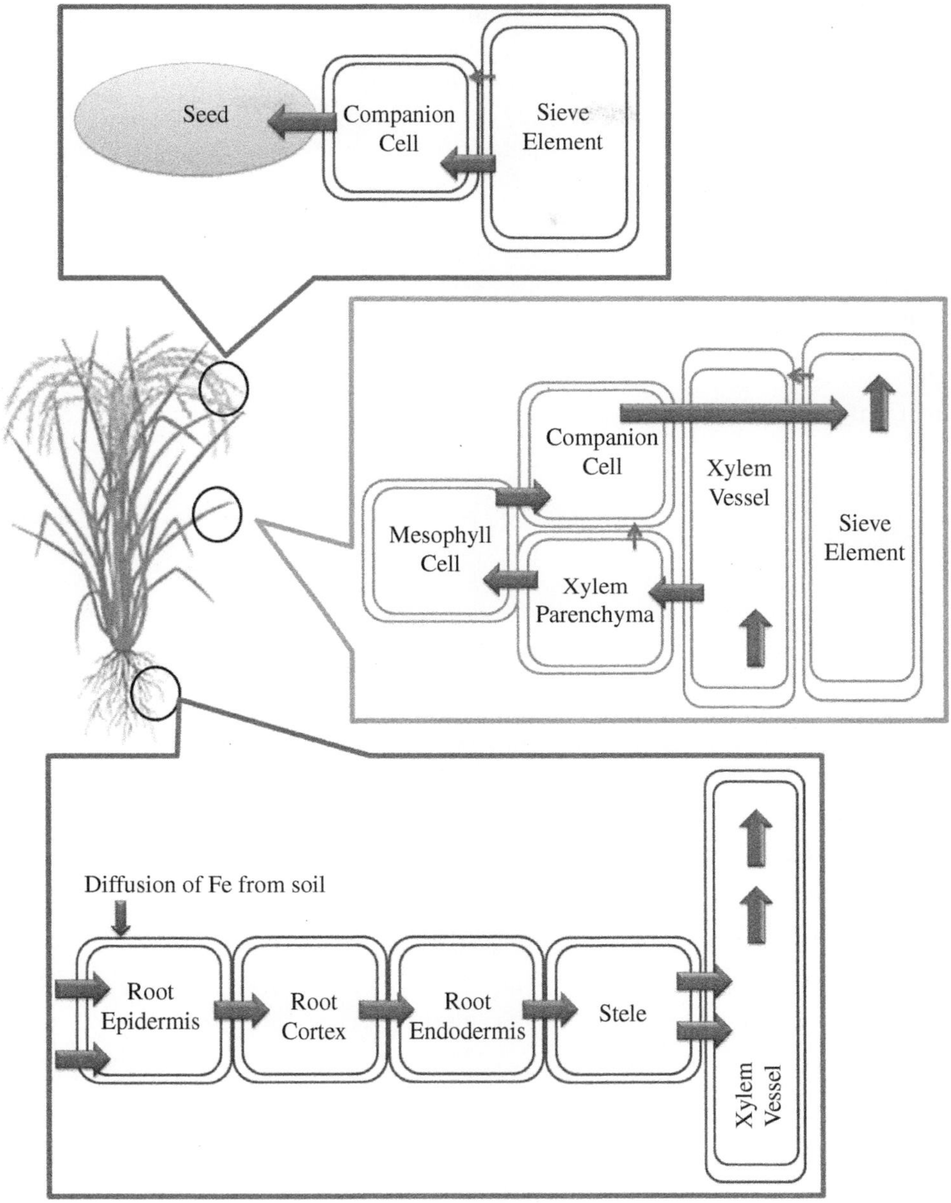

Figure 17.1 Schematic diagram for movement of Fe from soil to seed. *Note:* The red arrow depicts symplastic movement through transporters and blue arrow shows apoplastic movement.

Genes, Transporters, and Mechanisms of Fe Transport

Plants utilize two types of strategies for Fe uptake from soil namely, strategy I (reduction-based strategy) and strategy II (chelation-based strategy). Usually, non-leguminous plants use the reduction-based strategy, whereas leguminous plants use the chelation-based strategy. Non-leguminous graminaceous plants like maize and wheat

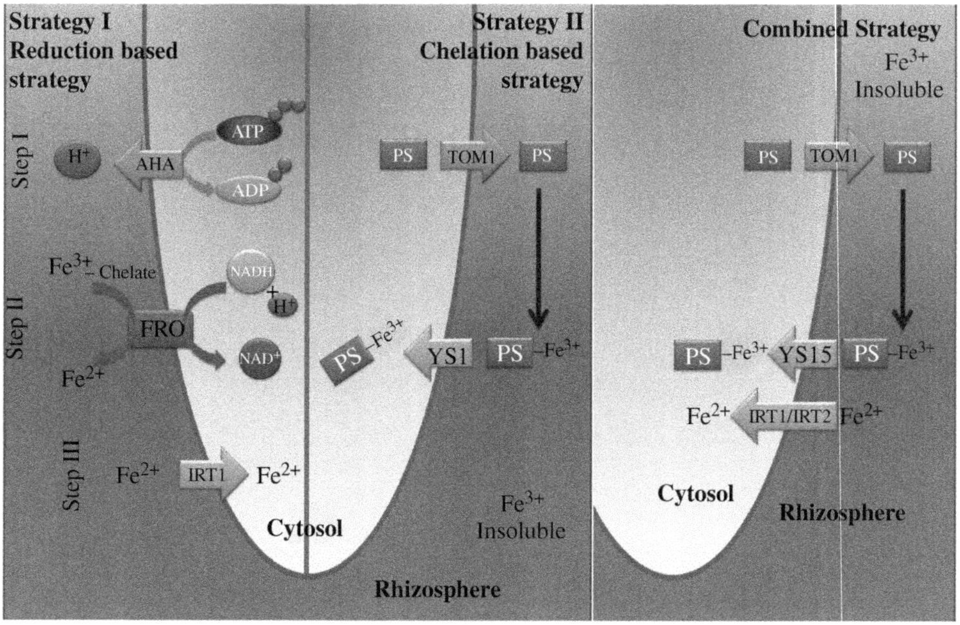

Figure 17.2 Strategies used by rice plants for Fe uptake from rhizosphere to roots

use strategy II, whereas rice has the ability to utilize both reduction-based as well as chelation-based strategies. A recent study by Wairich et al. (2019) through comparative transcription analysis revealed about the adoption of the combined strategy by *Oryza sativa* group containing AA genome only. The detailed mechanism of both pathways as well as the combined strategy has been depicted in Figure 17.2. In strategy I, Fe^{3+} (non-bioavailable) is reduced to Fe^{2+} (bioavailable) form of iron and then transported into the cytosol, whereas in strategy II, Fe^{3+} is attached to a chelating agent and then transported into cytosol in chelated form (Figure 17.2).

In strategy I, the rice plant releases protons to the surrounding rhizosphere by proton pumps, that is, autoinhibitory H+ ATPases (AHA) in order to increase the acidity of soil. Along with protons, many phenolics are released to the rhizosphere by the rice plant that help in increasing the solubility of Fe ions (Römheld and Marschner 1986; Rodríguez-Celma et al. 2013; Fourcroy et al. 2014; Schmidt and Eickhorst 2014). This process takes place under Fe-deficient conditions. Under low pH of soil, Fe^{3+} gets reduced to Fe^{2+} by activity of Ferric reductase oxidase (FRO) that is present in the epidermal layer of rice plant (Vert et al. 2002; Dinneny et al. 2008). Then Fe^{2+} is transported across the plasma membrane through Fe-transporters located on the membrane. OsIRT1, a divalent cation transporter belonging to the Zinc and Iron permease (ZIP) family plays a key role in the transport of Fe^{2+} through the plasma membrane (Eckhardt et al. 2001; Ishimaru et al. 2006). In addition to OsIRT1, many other Fe^{2+} transporters in the plasma membrane of rice roots, namely OsIRT2, OsNRAMP1 and OsNRAMP5, are involved in transport of Fe^{2+} into the cytosol (Ishimaru et al. 2006, 2012; Takahashi et al. 2011; Ogo et al. 2014). The rice plant has phenolics efflux transporters (PEZ1 and PEZ2) in order to secrete phenolics that have chemical properties for chelating and reducing Fe^{3+} to Fe^{2+} for its uptake under conditions of Fe deficiency in soil (Yoshino and Murakami 1998; Bashir et al. 2011; Ishimaru et al. 2011). PEZ2 located on the plasma membrane of the root epidermis helps in secretion of protocatechuic acid and caffeic acid into the rice root rhizosphere (Bashir et al. 2011; Ogo et al. 2014). A very low ferric-chelate reductase activity is observed in rice under Fe deficiency conditions (Ishimaru et al. 2006). The crop is adapted to anaerobic conditions, where Fe^{2+} is abundantly available, hence direct uptake of this form occurs without active ferric-chelate reduction.

In strategy II, chelating agents like mugineic acid (MA), nicotianamine (NA), malate, succinate, and citrate facilitate transportation of Fe (López-Millán et al. 2000, 2012). These Fe chelating molecules known as phytosiderophores (PS) are secreted by the roots to the rhizosphere, thereby freely available to bind to Fe^{3+} (Takagi 1976; Takagi et al. 1984, 1988). Many transporters facilitate the process of secretion of PS. For example, transporter of mugineic acid (TOM1) helps in secretion of phytosiderophore 2-deoxy-mugineic acid (DMA) through the plasma membrane (Nozoye et al. 2011). Then the Fe^{3+}-PS complex, formed in the rhizosphere, is transported into root cells by different tissue-specific transporters (Römheld and Marschner 1986; Von Wiren et al. 1994; Curie et al. 2009). These transporters include ITP (Iron transport protein), dehydrins (DHN), oligopeptide transporter 3 (OPT3), OsYSL2, OsYSL9, OsYSL15, OsYSL16, OsYSL18, AtYSL1, AtYSL2 and AtYSL3 (Hell and Stephan 2003; Wintz et al. 2003; Koike et al. 2004; Waters et al. 2006; Morrissey and Guerinot 2009; Aoyama et al. 2009; Inoue et al. 2009; Lee et al. 2009; Ishimaru et al. 2010; Kakei et al. 2012; Lee et al. 2012; Mendoza-Cózatl et al. 2014; Senoura et al. 2017; DiDonato et al. 2004). These transporters specifically transport the Fe^{3+}-PS complex through the plasma membrane but not Fe^{2+} (Hell and Stephan 2003; Morrissey and Guerinot 2009). All these transporters and enzymes for DMA-based Fe uptake are highly up-regulated under Fe-deficiency conditions to meet the Fe requirement by the plant. Specifically, Fe-nicotianamine and Fe-MAs are transported by Ferroportins (FPN1), Fe^{3+}-citrate complex by FRDL1 (ferric reductase defective 3 ortholog), and Fe^{3+}-DMA complex by OsYSL15, (Von Wiren et al. 1999; Inoue et al. 2003; Green and Rogers 2004; Durrett et al. 2007; Rellán-Álvarez et al. 2008; Rellán-Álvarez et al. 2009; Yokosho et al. 2009; Morrissey and Guerinot 2009; Kakei et al. 2012; Lee et al. 2012; Bonneau et al. 2016; Yokosho et al. 2016).

In the combined strategy, the rice plant uses both reduction- and chelation-based strategies for Fe uptake. The chelation-based strategy is enhanced under Fe deficiency condition by up-regulation of the transporters and increased secretion of chelating agents.

QTLs Involved for Enhancement of Fe in Rice Grain

Iron content in rice grain is governed by multiple genes and QTLs located on different chromosomes (Table 17.1). These QTLs were detected by bi-parental mapping (Gregorio et al. 2000; James et al. 2007; Stangoulis et al. 2007; Garcia-Oliveira et al. 2009; Norton 2010; Anuradha et al. 2012) as well as genome-wide association mapping or linkage disequilibrium mapping (Descalsota et al. 2018; Pradhan et al. 2020). The phenotypic variance of these QTLs individually varied up to 30% (Gregorio et al. 2000; James et al. 2007; Stangoulis et al. 2007; Garcia-Oliveira et al. 2009; Kaiyang et al. 2008; Norton 2010; Anuradha et al. 2012; Zhang et al. 2014; Nawaz et al. 2015; Descalsota et al. 2018; Pradhan et al.

2020). Gregorio et al. (2000) reported three QTLs located on chromosome 7, 8, and 9 with 19–30% phenotypic variance for grain Fe content. Another three QTLs on chromosomes 2, 8, and 12 were reported by James et al. (2007) that explained 17%, 18%, and 14% of the total PV, respectively. Two QTLs were detected on chromosome 2 by Stangoulis et al. (2007) and Garcia-Oliveira et al. (2009) in independent studies. The QTL, *qFe-1* explaining 25.81% phenotypic variance was reported by Norton (2010). Anuradha et al. (2012) identified 14 QTLs for grain Fe and Zn content in unpolished rice. Recent study by Dixit et al. (2019) reported two major QTLs for high grain Fe content in rice. However, comparative analysis for multi-season data revealed QTLs *qFe1.1, qFe1.2, qFe6.1*, and *qFe6.2* to be consistent in population. Genome-wide association mapping or linkage disequilibrium mapping could identify many QTLs namely, *qFe3:1, qFe3:2, qFe7:1, qFe9:1, qFe9:2, qFe10:1, Fe11:1, qFe3.3*, and *qFe7.3* controlling grain Fe-content (Descalsota et al. 2018; Pradhan et al. 2020). These studies showed the QTLs governing grain Fe content are scattered throughout the rice genome. Nawaz et al. (2015) had identified loci on chromosome 1, 2, 3, 4, 5, 6, 10, and 12 and reported that the genes, namely, *OsACA2, OsACA9, OsHMA4, OsZIP1, OsZIP2, OsZIP3, OsZIP5, OsZIP9, OsHma2*, ABC transporter, *OsNAS3,* heavy metal transporter, CHY zinc finger, and *OsCNGC,* present in those loci to be responsible for Fe content in rice (Table 17.1).

Approaches for Fe Biofortification in Rice

Biofortification strategies aim to produce nutrient-dense safe food sustainably and economically (Saltzman et al. 2014). Such biofortification can be achieved by three approaches, that is, through agronomic practices, transgenic/genetic engineering approaches, and/or through breeding approaches, which can be of two different types, namely, by conventional breeding and molecular breeding (Garg et al. 2018). Each approach has its own advantages and disadvantages. A higher number of crops including rice has been targeted for Fe biofortification through transgenic approaches, whereas the practical utility and success of biofortified lines has been achieved through breeding approaches.

Iron Biofortification in Rice Through Agronomic Practices

Iron biofortification in rice through agronomic practices includes direct physical application of Fe fertilizers/nutrients to enhance the nutritional status and health of crops, thereby helping in improving human nutritional status (Cakmak and Kutman 2017). This method improves the nutritional quality of the plant temporarily. Unlike other nutrients, application of Fe fertilizers like ferrous sulphate to the soil cannot be of much value, because the ferrous form of Fe soon gets converted to the ferric form. Hence, for Fe fortification, foliar application of Fe fertilizers is the preferred method (Prasad and Shivay 2020). Grain Fe content of rice could be increased by 20–40% by application of this method (Dhaliwal et al. 2010; Yadav et al. 2013; Prasad and Shivay 2018). Also, it is reported that supplementation of ferrous sulfate to rice plantlets increases Fe concentration in germinated brown rice up to 15.6 fold than control (Yuan et al. 2013).

Table 17.1 Genes/QTLs available for iron biofortification program in rice grain and their position and linked markers.

Sl. No	QTLs	Population used	Chromosome	Markers to be used for molecular breeding	References
1	Fe con		1	STSI163	Wan et al. (2003)
2	*qFe.2*	DHs of IR64/Azucena	2	RM53-RM300	James et al. (2007)
3	*qFe.8*	DHs IR64/Azucena	8	RM137-RM325A	James et al. (2007)
4	*qFe1*	RILs Zhenshan 97/Minghui 63	1	RM259-RM243	Kaiyang et al. (2008)
5	*qFe2–1*	ILs *O. rufipogon*/Teqing	2	RM6641	Garcia-Oliveira et al. (2009)
6	*qFe.11*	RILs Zhenshan 97/Minghui 63	11	RZ536-TEL3	Kaiyang et al. (2008)
7	*qFe12.2*	RILs Madhukar/Swarna	12	RM260-RM7102	Anuradha et al. (2012)
8	*qFe1.1*	RILs Madhukar/Swarna	1	RM243-RM488	Anuradha et al. (2012)
9	*qFe1.2*	RILs	1	RM488-RM490	Anuradha et al. (2012)
10	*qFe5.1*	RILs of Madhukar/Swarna	5	RM574-RM122	Anuradha et al. (2012)
11	*qFe7.1*	RILs	7	RM234-RM248	Anuradha et al. (2012)
12	*qFe.12*	DHs	12	RM270-RM17	Anuradha et al. (2012)
13	*qFe12.1*	RILs		RM17-RM260	Anuradha et al. (2012)
14	*OsHMA2* (Os06g0700700)		6	RM340-RM494	Du et al. (2013)
15	*qFe.2*	LT/TL-RILs Lemont/TeQing	2	RM452	Zhang et al. (2014)
16	*OSZIP1* (Os01g0972200)		1	RM431	Nawaz et al. (2015)
17	*OsZIP2* (Os03g0411800)		3	RM5864 and RM3400	Nawaz et al. (2015)
18	*OsCNGC14* (Os03g0758300)		3	RM8203	Nawaz et al. (2015)
19	*OsACA2* (Os02g0176700)		2	RM279 and RM555	Nawaz et al. (2015)
20	1. *OsNAS3* (Os07g0689600) 2. Peroxidase (0Os07g0677300)		6	RM162	Nawaz et al. (2015)
21	*OsZIP3* (Os04g0613000)		4	RM3217	Nawaz et al. (2015)

Table 17.1 (Continued)

Sl. No	QTLs	Population used	Chromosome	Markers to be used for molecular breeding	References
22	1. *OsZIP5* (Os05g0472700) 2. *OsZIP9* (Os05g0472400)		5	RM161	Nawaz et al. (2015)
23	Os10g0456800 CHY zinc finger family protein		10	RM271	Nawaz et al. (2015)
24	*OsACA9* (Os12g0136900)		12	RM19	Nawaz et al. (2015)
25	qFe3.1	MAGIC plus population	3	S3_13026098	Descalsota et al. (2018)
26	qFe3.1	MAGIC plus population	3	S3_29613916	Descalsota et al. (2018)
27	qFe7.1	MAGIC plus population	7	S7_29288449	Descalsota et al. (2018)
28	qFe9.1	MAGIC plus population	9	S9_11707990	Descalsota et al. (2018)
29	qFe9.2	MAGIC plus population	9	S9_22203582	Descalsota et al. (2018)
30	qFe10.1	MAGIC plus population	10	S10_13883426	Descalsota et al. (2018)
31	qFe11.1	MAGIC plus population	11	S11_7122972	Descalsota et al. (2018)
32	qFe3.3	GWAS	3	RM7	Pradhan et al. (2020)
33	qFe7.3	GWAS	7	RM1132	Pradhan et al. (2020)

Limitations of Agronomic Biofortification

Although the agronomic biofortification approach is simple and inexpensive, it is drastically affected by environment. It requires extra attention in terms of source of the micronutrient, application method, and environmental conditions, as well as the effect of the component on the environment. It must be applied in every crop season making it laborious and less economical. Hence, it is more practicable for developed countries than in developing ones.

Iron Biofortification in Rice Through Transgenics/Genetic Engineering

The transgenic approach seems to be a highly effective method for enhancement of Fe in rice, where there is no taxonomic constraint. The development of a transgenic biofortified crop takes a substantial amount of time, effort, and huge financial requirements at the initial stage, but when considered in the long run, it is a more cost-effective and sustainable approach than agronomic biofortification programs (Hefferon 2016; White and Broadley 2005). This method has proved its credibility in development of Fe-dense transgenic rice by using gene sources from wheat and soybean. The first transgenic rice having threefold increase in grain Fe content by soybean gene transfer was reported by Goto et al. (1999). There are many success stories of higher Fe uptake by rice plants, thereby increasing grain Fe content by a transgenic approach (Lucca et al. 2001; Higuchi et al. 2001; Kobayashi et al. 2001; Vasconcelos et al. 2003; Koike et al. 2004; Khalekuzzaman et al. 2006; Masuda et al. 2008; Suzuki et al. 2008; Masuda et al. 2009; Lee et al. 2009, 2010; Zhang et al. 2015; Ishimaru et al. 2010; Ogo et al. 2011; Johnson et al. 2011; Soumitra et al. 2012; Kobayashi and Nishizawa 2012; Lee et al. 2012; Zhang et al. 2012; Bashir et al. 2013; Masuda et al. 2013; Slamet-Loedin et al. 2015; Trijatmiko et al. 2016; Boonyaves et al. 2017). These transgenic rice lines with higher Fe content derived from other crops are available for successful crop improvement programs. These transgenic lines may also be useful as potential donors by rice breeders for enhancement of Fe content in popular rice varieties through molecular breeding approaches.

Limitations of Fe Biofortification through Genetic Engineering

One major limitation of this approach is very low acceptance by consumers and low adoption by farming communities (Al-Babili and Beyer 2004). Some consumers restrict genetically modified crops because they consider it "unnatural" (Manshardt 2004). Another drawback is that different countries have adopted different rules and regulation for commercialization of transgenic products. Unfortunately, these ongoing political and economic scenarios are not receptive to this approach (Inaba and Macer 2004). Moreover these regulatory processes take a long time and are expensive (Watanabe et al. 2005). For example, golden rice research published its first details in 2000 eight years after the project's initiation (Ye et al. 2000), and still it has not been approved for cultivation in many countries except for the United States. Despite all these issues, researchers are applying more efforts to the transgenic approach than to breeding-based approaches although the success rate is very low due to the time required for gene identification, gene modification, expression analysis, and studying possible effects on other life forms.

Iron Biofortification in Rice through Conventional Breeding

This is the most widely used method for development of nutrient-dense rice (Garg et al. 2018). Three rice varieties, namely, BRRI dhan 62, BRRI dhan 64, and BRRI dhan 72 have been developed by the Bangladesh Rice Research Institute, Bangladesh, and IR68144-3B-2-2-3 has been jointly developed by IRRI, Philippines and India, by using this approach. This method was popular, as earlier as there was not much information available on genes and QTLs governing Fe content in rice grain. Presently, the molecular breeding method is being preferred over this method.

Limitations of Fe Biofortification through Conventional Breeding

The strategy of conventional plant breeding to enhance micronutrient concentration has proved to be promising, and it is sustainable and cost-effective over long time periods; even so there are several restrictions, along with aspects to genetic variability in selected genotypes, and the time required to develop genotype. In some instances, these problems can be overcome by crossing diverse genome relatives and then introducing targeted traits into commercial cultivars; however, in many instances, it seems to be impossible to breed for the characteristics of interest to us using nonspecific conventional technique, and the efforts taken for long durations are somewhat impractical. Nonspecific conventional breeding occurs when two plants are crossed and several other traits are transferred in addition to the trait of interest to us, which potentially affects the targeted trait (DANIDA 2002).

Iron Biofortification in Rice through Molecular Breeding

The molecular breeding approach is the most reliable and suitable method for grain Fe content in rice. This approach is widely used for trait enhancement. Earlier this approach could not be used on a large scale due to a lack of suitable genes, major QTLs, and their associated markers to be deployed for Fe-content enhancement in rice through molecular breeding. Iron content in rice grain is a complex trait governed by multiple genes and QTLs that make trait enhancement very intricate. Currently, the genes, QTLs, and pathways involved in the process are well characterized (Kim and Guerinot 2007; Bashir et al. 2010; Conte and Walker 2011; Garg et al. 2018; Majumder et al. 2019). Many donor lines with high grain Fe content have now been screened and identified (Ravindra Babu 2013; Jahan et al. 2013; Mahender et al. 2016; Pradhan et al. 2019). In addition transgenic donor lines are also available to be used in molecular breeding programs. Development of high-Fe rice along with other essential micronutrients through molecular breeding is a priority research area of the ICAR-National Rice Research Institute, Cuttack, India, the ICAR-Indian Institute of Rice Research, Hyderabad, and the International Rice Research Institute, Philippines.

Strategies Involved in Molecular Breeding

Molecular breeding is a target-oriented, cost-effective, and less time-consuming approach for enhancing Fe content in rice. The detailed processes involved are outlined in Figure 17.3. In this process, the identification of source/donor lines and the identification of genes/QTLs are important steps for the process. Bi-parental mapping and genome-wide association mapping (GWAS) are two commonly used methods for identification of the

QTLs that are responsible for grain Fe content. Further characterization of these genomic locations and cloning leads identification and characterization of specific responsible genes. Gene mapping through bi-parental mapping is costly and time consuming and provides less resolution than GWAS (Pradhan et al. 2019; Flint-Garcia et al. 2003; Roy et al. 2006; Pradhan et al. 2016; Pandit et al. 2017), which can be overcome by the recent GWAS approach. Association mapping is carried out based on the linkage disequilibrium in an existing natural population or developed populations (Pradhan et al. 2019; Flint-Garcia et al. 2003; Pradhan et al. 2016; Pandit et al. 2017). Many QTLs have been reported for Fe content through bi-parental mapping (Wan et al. 2003; James et al. 2007; Kaiyang et al. 2008; Garcia-Oliveira et al. 2009; Norton 2010; Anuradha et al. 2012; Du et al. 2013; Zhang et al. 2014; Nawaz et al. 2015; Descalsota et al. 2018; Pradhan et al. 2020), whereas recently few robust QTLs have been identified by Zhao et al. (2011), Huang et al. (2015) and Pradhan et al. (2020). The major advantage of molecular breeding is that the presence of the target gene is checked through foreground selection by using linked or gene-specific molecular markers in each backcross generation. This confirms the presence of the trait in the end product along with highest genome recovery of the recurrent rice variety, which gets biofortified for Fe content. Pyramiding the major QTLs governing grain Fe content with *Ferritin*/IDS3/YSL/VIT genes can be a good strategy for enhancing grain Fe content. For this purpose, multiple donor lines having the target genes can be used.

Recent Molecular Breeding Approaches to Enhance Grain Fe Content

The genes and QTLs governing Fe content have been reported. In total, 21 QTLs and many genes are now available for development of Fe-fortified lines through molecular breeding approaches. The genes like *Ferritin* in soybean and *IDS3* in barley, which on transgenic incorporation into rice, have showed many-fold increase in grain Fe content in rice (Masuda et al. 2012; Aung et al. 2013). These developed lines can be used as donor lines to incorporate these genes into various high-yielding popular rice varieties. The genes—namely VIT (vacuolar iron transport), NAS (nicotianamine synthase), and YSL (yellow stripe-like)—that have been identified to enhance Fe content in rice can be transferred for high expression through molecular breeding approaches.

Through Ferritin Genes for Iron Storage Improvement

Ferritin is a protein that has a role in storing up to 4500 iron molecules in endosperms in a nontoxic form (Ford et al. 1984; Briat et al. 2006). Expression of *ferritin* is promoted by globulin, and glutenin and it can lead to increasing iron content by 2–3.7-fold in polished rice grain. Soybean *Ferritin* is bioavailable for the human system; so SoyferH1 and SoyferH2 have been introduced in rice. Compared to single-gene introduction strategies, introgression of multiple gene clusters has shown increases in iron folds. According to Masuda et al. (2012) and Aung et al. (2013), over-expression of these soybean *Ferritin* with barley *NAS1* and *OsYLS* construct increased iron content up to 3.4-fold and 6-fold in polished rice grains of Myanmar and Japanese genotypes. Through the genetic engineering approach, Boonyaves et al. (2017) have proved that the gene cassettes with *PvFerritin AtIRT1,* and *Atnas1* reached the highest iron content—up to 10.46 $\mu g\ g^{-1}$ DW. These developed rice lines can be subjected to gene-editing tools like CRISPR-Cas9 and TALENs to make it free from unwanted cassettes and then be used as donor lines for molecular

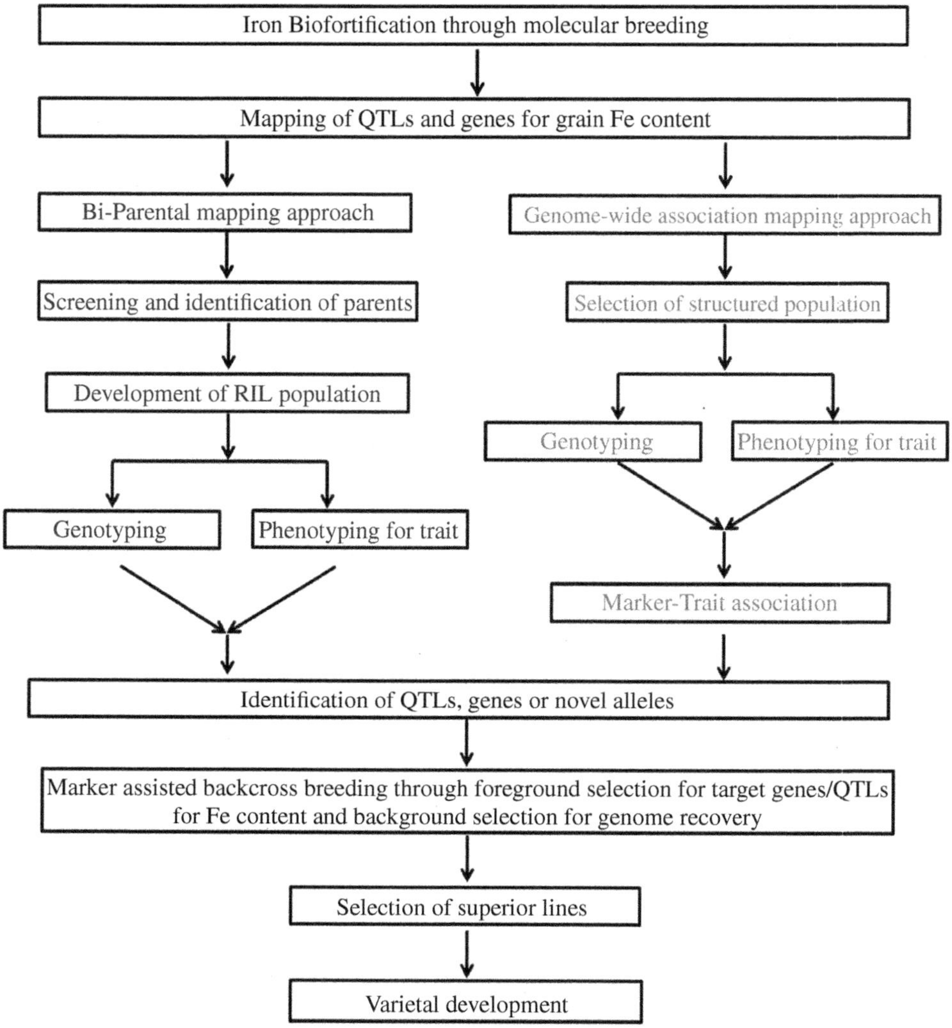

Figure 17.3 Steps involved in the process of molecular breeding for enhancing grain Fe content in rice.

breeding programs. Enhancement of Fe content in rice grain through the *ferritin* gene is summarized in Table 17.2.

Through VIT Genes

VIT was first identified in Aabidopsis by Kim et al. (2006), as a homologous of yeast Cross-complementer of CSG1 protein 1 (CCC1), which has a role of iron transport in yeast vacuoles (Li et al. 2001). Two VIT genes were reported in rice, namely, *OsVIT1* and *OsVIT2*, which show different expression in rice plants by maintaining iron translocation in different parts of plants. Specifically, expression of *OsVIT* was found in flag leaves by transportation of Fe^{2+} into vacuoles via tonoplast (Kim et al. 2006). Reported studies used modification and silencing approaches for *OsVIT* genes to increase Fe content in rice grains

Table 17.2 Enhancement of Fe in rice grain through *ferritin* gene.

Genes-promoter	Rice genotypes	Increased Fe folds	References
GlutelinA2 pro-Osfer2	Swarna-Pusa sugandhi 2	2.54 fold (polished rice grain)	Paul et al. (2012)
OsGluB1 pro-*SoyferH1*	*Indica* cv.-IR64	3.4 fold (polished rice grain)	Oliva et al. (2014)
OsGluA2 pro-*Osfer2*	*Indica* cv.-Pusa-Sugandh II	2.1 fold (polished rice grain)	Paul et al. (2012)
OsGluB1 pro-*SoyferH1*	*Japonica* cv.- Kitaake	1.5 fold (brown rice grain)	Qu et al. (2005)
OsGluB1 pro-*SoyferH1*	*Indica* cv.-IR68144	3.7 fold (polished rice grain)	Vasconcelos et al. (2003)
OsGluB1 pro-*SoyferH1*	*Japonica* cv.-Taipei 309	2.2 fold (brown rice grain)	Lucca et al. (2002)
OsGluB1 pro-*SoyferH1*	*Japonica* cv.- Kitaake	2 fold (polished rice grain)	Goto et al. (1999)

rather than flag leaves (Bashir et al. 2013; Zhang et al. 2012). According to Bashir et al. (2013), the knockdown approach of *OsVIT* genes in rice increases Fe concentration up to 8 $\mu g\, g^{-1}$ in polished grains; moreover, further study showed that expression of OsVIT2 and MIT (Mitochondrial iron transporter) influence the concentration and localization of iron in both brown and polished rice. Both the research groups—namely, Zhang et al. (2012) and Bashir et al. (2013)—have managed to achieve 1.4- and 1.8-fold increases in Fe concentration in rice grains through modification in *OsVIT* via T-DNA mutant lines development (Table 17.3). However, overexpression of *OsVIT* genes has not been tested, and its physiological influence has not been studied much (Ravet et al. 2009). Descalsota et al. (2018) studied genome-wide association mapping in 144 MAGIC Plus rice lines, and surprisingly this research group has found significant linking of *OsVIT* genes along with other candidate genes in their population. Besides, the silencing/knockdown approach of *OsVIT* is only applicable when mutated rice plants are cultivated in Cd^{2+} (Cadmium) uncontaminated soil because research has found accumulation of Cd^{2+} content in rice when it was cultivated in contaminated soil solution (Zhang et al. 2012). Thus, a great amount of investigation is needed to avoid absorption of Cd^{2+} like toxic compounds to reach health safety.

Through IDS3 Gene

Uptake of Fe by rice plants is dependent on soil redox potential and the pH of the soil solution. Fe gets oxidized in the insoluble form in aerobic soils at alkaline pH, while under acidic pH Fe is more available to plants in soluble form. Several genes are involved in the Fe uptake mechanism (Marschner 1995). The rice plant lacks IDS (iron deficiency specific) genes, namely, *IDS2* and *IDS3*. However, barley is rich in IDS genes (Kobayashi et al. 2001). This could be a reason why barley has more resistance to Fe deficiency than rice.

Table 17.3 Enhancement of Fe in rice grain through *OsVIT* genes.

Genes/promoter	Rice genotypes	Increased Fe folds/μg g^{-1}	References
OsVIT1 /*OsVIT2* T-DNA insertion line	*Japonica* cv. Zhonghua11 and *Japonica* cv. Dongjin	1.4 fold (brown rice grain)	Zhang et al. (2012)
OsVIT2 T-DNA insertion line	*Japonica* cv. Dongjin	1.8 fold (polished Rice grain)	Bashir et al. (2013)

Table 17.4 Enhancement of Fe in rice grain through *IDS3* gene.

Genes/promoter	Rice genotypes	Increased Fe folds	References
Fer-NAS-NAAT-IDS3	*Japonica* cv. Tsukinohikari	3 fold (polished rice grain)	Masuda et al. (2013)
35S pro-barley 20-kb *IDS3* genome fragment	*Japonica* cv. Tsukinohikari	1.21 fold (polished rice grain)	Masuda et al. (2008)
35S pro-barley 20-kb *IDS3* genome fragment	*Japonica* cv. Tsukinohikari	1.3 fold (brown rice grain)	Suzuki et al. (2008)

Barley contains several biosynthetic genes for mugeneic acid such as *HvNAS1*, *HvNAAT-A*, *HvNAAT-B*, *HvDMAS1*, *IDS2*, and *IDS3*, which are up-regulated in roots of barley under Fe-deficit condition (Takahashi et al. 1999; Higuchi et al. 1999; Nakanishi et al. 2000; and Bashir et al. 2006).

Barley IDS genes in rice showed up to 1.21-fold increased Fe accumulation in brown and polished rice grains (Masuda et al. 2008). These IDS lines, cultivated in Fe-deficit soil, showed 1.26 times higher Fe content in brown rice. Later Masuda et al. (2013) analyzed the introduction of three barley genes along with soybean *ferritin* gene in rice (Table 17.4). The effect of gene clusters (*Fer-NAS-NAAT-IDS3*) showed two to three times higher concentration of Fe in polished rice grains than control lines. Growing of these developed lines in calcareous soil exhibited the highest Fe concentration of 4.9 μg g^{-1} in polished grains.

Through NAS Genes

Synthesis of nicotianamine (NA), a non-protein amino acid, is achieved by trimerization of S-adenosyl L-methionine (SAM) via NAS enzyme (Noma and Noguchi 1976; Higuchi et al. 1994). Nicotianamine (NA) acts as a natural metal chelator for FeII and FeIII and is also involved in iron homeostasis and translocation by becoming converted into mugineic-acid-family phytosiderophores (MAs), such as 20-deoxymugineic acid (DMA) (Masuda et al. 2012, Hell and Stephan 2003, Koike et al. 2004). *Nicotianamine synthase* (*NAS*) catalyzes the synthesis of nicotianamine (NA) from *S*-adenosyl methionine. Rice

Table 17.5 Enhancement of Fe in rice grain through *NAS* and *YSL* genes.

Genes-promoter	Rice genotypes	Increased Fe folds	References
OsNAS2-SoyferH-1	IR64	6 fold (polished grain)	Trijatmiko et al. (2016)
35S pro-*OsNAS1, 2, 3*	*Japonica* cv. Nipponbare	4 fold (polished grain)	Johnson et al. (2011)
Maize *Ubiquitin* pro-*OsNAS2*	*Japonica* cv. Kitaake	2.9 fold (polished grain)	Lee et al. (2012)
OsSUT1 pro-*OsYSL2*	*Japonica* cv. Tsukinohikari	4.4 fold (polished grain)	Ishimaru et al. (2010)
Maize *Ubiquitin* pro-*OsNAS3*	*Japonica* cv. Dongjin	2.9 fold (polished grain)	Lee et al. (2009)
35S pro-*HvNAS1*	*Japonica* cv. Tsukinohikari	2.3 fold (polished grain)	Masuda et al. (2009)

contains three different *NAS* genes, namely, *OsNAS1, OsNAS2* and *OsNAS3*, which are involved in long-distance transport of Fe in the plant. The enhancement of grain Fe content through these NAS genes is summarized in Table 17.5.

Through YSL Genes

YSL genes encode Fe transporters involved in long-distance transport of iron in the form of DMA-FeIII/Na-FeII. To name a few, *OsYSL2, OsYSL15, OsYSL16,* and *OsYSL18* play important roles in Fe transport (Aoyama et al. 2009; Ishimaru et al. 2010; Zheng et al. 2012). Koike et al. (2004) reported 18 putative genes of the YS1 family, including the genes *OsYSL1* to *OsYSL17* and *ZmYS1*. Further, Koike et al. (2004) identified the role of other YS1 genes along with place of act in the plant. *OsYSL2,* a metal chelator transporter of iron, played an important role in translocation and accumulation of Fe in the endosperm of rice grain. *OsYSL6* was expressed in both roots and leaves under Fe-sufficient and Fe-deficient conditions, whereas *OsYSL13* was expressed in shoots only. Further, *OsYSL15* expression was induced in roots under Fe-deficiency situations. *OsYSL14* and *OsYSL16* expressed in both shoots and roots but with higher level of expression in roots. Ishimaru et al. (2010) also corroborated the findings of Koike et al. (2004).

Through qFe QTLs

More than 21 QTLs have been reported for high grain Fe content in rice (Stangoulis et al. 2007; Norton 2010; Anuradha et al. 2012; Pradhan et al. 2020). These QTLs are distributed throughout the genome, explaining up to 30% of phenotypic variance individually (Gregorio et al. 2000; James et al. 2007; Stangoulis et al. 2007; Garcia-Oliveira et al. 2009; Kaiyang et al. 2008; Norton 2010; Anuradha et al. 2012; Zhang et al. 2014; Nawaz et al. 2015; Descalsota et al. 2018; Dixit et al. 2019; Pradhan et al. 2020). These QTLs were mapped from either developed recombinant inbred lines, double haploid population, or structured natural populations through bi-parental mapping or association mapping approaches. The details regarding these QTLs including the population from which they were mapped, their position, the linked markers that can be utilized for marker-assisted breeding program, and so on are presented in Table 17.1. But there is no major QTL explaining >50% of phenotypic variance for grain Fe content is available to date that can be used in molecular breeding programs. Hence, combinations of these QTLs having cumulative effects can be carefully chosen for Fe biofortification in rice.

Global Success Towards Development of Fe Fortified Rice Lines

Transgenic rice lines with higher Fe content derived from other crops are already available for successful crop improvement programs (Goto et al. 1999; Lee et al. 2009; Ogo et al. 2011; Kobayashi and Nishizawa 2012; Lee et al. 2012; Soumitra et al. 2012; Bashir et al. 2013; Masuda et al. 2013; Slamet-Loedin et al. 2015; Trijatmiko et al. 2016; Boonyaves et al. 2017). But these transgenic lines could not be utilized at commercial level/farmer's field that would serve the purpose to combat Fe-deficiency-related problems in humans. These transgenic lines may be useful as potential donors to rice breeders for enhancement of Fe content in popular rice varieties through molecular-breeding approaches. However, there has been significant advancement of Fe-fortified lines through breeding approaches. The Bangladesh Rice Research Institute, Bangladesh, has released three rice varieties, namely, BRRI dhan62, BRRI dhan 72, and BRRI dhan64, which are reported to have high Fe as well as high Zn content. India and the Philippines have jointly developed an improved line namely, IR68144-3B-2-2-3 having high grain Fe content (~21 ppm in brown rice). Jalmagna, a deep water rice variety in India is rich in grain Fe content. The identified donor lines that can be used for Fe biofortification either in molecular breeding approaches or conventional breeding approaches are summarized in Table 17.6.

Conclusion and Future Perspective

Biofortification is a promising approach to find long-term and sustainable solutions to eradicate Fe-malnutrition-related problems from the world. The agronomic practice of enhancing Fe in rice grain by application of fertilizer may pollute the environment in addition to the fact there is no confirmation of achieving the target. The transgenic method of fortification faces many challenges related to ethical issues. Many countries do not allow consumption of transgenic food. Also, there is a need for more studies on the impact of the transgenic and antibiotic-resistance genes associated with the construct on human health and the environment. Recent gene editing tools such as TALENs and CRISPR-Cas9 could be used to remove the undesired cassette without affecting the plant. Again, gene-edited crops are subjected to different regulations and monitoring. Under such scenarios, the molecular plant breeders have much scope to develop eco-friendly Fe-biofortified rice varieties to combat Fe malnutrition. The cis-genic transfer of target genes through marker-assisted backcross breeding is economical, less time and labour consuming, and sustainable.

Table 17.6 List of donor germplasms lines for Fe biofortification and Fe-fortified lines developed through breeding approach.

Donor/developed lines	Names	Reference/source
Donor germplasm lines for Fe biofortification	MSE-9, Kalanamak, Kanchana, Karjat, Chittimutyalu, Udayagiri, Jyothi, VRM 7, Matta Triveni, Varsha	Ravindra Babu (2013)
	Enghi, Dakh Shail, Moina Moti, Hamai, Patnai, Khak Shail, Lal Gotal, Ful Kainja, Kakua Binni, Sada Binni, BRRI dhan5, BRRI dhan 28, and BRRI dhan 29.	Jahan et al. 2013
	Annada, ASD16, CH-45, Nagina 22, IR-29, Pusa Sugandha-1, IRGC-106187, IR68144-3B-2-2-3, IRGC-105320, IRGC-105320, IRGC-86476, CH-45, Jyoti, HKR-126, Jalmagna, Zuchem, Kalabath, Pusa Basmati, Noothipattu, Pitchavari, Thanu, TKM-9, NDR-6279, and Aghonibora.	Mahender et al. (2016)
	Chittimuthylu, BPT5204, Kalanamak, ET25450, IET 24775, IET 24760, IET 24316, BPT5204, IET 23832, IET25441, IET25465, IET 23829 and IET 24779	Pradhan et al. (2020)
Developed/ released Fe biofortified varieties	BRRI dhan62, BRRI dhan72, BRRI dhan 64 (Bangladesh)	CIAT, HarvestPlus (2014)
	IR68144-3B-2-2-3 (improved line) (India, Phillipines)	Gregorio et al. (2000)
	Jalmagna (India)	Gregorio et al. (2000)

Acknowledgments

The authors acknowledge the support from The Director, ICAR-National Rice Research Institute, Cuttack, India and Department of Science and Technology (WOS-A Program), Govt. of India.

References

Al-Babili, S. and Beyer, P. (2004). Golden rice on the road-five years to go? *Trends Plant Sci.* 10 (12): 565–573.

Anuradha, K., Agarwal, S., Rao, Y.V. et al. (2012). Mapping QTLs and candidate genes for iron and zinc concentrations in unpolished rice of Madhukar x Swarna RILs. *Gene* 508 (2): 233–240.

Aoyama, T., Kobayashi, T., Takahashi, M. et al. (2009). OsYSL18 is a rice iron (III)-deoxymugineic acid transporter specifically expressed in reproductive organs and phloem of lamina joints. *Plant Mol. Biol.* 70: 681–692.

Arcanjo, F.P.N., Santos, P.R., and Arcanjo, C.P.C. (2013). Daily and weekly iron supplementations are effective in increasing haemoglobin and reducing anemia in infants. *J. Trop. Pediatr.* 59: 175–179.

Aung, M.S., Masuda, H., Kobayashi, T. et al. (2013). Iron biofortification of Myanmar rice. *Front. Plant Sci.* 4: 158.

Bashir, K., Inoue, H., Nagasaka, S. et al. (2006). Cloning and characterization of deoxymugineic acid synthase genes from graminaceous plants. *J. Biol. Chem.* 281: 32395–32402.

Bashir, K., Ishimaru, Y., and Nishizawa, N.K. (2010). Iron uptake and loading into rice grains. *Rice* 3: 122–130. https://doi.org/10.1007/s12284-010-9042-y.

Bashir, K., Ishimaru, Y., Shimo, H. et al. (2011). The rice mitochondrial iron transporter is essential for plant growth. *Nat. Commun.* 2: 322–329.

Bashir, K., Takahashi, R., Akhtar, S. et al. (2013). The knockdown of *OsVIT2* and *MIT* affects iron localization in rice seed. *Rice* 6 (1): 31.

de Benoist, B., McLean, E., Egli, I., and Cogswell, M. (2008). *Worldwide Prevalence of Anaemia 1993–2005: WHO Global Database on Anaemia*. Geneva: World Health Organization.

Bonneau, J., Baumann, U., Beasley, J. et al. (2016). Identification and molecular characterization of the nicotianamine synthase gene family in bread wheat. *Plant Biotechnol. J.* 14: 2228–2239.

Boonyaves, K., Ting-Ying, W., Wilhelm, G., and Navreet, K. (2017). Enhanced grain iron levels in rice expressing an *iron-regulated metal transporter*, *nicotianamine synthase* and *FERRITIN* gene cassette. *Front. Plant Sci.* 8: 130.

Bouis, H.E., Hotz, C., McClafferty, B. et al. (2011). Biofortification: a new tool to reduce micronutrient malnutrition. *Food. Nutr. Bull.* 32 (1): S31–S40.

Briat, J.F. and Cellier, F. (2006). Ferritins and iron accumulation in plant tissues. In: *Iron Nutrition in Plants and Rhizospheric Microorganisms*, vol. 17 (eds. L.L. Barton and J. Abadia), 341–357. Dordrecht: Springer.

Brinch-Pedersen, H., Borg, S., Tauris, B., and Holm, P.B. (2007). Molecular genetic approaches to increasing mineral availability and vitamin content of cereals. *J. Cereal Sci.* 46: 308–326.

Cakmak, I. and Kutman, U.B. (2017). Agronomic biofortification of cereals with zinc: a review. *Eur. J. Soil Sci.* 69: 172–180.

Conte, S.S. and Walker, E.L. (2011). Transporters contributing to iron trafficking in plants. *Mol. Plant* 4 (3): 464–476.

Curie, C., Cassin, G., Couch, D. et al. (2009). Metal movement within the plant: contribution of nicotianamine and yellow stripe 1-like transporters. *Ann. Bot.* 103: 1–11.

Da Silva, J.G. (2013). Food losses mean hunger. The think. Eat. Save. Reduce your footprint-campaign of the save food initiative is a partnership between UNEP, FAO and Messe Dusseldorf.

DANIDA. (2002). Assessment of potentials and constraints for development and use of plant biotechnology in relation to plant breeding and crop production in developing countries. Working paper. Ministry of Foreign Affairs. Denmark.

Descalsota, G.I.L., Swamy, M., Zaw, H. et al. (2018). Genome-wide association mapping in a rice MAGIC plus population detects QTLs and genes useful for biofortification. *Front. Plant Sci.* 9: 1347.

Dhaliwal, S.S., Sadana, U.S., Khurana, M.P.S. et al. (2010). Enrichment of rice grains with zinc and iron through ferti–fortification. *Indian J. Fertil.* 6 (7): 28–35.

DiDonato, R.J., Roberts, L.A., Sanderson, T. et al. (2004). *Arabidopsis* yellow stripe-like2 (YSL2): a metal–regulated gene encoding a plasma membrane transporter of nicotianamine–metal complexes. *Plant J.* 39: 403–414.

Dinneny, J.R., Long, T.A., Wang, J.Y. et al. (2008). Cell identity mediates the response of Arabidopsis roots to abiotic stress. *Science* 320: 942–945. https://doi.org/10.1126/science.1153795.

Dixit, S., Singh, U.M., Abbai, R. et al. (2019). Identifcation of genomic region(s) responsible for high iron and zinc content in rice. *Sci. Rep.* 9: 8136.

Du, J., Zeng, D., Wang, B. et al. (2013). Environmental effects on mineral accumulation in rice grains and identification of ecological specific QTLs. *Environ. Geochem. Health* 35: 161–170.

Durrett, T.P., Gassmann, W., and Rogers, E.E. (2007). The FRD3–mediated efflux of citrate into the root vasculature is necessary for efficient iron translocation. *Plant Physiol.* 144: 197–205.

Eckhardt, U., Mas Marques, A., and Buckhout, T.J. (2001). Two iron– regulated cation transporters from tomato complement metal uptake–deficient yeast mutants. *Plant Mol. Biol.* 45: 437–448.

Flint-Garcia, S.A., Thornsberry, J.M., and Buckler, E.S. (2003). Structure of linkage disequilibrium in plants. *Annu. Rev. Plant Biol.* 54 (1): 357–374.

Ford, G.C., Harrison, P.M., Rice, D.W. et al. (1984). Ferritin: design and formation of an iron–storage molecule. *Philos. Trans. R. Soc. Lond., B Biol. Sci.* 304: 551–565.

Fourcroy, P., Siso-Terraza, P., Sudre, D. et al. (2014). Involvement of the ABCG37 transporter in secretion of scopoletin and derivatives by *Arabidopsis* roots in response to iron deficiency. *New Phytol.* 201: 155–167.

Freitas, B.A., Lima, L.M., Moreira, M.E. et al. (2016). Micronutrient supplementation adherence and influence on the prevalence of anemia and iron, zinc and vitamin A deficiencies in preemies with a corrected age of six months. *Clinics* 71 (8): 440–448.

Garcia-Oliveira, A.L., Tan, L., Fu, Y., and Sun, C. (2009). Genetic identification of quantitative trait loci for contents of mineral nutrients in rice grain. *J. Integr. Plant Biol.* 51 (1): 84–92.

Garg, M., Sharma, N., Sharma, S. et al. (2018). Biofortified crops generated by breeding, agronomy, and transgenic approaches are improving lives of millions of people around the world. *Front. Nutr.* 5: 12.

Goto, F., Yoshihara, T., Shigemoto, N. et al. (1999). Iron fortification of rice seed by the soybean ferritin gene. *Nat. Biotechnol.* 17 (3): 282–286.

Green, L.S. and Rogers, E.E. (2004). FRD3 controls iron localization in *Arabidopsis. Plant Physiol.* 136: 2523–2531.

Gregorio, G.B., Senadhira, D., Htut, H., and Graham, R.D. (2000). Breeding for trace mineral density in rice. *Food. Nutr. Bull.* 21: 382–386.

Hallberg, L. (2000). Iron. In: *Human Nutrition and Dietetics*, 177–192. Edinburgh: Churchill–Livingstone.

HarvestPlus. (2014) Biofortification progress briefs. http://www.harvestplus.org/sites/default/files/Biofortification_Progress_Briefs_August2014_WEB_0.pdf

Hefferon, K.L. (2016). Can biofortified crops help attain food security? *Curr. Mol. Biol. Rep.* 2 (4): 180–185.

Hell, R. and Stephan, U.W. (2003). Uptake, trafficking and homeostasis in plants. *Planta* 2216 (4): 541–551.

Higuchi, K., Kanazawa, K., Nishizawa, N.K. et al. (1994). Purification and characterization of nicotianamine synthase from Fe-deficient barley roots. *Plant Soil* 165: 173–179.

Higuchi, K., Suzuki, K., Nakanishi, H. et al. (1999). Cloning of nicotianamine synthase genes, novel genes involved in the biosynthesis of phytosiderophores. *Plant Physiol.* 119: 471–479.

Higuchi, K., Watanabe, S., and Takahashi, M. (2001). Nicotianamine synthase gene expression differs in barley and rice under Fe–deficient conditions. *Plant J.* 25: 159–167.

Huang, Y., Sun, C., Min, J. et al. (2015). Association mapping of quantitative trait loci for mineral element contents in whole grain rice (*Oryza sativa* L.). *J. Agric. Food. Chem.* 63: 10885–10892.

Inaba, M. and Macer, D. (2004). Policy, regulation and attitudes towards agricultural biotechnology in Japan. *J. Int. Biotechnol. Law* 1 (2): 45–53.

Inoue, H., Higuchi, K., Takahashi, M. et al. (2003). Three rice nicotianamine synthase genes, *OsNAS1*, *OsNAS2* and *OsNAS3* are expressed in cells involved in long-distance transport of iron and differentially regulated by iron. *Plant J.* 36: 366–381.

Inoue, H., Kobayashi, T., Nozoye, T. et al. (2009). Rice *OsYSL15* is an iron-regulated iron(III)-Deo–xymugineic acid transporter expressed in the roots and is essential for iron uptake in early growth of the seedlings. *J. Biol. Chem.* 284: 3470–3479.

Ishimaru, Y., Suzuki, M., Tsukamoto, T. et al. (2006). Rice plants take up iron as a Fe^{3+}-phytosiderophore and as Fe. *Plant J.* 45: 335–346.

Ishimaru, Y., Masuda, H., Bashir, K. et al. (2010). Rice metal-nicotianamine transporter, OsYSL2, is required for the long-distance transport of iron and manganese. *Plant J.* 62: 379–390.

Ishimaru, Y., Bashir, K., Nakanishi, H., and Nishizawa, N.K. (2011). The role of rice phenolics efflux transporter in solubilizing apo–plasmic iron. *Plant Signaling Behav.* 6: 1624–1626.

Ishimaru, Y., Takahashi, R., Bashir, K. et al. (2012). Characterizing the role of rice NRAMP5 in manganese, iron and cadmium transport. *Sci. Rep.* 2: 286.

Jahan, G.S., Hassan, L., Begum, S.N., and Islam, S.N. (2013). Identification of Iron rich rice genotypes in Bangladesh using chemicalanalysis. *J Bangladesh Agril. Univ.* 11 (1): 73–78.

James, C.R., Huynh, B.L., Welch, R.M. et al. (2007). Quantitative trait loci for phytate in rice grain and their relationship with grain micronutrient content. *Euphytica* 154 (3): 289–294.

Jiang, S., Wu, J., Thang, N. et al. (2008). Genotypic variation of mineral elements contents in rice (*Oryza sativa* L.). *Eur. Food Res. Technol.* 228: 115–122.

Johnson, A.A.T., Kyriacou, B., Callahan, D.L. et al. (2011). Constitutive overexpression of the *OsNAS* gene family reveals single-gene strategies for effective iron– and zinc-biofortification of rice endosperm. *PLoS One* 6 (9): e24476.

Kaiyang, L., Li, L., Zheng, X. et al. (2008). Quantitative trait loci controlling Cu, Ca, Zn, Mn and Fe content in rice grains. *J. Genet.* 87 (3): 305–310.

Kakei, Y., Ishimaru, Y., Kobayashi, T. et al. (2012). OsYSL16 plays a role in the allocation of iron. *Plant Mol. Biol.* 79: 583–594.

Khalekuzzaman, M., Datta, K., Oliva, N. et al. (2006). Stable integration, expression and inheritance of the ferritin gene in transgenic elite indica rice cultivar BR29 with enhanced iron level in the endosperm. *Indian J. Biotechnol.* 5: 26–31.

Kim, S.A. and Guerinot, M.L. (2007). Mining iron: iron uptake and transport in plants. *FEBS Lett.* 581: 2273–2280.

Kim, S.A., Punshon, T., Lanzirotti, A. et al. (2006). Localization of iron in *Arabidopsis* seed requires the vacuolar membrane transporter VIT1. *Science* 314 (80): 1295–1298.

Kobayashi, T. and Nishizawa, N.K. (2012). Iron uptake, translocation and regulation in higher plants. *Annu. Rev. Plant Biol.* 63: 131–152.

Kobayashi, T., Nakanishi, H., Takahashi, M. et al. (2001). In vivo evidence that IDS3 from *Hordeum vulgare* encodes a dioxygenase that converts 2-deoxymugineic acid to mugineic acid in transgenic rice. *Planta* 212: 864–871.

Koike, S., Inoue, H., Mizuno, D. et al. (2004). OsYSL2 is a rice metal-nicotianamine transporter that is regulated by iron and expressed in the phloem. *Plant J.* 39: 415–424.

Lee, S., Chiecko, J.C., Kim, S.A. et al. (2009). Disruption of *OsYSL15* leads to iron inefficiency in rice plants. *Plant Physiol.* 150: 786–800.

Lee, S., Kim, S.A., Lee, J. et al. (2010). OsZIP5 is a plasma membrane zinc transporter in rice. *Plant Mol. Biol.* 73: 507–517.

Lee, S., Ryoo, N., Jeon, J.S. et al. (2012). Activation of rice *Yellow Stripe 1-Like 16* (*OsYSL16*) enhances iron efficiency. *Mol. Cell* 33: 117–126.

Li, L., Chen, O.S., McVey Ward, D., and Kaplan, J. (2001). CCC1 is a transporter that mediates vacuolar iron storage in yeast. *J. Biol. Chem.* 276: 29515–29519.

López-Millán, A.F., Morales, F., Abadía, A., and Abadía, J. (2000). Effects of iron deficiency on the composition of the leaf apoplastic fluid and xylem sap in sugar beet implications for iron and carbon transport. *Plant Physiol.* 124: 873–884. https://doi.org/10.1104/pp.124.2.873.

López-Millán, A.F., Grusak, M.A., and Abadía, J. (2012). Carboxylate metabolism changes induced by Fe deficiency in barley, a strategy II plant species. *J. Plant Physiol.* 169: 1121–1124. https://doi.org/10.1016/j.jplph.2012.04.010.

Lucca, P., Hurrell, R., and Potrykus, I. (2001). Genetic engineering approaches to improve the bioavailability and the level of iron in rice grains. *Theor. Appl. Genet.* 102: 392–397.

Lucca, P., Hurrell, R., and Potrykus, I. (2002). Fighting iron deficiency anemia with iron-rich rice. *J. Am. Coll. Nutr.* 21: 184S–190S.

Lynch, S.R. (2003). Iron deficiency anemia. In: *Encyclopaedia of Food Science and Nutrition* (eds. C. Benjamin, F. Paul and T. Luiz), 215–320. Amsterdam: Academic Press.

Mahender, A., Anandan, A., Pradhan, S.K., and Pandit, E. (2016). Rice grain nutritional traits and their enhancement using relevant genes and QTLs through advanced approaches. *Springerplus* 5: 2086.

Majumder, S., Datta, K., and Datta, S.K. (2019). Rice biofortification: high iron, zinc, and vitamin-A to fight against "hidden hunger.". *Agronomy* 9: 803.

Manshardt, R. (2004). Crop improvement by conventional breeding or genetic engineering: how different are they?. Cooperative extension service bio. 5.

Marschner, H. (1995). *Mineral Nutrition of Higher Plants*. 2nd Edition, Boston: Academic Press.

Masuda, H., Suzuki, M., Morikawa, K.C. et al. (2008). Increase in iron and zinc concentrations in rice grains via the introduction of barley genes involved in phytosiderophore synthesis. *Rice* 1: 100–108.

Masuda, H., Usuda, K., Kobayashi, T. et al. (2009). Overexpression of the barley nicotianamine synthase gene *HvNAS1* increases iron and zinc concentrations in rice grains. *Rice* 2: 155–166.

Masuda, H., Ishimaru, Y., Aung, M.S. et al. (2012). Iron biofortification in rice by the introduction of multiple genes involved in iron nutrition. *Sci. Rep.* 2 (1): 543.

Masuda, H., Aun, M.S., and Nishizawa, N.K. (2013). Iron biofortification of rice using different transgenic approaches. *Rice* 6: 40.

McMahon, L.P. (2010). Iron deficiency in pregnancy. *Obstet. Med.* 3: 17–24. https://doi.org/10.1258/om.2010.100004.

Mendoza-Cózatl, D.G., Xie, Q., Akmakjian, G.Z. et al. (2014). OPT3 is a component of the iron-signaling network between leaves and roots and misregulation of OPT3 leads to an over-accumulation of cadmium in seeds. *Mol. Plant* 7: 1455–1469.

Morrissey, J. and Guerinot, M.L. (2009). Iron uptake and transport in plants: the good, the bad, and the ionome. *Chem. Rev.* 109: 4553–4567.

Nakandalage, N., Nicola, S.M., Norton, R.M. et al. (2016). Improving rice zinc biofortification success rates through genetic and crop management approaches in a changing environment. *Front. Plant Sci.* 7: 1–13.

Nakanishi, H., Yamaguchi, H., Sasakuma, T. et al. (2000). Two dioxygenase genes, *Ids3* and *Ids2*, from *Hordeum vulgare* are involved in the biosynthesis of mugineic acid family phytosiderophores. *Plant Mol. Biol.* 44: 199–207.

Nawaz, Z., Kakar, K.U., Li, X. et al. (2015). Genome-wide association mapping of quantitative trait loci (QTLs) for contents of eight elements in brown rice (*Oryza sativa* L.). *J. Agric. Food. Chem.* 63: 8008–8016.

Noma, M. and Noguchi, M. (1976). Occurrence of nicotianamine in higher plants. *Phytochemistry* 15: 1701–1702.

Norton, G.J. (2010). Genetic mapping of the rice ionome in leaves and grain: identifcation of QTLs for 17 elements including arsenic, cadmium, iron and selenium. *Plant Soil* 329: 139–153.

Nozoye, T., Nagasaka, S., Kobayashi, T. et al. (2011). Phytosiderophore efflux transporters are crucial for iron acquisition in graminaceous plants. *J. Biol. Chem.* 286: 5446–5454.

Ogo, Y., Itai, R.N., Kobayashi, T. et al. (2011). OsIRO2 is responsible for iron utilization in rice and improves growth and yield in calcareous soil. *Plant Mol. Biol.* 75: 593–605.

Ogo, Y., Kakei, Y., Itai, R.N. et al. (2014). Spatial transcriptomes of iron-deficient and cadmium-stressed rice. *New Phytol.* 201: 781–794.

Oliva, N., Chadha-Mohanty, P., Poletti, S. et al. (2014). Large–scale production and evaluation of marker-free indica rice IR64 expressing phytoferritin genes. *Mol. Breed.* 33 (1): 23–37.

Palmer, C.M. and Guerinot, M.L. (2009). Facing the challenges of Cu, Fe and Zn homeostasis in plants. *Nat. Chem. Biol.* 5: 333–340.

Pandav, C.S., Yadav, K., Srivastava, R. et al. (2013). Iodine deficiency disorders (IDD) control in India. *Indian J. Med. Res.* 138 (3): 418–433.

Pandit, E., Tasleem, S., Barik, S.R. et al. (2017). Genome-wide association mapping reveals multiple QTLs governing tolerance response for seedling stage chilling stress in indica rice. *Front. Plant Sci.* 8: 552.

Paul, S., Ali, N., Gayen, D. et al. (2012). Molecular breeding of *Osfer2* gene to increase iron nutrition in rice grain. *GM Crops Food* 3 (4): 310–316.

Pradhan, S.K., Barik, S.R., Sahoo, A. et al. (2016). Population structure, genetic diversity and molecular marker-trait association analysis for high temperature stress tolerance in rice. *PLoS One* 11: e0160027.

Pradhan, S.K., Pandit, E., Pawar, S. et al. (2019). Association mapping reveals multiple QTLs for grain protein content in rice useful for biofortification. *Mol. Genet. Genom.* 294 (4): 963–983.

Pradhan, S.K., Pandit, E., Pawar, S. et al. (2020). Linkage disequilibrium mapping for grain Fe and Zn enhancing QTLs useful for nutrient dense rice breeding. *BMC Plant Biol.* 20: 57.

Prasad, R. and Shivay, Y.S. (2018). Iron fertilization of cereal crops for human health—a new concept. *Indian J. Fertil.* 14 (5): 68–77.

Prasad, R. and Shivay, Y.S. (2020). Agronomic biofortification of plant foods with minerals, vitamins and metabolites with chemical fertilizers and liming. *J. Plant Nutr.* 43: 1534–1554. https://doi.org/10.1080/01904167.2020.1738464.

Prasad, R., Shivay, Y.S., and Kumar, D. (2017). Current status, challenges and opportunities in rice production. In: *Rice Production Worldwide* (eds. B. Chauhan, K. Jabran and G. Mahajan), 1–32. Switzerland, Chamok: Springer International Publishing AG.

Qu, L.Q., Yoshihara, T., Ooyama, A. et al. (2005). Iron accumulation does not parallel the high expression level of ferritin in transgenic rice seeds. *Planta* 222 (2): 225–233.

Ravet, K., Touraine, B., Kim, S.A. et al. (2009). Post-translational regulation of AtFER2 ferritin in response to intracellular iron trafficking during fruit development in Arabidopsis. *Mol. Plant* 2: 1095–1106.

Ravindra Babu, V. (2013). Importance and advantages of rice biofortification with iron and zinc. *J. SAT Agric. Res.* 11: 1–6.

Rellán-Álvarez, R., Abadia, J., and Álvarez-Fernández, A. (2008). Formation of metal-nicotianamine complexes as affected by pH, ligand exchange with citrate and metal exchange: a study by electrospray ionization time-offlight mass spectrometry. *Rapid Commun. Mass Sp.* 22: 1553–1562.

Rellán-Álvarez, R., Giner-Martinez-Sierra, J., Orduna, J. et al. (2009). Identification of a tri-iron(III), tri-citrate complex in the xylem sap of iron–deficient tomato resupplied with iron: new insights into plant iron long-distance transport. *Plant Cell Physiol.* 51: 91–102.

Rodríguez-Celma, J., Lin, W.D., Fu, G.M. et al. (2013). Mutually exclusive alterations in secondary metabolism are critical for the uptake of insoluble iron compounds by arabidopsis and *Medicago truncatula*. *Plant Physiol.* 162: 1473–1485.

Römheld, V. and Marschner, H. (1986). Evidence for a specific uptake system for iron phytosiderophore in roots of grasses. *Plant Physiol.* 80: 175–180.

Roy, J.K., Bandopadhyay, R., Rustgi, S. et al. (2006). Association analysis of agronomically important traits using SSR, SAMPL and AFLP markers in bread wheat. *Curr. Sci.* 90 (5): 683–689.

Saltzman, A., Birol, E., Bouis, H.E. et al. (2014). Biofortification: progress toward a more nourishing future. *Glob. Food Secur.* 2 (1): 9–17.

Santos, R.S., Junior, A.T.A., Pegoraro, C. et al. (2017). Dealing with iron metabolism in rice: from breeding for stress tolerance to biofortification. *Genet. Mol. Biol.* 40 (1): 312–325. https://doi.org/10.1590/1678-4685-GMB-2016-0036.

Schmidt, H. and Eickhorst, T. (2014). Detection and quantification of native microbial populations on soil–grown rice roots by catalyzed reporter deposition-fluorescence in situ hybridization. *FEMS Microbiol. Ecol.* 8: 390–402.

Senoura, T., Sakashita, E., Kobayashi, T. et al. (2017). The iron–chelate transporter OsYSL9 plays a role in iron distribution in developing rice grains. *Plant Mol. Biol.* 95: 375–387.

Seshadri, S. (1997). Prevalence of micronutrient deficiency particularly of iron, zinc and folic acid in pregnant women in South East Asia. *Br. J. Nutr.* 85: 87–92.

Shahzad, Z., Rouached, H., and Rakha, A. (2014). Combating mineral nutrition through iron and zinc biofortification of cereals. *Compr. Rev. Food Sci. Food Saf.* 13: 329–346.

Simbauranga, R.H., Kamugisha, E., Hokororo, A. et al. (2015). Prevalence and factors associated with severe anaemia amongst under–five children hospitalized at Bugando Medical Centre, Mwanza, Tanzania. *BMC Hematol.* 15: 13.

Slamet-Loedin, I.H., Johnson-Beebout, S.E., Impa, S. et al. (2015). Enriching rice with Zn and Fe while minimizing Cd risk. *Front. Plant Sci.* 6 (121): 1–9.

Soumitra, P., Nusrat, A., and Dipak, G. (2012). Molecular breeding of *Osfer2* gene to increase iron nutrition in rice grain. *GM Crops Food* 3 (4): 310–316.

Stangoulis, J.C., Huynh, B.L., Welch, R.M. et al. (2007). Quantitative trait loci for phytate in rice grain and their relationship with grain micronutrient content. *Euphytica* 154: 289–294.

Stevens, G.A., Finucane, M.M., De-Regil, L.M. et al. (2013). Global, regional and national trends in haemoglobin concentration and prevalence of severe anaemia in children and pregnant and non-pregnant women for 1995–2011: a systematic analysis of population-representative data. *Lancet Glob. Health* 1 (1): e16–e25. https://doi.org/10.1016/S2214-109X(13)70001-9.

Stoltzfus, R.J. and Dreyfuss, M.L. (1998). *Guidelines for the Use of Iron Supplements to Prevent and Treat Iron Deficiency Anemia*. Washington, DC: ILSI Press.

Suzuki, M., Morikawa, K.C., Nakanishi, H. et al. (2008). Transgenic rice lines that include barley genes have increased tolerance to low iron availability in a calcareous paddy soil. *Soil Sci. Plant Nutr.* 54 (1): 77–85.

Swamy, M.B.P., Rahman, M.A., Inabangan-Asilo, M. et al. (2016). Advances in breeding for high grain Zn in rice. *Rice* 9: 49–57.

Swamy, B.P.M., Kaladhar, K., Anuradha, K. et al. (2018). QTL analysis for grain iron and zinc concentrations in two *O. nivara* derived backcrossp populations. *Rice Sci.* 25 (4): 197–207.

Takagi, S. (1976). Naturally occurring iron-chelating compounds in oat-and rice-root washings. I. Activity measurement and preliminary characterization. *Soil Sci. Plant Nutr.* 22: 423–433.

Takagi, S., Nomoto, K., and Takemoto, T. (1984). Physiological aspect of mugineic acid, a possible phytosiderophore of graminaceous plants. *J. Plant Nutr.* 7: 469–477.

Takagi, S., Kamei, S., and Yu, M.H. (1988). Efficiency of iron extraction from soil by mugineic acid family phytosiderophores. *J. Plant Nutr.* 11: 643–651.

Takahashi, M., Yamaguchi, H., Nakanishi, H. et al. (1999). Cloning two genes for nicotianamine aminotransferase, a critical enzyme in iron acquisition (Strategy II) in graminaceous plants. *Plant Physiol.* 121: 947–956.

Takahashi, R., Ishimaru, Y., Senoura, T. et al. (2011). The OsNRAMP1 iron transporter is involved in Cd accumulation in rice. *J. Exp. Bot.* 62: 4843–4850.

Trijatmiko, K.R., Duenas, C., Tsakirpaloglou, N. et al. (2016). Biofortifiedindica rice attains iron and zinc nutrition dietary targets in the field. *Sci. Rep.* 6: 19792.

UNICEF (2017). The problem: about iron deficiency. Available at https://www.unicef.org/nutrition/23964_iron.html

Vasconcelos, M., Datta, K., Oliva, N. et al. (2003). Enhanced iron and zinc accumulation in transgenic rice with the ferritin gene. *Plant Sci.* 164 (3): 371–378.

Vert, G., Grotz, N., Dedaldechamp, F. et al. (2002). IRT1, an arabidopsis transporter essential for iron uptake from the soil and for plant growth. *Plant Cell* 14: 1223–1233.

Von Wiren, N., Mori, S., Marschner, H., and Romheld, V. (1994). Iron inefficiency in maize mutant ys1 (*Zea mays* L. cv yellow-Stripe) caused by a defect in uptake of iron phytosiderophores. *Plant Physiol.* 106: 71–77.

Von Wiren, N., Klair, S., Bansal, S. et al. (1999). Nicotianamine chelates both FeIII and FeII. Implications for metal transport in plants. *Plant Physiol.* 119: 1107–1114.

Wairich, A., De Oliveira, B.H., Arend, E.B. et al. (2019). The combined strategy for iron uptake is not exclusive to domesticated rice (*Oryza sativa*). *Sci. Rep.* 9: 16144.

Walker, E.L. and Connolly, E.L. (2008). Time to pump iron: iron deficiency–signaling mechanisms of higher plants. *Curr. Opin. Plant Biol.* 11: 530–535.

Wan, J.L., Zhai, H.Q., Wan, J.M., and Ikehashi, H. (2003). Detection and analysis of QTLs for ferrous iron toxicity tolerance in rice, *Oryza sativa* L. *Euphytica* 131: 201–206.

Watanabe, K.N., Sassa, Y., Suda, E. et al. (2005). Global political, economic, social and technological issues on transgenic crops–review. *Plant Biotechnol.* 22 (5): 515–522.

Waters, B.M., Chu, H.H., DiDonato, R.J. et al. (2006). Mutations in *Arabidopsis* yellow stripe-like1 and yellow stripe–like3 reveal their roles in metal ion homeostasis and loading of metal ions in seeds. *Plant Physiol.* 141: 1446–1458.

Welch, R.M. and Graham, R.D. (2004). Breeding for micronutrients in staple food crops from a human nutrition perspective. *J. Exp. Bot.* 55: 353–3640.

Wessells, K. and Brown, K. (2012). Estimating the global prevalence of zinc deficiency: results based on zinc availability in national food supplies and the prevalence of stunting. *PloS One* 7 (11): e50568.

White, J. and Broadley, M.R. (2005). Biofortifying crops with essential mineral elements. *Trends Plant Sci.* 10: 586–593.

WHO. (2016). Iron Deficiency Anaemia. Available at: http://www.who.int/nutrition/topics/ida/en/

Wintz, H., Fox, T., Wu, Y.Y. et al. (2003). Expression profiles of *Arabidopsis thaliana* in mineral deficiencies reveal novel transporters involved in metal homeostasis. *J. Biol. Chem.* 278: 47644–47653.

Yadav, G.S., Shivay, Y.S., Kumar, D., and Babu, S. (2013). Enhancing iron density and uptake in grain and straw of aerobic rice through mulching and rhizo–foliar fertilization of iron. *Afr. J. Agric. Res.* 8: 5447–5454.

Yang, X., Ye, Z., Shi, C. et al. (1998). Genotypic differences in concentrations of iron, manganese, copper, and zinc in polished rice grains. *J. Plant Nutr.* 21: 1453–1462.

Ye, X., Al-Babili, S., Kloti, A. et al. (2000). Engineering the provitamin A (β-carotene) biosynthetic pathway into (carotenoids-free) rice endosperm. *Science* 287: 303–305.

Yokosho, K., Yamaji, N., Ueno, D. et al. (2009). OsFRDL1 is a citrate transporter required for efficient translocation of iron in rice. *Plant Physiol.* 149: 297–305.

Yokosho, K., Yamaji, N., and Ma, J.F. (2016). OsFRDL1 expressed in nodes is required for distribution of iron to grains in rice. *J. Exp. Bot.* 67: 5485–5494.

Yoshino, M. and Murakami, K. (1998). Interaction of iron with polyphenolic compounds: application to antioxidant characterization. *Anal. Biochem.* 257: 40–44.

Yuan, L., Wu, L., Yang, C., and Quin, L.V. (2013). Effects of iron and zinc foliar applications on rice plants and their grain accumulation and grain nutritional quality. *J. Sci. Food Agric.* 93 (2): 254–261.

Zeng, Y., Zhang, H., Wang, L. et al. (2010). Genotypic variation in element concentrations in brown rice from Yunnan landraces in China. *Environ. Geochem. Health* 32: 165–177.

Zhang, Y., Xu, Y.H., Yi, H.Y., and Gong, J.M. (2012). Vacuolar membrane transporters OsVIT1 and OsVIT2 modulate iron translocation between flag leaves and seeds in rice. *Plant J.* 72: 400–410.

Zhang, M., Pinson, S.R.M., Tarpley, L. et al. (2014). Mapping and validation of quantitative trait loci associated with concentrations of 16 elements in unmilled rice grain. *Theor. Appl. Genet.* 127: 137–165.

Zhang, J., Liu, B., and Li, M. (2015). The bHLH transcription factor bHLH104 interacts with IAA-LEUCINE RESISTANT3 and modulates iron homeostasis in Arabidopsis. *Plant Cell* 27: 787–805.

Zhao, K., Tung, C.W., Eizenga, G.C. et al. (2011). Genome-wide association mapping reveals a rich genetic architecture of complex traits in *Oryza sativa*. *Nat. Commun.* 2: 467.

Zheng, L., Yamaji, N., Yokosho, K., and Ma, J.F. (2012). YSL16 is a phloem-localized transporter of the copper–nicotianamine complex that is responsible for copper distribution in rice. *Plant Cell* 24: 3767–3782.

Zimmermann, M.B. and Hurrell, R. (2002). Improving iron, zinc and vitamin A nutrition through plant biotechnology. *Curr. Opin. Biotechnol.* 13: 142–145.

18

Aromatic Rices: Evolution, Genetics and Improvement through Conventional Breeding and Biotechnological Methods

Lakshminarayana R. Vemireddy[1], *Bhaben Tanti*[2], *Lipika Lahkar*[2] *and Zina M. Shandilya*[2]

[1] *Department of Genetics and Plant Breeding, S.V. Agricultural College, Acharya NG Ranga Agricultural University, Tirupati 517502, Andhra Pradesh, India*
[2] *Department of Botany, Gauhati University, Guwahati 781014, Assam, India*

Aromatic Rices: Background

Throughout the world rice (*Oryza sativa* L.) is one of the oldest staple crops, endowed with phenomenal genetic resources. During the course of evolution, rice's genetic resources have diverged into a handful of sub-groups, with specific traits, which are largely cultivated and nurtured by the different ethnic groups in specific regions. The aromatic or scented or fragrant rice varieties are a special group, which carved out its niche owing to their unique aroma and other cooking quality traits. The first reference to the existence of aromatic rice can be traced to the documents of Susrutha (c. 400 BC), the great Indian pioneer in medicine and surgery (Siddiq et al. 2012). The scented rice genotypes are largely grouped into three categories: the basmati, jasmine, and other scented rice group. The basmati types of rice cultivars originated in India and Pakistan and are characterized by exquisite aroma and extra-long elongation after cooking (Siddiq et al. 2012). Jasmine rice has a slightly larger kernel length than basmati and originated in Thailand (Juliano and Villarreal 1993). Jasmine rice, also known as "Thai Hom Mali Rice." was collected from Chachoengsao, a central province of Thailand, and after pure-line selection was officially named Khao Dawk Mali 105, (KDML105). Because of its superb fragrance and cooking quality, KDML105 has been a model variety for studying genes controlling grain quality and aroma. In addition to these two large aromatic groups, many more small aromatic groups can be found that are specific to particular regions in the world, including Sadri from Iran and Della from the United States, to name a few (Table 18.1).

The aromatic rices occupy a special place among the diverse subspecies of rice owing to their exquisite aroma. Aromatic rice includes basmati from India and Pakistan, jasmine from Thailand and sadri from Iran (Figure 18.1). Aromatic rices fetch a premium price in the domestic as well as overseas markets due to their unique special quality characteristics and their being a rich source of nutrients. The basmati varieties, often described as the "prince of rice" are reported to have low glycemic index and be a rich source of Fe and Zn.

Molecular Breeding for Rice Abiotic Stress Tolerance and Nutritional Quality, First Edition.
Edited by Mohammad Anwar Hossain, Lutful Hassan, Khandakar Md. Iftekharuddaula, Arvind Kumar and Robert Henry.

Table 18.1 List of aromatic rice varieties in the world.

S. No.	Aromatic variety	Region	Characteristics
1	Basmati	India and Pakistan	Extremely fragrant variety
2	Jasmine rice	Thailand	Floral aroma
3	Bahra	Afghanistan	
4	Sadri, Domsiah, Hashemi	Iran	Strong aroma
5	Della, Texamati and Kasmati	USA	Combines the qualities of basmati with that of traditional American long-grain rice
6	MRQ74	Malaysia	Popular aromatic landrace
7	Ambemohar	Maharashtra, India	Popular for its quick-cooking characteristics and its beautiful aroma
8	Mullan Kazhama	Wayanad, Kerala, India	Unconventional taste and aroma
9	Gobindo Bhog	West Bengal, India	Small-grained fragrant rice
10	Seeraga Samba	Tamil Nadu, India	Slender rice with a sublime scent
11	Mushk Budji	Kashmir, India	Short-grain rice with a resplendent aroma
12	Radhuni Pagol	West Bengal, India	Fragrant rice
13	Chak Hao Amubi	Manipur, India	Fragrant variety of sticky black rice
14	Kalijira and Chiniguri	West Bengal, India	• Tiny aromatic rice cooks in only 10 min • Potential source of antitumor agent
15	Joha rice	Assam, India	Unique aroma, superfine kernel, good cooking qualities and excellent palatability
16	Gandhakasala rice Jeerakasala rice	Wayanad, Kerala, India	Disease-resistant properties, high nutritional value, fine taste and cooking properties
17	Tulaipanji	West Bengal, India	Non-sticky and friable due to high amylose content
18	BR5, BRRI dhan34, Binadhan-13	Bangladesh	Small-grained fragrant rice
19	BRRI dhan37, BRRI dhan38, BRRI dhan50, BRRI dhan70, BRRI dhan80	Bangladesh	Long, slender-grained fragrant rice

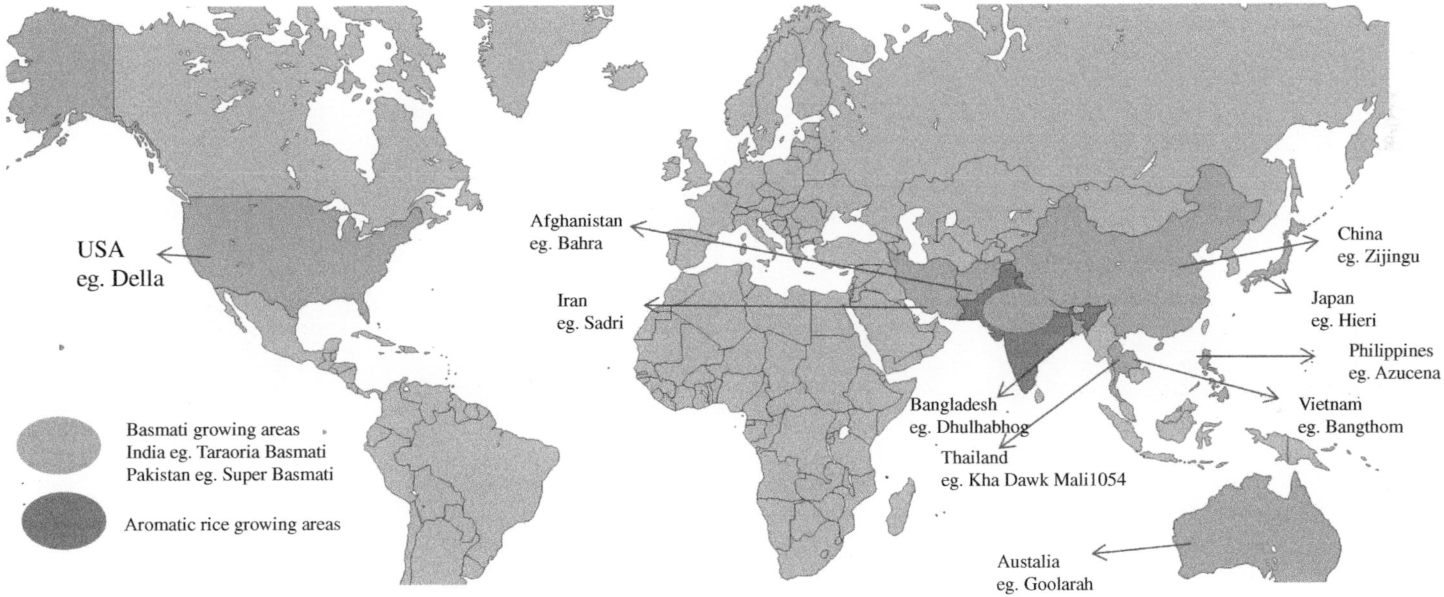

Figure 18.1 Global distribution of aromatic rice varieties.

Aromatic rice is preferred in the Middle East, Asia, Europe, the United States, and Australia. However, aromatic rices are low yielding, tall, and photosensitive, which has prompted breeders to develop high-yielding and photo-insensitive varieties with the least trade off on their quality traits. In the following sections, the genetic, molecular, and biochemical basis of aroma is discussed, and the further development of aromatic rice in breeding is described.

Genetics, Origin, and Evolution of Aromatic Rices

Aroma development is influenced by both genetic and environmental factors. The unprecedented popularity of aromatic rices has stimulated many researchers in the world to explore an in depth understanding of the inheritance of the aroma trait and also to trace back their origin and evolution in order to deploy this knowledge to develop customized high-yielding aromatic varieties.

Genetics of Aroma

The aroma of rice is the most sought-after quality trait for scientists (as a subject of research), farmers (for their higher price), and customers (for their fragrance and taste). There are conflicting reports on the inheritance patterns and the number of genes controlling aroma. For instance, Reddy and Reddy (1987) reported that aroma is controlled by two to three recessive or dominant genes, two recessive (Vivekanandan and Giridharan 1995), a single dominant gene (Kuo et al. 2005), a dominant epistasis (Chen et al. 2010), and complementary gene action (Chen et al. 2010). These contradictions about aroma inheritance can be attributed to employment of various phenotyping methods for aroma estimation and the use of different parents in the breeding programs. However, for the most part aroma is reported to be inherited as a single recessive nuclear gene (Sood and Siddiq 1978). However, with advances in molecular marker technology, the genomic regions or quantitative traits loci (QTLs) governing fragrance (*fgr*) have been found to be located on chromosome 8 with a restriction fragment length polymorphism (RFLP) marker RG28 (Ahn et al. 1992). In 2004, after completion of rice whole genome reference sequence and its availability in public databases many molecular markers and high-density linkage maps became available. The Robert Henry group from Southern Cross University, Australia, uncovered a candidate gene underlying the aroma QTL on chromosome 8 and reported that the gene encodes the betaine aldehyde dehydrogenase (BADH) enzyme (Bradbury et al. 2005). They reported that an 8 bp deletion (GATTATGG) in exon 7 introduced a stop codon at position 734 and rendering BADH non-functional in aromatic varieties including basmati and jasmine. Later, Chen et al. (2008) demonstrated that Badh2 inhibits the biosynthesis of 2-acetyl-1-pyrroline (2-AP), a potent flavor compound in aromatic cultivars.

Natural Variability in the Aroma (fgr) Gene

The DNA sequence of the BADH2 gene is 1509 bp long comprising 15 exons and 14 introns and encodes 503 amino acids (Chen et al. 2008). However, comparison of aromatic

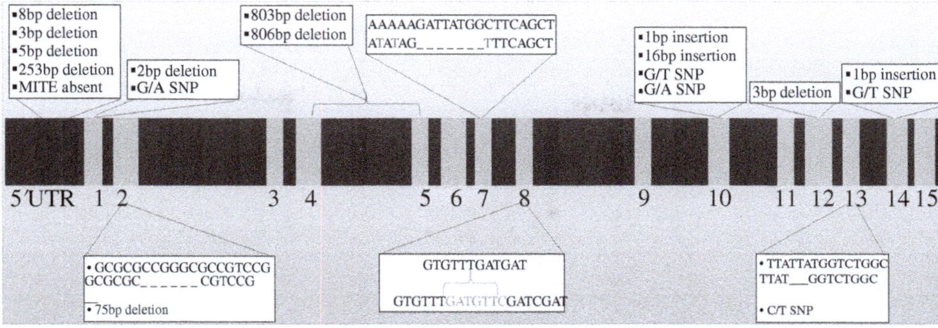

Figure 18.2 Overall distributions of alleles in betain aldehyde dehydragenase (*Badh2*) gene. *Note:* Yellow color lines indicate exons. The dashes (----) indicates deletion, red color nucleotides are SNPs and green color nucleotides are insertion.

and non-aromatic varieties has revealed several alleles of the *Badh2* gene sequence (Figure 18.2). Among all of these alleles, an 8 bp deletion and 3 SNPs (single nucleotide polymorphism) in the 7th exon of the gene is the most predominant allele found in the majority of the aromatic varieties. However, many exceptions have also been reported across rice germplasm for this allele. As of now, there are at least 23 sequence variations that have been found in the *Badh2* gene, and these sites are located in the 5′UTR (Untranslated region), promoter, exon1, exon1 and intron1 junction, exon2, exons 7,8,10,12,13, and 14 (He and park 2015) (Table 18.2). Of the 23 alleles of *Badh2*, two alleles, that is, *Badh7.1* (8 bp deletion in exon 7) and *Badh-13.2* (*C/T* SNP in exon 13) have been shown to explain as much as 97% of the aroma variation in rice germplasm (He et al. 2015). These *fgr* alleles provide insights for the origin of the genes and can be exploited to develop varieties with different aroma intensities. All these sequence variations lead to truncated proteins, thereby resulting in non-functioning of the *Badh2* gene.

Additional QTLs for Aroma

Although *fgr* located on chromosome 8 is the major QTL explaining more than 97% of the phenotypic variation, there are many minor QTLs that also been identified (Table 18.3). For instance, assessment of 478 aromatic rice accessions revealed that the majority of the lines contained the *Badh7.1* allele; however, approximately 10% of the varieties accumulated a significant amount of 2-AP even though the *Badh7.1* allele was absent (Fitzgerald et al. 2008). Initially, Lorieux et al. (1996) identified two additional QTLs on chromosome 4 and 12 using RFLP marker RG1. Subsequently, two minor QTLs on chromosomes 3 and 4, besides a major QTL on chromosome 8, were identified (Amarawathi et al. 2008). Using an F_2 population derived from a cross between Basmati370 and Jaya, six QTLs designated as *qARM1.1, qARM2.1, qARM8.1, qARM8.2, qARM8.3,* and *qARM12.1* influencing aroma were identified on chromosomes 1, 2, 8, and 12 (Vemireddy et al. 2015). Analysis of the rice genome database revealed that in addition to badh2 on chromosome 8, badh1 (with 92% homology to badh2) has also been located on chromosome 4; however, it has a major role in salt stress tolerance rather than aroma. These QTLs are specific to either the environment

Table 18.2 Summary of sequence variations in *Badh2* gene.

S.No.	Allele	Sequence variation	Location	Genotype	References
1	*Badh-p*	8 bp insertion	Promoter	Seeraga samba	Bindusree et al. (2017)
2	*Badh-5′UTR-1*	8 bp insertion	5′ UTR	Nankai 138	Shi et al. (2014)
3	*Badh-5′UTR-2*	3 bp deletion	5′ UTR	Nankai 138	Shi et al. (2014)
4	*Badh-5′UTR-3*	5 bp deletion	5′ UTR	Velchi	Khandagale et al. (2017)
5	*Badh-5′UTR-4*	253 bp deletion	5′ UTR	Velchi	Khandagale et al. (2017)
6	*Badh-5′UTR-5*	MITE absent	5′ UTR	Azucena	Bourgis et al. (2008)
7	*Badh-1.1*	2 bp deletion	Exon 1		Kovach et al. (2009)
8	*Badh-1.2*	G/A SNP	Exon 1 and Intron 1 junction	Nioi-mochi 2	Ootsuka et al. (2014)
9	*Badh-2.1*	7 bp deletion	Exon 2		Shi et al. (2008)
10	*Badh-2.2*	75 bp deletion	Exon 2		Shao et al. (2011)
11	*Badh-4.1*	803 bp deletion	Exon 4 to Exon 5		Shao et al. (2011)
12	*Badh-4.2*	806 bp deletion	Exon 4 to Exon 5		Shao et al. (2013)
13	*Badh-7.1*	8 bp deletion and 3 SNPs	Exon 7	Basmati Jasmine	Bradbury et al. (2005)
14	*Badh-8.1*	7 bp deletion	Exon 8	Pusa 1121	Amaravathi et al. (2008)
15	*Badh-10.1*	1 bp insertion	Exon 10		Kovach et al. (2009)
16	*Badh-10.2*	1 bp deletion	Exon 10		Kovach et al. (2009)
17	*Badh-10.3*	G/T SNP	Exon 10		Kovach et al. (2009)
18	*Badh-10.4*	G/A SNP	Exon 10		Shao et al. (2013)
19	*Badh-12*	3 bp deletion	Exon 12		He and Park (2015)
20	*Badh-13.1*	3 bp insertion	Exon 13		Kovach et al. (2009)
21	*Badh-13.2*	C/T SNP	Exon 13	Shiratama	Kovach et al. (2009)
22	*Badh-14.1*	1 bp insertion	Exon 14		Kovach et al. (2009)
23	*Badh-14.2*	G/T SNP	Exon 14		Kovach et al. (2009)

or the genotype. However, the precise role of these additional QTLs in aroma development has not been demonstrated yet.

Origin and Evolution of Aromatic Varieties

The understanding of origin and evolution of aromatic rice is quite interesting and complex. The common ancestor of rice is believed to have been domesticated separately in Southeast

Table 18.3 List of QTLs governing for aroma using diverse mapping population in rice.

Trait	QTL	Marker interval	Chr.	LOD	PVE	Population	Reference
Aroma sensory score	*aro3-1*	CHR3_22-CHR3_24	3	2.92	11	F6 RIL –Pusa1121/Pusa1342	Singh et al. (2007)
Aroma sensory score	*aro8-1*	CHR8_49-RM44	8	3.35	8.6	F6 RIL –Pusa1121/Pusa1342	Singh et al. (2007)
Aroma sensory score	*aro4-1*	RM5633-RM273	4	3.3	6.1	F6 RIL –Pusa1121/Pusa1342	Singh et al. (2007)
2-AP conc	*aro8-1*	CHR8_49-RM44	8	2.3	5	F6 RIL –Pusa1121/Pusa1342	Singh et al. (2007)
2-AP conc	*aro3-1*	CHR3_22-CHR3_24	3	5.08	15	F6 RIL –Pusa1121/Pusa1342	Singh et al. (2007)
Aroma sensory score	*aro_cim5-2*	RM169-RM430	5	3.296	3.08	F2-Ranjit × Kola Joha	Talukdar et al. (2017)
Aroma sensory score	*qARM1.1*	RM8278-RM582	1	6.735	1.8	F2-Basmati370 × Jaya	Vemireddy et al. (2015)
Aroma sensory score	*qARM2.1*	RM138-RM475	2	7.59	0.13	F2-Basmati370 × Jaya	Vemireddy et al. (2015)
Aroma sensory score	*qARM8.1*	RM502-RM310	8	6013	0.21	F2-Basmati370 × Jaya	Vemireddy et al. (2015)
Aroma sensory score	*qARM8.2*	RM152-RM42	8	4.99	3.16	F2-Basmati370 × Jaya	Vemireddy et al. (2015)
Aroma sensory score	*qAR-8-1*	RM42 – RM223	8	23.28	24	F2-Basmati 370 × MR 84	Cheng et al. (2014)
Aroma sensory score	*qAR-8-2*	RM223– AROE	8	24.17	28	F2-Basmati 370 × MR 84	Cheng et al. (2014)
Aroma sensory score	*qAR-8-3*	AROE– RM342	8	13.37	24	F2-Basmati 370 × MR 84	Cheng et al. (2014)
Aroma sensory score	*qAR-8-4*	RM515–RM210	8	6.48	20	F2-Basmati 370 × MR 84	Cheng et al. (2014)

Chr., Chromosome; PVE, Phenotypic variance explained.

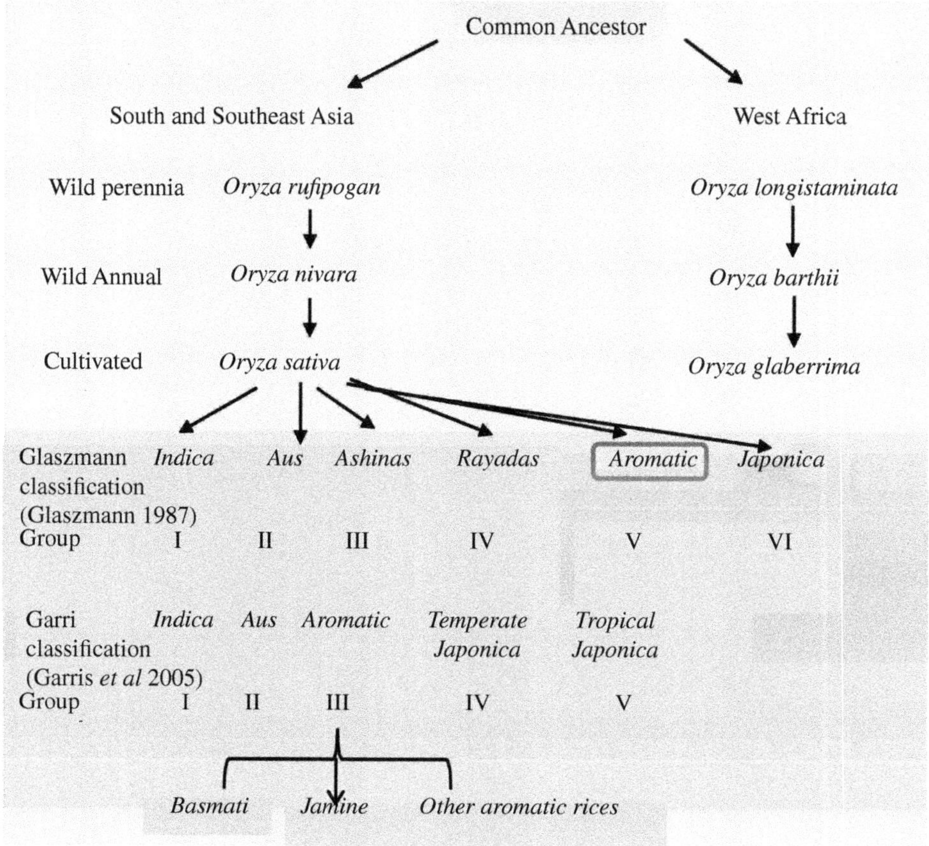

Figure 18.3 Evolution and origin of aromatic rice.

Asia and West Africa as wild perennials, *Oryza rufipogan* and *Oryza longistaminata,* respectively. From these wild annuals *Oryza nivara* and *Oryza barthii* may have originated, which eventually led to two cultivated species *O. sativa* and *Oryza glaberrima,* respectively. Using chloroplast DNA analysis, however, it has been demonstrated that *Oryza barhii* is closely related to *O. rufipogan* than to *O. longistaminata* (Wambugu et al. 2015). Cultivated *O. sativa* has been further subdivided based on their unique characters. Aromatic rice is one of the sub-genotypes of *O. sativa* (Figure 18.3). Its origin and evolution are not known. Many scientists have attempted to understand the origin of aromatic rice using isozyme, simple sequence repeats (SSR), and SNP markers. Initially, Glaszmann (1987) classified the aromatic varieties into three isozyme groups: Group-I (indica-jasmine), Group-V (basmati), and GroupVI (tropical *japonica*-azucena). For the most part, aromatic varieties can be subdivided into two large groups: *basmati* and *jasmine* and the small group of other aromatic rices comprising aromatic short grain, joha, sadri, and so on. However, molecular-marker-based genetic analysis of rice accessions has provided great insights into the origin and evolution of aromatic rices. Kovach et al. (2009) reported that aromatic varieties are closely associated with the *japonica* varietal group. This finding was also supported by Garris et al. (2005). However, very recently, it was demonstrated that the aromatic varieties originated from the Indian subcontinent by hybridization between local and wild populations. They also showed that present-day aromatic accessions inherited their cytoplasm with only 29–47% and their nuclear genome from local Indian rice (Civáň et al. 2019). However, the survey of diverse fragrant rice accessions revealed that all have been shown to carry the *badh2.1* allele, obviously suggesting that this allele is common by descent in all aromatic rices (Kovach et al. 2009). It is reasonable to presume that the source of aroma in the aromatic accessions originated from the foothills of Himalayas, where the basmati accessions originated. The low temperatures in the Himalayas may have caused the increased expression of aroma (Bourgis et al. 2008). Also it can be speculated that the jasmine and sadri group of aromatic varieties also may have been derived from basmati cultivars; later they spread east and west, respectively. Moreover, the aromatic trait appears to have originated based on the geographic area and its environment rather than by the human selection from the wild species as the 8 bp deletion of *Badh2* gene was found to be in very low frequency in wild species (Prathepa 2009). However, further detailed investigations are warranted to better understand the origin of the aromatic varieties.

Biochemical Basis of Aroma

The aroma of aromatic rice varieties is present in all parts of the crop except the roots and almost all stages of the crop, especially at the seedling, booting, and flowering stage of the plant. The aroma is due to more than 300 volatile and semi-volatile compounds such as hydro carbons, alcohols, aldehydes, ketones, acids, esters, phenols, and other compounds. The most significant volatile compounds among 64 recognized odors contributing to the aroma of cooked rice are hexanol, (E,E), decanal, 2-Acetyl-1-pyrroline (2-AP), nonanal, decanol, (E)-non-2-enol, 4-vinylphenol, octanol and 4-vinyl-2-methoxy phenol (Buttery et al. 1988). 2-AP has been identified as the primary aromatic compound in aromatic varieties especially basmati (Buttery et al. 1988), black rice (Yang et al. 2008), brown rice (Jezussek et al. 2002) and Thai jasmine rice KDML105 (Mahattanatawee and Rouseff 2014). Aromatic rice varieties include 0.04–0.09 ppm 2AP while in non-aromatic rice; it is 10 times lower (Champagne 2008). The concentration of 2AP is 6–90 ppb in milled rice, while it is 100–200 ppb in brown rice. Besides the presence in fragrant rices, the high amount of 2AP compound can also be found in popcorn, mung bean, pandanas leaves, tortillas, beguettes, ham, cheese, green tea, bread flowers, and wines and also in some microbes such as Bacillus cereus, *Lactobacillus hilgardii* and in some fungi (Bradbury et al. 2008; Wakte et al. 2017). Keeping in mind the importance of aroma, many researchers have attempted to elucidate the precise mechanism of biosynthesis of 2-AP, employing both biochemical and molecular approaches. It has been reported that the BADH2 enzyme inhibits the synthesis of 2AP (Chen et al. 2008) and also demonstrated that inactivation of an aminoaldehyde dehydrogenase is responsible for fragrance in rice (Bradbury et al. 2008). Based on the accumulated evidence, two hypothesis have been formulated to explain the 2AP synthesis, that is, BADH2-dependant and BADH2-independent synthesis. In BADH2-dependant synthesis, L-Proline and ornithine are believed to be the precursors for 2AP through gamma-butyric acid (GABA) (Vanavichit et al. 2005) (Figure 18.4). The oxidation of gamma-aminobutyraldehyde (GAB-ald) is catalyzed by the functional BADH2 enzyme

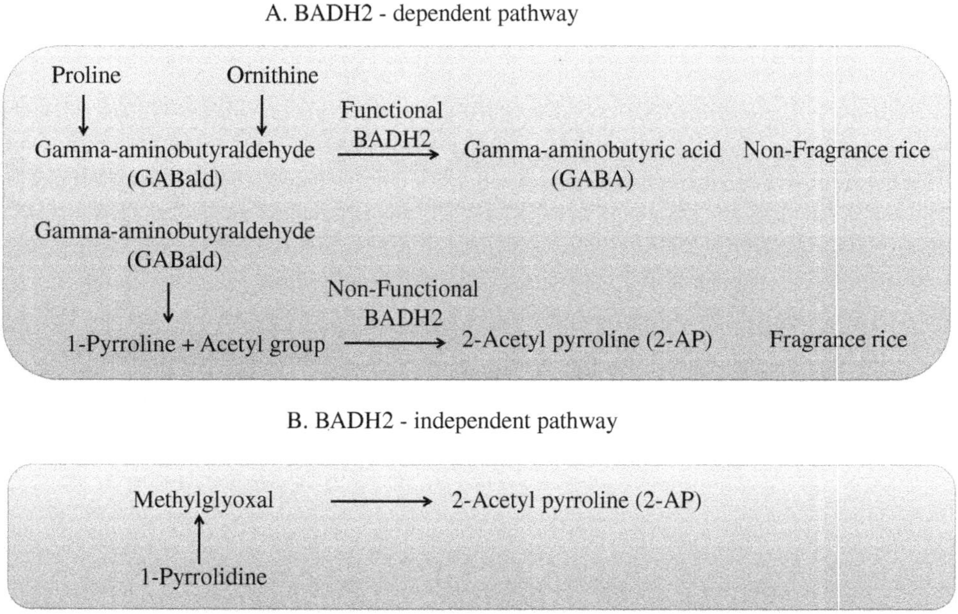

Figure 18.4 Biochemical pathway of the synthesis of 2-acetyl pyrrolidine.

and is converted to GABA while inhibiting the 1-pyrroline (2-AP precursor) (Fitzgerald et al. 2009). Hence, the content of GABA is much higher in non-fragrant varieties. Whereas the non-functional BADH2 enzyme leads to accumulation of both AB-ald, which in turn transformed the 1-pyrroline, which by reacting with acetyl group ultimately synthesizes 2-AP in aromatic rices (Kovach et al. 2009; Fritzerald et al. 2010). However, the discovery of key fragrance (*fgr*) gene, that is, BADH2 – betain aldehyde dehydrogensae enzyme that is located on chromosome 8 involved in 2AP synthesis enhanced our understanding of the 2AP synthesis.

Due to its unstable and thermo-liable nature, 2-AP concentration is affected by several environmental and crop growth factors including post-harvest storage time, effect of planting density, harvesting time (early harvesting and low planting density increases aroma), humidity (70–80% required during grain filling for increased aroma), and temperature (day temperature of 20–32 °C and night temperature of 20–25 °C enhances aroma). Interestingly, the level of 2AP is higher in plants exposed to water stress because of the association of BADH2 with stress tolerance (Bradbury et al. 2005).

Improvement of Aromatic Rice Varieties

Conventional Breeding

Improvement of aromatic varieties or transfer of the aroma trait through conventional breeding methods is indeed a challenging task because of the large effects of the environment and low narrow-sense heritability of the trait. Besides, all aromatic varieties are believed to possess excellent cooking quality but are considerably lower yielding. Attempts to combine both the quality with high-yield through conventional methods have more or less resulted in high-yielding varieties that are short of the quality traits of the traditional basmati varieties. However, the first breakthrough came in 1989 with the release of "Pusa Basmati-1" (PB-1), the first semidwarf, photoperiod-insensitive and high-yielding basmati variety from the Indian Agricultural Research Institute (IARI), New Delhi, India. The PB-1 variety revolutionized basmati rice production, especially during 1995–2007, which contributed nearly 60% of the total basmati rice exports (Siddiq et al. 2009). Subsequently, several high-yielding basmati varieties were released (Siddiq et al. 2012). Among them, one iconic basmati variety having extra-long slender grains, which stands out with exceptional kernel elongation after cooking and has good market value earns more than US$ 20.8 billion of foreign exchange to India is Pusa 1121. Currently, it accounts for 70% of the total basmati grown area in India (Singh et al. 2018). Further, IARI has developed a series of early-maturing and extra-long high-yielding varieties such as Pusa sugandh 2, Pusa sugandh 3, and Pusa sugandh 5, which are also excellent restorers of WA (wild abortive) cytoplasm. IARI also made remarkable achievements in developing the world's first semidwarf, high-yielding, super-fine-grain aromatic rice hybrid, Pusa RH10 by crossing Pusa 6A with PRR78.

As far as jasmine rice is concerned, the first variety released was "Khao Dawk Mali 105" or "KDML105" for its superior physical appearance, cooking quality, and grain aroma from a local landrace Hom Mali in 1959. Later, the RD6, waxy variety, and RD15, early-maturing varieties were released in 1977 and 1978, respectively through mutagenesis of KDML105. These two varieties now occupy more than 70% of the rice area in Thailand due to their tolerance to drought, salinity, and acid sulfate soils (Vanavichit et al. 2018). The first blast-tolerant variety with a jasmine-like quality RD33, photoperiod-insensitive jasmine rice was released in 2006, as an alternative aromatic variety with blast resistance for northeast and northern Thailand (Rice Today 2006).

Molecular Breeding of Aromatic Rice

With the availability of precise rice genome sequences, many candidate genes for diverse traits have been cloned and characterized. For many traits such as biotic and abiotic stress-tolerance traits, functional or closely linked molecular markers are available. This has facilitated the precise transfer of key genes through marker-assisted backcross breeding. Attracted by the key features of aromatic rice varieties, several breeders have attempted to transfer the aroma trait into non-aromatic varieties through conventional methods. However, technically challenging, inconsistent, and labor-intensive measurements of aroma trait have prompted breeders to look for easy and inexpensive assays to select aromatic plants in the segregating populations. Hence, molecular-marker technology has the potential to supplement traditional phenotyping of aroma. To achieve this, many markers linked to aroma gene have been developed. At the outset, an RFLP marker linked to fragrance, RG28 located on chromosome 8 was identified. Due to the laborious nature of RFLP marker technology, the RG28 marker has been converted to PCR-based STS (sequence tagged site) marker (Kottearachchi et al. 2010). However, this STS marker was found to be ineffective as the location of the marker is far from the aroma gene.

The functional markers, FME2-7, FME7, FME12-3, FME13, and FME14 of *Badh2* gene were used to identify the mutation types of the fragrant gene in the exons 2, 7, 12, 13, and 14, respectively.

Molecular breeding has been employed to improve biotic and abioitc stress resistance in basmati and jasmine rice. The PB-1 variety has been improved for bacterial blight resistance by incorporating *Xa13* and *Xa21* genes from IRBB55 and released as Pusa1460 (Improved Pusa Basmati 1) in 2007. Likewise, Pusa1121 also improved for bacterial blight resistance using the same genes and released as PB1718 (Singh et al. 2018). The popular jasmine variety KDML105 also has been improved by incorporating many bacterial-blight, brown-plant-hopper (BPH), and abiotic-stress-tolerance genes and released varieties such as HM80, HM812, and HM84 (Vanavichit et al. 2018) (Table 18.4).

Genetic Engineering

Ever since the *Badh2* gene was discovered, there have been many attempts to understand the gene function and to enhance the 2-AP content employing various genetic engineering

Table 18.4 List of QTLs/genes transferred to improve various aromatic varieties using marker-assisted breeding.

Trait	QTL/gene	Donor	Recipient	Improved variety	Reference
Bacterial blight (BB) resistance	*xa13* and *Xa21*	IRBB55	Pusa Basmati1	Improved Pusa Basmati 1 (Pusa 1460)	Gopalakrishnan et al. (2008)
Bacterial blight (BB) resistance	*Xa21, xa13* and *xa5*	IRBB-60	CSR-30	Advanced lines (IC-R28, IC-R68, IC-R32, and IC-R42)	Baliyan et al. (2017)
Bacterial blight resistant	*xa13* and *Xa21*	SPS97	PB 1121	Pusa Basmati 1718	Singh et al. (2018)
Semi-dwarf	*Sd1*	CSR10	Kalanamak		Srivastava et al. (2019)
Semi-dwarf and blast resistance	*Pi9* and *sd1*	Pusa Basmati 1637	Ranbir Basmati	Two superior BC_2F_1 recombinants	Samal et al. (2019)
Submergence tolerance	*Sub1*	IR49830–7–1-2-2	KDML105	HM80	Siangliw et al. (2003)
Bacterial blight resistant	*Xa21, xa5, xa33 (t), xa34(t), and qBB11*	IR1188	KDML105	HM812	Korinsak (2009)
BPH, BLB and blast-resistant and salt and drought tolerance	*Sub1A, Xa21, xa5,BphQ12, Bph32, Bph3, BLQ1* and *BLQ1*	Abhaya and Rathu Heenati	KDML105	HM84	Vanavichit et al. (2018)

approaches. Using the RNAi approach, 2-AP content has been increased twenty-fold in the Nipponbare variety (Vanavichit et al. 2005; Niu et al. 2008). Chen et al. (2015) enhanced the 2-AP level to 0.35–0.75 mg kg^{-1} by knocking out the *Badh2* gene using transcription activator-like effector nucleases (TALENs). Very recently, Ashokkumar et al. (2020) have created novel alleles of fragrance gene *OsBADH2* in rice through CRISPR/Cas9 mediated gene editing. Further, over-expression of the P5CS gene, a key enzyme in proline synthesis, was demonstrated to increase 2-AP content (Yoshihashi et al. 2002; Thimmaraju et al. 2005; Kayghobad et al. 2015).

Trade and Geographical Indication (GI)

The aromatic rice varieties, owing to their extraordinary quality traits, fetch premium prices in both domestic and overseas markers. Especially, basmati from India and jasmine rice varieties from Thailand exports account for more than 50% of their total exports. However, the obvious price variation among different basmati classes, such as traditional basmati (TB), evolved basmati (EB), and non-basmati (NB) varieties in the international markets, prompted some traders to mix the substandard quality look-a-like evolved or non-basmati varieties with premium-quality traditional basmati varieties. To curb this adulteration, a capillary electrophoresis-based microsatellite marker system has been developed, which is able to detect even a 1% level of the adulteration in the export samples (Vemireddy et al. 2007). Based on this technology, the APEDA-CDFD Centre for Basmati DNA analysis housed in the Centre for DNA Fingerprinting and Diagnostics (CDFD), Hyderabad, has been established to provide authenticity certificates for the export samples.

For the past 10 years, many attempts have been made to enhance the value of the aromatic varieties by recognizing them with geographical indications (GI). The GI tag is given generally to products that have a specific geographical origin and possess qualities and reputation that are due to that origin. The unique quality traits of basmati rice varieties are due to the unique blend of soil, water, and climate of the Himalayan hills where it was originated. In addition to basmati, the other aromatic rice varieties of India such as Kala namak from Uttar Pradesh, Gobind bhog of west Bengal, Joha rice from Assam are among the most important. Till now, the GI tag has been given to basmati varieties grown in seven states of India – Punjab, Haryana, Himachal Pradesh, Uttarakhand, Delhi, Western Uttar Pradesh – and two districts of Jammu and Kashmir. In the case of jasmine rice, Thung Kula Rang-Hai Thai Hom Mali rice has been registered as GI in Thailand. It was the first Asian rice product that was certified in the European Union with GI registration.

Future Prospects of Aromatic Rices

Among all specialty rices, aromatic rice varieties occupy a special place in domestic and overseas markets owing to their exceptional quality features for which they fetch almost three times higher prices compared to non-aromatic varieties. Moreover, the demand for aromatic rice is on the rise every passing year due to the preference for aromatic rice varieties increasing and expanding to almost all corners of the world. In addition to the aromatic long slender varieties, there are several short grain aromatic varieties, which are unique to different states of India; however, their export market share is very meager, and their origin and evolution are not known with certainty. However, with the exploitation of advanced tools such as marker-assisted breeding and genome-editing techniques, it is now easy to combine high yield with quality traits.

References

Ahn, S.H., Bollich, C.N., and Tanksley, S.D. (1992). RFLP tagging of a gene for aroma in rice. *Theor. Appl. Genet.* 84: 825–828.

Amarawathi, Y., Singh, R., Singh, A.K. et al. (2008). Mapping of quantitative trait loci for basmati quality traits in rice (*Oryza sativa* L.). *Mol. Breed.* 21: 49–65.

Ashokkumar, S, Jaganathan, D., Ramanathan, V., Rahman, H, Palaniswamy, R, Kambale, R, et al. (2020) Creation of novel alleles of fragrance gene OsBADH2 in rice through CRISPR/Cas9 mediated gene editing. *PLoS One*, 15(8): e0237018. https://doi.org/10.1371/journal.pone.0237018

Baliyan, N., Malik, R., Rani, R. et al. (2017). Integrating marker-assisted background analysis with foreground selection for pyramiding bacterial blight resistance genes into basmati rice. *C. R. Biol.* 341: 1–8.

Bindusree, G., Natarajan, P., Kalva, S., and Madasamy, P. (2017). Whole genome sequencing of *Oryza sativa* L. cv. Seeragasamba identifies a new fragrance allele in rice. *PLoS One* 12 (11): 1–9.

Bourgis, F., Guyot, R., Gherbi, H. et al. (2008). Characterization of the major fragance gene from an aromatic japonica rice and analysis of its diversity in Asian cultivated rice. *Theor. Appl. Genet.* 117: 353–368.

Bradbury, L.M.T., Fitzgerald, T.L., Henry, R.J. et al. (2005). The gene for fragrance in rice. *Plant Biotechnol. J.* 3: 363–370.

Bradbury, L.M.T., Gillies, S.A., Brushett, D.J. et al. (2008). Inactivation of an aminoaldehyde dehydrogenase is responsible for fragrance in rice Plant Mol. *Biol.* 68: 439–449. https://doi.org/10.1007/s11103-008-9381-x.

Buttery, R.G., Turnbaugh, J.G., and Ling, L.C. (1988). Contribution of volatiles to rice aroma. *J. Agric. Food Chem.* 36 (5): 1006–1009. https://doi.org/10.1021/jf00083a025.

Champagne, E.T. (2008). Rice aroma and flavor: a literature review. *Cereal Chem.* 85 (4): 445–454.

Chen, S.H., Yang, Y., Shi, W.W. et al. (2008). Badh2, encoding betaine aldehyde dehydrogenase, inhibits the biosynthesis of 2-acetyl-1-pyrroline, a major component in rice fragrance. *Plant Cell* 20 (7): 1850–1861.

Chen, Z., Yang, Z., and Gu, Z. (2010, 2013). Determination of volatile flavor compounds in Ganoderma lucidum by HSSPME-GC-MS. *Afr. J. Tradit. Complement. Altern. Med.* 10 (2): 353–355.

Chen, S., Zhang, R., Hao, L. et al. (2015). Profiling of volatile compounds and associated gene expression and enzyme activity during fruit development in two cucumber cultivars. *PLoS One* 10: 1–22.

Cheng, A., Massawe, F., Ismail, I. et al. (2014). High resolution agarose-based system for single-tube genotyping of fgr and Waxy genes in rice: MAGE to displace PAGE? *Plant Omics* J8 (4): 348–352.

Civáň, P., Ali, S., Batista-Navarro, R. et al. (2019). Origin of the aromatic group of cultivated rice (*Oryza sativa* L.) traced to the Indian subcontinent. *Genome Biol. Evol.* 11 (3): 832–843.

Fitzgerald, M.A., Hamilton, N.R.S., Calingacion, M.N. et al. (2008). Is there a second fragrance gene in rice? *Plant Biotechnol. J.* 6: 416–423.

Fitzgerald, M.A., McCouch, S.R., and Hall, R.D. (2009). Not just a grain of rice: the quest for quality. *Trends Plant Sci.* 14: 133–139.

Fitzgerald, T.L., Waters, D.L.E., Brooks, L.O., and Henry, R.J. (2010). Fragrance in rice (*Oryza sativa*) is associated with reduced yield under salt treatment. *Environ. Exp. Bot.* 68 (3): 292–300.

Garris, A.J., Tai, T.H., Coburn, J. et al. (2005). Genetic structure and diversity in *Oryza sativa* L. *Genetics* 169 (3): 1631–1638. https://doi.org/10.1534/genetics.104.035642.

Glaszmann, J.C. (1987). Isozymes and classification of Asian rice varieties. *Theoret. Appl. Genetics* 74: 21–30. https://doi.org/10.1007/BF00290078.

Gopalakrishnan, S., Sharma, R.K., Rajkumar, K.A. et al. (2008). Integrating marker assisted background analysis with foreground selection for identification of superior bacterial blight resistant recombinants in basmati rice. *Plant Breed.* 127: 131–139.

He, Q. and Park, Y.J. (2015). Discovery of a novel fragrant allele and development of functional markers for fragrance in rice. *Mol. Breed.* 35 (217): 1–10.

He, Q., Yu, J., Kim, T.S. et al. (2015). Resequencing reveals different domestication rate for BADH1 and BADH2 in rice (*Oryza Sativa*). *PLoS One* 10 (8): 1–12.

Jezussek, M., Juliano, B.O., and Schieberle, P. (2002). Comparison of key aroma compounds in cooked brown rice varieties based on aroma extract dilution analyses. *J. Agric. Food Chem.* 50 (5): 1101–1105.

Juliano, B.O. and Villarreal, C.P. (1993). *Grain Quality Evaluation of World Rice*, 22–31. Los Banos, Philippines: International Rice Research Institute.

Kayghobad, K., Kad, T.D., Zanan, R.L., and Nadaf, A.B. (2015). 2-acetyl-1-pyrroline augmentation in scented indica rice (*Oryza sativa* L.) varieties through Δ1-pyrroline-5-carboxylate synthetase (P5CS) gene transformation. *Appl. Biochem. Biotechnol.* 177 (7): 1466–1479.

Khandagale, K.S., Chavan, R., and Nadaf, A.B. (2017). RNAi mediated down regulation of BADH2 gene for expression of 2-acetyl-1-pyrroline in non-scented indica rice IR-64 (*Oryza sativa* L.). *Can. J. Biotech.* 1: 169.

Korinsak, S., Sriprakhon, S., Sirithanya, P. et al. (2009). Identification of microsatellite markers (SSR) linked to a new bacterial blight resistance gene xa33(t) in rice cultivar "Ba7.". *Maejo Int. J. Sci. Technol.* 3 (2): 235–247.

Kottearachchi, N.S., Priyangani, E.G.D., and Attanayaka, D.P.S.T.G. (2010). Identification of fragrant gene, fgr, in traditional rice varieties of Sri Lanka. *J. Natl. Sci. Found.* 38 (2): 139–143.

Kovach, M.J., Calingacion, M.N., Fitzgerald, M.A., and McCouch, S.R. (2009). The origin and evolution of fragrance in rice (*Oryza sativa* L.). *Proc. Natl Acad. Sci. USA* 106 (34): 14444–14449.

Kuo, S.M., Chou, S.Y., Wang, A.Z., Tseng, T.H., Chueh, F.S., Yen, H.E., and Wang, C.S. (2005). The betaine aldehyde dehydrogenase (BAD2) gene is not responsible for the aroma trait of

SA0420 rice mutant derived by sodium azide mutagenesis. 5th International Rice Genetics Symposium and 3rd International Rice Functional Genomics Symposium, IRRI. Manila. Abstract n. 224.

Lorieux, M., Petrov, M., Huang, N. et al. (1996). Aroma in rice: Genetic analysis of a quantitative trait. *Theor. Appl. Genet.* 93 (7): 1145–1151.

Mahattanatawee, K. and Rouseff, R.L. (2014). Comparison of aroma active and sulfur volatiles in three fragrant rice cultivars using GC-Olfactometry and GC-PFPD. *Food Chem.* 154: 1–6.

Niu, X.L., Tang, W., Huang, W.Z. et al. (2008). RNAi-directed downregulation of OsBADH2 results in aroma (2-acetyl-1-pyrroline) production in rice (*Oryza sativa* L.). *BMC Plant Biol.* 8 (100): 1–10.

Ootsuka, K., Takahashi, I., Tanaka, K. et al. (2014). Genetic polymorphisms in Japanese fragrant landraces and novel fragrant allele domesticated in northern Japan. *Breed. Sci.* 64: 115–124.

Prathepha, P. (2009). The fragrance (fgr) gene in natural populations of wild rice (*Oryza rufipogon* Griff.). *Genet. Resour. Crop. Evol.* 56 (1): 13–18.

Reddy, V.D. and Reddy, G.M. (1987). Genetics and biochemical basis of scent in rice (*Oryza sativa* L.). *Theor. Appl. Genet.* 73: 699–700.

Rice Today (2006). Improving the sacred. January–March, 2006. https://www.scribd.com/document/98480901/RT-Vol-5-No-1-All-in-the-genes

Samal, P., Pote, T.D., Krishnan, S.G. et al. (2019). Integrating marker-assisted selection and doubled haploidy for rapid introgression of semi-dwarfing and blast resistance genes into a basmati rice variety "Ranbir Basmati.". *Euphytica* 215 (149): 1–13.

Shao, G., Tang, A., Tang, S. et al. (2011). A new deletion mutation of fragrant gene and the development of three molecular markers for fragrance in rice. *Plant Breed.* 130 (2): 172–176.

Shao, G., Tang, S., Chen, M. et al. (2013). Haplotype variation at Badh2, the gene determining fragrance in rice. *Genomics* 101 (2): 157–162.

Shi, W., Yang, Y., Chen, S., and Xu, M. (2008). Discovery of a new frangance allele and the development of functional marker for the breeding of fragrant rice varieties. *Mol. Breed.* 22: 185–192.

Shi, Y., Zhao, G., Xu, X., and Li, J. (2014). Discovery of a new fragrance allele and development of functional markers for identifying diverse fragrant genotypes in rice. *Mol. Breed.* 33 (3): 701–708.

Siangliw, M., Toojinda, T., Tragoonrung, S., and Vanavichit, A. (2003). Thai jasmine rice carrying QTLch9 (SubQTL) is submergence tolerant. *Ann. Bot.* 91 (2): 255–261.

Siddiq, E.A., Sigh, V.P., Zaman, F.U. et al. (2009). *Development of High Yielding Basmati Quality Rice Varieties: A Success Story*, 13–17. Indian Farming.

Siddiq, E.A., Vemireddy, L.R., and Nagaraju, J. (2012). Basmati rices: genetics, breeding and trait. *Agric. Res.* 1 (1): 25–36.

Singh, R., Singh, A.K., Sharma, T.R. et al. (2007). Fine mapping of aroma QTLs in basmati rice (*Oryza sativa* L) on chromosomes 3, 4 and 8. *J. Plant Biochem. Biotechnol.* 16 (2): 75–82.

Singh, V., Singh, A.K., Mohapatra, T. et al. (2018). Pusa basmati 1121 – a rice variety with exceptional kernel elongation and volume expansion after cooking. *Rice* 11 (19): 1–10.

Sood, B.G. and Siddiq, E.A. (1978). A rapid technique for scent determination in rice. *Indian J. Genet. Plant Breed.* 38: 268–271.

Srivastava, D., Shamim, M., Mishra, A. et al. (2019). Introgression of semi-dwarf gene in Kalanamak rice using marker-assisted selection breeding. *Curr. Sci.* 116: 4.

Talukdar, P.R., Rathi, S., Pathak, K. et al. (2017). QTL analysis in aromatic rice of Assam. *India. J. Rice Res.* 5 (4): 1–7.

Thimmaraju, R., Bhagyalakshmi, N., and Narayan, M.S. (2005). In vitro culture of *Pandanus amaryllifolius* and enhancement of 2-acetyl-1-pyrroline, the major flavouring compound of aromatic rice, by precursor feeding of L-proline. *J. Sci. Food Agric.* 85: 2527–2534.

Vanavichit, A., Tragoonrung, S., Theerayut, T., Wanchana, S., and Kamolsukyunyong, W. (2005). Transgenic rice plants with reduced expression of Os2AP and elevated levels of 2-acetyl-1-pyrroline. United States Patent, Patent No. US 7,319,181 B2.

Vanavichit, A., Kamolsukyeunyong, W., Siangliw, M. et al. (2018). Thai hom mali rice: origin and breeding for subsistence rainfed lowland rice system. *Rice* 11 (20): 1–12.

Vemireddy, L.R., Archak, S., and Nagaraju, J. (2007). Capillary electrophoresis is essential for microsatellite marker based detection and quantification of adulteration of Basmati rice (*Oryza sativa*). *J. Agric. Food Chem.* 55 (20): 8112–8117. https://doi.org/10.1021/jf0714517.

Vemireddy, L.R., Noor, S., Satyavathi, V.V. et al. (2015). Discovery and mapping of genomic regions governing economically important traits of basmati rice. *BMC Plant Biol.* 15 (207): 1–19.

Vivekanandan, P. and Giridharan, S. (1995). Inheritance of aroma and breadthwise grain expansion in basmati and non-basmati rices. *AGRIS* 19 (2): 4–5.

Wakte, K., Zanan, R., Hinge, V. et al. (2017). Thirty-three years of 2-acetyl-1-pyrroline, a principal basmati aroma compound in scented rice (*Oryza sativa* L.): a status review. *J. Sci. Food Agric.* 97 (2): 384–395. https://doi.org/10.1002/jsfa.7875.Epub.

Wambugu, P.W., Brozynska, M., Furtado, A. et al. (2015). Relationships of wild and domesticated rices (*Oryza* AA genome species) based upon whole chloroplast genome sequences. *Sci. Rep.* 5: 13957. https://doi.org/10.1038/srep13957.

Yang, D.S., Lee, K.S., Jeong, O.Y., Kim, K.J., and Kays, S.J. (2008). Characterization of volatile aroma compounds in cooked black rice. *Journal of Agricultural and Food Chemistry* 56: 235–240.

Yoshihashi, T., Huong, N.T.T., and Inatomi, H. (2002). Precursors of 2-acetyl-1-pyrroline, a potent flavor compound of an aromatic rice variety. *J. Agric. Food Chem.* 50: 2001–2004.

19

Genetic Engineering for Increasing Antioxidant Content in Rice: Recent Progress and Future Perspectives

Qinlong Zhu[1,2], Jiantao Tan[1,2], Bin Wang[1,2] and Yao-Guang Liu[1,2]

[1] *State Key Laboratory for Conservation and Utilization of Subtropical Agro-Bioresources, Guangzhou 510642, China*
[2] *College of Life Sciences, South China Agricultural University, Guangzhou 510642, China*

Introduction

The twenty-first century is an era with rapid development of the global economy and industry, but it has been accompanied by an increase in the incidence and rejuvenation of various diseases. One of the main reasons is the pathway for producing reactive oxygen species (ROS) and substances capable of producing ROS increase rapidly, which cause harm to human health (Lobo et al. 2010). ROS are excited state molecules with several forms in the body, including oxygenated free radicals (O_2^-, $\cdot OH$, and HOO^-) and non-radical forms of oxygenates (H_2O_2, ROOH, and NO) (Winterbourn 2008). ROS can be created by normal metabolism in the human body, and are produced mainly in the mitochondria (Tomycz et al. 2011). In addition, unhealthy living habits (pressure and smoking, etc.), radiation (including ultraviolet light and electromagnetic waves from mobile phones or computers), and chemical factors (antibiotics, insecticides and metal ions, etc.) will cause excessive production of ROS in the human body (Mailloux 2015). Irreversible oxidative forms of damage to biological macromolecules, such as nucleic acids, lipids, and proteins are often connected with these unbalanced high concentrations of active oxygen (Gülçin et al. 2012), which induce programmed cell death (PCD), loose skin, and pigmentation, and accelerate the aging of body, even death (Frankel and German 2006; Liochev 2013). In order to maintain the balance of active oxygen in vivo, excess ROS can be removed by enzymatic (superoxide dismutase, peroxidase and catalase, etc.) and non-enzymatic (antioxidant substances) antioxidant systems (Ahmad et al. 2010). However, these antioxidant substances produced by the human body are not enough to meet the demand, so it is necessary to obtain more effective antioxidant substances by eating food.

Human can get needed antioxidants in two ways: one is to eat staple foods with essential nutrients that happen to have antioxidant properties (such as carotenoids, folates, and tocopherols) and the other is to take luxury foods adding non-essential but health-promoting antioxidants (such as flavonoids and melatonin) (Zhu et al. 2013). Obviously, it is impractical for low-income groups in developing countries with dense populations to take the above antioxidants by eating luxury foods (Steiger et al. 2014). Therefore, fortifying and

Molecular Breeding for Rice Abiotic Stress Tolerance and Nutritional Quality, First Edition.
Edited by Mohammad Anwar Hossain, Lutful Hassan, Khandakar Md. Iftekharuddaula, Arvind Kumar and Robert Henry.

improving the types and contents of these antioxidant compounds in staple foods have become the key to promoting human health and solving diseases related to oxidative damage. The addition of exogenous antioxidant active substances in staple foods by processing and crops-improvement by genetic engineering for producing natural antioxidants or antioxidant precursors can achieve this goal (Fatima et al. 2013). But the latter is more suitable for large-scale promotion, because of autosynthesis of antioxidants by optimized metabolic pathways, which save labor costs and time (Mattoo et al. 2010). Furthermore, genetically modified crops can not only improve the content of endogenous antioxidants, but also add or change the flow of metabolites to produce non-essential antioxidants (such as anthocyanins and astaxanthins) (Dixon et al. 2013; Bai et al. 2017; Zhu et al. 2017, 2018) or their scarce precursors (proanthocyanidins) (Dixon et al. 2013) that the species should not be able to synthesize. Imagine that a cheap staple food can provide human-rich and high-level antioxidant-active substances, which will fundamentally overcome the problems caused by premature aging of the body and oxidative-damage–related diseases. It will make foods advance from simply eating full to healthy eating.

Rice, one of the most important food crops in the world, provides energy and nutrients to more than half of the population on earth (Zhou et al. 2002). Especially in Southeast Asia and Northeast Asia, where rice is the staple food, the name of rice is basically the same as food (Steiger et al. 2014). In unprocessed brown rice, it contains a certain amount of antioxidants such as flavonoids, anthocyanins, proanthocyanidins, tocopherols, and folates (Goufo and Trindade 2014; Wang et al. 2018). Their content in pigmented rice (such as red rice and black rice, which is only 15% of the human consumption of rice) is generally higher than that in white rice (Deng et al. 2013). The distribution of these secondary metabolites in the rice grains is tissue-specific, which mainly present in the pericarp and embryo and is extremely low or absent in the starch-rich endosperm (Shen et al. 2009; Shao et al. 2011; Min et al. 2012; Goufo and Trindade 2014). Because the taste quality of brown rice is poor and the taste of rice is negatively correlated with the proportion of protein contained in it, polished rice (the starchy endosperm), from which the rough pericarp and protein-rich embryos have been removed, has become the rice consumption habit of most countries that consume rice (Tran et al. 2004). However, this greatly reduces the nutritional value of rice and the amount of antioxidants that can be absorbed by humans. Therefore, establishing the optimal synthetic route of these antioxidant active substances in the endosperm by biofortified methods of genetic engineering can effectively improve the edible value of polished rice, so that humans can obtain the necessary antioxidant compounds of the body by eating this staple food (Fang et al. 2019). In addition, the synthesis of certain antioxidant compounds that are non-essential but health-promoting in the rice endosperm can, to a certain extent, eliminate the dependence on luxury health products and medicines (Zhu et al. 2013; Martin and Li 2017). However, this process is extremely challenging. It needs to increase the content of antioxidant compound-synthesis precursors in the pathway without inhibiting the growth and development of plants. At the same time, it is necessary to inhibit other metabolic pathways that consume common precursors and reduce the consumption of antioxidants by adjusting related enzyme activity. Last but not least, it is necessary to increase the capacity of a reservoir suitable for storing antioxidants and ensure that it is not destroyed before being ingested by humans (Farré et al. 2014; Giller 2016). Two or more of the above strategies are often required to run in parallel for increasing and stabilizing antioxidant substances in the endosperm. However, the metabolic pathways in plants are complex; fully understanding the enzymatic reaction of each node in the metabolic pathway and the potential flow of key metabolic intermediates are both important premises for transforming biofortification (Farré et al. 2014; Jeffryes et al. 2018). It often requires many attempts and experiences.

In this chapter, we first introduce the characteristics of some rice-specific antioxidant compounds based on the descriptions in the previous reports: we then summarize their antioxidant activity characteristics and contributions. Then the current status of these compounds in the biofortified enhancement of rice by genetic engineering are reviewed one by one. Finally, the existing challenges are summarized, and the prospects for future development are discussed.

Antioxidant Compounds of Rice

Flavonoids, Anthocyanins, and Proanthocyanidins Composition of Rice

Flavonoids are one class of 15-carbon skeleton polyphenol compounds in which a plurality of hydroxyl (OH) or methoxyl (OCH_3) groups are attached to benzene rings and two aromatic rings (A- and B-rings) are interlinked by a three-carbon chain (structure C6—C3—C6). The flavonoids are synthesized by the phenylpropanoid metabolic pathway in rice (Grotewold 2006). According to the binding positions of OH groups on the three-carbon chain, the presence or absence of unsaturated double bonds in the B-ring, the number and position of the OH groups bound to each ring, flavonoids can be classified into flavones, flavonols, flavanones, flavanols (proanthocyanidins can also be summarized in this category), isoflavones, and anthocyanins (Wang et al. 2018). Flavonoids are mostly present as glycoside ligands in rice seeds, and most of them have absorption maximum at 370 nm (Shen et al. 2009). Flavones are the main common flavonoids in white rice, which are mainly found in endosperm. Otherwise, tricin is the most abundant flavonoid compound in the pericarp (about 77%) (Shao et al. 2011; Goufo and Trindade 2014; Zhu et al. 2017).

Many reports have show that flavonoids have strong antioxidant activity, including capturing superoxide radicals, chelation with metal ions, and blocking lipid peroxidation chain reactions (Shao et al. 2011; Min et al. 2012; Goufo and Trindade 2014). The strength of these activities depends on the number and position of the phenolic hydroxyl groups on the structure of flavonoids. The 3′OH and 4′OH of the three carbon chains contribute the most to antioxidant capacity, while the OH groups-lacking flavones and flavanones have no or only weak antioxidant activity (Shao et al. 2011; Wang et al. 2018). So it is necessary to convert the flavones into other flavonoid derivatives (such as anthocyanins and proanthocyanidins) with stronger antioxidant properties in rice (Min et al. 2012).

Anthocyanins, another class of flavonoids, mainly exist as two forms in rice: cyanidin-3-O-glucoside and paeoniflorin-3-O-glucoside (Finocchiaro et al. 2007; Wang et al. 2018). The anthocyanins, existing in monomeric form, are highly water-soluble and appear black, and are mainly contained in the peel of black rice. Proanthocyanidins are flavanol polymers that have been polymerized 4–11 times. The weakly water-soluble proanthocyanidins are easily oxidized to complex compounds and become red (mainly present in the peel of red rice) (Finocchiaro et al. 2007). Red rice contains proanthocyanidins, which have three times the antioxidant capacity of white rice, but are much weaker than that with anthocyanins (that has 15- and 20-fold the antioxidant capacity of tocopherols and ascorbic acid respectively) (Goufo and Trindade 2014; Martin and Li 2017). In rice grains, the synthesis of anthocyanins is divided into three steps, namely the formation of an anthocyanin skeleton, the formation of anthocyanin precursors, and the formation of various pigments (anthocyanins) (Figure 19.1) (Hong et al. 2012; Zhu et al. 2017; Wang et al. 2018). Phenylalanine ammonia lyase (PAL), chalcone synthesis (CHS), chalcone isomerase (CHI) (the above three proteins are translated in all rice), flavanone 3-hydroxylase (F3H), dihydroflavonol reductase (DFR) (the above two proteins are translated in pigmented rice), and anthocyanidin synthesis (ANS) (only exists in black rice) are key enzymes in this metabolic pathway, which is regulated by multiple transcription factors (Grotewold 2006; Zhang et al. 2014; Fang et al. 2019). Proanthocyanidins are formed by catalytic polymerization of anthocyanin reductase (ANR) and leucoanthocyanidin reductase (LAR) on the basis of anthocyanin synthesis (Peng et al. 2012). The genes that encode these pigment metabolism synthesis enzymes are expressed in all rice (white rice and pigmented rice). In white rice, there are no significant differences at the expression levels of these genes in the pericarp, embryo, and endosperm of white rice, which express at a relative low level (Chen et al. 2013). On the other hand, the expression levels of these genes in the pericarp and embryos of pigmented rice are much higher than those in the endosperm, which may be related to the activity of the gene promoter or the difference in transcription factors between different varieties of rice (Chen et al. 2013; Lim and Ha 2013). It has been reported that the eight genes related to anthocyanin synthesis are specifically expressed in rice endosperm by using a polygenic assembly and expression system (transgene stacking II (TGSII) system) and obtain high anthocyanin content in the endosperm (purple endosperm rice [PER]) (Zhu et al. 2017). To our knowledge, eating the anthocyanin-rich rice can effectively improve blood circulation and prevent cardiovascular disease, anti-tumor, anti-radiation, and anti-platelet condensation (Wang and Stoner 2008; Zhang et al. 2014).

Carotenoids Composition of Rice

"Carotenoids" is a general term for a class of pigment compounds having C_{40} hydrocarbons (carotenes) and their oxygen-containing derivatives thereof (lutein) (Sankari et al. 2018). It is formed by a condensation polymerization of eight isoprenoid units (having nine double bonds) and has a β-ionrone ring at each end (Giuliano 2014). Different carotenoids have various heterogeneous, substituted, and ring-opened forms on the β-ionrone ring, which are classified into lycopene, carotene, lutein, and zeaxanthin according to these characteristics (Giuliano 2014; Sankari et al. 2018). The most important part of determining the biological function of carotenoids is the conjugated double-bond system (which has a strong absorption peak between 400 and 500 mm), and their antioxidant capacities depend on the number of double bonds on the ring (lycopene is perceived as the most potent) (Zhu et al. 2013; Martin and Li 2017). The human body cannot synthesize carotenoids by themselves, and so they must be ingested by eating (Chen et al. 2015). In higher plants, carotenoids are

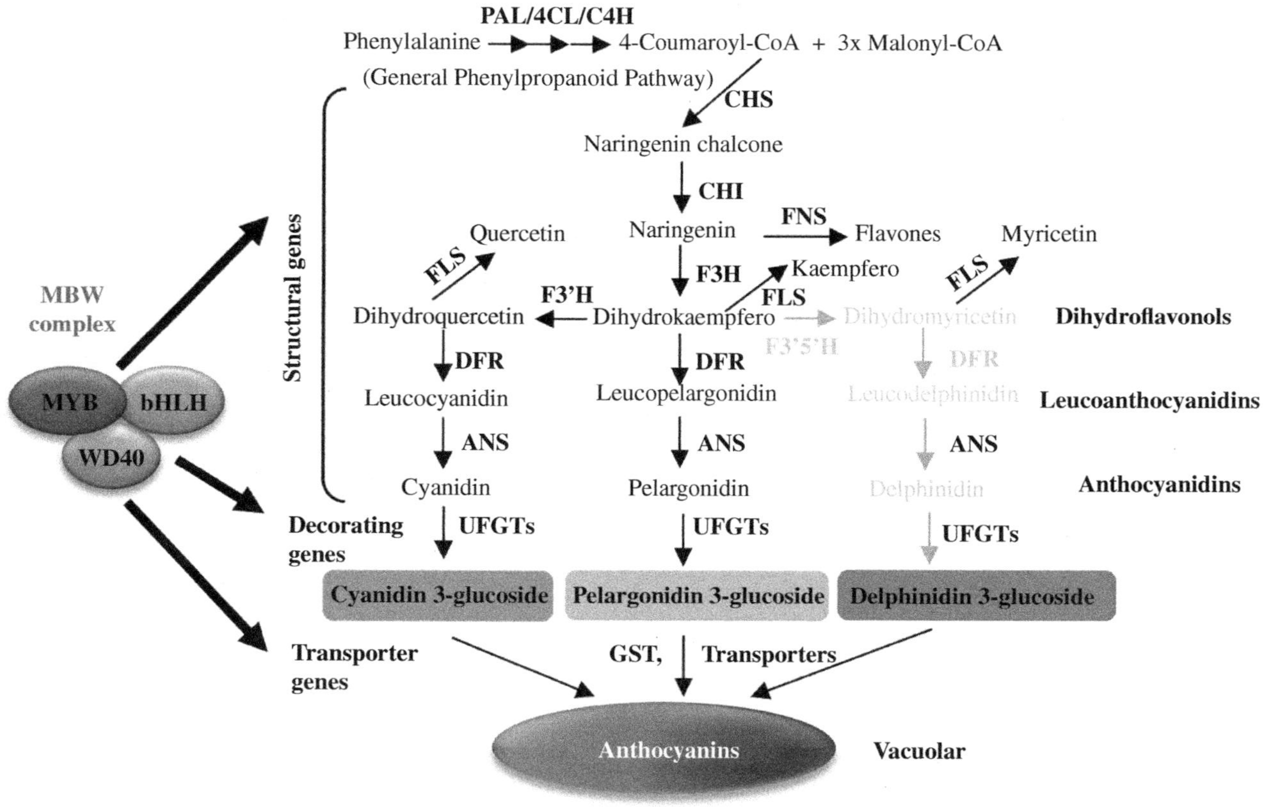

Figure 19.1 Schematic representation of the anthocyanin biosynthetic pathway in plant (rice). *Note:* The general phenylpropanoid pathway includes PAL (phenylalanine ammonia lyase), C4H (cinnamate 4-hydroxylase), and 4CL (4-coumaryol CoA ligase). CHS, chalcone synthase; CHI, chalcone isomerase; F3H, flavanone 3-hydroxylase; F3′H, flavonoid 3′-hydroxylase; F3′5′H, flavonoid 3′,5′-hydroxylase; DFR, dihydroflavonol 4-reductase; ANS, anthocyanidin synthase; UFGT, flavonoid 3-O-glucosyltransferase; FLS, flavonol synthase; FNS, flavone synthase. Gray fonts represent pathways that do not exist in rice due to F3′5′H and DFR no function. (*See color plate section for color representation of this figure*).

synthesized in the plastids via the isoprenoid pathway while rice as a staple food lacks the key enzymes and precursors of the carotenoid anabolic pathway, which make it unable to synthesize downstream carotenoids (Martin and Li 2017).

Both *in vivo* and *in vitro* studies have shown that carotenoids can protect cellular genetic material (DNA and RNA) and cell membranes (such as proteins, lipids, and carbohydrates) by quenching singlet oxygen ($^1O_2$) and scavenging free radicals (Zhu et al. 2013; Jeffryes et al. 2018). Antioxidants in the human body are functionally divided into two types. One is a preventive antioxidant, including riboflavin and retinol, and the other is a chain-breaking antioxidant, including water-soluble ascorbic acid, glutathione (GSH), and fat-soluble α-tocopherol, β-tocopherol, α-carotene, β-carotene, lycopene, zeaxanthin, and lutein. Among them, carotenoids have a significantly higher ability to quench $^1O_2$ than α-tocopherol and ascorbic acid, which is a benefit by the simultaneous elimination of $^1O_2$ by physical quenching and chemical reaction (Martin and Li 2017). In addition, superoxide radicals can also be removed by carotenoids through electron transfer and free radical addition. Therefore, the synthesis of a variety of carotenoids with strong antioxidant activity in rice (especially in endosperm) can greatly improve the edible value of rice (Sankari et al. 2018).

α-carotene and β-carotene, two members of the carotenoid family, are intermediates in the synthesis of many important carotenoids. Their biosynthesis includes a series of reactions such as condensation, dehydrogenation, cyclization, hydroxylation, and epoxidation (Figure 19.2) (Grotewold 2006; Sankari et al. 2018). Isopentenyl pyrophosphate (IPP) is a co-synthesis precursor of all isoprenoids, which produces dimethyl allyl diphosphate (DMAPP) by catalyzing of IPP isomerase. Then one part of IPP condenses with DMAPP to form geranyl pyrophosphate (C_{10}). Catalyzed by geranylgeranyl pyrophosphate synthase (GGPS), geranylgeranyl diphosphate (GGPP) was synthesized by geranyl pyrophosphate and two IPPs. Subsequently, two molecules of GGPPs (C_{20}) are catalyzed by phytoene synthase (PSY) to form a nonpigmented phytoene (C_{40}) in the plastid. Under the action of phytoene dehydrogenase (PDS) and ζ-carotene dehydrogenase (ZDS), phytoene is dehydrogenated to form pink lycopene. β-lycopene cyclase (LCYB) and ε-lycopene cyclase (LCYE) compete for the lycopene (as a common substrate) and catalyze the formation of β-carotene (with two β-rings) and α-carotene (with one β- and one ε-ring). This reaction is a key point of the carotenoid synthesis pathway (Giuliano 2014; Sankari et al. 2018; Fang et al. 2019). The Golden Rice Project was first introduced in 1999 (NYU n.d.) with the aim of accumulating β-carotene in the buttery endosperm of rice. Lutein can be formed from α-carotene by a continuous hydroxylation reaction (Potrykus 2001). While β-carotene can produce zeaxanthin under the action of β-Carotene hydroxylase (BHY) or produce astaxanthin catalyzed by BHY and β-carotene ketolase (BKT) (Farré et al. 2014).

Astaxanthin is a red ketone carotenoid and can only be found in some algae, bacteria and yeast, and its antioxidant activity is 550 times that of tocopherol (Higuera-Ciapara et al. 2006; Nogueira et al. 2017). In higher plants, ketone carotenoids cannot be produced because of the lack of a gene to encode BKT, a key enzyme in this synthetic pathway (Cunningham and Gantt 2005). In order to synthesize astaxanthin in rice endosperm, researchers in many countries have conducted long-term experiments. But due to the lack of profound understanding of synthetic pathways in the early years of these attempts

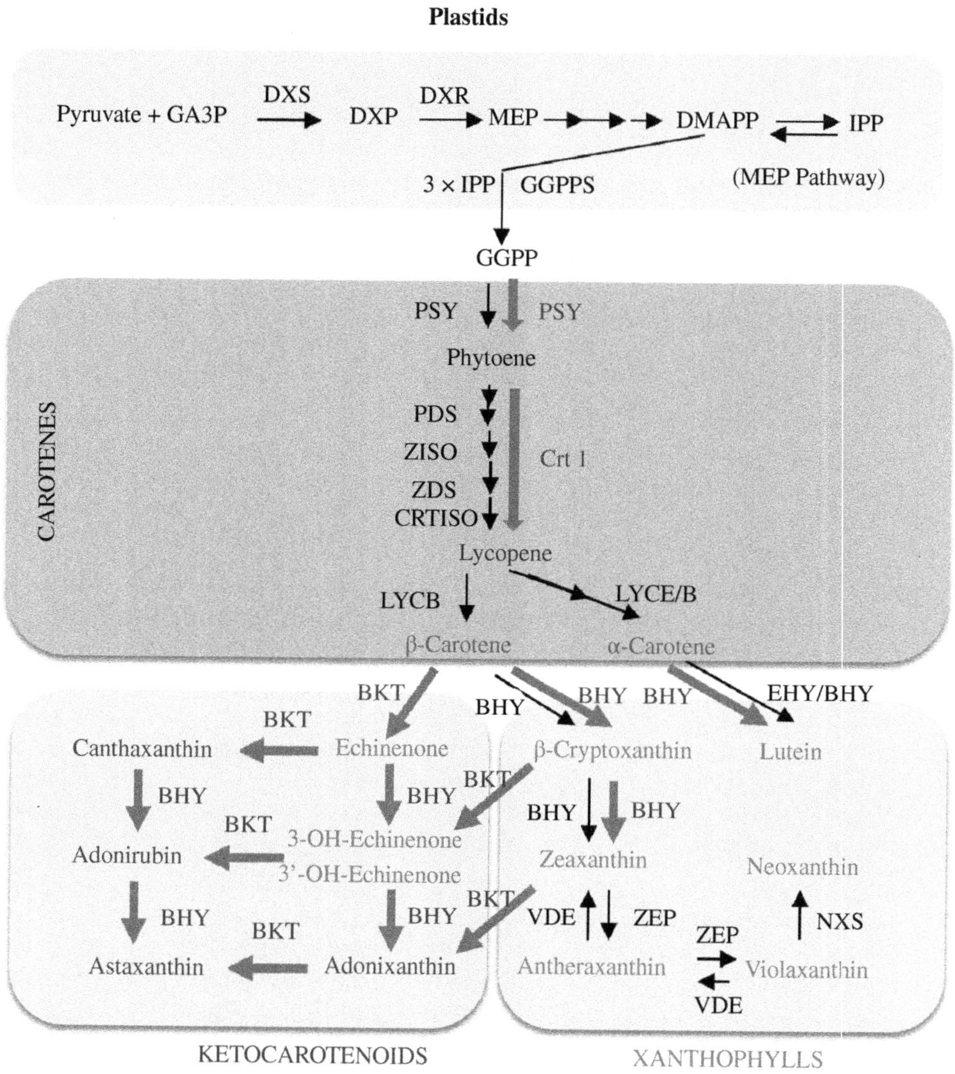

Figure 19.2 Schematic diagram of the carotenoid biosynthetic pathway in plants and in transgenic rice that accumulate ketocarotinoids. *Note:* The precursors of carotenoid biosynthesis are from the upstream mevalonate pathway (MEP). The transgene-encoded enzymes (in red) are phytoene synthase (PSY), bacterial phytoene desaturase (CrtI), β-carotene hydroxylase (BHY), and β-carotene ketolase (BKT), which can catalyze the geranylgeranyl diphosphate (GGPP) to the target end-product astaxanthin in transgenic plants. DXS, 1-deoxy-D-xylulose 5-phosphate synthase (DXP); DXR, DXP reductoisomerase; GGPPS, geranylgeranyl diphosphate synthase; PDS, phytoene desaturase; ZISO, ζ-carotene isomerase; ZDS, ζ-carotene desaturase; CRTISO, carotenoid isomerase; LCYE, lycopene ε-cyclase; LCYB, lycopene β-cyclase; EHY, ε-carotene hydroxylase; VDE, violaxanthin de-epoxidase; ZEP, zeaxanthin epoxidase; NXS, neoxanthin synthase.

and the complex metabolic processes in endosperm, they have not been successful after many years of trials (Bai et al. 2017). Recently, Zhu et al. (2018) successfully obtained astaxanthin-enriched functional rice (aSTARice) by expressing four astaxanthin-related foreign genes in rice endosperm (see section "Greenhouse Gas Emissions under Aerobic Conditions"). By eating ketocarotenoids-enriched food, it can remove excess free radicals in the human body, enhance human immunity, and prevent cardiovascular diseases and promote the fight against cancer.

Tocopherols and Tocotrienols Composition of Rice

Vitamin E is a derivative of benzo-hydropyran, which is classified into tocopherols and tocotrienols according to the saturating degree of the side chain (Wang et al. 2018). Both of them have an isoprenoid side chain; the side chains of the tocopherols are saturated fatty acid while the tocotrienols contain unsaturated fatty acid side chains with three double bonds (Wang et al. 2018). There are eight isomers of vitamin E in plants, including four tocopherols and four tocotrienols (α, β, γ, and δ, respectively) (Goufo and Trindade 2014). Among them, α-tocopherol has the highest physiological activity, followed by β-tocopherol, and tocotrienols are considered to have no vitamin activity. But tocotrienols have better performance in antioxidation than tocopherols (Martin and Li 2017). Vitamin E is very high in rice seed (up to 300~500 $\mu g\,g^{-1}$) and mainly forms into α-tocopherol or γ-tocotrienol (Hess et al. 2001). The contents of these isomers in different varieties of rice are significantly different, the contents of them in *japonica* rice are generally higher than that in *indica* rice. Otherwise, the ratio of tocotrienols to tocopherols in *indica* rice seeds (1.61) is significantly higher than that in *japonica* rice (0.95) (Britz et al. 2007; Heinemann et al. 2008). The total vitamin E concentration in brown rice can reach 443 $\mu g\,g^{-1}$, which is much higher than that in polished rice (Britz et al. 2007; Ng et al. 2013). Therefore, finding how to increase the content of vitamin E in polished rice (starch endosperm) becomes one of the key points to enhance the antioxidant value of rice.

Tocopherols and tocotrienols, as fat-soluble chain-breaking antioxidants, are mainly in the cell membrane. Both of them can react with free radicals and have anti-lipid peroxidation properties (Xu et al. 2001; Martin and Li 2017). It has been reported that α-tocopherol and α-tocotrienol have no difference in the ability to scavenge free radicals. But in liposomes, the ability of α-tocotrienol to scavenge lipid peroxides is 1.5 times to α-tocopherol and its effect of protecting cytochrome P-450 against suffering oxidative damage is also higher (6.5 times) than that of α-tocopherol (Packer et al. 2001). In the rat liver lipid peroxidation experiment, the antioxidant activity of α-tocotrienol can reach 40–60 times that of α-tocopherol, which may be related to its structural characteristics (Hermizi et al. 2009). First, the α-tocotrienol molecule can be more efficiently reduced by the action of reducing agents. Secondly, α-tocopherol can be better dissolved in membrane lipids, which make the lipid structure of cell membrane become disordered, so it is more conducive for α-tocopherol to interact with the free radical in lipids. Recent studies have shown that γ-tocopherol may have superior antioxidant properties to those of α-tocopherol, but its working mechanism remains to be further elucidated. In rice, homogentisate phytyltransferase (HPT) and homogentisate geranylgeranyl transferase (HGGT) are two of the key enzymes in the tocopherols and tocotrienols synthesis pathways, respectively (Figure 19.3)

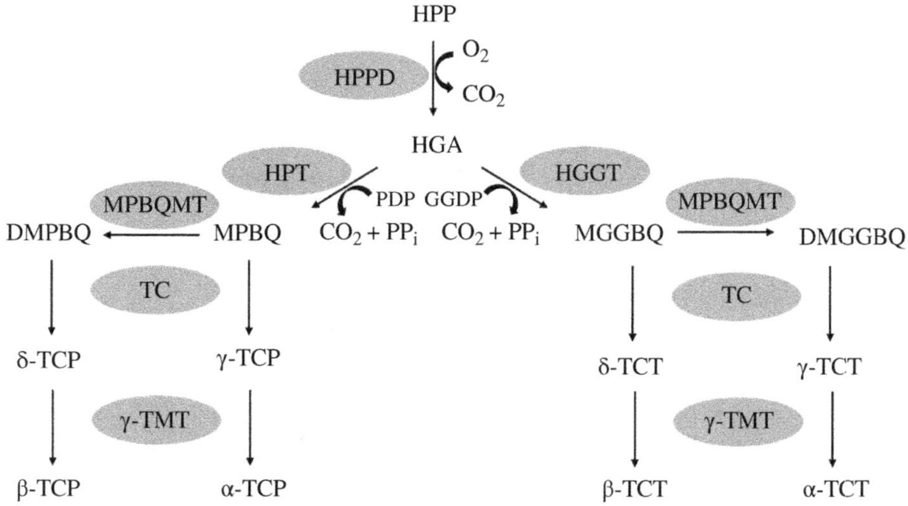

Figure 19.3 The tocotrienol and tocopherol biosynthesis pathway in rice. *Note:* Enzymes (represented by ovals): HPPD, 4-hydroxyphenylpyruvate dioxygenase; HPT, homogentisate phytyltransferase; HGGT, homogentisate geranylgeranyl transferase; MPBQMT, 2-methyl-6-phytylbenzoquinone methyltransferase; TC, tocopherol cyclase; γ-TMT, γ-tocopheryl methyltransferase. Chemicals: HPP, p-hydroxyphenyl pyruvate; HGA, homogentisate; MPBQ, 2-methyl-6-phytyl-1,4-benzoquinol; DMPBQ, 2,3-methyl-6-phytyl-1,4-benzoquinol; MGGBQ, 2-methyl-6-geranylgeranyl-1,4-benzoquinol; DMGGBQ, 2,3-methyl-6-geranylgeranyl-1,4-benzoquinol; α-, β-, γ-, and δ-TCP, α-, β-, γ-, and δ-tocopherols; α-, β-, γ-, and δ-TCT, α-, β-, γ-, and δ-tocotrienols.

(Hunter and Cahoon 2007). In the final step of this synthetic route, the γ and δ isoforms can be catalytically converted to α and β isomers by γ-tocopherol methyltransferase (γ-TMT) (Chaudhary and Khurana 2009). Specific activation or inhibition of these enzyme activities in rice endosperm can shift the type and content of vitamin E in a direction that favors humans (Zhang et al. 2012).

Folate Composition of Rice

As a water-soluble vitamin B, folates (pteroylglutamic acid) are a mixture of tetrahydrofolic acid and its derivatives, which are indispensable nutrients for fetal development (Jabrin et al. 2003). Folates have a strong scavenging effect on ·OH and O_2^- and this effect is affected by the concentration. It has been reported that the clearance rate of folates to ·OH at a concentration of 0.062 mg ml^{-1} is 90.73% while the clearance rate for O_2^- at the same concentration is 66.54% (Chi et al. 2011). The content of folates in rice is low and is extremely prone to loss during storage and cooking. The average content of folates in brown rice is 488 ng g^{-1}, which will cause it to lose about 50% (only left 246 ng g^{-1}) during fashioning (Dong et al. 2011). It further loses nearly 50% after cooking, which is far below the international recommended daily intake for adults (400 μg d^{-1}) (Blancquaert et al. 2010). The content difference of folates in different varieties can reach 8.4 times, which in *indica* rice is generally higher than that in *japonica* rice. Studies show that this difference is mainly due to the different content of 5-methyltetrahydrofolate (Dong et al. 2011). In order to reduce disease caused by the lack of folates, it is necessary to increase the content of folates in the endosperm of rice by biofortification.

The molecular structure of folates consists of pteridine, para-aminobenzoic acid (PABA), and glutamine (Hanson and Gregory 2011). They are synthesized in plastids, mitochondria, and cytoplasm, respectively, in rice and transported to vacuoles for storage (Figure 19.4). First, guanosine triphosphate cyclohydrolase I (GTPCHI) catalyzes the synthesis of pteridine by using guanosine triphosphate (GTP) as a precursor, and the part of PABA is synthesized from a chorismic acid salt by aminodeoxy chorismate synthase (ADCS) and aminodeoxy chorismate lyase (ADCL). Subsequently, pteridine and PABA are transported to the mitochondria and reduced to tetrahydrofolate with glutamate, which requires the participation of 6-hydroxymethyldihydrofuran pyrophosphokinase (HHPK) and dihydropteroate synthase (DHPS). Finally, the γ-position of glutamic acid in tetrahydrofolate is catalyzed by folylpolyglutamate synthase (FPGS) to bind a plurality of glutamic acid

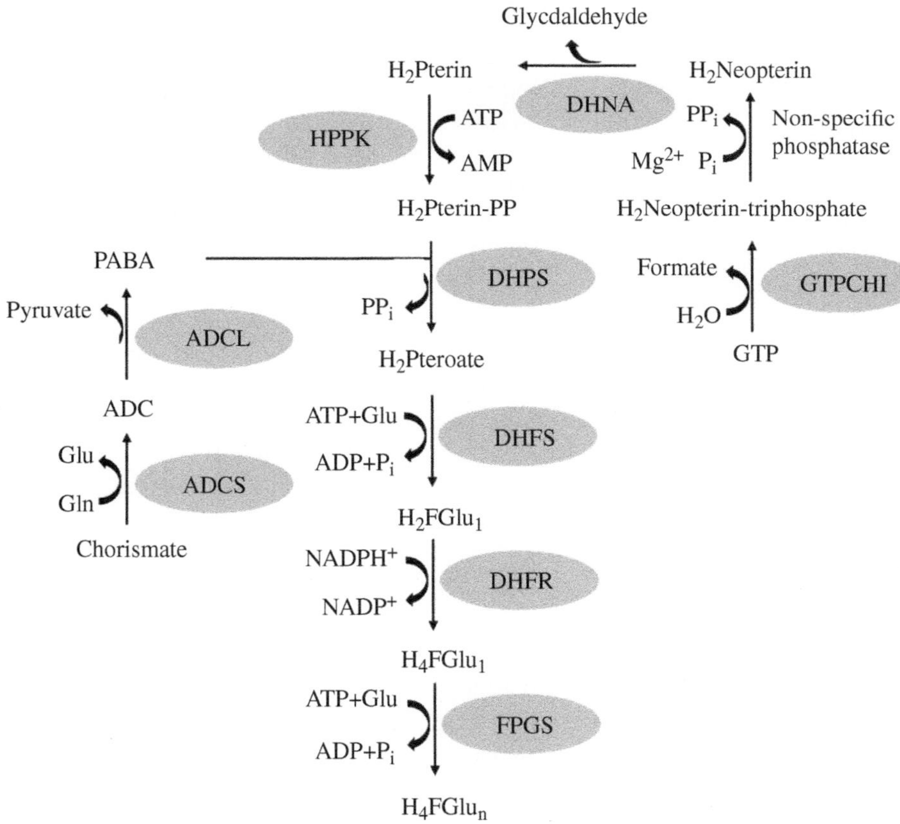

Figure 19.4 The folate biosynthesis pathway in rice. *Note:* Enzymes (represented by ovals): GTPCHI, guanosine triphosphate cyclohydrolase I; DHNA, dihydroneopterin aldolase; HPPK, 6-hydroxymethyldihydrofuran pyrophosphokinase; ADCS, aminodeoxy chorismate synthase; ADCL, aminodeoxy chorismate lyase; DHPS, dihydropteroate synthase; DHFS, dihydrofolate synthetase; DHFR, dihydrofolate reductase; FPGS, folylpolyglutamate synthase. Chemicals: GTP, guanosine triphosphate; H_2Neopterin, dihydroneopterin; H_2Pteroate, dihydropteroate; H_2Pterin-PP, hydroxymethyldihydropterin; ADC, minodeoxychorismate; PABA, para-aminobenzoate; Glu, glutamate; H_2Pterin, dihydropterin; H_2FGlu_1, dihydrofolate; H4FGlu1, tetrahydrofolate.

residues, which form folates having a polyglutamic acid tail (Bekaert et al. 2008; Hanson and Gregory 2011). At present, based on the low content and easy degradation of folates in rice, two strategies have been adopted for biofortification of folates. One is to increase the absolute content of folates by improving the expression level of key enzymes in the folate synthesis pathway (Storozhenko et al. 2007; Blancquaert et al. 2015), while the other is to enhance the stability of folates by folate-binding protein (FBP) (Goto et al. 1999).

Melatonin Composition of Rice

Melatonin (N-acetyl-5-methoxytryptamine) is a low molecular-weight (232.27) indoleamine that not only has the function for regulating the biological circadian rhythm, but also is a highly effective antioxidant compound (Zhu et al. 2013; Zhang and Zhang 2014). The chemical structure of melatonin is stable and has no redox activity, which is quite different from other antioxidants (Bonnefont-Rousselot and Collin 2010). It has an electron-rich indole ring, which can be used as a strong electron donor to quench high-activity oxygen radicals ($^1O_2$, ·OH, and HOO—) by electron-donating reaction. But there is no direct reaction between melatonin and low oxidation potential electrophiles (H_2O_2 and O_2^-) (Zhang and Zhang 2014). Studies have shown that melatonin is 4 times more capable of clearing ·OH than glutathione (GSH) and twice as effective as vitamin E (Tan 1993). Melatonin, an ancient antioxidant, is widely found in unicellular algae, plants, and animals (Tsia and Hu 2003). The content of melatonin has been measured in 25 types of cultivated rice, ranging from 11 to 264 ng g^{-1} (average at 16.0 ng g^{-1}) (Wang et al. 2009). Some studies suggest that melatonin may be mainly distributed in the colored pericarp of pigmented rice (Zhu et al. 2013).

In rice, melatonin is synthesized using tryptamine as a precursor and catalyzed by tryptophan decarboxylase (TDC), tryptamine 5-hydroxylase (T5H), serotonin N-acetyltransferase (SNAT), and acetyl serotonin methyltransferase (ASMT) (Figure 19.5) (Kang et al. 2011;

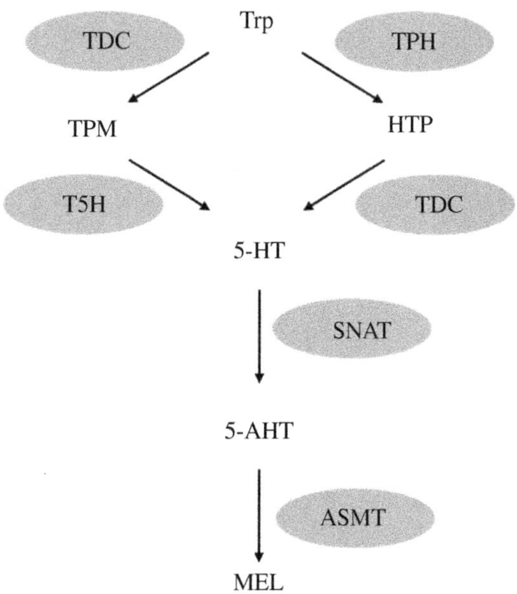

Figure 19.5 The melatonin biosynthesis pathway in rice. *Note:* Enzymes (represented by ovals): TDC, tryptophan decarboxylase; T5H, tryptamine 5-hydroxylase; TPH, tryptophan hydroxylase; SNAT, serotonin N-acetyltransferase; ASMT, acetyl serotonin methyltransferase. Chemicals: Trp, tryptophan; TPM, tryptamine; HTP, 5-hydroxy tryptophan; 5-HT, serotonin; 5-AHT, N-acetylserotonin; MEL, melatonin.

Zhu et al. 2013). The production of melatonin in plants is related to circadian rhythm and developmental stage. The rice seedlings can be promoted to produce melatonin in the dark and in high-temperature conditions (Byeon and Back 2014). The tomato fruit has the lowest melatonin content at the beginning of the fruiting stage (green ripening period), and it increases with the ripening process until it reaches the highest level (full ripening period) (Van Tassel et al. 2001). These results provide a theoretical basis and practical guidance for melatonin biofortification in rice.

Genetic Engineering for Biofortification with Antioxidants

The nutrients in the grain, such as fiber, phenolic acid, vitamins, phytonutrients, and micronutrients, mainly exist in the pericarp and the embryo, but in the process of refining, the outer layer of the grain (pericarp and aleurone) is often removed and only endosperm remains, which greatly reduce the value, especially for rice. In polished rice, the nutrient components and antioxidant-active substances of the endosperm are even rarer (Boers et al. 2015). Therefore, the endosperm nutrition enhancement by genetic engineering can effectively improve its nutritional value and increase antioxidant activity.

Flavonoids, Anthocyanins, Proanthocyanidins

Flavonoids are a large class of phytonutrients, including intermediates (e.g., flavonols and flavones) and downstream products (e.g., anthocyanins and proanthocyanidins), which have strong antioxidant activity and are important health-promoting bioactive compounds (Deng et al. 2013). They are widely found in fruits and vegetables. In rice, anthocyanins and proanthocyanidins mainly exist in the pericarp of specific varieties such as black rice and red rice. The pericarp and embryo contain a small amount of flavonoids, but they are not present in the endosperm (Shao et al. 2014). Due to the habit of eating refined rice (polished to remove pericarp and embryo) in East Asia, the nutritional imbalance is even more pronounced (Boers et al. 2015), and there is a lack of antioxidants such as anthocyanins. Because of the restriction of wild-type rice genotypes, traditional breeding methods cannot obtain new rice germplasm with endosperm accumulation of flavonoids or anthocyanins. Therefore, the use of biotechnology, especially metabolic engineering, can quickly complete the above breeding needs.

Anthocyanin biosynthesis involves a series of genes, nearly 20, such as structural genes, decorating genes, transporters, and transcriptional regulatory genes (Zhang et al. 2014; Yuan and Grotewold 2015). In nature black rice and red rice mainly contain anthocyanins and proanthocyanidins in pericarp, respectively, which distribution of pigments is mainly controlled by a complex of transcriptional regulatory factors (MYB-bHLH-WD40, MBW), especially basic helix-loop-helix (bHLH)-type regulatory factors. Black rice and red rice are mainly regulated by *OsB1/Kala4* (Oikawa et al. 2015), and *Rc* regulators (Sweeney et al. 2006), respectively. In common white rice, many key genes (such as the *DFR*) and transcription factors (such as the *OsB1/OsB2*, *Rc*) related with the anthocyanin synthesis pathway are functional deletion mutations, resulting in the common white rice pericarp, which is usually light brown. Therefore, it is difficult to synthesize anthocyanins in white rice, especially in the endosperm.

Many attempts have been made to obtain a new rice germplasm with anthocyanins in the endosperm. For example, Shin et al. (2006) transformed two maize anthocyanin regulators, ZmR-S and ZmC1, driven by endosperm-specific promoters into rice, but only the brown seeds were obtained, which accumulated flavonoids (mainly dihydroquercetin, 3′-*O*-methyl dihydroquercetin, and 3′-*O*-methyl quercetin) in the pericarp and didn't produce anthocyanin in endosperm. Ogo et al. (2013) obtained a flavonoid-enhanced germplasm that accumulated various flavonoid intermediates (mainly naringenin, kaempferol, genistein, apigenin, and tricin) in the endosperm by co-expressing different combinations of several anthocyanin biosynthetic key enzyme genes (e.g., *PAL*, *CHS*, *F3H*, *FLS*, and *FNS*) controlled by endosperm-specific promoters, although no anthocyanins are synthesized.

By analyzing the genomic and transcriptional data of rice anthocyanin biosynthesis-related key genes and transcription factors, Zhu et al. (2017) use a new strategy, anthocyanin pathway-related regulatory factors together with complete structural genes, to successfully synthesize anthocyanins in rice endosperm. By using a robust TGSII system, two maize anthocyanin biosynthesis regulatory factors (*ZmLc* and *ZmPl*), and six coleus anthocyanin synthetic structural genes (*SsCHS*, *SsCHI*, *SsF3H*, *SsF3′H*, *SsDFR*, and *SsANS*) driven by endosperm-specific promoters, are assembled and transformed into rice to produce the first PER (called "Zijingmi" in Chinese) (Zhu et al. 2017). The pericarp color of PER rice is the same as the wild type white rice, while the aleurone layer and endosperm specifically accumulate purple anthocyanins. The total content of anthocyanins in polished rice is up to around 1 $mg\,g^{-1}$ DW (dry weight), and its main components are cyanidin 3-*O*-glucoside and peonidin 3-*O*-glucoside. By the trolox equivalent antioxidant capacity (TEAC) assay, the unpolished and polished grains of PER show much higher antioxidant activity than those of white rice grains (Zhu et al. 2017). The above studies demonstrate that the metabolic engineering method can effectively achieve nutrient biofortification and enhance the antioxidant activities of refined rice.

Carotenoids

Carotenoid is a large class of fat-soluble phytonutrients with important biological activities for human health. For example, β-carotene is the precursor of vitamin A. Furthermore, as an expensive nutritional health product, the advanced ketocarotenoids, such as canthaxanthin or astaxanthin, have strong antioxidant activity, and significant benefit for human health. The diterpenoid GGPP is the carotenoid biosynthetic precursor and is derived from the mevalonate pathway (MEP), which is widespread in plants (Chen and Lu 2019). But rice endosperm does not contain any carotenoids, due to low or no expression of carotenoid-biosynthesis-related key enzyme genes. Through the endosperm-specific expression of *PSY* (encoding phytoene synthase) and *CrtI* (encoding phytoene desaturase) genes in rice, the famous β-carotene- enriched "golden rice" was developed (Ye et al. 2000; Paine et al. 2005), which has important nutritional value for impoverished populations in developing countries.

Because they lack the *BKT* gene (encoding β-carotene ketolase), most higher plants do not synthesize ketocarotenoids such as canthaxanthin and astaxanthin. The synthesis of astaxanthin is mainly found in microalgae and yeast. In animals such as shrimp, it is derived from ingesting algae. As the strongest antioxidant, it is beneficial for human health, such as preventing certain types of cancer, inhibiting the oxidation of low-density lipoproteins, and boosting the immune system (Chew et al. 1999; Iwamoto et al. 2000; Tatsuzawa et al. 2000). Astaxanthin is biosynthesized from β-carotene after two steps of the enzymes reaction by β-carotene ketolase (*BKT*) and β-carotene hydroxylase (*BHY*). Generally, the *BHY* gene widely exists in plant green tissues and fruits that have abundant carotenoid precursors. Therefore, astaxanthin can be produced only by transferring the *BKT* gene into these parts. However, because there are no carotenoid precursors in rice endosperm, the synthesis of astaxanthin in endosperm is more complicated and involves multiple steps: it needs to start from the GGPP and be catalyzed by four enzymes PSY, CrtI, BHY, and BKT in order.

By using the transgene stacking system TGSII, Zhu et al. (2018) engineered astaxanthin biosynthesis in rice endosperm by delivering different combinations of four synthetic rice codon-optimized key genes, *sZmPSY1*, *sPaCrtI*, *sCrBKT*, and *sHpBHY*, under the control of rice endosperm-specific promoters. Though over-expression of two (*sZmPSY1* and *sPaCrtI*), three (*sZmPSY1*, *sPaCrtI* and *sCrBKT*), and all these four genes, the carotenoid, ketocarotenoids, and astaxanthin biosynthetic pathways are rebuilt in endosperm, and various types of biofortified germplasm are produced from yellow golden rice (β-carotene, ~25 $\mu g\,g^{-1}$ DW, accounting for 90% of total carotenoids) to orange–red canthaxanthin rice (canthaxanthin, ~26 $\mu g\,g^{-1}$ DW, accounting for 82% of total carotenoids) and astaxanthin rice (astaxanthin, ~16 $\mu g\,g^{-1}$ DW, accounting for 74% of total carotenoids; called "Chiijingmi" in Chinese), respectively. In addition, by using a particle-bombardment method, three genes (*PSY*, *CrtI*, and *BKT*) were co-expressed in rice to mainly produce canthaxanthin (4 $\mu g\,g^{-1}$ DW) (Bai et al. 2017). And Ha et al. (2019) transformed similar four genes, pepper *PSY*, *PaCrtI*, *CrBKT*, and pepper *BHY* into rice to produce low content of astaxanthin in endosperm (1.1 $\mu g\,g^{-1}$ DW, accounting for 61% of total carotenoids), by using the polycistronic transgene method that fused genes linking with self-cleavage peptide 2A.

However, compared with other horticultural crops that biosynthesize astaxanthin, such as astaxanthin-enriched tomato (up to 16 $mg\,g^{-1}$ DW) (Huang et al. 2013), the contents of astaxanthin rice are still low. Therefore, in the future, it is expected to further increase astaxanthin production by increasing upstream precursors through continuously stacking and co-expressing the key enzyme genes of the upstream pathway (e.g., the *DXS* gene) and regulatory genes (e.g., *OR*). It is an effective method to enhance carotenoid biofortification by means of metabolic engineering.

Tocopherols, Tocotrienols

As an essential antioxidant, vitamin E is very important for human diet and health. Plants are the main source of vitamin E intake in humans. Vitamin E has eight isomers with four tocopherols (α, β, γ, and δ) and four tocotrienols (α, β, γ, and δ), and is synthesized in plant plastids (Falk and Munne-Bosch 2010; Yang et al. 2011). Vitamin E synthesis requires two substrates: homogentisic acid (HGA) and phytyl diphosphate (PDP). HGA is produced by the shikimate pathway. Firstly, ρ-hydroxyphenylpyruvic acid (HPP) is produced through deamination of tyrosine, and then ρ-hydroxyphenylpyruvic acid dioxygenase (HPPD) brings about the oxidation of HPP to produce HGA. The second substrate, PDP, is produced by the non-MEP. PDP has two forms: PDP or geranylgeranyldiphosphate (GGDP). GGDP can be used directly in tocotrienol synthesis or reduced to PDP for tocopherol biosynthesis (Hunter and Cahoon 2007). The initial precursor of vitamin E synthesis was obtained from prenylation of HGA. MBPQ and MGGBQ, from HGA and PDP/GGDP by HPT/HGGT, are common precursors for all tocopherols and tocotrienols, respectively. MPBQ can produce δ-tocopherol under the action of tocopherol cyclase (TC); it can also produce γ-tocopherol under the action of 2-methyl-6-phenylphenylhydrazine methyltransferase (MPBQ-MT) and TC. Under the action of tocopherol methyltransferase (γ-TMT), δ-tocopherol, and γ-tocopherol were converted into β-tocopherol and α-tocopherol, respectively. Studies have shown that MGGBQ also produces δ-tocotrienol, γ-tocotrienol, β-tocotrienol, and α-tocotrienol in the same manner. Among them, α-tocotrienol has the highest physiological activity.

In general, the main way to increase vitamin E is to enhance the activity of the enzymes related with the vitamin E pathway, by over-expression of one or multiple genes (Mène-Saffrané 2017). Therefore, genetic engineering can be used, not only to increase the total vitamin E content, but also to change the relative levels of different forms of vitamin E (Jiang et al. 2017). Over-expression of the key upstream enzymes coding genes of vitamin E biosynthesis (e.g., *TYRA*, *HPPD*, *HPT*, and *HGGT*) can increase tocopherol content in *Arabidopsis*, maize, rice, lettuce, and soybean (Jiang et al. 2017). Expression of downstream enzymes, such as the 2-methyl-6-phenylphenylhydrazine methyltransferase gene (*MPBQ-MT*), and the tocopherol methyltransferase gene (*γ-TMT*) in lettuce and *Arabidopsis* can alter the proportion of different forms of tocopherol. Over-expression of the *Arabidopsis* HPPD gene changed tocopherol synthesis from the α to the γ isoform but not increasing the total tocopherol levels (Farre et al. 2012). Over-expression of *HGGT* or *HPT* genes showed no significant changes in the content of tocopherols and tocotrienols in brown rice, but the total tocopherol content in polished rice was increased by 1.3 times and 1.5 times, while the content of tocotrienol was not affected. In brown rice that over-expresses the *Arabidopsis* γ-TMT gene, the ratio of α-tocopherol to total tocopherol and tocotrienol was increased to over 97%, but the content of δ-tocotrienol was almost none (Zhang et al. 2013).

Folates

The folate biosynthesis pathway in plants consists of three pathways, which are localized in the cytoplasm, mitochondria, and plastids. The synthesis of the pteridine ring component, 6-hydroxymethylindoline phosphate, starts from GTP, and a group of pteridine complexes are synthesized from GTP by GTP cyclohydrolase. This series of reactions is carried out in cytoplasm. The other substrate for folate, PABA, is synthesized from methylidene benzoic acid by m-benzoate synthase and methylidene cyclase in the plastid (Hanson and Gregory 2011; Dong et al. 2014). The pteridine compound is catalyzed by HPPK to form 6-hydroxymethylindoline phosphate, and the latter combines with PABA from plastids to form polyglutamyltetrahydrofolate by DHPS, dihydrofolate synthetase (DHFS), dihydrofolate reductase (DHFR), and folate polyglutamate synthase (FPGS). This form of the folate molecule is a preferred form for folate-dependent enzyme involved in one-carbon metabolism (Teplyakov et al. 2004). In contrast, folate transporters prefer the monoglutamic folate (Hamid et al. 2009). The expression of folate biosynthesis-related genes *GTPCHI*, *ADCS*, *HPPK*, *DHPS*, and folate polyglutamylase gene *FPGS* alone or in combination can significantly increase the folate content in tomatoes (Díaz de la Garza et al. 2004), beans (Ramírez Rivera et al. 2016), corn (Naqvi et al. 2009), and lettuce (Nunes et al. 2009). The use of the same strategy to over-express *ADCS* and *GCHI* genes in rice endosperm results in 100-fold folate enhancement (up to 38.3 nmol g^{-1}) (Storozhenko et al. 2007). Further research shows that simultaneously expressing three genes (*GTPCHI*, *ADCS*, and *FPGS*) can increase the folate level up to 100 times in rice grains (Blancquaert et al. 2015).

In nature, folate is very unstable, especially the pteridine and PABA. Light, high temperature, active oxygen, and degradation enzymes can easily reduce the content of folate. In contrast, folate in plants is protected by antioxidants (such as ascorbic acid and GSH) or binding protein that forms relatively stable complexes. Therefore, enhancing the stability of folate by genetic engineering is an effective means to increase its content. Expression of the folate-binding protein gene *FBP* together with three genes (*GTPCHI*, *ADCS*, and *FPGS*) in rice for forming a complex with folate improves folate stability, obtaining a folate concentration 150 times higher than that of wild-type rice. This way enables long-term storage of biofortified high-folate rice seeds (Blancquaert et al. 2015).

Melatonin

Melatonin is an important antioxidant. It was first identified in plants in 1995 (Dubbels et al. 1995; Hattori et al. 1995). The melatonin content of rice grains from various rice varieties was detected. The results show that unpolished pigmented rice (black and red grains) have high levels of melatonin, in which the red *indica* rice grains is the highest up to 212 μg kg^{-1}, but the content of polished grains is low (average around 32 μg kg^{-1}) (Setyaningsih et al. 2015).

Tryptophan is a precursor for the synthesis of melatonin. The synthesis of melatonin by tryptophan has undergone a four-step reaction. There are two main synthetic pathways for melatonin: one is the tryptophan/tryptamine/serotonin/N-acetylserotonin/melatonin pathway, which may occur under normal growth conditions; the other is the tryptophan/tryptamine/serotonin/5-methoxytryptamine/melatonin pathway, which may occur when plants produce large amounts of serotonin (Back et al. 2016). Plant melatonin synthesis is related to six enzymes: tryptophan decarboxylase (TDC), tryptophan hydroxylase (TPH), tryptophan 5-hydroxylase (T5H), serotonin n-acetyltransferase (SNAT), n-Acetyl serotonin methyltransferase (ASMT), and caffeic acid o-methyltransferase (COMT). Almost all key enzymes responsible for melatonin biosynthesis have been cloned. The combination expression of the three enzyme genes *TDC*, *T5H*, and *TPH* leads to the synthesis of serotonin, the key intermediate of melatonin synthesis. The synthesis of melatonin from serotonin is a two-step reaction involving three different enzymes, SNAT, ASMT, and COMT (Back et al. 2016). SNAT catalyzes the production of serotonin to N-acetyl serotonin, which is then converted to melatonin by ASMT or COMT enzyme. The rate of synthesis from serotonin to melatonin is lower than that of serotonin's synthesis. The inefficient enzymatic reaction of melatonin produced by serotonin limits the biosynthesis of melatonin. Therefore, the regulation of synthesis from serotonin to melatonin provides an opportunity to increase the melatonin content.

Several enzymes in the melatonin synthesis pathway have been expressed in transgenic plants, including SNAT, ASMT, and TDC. For example, mini-tomato with algae arylalkylamine *N*-acetyltransferase (Okazaki et al. 2009) and rice with human serotonin *N*-acetyltransferase (Kang et al. 2010) result in increasing melatonin levels in transgenic plants. In rice, there are three TDC genes (Kang et al. 2007; Kanjanaphachoat et al. 2012). Over-expressing one of the three endogeneous TDC genes all can increase the melatonin content in rice (Byeon et al. 2014). The over-expression of *TDC3-1* leads to 31-fold melatonin concentration higher than those of the wild type grains in transgenic rice plants. But the content of melatonin in transgenic seedlings only increases two times. In addition, the amount of melatonin intermediates is produced in transgenic rice. Therefore, simultaneous expression of multiple key enzyme genes by multigene vectors is expected to further increase the content of melatonin in rice seeds.

Conclusions and Perspectives

With the rapid development of biotechnology in recent years, especially the widespread application of metabolic engineering and synthetic biology, more and more genetically modified and health-promoting foods have come into our sight. There are various natural antioxidants in rice, but their contents are low or mainly in the pericarp or embryo, which are almost completely lost in the process of processing into polished rice (Goufo and Trindade 2014; Steiger et al. 2014; Wang et al. 2018). Humans are eager to obtain essential (or nonessential but health-promoting) antioxidants from staple foods (such as rice) (Zhu et al. 2013). It is clear that traditional breeding methods are already weak and unable to make this goal come true. The biofortification of rice by genetic engineering, which can produce functional rice with abundant kinds and high content of antioxidants in the endosperm, provide new ideas and directions for humans to prevent diseases and resist aging (Farré et al. 2014).

The biosynthesis regulation network of rice secondary metabolites is very complicated. The synthesis, transportation, and storage methods of many rice antioxidant compounds remain unresolved (Goufo and Trindade 2014; Steiger et al. 2014). For example, at least three folate transporters (working in mitochondria, vacuoles, and plasma membranes, respectively) have yet to be discovered, and how transcription factors regulate the coordinated use of antioxidants biosynthetic genes remains unclear (Wang et al. 2018).

In order to separate and identify more unknown antioxidants in rice, various analytical techniques such as microwave extraction (Setyaningsih et al. 2015), diode array spectroscopy, and mass spectroscopy (Lee 2010) have been rapidly developed and widely used. In recent years, the combined application of bioinformatics (Vasconcelos et al. 2017), transcriptomics (Gao et al. 2013), proteomics (Wang et al. 2013), metabolomics (Kusano et al. 2015), and compound prediction based on biosynthetic gene clusters (Yamane 2013) has become a powerful driving force for biofortification of rice antioxidant nutrients. However, due to the fact that research in the area of rice antioxidants is still in its infancy, many antioxidant active ingredients are poorly known (such as the identity of phenolic compounds) or lack of standardized-extraction methods (Giller 2016), the cooperation between international laboratories may be a key for promoting the rice antioxidant biofortification (Goufo and Trindade 2014). With the combination of traditional breeding methods and metabolic engineering, crossing the engineered rice with carotenoid or anthocyanin-enriched endosperm with the high antioxidant-activity wild-type black rice or red rice (containing anthocyanins or proanthocyanidins in the pericarp) is expected to produce a new multi-functional rice germplasm with higher antioxidant activity in grains. The use of synthetic biology and metabolic engineering methods to enhance rice antioxidant nutrients is a promising development direction, which will provide people with more and better quality new rice germplasm resources.

Acknowledgments

This work was supported by grants from the National Natural Science Foundation of China (31971915; 31771740), and the Guangdong Special Support Program of Young Top-Notch Talent in Science and Technology Innovation (2019TQ05N147).

References

Ahmad, P., Jaleel, C.A., Salem, M.A. et al. (2010). Roles of enzymatic and nonenzymatic antioxidants in plants during abiotic stress. *Crit. Rev. Biotechnol.* 30: 161–175.

Back, K., Tan, D.X., and Reiter, R.J. (2016). Melatonin biosynthesis in plants: multiple pathways catalyze tryptophan to melatonin in the cytoplasm or chloroplasts. *J. Pineal Res.* 61: 426–437.

Bai, C., Berman, J., Farre, G. et al. (2017). Reconstruction of the astaxanthin biosynthesis pathway in rice endosperm reveals a metabolic bottleneck at the level of endogenous β-carotene hydroxylase activity. *Transgenic Res.* 26: 13–23.

Bekaert, S., Storozhenko, S., Mehrshahi, P. et al. (2008). Folate biofortification in food plants. *Trends Plant Sci.* 13: 28–35.

Blancquaert, D., Storozhenko, S., Loizeau, K. et al. (2010). Folates and folic acid: from fundamental research toward sustainable health. *Crit. Rev. Plant Sci.* 29: 14–35.

Blancquaert, D., Van Daele, J., Strobbe, S. et al. (2015). Improving folate (vitamin B9) stability in biofortified rice through metabolic engineering. *Nat. Biotechnol.* 33: 1076–1078.

Boers, H.M., Seijen Ten Hoorn, J., and Mela, D.J. (2015). A systematic review of the influence of rice characteristics and processing methods on postprandial glycaemic and insulinaemic responses. *Br. J. Nutr.* 114: 1035–1045.

Bonnefont-Rousselot, D. and Collin, F. (2010). Melatonin: action as antioxidant and potential applications in human disease and aging. *Toxicology* 278: 55–67.

Britz, S.J., Prasad, P.V., Moreau, R.A. et al. (2007). Influence of growth temperature on the amounts of tocopherols, tocotrienols, and gamma-oryzanol in brown rice. *J. Agric. Food. Chem.* 55: 7559–7565.

Byeon, Y. and Back, K. (2014). Melatonin synthesis in rice seedlings in vivo is enhanced at high temperatures and under dark conditions due to increased serotonin N-acetyltransferase and N-acetylserotonin methyltransferase activities. *J. Pineal Res.* 56: 189–195.

Byeon, Y., Park, S., Lee, H.Y. et al. (2014). Elevated production of melatonin in transgenic rice seeds expressing rice tryptophan decarboxylase. *J. Pineal Res.* 56: 275–282.

Chaudhary, N. and Khurana, P. (2009). Vitamin E biosynthesis genes in rice: molecular characterization, expression profiling and comparative phylogenetic analysis. *Plant Sci.* 177: 479–491.

Chen, W.C. and Lu, S. (2019). From golden rice to aSTARice, more than just two steps forward in a pathway. *Sci. China Life Sci.* 62: 280–281.

Chen, X., Itani, T., Wu, X. et al. (2013). Physiological factors affecting transcription of genes involved in the flavonoid biosynthetic pathway in different rice varieties. *Plant Signal. Behav.* 8: e27555.

Chen, G., Wang, B., Han, D. et al. (2015). Molecular mechanisms of the coordination between astaxanthin and fatty acid biosynthesis in *Haematococcus pluvialis* (Chlorophyceae). *Plant J.* 81: 95–107.

Chew, B.P., Park, J.S., Wong, M.W., and Wong, T.S. (1999). A comparison of the anticancer activities of dietary beta-carotene, canthaxanthin and astaxanthin in mice in vivo. *Anticancer Res.* 19: 1849–1853.

Chi, X., Zhang, T., Zhao, J., and Feng, J. (2011). Study on extraction conditions and antioxygenic activity of folic acid. *Modern Food Sci. Technol.* 27: 1234–1237.

Cunningham, F.X. Jr., and Gantt, E. (2005). A study in scarlet: enzymes of ketocarotenoid biosynthesis in the flowers of *Adonis aestivalis*. *Plant J.* 41: 478–492.

Deng, G.F., Xu, X.R., Zhang, Y. et al. (2013). Phenolic compounds and bioactivities of pigmented rice. *Crit. Rev. Food Sci. Nutr.* 53: 296–306.

Díaz de la Garza, R., Quinlivan, E.P., Klaus, S.M. et al. (2004). Folate biofortification in tomatoes by engineering the pteridine branch of folate synthesis. *Proc. Natl. Acad. Sci. U.S.A.* 101: 13720–13725.

Dixon, R.A., Liu, C., and Jun, J.H. (2013). Metabolic engineering of anthocyanins and condensed tannins in plants. *Curr. Opin. Biotechnol.* 24: 329–335.

Dong, W., Cheng, Z., Wang, X. et al. (2011). Determination of folate content in rice germplasm (*Oryza sativa* L.) using tri-enzyme extraction and microbiological assays. *Int. J. Food Sci. Nutr.* 62: 537–543.

Dong, W., Cheng, Z.J., Lei, C.L. et al. (2014). Overexpression of folate biosynthesis genes in rice (*Oryza sativa* L.) and evaluation of their impact on seed folate content. *Plant Foods Hum. Nutr.* 69: 379–385.

Dubbels, R., Reiter, R.J., Klenke, E. et al. (1995). Melatonin in edible plants identified by radioimmunoassay and by high performance liquid chromatography-mass spectrometry. *J. Pineal Res.* 18: 28–31.

Falk, J. and Munne-Bosch, S. (2010). Tocochromanol functions in plants: antioxidation and beyond. *J. Exp. Bot.* 61: 1549–1566.

Fang, C., Luo, J., and Wang, S. (2019). The diversity of nutritional metabolites: origin, dissection, and application in crop breeding. *Front. Plant Sci.* 10: 1028.

Farre, G., Sudhakar, D., Naqvi, S. et al. (2012). Transgenic rice grains expressing a heterologous rho-hydroxyphenylpyruvate dioxygenase shift tocopherol synthesis from the gamma to the alpha isoform without increasing absolute tocopherol levels. *Transgenic Res.* 21: 1093–1097.

Farré, G., Blancquaert, D., Capell, T. et al. (2014). Engineering complex metabolic pathways in plants. *Annu. Rev. Plant Biol.*. 65: 187–223.

Fatima, T., Handa, A.K., and Mattoo, A.K. (2013). Functional foods: genetics, metabolome, and engineering phytonutrient levels. In: *Natural Products: Phytochemistry, Botany and Metabolism of Alkaloids, Phenolics and Terpenes* (eds. K.G. Ramawat and J.M. Merillon), 1715–1749. Berlin, Heidelberg: Springer-Verlag.

Finocchiar, F., Ferrari, B., Gianinetti, A. et al. (2007). Characterization of antioxidant compounds of red and white rice and changes in total antioxidant capacity during processing. *Mol. Nutr. Food Res.* 51: 1006–1019.

Frankel, E.N. and German, J.B. (2006). Antioxidants in foods and health: problems and fallacies in the field. *J. Sci. Food Agric.* 86: 1999–2001.

Gao, Y., Xu, H., Shen, Y., and Wang, J. (2013). Transcriptomic analysis of rice (*Oryza sativa*) endosperm using the RNA-Seq technique. *Plant Mol. Biol.* 81: 363–378.

Giller, K.E. (2016). *Genetically Engineered Crops: Experiences and Prospects*. National Academies of Sciences, Engineering, and Medicine. Washington, DC: National Academies Press https://doi.org/10.17226/23395.

Giuliano, G. (2014). Plant carotenoids: genomics meets multi-gene engineering. *Curr. Opin. Plant Biol.* 19: 111–117.

Goto, F., Yoshihara, T., Shigemoto, N. et al. (1999). Iron fortification of rice seed by the soybean ferritin gene. *Nat. Biotechnol.* 17: 282–286.

Goufo, P. and Trindade, H. (2014). Rice antioxidants: phenolic acids., flavonoids, anthocyanins, proanthocyanidins, tocopherols, tocotrienols, γ-oryzanol, and phytic acid. *Food Sci. Nutr.* 2: 75–104.

Grotewold, E. (2006). The genetics and biochemistry of floralpigments. *Annu. Rev. Plant Biol.* 57: 761–780.

Gülçin, İ., Elmastaş, M., and Aboul-Enein, H.Y. (2012). Antioxidant activity of clove oil–a powerful antioxidant source. *Arabian J. Chem.* 5: 489–499.

Ha, S.H., Kim, J.K., Jeong, Y.S. et al. (2019). Stepwise pathway engineering to the biosynthesis of zeaxanthin, astaxanthin and capsanthin in rice endosperm. *Metab. Eng.* 52: 178–189.

Hamid, A., Wani, N.A., and Kaur, J. (2009). New perspectives on folate transport in relation to alcoholism-induced folate malabsorption–association with epigenome stability and cancer development. *FEBS J.* 276: 2175–2191.

Hanson, A.D. and Gregory, J.F. 3rd, (2011). Folate biosynthesis, turnover, and transport in plants. *Annu. Rev. Plant Biol.* 62: 105–125.

Hattori, A., Migitaka, H., Iigo, M. et al. (1995). Identification of melatonin in plants and its effects on plasma melatonin levels and binding to melatonin receptors in vertebrates. *Biochem. Mol. Biol. Int.* 35: 627–634.

Heinemann, R.J., Xu, Z., Godber, J.S., and Lanfer-Marquez, U.M. (2008). Tocopherols. tocotrienols, and γ-oryzanol contents in japonica and indica subspecies of rice (*Oryza sativa* L.) cultivated in Brazil. *Cereal Chem.* 85: 243–247.

Hermizi, H., Faizah, O., Ima-Nirwana, S. et al. (2009). Beneficial effects of tocotrienol and tocopherol on bone histomorphometric parameters in Sprague-Dawley male rats after nicotine cessation. *Calcif. Tissue Int.* 84: 65–74.

Hess, K.R., Zhang, W., Baggerly, K.A. et al. (2001). Microarrays: handling the deluge of data and extracting reliable information. *Trends Biotechnol.* 19: 463–468.

Higuera-Ciapara, I., Felix-Valenzuela, L., and Goycoolea, F.M. (2006). Astaxanthin: a review of its chemistry and applications. *Crit. Rev. Food Sci. Nutr.* 46: 185–196.

Hong, L., Qian, Q., Tang, D. et al. (2012). A mutation in the rice chalcone isomerase gene causes the golden hull and internode 1 phenotype. *Planta* 236: 141–151.

Huang, J.C., Zhong, Y.J., Liu, J. et al. (2013). Metabolic engineering of tomato for high-yield production of astaxanthin. *Metab. Eng.* 17: 59–67.

Hunter, S.C. and Cahoon, E.B. (2007). Enhancing vitamin E in oilseeds: unraveling tocopherol and tocotrienol biosynthesis. *Lipids* 42: 97–108.

Iwamoto, T., Hosoda, K., Hirano, R. et al. (2000). Inhibition of low-density lipoprotein oxidation by astaxanthin. *J. Atheroscler. Thromb.* 7: 216–222.

Jabrin, S., Ravanel, S., Gambonnet, B. et al. (2003). One-carbon metabolism in plants: regulation of tetrahydrofolate synthesis during germination and seedling development. *Plant Physiol.* 131: 1431–1439.

Jeffryes, J.G., Seaver, S.M.D., Faria, J.P., and Henry, C.S. (2018). A pathway for every product? Tools to discover and design plant metabolism. *Plant Sci.* 273: 61–70.

Jiang, L., Wang, W., Lian, T., and Zhang, C. (2017). Manipulation of metabolic pathways to develop vitamin-enriched crops for human health. *Front. Plant Sci.* 8: 937.

Kang, S., Kang, K., Lee, K., and Back, K. (2007). Characterization of rice tryptophan decarboxylases and their direct involvement in serotonin biosynthesis in transgenic rice. *Planta* 227: 263–272.

Kang, K., Lee, K., Park, S. et al. (2010). Enhanced production of melatonin by ectopic overexpression of human serotonin N-acetyltransferase plays a role in cold resistance in transgenic rice seedlings. *J. Pineal Res.* 49: 176–182.

Kang, K., Kong, K., Park, S. et al. (2011). Molecular cloning of a plant N-acetylserotonin methyltransferase and its expression characteristics in rice. *J. Pineal Res.* 50: 304–309.

Kanjanaphachoat, P., Wei, B.Y., Lo, S.F. et al. (2012). Serotonin accumulation in transgenic rice by over-expressing tryptophan decarboxylase results in a dark brown phenotype and stunted growth. *Plant Mol. Biol.* 78: 525–543.

Kusano, M., Yang, Z., Okazaki, Y. et al. (2015). Using metabolomic approaches to explore chemical diversity in rice. *Mol. Plant* 8: 58–67.

Lee, J.H. (2010). Identification and quantification of anthocyanins from the grains of black rice (*Oryza sativa* L.) varieties. *Food Sci. Biotechnol.* 19: 391–397.

Lim, S.H. and Ha, S.H. (2013). Marker development for the identification of rice seed color. *Plant Biotechnol. Rep.* 7: 391–398.

Liochev, S.I. (2013). Reactive oxygen species and the free radical theory of aging. *Free Radic. Biol. Med.* 60: 1–4.

Lobo, V., Patil, A., Phatak, A., and Chandra, N. (2010). Free radicals, antioxidants and functional foods: impact on human health. *Pharmacogn. Rev.* 4: 118.

Mailloux, R.J. (2015). Teaching the fundamentals of electron transfer reactions in mitochondria and the production and detection of reactive oxygen species. *Redox Biol.* 4: 381–398.

Martin, C. and Li, J. (2017). Medicine is not health care, food is health care: plant metabolic engineering, diet and human health. *New Phytol.* 216: 699–719.

Mattoo, A.K., Shukla, V., Fatima, T. et al. (2010). Genetic engineering to enhance crop-based phytonutrients (nutraceuticals) to alleviate diet-related diseases. *Adv. Exp. Med. Biol.* 698: 122–143.

Mène-Saffrané, L. (2017). Vitamin E biosynthesis and its regulation in plants. *Antioxidants (Basel)* 7 (1): pii: E2.

Min, B., Gu, L., McClung, A.M. et al. (2012). Free and bound total phenolic concentrations, antioxidant capacities, and profiles of proanthocyanidins and anthocyanins in whole grain rice (*Oryza sativa* L.) of different bran colours. *Food Chem.* 133: 715–722.

Naqvi, S., Zhu, C., Farre, G. et al. (2009). Transgenic multivitamin corn through biofortification of endosperm with three vitamins representing three distinct metabolic pathways. *Proc. Natl. Acad. Sci. U.S.A.* 106 (19): 7762–7767.

Ng, L.T., Huang, S.H., Chen, Y.T., and Su, C.H. (2013). Changes of tocopherols, tocotrienols, γ-oryzanol, and γ-aminobutyric acid levels in the germinated brown rice of pigmented and nonpigmented cultivars. *J. Agric. Food. Chem.* 61: 12604–12611.

Nogueira, M., Enfissi, E.M.A., Martínez Valenzuela, M.E. et al. (2017). Engineering of tomato for the sustainable production of ketocarotenoids and its evaluation in aquaculture feed. *Proc. Natl. Acad. Sci. U.S.A.* 114: 10876–10881.

Nunes, A.C., Kalkmann, D.C., and Aragao, F.J. (2009). Folate biofortification of lettuce by expression of a codon optimized chicken GTP cyclohydrolase I gene. *Transgenic Res.* 18 (5): 661.

NYU. (n.d.). Genetically modified organisms: the "golden rice" debate. NYU Langone Health. https://med.nyu.edu/highschoolbioethics/genetically-modified-organisms-%E2%80%9Cgolden-rice%E2%80%9D-debate

Ogo, Y., Ozawa, K., Ishimaru, T. et al. (2013). Transgenic rice seed synthesizing diverse flavonoids at high levels: a new platform for flavonoid production with associated health benefits. *Plant Biotechnol. J.* 11: 734–746.

Oikawa, T., Maeda, H., Oguchi, T. et al. (2015). The birth of a black rice gene and its local spread by introgression. *Plant Cell* 27: 2401–2414.

Okazaki, M., Higuchi, K., Hanawa, Y. et al. (2009). Cloning and characterization of a *Chlamydomonas reinhardtii* cDNA arylalkylamine N-acetyltransferase and its use in the genetic engineering of melatonin content in the micro-tom tomato. *J. Pineal Res.* 46: 373–382.

Packer, L., Weber, S.U., and Rimbach, G. (2001). Molecular aspects of alpha-tocotrienol antioxidant action and cell signalling. *J. Nutr.* 131: 369S–373S.

Paine, J.A., Shipton, C.A., Chaggar, S. et al. (2005). Improving the nutritional value of golden rice through increased pro-vitamin A content. *Nat. Biotechnol.* 23: 482–487.

Peng, Q.Z., Zhu, Y., Liu, Z. et al. (2012). An integrated approach to demonstrating the ANR pathway of proanthocyanidin biosynthesis in plants. *Planta* 236: 901–918.

Potrykus, I. (2001). Golden rice and beyond. *Plant Physiol.* 125: 1157–1161.

Ramírez Rivera, N.G., García-Salinas, C., Aragão, F.J., and Díaz de la Garza, R.I. (2016). Metabolic engineering of folate and its precursors in Mexican common bean (*Phaseolus vulgaris L.*). *Plant Biotechnol. J.* 14 (10): 2021–2032/.

Zhu, C., Sanahuja, G., Yuan, D. et al. (2013). Biofortification of plants with altered antioxidant content and composition: genetic engineering strategies. *Plant Biotechnol. J.* 11: 129–141.

Sankari, M., Rao, P.R., Hemachandran, H. et al. (2018). Prospects and progress in the production of valuable carotenoids: insights from metabolic engineering, synthetic biology, and computational approaches. *J. Biotechnol.* 266: 89–101.

Setyaningsih, W., Saputro, I.E., Barbero, G.F. et al. (2015). Determination of melatonin in rice (*Oryza sativa*) grains by pressurized liquid extraction. *J. Agric. Food. Chem.* 63: 1107–1115.

Shao, Y., Jin, L., Zhang, G. et al. (2011). Association mapping of grain color, phenolic content, flavonoid content and antioxidant capacity in dehulled rice. *Theor. Appl. Genet.* 122: 1005–1016.

Shao, Y., Xu, F., Sun, X. et al. (2014). Identification and quantification of phenolic acids and anthocyanins as antioxidants in bran, embryo and endosperm of white, red and black rice kernels (*Oryza sativa* L.). *J. Cereal Sci.* 59: 211–218.

Shen, Y., Jin, L., Xiao, P. et al. (2009). Total phenolics, flavonoids, antioxidant capacity in rice grain and their relations to grain color, size and weight. *J. Cereal Sci.* 49: 106–111.

Shin, Y.M., Park, H.J., Yim, S.D. et al. (2006). Transgenic rice lines expressing maize C1 and R-S regulatory genes produce various flavonoids in the endosperm. *Plant Biotechnol. J.* 4: 303–315.

Steiger, G., Müller-Fischer, N., Cori, H., and Conde-Petit, B. (2014). Fortification of rice: technologies and nutrients. *Ann. N.Y. Acad. Sci.* 1324: 29–39.

Storozhenko, S., De Brouwer, V., Volckaert, M. et al. (2007). Folate fortification of rice by metabolic engineering. *Nat. Biotechnol.* 25: 1277–1279.

Sweeney, M.T., Thomson, M.J., Pfeil, B.E., and McCouch, S. (2006). Caught red-handed: Rc encodes a basic helix-loop-helix protein conditioning red pericarp in rice. *Plant Cell* 18: 283–294.

Tan, D.X. (1993). Melatonin: a potent, endogenous hydroxyl radical scavenger. *Endocr. J.* 1: 57–60.

Tatsuzawa, H., Maruyama, T., Misawa, N. et al. (2000). Quenching of singlet oxygen by carotenoids produced in *Escherichia coli* - attenuation of singlet oxygen-mediated bacterial killing by carotenoids. *FEBS Lett.* 484: 280–284.

Teplyakov, A., Obmolova, G., Sarikaya, E. et al. (2004). Crystal structure of the YgfZ protein from Escherichia coli suggests a folate-dependent regulatory role in one-carbon metabolism. *J. Bacteriol.* 186 (21): 7134–7140.

Tomycz, N.D., Richardson, R.M., and Friedlander, R.M. (2011). Seizures and free radicals: the antioxidant may become the next anticonvulsant. *Neurosurgery* 69: N12–N13.

Tran, T.U., Suzuki, K., Okadome, H. et al. (2004). Analysis of the tastes of brown rice and milled rice with different milling yields using a taste sensing system. *Food Chem.* 88: 557–566.

Tsia, P.L. and Hu, M.K. (2003). Free radical scavenging and antioxidative activity of melatonin derivatives. *J. Pharm. Pharmacol.* 55: 1655–1660.

Van Tassel, D.L., Roberts, N., Lewy, A., and O'Neill, S.D. (2001). Melatonin in plant organs. *J. Pineal Res.* 31: 8–15.

Vasconcelos, M.W., Gruissem, W., and Bhullar, N.K. (2017). Iron biofortification in the 21st century: setting realistic targets, overcoming obstacles, and new strategies for healthy nutrition. *Curr. Opin. Biotechnol.* 44: 8–15.

Wang, L.S. and Stoner, G.D. (2008). Anthocyanins and their role in cancer prevention. *Cancer Lett.* 269: 281–290.

Wang, J., Liang, C., Li, S., and Zheng, J. (2009). Study on analysis method of melatonin and melatonin content in corn and rice seeds. *Chinese Agric. Sci. Bull.* 17: 6.

Wang, Y.D., Wang, X., Ngai, S.M., and Wong, Y.S. (2013). Comparative proteomics analysis of selenium responses in selenium-enriched rice grains. *J. Proteome Res.* 12: 808–820.

Wang, W., Li, Y., Dang, P. et al. (2018). Rice secondary metabolites: structures, roles, biosynthesis, and metabolic regulation. *Molecules* 23: 3098.

Winterbourn, C.C. (2008). Reconciling the chemistry and biology of reactive oxygen species. *Nat. Chem. Biol.* 4: 278–286.

Xu, Z., Hua, N., and Godber, J.S. (2001). Antioxidant activity of tocopherols, tocotrienols, and gamma-oryzanol components from rice bran against cholesterol oxidation accelerated by 2,2′-azobis(2-methylpropionamidine) dihydrochloride. *J. Agric. Food. Chem.* 49: 2077–2081.

Yamane, H. (2013). Biosynthesis of phytoalexins and regulatory mechanisms of it in rice. *Biosci. Biotechnol., Biochem.* 77: 1141–1148.

Yang, W., Cahoon, R.E., Hunter, S.C. et al. (2011). Vitamin E biosynthesis: functional characterization of the monocot homogentisate geranylgeranyl transferase. *Plant J.* 65 (2): 206–217.

Ye, X., Al-Babili, S., Klöti, A. et al. (2000). Engineering the provitamin A (beta-carotene) biosynthetic pathway into (carotenoid-free) rice endosperm. *Science* 287: 303–305.

Yuan, L. and Grotewold, E. (2015). Metabolic engineering to enhance the value of plants as green factories. *Metab. Eng.* 27: 83–91.

Zhang, H.M. and Zhang, Y. (2014). Melatonin: a well-documented antioxidant with conditional pro-oxidant actions. *J. Pineal Res.* 57: 131–146.

Zhang, G., Liu, R., Zhang, P. et al. (2012). Increase of tocopherol content in endosperm by overexpression of homogentisate phytyltransferase in transgenic rice. *Mol. Plant Breed.* 10: 528–534.

Zhang, G.Y., Liu, R.R., Xu, G. et al. (2013). Increased α-tocotrienol content in seeds of transgenic rice overexpressing *Arabidopsis* γ-tocopherol methyltransferase. *Transgenic Res.* 22 (1): 89–99.

Zhang, Y., Butelli, E., and Martin, C. (2014). Engineering anthocyanin biosynthesis in plants. *Curr. Opin. Plant Biol.* 19: 81–90.

Zhou, Z., Robards, K., Helliwell, S., and Blanchard, C. (2002). Composition and functional properties of rice. *Int. J. Food Sci. Technol.* 37: 849–868.

Zhu, Q., Yu, S., Zeng, D. et al. (2017). Development of "purple endosperm rice" by engineering anthocyanin biosynthesis in the endosperm with a high-efficiency transgene stacking system. *Mol. Plant* 10: 918–929.

Zhu, Q., Zeng, D., Yu, S. et al. (2018). From Golden Rice to aSTARice: bioengineering astaxanthin biosynthesis in rice endosperm. *Mol. Plant* 11: 1440–1448.

20

Molecular Breeding Approaches for Improvement and Development of Water Saving Aerobic Rice

Rahul K. Meena[1], Kuldeep Kumar[2], Saurabh K. Dubey[3], Ashish K. Singh[4], Adarsh Kumar[5], Deepanshu Jayaswal[2], Badal Singh[6], Rajinder Jain[7] and Sunita Jain[7]

[1] *Department of Molecular Biology, Biotechnology and Bioinformatics, CCS Haryana Agricultural University, Hisar 125004, Haryana, India*
[2] *National Institute for Plant Biotechnology, New Delhi, India*
[3] *Division of Plant Pathology, Indian Agricultural Research institute, New Delhi, India*
[4] *ICAR-VPKAS, Uttarakhand, India*
[5] *ICAR-NBAIM, Uttar Pradesh, India*
[6] *Regional station Shimla, ICAR-NBPGR, Himanchal Pradesh, India*
[7] *Department of Molecular Biology and Biotechnology, CCSHAU, Haryana, India*

Introduction

Rice (*O. sativa* L. 2n = 24) is a member of the family *poaceae*, and the primary source of food for about three billion people across the world. It is widely grown in tropical and subtropical regions. "Rice is life" aptly describes the importance of rice in ensuring food and nutritional security in Asian countries. It ranks first among the cereal crops globally for the area under cultivation and total production, which, makes it one of the most diverse crops, since it is grown as far north as Manchuria (China) and far south as Uruguay and New South Wales in Australia.

Rice is a "model cereal crop" amenable to genetic manipulation due to its small genome size, enriched genetic map, available genome sequence data, large collection of germplasm, and so on. It is a heavy consumer of water; irrigated lowland rice receives some 34–43% of the total world's irrigation water, or 24–30% of the world's total freshwater withdrawals. Rice is the single biggest user of fresh water; it takes 3000–5000 l of water to produce 1 kg of rice (Bouman 2002). The increasing scarcity of water has threatened the traditional way of irrigated rice production. Worldwide, water for agriculture is becoming increasingly scarce.

Ecosystems of rice are generally classified into four types: irrigated, rain-fed lowland, deep-water, and rain-fed upland. Irrigated rice is most common, comprising 55% of the global production area. Rain-fed lowland rice represents 25%, upland rice 12%, and deep-water 8% of global production area. Rice is a profligate user of water. A significant portion of the total water in rice production is used for land preparation alone. In Asia, 17 million ha of irrigated rice areas may experience "physical water scarcity" and 22 million ha "economic water scarcity" by 2025 (Bouman et al. 2005).

Molecular Breeding for Rice Abiotic Stress Tolerance and Nutritional Quality, First Edition.
Edited by Mohammad Anwar Hossain, Lutful Hassan, Khandakar Md. Iftekharuddaula, Arvind Kumar and Robert Henry.

In the last 20 years, rapid progress has been made in the area of rice biotechnology and molecular biology, resulting in a paradigm shift in rice breeding programs. Useful knowledge based on rice molecular genetics, tissue culture, and cellular biology has been established. Farmers have been selecting those plants that survived drought events for centuries, and there is a wealth of genetic variation for response to water deficit among traditional cultivars. One technology that enables rice to be grown in dry land without flooding, and help farmers cope with water scarcity is the aerobic rice system. Research has been undertaken to develop "aerobic rice varieties" that require less water but give optimum yield. Efforts are already underway in many countries, including China and India, to develop "aerobic rice" varieties that combine drought-resistant characteristics with high-yielding traits.

Development of rice varieties for aerobic situations or non-submerged situations in India was initiated at University of Agricultural Sciences, Bangalore, in early 1980, using the upland paddy. A new aerobic rice variety, MAS 946-1, was formally released by the University of Agricultural Sciences, Bangalore, in 2006–2007, which yields 6.5–7.0 tons ha^{-1} of grain yield (GY) and 7 tons of fodder yield and consumes only 40% of the water as against the submerged condition.

Aerobic rice is a production system in which specially developed varieties are grown in well-drained, non-puddled, and non-saturated soils. The soil is therefore "aerobic" or with oxygen throughout the growing season, as compared to traditional flooded fields, which are "anaerobic." Aerobic rice entails the use of nutrient-responsive cultivars that are adapted to aerobic soils (Bouman and Tuong 2001), aiming at yields of 70–80% of high-input flooded rice.

Rice Facts and Figure

Globally rice is grown over an area of about 160 million ha, which accounts for approximately 11% of the world's cultivated land to produce 503.8 million tons (http://www.fao.org/fileadmin/templates/est/COMM_MARKETS_MONITORING/Rice/Images/RMM/RMM_APR17_H.pdf). The percent contribution of the area for rice production is presented in Figure 20.1. According to the UN Convention to Combat Desertification (UNCCD), with the existing climate change scenario, almost half the world's population will be living in areas of water scarcity by 2030. As per International Water Management Institute (IWMI), agriculture, which accounts for about 70% of global water withdrawals, is constantly competing with industrial and environmental uses for water supply. India is the largest user of groundwater in the world, and over a quarter of the global total and about 60% of aquifers in India will be in critical condition in another 15 years if the trend of indiscriminate exploitation of ground water continues. As water scarcity increases, improved water management in agriculture could increase global water availability, reduce soil erosion, and lead to increased and diversified agricultural yields, augmenting our ability to feed a population projected to reach 9 billion by 2050.

As expected, Asia accounts for over 90% of the world's production of rice, with China, India, and Indonesia producing the most. Currently, total rice is cultivated in 167.25 mha area with the production of 769.66 mt in the world. Whereas, in Asia it occupies 145.54 mha area for the production of 692.59 mt, which is approximately 90% of

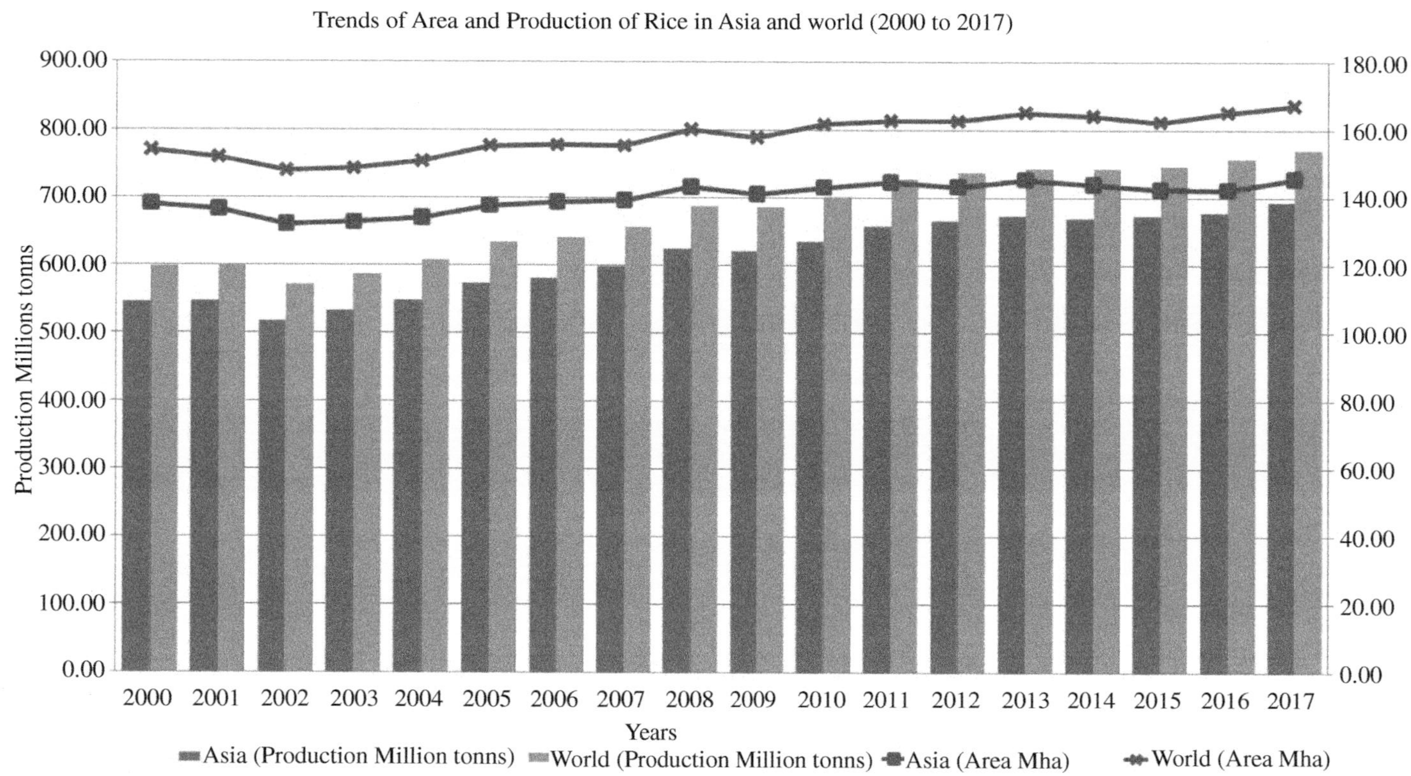

Figure 20.1 Trends of area and production of rice in Asia and World (2000 to 2017). *Source:* Food and Agriculture Organization of the United States. http://www.faostat.fao.org 2019 © FAO.

the total rice production. The trends of 2000 to 2017 showed increments in rice area and production with average 0.471% and 1.44%, respectively throughout the world. However, in Asia increasing growth rate of area is 0.297% and production is 1.370% (Figure 20.1).

India is the second-largest producer with a production of 110.15 million tons from 43.25 million ha during the crop season 2016–2017. Approximately, an area of 60 mha is irrigated, which supplies more than 75% of the annual rice. The trends of rice production during 2000–2001 to 2016–2017 showed an average 1.09% increase annually, while the area under rice production has become stable and is expected to decrease by the upcoming cropping seasons (Figure 20.2).

Concept of Aerobic Rice

Aerobic rice is a production system in which specially developed varieties are grown in well-drained, non-puddled, and non-saturated soils. The soil is therefore "aerobic" or with oxygen throughout the growing season, as compared to traditional flooded fields, which are "anaerobic." The term "aerobic rice" was coined by the International Rice Research Institute (IRRI), for high-yielding rice grown under non-flooded conditions in non-puddled and unsaturated (aerobic) soil, which is responsive to nutrient supply, can be rain-fed or irrigated, and tolerates (occasional) flooding (Bouman and Tuong 2001).

Aerobic rice cultivation system is the method of cultivation, where the rice crop is established by direct seeding (dry or water-soaked seed) in non-puddle field and un-flooded field conditions. It is the most promising approaches for saving water and labor. The cultivars used are adapted to aerobic soils and have higher yield potential than traditional upland cultivars (Atlin et al. 2006). Compared with lowland rice breeding, the aerobic rice breeding program is extremely small, and the genetic basis is very narrow. Only a few successful aerobic rice cultivars with high-yield potential and broad biotic and abiotic stress tolerance are commercially grown.

A group of temperate aerobic rice cultivars, called Han Dao, has been developed by the breeders from China Agricultural University and have been commercially grown by farmers in northern China since the early 1990s. To achieve high yields under aerobic conditions, aerobic rice cultivars should combine the drought-tolerance characteristics of upland cultivars with the high-yielding characteristics of lowland cultivars (Lafitte et al. 2002).

Effect of Aerobic Rice on Water Saving and Water Productivity

The conventional transplanting method of rice used higher quantities of water (16 200 m^3 $ha^{-1,}$) whereas aerobic rice used minimum quantity (9687 m^3 ha^{-1}) and observed a water saving of 32.9% to 43.9% over transplanted rice (Geethalakshmi et al. 2009). Aerobic ways of growing rice save water by eliminating continuous seepage, percolation, and by reducing evaporation (Castaneda et al. 2002). The direct seeded aerobic rice is a typical technology, wherein an additional 130–150 mm of water input can be saved by foregoing the wet land preparation (Bouman et al. 2005). Patel et al. (2010) reported the higher water use efficiencies (WUE) of aerobic rice compared to flooded condition. The WUE of aerobic varieties under aerobic conditions were 164–188% higher than that of lowland varieties under lowland conditions (Wang and Tang 2000).

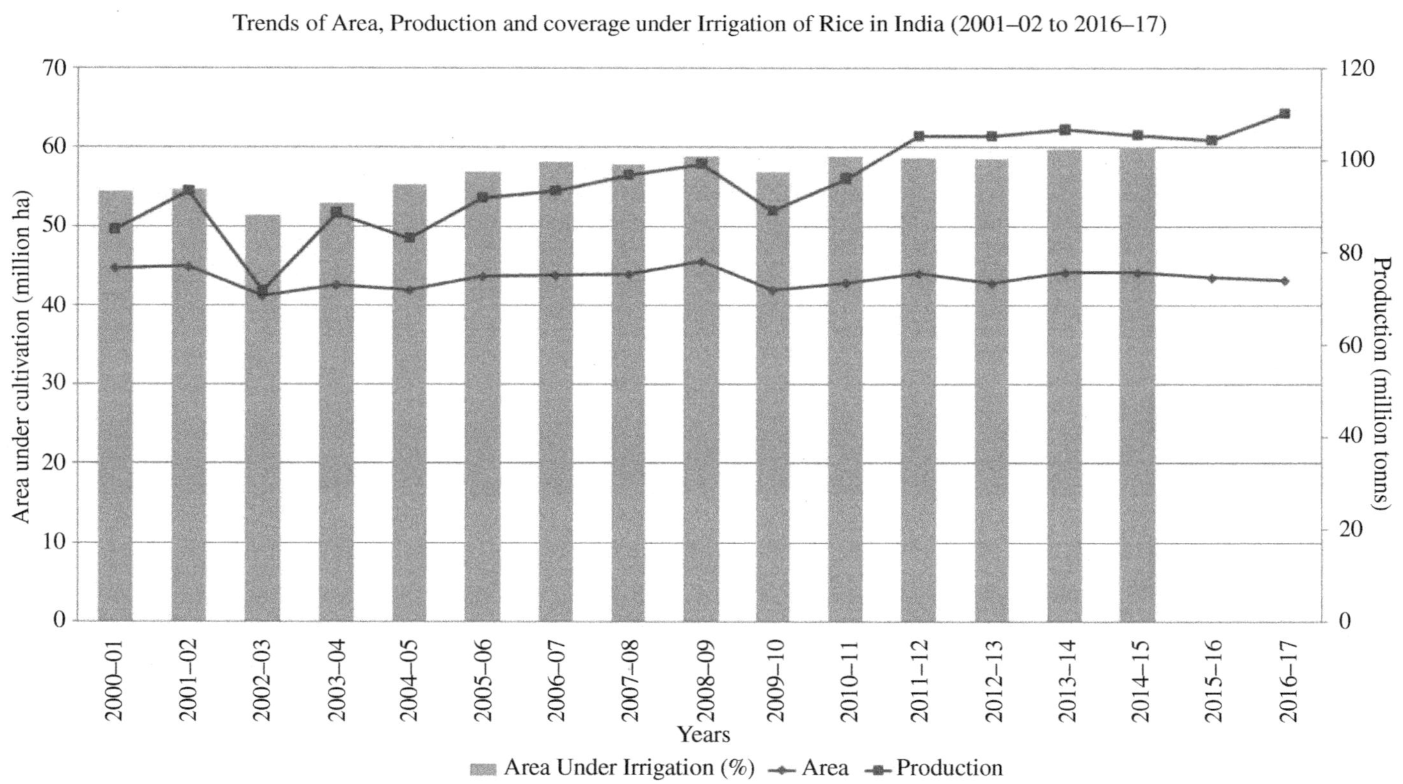

Figure 20.2 Trends of area, production, and coverage under irrigation of rice in India (2001–2002 to 2016–2017). *Source:* Agriculture, Cooperation & Farmer Welfare, www.agricoop.nic.in.

Greenhouse Gas Emissions under Aerobic Conditions

Globally, anthropogenic sources of methane (CH_4) and nitrous oxide (N_2O) are dominated by agriculture, and agricultural CH_4 and N_2O emissions have increased significantly. Traditional flooded paddy fields have been identified as a major source of increasing atmospheric CH_4, accounting for approximately 5–19% of the annual global CH_4 emissions to the atmosphere. Methane is produced by anaerobic (without oxygen) decomposition of organic matter in the soil under flooded rice cultivation. The absence of standing water drastically reduces emissions of methane to the atmosphere. Adopting aerobic rice will help to minimize both methane and nitrous oxide emission rates from rice fields without affecting productivity.

Achieving high yields under relatively favorable aerobic soil conditions requires new varieties of aerobic rice that combine the drought-resistant characteristics of upland varieties (grown on non-flooded sloping uplands) with the high-yielding characteristics of lowland varieties (grown in irrigated, flooded fields). Consequently several aerobic rice genotypes (*indica, tropical japonica, indica/tropical japonica, aus/tropical japonica, and indica/tropical japonica/aus*) have been developed using conventional breeding and marker-assisted selection techniques in combination. Molecular markers associated with quantitative trait loci and monogenic trait associated with roots, osmotic adjustment, and grain yield have been used to facilitate this task.

Diversity in Rice Germplasm

Rice (*O. sativa* L.) was domesticated from the wild grass *Oryza rufipogon* around 10,000–14,000 years ago. The two main subspecies of rice – *indica* (tropical regions) and *japonica* (East Asia) are not believed to have been derived from independent domestication events. Another cultivated species, *Oryza glaberrima*, was domesticated later on in West Africa. It has been believed that *indica* variety of rice was first domesticated in the area of the eastern Himalayas (i.e. northeastern India), stretching through Burma, Thailand, Laos, Vietnam, and southern China. The *japonica* variety was domesticated from wild rice in southern China and was later introduced to India. It is also believed that rice may have originated in southern India, then spread to the north of the country and to China and transmitted to Korea, Philippines (about 2000 BC) and then Indonesia and Japan (about 1000 BC).

There are altogether 24 species in the genus *Oryza* of which 22 are wild species and two (*O. sativa* and *O. glaberrima*) are cultivated. *O. sativa* is grown in all rice-growing areas, but *O. glaberrima* is confined to the West Africa only. The International Rice Gene Bank holds more than 127,000 rice accessions and wild relatives, which is the biggest collection of rice genetic diversity in the world. These subspecies are adapted to diverse environmental conditions ranging from upland to lowland and from tropical to temperate zones.

Root Morphology for Aerobic Traits

Aerobic rice varieties developed so far have an upland rice root system, which is composed of deep and thickened roots with large xylem vessels resulting in higher yield potential under water-scarce conditions. Varieties adapted to anaerobic (flooded) systems tend to stop growing as soil moisture declines, and leaves roll or senesce, effectively shedding excess radiation. This is associated with limited deep rooting and water extraction in many lowland varieties. Rice extracts little or no water from below 60 cm, so aerobic rice plants with a deep root system are able to avoid water stress by absorbing water from deep soil layers.

Mapping of Aerobic Traits

Molecular marker technology, linkage mapping, and marker-assisted selection of the target traits can greatly improve the selection efficiency and precision in rice breeding programs including the development of new water-efficient cultivars with improved root traits. Rice genome sequencing and comparison of sequence databases of *indica* and *japonica* rice genomes have provided an almost unlimited number of DNA markers (SSR, SNP, InDel, etc.) for high-resolution genetic analysis. Several marker systems like RAPD, RFLP, AFLP, SSR, and SNPs have been used for mapping traits promoting aerobic adaptation in rice. But due to their genome wide abundance, co-dominant nature, high polymorphism rate, and reproducibility SNPs and SSRs are the best marker system to be used for marker trait linkage through any approach. A saturated map of SSR (simple sequence repeats) markers in rice with 2240 validated markers and another with 52,845 SSR markers is already available, which shows the presence of one SSR marker every 157 bp (McCouch et al. 1997).

Linking Mapping and Marker-Assisted Selection

Mapping of QTLs for traits promoting aerobic cultivation of rice has gained momentum in the last decade. A list of a few QTLs identified has been provided in Table 20.1. Vikram et al. (2011) studied genetic control of grain yield (GY) under reproductive-stage drought stress (RS) in elite genetic background in three $F_{3:4}$ mapping populations derived from crosses of N22, a drought-tolerance aus cultivar, with Swarna, IR64, and MTU1010, three high yielding varieties. A major QTL for grain yield was identified on rice chromosome 1 flanked by RM11943 and RM431 in all the three populations. Using four different backcross-derived populations, Dixit et al. (2012b) mapped QTLs ($qDTY_{2.1}$, $qDTY_{2.2}$, $qDTY_{9.1}$, and $qDTY_{12.1}$), which increased GY under stress conditions and did not have any effect on GY under non-stress. Sandhu et al. (2014) reported that development of rice varieties for dry direct-seeded conditions can be accelerated by selecting for suitable agronomic aerobic traits such as plant height, effective number of tillers, grain yield/plant, and root parameters. Sandhu et al. (2013) reported the mapping of >40 QTL associated for realizing higher grain yield under aerobic conditions using MASARB25 × Pusa Basmati 1460 and HKR47 × MAS26 derived populations.

Zheng et al. (2006) studied the role of root systems in drought resistance in rice using doubled haploid population derived from the cross between "IR64" and "Azucena" and reported that 12 putative QTLs were detected by interval mapping comprising four QTLs for root-penetration ability, four QTLs for root thickness, two QTLs for penetrated root number, and two QTLs for total root number. Sandhu et al. (2014) conducted studies for mapping QTLs for yield, root-related traits, and agronomic traits and identified 35 QTL associated

Table 20.1 Summary of major QTL reported earlier for traits promoting aerobic adaptation.

QTL	Chr	Interval/markers name	References
$qDTY_{1.1}$	1	RM431-RM104	Ghimire et al. (2012)
$qDTY_{1.2}$	1	RM259–RM315	Sandhu et al. (2014), Bernier et al. (2009)
$qPH_{1.1}$	1	RM315–RM431	Sandhu et al. (2014), Bernier et al. (2009)
$qDTY_{1.3}$	1	RM263–RM573	Sandhu et al. (2014), Bernier et al. (2009)
$qPL_{1.1}$	1	RM1282-RM220	Marathi et al. (2012)
$qDTY_{2.1}$	2	RM327-RM262	Venuprasad et al. (2008)
$qPL_{2.1}$	2	RM3732-RM324	Marathi et al. (2012)
$qDTY_{2.2}$	2	RM236-RM279	Yadaw et al. (2013)
$qDTY_{2.2}$	2	RM211-RM263	Sandhu et al. (2013)
$qDTY_{2.1}$	2	RM263-RM573	Sandhu et al. (2014)
$qDTY_{2.3}$	2	RM211-RM250	Palanog et al. (2014)
$qDTY_{2.3}$	2	RM263–RM573	Sandhu et al. (2014), Venuprasad et al. (2009), Bernier et al. (2009)
$qDTY_{2.2}$	2	RM211–RM263	Sandhu et al. (2014), Venuprasad et al. (2009), Bernier et al. (2009)
$qDTY_{2.1}$	2	RM324–RM263	Sandhu et al. (2014), Venuprasad et al. (2009), Bernier et al. (2009)
$qDTF_{2.3}$	2	RM450–RM250	Sandhu et al. (2014), Venuprasad et al. (2009), Bernier et al. (2009)
$qYLD_{2.1}$	2	RM3732-RM7144	Marathi et al. (2012)
$qDTY_{3.1}$	3	RM520-RM16030	Venuprasad et al. (2009)
$qPL_{3.1}$	3	RM3766-RM157A	Marathi et al. (2012)
$qDTY_{3.1}$	3	RM168-RM468	Dixit et al. (2014)
$qDTY_{3.2}$	3	RM569-RM517	Yadaw et al. (2013)
$qDTY_{3.2}$	3	RM20-RM22	Vikram et al. (2011)
$qGL_{3.1}$	3	RM6283–RM16	Lou et al. (2009)
qGL3a	3	RMw357–RMw353	Wan et al. (2006)
$qGY_{3.1}$	3	RG100	Martinez and Thome (2001)
qGL_{4b}	4	RM5586–RM3524	Kato and Okami (2011)
$qGY_{5.1}$	5	RM13	Martinez and Thome (2001)
$qGY_{6.1}$	6	RM217	Martinez and Thome (2001)
$qDTY_{6.1}$	6	RM589-RM204	Venuprasad et al. (2012)
$qGL_{6.1}$	6	RM588–RM540	Lou et al. (2009)
$qDTY_{6.1}$	6	RM586-RM217	Dixit et al. (2014)
$qDTY_{6.1}$	6	RM212-RM541	Dixit et al. (2014)
$qDTY_{6.2}$	6	RM121-RM541	Dixit et al. (2014)
$qLB_{7.1}$	7	SSR	Redoña and Mackill (1998), Rabiei et al. (2004)
$qGL_{7.1,}$	7	SSR	Wan et al. (2006), Amarawathi et al. (2008)
qGL_{7}	7	RID711–RM6389	Bai et al. (2010)

Table 20.1 (Continued)

QTL	Chr	Interval/markers name	References
$qPH_{8.1}$	8	RZ333-RZ562	Xiao et al. (1995)
$qYPP_{8.1}$	8	RM547-RM72	Kharb et al. (2016)
$qPH_{9.1}$	9	RZ9106-RZ777	Li et al. (1995)
$qPH_{9.1}$	9	17141056-18060875	Lu et al. (2009)
$qDTY_{9.1}$	9	RM105-RM434	Swamy et al. (2013)
$qGY_{9.1}$	9	RM215	Martinez and Thome (2001)
$qGY_{9.2}$	9	RM242	Martinez and Thome (2001)
$qGY_{10.1}$	10	RM10030-RM228	Zhao et al. (2010)
$qDTY_{10.1}$	10	RM216-RM304	Vikram et al. (2011)
$qPH_{10.1}$	10	G1082-G291	Lu et al. (1996)
$qDTY_{10.2}$	10	RM269-G2155	Swamy et al. (2013)
$qPH_{11.1}$	11	RG118	Zhuang et al. (1997)
$qTGW_{11.1}$	11	RM287-RM21	Kharb et al. (2016)
$qENT_{11.1}$	11	RM26245-RM202	Kharb et al. (2016)
$qGL_{11.1}$	11	SSR	Lin et al. (1995), Wan et al. (2006), Amarawathi et al. (2008)
qGY_{11}	11	RZ537-RZ900	Martinez and Thome (2001)
$qGY_{11.1}$	11	RM287-RM26775	Zhao et al. (2010)
$qTGW_{11}$	11	RM287-RM26775	Zhao et al. (2010)
$qDTY_{12.1}$	12	RM28166-M28199	Mishra et al. (2013)
$qDTY_{12.1}$	12	RM280-RM28166	Bernier et al. (2007)
$qGY_{12.1}$	12	R901	Martinez and Thome (2001)
$qGY_{12.2}$	12	G1112	Martinez and Thome (2001)
Badh2	8	nksbadh2	Amrawathi et al. (2008), Bradbury et al. (2005), Kovach et al. (2009)

Note: Abbreviations: DTY- Drought yield, GY- Grain yield, PH- Plant height, ENT- Effective number of tillers, PL-Panicle length, TGW- 1000 grain weight, YPP- Yield per plant, LB- Grain length/breadth ratio, GL- Grain length.
Source: Meena et al. (2019) © Springer Nature

with 14 aerobic adaptation-promoting traits. Kharb et al. (2015) reported the major effect QTL for traits promoting aerobic adaptation using molecular markers and breeding programs to develop high-yielding, direct-seeded, water-efficient rice varieties.

Problems in Aerobic Rice

Weed Problems in Aerobic Rice

In rice cultivation, weeds pose a serious threat in upland and aerobic rice as compared to irrigated/lowland rice. In direct-seeded rice or upland rice, the field conditions are more conducive for weeds due to the dry tillage and alternate wetting and drying conditions promoting their germination and growth ultimately causing grain-yield loss of 50–91%. Mahajan et al. (2010) reported significant increase in the density and biomass of weeds in aerobic rice field when compared with conventional transplanted rice at 35 and 75 days after sowing/transplanting, and recommended a higher seed rate in aerobic rice cultivation to reduce the weed biomass. Thus, it is apparent that effective weed management in the aerobic rice cultivation system holds a key place. The critical period has two components: (i) the length of time weed control is required to prevent crop yield losses, and (ii) the length of time crops can tolerate weeds before resulting in yield losses.

Approximately 51 weed species have been reported to be associated with aerobic rice. Some of the important weed flora that pose great threats to aerobic rice cultivation are mentioned here, such as *Aegeratum conyzoides, Mimosa pudica, Aeschynomene indica, Celosia argentia, Portulaca oleraceae, Cynodon dactylon, Echinochloa colonum, Eleusine indica, Cyperus iria, Cyperus rotundus*, and so on.

Nematode Problems in Aerobic Rice

Nematodes were usually believed to be the primary factor among biotic constraints responsible for incessant cropping hindrance in upland rice. Under partially or fully aerobic conditions, nematodes especially root knot nematodes (RKN), that is, *Meloidogyne graminicola* emerge as a major problem for successful cultivation. However, the role of nematodes causing sickness of soil under continuous upland rice cropping is denied by some researchers as they could not observe any reduction in yield when nematodes were added in huge inoculum in the soil. *Pyrenochaeta* sp. affected the growth of aerobic rice by producing some substances inhibiting the growth of upland rice seedlings. Moreover Christie (1960) had already reported the effect of combined interaction between nematodes and fungi in causing plant disease in monoculture. Mountain (1960) had also noted the positive impact of nematode infection on greater pathogenicity of other pathogens in continuous cropping systems because continuous cropping can change the component, reproduction, and activities of microorganisms in soil and rhizosphere.

The most prominent plant pathogenic nematodes are *M. graminicola* (most severe in upland rice), *Aphelencoides besseyi, Ditylenchus angustus* (in deep water rice), and *Hirschmanniella* spp. The symptoms produced due to *M. graminicola* under nursery conditions can easily be recognized as severe stunting, yellowing on above-ground parts of the plants, and terminal hook shape gall (swelling) on the tip of the root. The host ranges are mostly confined to rice, wheat, and weeds like *Echinochloa crusgalli* and *Echinochloa colona*.

In order to avoid the losses caused by *M. graminicolai* various interventions have been taken to bring down damage level. Resistance sources against *M. graminicola* have been identified in *O. glaberrima* (Plowright et al. 1999). Use of bioagents like *Trichoderma* species has shown promise in the form of culture filtrate, soil application @4–8 g kg^{-1} of soil to reduce gall number, significantly (Le et al. 2009). Application of organic amendments (buckwheat (*Fagopyrum esculentum*) biomass, sesame (*Sesamum indicum*) biomass, neem (*Azadirachta indica*) leaves, chinaberry (*Melia azedarch*) leaves and poultry manure

@ 1, 2, and 3 t ha^{-1}) has been found effective in reducing severity of *M. graminicola* in rice nurseries by releasing ammonia and nematicidal compounds upon decomposition and increasing abundant antagonistic fauna in soil environment.

Disease and Insect Pests Constraints in Aerobic Rice Cultivation

Some of the most threatening diseases of the conventional lowland rice cropping also raise serious concerns over aerobic rice cultivation. Some major diseases such as rice blast (caused by *Magnoporthe oryzae*), sheath blight of rice (caused by *Rhizoctonia solani*), bakane or foolish seedling disease of rice (caused by *Gibberella fujikuroi*), brown spot of rice (caused by *Helminthosporium oryzae*), crown sheath rot of rice (caused by *Gaeumannomyces graminis var. graminis*), and bacterial blight of rice (caused by *Xanthomonas oryzae pv. Oryzae*) cause significant yield loss, with rice blast being the most disastrous. Rice blast causes more loss under water-stress conditions, because many critical processes in the infection cycle of *M. oryzae* such as liberation and germination of spores and infection by it are negatively regulated by adequate water levels. Breeding for blast resistance is one of the most important objectives in breeding programs of various countries. During the vegetative stage, stem borers (*Scirpophaga incertulas*) become a more serious problem for rice plants under aerobic conditions than under flooded conditions, and varieties with thick stems (traditional upland rice varieties) seem to be more susceptible to these pests. Damage by invertebrates such as armyworm larvae (*Mythimna separata*), root aphids (*Tetraneura nigriabdominalis*), and mole crickets (*Gryllotalpa orientalis* Burmeister) is a major problem for crop establishment, as they attack the stem base and roots (Saito et al. 2006). Likewise there are many insect pests that cause hindrance in successful aerobic rice cultivation among which plant hoppers (brown plant hopper- *Nilaparvata lugens*) are most important.

Integrated pest management (IPM) is a holistic approach of managing these harmful pests by using combinations of all available and possible techniques that are socially acceptable, economically viable, farmer-friendly, and environmentally sound. Some of the possible management of biotic constraints are listed here:

1) Integrated weed management by applying all the available and possible techniques/practices in a cohesive and synergistic manner.
2) Use of clean cultivation by following rouging, weeding, and burning of stubbles, etc.
3) Developing systematic monitoring programs for weeds.
4) Use of different chemicals like herbicides, nematicides, fungicides, antibiotics, insecticides, etc. which are of low risk to resistance.
5) Proper nutrient management by using slow release N fertilizers, split application of N, and use of nitrification inhibitors, judicious application of P and K fertilizers.
6) Delivering micronutrients in the best and most judicious way such as foliar application of Fe and soil application of Zn.
7) Use of well-established bio-control agents such as *Trichoderma sp.* and *Pseaudomonas sp.*, etc.
8) Breeding for resistance to these pests.
9) Use of biotechnological and genetic engineering approaches to develop resistant lines.

Other Problems in Aerobic Rice

Problems similar to those in traditional upland rice cultivation also occur in aerobic culture. Since rice is directly sown on an unsaturated soil and grows aerobically, it is difficult to fully protect rice plants against pest damage and weed infestations. Diseases such as seedling blight (caused by *Pseudomonas plantarii*) and foot rot (caused by *Erwinia chrysanthemi pv. zeae*) are also observed at higher levels on rice seedlings grown under aerobic conditions than under flooded conditions. If we sow rice seeds at a shallow depth to promote germination and tillering, the risk of damage by rodents and birds increases sharply. Nutrient availability is also a critical issue for growing rice under aerobic conditions. Problems include the immobilization of P in acidic soils (Kirk et al. 1998) and increased N leaching and volatilization. In addition, rice plants are more likely to suffer from deficiencies of micronutrients such as Fe (Fan et al. 2012), Mn (Kreye et al. 2009), and Zn (Gao et al. 2012) under aerobic conditions when the soil pH is higher than 6.0.

Table 20.2 List of various water-saving varieties of rice released throughout the World.

Variety	Country	Grain yield (t ha^{-1})	Irrigation method	Year
Lemont	Australia	8.6	Sprinkler	1989
Suweon 290	Japan	8.8	Rain-fed	1994
9516	China	8.4	Flush irrigation	1999
Apo (IR55423-0)	Philippines	5.7	Aerobic	2001
HD 502	China	5	Aerobic	2003
HD 297	China	4.8	Aerobic	2003
HD 277	China	4.5	Aerobic	2003
MAS 946-1	India	5.5–6	Aerobic	2007
Takanari	Japan	11.4	Sprinkler	2007
NERICA5	Japan	8.1	Rain-fed	2007
MAS 26	India	5.5–6	Aerobic	2008
MAS ARB25	Philippines	5.5	Aerobic	2008
MAS 25	India	5.5	Aerobic	2008
YTH323	Japan	5.1	Sprinkler irrigation	2009
CR Dhan 200	India	4.5–5	Aerobic	2012
CR Dhan 201	India	4	Aerobic	2014
CR Dhan 202	India	4	Aerobic	2014
CR Dhan 203	India	4.05	Aerobic	2014
CR Dhan 204	India	4.8–5.6	Aerobic	2014
CR Dhan 205	India	4.5	Aerobic	2014
CR Dhan 206	India	3.95	Aerobic	2014
CR Dhan 207	India	3.7	Aerobic	2016
CR Dhan 209	India	4.07	Aerobic	2016

Source: Adapted from Meena et al. (2019). © Springer Nature.

Success Achieved

Since the initiation of research on aerobic rice, significant success has been achieved. Many QTLs and genomic intervals regulating these traits have been mapped as provided in (Table 20.1). Besides these successful screening has been done in order to find the donor lines. Many varieties have been released throughout the world, which saves water consumption as compared to conventional methods. Brief descriptions of such varieties are provided in Table 20.2.

Conclusion and Ways Forward

Available literature related to aerobic rice production, linkage mapping, and molecular breeding for WUE and yield related attributes have been reviewed and presented here. There is an urgent necessity to develop and implement solutions that have the potential of increasing the efficiency, equity, and sustainability of water use. This requires a shift from the focus on pure "land productivity" without concern for water use to "water productivity," that is, getting the highest yield out of every drop of water used in agriculture. To cope with water scarcity, agricultural scientists are now taking on the challenging task of developing water-efficient crops including rice. Several strategies are being used for reducing water consumption in rice cultivation, such as saturated soil culture on raised beds, alternate wetting and drying, ground cover systems, systems of rice intensification, and aerobic rice cultivation. Among these strategies, aerobic rice is considered to be one of the most promising strategies in terms of water-use efficiency.

References

Amrawathi, Y., Singh, R., Singh, A. et al. (2008). Mapping of quantities trait loci for Basmati rice quality traits in rice (*Oryza sativa* L.). *Mol. Breed.* 21: 49–65.

Atlin, G.N., Lafitte, H.R., Tao, D. et al. (2006). Developing rice cultivars for high fertility upland systems in the Asian tropics. *Field Crops Res.* 97: 43–52.

Bai, X., Luo, L., Yan, W. et al. (2010). Genetic dissection of rice grain shape using a recombinant inbred line population derived from two contrasting parents and fine mapping a pleiotropic quantitative trait locus qGL7. *BMC Genet.* 11: 16.

Bernier, J., Kumar, A., Ramaiah, V. et al. (2007). A large effect QTL for grain yield under reproductive-stage drought stress in upland rice. *Crop Sci.* 47: 505–516.

Bernier, J., Kumar, A., Venuprasad, R. et al. (2009). Characterization of the effect of a QTL for drought resistance in rice, qtl12.1, over a range of environments in the Philippines and eastern India. *Euphytica* 166: 207–217.

Bouman, B.A. (ed.) (2002). *Water-wise Rice Production*, vol. 1. Int. Rice Res. Inst.

Bouman, B.A.M. and Tuong, T.P. (2001). Field water management to save water and increase its productivity in irrigated rice. *Agric. Water Manag.* 49: 11–30.

Bouman, B.A.M., Peng, S., Castaneda, A.R., and Visperas, R.M. (2005). Yield and water use of irrigated tropical aerobic rice systems. *Agric. Water Manag.* 74 (2): 87–105.

Bradbury, L.M.T., Henry, R.J., Jin, Q. et al. (2005). A perfect marker for fragrance genotyping in rice. *Mol. Breed.* 16: 279–283.

Castaneda, A.R., Bouman, B.A.M., Peng, S., and Visperas, R.M. (2002). The potential of aerobic rice to reduce water use in water-scarce irrigated lowlands in the tropics. In: *Proceedings of the International Work-shop on Water-wise Rice Production* (eds. B.A.M. Bouman, H. Hengsdijk, B. Hardy, et al.), 165–176. Manila: IRRI.

Christie, J.R. (1960). Some international relationships between nematodes and other soil-borne pathogens. In: *Nematology, Fundamentals and Recent Advances with Emphasis on Plant Parasitic and Soil Forms* (eds. J.N. Sasser and W.R. Jenkins), 432–436. Raleigh: University of North Carolina Press.

Dixit, S., Swamy, B.P.M., Vikram, P. et al. (2012b). Increased drought tolerance and wider adaptability of qDTY12.1 conferred by its interaction with qDTY2.3 and qDTY3.2. *Mol. Breed.* 30: 1767–1779.

Dixit, S., Singh, A., Sta Cruz, M.T. et al. (2014). Multiple major QTL lead to stable yield performance of rice cultivars across varying drought intensities. *BMC Genet.* 15: 16.

Fan, X., Karim, M.R., Chen, X. et al. (2012). Growth and iron uptake of lowland and aerobic rice genotypes under flooded and aerobic cultivation. *Commun. Soil Sci. Plant Anal.* 43: 1811–1822.

Gao, C., Shen, Q., and Guo, S. (2012). Drought-induced root aerenchyma formation restricts water uptake in rice seedlings supplied with nitrate. *Plant Cell Physiol.* 53: 495–504.

Geethalakshmi, V. Ramesh, T. Muthisolai A, and Lakshmanan, A. (2009). Agronomic evaluation of rice cultivation system for water and grain productivity. www.scribd.com/doc/4991320/agronomicevolution

Ghimire, K.H., Quiatchon, L.A., Vikram, P. et al. (2012). Identification and mapping of a QTL (qDTY1.1) with a consistent effect on grain yield under drought. *Field Crops Res.* 131: 88–96.

Kato, Y. and Okami, M. (2011). Root morphology, hydraulic conductivity and plant water relations of high-yielding rice grown under aerobic conditions. *Ann. Bot.* 108 (3): 575–583.

Kharb, A., Sandhu, N., Jain, S., and Jain, R.K. (2015). Linkage mapping of quantitativetrait loci for traits promoting aerobic adaptation on chromosome 8 in Indica rice (*Oryza sativa* L.). *Rice Genom. Genet.* 6: 1–5.

Kharb, A., Jain, S., and Jain, R.K. (2016). Phenotypic analysis, correlation studies and linkage mapping of QTL for traits promoting cultivation under dry direct seeded aerobic conditions for the development of water-efficient high yielding rice lines. *Cereal Res. Commun.* 44: 658–668.

Kirk, G.J.D., George, T., Courtois, B., and Senadhira, D. (1998). Opportunities to improve phosphorus efficiency and soil fertility in rainfed lowland and upland rice ecosystems. *Field Crops Res.* 56: 73–92.

Kovach, M.J., Calingacion, M.N., Fitzgerald, M.A., and McCouch, S.R. (2009). The origin and evolution of fragrance in rice (*Oryza sativa* L.). *Proc. Natl. Acad. Sci. U.S.A.* 106 (34): 14444–14449.

Kreye, C., Bouman, B.A.M., Castañeda, A. et al. (2009). Possible causes of yield failure in tropical aerobic rice. *Field Crops Res.* 111: 197–206.

Lafitte, H.R., Courtois, B., and Arraudeau, M. (2002). Genetic improvement of rice in aerobic systems: progress from yield to genes. *Field Crops Res.* 75: 171190.

Le, H.T., Padgham, J.L., and Sikora, R.A. (2009). Biological control of the rice root-knot nematode Meloidogyne graminicola on rice, using endophytic and rhizosphere fungi. *Int. J. Pest. Manag.* 55: 31–36.

Li, Z., Pinson, S.R.M., Stansel, J.W., and Park, W.D. (1995). Identification of quantitative trait loci (QTLs) for heading date and plant height in cultivated rice (*Oryza sativa* L.). *Theor. Appl. Genet.* 91: 374–381.

Lin, H.X., Min-Shao, K., Xiong, Z.M. et al. (1995). RFLP mapping of QTLs for grain shape traits in indica rice (*Oryza sativa* L. subsp. indica). *Sci. Agric. Sin.* 28: 1–7.

Lou, J., Chen, L., Yue, G. et al. (2009). QTL mapping of grain quality traits in rice. *J. Cereal Sci.* 50 (2): 145–151.

Lu, C., Shen, L., Tan, Z. et al. (1996). Comparative mapping of QTLs for agronomic traits of rice across environments using a doubled haploid population. *Theor. Appl. Genet.* 93: 1211–1217.

Lu, F., Ammiraju, J.S., Sanyal, A. et al. (2009). Comparative sequence analysis of *MONOCULM1*-orthologous regions in 14 *Oryza* genomes. *Proc. Natl. Acad. Sci. U.S.A.* 106: 2071–2076.

Mahajan, G., Mugalodi, S., and Ramesha Chauhan, B.S. (2010). Optimizing seed rate for weed suppression and higher yield in aerobic direct seeded rice in North Western Indo Gangetic Plains. *J. New Seeds* 11: 225–238.

Marathi, B., Guleria, S., Mohapatra, T. et al. (2012). QTL analysis of novel genomic regions associated with yield and yield related traits in new plant type based recombinant inbred lines of rice (*Oryza sativa* L.). *BMC Plant Biol.* 12 (1): 137.

McCouch, S.R., Chen, X., and Panaud, O. (1997). Microsatellite marker development, mapping and applications in rice genetics and breeding. *Plant Mol. Biol.* 35 (2): 89–99.

Meena, R.K., Bhusal, N., Kumar, K. et al. (2019). Intervention of molecular breeding in water saving rice production system: aerobic rice. *3 Biotech* 9 (4): 133.

Mishra, K.K., Vikram, P., Yadaw, R.B. et al. (2013). *qDTY 12.1*: a locus with a consistent effect on grain yield under drought in rice. *BMC Genet.* 14 (1): 12.

Mountain, W.B. (1960). Theoretical considerations of plant–nematode relationships. In: *Nematology, Fundamentals, Recent Advances with Emphasis on Plant Parasitic and Soil Forms* (eds. J.N. Sasser and W.R. Jenkins), 419–421. Raleigh: University of North Carolina Press.

Palanog, A.D., Swamy, B.P.M., Shamsudin, N.A.A. et al. (2014). Grain yield QTLs with consistent-effect under reproductive-stage drought stress in rice. *Field Crops Res.* 161: 46–54.

Patel, D.P., Das, A., Munda, G.C. et al. (2010). Evaluation of yield and physiological attributes of high-yielding rice varieties under aerobic and flooded-irrigated management practices in mid-hills eco system. *Agric. Water Manag.* 97 (9): 1269–1276.

Plowright, R.A., Coyne, D.L., Nash, P., and Jones, M.P. (1999). Resistance to the rice nematodes *Heterodera sacchari, Meloidogyne graminicola* and *M. incognita* in *Oryza glaberrima* and *O. glaberrima* × *O. sativa* interspecific hybrids. *Nematology* 1: 745–751.

Rabiei, B., Valizadeh, M., Ghareyazie, B. et al. (2004). Identification of QTLs for rice grain size and shape of Iranian cultivars using SSR markers. *Euphytica* 137: 325–332.

Redoña, E.D. and Mackill, D.J. (1998). Quantitative trait locus analysis for rice panicle and grain characteristics. *Theor. Appl. Genet.* 96: 957–963.

Saito, K., Linquist, B., and Atlin, G.N. (2006). Response of traditional and improved rice cultivars to N and P fertilizer in northern Laos. *Field Crops Res.* 96: 216–223.

Sandhu, N., Jain, S., Kumar, A. et al. (2013). Genetic variation, linkage mapping of QTL and correlation studies for yield, root, and agronomic traits for aerobic adaptation. *BMC Genet.* 14: 104–119.

Sandhu, N., Torres, R.O., Sta Cruz, M.T. et al. (2014). Traits and QTLs for development of dry direct-seeded rainfed rice varieties. *J. Exp. Bot.* 66 (1): 225–244.

Swamy, B.P.M., Kaladhar, K., Reddy, G.A. et al. (2013). Mapping and introgression of QTL for yield and related traits in two backcross populations derived from *Oryza sativa* cv: swarna and two accessions of *O. nivara*. *J. Genet.* 93: 643–654.

Venuprasad, R., StaCruz, M.T., Amante, M. et al. (2008). Response to two cycles of divergent selection for grain yield under drought stress in four rice breeding populations. *Field Crops Res.* 107: 232–244.

Venuprasad, R., Dalid, C.O., Del, V.M. et al. (2009). Identification and characterization of large-effect quantitative trait loci for grain yield under lowland drought stress in rice using bulk-segregant analysis. *Theor. Appl. Genet.* 120: 177–190.

Venuprasad, R., Bool, M.E., Quiatchon, L. et al. (2012). A large-effect QTL for rice grain yield under upland drought stress on chromosome 1. *Mol. Breed.* 30: 535–547.

Vikram, P., Swamy, B.P.M., Dixit, S. et al. (2011). qDTY1.1, a major QTL for rice grain yield under reproductive stage drought stress with a consistent effect in multiple elite genetic backgrounds. *BMC Genet.* 12: 89–93.

Wan, X.Y., Wan, J.M., Jiang, L. et al. (2006). QTL analysis for rice grain length and fine mapping of an identified QTL with stable and major effects. *Theor. Appl. Genet.* 112: 1258–1270.

Wang, H.Q., and Tang, S.X. (2000). Upland rice production and breeding in China: it's past, today and future. In: Upland Rice Research Consortium Review and Synthesis Meeting and Aerobic Rice Workshop. International Rice Research Institute. Los Baños, Philippines. 7–8 September 2000.

Xiao, J., Li, J., Yuan, L., and Tanksley, S.D. (1995). Dominance is the major genetic basis of heterosis in rice as revealed by QTL analysis using molecular markers. *Genetics* 140: 745–754.

Yadaw, R.B., Dixit, S., Raman, A. et al. (2013). A QTL for high grain yield under lowland drought in the background of popular rice variety Sabitri from Nepal. *Field Crops Res.* 144: 281–287.

Zhao, X., Qin, Y., and Sohn, J.K. (2010). Identification of main effects, epistatic effects and their environmental interactions of QTLs for yield traits in rice. *Genes Genom.* 32: 37–45.

Zheng, B., Yang, L., Mao, C. et al. (2006). QTLs and candidate genes for rice root growth under flooding and upland conditions. *Acta Genet. Sinic.* 33: 141–151.

Zhuang, J.Y., Lin, H.X., Lu, J. et al. (1997). Analysis of QTL x environment interaction for yield components and plant height in rice. *Theor. Appl. Genet.* 95: 799–808.

21

Targeting the Ascorbate-Glutathione Pathway and the Glyoxalase Pathway for Genetic Engineering of Abiotic Stress-Tolerance in Rice

Mohammad A. Hossain[1], *Tahsina S. Hoque*[2], *Abbu Zaid*[3], *Shabir H. Wani*[4], *Mohammad G. Mostofa*[5] *and Robert Henry*[6]

[1] *Department of Genetics and Plant Breeding, Bangladesh Agricultural University, Mymensingh 2202, Bangladesh*
[2] *Department of Soil Science, Bangladesh Agricultural University, Mymensingh 2202, Bangladesh*
[3] *Plant Physiology and Biochemistry Section, Department of Botany, Aligarh Muslim University, Aligarh 202002, India*
[4] *Mountain Research Centre for Field Crops, Khudwani, Anantnag 192 101, Sher-e-Kashmir University of Agricultural Sciences and Technology of Kashmir, J&K, India*
[5] *Department of Biochemistry and Molecular Biology, Bangabandhu Shiekh Mujibur Rahman Agricultural University, Gazipur 1706, Bangladesh*
[6] *Queensland Alliance for Agriculture and Food Innovation, University of Queensland, Brisbane 4072, Australia*

Introduction

Currently, about 200,000 individuals are being added to the existing world population every day, and it is presumed that the global population will surpass 9 billion by 2050. To quench the increasing demand for food, clothing, and shelter for such an increasing population, the world's agricultural resources have been put under massive pressure that is causing reduced agricultural productivity. Abiotic stresses, such as salt, drought, high and low temperatures, nutrient deficiency and toxicity, heavy metals, reduce photosynthesis, growth, and finally the yield of plants (Ahmad et al. 2019). Rising anthropogenic emissions and the population explosion impose multifarious effects on crop yields and quality (Wani et al. 2018a). Plants are sessile and sensitive creatures, and are inevitably exposed to a variety of abiotic stresses. During millennia of evolution, plants have developed various strategies to survive under unfavorable conditions. Under sub-optimal conditions, it has been reported that losses in terms of yields were in the range of 60–90%, with two-thirds of the decline associated with abiotic stresses alone (Zurbriggen et al. 2010). The perception of abiotic stresses in plants triggers a complex set of responses that is regulated by cascades of signaling elicitors, transcriptional factors, and a variety of upstream and downstream responsive genes (Vinocur and Altman 2005; Wani et al. 2018b). Various abiotic stresses elicit idiosyncratic responses in plants. The significant cross-talk that exists amongst plants during abiotic stress conditions may be synergistic or antagonistic in nature (Mittler 2006).

The rise in global population, is accelerating the demands for food produce in general and cereals in particular, requiring increased crop productivity. However, plants encounter biotic and abiotic stresses that impact their growth and productivity. There is ample scope for improvement of yields and quality traits in agricultural crops in response to abiotic

Molecular Breeding for Rice Abiotic Stress Tolerance and Nutritional Quality, First Edition.
Edited by Mohammad Anwar Hossain, Lutful Hassan, Khandakar Md. Iftekharuddaula, Arvind Kumar and Robert Henry.

stresses by genetic engineering tolerance in plants. The selection of crop plants with engineered agricultural productivity and other traits associated with tolerance to stress has become particularly important.

Rice (*Oryza sativa* L.) is a major cereal crop and is considered a staple food for 50% of the global population, and 2.5–3 billion population supply their main dietary intake from rice (Khush 2005; Arif et al. 2012; FAO 2019; Hanif et al. 2020). Moreover, about 23% of the global human per capita energy consumption and 16% of the per capita protein consumption is provided by rice (IRRI 1997). The principal rice-growing countries are in Asian, American, and African regions; Thailand, the United States, Vietnam, Pakistan, and India lead in rice exports. With a burgeoning human population by 2050, there is an urgent need to accelerate the production of rice by an estimated 87% (Kromdijk and Long 2016). This is a challenging task as the arable land available to meet this demand for increased rice production is very low if current agricultural practices are considered. In addition, the global production of food will have to face dynamic environmental fluctuations and associated abiotic pressures under a more challenging environment (FAO 2019; Arif et al. 2020). Therefore, plant biotechnologists and breeders are making efforts to unravel the underlying genetic mechanisms in various metabolic pathways, which could help elucidate regulatory interactions to improve tolerance of rice to various abiotic stresses. It is well established that an abrupt increase in reactive oxygen species (ROS) and methylglyoxal (MG) is a common consequence of abiotic stress in plants and coordinated induction of the ROS detoxifying Ascorbate-Glutathione (AsA-GSH) pathway and the MG detoxifying glyoxalase pathway is essential in modulating abiotic stress tolerance in plants including rice (Yadav et al. 2005a,b; El-Shabrawi et al. 2010; Hossain et al. 2009, 2010, 2011a,b, 2015; Hossain and Fujita 2012; Pandey et al. 2015;Rahman et al. 2016a,b ; Hoque et al. 2016, 2018; Gupta et al. 2018; Mostofa et al. 2018; Hasanuzzaman et al. 2019; Singla-Pareek et al. 2020). Additionally, genetic engineering of the genes in these pathways has shown promising results in various crop plants including rice (Yadav et al. 2005a,b; Gupta et al. 2018). Therefore, the present chapter focuses on the roles of the AsA-GSH pathway and glyoxalase pathways and their possible interactions in modulating abiotic stress tolerance in rice. This chapter also highlights the possibility of manipulation of these genes for developing climate-resilient rice to ensure food security.

Accumulation of ROS and MG in Plants in Response to Abiotic Stress

Accumulation of ROS in Plants in Response to Abiotic Stresses

In plants, ROS are normally produced during various metabolic processes under non-stressful environmental conditions, as by-products of metabolic pathways, such as photosynthesis, respiration, and photorespiration. Abiotic stresses disrupt cellular homeostasis, leading to the generation of ROS including superoxide ($O_2^{\bullet-}$), hydrogen peroxide (H_2O_2), the hydroxyl radical ($OH^{\bullet}$), and singlet oxygen ($^1O_2$) (Hossain et al. 2011a; Czarnocka and Karpiński 2018; Gómez et al. 2019). Each type of ROS in plants possesses a different oxidative capacity and affects various physiological and biochemical

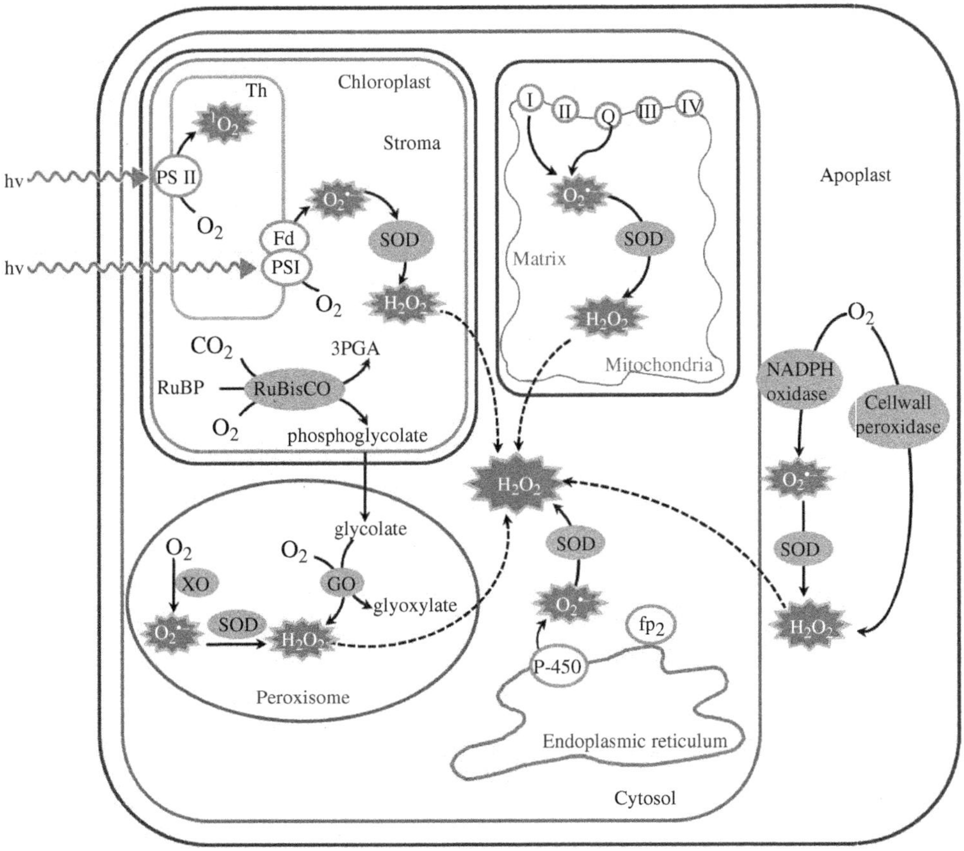

Figure 21.1 Major sites and sources of ROS production in plant cells. *Source:* Adapted from Hossain, M.A., Teixeira da Silva, J.A., and Fujita, M. (2011b). (*See color plate section for color representation of this figure*).

reactions in different ways. Organelles with a high oxidizing metabolic activity or with an intense rate of electron flow, such as the chloroplast, mitochondria, and peroxisomes are the major sources of ROS production in plant cells (for detailed description, see Hossain et al. 2011a; Czarnocka and Karpiński 2018; Foyer 2018; Zaid and Wani 2019) (Figure 21.1). Of the ROS, H_2O_2 is an important redox regulatory molecule, and has specific physical and chemical properties, owing to its stability (half life of 10^{-3} s), and ability to cause oxidation of proteins (Mhamdi and Van Breusegem 2018). In plants, H_2O_2 participates in various processes such as cell differentiation, senescence, and formation of cell walls (Moller et al. 2007; Waszczak et al. 2016; Zeng et al. 2017). In addition, H_2O_2 acts as a signaling molecule and interplays with other plant hormones to regulate various plant developmental processes under stress conditions (Hossain et al. 2015; Nazir et al. 2020). Many recent studies of rice showed an increase in ROS specially H_2O_2 and $O_2^{\cdot-}$ in response to salt stress (Hossain et al. 2011a; Mostofa et al. 2015c,d,e; Rahman et al. 2016b,c), drought stress (Xiong et al. 2018; Wang et al. 2019; El-Esawi and Alayafi 2019; Sohag et al. 2020), heat stress (Mostofa et al. 2014a; El-Esawi and Alayafi 2019), cold stress (Xu et al. 2020), copper (Cu) stress (Mostofa and Fujita 2013; Mostofa et al. 2014b,2015b), cadmium (Cd) stress (Mostofa et al. 2015a; Rahman et al. 2016a), and arsenic (As) stress (Rahman et al. 2015). Singh et al. (2020) applied 50 μM Cd to rice seedlings and found that the ROS (H_2O_2 and $O_2^{\cdot-}$) increased linearly with increasing days of growth. The accumulation of ROS (H_2O_2 and $O_2^{\cdot-}$) in leaves of rice plants under stress conditions has also been observed by nitroblue tetrazolium (NBT) and 3,3′-diaminobenzidine (DAB) staining techniques. Ji et al. (2020) applied various doses of NaCl (0, 100, and 200 mmolL^{-1}) in rice seedlings. The intensity of staining increased with increasing doses of salt stress, indicating the production of ROS (H_2O_2 and $O_2^{\cdot-}$) under salt stress conditions. Ueda et al. (2013) analyzed the impacts of acute ozone (O_3) stress on superoxide dismutase (SOD) expression, lipid peroxidation, NADPH oxidase activity, and *in vivo* H_2O_2 formation in rice leaves. Under oxygen-stress conditions, genes encoding calcium-dependent protein kinases stimulated a respiratory burst oxidase homolog (RBOH), which mediated the ROS production required for aerenchyma formation in rice (Yamauchi et al. 2017). In rice, modulation of ROS levels functions in late anther development and pollen formation through the *MADS3* transcriptional regulator (Hu et al. 2011). These results suggest the important role of ROS in plant development.

Accumulation of MG in Plants in Response to Abiotic Stresses

MG was discovered by Von Pechmann in 1887 (Von Pechmann 1887). The glyoxalases were discovered in 1913 as a single enzyme and are considered as principal enzymes of the glycolysis pathway that converts MG to lactate. However, in 1951 it was shown that two enzymes were needed and that glutathione (GSH) was the required catalytic co-factor. Szent-Georgyi (1965) elucidated the role of glyoxalases in cell division. MG is a highly reactive oxygenated short-aldehyde-containing compound that can react with lipids, nucleic acids, and proteins. MG is generated as a by-product in various metabolic reactions, such as glycolysis, per-oxidation of proteins and lipids, and degradation of sugars (for more detailed description, see Hossain et al. 2011b; Hoque et al. 2016; Singla-Pareek et al. 2020). Being a potent cytotoxin, MG once generated inside the cell can disrupt normal processes. Under adverse environmental conditions, MG accumulates to higher concentrations in all living systems. In plants, MG metabolism is linked with the glyoxalase pathway comprising two enzymes, glyoxalase I (Gly I) and glyoxalase II (Gly II), and this is regarded as the principal pathway associated with MG detoxification (Figure 21.2).

In plants, MG could play a role in signaling mechanisms via Ca^{2+}, ROS, K^+, and abscisic acid linkage. MG also initiates signaling cascade events, and it is presumed that in plants a tolerance trade-off exists between MG-responsive and stress-responsive signaling events (Hoque et al. 2016; Mostofa et al. 2018). As MG over-accumulation in cells is detrimental, by generating excess ROS, glycation end-product formation and inactivating antioxidative defense systems, plants are equipped with detoxification systems. The detoxification of MG via the glyoxalase system ameliorates MG-induced oxidative stress in plants. In humans, high MG levels are associated with various disorders, such as diabetes and obesity (Rabbani and Thornalley 2011). It has been reported that the levels of MG are usually 30–75 μM under normal metabolic conditions in plants (Yadav et al. 2005a; Hossain et al. 2009). On the onset of abiotic pressures, an increase has been observed (Yadav et al. 2005a; Hasanuzzaman and Fujita 2011; Hasanuzzaman et al. 2011; 2012; Li et al. 2018;

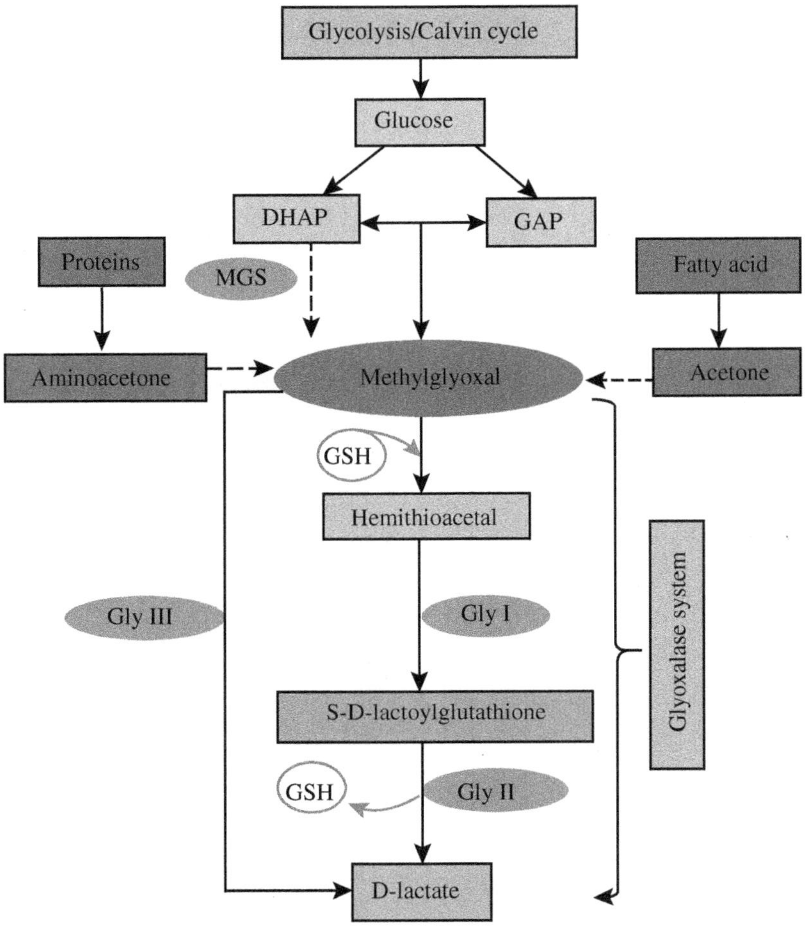

Figure 21.2 Schematic representation of the glyoxalase system in plants. *Source:* Modified from Hossain, M.A., Teixeira da Silva, J.A. and Fujita, M. (2011b) and Bhowal, B., Singla-Pareek, S.L., Sopory, S.K., et al. (2020). *Note:* MG is mainly produced during glycolysis and the Calvin cycle as a by-product of carbohydrate metabolism from triose sugar phosphates namely glyceraldehyde 3-phosphate (GAP) and dihydroxyacetone phosphate (DHAP). A small amount of MG is synthesized from aminoacetone and acetone in protein and lipid metabolism, respectively. Cytotoxic MG is principally detoxified via the glyoxalase system where two enzymes, namely, glyoxalase I (Gly I) and glyoxalase II (Gly II), are associated to transform MG. In this pathway, MG spontaneously reacts with glutathione (GSH) to form hemithioacetal, which is converted to S-D-lactoylglutathione (SLG) by Gly I. Subsequently, this S-D-lactoylglutathione is transformed into D-lactate with regeneration of GSH by the catalytic function of Gly II. Apart from conventional glyoxalase system, glyoxalase III (Gly III) irreversibly converts MG to D-lactate without requiring GSH. The dotted lines indicate the pathways that have not yet been confirmed in the plant system. (*See color plate section for color representation of this figure*).

Zaid et al. 2019; Majláth et al. 2020). MG production in response to environmental stresses and its detoxification by the glyoxalase pathway and its ability in eliciting widespread plant responses, make these events suitable candidates for depicting plant stress tolerance (Kaur et al. 2014b; Hoque et al. 2016; Mostofa et al. 2018). A large body of literature has revealed that MG is a central metabolite that controls transduction of signals, expression of genes, and protein modification (Hoque et al. 2016; Kaur et al. 2015; Banerjee et al. 2016; Bhowal et al. 2020). Various abiotic stresses in diverse crop plants increase MG production. For example, the salt-stress-mediated increase in MG levels in *Cucurbita maxima* were found to be 77% (Hossain et al. 2009) as compared to control plants. In *Nicotiana tabacum* an increase of 67% (Banu et al. 2010) was found, and in *Solanum tuberosum* a 50% increase was noticed as compared with the respective control plants (Upadhyaya et al. 2011). Increased MG levels were also found in *Brassica juncea* plants exposed to Ni stress (Zaid et al. 2019). Ojeda et al. (2013) reported a dose-dependent increment in MG levels in *Lepidium sativum* plants by subjecting them to excessive Cd and Se stresses. In *Vigna radiata* drought increased MG by 90–107% (Nahar et al. 2015) as compared with the control treatment. In rice excessive Cd increased MG by 60–260%, with respect to control plants (Mostofa et al. 2015a). Recently, Gupta et al. (2018) showed that rice plants subjected to salt and drought stress showed a 70–90% increase in MG as compared to control plants. These reports indicate that the increment in MG biosynthesis in diverse crop plants including rice is a common phenomenon on subjecting them to myriads of abiotic pressures. From the above discussion it can be inferred that the abiotic stress induced by MG production could serve as a potential signaling agent.

Importance of Glyoxalase and AsA-GSH Pathways in Modulating Abiotic Stress Tolerance

Importance of Glyoxalase Pathway in Modulating Abiotic Stress Tolerance

In plants, excess production of MG and ROS is a general phenomenon during abiotic stresses (Yadav et al. 2005a,b; Singla-Pareek et al. 2006; Hossain and Fujita 2009, 2010; Banu et al. 2010; El-Shabrawi et al. 2010; Hossain et al. 2009, 2010, 2011a,b; Zaid and Wani 2019), and these molecules are lethal to plant life, as they react with metabolic compounds leading to inactivation of the vital defense system. In order to counter the deleterious influence of MG and ROS, plants possess a complex and robust network of detoxification and scavenging pathways that act as an efficient defense system. As abiotic stress induces ROS and MG formation, plants up-regulate both antioxidant defense and glyoxalase pathways to minimize stress-induced damage (Bose et al. 2014; Gupta et al. 2018; Zaid et al. 2019).

MG, as a by-product of glycolysis, is principally detoxified via the glyoxalase system, which is an integral component of cellular MG metabolism in living cells. In plant cells, the glyoxalase pathway is present in the cytosol and organelles, with high levels of glyoxalase enzyme activity found in chloroplasts and mitochondria (Yadav et al. 2008; Rabbani and Thornalley 2012). The ubiquitous glyoxalase pathway consists of two thiol-dependent enzymes (Figure 21.2) namely glyoxalase I (Gly I; lactoyl glutathionelyase; EC 4.4.1.5) and glyoxalase II (GlyII; hydroxyacylglutathione hydrolase; EC 3.1.2.6) (Kaur et al. 2014a,b; Mustafiz et al. 2011), both of which function in coordination to transform MG and other 2-oxoaldehydes to 2-hydroxyacids with release of reduced glutathione (GSH) as a cofactor in a two-step irreversible reaction (Thornalley 1990) (Figure 21.2). In the first step, MG reacts with GSH, resulting in hemithioacetal (HTA) generation followed by conversion of HTA to

S-D-lactoyl-glutathione (SLG) in a reaction catalyzed by Gly I whereas in the second step, one molecule of GSH is recycled back to the system and D-lactate is formed through SLG hydrolysis, which is catalyzed by GlyII (Thornalley 1993; Kaur et al. 2014a; Li 2016). Thus, MG detoxification is sharply conditional on cellular GSH, the deficiency of which limits HTA generation resulting in MG accumulation (Hossain et al. 2011b). In addition to MG detoxification, the glyoxalase system takes part in oxidative stress tolerance by maintaining glutathione homeostasis via recycling GSH to be trapped non-enzymatically by MG to form HTA. Moreover, some recent reports have confirmed the presence of another unique glyoxalase member, namely, Glyoxalase III (GlyIII) (Lee et al. 2012; Kwon et al. 2013; Ghosh et al. 2016), which is able to irreversibly convert MG to D-lactate in a single step, without the need for GSH. The glyoxalase system plays a pivotal role in the cellular defense against glycation and oxidative stress, where glyoxalase and MG are considered to be potential biomarkers associated with plant stress responses and physiology (Thornalley 1993; Kalapos 2008; Kaur et al. 2014b).

Over-expression of the glyoxalase pathway genes causes better glutathione homeostasis and retains higher activities of antioxidant as well as glyoxalase enzymes that contribute to retain optimum cellular level of MG and ROS preventing cellular damage during stressed conditions (Yadav et al. 2005a,b). In recent years, various transgenic, transcriptomic, and proteomic studies have highlighted the novel role of the glyoxalase pathway in abiotic stress tolerance in plants (Singla-Pareek et al. 2003, 2006; Hossain et al. 2009; Lin et al. 2010; Mustafiz et al. 2011; Gupta et al. 2018). Considering their protective and regulatory roles, glyoxalase genes have been targeted to provide significant resistance against different clinical disorders (Rabbani and Thornalley 2011; Xue et al. 2011) and tolerance to multiple abiotic stresses including salinity, drought, heavy metal toxicity, nutrient deficiency, and others. in different plant species (Singla-Pareek et al. 2003, 2006; Saxena et al. 2011; Alvarez Viveros et al. 2013; Ghosh et al. 2014; Mustafiz et al. 2014; Kaur et al. 2014a; Zeng et al. 2016; Gupta et al. 2018; Batth et al. 2017).

Importance of AsA-GSH Pathway in Modulating Abiotic Stress Tolerance

Since excess production of ROS causes detrimental effects in plants, its detoxification is crucial to protect cellular integrity and to maintain redox homeostasis. To prevent such stress-induced damage, plants activate an antioxidant defense system or machinery, which is considered to be a unique and remarkable mechanism to regulate ROS levels according to the cellular needs at a certain period. This intrinsic scavenging or detoxification system comprises both enzymatic and non-enzymatic pathways to regulate oxidative stress. Some of the major enzymatic antioxidants include ascorbate peroxidase (APX; EC 1.11.1.11), monodehydroascorbate reductase (MDHAR; EC 1.6.5.4), dehydroascorbate reductase (DHAR; EC 1.8.5.1), glutathione reductase (GR; EC 1.6.4.2) and so on, whereas the most abundant non-enzymatic antioxidants include water-soluble compounds such as ascorbate (AsA) and glutathione (GSH) (Gill and Tuteja 2010; Miller et al. 2010; Gill et al. 2011; Zaid and Wani 2019). These antioxidants are found in almost all the cellular compartments in plants such as mitochondria, chloroplasts, peroxisomes, and cytosol, which function in coordination to control the redox signaling cascades and to scavenge ROS for protecting cells from oxidative damage (Mittler et al. 2004; Gill and Tuteja 2010).

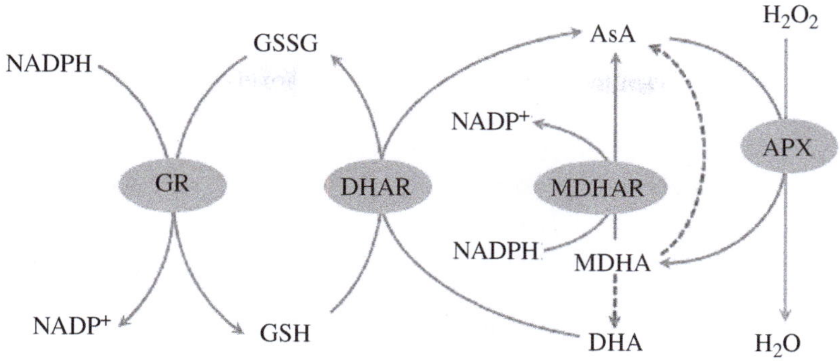

Figure 21.3 Diagram showing ascorbate-glutathione (AsA-GSH) pathway in plants. *Note:* Reactive oxygen species (ROS) namely hydrogen peroxide (H_2O_2) are detoxified forming water (H_2O) by catalytic action of ascorbate peroxidase (APX). In this detoxification reaction, AsA is oxidized to form an unstable compound monodehydroascorbate (MDHA), which is either reduced by monodehydroascorbate reductase (MDHAR) to form AsA or, reacts to produce dehydroascorbate (DHA), which is further reduced by dehydroascorbate reductase (DHAR) to form AsA. In this conversion, glutathione (GSH) is oxidized to glutathione disulfide (GSSG), which is again reduced by glutathione reductase (GR) to re-synthesize GSH. At the time of reduction of MDHA and GSSG by the respective enzymes in this pathway, the electron acceptor nicotinamide adenine dinucleotide phosphate (NADP) is regenerated using NADPH as a reducing agent. Since GSH, AsA, and NADPH are present in high concentrations in plant cells, it is obvious that the AsA-GSH cycle plays a key role for H_2O_2 detoxification. Dotted lines represent non-enzymatic reactions.

To combat abiotic stress conditions, the redox buffer capacity is principally represented by AsA and GSH pools where the antioxidants are closely associated in a set of reactions known as the AsA-GSH cycle (Bartoli et al. 2017). The AsA-GSH pathway (known as the Asada–Halliwell pathway) consisting of AsA, GSH, and four enzymes, namely, APX, MDHAR, DHAR, and GR, is one of the most important antioxidant systems in plants and greatly contributes in detoxifying ROS (Figure 21.3). Additionally, the components of this cycle participate in optimizing photosynthesis and cell signaling to stimulate adaptive stress responses in plants.

For example, APX reduces H_2O_2 into H_2O, employing AsA as an electron donor in the AsA-GSH cycle, and the balance between GSH and GSSG is of paramount importance in maintaining a favorable redox status to detoxify H_2O_2. The oxidation of AsA leads to the generation of monodehydroascorbate (MDHA), which is either reduced by MDHAR requiring NADPH or disproportionate into AsA and dehydroascorbate (DHA) (oxidized state) where DHAR again reduces DHA to AsA using GSH and thus regenerates AsA in the system (Chen et al. 2003; Martínez and Araya 2010). Thus, MDHAR and DHAR play important roles in regulating the level and redox state of cellular AsA under oxidative stress (Eltayeb et al. 2006, 2007). In addition, GR, the rate-limiting enzyme of AsA–GSH cycle, catalyzes the reduction of AsA by GSH generated from GSSG at the expense of NADPH, keeping the GSH/GSSG ratio favorable for AsA reduction (Foyer and Noctor 2005). Acting as cofactors of enzymes in glyoxalase and antioxidant pathways, both AsA and GSH directly mitigate ROS and regulate gene expression associated with stress responses to improve plant defense. In summary, antioxidants of the AsA-GSH pathway have a vital role in maintaining the ROS concentration at an optimum level that can enhance plant growth and establish plant tolerance to abiotic stresses via modulation of defense gene expression and redox signaling cascades (Neill et al. 2002).

There are many recent reports of higher activities of these antioxidants (enzymatic and non-enzymatic) and of induced expression of antioxidant genes in several plant species grown in stressed conditions, including salinity, drought, heat, chilling, nutrient deficiency, heavy metal, or other (Mittova et al. 2004; Hernández et al. 2000, 2010; Chaparzadeh et al. 2004; Gao et al. 2008 ; Sairam et al. 2011; Almeselmani et al. 2006; Kusvuran et al. 2007, 2012; Sevengor et al. 2011; Vardharajula et al. 2011; Wu et al. 2012; Abbas et al. 2015; El-Tayeb 2006; Basu et al. 2010; Kiran et al. 2014; Lee and Lee 2000; Dong et al. 2006; Tanyolac et al. 2007; Ekmekci et al. 2008). According to the findings of these reports higher levels of antioxidants, whether constitutive or induced, showed greater resistance to abiotic stresses in plants including rice.

Interconnection of the Antioxidant Pathway and the Glyoxalase Pathway in Modulating Abiotic Stress Tolerance

Over the past decades the diversified roles of GSH has greatly advanced understanding related to ROS metabolism, redox regulation, and more recently in redox signaling. The MG detoxifying glyoxalase pathway and the H_2O_2 detoxifying AsA-GSH pathway are interconnected through GSH, which regulates the ROS and MG levels in plants and their signaling to counteract the negative consequence of ROS and MG. In conferring abiotic stress tolerance the interconnection of the AsA-GSH pathway and the glyoxalase pathway (Figure 21.4) was evident from a series of studies with stress tolerant and susceptible plants, from chemical-priming-induced abiotic stress tolerance as well as from transgenic plants. The first experimental evidence of Yadav et al. (2005b) signifies the interaction of the glyoxalase pathway and the AsA-GSH pathway in modulating salt tress tolerance in transgenic tobacco plants. Transgenic plants overexpressing both Gly I and II genes showed higher Gly I and II activities and reflect better antioxidative protection under salt stress (200 mM NaCl), while wild-type (WT) plants suffered from serious oxidative stress. Importantly transgenic plants showed higher constitutive AsA-GSH pathway enzymes (APX, GR) and GSH dependent antioxidant enzyme (GST, Glutathione *S*-transferase and GPX, Glutathione peroxidase) activities, but a sharp increase of these enzymatic activities was observed in response to salt stress. They concluded that the transgenic plants showed better salt stress tolerance by regulating multiple biochemical pathways and functions by (i) maintaining higher activities of AsA-GSH cycle enzymes that control the H_2O_2 and GSH-AsA redox ratios; (ii) maintaining non-destructive levels of MG that usually deplete GSH; (iii) maintaining higher activities of GPX and GST that utilize GSH in degrading organic hydroperoxide, lipid peroxide, and H_2O_2; (iv) protecting non-enzymatic antioxidants from oxidative and antioxidative enzymes from inactivation through advanced glycation. These four mechanisms hold true for various abiotic stresses, and recent studies with exogenous chemical priming also demonstrated that co-ordinated induction of both detoxification pathways showed enhanced abiotic stress tolerance in different plant species and cultured cells.

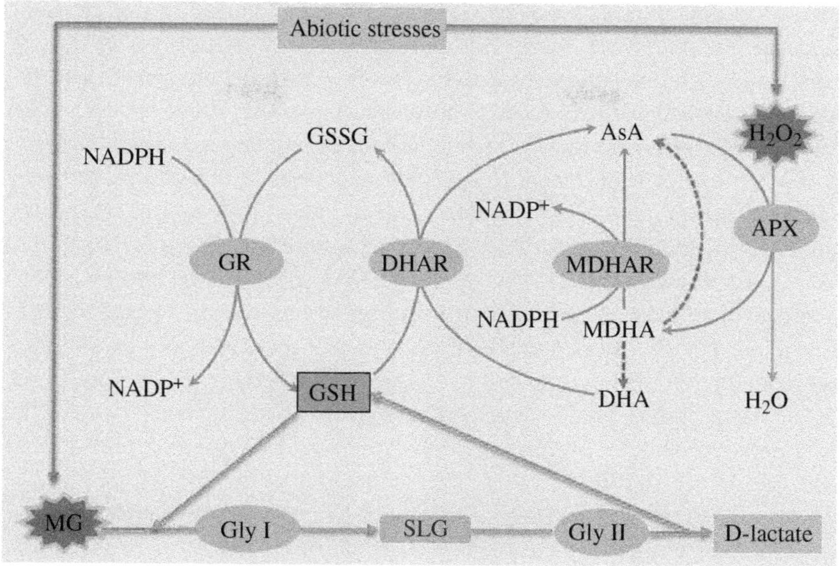

Figure 21.4 Inter-connection of the AsA-GSH pathway and the glyoxalase pathway through GSH. *Note:* For further discussion, see Section 21.4 in the main text. (*See color plate section for color representation of this figure*).

El-Shabrawi et al. (2010) provided further evidence that simultaneous induction of AsA-GSH cycle and glyoxalase cycle is very important in modulating salt stress tolerance using susceptible and tolerant rice plants. In addition with other antioxidative enzymes activities (SOD, catalase [CAT], GPX, and peroxidase [POX]) they found higher activities of AsA-GSH cycle enzymes (APX and GR) and isozyme expression in the salt-tolerant cultivar (Pokkali), and lower oxidative damage (in terms of lower H_2O_2 and oxidative DNA damage) as compared to the salt-sensitive cultivar (IR64). The tolerant cultivar also showed a higher GSH/GSSG ratio and a higher AsA/DHA ratio. They concluded that fine modulation of glyoxalase cycle and AsA-GSH cycle allowed the tolerant plant to show a tolerance reaction as compared to the susceptible IR64. The above funding further confirms that ROS and MG metabolism are tightly correlated and that a plant induces both pathways in response to abiotic stress tolerance.

Recently, a series of our studies with exogenous chemical priming induced abiotic stress tolerance in rice further pointed out that the co-ordinated up-regulation of both of the pathways confers abiotic stress oxidative stress tolerance in rice. Mostofa and Fujita (2013) showed that copper (Cu) stress in rice seedlings caused toxicity symptoms (chlorosis, necrosis, and rolling in leaves), sharp increases in malondialdehyde (MDA), H_2O_2 content and lipoxygenase (LOX) activity with concomitant reductions of chlorophyll (Chl), and relative water content (RWC). Salicylic acid (SA) supplemented seedlings showed a significant decrease in ROS and MDA levels and the activation of AsA-GSH cycle enzymes (APX, GR) and glyoxalase pathway enzyme activities (Gly I and II). These findings indicated that coactivation of ROS and MG detoxification improved the plants to Cu stress. Similarly, Mostofa et al. (2014a) showed that heat stress (42 °C, 48 hours) significantly increases the ROS and MDA levels in rice seeding. Heat-induced damage was also evident in decreased levels of AsA, GSH, and AsA and GSH redox ratios. In response to heat stress, the activities of APX, DHAR, MDHAR, and GR decreased along with the level of AsA and GSH. Heat stress also increases the activities of glyoxalase pathway enzymes. However, spermidine-pretreated heat-exposed seedlings displayed higher activities of SOD, CAT, GPX, GST APX, DHAR, and GR as well as of Gly I and Gly II and maintained markedly lower ROS and MDA levels. They conclude that modulation of the activities of enzymes of both pathways improved heat-stress tolerance. Consequently, Mostofa et al. (2014b) showed that nitric oxide (NO) donor or GSH induces Cu stress tolerance in rice seedlings. Hydroponically grown rice seedlings (12 days) were subjected to Cu stress (100 μM $CuSO_4$) alone or in combination with 200 μM sodium nitroprusside (SNP) (an NO donor) and 200 μM GSH. Cu stress resulted in a significant growth reduction and chlorosis, as well as leaf rolling. The seedlings also showed higher oxidative damage as indicated by higher ROS and MDA levels as well as LOX activity and decreased AsA content. The application of GSH, SNP, or SNP + GSH reduced the oxidative burden of the seedlings as indicated by lower ROS and MDA levels as well as LOX activity. The activities of SOD, APX, MDHAR, and GR, as well as glyoxalase pathway enzyme (Gly I and Gly II), increased in response Cu stress, but a greater increase in other enzymes of these pathways (DHAR, GST, GPX, and CAT) and AsA and phytochelatins content were increased in GSH and NO supplemented seedlings. Their results further proved that modulation of ROS and MG detoxification pathways improved oxidative stress tolerance in rice.

Later, Mostofa et al. (2015a) reported that the Cd stress in rice resulted in growth inhibition and biomass reduction, which were correlated with the increased uptake of Cd and depletion of the photosynthetic pigments, leaf water contents, essential minerals, water-soluble proteins, and enzymatic and non-enzymatic antioxidants. Hydrogen sulfide (H_2S) supplemented seedlings showed a significant increase in AsA-GSH cycle enzymes (MDHAR, DHAR, GR) and other antioxidant enzymes (SOD, CAT, GST, GPX) and glyoxalase pathway enzyme (Gly II), and maintained a lower level of MG as compared to seedlings treated with Cd stress only. H_2S supplemented Cd stress seedlings maintained higher AsA/DHA and GSH/GSSG ratios. Their findings indicated that simultaneous induction of ROS and the MG detoxification system by H_2S improved Cd stress tolerance.

Mostofa et al. (2015b) showed that copper (Cu) stress in rice seedlings significantly increases ROS and MDA levels as compared to the control plants. Trehalose (Tre) supplemented Cu-stressed seedlings maintained higher AsA/DHA and GSH/GSSG ratios and maintained higher AsA-GSH cycle enzyme (APX and DHAR) and glyoxalase pathway enzyme activities (Gly I and Gly II). Cu stress significantly increases ROS and MG levels whereas supplementation of AsA reversed the negative effect of Cu stress. These results indicate that modifying the endogenous Tre content by Tre pretreatment improved Cu tolerance in rice plants by inhibiting Cu uptake and regulating the antioxidant and glyoxalase systems, and thereby demonstrated the important role of Tre in mitigating heavy metal toxicity.

Mostofa et al. (2015c) studied the effects of H_2O_2 and SA on salt stress tolerance in rice seedlings. Seedlings were subjected to salt stress (200 mM NaCl, 72 hours) with or without H_2O_2 and SA pre-treatment for 24 hours. Salt stress resulted in a significant increase in ROS ($O_2^{\bullet -}$ and H_2O_2) and a significant decrease in AsA/DHA and GSH/GSSG ratio, whereas pre-treated salt stress seedlings showed a significantly lower level of ROS and an increase in AsA and GSH redox ratio. Importantly, H_2O_2 pre-treated seedlings showed significantly higher AsA-GSH cycle enzyme activities (APX and DHAR) along with other antioxidant enzymes (CAT, GPX, and GST) and glyoxalase pathway enzyme (Gly II), as compared to the seedlings subjected to salt stress without H_2O_2 pre-treatment. They finally concluded that H_2O_2 and SA mediated salt stress tolerance is due to increase in endogenous NO level that finally modulated the glyoxalase cycle and AsA-GSH cycle.

Mostofa et al. (2015d) showed that rice seedlings pre-treated with H_2S had improved salinity tolerance by modulating ROS and the MG detoxification system as well as by decreasing Na^+ uptake. Salt stress resulted in a significant increase in ROS, MG, and MDA levels and decreased the levels of AsA/DHA and GSH/GSSG ratios. Salt stress increases the activities of antioxidant enzymes, whereas a greater increase was observed in the GR activities in H_2S pre-treated seedlings. The activities of GPX activity were also found to increase in response to H_2S treatment. H_2S pre-treated seedlings maintained significantly lower levels of ROS, MG, and MDA due to synergistic coactivation of ROS and the MG detoxification system.

Mostofa et al. (2015e) reported that salt stress in rice seedlings induced significant oxidative stress as indicated by higher ROS ($O_2^{\cdot-}$ and H_2O_2) and MDA accumulation whereas total Chl and RWC decreased. Salt stress caused an imbalance in non-enzymatic antioxidants, that is, AsA content, AsA/DHA ratio, and GSH/GSSG ratio decreased, but GSH content increased significantly. In contrast, Tre pretreatment (10 mM, 48 hours) significantly addressed salt-induced toxicity symptoms and dramatically depressed LOX activity, ROS, MDA, and Pro accumulation whereas AsA, GSH, RWC, Chl contents, and redox status improved considerably. Salt stress stimulated the activities of SOD, GPX, APX, MDHAR, DHAR, and GR but decreased the activities of CAT and GST. However, Tre-pretreated salt-stressed seedlings counteracted SOD and MDHAR activities, elevated CAT and GST activities, further enhanced APX and DHAR activities, and maintained GPX and GR activities similar to the seedlings stressed with salt alone.

Rahman et al. (2016a) reported that rice seedlings exposed to Cd stress (0.25 and 0.5 mM $CdCl_2$) resulted in enhanced MG and ROS accumulation through the inhibition of the antioxidant defense system and glyoxalase system. Exogenous application of calcium (Ca) in the growth medium of rice seedlings reverses negative effect of Cd stress by increasing the level of non-enzymatic antioxidant (AsA) as well as the antioxidant enzymes activities including AsA-GSH cycle (SOD, CAT, GST, MDHAR, and DHAR) and glyoxalase cycle (Gly I and Gly II) along with the decreased levels of ROS and MG and reduces the negative effect of Cd stress.

Rahman et al. (2015) reported that As stress in rice seedlings resulted in a significant increase in growth inhibition, chlorosis, reduced leaf RWC, and increased As-accumulation. As-stress significantly increased the level of ROS and inactivated the antioxidant defense system and glyoxalase system and made the plant more susceptible to As stress. Calcium supplementation increases the AsA-GSH cycle enzymes (MDHAR and DHAR) and other antioxidant enzymes (SOD, GPX, CAT) as well as glyoxalase pathway enzymes. These results suggest that Ca supplementation improves rice seedlings' tolerance to As-induced oxidative stress by reducing As uptake, enhancing their antioxidant defense and glyoxalase systems, and also improving growth and physiological condition.

Rahman et al. (2016c) showed that salt stress (150 mM) in rice (*O. sativa* L. cv. BRRI dhan47) seedlings resulted in a significant growth retardation, osmotic stress, and leaf chlorosis. Salt-induced ionic toxicity and osmotic stress consequently resulted in oxidative stress by disrupting the antioxidant defense and glyoxalase systems through overproduction of ROS and MG. Importantly, exogenous application of manganese (Mn) reinforced the ROS and MG detoxification system as indicated by higher levels of non-enzymatic antioxidants like AsA and flavonoids and the activities of MDHAR, DHAR, SOD, CAT, and glyoxalase pathway enzyme (Gly I and II). They concluded that coordinated action of ion homeostasis and the antioxidant defense and glyoxalase systems in the salt-affected seedlings resulted in better salt stress tolerance.

Rahman et al. (2016b) showed that exogenous calcium (Ca) improves salinity tolerance in rice through induction of the antioxidant defense pathway and the glyoxalase pathway. Thirteen-day-old rice (*O. sativa* L. cv. BRRI dhan47) seedlings were subjected to salt stress (200 mM NaCl) alone and combined with 2 mM $CaCl_2$ and 2 mM ethylene glycol tetraacetic acid (EGTA, a Ca scavenger) for three days. Salt stress led to a significant reduction in growth inhibition, chlorosis, and water shortage in the rice seedlings. Salt stress resulted in a reduction in nutrient uptake and disruption of ion homeostasis. Salinity also induces oxidative stress as indicated by higher lipid peroxidation, ROS, and MG levels. Importantly, seedlings that have grown through Ca supplementation showed less growth reduction, chlorosis, and water loss. Ca-supplemented seedlings also improved ion homeostasis and modulated the activities of antioxidant and glyoxalase pathway enzymes. However, EGTA application reversed the positive of Ca under salt stress. Their results finally proved that Ca can modulate the different biochemical pathways to improve salt stress tolerance in rice.

The above study clearly indicates that co-regulation of the glyoxalase cycle and the AsA-GSH cycle is essential to maintain an optimum balance of MG and ROS to protect the plants from their overproduction as well as to perform their signaling function in stress-responsive gene expression that finally forces the plant to show a resistant reaction.

Manipulation of Glyoxalase Pathway Genes for Abiotic Stress Tolerance in Rice

Modification of the glyoxalase system can potentially induce abiotic stress tolerance in genetically engineered plants (Kaur et al. 2014a,b; Gupta et al. 2018). For example, transgenic plants over-expressing *glyI* and *glyII* genes individually or together are able to survive, grow, and show better tolerance towards multiple abiotic stresses including salinity, drought, metal toxicity, and extreme temperature, due to the check on MG level, and regulation of glutathione homeostasis under stress conditions (Mustafiz et al. 2011).

Transgenic Plants Over-expressing *gly I* Gene and Abiotic Stress Tolerance in Rice

Glyoxalase I is a primary enzyme of the glyoxalase pathway; its activity has been detected in various monocots and dicots such as *Lycopesicon esculentum* (Espartero et al. 1995), *Brassica oleracea* (Clugston et al. 1998), *B. juncea* (Veena et al. 1999), *O. sativa* (Usui et al. 2001), *Triticum aestivum* (Lin et al. 2010), and *Arabidopsis thaliana* (Jain et al. 2016). As gly I is involved in detoxification of MG in the glyoxalase pathway, the gene serves as a good candidate to produce transgenic lines with enhanced abiotic stress tolerance in several plant species, including rice. Verma et al. (2005) biolistically transferred the gly I gene of *B. juncea* (Veena et al. 1999), driven by a stress-inducible ABRC1-Act1-100P-HVA22 (I) promoter complex and flanked by tobacco Rb7 MAR sequences into scutellum-derived, three-week-old calli of *japonica* rice (*O. sativa* L.) variety TNG67 and *indica* rice variety Pusa Basmati 1. They showed that transgenic *japonica* and *indica* rice plants over-expressing *B. juncea glyI* gene had salinity tolerance. Recently, Zeng et al. (2016) cloned and characterized a gene of rice cv. Nipponbare encoding gly I that was ubiquitously expressed in various plant organs: namely, roots, stems, leaves, leaf sheathes, and spikelets in varying proportions, and found that the gene (*Osgly I*) was markedly up-regulated in response to NaCl, $ZnCl_2$, and mannitol stress in rice seedlings. They investigated the over-expression of *Osgly I* using Agrobacterium-mediated transformation and observed that transgenic rice plants had increased glyoxalase enzyme activity, a decreased MG level, better tolerance to NaCl, $ZnCl_2$ and mannitol, higher seed-setting rate, and yield compared to wild-type plants. Improved stress tolerance in transgenic lines was related to low malondialdehyde content produced from cellular lipid peroxidation. The results of the study highlighted the potential of the *Osgly I* gene to be a useful target for genetic engineering in cereals to withstand abiotic stresses. Again, enhanced tolerance to MG and salt stress was also observed by ectopic expression of the *OsglyI* gene in tobacco (Mustafiz et al. 2014).

Transgenic Plants Over-expressing *glyII* Gene and Abiotic Stress Tolerance in Rice

Glyoxalase II is another novel target of genetic engineers for developing stress resistant crops using transgenic approaches. In the last two decades, purification, cloning, and characterization of the *glyII* gene have been reported in many plant species, such as *Aloe vera*, spinach, and *A. thaliana* (Norton et al. 1990; Talesa et al. 1990; Ridderstrom and Mannervik 1997). Yadav et al. (2007) successfully performed characterization and functional validation of glyoxalase II from *O. sativa,* and transcript analysis in rice showed that *OsglyII* gene expression was stimulated within 15 minutes in response to different abiotic stresses. To investigate the role of glyoxalase II, Singla-Pareek et al. (2008) cloned rice glyoxalase II in pCAMBIA1304 followed by transformation into rice cv. PB1 via Agrobacterium and found higher constitutive activity of glyoxalase II in transgenic plants under salinity stress. Compared to the non-transgenic lines, the transgenic rice plants up-regulated the enzyme activity and exhibited sustained growth and more favorable ion balance conferring significant tolerance to salt stress. A possible mechanism behind stress tolerance in *OsglyII* over-expressing transgenic rice was attributed to detoxification of MG as well as SLG and maintaining GSH homeostasis. Furthermore, Wani and Gosal (2011) introduced the *OsglyII* gene into *indica* rice cv. PAU201 using a particle bombardment transformation technique and indicated that *OsglyII* transgenic plants showed tolerance to salt stress both at the calli and plantlet level. *In vitro* screening for salt tolerance showed that both the *OsglyII* putative transgenic calli and plantlets had increased fresh mass and growth under NaCl stress compared to non-transgenic calli and plantlets. Furthermore, ectopic expression of rice *glyII* gene in brassica and tobacco has been shown to confer tolerance against salinity and exogenous MG stress (Ghosh et al. 2014; Saxena et al. 2011).

Transgenic Plants Over-expressing *glyI* and *glyII* Gene and Abiotic Stress Tolerance in Rice

Under abiotic stresses, the presence and characterization of both glyoxalase I and II in the same plant has been reported where the genes encoding these have been cloned and found to be up-regulated (Yadav et al. 2008). Genome-wide analysis for *glyI* and *glyII* genes was carried out by Mustafiz et al. (2011) confirming the existence of 11 *glyI* genes and 3 *glyII* genes in rice, which were highly expressed under abiotic stresses. Over-expression of both *glyI* and *glyII* in double transgenic plants exhibited a better response than single-gene transformed lines as well as untransformed plants under salinity and heavy metal (Zn) stress in tobacco, tomato, and citrus (Singla-Pareek et al. 2003, 2006; Yadav et al. 2005b; Alvarez Viveros et al. 2013; Alvarez-Gerding et al. 2015). Recently, Gupta et al. (2018) suggested that manipulating both the enzymes of the glyoxalase pathway in rice plant confers tolerance to multiple abiotic stresses such as salinity, drought, and heat. They generated transgenic plants over-expressing the entire glyoxalase pathway by cloning the *glyI* gene from *Brassica* and the *glyII* gene from *O. sativa* together in a single vector pCAMBIA1304 and performing Agrobacterium-mediated transformation. Consistent with higher accumulation of gly I and gly II proteins, a threefold to fourfold increase in enzyme activities of both gly I and gly II and lower accumulation of MG in the transgenic lines were observed as compared with the WT plants. Transgenic plants regulated higher glyoxalase and triose phosphate isomerase activity that restricts MG accumulation, maintained organellar structure and a higher level of reduced glutathione and membrane stability index, and showed lower lipid peroxidation leading to better tolerance under salinity, drought, and heat-stress conditions. Taken together, in double transgenic rice plants, the possible mechanism of abiotic stress tolerance involves restricted MG accumulation for higher glyoxalase activity, leading to reduced cytotoxic effects, better protection of chloroplast and mitochondrial ultra-structure, higher chlorophyll content and photosynthetic capability, and better antioxidant capacity.

Transgenic Plants Over-expressing *glyIII* Gene and Abiotic Stress Tolerance in Rice

The existence of DJ-1 genes in plants was revealed by a sequence analysis study of DJ-1/PfpI member proteins in various plant species. Ghosh et al. (2016) performed in-depth investigation of this protein confirming the presence of Gly III protein in rice that can utilize MG as a substrate to produce D-lactate in a GSH-independent manner. By *in silico* analysis of rice genomes, they identified 12 DJ-1 proteins encoded by six genes where OsDJ-1 was up-regulated in response to abiotic stresses like cold, heat, drought, and heavy metals, and could interact with a total of 503 proteins of rice. Transcript profiling under multiple stress conditions showed high consecutive expression of OsDJ-1C (a member of the OsDJ-1 family) at all developmental stages in rice tissues suggesting the protective role of *OsglyIII* in abiotic stress tolerance in plant. In comparison with the conventional glyoxalase (glyI/II) system, the efficiency of *OsglyIII* has been found to be relatively low in terms of kinetic parameters, but its over-expression still has the potential to confer tolerance towards multiple abiotic stresses.

Genetic Engineering of the AsA-GSH Pathway Genes for Abiotic Stress Tolerance in Rice

Plenty of research results showed that the transgenic plants over-expressing AsA-GSH pathway genes showed better abiotic stress tolerance by performing specific function as well as by regulating multiple pathways of stress tolerance.

Transgenic Plants Over-expressing APX Gene and Abiotic Stress Tolerance in Rice

APX, the most important H_2O_2 detoxifying enzyme, is one of the main components of the AsA-GSH pathway. APX differs from other peroxidase due to its dependency on AsA as a source of reducing power (Shigeoka et al. 2002). *O. sativa* has eight isozymes (*Os*APX1-*Os*APX8), having two APX each in cytosol, peroxisome, chloroplast, and mitochondria (reviewed in Pandey et al. 2017). The existence of multiple molecular forms of APX within the cells and organelles signifies the important role played by them in developmental process and stress tolerance (Shigeoka et al. 2002; Pandey et al. 2017). Due to its profound effects on plant stress tolerance, a large number of APX transgenic plants have been produced in various crop plants including rice. Sato et al. (2011) showed that transgenic rice plants over-expressing *OsAPXa* gene displayed higher oxidative stress tolerance induced by cold stress at the booting stage. Rice plants subjected to cold stress (12 °C treatment for six days) showed 1.5-fold increase in H_2O_2 and twofold increase in MDA levels. Importantly, transgenic plants maintained a significantly lower level of ROS and MDA levels as compared to non-transgenic control plants. Transgenic plants showed significantly higher spikelet fertility. Their reports indicated higher ROS detoxification capacity of transgenic plants that make the plant more tolerant to cold-induced oxidative stress. Later on, Chiang et al. (2014) reported that transgenic rice plants over-expressing *SmAPX* showed higher tolerance to flooding stress as indicated by faster growth rates and less oxidative injury as compared to WT plants. Similarly, Zhang et al. (2015) showed that transgenic rice plants over-expressing AsA biosynthetic genes GDP-L-galactose phosphorylase (GGP) showed higher AsA synthesis and salt stress tolerance. Transgenic homozygous line showed 2.5-fold higher AsA and showed higher plant height, root length, and the relative growth rate of fresh weight as compared to WT plants. Their results showed that GGP may be a key rate-limiting step in rice AsA biosynthesis, and the plants with elevated AsA contents demonstrated enhanced tolerance for salt stress. Recently, Feng et al. (2017) reported that transgenic rice plants over-expressing *OsAPX2* showed enhanced drought tolerance in the laboratory as well as in natural field conditions. Transgenic plants at the seedling stage showed resistance when subjected to drought stress in hydroponic solution by 20% PEG-4000. Transgenic plants showed less wilting as compared to the WT type, and transgenic plants showed higher survival percentage after rewatering. The H_2O_2 level after 10 days of drought stress showed a threefold increase, whereas transgenic plants showed a 1.8-fold increase. Additionally, the transgenic line also showed a lower MDA level as compared to non-transgenic plants. The transgenic plants were also subjected to drought stress at the flowering stage. The transgenic line showed a higher seed-setting rate, total number of seeds per panicle, and yield per panicle than did the wild type. The seed-setting rate for line 1 was 70%, whereas that of the wild type was 35%. This finding suggested the potential use of line 1 in producing marker-free rice material for improving grain yield under drought-stressed conditions. The roles of APX were also studied by producing knockout lines. By producing knockout and transgenic plants of rice, Zhang et al. (2013) showed that *OsAPX2* plays a critical role in growth and reproduction under drought, salt, and cold stresses in rice. Loss of function in *OsAPX2* affected the growth and development of rice seedlings, resulting in semi-dwarf seedlings, yellow-green leaves, leaf lesion mimic, and seed sterility. The knockout seedlings showed lower APX activity and higher sensitivity to abiotic stresses as indicated by higher H_2O_2 and MDA levels as compared to the over-expressed lines. They concluded that the cytosolic APX *OsAPX2* plays an important role in rice growth and development by protecting the seedlings from abiotic stresses through scavenging ROS.

Transgenic Plants Over-expressing the MDHAR Gene and Abiotic Stress Tolerance in Rice

MDHAR, the second enzyme of the AsA-GSH pathway, has been extensively studied in plants. This enzyme uses NADPH as an electron donor to reduce MDHA to AsA. A large number of transgenic plants, including rice, were produced for functional characterization of the enzyme. Sultana et al. (2012) showed that rice plants over-expressing *AeMDHAR* showed significantly higher MDHAR activity as compared to WT plants under control and salt stress conditions. The transgenic plants under salt-stress conditions showed higher root and shoot length at the seedling stage as compared to WT plants. The transgenic plants also showed a higher tiller number and 100-seed weight as compared to non-transgenic plants. The transgenic plants also showed lower oxidative damage as indicated by higher total chlorophyll content as compared to WT plants. More recently, the functional roles of MDHAR were also assessed by Kim et al. (2017) by producing an *OsMDHAR3* suppression line. Transgenic *(mdhar3)* rice plants were more sensitive to salt stress than the wild-type (WT) plants. In addition, the *mdhar3* rice plants showed decreased ability for environmental adaptation because of an imbalance in redox homeostasis and reduced AsA pool. The transgenic line also showed higher hydroperoxide level and decreased chlorophyll content and AsA/DHA ratio under paddy field conditions; they also exhibited a reduction in the total biomass and grain yield. These results suggest that *OsMDHAR3* plays a critical role in intrinsic resistance, as well as in the sensitivity of seed maturation and productivity, of plants to environmental stresses, thereby indicating the functional importance of MDHAR gene in rice plants.

Transgenic Plants Over-expressing DHAR Gene and Abiotic Stress Tolerance in Rice

Oxidation of AsA by APX leads to the formation of a short-lived MDHA radical, which is converted to AsA by MDHAR or disproportionates non-enzymatically to AsA and DHA. DHA is recycled to AsA by DHAR, which requires GSH as the reductant (Chen et al. 2003). The functions of DHAR in regulating AsA/DHA and GSH/GSSG ratios in cell signaling have been well documented in plants. Kim et al. (2013) showed that transgenic lines over-expressing OsDHAR1 showed higher antioxidant enzyme activities, which led to an improved AsA pool and redox homeostasis and higher phostosynthetic capacity under paddy field conations as compared to WT plants. Transgenic lines also showed higher biomass and yield due to enhanced panicle and spikelet numbers. Later on, Kim et al. (2014) also showed that transgenic rice plants over-expressing *OsDHAR1* greatly increased DHAR activity and the AsA/DHA ratio, following increase in AsA content and decrease in DHA content. The activity of antioxidant enzymes, such as APX, MDHPR, and GR, was also enhanced under control and salt-stress conditions in transgenic plants. The transgenic plants also maintained higher redox states by reducing the ROS and MDA levels under salt and methyl viologen (MV) stress conditions, which led to better plant growth, ion leakage, and quantum yield (Fv/Fm). They concluded that over-expression of *OsDHAR1* increases the adaptation of rice plants to salt stress by maintaining the AsA pool, ion homoeostasis, and redox homoeostasis, and plays an important role in attenuating the deleterious effects of various abiotic stresses.

Transgenic Plants Over-expressing the GR Gene and Abiotic Stress Tolerance in Rice

GR is a flavoprotein oxidoreductase that catalyzes the reduction of GSSG to GSH with the accompanying oxidation of NADPH, which plays a pivotal role in maintaining the cellular redox balance of GSH/GSSG. Recently, GR was found to play a positive role in tolerance to abiotic stress. Studies of transgenic plants have shown that GR plays important role in providing resistance to oxidative stress caused by different environmental cues. Transgenic tobacco over-expressing chloroplastic or cytosolic GR showed enhanced tolerance to photo-oxidative stress caused by paraquat, the air pollutant sulfur dioxide, ozone, MV, salt, and cold stresses (Aono et al. 1991; Broadbent et al. 1995; Le Martret et al. 2011). Kouril et al. (2003) reported that the genetically engineered rice plants showed improved protection against photo-bleaching of chlorophyll and photo-oxidative action of MV in thylakoid membranes at 25 °C.

Transgenic Plants Over-expressing GSH Biosynthetic Genes and Abiotic Stress Tolerance in Rice

An important component of the AsA-GSH pathway, GSH, is used to detoxify ROS in plant cells and is mainly produced by two enzymes: gamma-glutamylcysteine synthetase (γ-ECS) and glutathione synthetase (GS). Due to its multiple roles in plant growth, development, and stress tolerance, GSH is now a hot research topic. Choe et al. (2013) showed that transgenic rice plants over-expressing glutathione biosynthetic gene *γ-ECS* (gamma-glutamylcysteine synthetase) displaying higher salinity stress tolerance at various phases of plant growth. Transgenic plants grown under salt stress (100 mM NaCl) at the vegetative-stage of growth showed lower oxidative damage as indicated by bright green phenotypic appearance as compared to WT plants. Transgenic lines also showed higher oxidative damage as indicated by lower ion leakage and higher chlorophyll-fluorescence in response to MV-induced oxidative stress. Transgenic plants showed enhanced redox homeostasis and higher GSH/GSSG ratio and higher germination rate under salt-stress conditions (200 mM NaCl). Transgenic plants also showed higher grain yield, 1000 grain weight, and improved biomass under paddy field conditions. Finally, they concluded that over-expression of GSH biosynthetic gene could be a potential option for use in developing abiotic stress tolerance in plants. Similarly, Bae et al. (2013) also showed that the genetically engineered rice plants over-expressing the *Brassica juncea* L. *ECS* (*BrECS*) gene showed higher adaptation to abiotic stresses. Transgenic rice plants showed higher oxidative stress tolerance induced by MV and salt stress as indicated by lower ion leakage and higher chlorophyll-fluorescence than WT rice plants. Transgenic plants showed a moderate increase in biomass and grain yield under general paddy field conditions as compared to WT plants. They concluded that transgenic plants showed better cellular defense against ROS attacks produced by MV and salt stress by promoting the metabolic process involved in environmental stress tolerance and showed better growth and grain yield. Recently, Park et al. (2017) showed that transgenic plants over-expressing glutathione synthetase gene (*OsGs*) showed better plant growth yield as compared to WT plants. Transgenic plants grown under natural paddy field conditions showed enhanced plant growth, higher leaf chlorophyll content, and higher GSH/GSSG ratio as compared to WT plants. Transgenic rice plants showed better oxidative stress tolerance induced by MV.

Transgenic rice plants also showed an increase in panicle number, number of spikelets per panicle, and total biomass as compared to WT plants. They conclude that that *OsGS* over-expression improved redox homeostasis by enhancing the glutathione pool, which resulted in greater tolerance to environmental stresses in the paddy fields.

Transgenic Plants Over-expressing AsA Biosynthetic Genes and Abiotic Stress Tolerance in Rice

Ascorbic acid is one of the most abundant and powerful antioxidants in plant cells. It is an integral weapon in the defense against ROS generated by various abiotic and biotic stresses in different subcellular organelles and the apoplast. AsA has the ability to donate electrons in a number of cellular redox reactions and serves as a major cellular redox regulatory antioxidant (Smirnoff 2000). It is the substrate of APX, which is a critical component of the AsA-GSH cycle for H_2O_2 detoxification. A wealth of evidence suggests that stress-resistant plants are up-regulated or maintain higher AsA levels than stress-sensitive plants (Shalata et al. 2001; Mittova et al. 2003). Liu et al. (2011) reported that L-galactono-1,4-lactone dehydrogenase (GLDH) over-expressing or suppressed plants showed an altered plant growth and seed set. In rice, suppression of GLDH expression resulted in a loss of chlorophyll, a lower Rubisco protein content, and a lower rate of CO_2 assimilation, which was accompanied with reduced growth rate (as measured by plant height, root length, leaf weight, and root weight) and seed setting in rice. Importantly, GLDH over-expressed lines maintained high levels of chlorophyll, Rubisco protein, and a higher rate of net photosynthesis, resulting in higher seed set. Increasing the level of GLDH expression correlated with reduced lipid peroxidation. Their work suggests that the level of AsA in transgenic rice for GLDH is associated with plant growth and seed set. Similarly, Zhang et al. (2015) reported that transgenic plants over-expressing GDP-mannose-3′, 5′-epimerase (GME), GDP-L-galactose phosphorylase (GGP) genes of AsA biosynthesis in plants showed higher salt-stress tolerance in rice. Transgenic plants showed higher plant height, root length, and fresh weight as compared to the WT plants. Their results suggest that GGP may be a key rate-limiting step in rice AsA biosynthesis, and the plants with elevated AsA contents demonstrated enhanced tolerance for salt stress. Recently, Qin et al. (2016) showed that the suppression of *OsVTC1-1* expression in rice reduced the grain yield of rice under salt-stress conditions. *OsVTC1-1* RNAi rice plants showed a significantly reduced growth rate and lower level of grain yield as compared to control plants. Exogenous application of AsA restored the salt tolerance by reducing the ROS levels. Their results indicate that the rice GMPase gene *OsVTC1-1* plays a critical role in salt tolerance of rice at both vegetative and reproductive stages through AsA scavenging of excess ROS.

Conclusion and Future Perspectives

Drastic fluctuations in global climate have added new dimension of uncertainty to world rice production. As abiotic stresses are a potential threat to plant growth and survival due to over-accumulation of cytotoxic MG and ROS, it becomes vital for plants to detoxify these metabolites by efficient scavenging mechanisms to perform their signaling function. Plants maintain a dynamic balance between generation and detoxification of MG and ROS by the glyoxalase pathway and antioxidant defense system including the AsA-GSH pathway. Considerable progress has been achieved by using stress-tolerant varieties, through chemical priming as well as through genetic engineering to reveal the mechanistic interaction of these stress-responsive metabolic pathways. Additionally, different genes of these pathways have been isolated, cloned, characterized, and exploited to construct transgenic lines for dissecting the individual roles or the pleiotropic roles of the genes in the individual pathways or other pathways. So far, the focus of such studies has proved that the manipulation of single gene may not be very effective. Therefore, it would be more advantageous to engineer plants with multiple genes of the same or different metabolic pathways to generate robust abiotic stress tolerant rice plants.

References

Abbas, T., Balal, R.M., Shahid, M.A. et al. (2015). Silicon-induced alleviation of NaCl toxicity in okra (*Abelmoschus esculentus*) is associated with enhanced photosynthesis, osmoprotectants and antioxidant metabolism. *Acta Physiol. Plant.* 37 (2): 1–15.

Ahmad, B., Zaid, A., Sadiq, Y. et al. (2019). Role of selective exogenous elicitors in plant responses to abiotic stress tolerance. In: *Plant Abiotic Stress Tolerance* (eds. M. Hasanuzzaman, K.R. Hakeem, K. Nahar, et al.), 273–290. Cham: Springer.

Almeselmani, M., Deshmukh, P.S., Sairam, R.K. et al. (2006). Protective role of antioxidant enzymes under high temperature stress. *Plant Sci.* 171: 382–388.

Alvarez Viveros, M.F., Inostroza-Blancheteau, C., Timmermann, T. et al. (2013). Overexpression of *GlyI* and *GlyII* genes in transgenic tomato (*Solanum lycopersicum* Mill.) plants confers salt tolerance by decreasing oxidative stress. *Mol. Biol. Rep.* 40: 3281–3290.

Alvarez-Gerding, X., Cortés-Bullemore, R., Medina, C. et al. (2015). Improved salinity tolerance in Carrizo Citrange rootstock through overexpression of glyoxalase system genes. *Biomed. Res. Int.* 2015: e827951.

Aono, M., Kubo, A., Saji, H. et al. (1991). Resistance to active oxygen toxicity of transgenic *Nicotiana tabacum* that expresses the gene for glutathione reductase from *Escherichia coli*. *Plant Cell Physiol.* 32 (5): 691–697.

Arif, M., Shehzad, M.A., Bashir, F. et al. (2012). Boron, zinc and microtone effects on growth, chlorophyll contents and yield attributes in rice (*Oryza sativa* L.) cultivar. *Afr. J. Biotechnol.* 11 (48): 10851–10858.

Arif, Y., Sami, F., Siddiqui, H. et al. (2020). Salicylic acid in relation to other phytohormones in plant: a study towards physiology and signal transduction under challenging environment. *Environ. Exp. Bot.* 175: 104040. https://doi.org/10.1016/j.envexpbot.2020.104040.

Bae, M.J., Kim, Y.S., Kim, I.S. et al. (2013). Transgenic rice overexpressing the *Brassica juncea* gamma glutamylcysteine synthetase gene enhances tolerance to abiotic stress and improves grain yield under paddy field conditions. *Mol. Breed.* 31: 931–945.

Banerjee, S., Maity, S., and Chakraborti, A.S. (2016). Methylglyoxal-induced modification causes aggregation of myoglobin. *Spectrochim. Acta A Mol. Biomol. Spectrosc.* 155: 1–10.

Banu, M.N.A., Hoque, M.A., Watanabe-Sugimoto, M. et al. (2010). Proline and glycinebetaine ameliorated NaCl stress via scavenging of hydrogen peroxide and methylglyoxal but not superoxide or nitric oxide in tobacco cultured cells. *Biosci. Biotechnol., Biochem.* 74: 2043–2049.

Bartoli, C.G., Buet, A., Gergoff Grozeff, G. et al. (2017). Ascorbate-glutathione cycle and abiotic stress tolerance in plants. In: *Ascorbic Acid in Plant Growth, Development and Stress Tolerance* (eds. M.A. Hossain, S. Munné-Bosch, D.J. Burritt, et al.), 177–200. Cham: Springer.

Basu, S., Roychoudhury, A., Saha, P.P. et al. (2010). Differential antioxidative responses of indica rice cultivars to drought stress. *Plant Growth Regul.* 60: 51–59.

Batth, R., Jain, M., Kumari, S. et al. (2017). Glyoxalase system: a glutathione-dependent pathway for abiotic stress tolerance in plants. In: *Glutathione in Plant Growth, Development and Stress Tolerance* (eds. M.A. Hossain, M.G. Mostofa, P. Diaz-Vivancos, et al.), 235–263. Cham: Springer.

Bhowal, B., Singla-Pareek, S.L., Sopory, S.K. et al. (2020). From methylglyoxal to pyruvate: a genome-wide study for the identification of glyoxalases and D-lactate dehydrogenases in *Sorghum bicolor*. *BMC Genomics* 21 (1): 145.

Bose, J., Rodrigo-Moreno, A., and Shabala, S. (2014). ROS homeostasis in halophytes in the context of salinity stress tolerance. *J. Exp. Bot.* 65: 1241–1257.

Broadbent, P., Creissen, G.P., Kular, B. et al. (1995). Oxidative stress responses in transgenic tobacco containing altered levels of glutathione reductase activity. *Plant J.* 8 (2): 247–255.

Chaparzadeh, N., D'Amico, M.L., Khavari-Nejad, R.A. et al. (2004). Antioxidative responses of *Calendula officinalis* under salinity conditions. *Plant Physiol. Biochem.* 42: 695–701.

Chen, Z., Young, T.E., Ling, J. et al. (2003). Increasing vitamin C content of plants through enhanced ascorbate recycling. *Proc. Natl. Acad. Sci. U.S.A.* 100: 3525–3530.

Chiang, C.M., Chen, L.F.O., Shih, S.W. et al. (2014). Expression of eggplant ascorbate peroxidase increases the tolerance of transgenic rice plants to flooding stress. *J. Plant Biochem. Biotechnol.* 24: 257–267.

Choe, Y.H., Kim, Y.S., Kim, I.S. et al. (2013). Homologous expression of γ-glutamylcysteine synthetase increases grain yield and tolerance of transgenic rice plants to environmental stresses. *J. Plant Physiol.* 170: 610–618.

Clugston, S.L., Daub, E., and Honek, J.F. (1998). Identification of glyoxalase I sequences in *Brassica oleracea* and *Sporobolus stapfianus*: evidence for gene duplication events. *J. Mol. Evol.* 47: 230–234.

Czarnocka, W. and Karpiński, S. (2018). Friend or foe? Reactive oxygen species production, scavenging and signaling in plant response to environmental stresses. *Free Radic. Biol. Med.* 122: 4–20.

Dong, J., Wu, F., and Zhang, G. (2006). Influence of cadmium on antioxidant capacity and four microelement concentrations in tomato seedlings (*Lycopersicon esculentum*). *Chemosphere* 64: 1659–1666.

Ekmekci, Y., Tanyolac, D., and Ayhan, B. (2008). Effects of cadmium on antioxidant enzyme and photosynthetic activities in leaves of two maize cultivars. *J. Plant Physiol.* 165: 600–611.

El-Esawi, M.A. and Alayafi, A.A. (2019). Overexpression of rice *Rab7* gene improves drought and heat tolerance and increases grain yield in rice (*Oryza sativa* L.). *Genes* 10 (1): 56.

El-Shabrawi, H., Kumar, B., Kaul, T. et al. (2010). Redox homeostasis, antioxidant defense, and methylglyoxal detoxification as markers for salt tolerance in Pokkali rice. *Protoplasma* 245: 85–96.

El-Tayeb, M.A. (2006). Differential response of two *Viciafaba* cultivars to drought: growth, pigments, lipid peroxidation, organic solutes, catalase and peroxidase activity. *Acta Agron. Hung.* 54: 25–37.

Eltayeb, A.E., Kawano, N., Badawi, G.H. et al. (2006). Enhanced tolerance to ozone and drought in transgenic tobacco overexpressing dehydroascorbate reductase in cytosol. *Physiol. Plant.* 127: 57–65.

Eltayeb, A.E., Kawano, N., Badawi, G.H. et al. (2007). Overexpression of monodehydro ascorbate reductase in transgenic tobacco confers enhanced tolerance to ozone, salt and polyethylene glycol stresses. *Planta* 225: 1255–1264.

Espartero, J., Aguayo, I.S., and Pardo, J.M. (1995). Molecular characterization of glyoxalase I from a higher plant; upregulation by stress. *Plant Mol. Biol.* 29: 1223–1233.

FAO (2019). The state of the world's biodiversity for food and agriculture: a call to action? *Environ. Policy Law* 49: 110–112.

Feng, D., Wang, Y., Wu, J. et al. (2017). Development and drought tolerance assay of marker-free transgenic rice with *OsAPX2* using biolistic particle-mediated co-transformation. *Crop J.* 5: 271–281.

Foyer, C.H. (2018). Reactive oxygen species, oxidative signaling and the regulation of photosynthesis. *Environ. Exp. Bot.* 154: 134–142.

Foyer, C.H. and Noctor, G. (2005). Redox homeostasis and antioxidant signaling: a metabolic interface between stress perception and physiological responses. *Plant Cell* 17: 1866–1875.

Gao, S., Ouyang, C., Wang, S. et al. (2008). Effects of salt stress on growth, antioxidant enzyme and phenylalanine ammonia-lyase activities in *Jatropha curcas* L. seedlings. *Plant Soil Environ.* 54: 374–381.

Ghosh, A., Pareek, A., Sopory, S.K. et al. (2014). A glutathione responsive rice glyoxalaseII, OsGLYII-2, functions in salinity adaptation by maintaining better photosynthesis efficiency and anti-oxidant pool. *Plant J.* 80: 93–105.

Ghosh, A., Kushwaha, H.R., Hasan, M.R. et al. (2016). Presence of unique glyoxalase III proteins in plants indicates the existence of shorter route for methylglyoxal detoxification. *Sci. Rep.* 6: 18358.

Gill, S.S. and Tuteja, N. (2010). Reactive oxygen species and antioxidant machinery in abiotic stress tolerance in crop plants. *Plant Physiol. Biochem.* 48: 909–930.

Gill, S.S., Khan, N.A., Anjum, N.A. et al. (2011). Amelioration of cadmium stress in crop plants by nutrients management: morphological, physiological and biochemical aspects. In: *Plant Nutrition and Abiotic Stress Tolerance III*, Plant Stress 5 (Special Issue 1) (eds. N.A. Anjum and F. Lopez-Lauri), 1–23. Ikenobe, Japan: Global Science Books Ltd.

Gómez, R., Vicino, P., Carrillo, N. et al. (2019). Manipulation of oxidative stress responses as a strategy to generate stress-tolerant crops: from damage to signaling to tolerance. *Crit. Rev. Biotechnol.* 39: 693–708.

Gupta, B.K., Sahoo, K.K., Ghosh, A. et al. (2018). Manipulation of glyoxalase pathway confers tolerance to multiple stresses in rice. *Plant Cell Environ.* 41 (5): 1186–1200.

Hanif, S., Saleem, M.F., Sarwar, M. et al. (2020). Biochemically triggered heat and drought stress tolerance in rice by proline application. *J. Plant Growth Regul.* https://doi.org/10.1007/s00344-020-10095-3.

Hasanuzzaman, M. and Fujita, M. (2011). Selenium pretreatment up-regulates the antioxidant defense and methylglyoxal detoxification system and confers enhanced tolerance to drought stress in rapeseed seedlings. *Biol. Trace Elem. Res.* 143: 1758–1776.

Hasanuzzaman, M., Hossain, M.A., and Fujita, M. (2011). Nitric oxide modulates antioxidant defense and the methylglyoxal detoxification system and reduces salinity-induced damage of wheat seedlings. *Plant Biotechnol. Rep.* 5 (4): 353.

Hasanuzzaman, M., Hossain, M.A., and Fujita, M. (2012). Exogenous selenium pretreatment protects rapeseed seedlings from cadmium-induced oxidative stress by up-regulating antioxidant defense and methylglyoxal detoxification systems. *Biol. Trace Elem. Res.* 149: 248–261.

Hasanuzzaman, M., Bhuyan, M.H.M.B., Anee, T.I. et al. (2019). Regulation of ascorbate-glutathione pathway in mitigating oxidative damage in plants under abiotic stress. *Antioxidants (Basel).* 8 (9): 384.

Hernández, J.A., Jiménez, A., Mullineaux, P. et al. (2000). Tolerance of pea (*Pisum sativum* L.) to long-term salt stress is associated with induction of antioxidant defences. *Plant Cell Environ.* 23: 853–862.

Hernández, M., Fernandez-Garcia, N., Diaz-Vivancos, P. et al. (2010). A different role for hydrogen peroxide and the antioxidative system under short and long salt stress in *Brassica oleracea* roots. *J. Exp. Bot.* 61: 521–535.

Hoque, T.S., Hossain, M.A., Mostofa, M.G. et al. (2016). Methylglyoxal: an emerging signaling molecule in plant abiotic stress responses and tolerance. *Front. Plant Sci.* 7: 1341. https://doi.org/10.3389/fpls.2016.01341.

Hoque, T.S., Burritt, D.J., and Hossain, M.A. (2018). Glyoxalase system and salt stress tolerance in plants. In: *Novel Perspectives for Salinity Responses and Tolerance in Plants: Targeting Sensory, Ion-Transport and Signalling Mechanisms* (eds. V. Kumar, S.H. Wani, P. Suprasanna, et al.), 257–281. Cham: Springer.

Hossain, M.A. and Fujita, M. (2009). Purification of glyoxalase I from onion bulbs and molecular cloningof its cDNA. *Biosci. Biotechnol., Biochem.* 73: 2007–2013.

Hossain, M.A. and Fujita, M. (2010). Evidence for a role of exogenous glycinebetaine and proline in antioxidant defense and methylglyoxal detoxification systems in mung bean seedlings under salt stress. *Physiol. Mol. Biol. Plants* 16: 19–29.

Hossain, M.A. and Fujita, M. (2012). Regulatory role of components of ascorbate-glutathione (AsA-GSH) pathway in plant tolerance to oxidative stress. In: *Oxidative Stress in Plants: Causes, Consequences and Tolerance* (eds. N.A. Anjum, S. Umar and A. Ahmad), 81–147. New Delhi: IK International Publishing House Pvt. Ltd.

Hossain, M.A., Hossain, M.Z., and Fujita, M. (2009). Stress-induced changes of methylglyoxal level and glyoxalase I activity in pumpkin seedlings and cDNA cloning of glyoxalase I gene. *Aust. J. Crop Sci.* 3: 53–64.

Hossain, M.A., Hasanuzzaman, M., and Fujita, M. (2010). Up-regulation of antioxidant and glyoxalase systems by exogenous glycinebetaine and proline in mung bean confer tolerance to cadmium stress. *Physiol. Mol. Biol. Plants* 16: 259–272.

Hossain, M.A., Hasanuzzaman, M., and Fujita, M. (2011a). Coordinate induction of antioxidant defense and glyoxalase system by exogenous proline and glycinebetaine is correlated with salt tolerance in mung bean. *Front. Agric. China* 5 (1): 1–14.

Hossain, M.A., Teixeira da Silva, J.A., and Fujita, M. (2011b). Glyoxalase system and reactive oxygen species detoxification system in plant abiotic stress response and tolerance: an intimate relationship. In: *Abiotic Stress in Plants—Mechanisms and Adaptations* (eds. A. Shanker and B. Venkateswarlu), 235–266. Rijeka, Croatia: INTECH-Open Access Publisher.

Hossain, M.A., Bhattacharjee, S., and Armin, S.M. (2015). Hydrogen peroxide-priming modulates abiotic oxidative stress tolerance: insights from ROS detoxification and scavenging. *Front. Plant Sci.* 6: 420.

Hu, L., Liang, W., Yin, C. et al. (2011). Rice MADS3 regulates ROS homeostasis during late anther development. *Plant Cell* 23: 515–533.

IRRI (1997). *Almanac*, 2e. Manila, Philippines: International Rice Research Institute , Los banos, Laguna, Philippines.

Jain, M., Batth, R., Kumari, S. et al. (2016). *Arabidopsis thaliana* contains both Ni^{2+} and Zn^{2+} dependent glyoxalase I enzymes and ectopic expression of the latter contributes more towards abiotic stress tolerance in *E. coli*. *PLoS One* 11: e0159348.

Ji, J., Yuan, D., Jin, C. et al. (2020). Enhancement of growth and salt tolerance of rice seedlings (Oryza sativa L.) by regulating ethylene production with a novel halotolerant PGPR strain Glutamicibacter sp. YD01 containing ACC deaminase activity. *Acta Physiol. Plant.* 42: 1–17.

Kalapos, M.P. (2008). The tandem of free radicals and methylglyoxal. *Chem. Biol. Interact.* 171: 251–271.

Kaur, C., Ghosh, A., Pareek, A. et al. (2014a). Glyoxalases and stress tolerance in plants. *Biochem. Soc. Trans.* 42: 485–490.

Kaur, C., Singla-Pareek, S.L., and Sopory, S.K. (2014b). Glyoxalase and methylglyoxal as biomarkers for plant stress tolerance. *Crit. Rev. Plant Sci.* 33: 429–456.

Kaur, C., Sharma, S., Singla-Pareek, S.L. et al. (2015). Methylglyoxal, triose phosphate isomerase, and glyoxalase pathway: implications in abiotic stress and signaling in plants. In: *Elucidation of Abiotic Stress Signaling in Plants* (ed. G.K. Pandey), 347–366. New York: Springer.

Khush, G.S. (2005). What it will take to feed 5.0 billion rice consumers in 2030. *Plant Mol. Biol.* 59: 1–6.

Kim, Y.S., Kim, I.S., Bae, M.J. et al. (2013). Homologous expression of cytosolic dehydroascorbate reductase increases grain yield and biomass under paddy field conditions in transgenic rice (*Oryza sativa* L. japonica). *Planta* 237: 1613–1625.

Kim, Y.S., Kim, I.S., Shin, S.Y. et al. (2014). Overexpression of dehydroascorbate reductase confers enhanced tolerance to salt stress in rice plants (*Oryza sativa* L. japonica). *J. Agron. Crop Sci.* 200: 444–456.

Kim, J.J., Kim, Y.S., Park, S.I. et al. (2017). Cytosolic monodehydroascorbate reductase gene affects stress adaptation and grain yield under paddy field conditions in *Oryza sativa* L. *japonica*. *Mol. Breed.* 37: 118.

Kiran, S., Kusvuran, S., Talhouni, M. et al. (2014). The studies on some biochemical changes and ion regulation in the tomato genotypes exposed to drought stress. Sixth Balkan Symposium on Vegetables and Potatoes. Zagreb, Croatia. 29 September–2 October 2014, 126.

Kouril, R., Lazar, D., Lee, H. et al. (2003). Moderately elevated temperature eliminates resistance of rice plants with enhanced expression of glutathione reductase to intensive photooxidative stress. *Photosynthetica* 41: 571–578.

Kromdijk, J. and Long, S.P. (2016). One crop breeding cycle from starvation? How engineering crop photosynthesis for rising CO_2 and temperature could be one important route to alleviation. *Proc. Biol. Sci.* 283: 20152578.

Kusvuran, S., Ellialtioglu, S., Yasar, F. et al. (2007). Effects of salt stress on ion accumulations and some of the antioxidant enzymes activities in melon (*Cucumis melo* L.). *J. Food Agric. Environ.* 2 (5): 351–354.

Kusvuran, S., Ellialtioglu, S., Yasar, F. et al. (2012). Antioxidative enzyme activities in the leaves and callus tissues of salt-tolerant and salt-susceptible melon varieties under salinity. *Afr. J. Biotechnol.* 11 (3): 635–641.

Kwon, K., Choi, D., Hyun, J.K. et al. (2013). Novel glyoxalases from *Arabidopsis thaliana*. *FEBS J.* 280 (14): 3328–3339.

Le Martret, B., Poage, M., Shiel, K. et al. (2011). Tobacco chloroplast transformants expressing genes encoding dehydroascorbate reductase, glutathione reductase, and glutathione-S-transferase, exhibit altered anti-oxidant metabolism and improved abiotic stress tolerance. *Plant Biotechnol. J.* 9 (6): 661–673.

Lee, D.H. and Lee, C.B. (2000). Chilling stress-induced changes of antioxidant enzymes in the leaves of cucumber: in gel enzyme activity assays. *Plant Sci.* 159 (1): 75–85.

Lee, J.Y., Song, J., Kwon, K. et al. (2012). Human DJ-1 and its homologs are novel glyoxalases. *Hum. Mol. Genet.* 21 (14): 3215–3225.

Li, Z.G. (2016). Methylglyoxal and glyoxalase system in plants: old players, new concepts. *Bot. Rev.* 82: 183–203.

Li, Z.G., Nie, Q., Yang, C.L. et al. (2018). Signaling molecule methylglyoxal ameliorates cadmium injury in wheat (*Triticum aestivum* L.) by a coordinated induction of glutathione pool and glyoxalase system. *Ecotoxicol. Environ. Saf.* 149: 101–107.

Lin, F., Xu, J., Shi, J. et al. (2010). Molecular cloning and characterization of a novel glyoxalaseI gene TaGly I in wheat (*Triticum aestivum* L.). *Mol. Biol. Rep.* 37 (2): 729–735.

Liu, Y., Yu, L., and Wang, R. (2011). Level of ascorbic acid in transgenic rice for L-galactono-1, 4-lactone dehydrogenase overexpressing or suppressed is associated with plant growth and seed set. *Acta Physiol. Plant.* 33: 1353–1363.

Majláth, I., Éva, C., Tajti, J. et al. (2020). Exogenous methylglyoxal enhances the reactive aldehyde detoxification capability and frost-hardiness of wheat. *Plant Physiol. Biochem.* 149: 75–85.

Martínez, J.P. and Araya, H. (2010). Ascorbate–glutathione cycle: enzymatic and non-enzymatic integratedmechanisms and its biomolecular regulation. In: *Ascorbate Glutathione Pathway and Stress Tolerance in Plants* (eds. N.A. Anjum, M.T. Chan and S. Umar), 303–322. Dordrecht: Springer.

Mhamdi, A. and Van Breusegem, F. (2018). Reactive oxygen species in plant development. *Development* 145 (15): dev164376.

Miller, G., Suzuki, N., Ciftci-Yilmaz, S. et al. (2010). Reactive oxygen species homeostasis and signalling during drought and salinity stresses. *Plant Cell Environ.* 33: 453–467.

Mittler, R. (2006). Abiotic stress, the field environment and stress combination. *Trends Plant Sci.* 11 (1): 15–19.

Mittler, R., Vanderauwera, S., Gollery, M. et al. (2004). Reactive oxygen gene network of plants. *Trends Plant Sci.* 9: 490–498.

Mittova, V., Tal, M., Volokita, M. et al. (2003). Up-regulation of the leaf mitochondrial and peroxisomal antioxidative systems in response to salt-induced oxidative stress in the wild salt-tolerant tomato species *Lycopersicon pennellii*. *Plant Cell Environ.* 26: 845–856.

Mittova, V., Guy, M., Tal, M. et al. (2004). Salinity up-regulates the antioxidative system in root mitochondria and peroxisomes of the wild salt-tolerant tomato species *Lycopersicon pennellii*. *J. Exp. Bot.* 55: 1105–1113.

Moller, I.M., Jensen, P.E., and Hansson, A. (2007). Oxidative modifications to cellular components in plants. *Annu. Rev. Plant Biol.* 58: 459–481.

Mostofa, M.G. and Fujita, M. (2013). Salicylic acid alleviates copper toxicity in rice (*Oryza sativa* L.) seedlings by up-regulating antioxidative and glyoxalase systems. *Ecotoxicology* 22 (6): 959–973.

Mostofa, M.G., Yoshida, N., and Fujita, M. (2014a). Spermidine pretreatment enhances heat tolerance in rice seedlings through modulating antioxidative and glyoxalase systems. *Plant Growth Regul.* 73 (1): 31–44.

Mostofa, M.G., Seraj, Z.I., and Fujita, M. (2014b). Exogenous sodium nitroprusside and glutathione alleviate copper toxicity by reducing copper uptake and oxidative damage in rice (*Oryza sativa* L.) seedlings. *Protoplasma* 251: 1373–1386.

Mostofa, M.G., Rahman, A., Ansary, M.M. et al. (2015a). Hydrogen sulfide modulates cadmium-induced physiological and biochemical responses to alleviate cadmium toxicity in rice. *Sci. Rep.* 5: 14078.

Mostofa, M.G., Hossain, M.A., Fujita, M. et al. (2015b). Physiological and biochemical mechanisms associated with trehalose-induced copper-stress tolerance in rice. *Sci. Rep.* 5: 11433.

Mostofa, M.G., Fujita, M., and Tran, L.S.P. (2015c). Nitric oxide mediates hydrogen peroxide-and salicylic acid-induced salt tolerance in rice (*Oryza sativa* L.) seedlings. *Plant Growth Regul.* 77 (3): 265–277.

Mostofa, M.G., Saegusa, D., Fujita, M. et al. (2015d). Hydrogen sulfide regulates salt tolerance in rice by maintaining Na^+/K^+ balance, mineral homeostasis and oxidative metabolism under excessive salt stress. *Front. Plant Sci.* 6: 1055.

Mostofa, M.G., Hossain, M.A., and Fujita, M. (2015e). Trehalose pretreatment induces salt tolerance in rice (*Oryza sativa* L.) seedlings: oxidative damage and co-induction of antioxidant defense and glyoxalase systems. *Protoplasma* 252 (2): 461–475.

Mostofa, M.G., Ghosh, A., Li, Z.G. et al. (2018). Methylglyoxal–a signaling molecule in plant abiotic stress responses. *Free Radic. Biol. Med.* 122: 96–109.

Mustafiz, A., Singh, A.K., Pareek, A. et al. (2011). Genome-wide analysis of rice and *Arabidopsis* identifies two glyoxalase genes that are highly expressed in abiotic stresses. *Funct. Integr. Genomics* 11: 293–305.

Mustafiz, A., Ghosh, A., Tripathi, A.K. et al. (2014). A unique Ni2C-dependent and methylglyoxal-inducible rice glyoxalase I possesses a single active site and functions in abiotic stress response. *Plant J.* 78: 951–963.

Nahar, K., Hasanuzzaman, M., Alam, M.M. et al. (2015). Exogenous glutathione induced drought stress tolerance in mung bean (*Vigna radiata* L.) seedlings: coordinated roles of the antioxidant defense and methylglyoxal detoxification systems. *AoB Plants* 7: plv069.

Nazir, F., Fariduddin, Q., and Khan, T.A. (2020). Hydrogen peroxide as a signalling molecule inplants and its crosstalk with other plant growth regulators under heavy metal stress. *Chemosphere* 252: 126486.

Neill, S., Desikan, R., and Hancock, J. (2002). Hydrogen peroxide signalling. *Curr. Opin. Plant Biol.* 5: 388–395.

Norton, S.J., Talesa, V., Yuan, W.J. et al. (1990). Glyoxalase I and glyoxalase II from *Aloe vera*: purification, characterization and comparison with animal glyoxalases. *Biochem. Int.* 22 (3): 411–418.

Ojeda, A.G., Escobosa, A.R.C., Wrobel, K. et al. (2013). Effect of Cd(II) and Se(IV) exposure on cellular distribution of both elements and concentration levels of glyoxal and methylglyoxal in *Lepidium sativum*. *Metallomics* 5 (9): 1254–1261.

Pandey, S., Fartyal, D., Agarwal, A. et al. (2017). Abiotic stress tolerance in plants: myriad roles of ascorbate peroxidase. *Front Plant Sci* 8: 581.

Pandey, P., Singh, J., and Achary, V.M.M. (2015). Redox homeostasis via gene families of ascorbate-glutathione pathway. *Front. Environ. Sci.* 3: 25.

Park, S.I., Kim, Y.S., Kim, J.J. et al. (2017). Improved stress tolerance and productivity in transgenic rice plants constitutively expressing the *Oryza sativa* glutathione synthetase OsGS under paddy field conditions. *J. Plant Physiol.* 215: 39–47.

Qin, H., Wang, Y., Wang, J. et al. (2016). Knocking down the expression of GMPase gene *OsVTC1-1* decreases salt tolerance of rice at seedling and reproductive stages. *PLoS One* 11 (12): e0168650.

Rabbani, N. and Thornalley, P.J. (2011). Glyoxalase in diabetes, obesity and related disorders. *Semin. Cell Dev. Biol.* 22: 309–317.

Rabbani, N. and Thornalley, P.J. (2012). Methylglyoxal, glyoxalase I and dicarbonyl proteome. *Amino Acids* 42: 1133–1142.

Rahman, A., Mostofa, M.G., Nahar, K. et al. (2015). Calcium mitigates arsenic toxicity in rice seedlings by reducing arsenic uptake and modulating the antioxidant defense and glyoxalase systems and stress markers. *Biomed. Res. Int.* 2015: 340812.

Rahman, M.G., Mostofa, K., Nahar, M. et al. (2016a). Exogenous calcium alleviates cadmium-induced oxidative stress in rice (*Oryza sativa* L.) seedlings by regulating the antioxidant defense and glyoxalase systems. *Braz. J. Bot.* 39 (2): 393–407.

Rahman, A., Nahar, K., Hasanuzzaman, M. et al. (2016b). Calcium supplementation improves Na^+/K^+ ratio, antioxidant defense and glyoxalase systems in salt-stressed rice seedlings. *Front. Plant Sci.* 7: 609.

Rahman, A., Hossain, M.S., Mahmud, J.A. et al. (2016c). Manganese-induced salt stress tolerance in rice seedlings: regulation of ion homeostasis, antioxidant defense and glyoxalase systems. *Physiol. Mol. Biol. Plants* 22: 291–306.

Ridderstrom, M. and Mannervik, B. (1997). Molecular cloning and characterization of the thiol esterase glyoxalase II from *Arabidopsis thaliana*. *Biochem. J.* 322: 449–454.

Sairam, R.K., Dharmar, K., and Lekshmy, S. (2011). Expression of antioxidant defense genes in mung bean (*Vigna radiata* L.) roots under water-logging is associated with hypoxia tolerance. *Acta Physiol. Plant.* 33: 735–744.

Sato, Y., Masuta, Y., Saito, K. et al. (2011). Enhanced chilling tolerance at the booting stage in rice by transgenic overexpression of the ascorbate peroxidase gene, OsAPXa. *Plant Cell Rep.* 30: 399–406.

Saxena, M., Roy, S.D., Singla-Pareek, S.L. et al. (2011). Overexpression of the glyoxalase II gene leads to enhanced salinity tolerance in *Brassica juncea*. *Open Plant Sci. J.* 5: 23–28.

Sevengor, S., Yasar, F., Kusvuran, S. et al. (2011). The effect of salt stress on growth, chlorophyll content, lipid peroxidation and antioxidative enzymes of pumpkin seed-ling. *Afr. J. Agric. Res.* 6 (21): 4920–4924.

Shalata, A., Mittova, V., Volokita, M. et al. (2001). Response of the cultivated tomato and its wild salt-tolerant relative *Lycopersicon pennellii* to salt-dependent oxidative stress: the root antioxidative system. *Physiol. Plant.* 112: 487–494.

Shigeoka, S., Ishikawa, T., Tamoi, M. et al. (2002). Regulation and function of ascorbate peroxidase isoenzymes. *J. Exp. Bot.* 53: 1305–1319.

Singh, P., Singh, I., and Shah, K. (2020). Alterations in antioxidative machinery and growth parameters upon application of nitric oxide donor that reduces detrimental effects of cadmium in rice seedlings with increasing days of growth. *S. Afr. J. Bot.* 131: 283–294.

Singla-Pareek, S.L., Reddy, M.K., and Sopory, S.K. (2003). Genetic engineering of the glyoxalase pathway in tobacco leads to enhanced salinity tolerance. *Proc. Natl. Acad. Sci. U.S.A.* 100: 14672–14677.

Singla-Pareek, S.L., Yadav, S.K., Pareek, A. et al. (2006). Transgenic tobacco overexpressing glyoxalase pathway enzymes grow and set viable seeds in zinc-spiked soils. *Plant Physiol.* 140: 613–623.

Singla-Pareek, S.L., Yadav, S.K., Pareek, A. et al. (2008). Enhancing salt tolerance in a crop plant by overexpression of glyoxalase II. *Transgenic Res.* 17: 171–180.

Singla-Pareek, S.L., Kaur, C., Kumar, B. et al. (2020). Reassessing plant glyoxalases: large family and expanding functions. *New Phytol.* 227: 714–721.

Smirnoff, N. (2000). Ascorbic acid: metabolism and functions of a multi-facetted molecule. *Curr. Opin. Plant Biol.* 3: 229–235.

Sohag, A.A.M., Tahjib-Ul-Arif, M., Brestic, M. et al. (2020). Exogenous salicylic acid and hydrogen peroxide attenuate drought stress in rice. *Plant Soil Environ.* 66: 7–13.

Sultana, S., Khew, C.Y., Morshed, M.M. et al. (2012). Over-expression of monodehydroascorbate reductase from a mangrove plant (*AeMDHAR*) confers salt tolerance on rice. *J. Plant Physiol.* 169: 311–318.

Szent-Georgyi, A. (1965). Cell division and cancer. *Science* 149: 34–37.

Talesa, V., Rosi, G., Contenti, S. et al. (1990). Presence of glyoxalase II in mitochondria from spinach leaves: comparison with the enzyme from cytosol. *Biochem. Int.* 22: 1115–1120.

Tanyolac, D., Ekmekci, Y., and Unalan, S. (2007). Changes in photochemical and antioxidant en-zyme activities in maize (*Zea mays* L.) leaves exposed to excess copper. *Chemosphere* 67: 89–98.

Thornalley, P.J. (1990). The glyoxalase system: new developments towards functional characterization of a metabolic pathway fundamental to biological life. *Biochem. J.* 269: 1–11.

Thornalley, P.J. (1993). The glyoxalase system in health and disease. *Mol. Aspects Med.* 14: 287–371.

Ueda, Y., Uehara, N., Sasaki, H. et al. (2013). Impacts of acute ozone stress on superoxide dismutase (SOD) expression and reactive oxygen species (ROS) formation in rice leaves. *Plant Physiol. Biochem.* 70: 396–402.

Upadhyaya, C.P., Venkatesh, J., Gururani, M.A. et al. (2011). Transgenic potato overproducing L-ascorbic acid resisted an increase in methylglyoxal under salinity stress via maintaining higher reduced glutathione level and glyoxalase enzyme activity. *Biotechnol. Lett.* 33 (11): 2297–2307.

Usui, Y., Nakase, M., Hotta, H. et al. (2001). A 33-kDa allergen from rice (*Oryza sativa* L. *Japonica*) cDNA cloning, expression, and identification as a novel glyoxalase I. *J. Biol. Chem.* 276: 11376–11381.

Vardharajula, S., Ali, S.Z., Grover, M. et al. (2011). Drought-tolerant plant growth promoting *Bacillus* spp. effect on growth, osmolytes, and antioxidant status of maize under drought stress. *J. Plant Interact.* 6: 1–14.

Veena, Reddy, V.S., and Sopory, S.K. (1999). Glyoxalase I from *Brassica juncea*: molecular cloning, regulation and its over-expression confer tolerance in transgenic tobacco under stress. *Plant J.* 17: 385–395.

Verma, M., Verma, D., and Jain, R.K. (2005). Overexpression of glyoxalase I gene confers salinity tolerance in transgenic *japonica* and *indica* rice plants. *Rice Genet. Newslett.* 22: 58–62.

Vinocur, B. and Altman, A. (2005). Recent advances in engineering plant tolerance to abiotic stress: achievements and limitations. *Curr. Opin. Biotechnol.* 16 (2): 123–132.

Von Pechmann, H. (1887). Über die Spaltung der Nitrosoketone. *Ber. Dtsch. Chem. Gesells.* 20: 3213–3214.

Wang, X., Liu, H., Yu, F. et al. (2019). Differential activity of the antioxidant defence system and alterations in the accumulation of osmolyte and reactive oxygen species under drought stress and recovery in rice (*Oryza sativa* L.) tillering. *Sci. Rep.* 9: 8543.

Wani, S.H. and Gosal, S.S. (2011). Introduction of OsglyII gene into *Oryza sativa* for increasing salinity tolerance. *Biol. Plant.* 55: 536–540.

Wani, W., Masoodi, K.Z., Zaid, A. et al. (2018a). Engineering plants for heavy metal stress tolerance. *Rend. Lincei. Sci. Fis.* 29 (3): 709–723.

Wani, S.H., Tripathi, P., Zaid, A. et al. (2018b). Transcriptional regulation of osmotic stress tolerance in wheat (*Triticum aestivum* L.). *Plant Mol. Biol.* 97 (6): 469–487.

Waszczak, C., Kerchev, P.I., Mühlenbock, P. et al. (2016). SHORT-ROOT deficiency alleviates the cell death phenotype of the Arabidopsis catalase2 mutant under photorespiration-promoting conditions. *Plant Cell* 28 (8): 1844–1859.

Wu, X., Zhu, Z., and Li, X. (2012). Effects of cytokinin on photosynthetic gas exchange, chlorophyll fluorescence parameters and antioxidative system in seedlings of eggplant (*Solanum melongena* L.) under salinity stress. *Acta Physiol. Plant.* 34: 2105–2114.

Xiong, H., Yu, J., Miao, J. et al. (2018). Natural variation in OsLG3 increases drought tolerance in rice by inducing ROS scavenging. *Plant Physiol.* 178: 451–467.

Xu, Y., Wang, R., Wang, Y. et al. (2020). A point mutation in *LTT1* enhances cold tolerance at the booting stage in rice. *Plant Cell Environ.* 43: 992–1007.

Xue, M., Rabbani, N., and Thornalley, P. (2011). Glyoxalase in ageing. *Semin. Cell Dev. Biol.* 22: 293–301.

Yadav, S.K., Singla-Pareek, S.L., Ray, M. et al. (2005a). Methylglyoxal levels in plants under salinity stress are dependent on glyoxalase I and glutathione. *Biochem. Biophys. Res. Commun.* 337: 61–67.

Yadav, S.K., Singla-Pareek, S.L., Ray, M. et al. (2005b). Transgenic tobacco plants overexpressing glyoxalase enzymes resist an increase in methylglyoxal and maintain higher reduced glutathione levels under salinity stress. *FEBS Lett.* 579: 6265–6271.

Yadav, S.K., Singla-Pareek, S.L., Kumar, M. et al. (2007). Characterization and functional validation of glyoxalase II from rice. *Protein Expr Purif.* 51: 126–132.

Yadav, S.K., Singla-Pareek, S.L., and Sopory, S.K. (2008). An overview on the role of methylglyoxal and glyoxalases in plants. *Drug Metabol. Drug Interact.* 23: 51–68.

Yamauchi, T., Yoshioka, M., Fukazawa, A. et al. (2017). An NADPH oxidase RBOH functions in rice roots during lysigenous aerenchyma formation under oxygen-deficient conditions. *Plant Cell* 29: 775–790.

Zaid, A. and Wani, S.H. (2019). Reactive oxygen species generation, scavenging and signaling in plant defense responses. In: *Bioactive Molecules in Plant Defense* (eds. S. Jogaiah and M. Abdelrahman), 111–132. Cham: Springer.

Zaid, A., Mohammad, F., Wani, S.H. et al. (2019). Salicylic acid enhances nickel stress tolerance by up-regulating antioxidant defense and glyoxalase systems in mustard plants. *Ecotoxicol. Environ. Saf.* 180: 575–587.

Zeng, Z., Xiong, F., Yu, X. et al. (2016). Overexpression of a glyoxalase gene, *OsGly I*, improves abiotic stress tolerance and grain yield in rice (*Oryza sativa* L.). *Plant Physiol. Biochem.* 109: 62–71.

Zeng, J., Dong, Z., Wu, H. et al. (2017). Redox regulation of plant stem cell fate. *EMBO J.* 36 (19): 2844–2855.

Zhang, Z., Zhang, Q., Wu, J. et al. (2013). Gene knockout study reveals that cytosolic ascorbate peroxidase 2 (OsAPX2) plays a critical role in growth and reproduction in rice under drought, salt and cold stresses. *PLoS One* 8 (2): e57472.

Zhang, G.Y., Liu, R.R., Zhang, C.Q. et al. (2015). Manipulation of the rice L-galactose pathway: evaluation of the effects of transgene overexpression on ascorbate accumulation and abiotic stress tolerance. *PLoS One* 10 (5): e0125870.

Zurbriggen, M.D., Hajirezaei, M.R., and Carrillo, N. (2010). Engineering the future. Development of transgenic plants with enhanced tolerance to adverse environments. *Biotechnol. Genet. Eng.* 27 (1): 33–56.

Index

a

abiotic stresses 2
abiotic stress-tolerance in rice
  AsA-GSH pathway
    APX gene 413–414
    AsA biosynthetic genes 416–417
    DHAR gene 414–415
    GR gene 415
    GSH biosynthetic genes 415–416
    MDHAR gene 414
    modulating 404–406
  future perspectives 417
  glyoxalase pathway 406–410
    antioxidant pathway 406–410
    glyI gene 410–412
    glyII gene 411, 412
    glyIII gene 412
    modulating 403–404
  methylglyoxal 401–403
  reactive oxygen species 399–401
abscisic acid (ABA) 186, 252–253
  signaling pathways 29
2-acetyl pyrrolidine 349–350
adenosine triphosphate (ATP) synthesis 253
ADP-glucose pyrophosphorylase (AGPase) 94
aerobic rice, water saving
  greenhouse gas emissions 387
  insect pests constraints 392
  marker-assisted selection 388–390
  model cereal crop 382
  nematode problems in 391–392
  problems in 393
  rice facts and figure 383–385
  root morphology 387–388
  suitable alleles 387
  water productivity 385–386
  weed problems in 390–391
agglomeration 1
Agrobacterium-mediated transformation 290
agronomical-superior genotypes 79
agronomic Fe biofortification approach
  description 320
  foliar application of Fe fertilizers 320
  limitations 323
alleles
  aerobic rice, water saving 387
  mining 35–36
1-aminocyclopropane-1-carboxylic acid (ACC) 222
anaerobic germination1 (AG1) 207
anaerobic germination potential (AGP) 80
anthocyanins composition, antioxidant 360–361, 369–370
anthropocene 1
antioxidant content in rice
  biofortification
    carotenoids 370–371
    flavonoids, anthocyanins, proanthocyanidins 369
    folates 372–373
    melatonin 373–374
    tocopherols, tocotrienols 371–372
  biosynthetic pathway in 362
  compounds of
    carotenoids composition 361–365
    flavonoids, anthocyanins, and proanthocyanidins composition of 360–361
    folate composition 366–368
    melatonin composition 368–369
    tocopherols and tocotrienols composition 365–366
  perspectives 374–375
antioxidant pathway, abiotic stress
  AsA-GSH pathway 406–407
  dehydroascorbate reductase 408–409
  malondialdehyde 407
apoplastic Fe transport 316
application gap 86
Arabidopsis thaliana, 187

Molecular Breeding for Rice Abiotic Stress Tolerance and Nutritional Quality, First Edition.
Edited by Mohammad Anwar Hossain, Lutful Hassan, Khandakar Md. Iftekharuddaula, Arvind Kumar and Robert Henry.

arbuscularmycorrhizal fungi (AMF) 141
aromatic rices
  Badh2 gene 344–346
  biochemical basis of 349–350
  conventional breeding 350–351
  future prospects of 353–354
  genetic engineering 352–353
  genetics of 344
  global distribution of 343
  molecular breeding of 351–352
  natural variability in 344–345
  origin and evolution of 346–349
  QTLs 345–347
  trade and geographical indication 353
  varieties list 341–342
arsenic stress tolerance, in rice
  genetic bases of 166–169
  genetic diversity 165–166
  GS 170
  MARS 169
  phytotoxic effects, of arsenic 164–165
  rice defense response 164–165
ascorbate (AsA) biosynthetic genes 416–417
ascorbate-glutathione (AsA-GSH) pathway, abiotic stress
  APX gene 413–414
  AsA biosynthetic genes 416–417
  DHAR gene 414–415
  glyoxalase pathway 403–404
  GR gene 415
  GSH biosynthetic genes 415–416
  MDHAR gene 414
  modulating 404–406
ascorbate oxidase (AO) proteins 191
ascorbate peroxidase (APX) gene 413–414
association mapping (AM) 239–240
atmospheric Co_2 concentration 282–283
ATP-binding cassette (ABC) transporter 141–142

b

bacterial blight (BB) 3
Bayesian regression BayesA 171
β–carotene synthesis
  golden rice
    endosperm-specific expression 289–290
    genes 290, 291
    rice genetic transformation system 290, 292
  in plant 287–289
betain aldehyde dehydragenase (Badh2) gene 344–346, 349
biofortification
  antioxidant genetic engineering
    carotenoids 370–371
    flavonoids, anthocyanins, proanthocyanidins 369
    folates 372–373
    melatonin 373–374
    tocopherols, tocotrienols 371–372
  defined 315–316
  Fe-biofortification in rice (*see* iron (Fe) biofortification)
biotic stresses 2
brown (cargo) grain (CG) 164
brown plant hopper (BPH) 15, 83
bulked segregant analysis (BSA) 40
bundle sheath cells (BSCs) 277

c

Ca^{2+} concentration 124
cadmium (Cd) pollution 133
calcium:cation (CaCA) 142
calcium messenger 125
calmodulin (CAM) 125
candidate genes 35–36
  nitrogen use efficient 240–242
  and pathways imparting tolerance
    cross talk b/w stresses 207–208
    low light perception 203–205
    molecular basis, of delayed senescence 205–207
    molecular targets, for erect lamina 205
    "omics" and mapping studies 208–210
capricious rainfall 11
carbon-concentrating mechanisms (CCMs) 277
carbon fertilization 4, 6
carbon monoxide (CO) 180
carotenoids 287
  antioxidant composition 361, 363–365, 370–371
cation diffusion facilitator (CDF) 142
CBF regulatory pathway 125–126
cell-level compartmentation 27–29
cellular elasticity 10
cellular metabolisms 13
Centre Français du Riz (CFR) 170
Chakhao Poreiton (CP) 226
chelating agents, for Fe transport 319
chelation-based strategy 316
chilling injury 120
chlorophyll florescence phenotyping 39
chromosome segment substitution lines (CSSLs) 33, 190
cleaved amplified polymorphic sequence (CAPS) 56
cloning 34
CO_2 fertilization effect 4
cold susceptibility 13

cold tolerance, in rice
map-based cloning 124
molecular mechanism, of cold-resistant QTLs 124
molecular regulatory networks
CBF regulatory pathway 125–126
MYB regulatory pathway 126
preliminary mapping 121–123
conventional and molecular breeding 82–85
conventional breeding method
aromatic rices 350–351
Fe biofortification 324
salinity tolerance 29–31
convergent breeding approach 83
C3 photosynthesis 279
C4 photosynthesis 278
crassulacean acid metabolism (CAM) 277–278
CRISPR-associated (Cas) nucleases 67
Crispr-Cas9
drought stress tolerance 258
golden rice 298–299
crop species, genetic variability of 275
CRT/DRE regulatory element 125
cryptochromes (CRY) 203
cultivar development 7
cytosol-specific AGPase 95

d
dehydration-responsive elements (DRE) 263
dehydroascorbate reductase (DHAR) 187, 414–415
deposition of ozone and stomatal exchange (DO_3SE) 182
dimethylarsinic acid (DMA) 164
direct seeding 75
DJ123 227
DNA double-strand break (DSB) 66
donor germplasm lines, for Fe biofortification 331
doubled haploid (DH) 32
double-strand break (DSB) 66
drought screening protocols 54
drought stress 54
drought tolerance
antioxidative defenses 255–256
challenges and its distribution 53
climate change and 53
defense mechanisms 251–252
defined 249
effect of 252–254
genetic basis of
development, of NILs and PLs 57–60
fine mapping 57
mapping and deployment, of drought QTLs 57
MAS introgression, of multiple traits 60–61
high-throughput technologies
CRISPR-Cas9 technology 258
expressed sequence tags 258
gene networks 258–259
Micro-RNA technology 259–260
quantitative trait loci 262
RNA-Seq technology 259
targeting induced local lesions in genomes 261–262
virus-induced gene-silencing machinery 260–261
impacts, on rice plant 54
molecular basis of, epistatic interactions 61–64
phenotyping platforms
design and management 54–55
direct selection, for yield under stress 55
drought screening protocols 54
field-based phenotyping 54
trait-based selection, for drought resistance 55
redox homeostatic mechanisms 255–256
ROS
production 254–255
signaling 256–257
tolerant genotype 262–265
drought-tolerant QTL (DT-QTL8) 33
dry-heat treatment 95

e
early-morning flowering (EMF) 96
embryo abortion 10
epigenetics 109
ethyl-methansulfonate (EMS) 31
exogenous abscisic acid (ABA) 108
expressed sequence tags (ESTs) 241, 258

f
ferric-chelate reductase activity 319
ferritin gene, Fe rice grain 325–327
field-based phenotyping 54
fine-mapping pyramiding 34
flag leaf *As* content (FL-*As*) 165
flavonoids composition, antioxidant 360–361, 369–370
flooding 75
folate composition, antioxidant 366–368, 372–373
foliar application, of Fe fertilizers 320
Food and Agriculture Organization (FAO) 163
free-air concentration-enrichment with ozone ($FACE\text{-}O_3$) 188
funRiceGenes 9
Fusarium fujikuroi, 14

g

gall midge (GM) infestation 16
gas chromatography–mass spectrometry 95
gene networks 258–259
genes
  aromatic rices 344
  for Fe transport 317–319
  golden rice 290, 291
  grain zinc concentration in rice 305–306
  for iron biofortification program 321–322
genetically modified organisms (GMOs) 3
genetic engineering 26
genetic variation, nitrogen use efficient 235–237
genome-deciphering techniques 3
genome editing 3, 40–41, 110
  deployment of 66–67
genome-wide association mapping 191, 282
genome-wide association studies (GWAS) 3, 33, 56, 109–110, 239
genome-wide scans, with microsatellite markers 56
genomic best linear unbiased prediction (GBLUP) 171
genomic estimated breeding value (GEBV) 38, 170
genomics-assisted breeding (GAB) 3, 9
genomic selection (GS) 3, 56
  breeding, implication for 172–173
  cross-validation, within reference population 171
  genomic prediction, across populations 171–172
  methods 170–171
genotype, drought stress tolerance 262–265
genotyping-by-sequencing (GBS) 33, 40, 109, 171, 282
geographical indication (GI) 353
geranylgeranyl diphosphate synthase (GGPPS) 288
germination stage oxygen deficiency (GSOD) 76, 79–82
Gihobyeo *(japonica)*, 96
global food security 274
glutamine oxoglutarate aminotransferase (GOGAT) 240
glutathione reductase (GR) gene 415
glutathione (GSH) 146
glyoxalase I (Gly I) gene 410–412
glyoxalase II (Gly II) gene 411, 412
glyoxalase III (Gly III) gene 412
glyoxalase pathway, abiotic stress 406–410
  antioxidant pathway 406–410
  glyI gene 410–412
  glyII gene 411, 412
  glyIII gene 412
  modulating 403–404
golden rice (GR)
  agronomic performance 295–296
  β–carotene synthesis pathway
    endosperm-specific expression 289–290
    genes 290, 291
    in plant 287–289
    rice genetic transformation system 290, 292
  confined field trial and regulation of 298
  CRISPR/Cas for 298–299
  defined 287
  development of 291–294
  economics 299–300
  food safety analysis of 297
  homozygous dihaploid 294
  introgression breeding programs 294–295
  lipoxygenase 296
  long-term storage of 296
  metabolic analysis of 297
  phenotypic performance 295–296
  substantial equivalence analysis of 296–297
grain filling 94–95
grain-filling rate (GFR) 93
grain weight (GW) 207
grain zinc concentration in rice
  genes control 305–306
  in human nutrition 304–305
  natural variations 306
  transgenic approaches 306
  wild relatives 307–308
  with yield 308–311
gramene 8
greenhouse gas emissions, aerobic rice 387
green leaf hopper (GLH) 15, 83
Green Revolution 274–275
GSH biosynthetic genes 415–416

h

haplotypes, phosphorus uptake1 224–226
heat shock proteins (HSPs) 97
heat stress (HS) tolerance 93–94, 108
heat-tolerant genes (HT) 93
heavy metal ATPase (HMA) 142
high-Cd *japonica* cultivars 148
high night temperatures (HNTs) 97
high resolution mapping 103
high-resolution melting (HRM) curve analysis 56
HKT-type transporter 36
homing endonucleases (HEs) 41
homogentisate geranylgeranyl transferase (HGGT) 365
homogentisate phytyltransferase (HPT) 365
homology-dependent recombination (HDR) 41

homozygous dihaploid golden rice 294
honestly significant difference (HSD) 101
human nutrition, grain zinc 304–305

i

IDS3 gene, Fe rice grain 327–328
Indian Agricultural Research Institute (IARI) 351
Indica golden rice 295
indica/japonica or *indica/indica*, 101
inositol polyphosphate-signaling molecules (InPs) 218
insect pests constraints, aerobic rice 392
insertions/deletions (InDels) 8, 39
integrated pest management (IPM) 392
Intergovernmental Panel on Climatic Change (IPCC) 92
International Rice Research Institute (IRRI) 31, 83, 92
introgression lines (IL) mapping population 32
In vitro mutagenesis 32
ionizing radiation 31
iron (Fe)
  deficiency, in human body 315
  fortified rice line development 330
  through agronomic practices
    description 320
    foliar application of Fe fertilizers 320
    limitations 323
  through conventional breeding 324
  through molecular breeding method 324
    advantage of 325
    bi-parental mapping 324, 325
    donor germplasms lines for 331
    enhancing grain Fe content 325–330
    genome-wide association mapping (GWAS) 324–325
  through transgenics/genetic engineering 323
  transport from soil to seeds 316, 317
  uptake strategies by plants
    chelation-based strategy 316, 317, 319
    reduction-based strategy 316–319
iron (Fe)-containing rice germplasm lines 316
isopentenyl diphosphate (IPP) 288
isopentenyl pyrophosphate (IPP) 363
"*Itai Itai*" disease 131

j

japonica, 7, 29, 165
jeopardize cropping pattern 7

k

KEGG 9
Khao Dawk Mali 105 (KDML105) 33, 341, 351
K^+ homeostasis (SKC1) 36
Kompetitive allele specific PCR (KASPTM) assay 56
Kyoto Encyclopedia of Genes and Genomes (KEGG) 241

l

late planting 54
leaf senescence 205
Least Absolute Shrinkage and Selection Operator (LASSO) 66
L-galactono-1,4-lactone dehydrogenase (GLDH) 416
light-harvesting complex (LHC) 208, 254
Light harvesting Complex II (LHCII) 205
linkage disequilibrium (LD) 239
linkage mapping 3
lipoxygenase (LOX) 296
logarithm of odds (LOD) 97
low-affinity cation transporter (LCT) 146
lower cadmium accumlation
  Cd contamination, in soils 132–133
  cd toxicity, to human health 131–132
  Cd uptake and distribution patterns 133
  diversity of, Cd accumulation trait 133–134
  environmental factors
    influence of, soil water 134–139
    ions and chemical compounds 139–141
    ph values in soil and irrigation water 139
    rhizosphere microorganisms 141
  genetic basis for 134
  genetic factors
    genes in, Cd chelation 146–147
    regulating genes, to mediate grain 147
    transporters for, Cd Uptake and translocation 142–146
  molecular breeding
    gene editing-based molecular breeding 148–149
    molecular marker-assisted low-Cd gene aggregation breeding 148
    screening and identification, of low-Cd varieties 148
low light, in rice physiology 202
low light tolerant/light use efficient (LUE) genotypes 201
low phosphorus in rice
  development and growth 215
  impact of 221
  molecular breeding
    haplotypes report 224–226
    PSTOL1 226–227
    PUP1 224
  P deficiency tolerance, rice 215, 217, 220
  reproductive stage 220–221
  signaling in 218
  tolerance

potential candidates 222–224
quantitative traits loci 221–222
uptake and transport of 215–218
vegetative stage 220
lycopene β-cyclase (LYCB) 289
lycopene ε-cyclase (LYCE) 289

m

magnetic resonance imaging (MRI) 39
MALDI-TOF/MSbased proteomics 108
malondialdehyde (MDA) 13
map-based cloning 124
marker-aided selection (MAS) 83
marker-assisted backcross breeding (MABB) 3
marker-assisted backcrossing (MABC) 35, 82
marker-assisted breeding (MAB) 3, 26
marker-assisted recurrent selection (MARS) 9, 169
marker-assisted selection (MAS) 3, 34–35, 103
aerobic rice 388–390
heat stress tolerance 105–107
salinity tolerance 56
Markov chain Monte Carlo (MCMC) algorithm 171
melatonin composition, antioxidant 368–369, 373–374
Meloidogyne graminicola, 16
membrane lipid composition 13
mesophyll cells (MCs) 277
methane (CH_4) 180
methionine 124
methylerythritol 4-phosphate (MEP) 288
methylglyoxal (MG), abiotic stress 401–403
MicroRNAs (miRNAs) 109, 259–260
Milyang 23 *(indica/japonica)*, 96
minor allele frequency (MAF) 171
mitochondrial electron transport chain (METC) 255
mitogen-activated protein kinase (MAPK) 191, 263
modern breeding method
novel breeding strategies 65–66
rapid breeding cycles 64–65
molecular breeding method
aromatic rices 351–352
for biotic stress tolerance
diseases 14–15
pests 15–16
weed competitiveness 16–17
for climate-resilient rice 7–9
gene editing-based molecular breeding 148–149
iron biofortification 324
advantage of 325
bi-parental mapping 324, 325
donor germplasms lines for 331
enhancing grain Fe content 325–330
genome-wide association mapping (GWAS) 324–325
molecular marker-assisted low-Cd gene aggregation breeding 148
novel tools
genomic selection 38
MAGIC 37–38
phenomics 38–39
in rice
abiotic stress tolerance 9
drought 9–10
high temperature 12
low temperature 13
salinity 10–11
submergence 11–12
screening and identification, of low-Cd varieties 148
molecular regulatory networks
CBF regulatory pathway 125–126
MYB regulatory pathway 126
mono-dehydroascorbate reductase (MDHAR) 187, 414
multidrug and toxic compound extrusion (MATE) 34
multi-parental advanced generation intercross (MAGIC) 37–38
mutation breeding 31
photosynthesis crop yield 280–281
MYB (v-myb avian myeloblastosis viral oncogene homolog) transcription factor 125, 126

n

N-acetyl-5-methoxytryptamine 368
NAS gene, Fe rice grain 328–329
National Agricultural Research System (NARS) 83
natural resistance-associated macrophage protein (NRAMP) 140, 141
natural variations, grain zinc 306
NCBI Genome Data Viewer 8
Neanastatus grallarius, 16
near-isogenic line (NIL) 97
nematode problems, aerobic rice 391–392
net photosynthesis rate 94
New Rice for Africa ("NERICA") 225
next-generation sequencing (NGS) 3, 38–40, 56, 109
5'-NGG protospacer adjacent motif (PAM) 41
NGS-assisted high-throughput genotyping 40
nicotinamide adenine dinucleotide (NADH) 255
nicotinamide adenine dinucleotide phosphate (NADPH) 255
nitrogen oxides (NOx) 180

nitrogen use efficient (NUE)
  candidate genes 240–242
  defined 234
  genetic variation 235–237
  genomic regions
    association mapping 239–240
    quantitative trait loci 237–239
  rice varieties 242–243
N-methyl-N-nitrosourea 31
nonhomologous end-joining (NHEJ) 41
non-inbred progeny 64
nonspecific conventional breeding 324
nonstructural carbohydrate (NSC) 97
novel breeding strategies 65–66
NUE. *see* nitrogen use efficient (NUE)

o
OMAP 9
Open Stomata 1 (OST1) gene 187
open-top chambers (OTC) 187
optimum light intensity 201
OryGenes 8
Oryzabase 8
Oryza rufipogon, 307
Oryza spp.
  O. glaberrima, 2
  O. nivara, 12, 29
  O. officinalis, 96
  O. rufipogon, 29
  O. sativa, 2, 120
OsGDB 8
OsIRT1 318
osmotic adaptation 10
osmotic tolerance 29
OsVIT (vacuolar iron transporter) genes 305
Overview of functionally characterized Genes in Rice Online (OGRO) 208
oxygen-evolving complex (OEC) 253
ozone layer 180
ozone pollution 181
ozone precursor gases 182
Ozone Responsive Apoplastic Protein1 (OsORAP1) 191
ozone tolerance, in rice
  breeding progress 192
  mechanisms 186–187
  ozone damage, mechanisms of 184–186
  ozone-impact assessments 187–190
  strategies 190–192
  tropospheric ozone 182–184

p
parasitoids 16
P deficiency tolerance, rice 215, 217, 220
phenolics efflux transporters (PEZ1 and PEZ2) 318–319
phenomics 38–39
Phosphorus Starvation Tolerance 1 (PSTOL1) 226–227
phosphorus uptake1 (Pup1)
  haplotypes 224–226
  history 224
phosphorus uptake efficiency (PUP) 220, 221
phosphorus utilization efficiency (PUE) 220, 222
photon flux density (PFD) 201
photosynthesis crop yield
  atmospheric Co_2 concentration 282–283
  C_4 pathway into C_3 crops 279–280
  marker assisted selection 281–282
  mode of 277–279
  mutation breeding 280–281
  rate of 276
photosynthetic fluorescence 39
phototropin (PHOT) 203
phytochelatins (PCs) 146
phytochelatin synthase (PCS) 146
phytochrome interacting factors (PIFs) 203
phytoene synthase (PSY) 288
phytosiderophores (PS) 319
phytotoxic ozone dose (POD_Y) 182
Phytozome 8
Pi-transporter (PT) 215–216
PlantCV 39
plant-level compartmentation 27
plant physiology, on heat-stress response
  amylose content response 97
  anthesis response 96–97
  germination response 95–96
  grain filling 94–95
  heat-induced quality decline, of rice 97
  net photosynthesis rate 94
  source sink dynamics 97–98
plants
  carotenoid biosynthesis pathway in 288
  desaturation reactions 289
plasma-membrane slow-anion channel (SLAC1) 186
Platygaster oryzae, 16
Poaceae, 2
polyamines 108
polymerase chain reaction (PCR)-based fluorescently-labeled high-throughput methods 56
post-transcriptional gene silencing (PTGS) 261
proanthocyanidins composition, antioxidant 360–361, 369–370
programmed cell death (PCD) 186
proteomics 110
protoplasmic resistance 10
purple acid phosphatases (PAPs) 218
46 pyramided lines (PLs) 55

q
Q-TARO 8
QTLtoGene 9
quantitative trait nucleotides (QTNs) 34
quantitative traits loci (QTLs) 3, 26, 55, 79
aerobic rice, water saving 388–390
aromatic rices 345–347
bi-parental mapping 324, 325
drought stress tolerance 262
genome-wide association mapping 324, 325
grain Zn concentration 306–309
for iron biofortification program 321–322
for iron (Fe) enhancement 319–320
mapping 32–34
nitrogen use efficient 237–239
phosphorus uptake1 221–222
photosynthesis crop yield 281–282
qFe QTLs 329–330

r
rainfed-shallow-drought and submergence-prone environments 55
rainfed-shallow-drought-prone 55
rainfed-shallow-favorable environments 55
Randomized Complete Block (RCB) 55
rapid generation advance (RGA) 64–65
rapid generation advance (RGA) 84
reactive oxygen species (ROS) 11, 358
abiotic stress 399–401
production 254–255
signaling 256–257
reactive oxygen species (ROS)-scavenging enzymes 103
recombinant inbred lines (RIL) 32, 96, 190
recommended daily intake (RDI) 297
redox homeostasis, drought tolerance 255–256
reduction-based strategy 316
reference population (RP) 170
reproducing kernel Hilbert space regressions (RKHS) 66, 171
respiratory burst oxidase homolog (RBOH) 255
restriction-enzyme-based genotyping by sequencing (GBS) 56
restriction fragment length polymorphism (RFLP) 98
resulting in hemithioacetal (HTA) 403–404
Rhizoctonia solani, 14
rhizosphere microorganisms 141
ribulose-1,5-bisphosphate carboxylase/oxygenase (RUBISCO) 93, 205
rice adaptation
rice breeding materials 99
rice breeding, progress on
genetic background
breeding materials (donors) 98–99
genetic diversity analysis 99–100
phenotyping 100–101
proteomics 107–109
putative and candidate genes/QTLs 101–105
plant physiology, on heat-stress response
amylose content response 97
anthesis response 96–97
germination response 95–96
grain filling 94–95
heat-induced quality decline, of rice 97
net photosynthesis rate 94
source sink dynamics 97–98
rice ecosystem 2, 5
rice genetic transformation system 290, 292
rice genome 7
rice glutelin genes 289
rice landraces 30
rice production and climate change 4–7
rice tungro bacilliform virus (RTBV) 15
rice tungro spherical virus (RTSV) 15
RNA-Seq technology 259
root distribution pattern 55
root knot nematodes (RKN) 16, 391
root-level ion exclusion 27
ROOTomics 9
RUBISCO activase (RCA) 93

s
Sahbhagi Dhan (SD) 226
salinity tolerance
conventional tools
conventional breeding 29–31
mutation breeding 31
tissue culture 32
molecular tools
candidate genes and allele mining 35–36
genetic engineering, to improve salt tolerance 36–37
marker-assisted selection 34–35
QTL mapping 32–34
Saltol, 11
salt-responsive protein 3-1 (SaSRP3-1) 36
salt stress 26
salt tolerance
mechanisms
cell-level compartmentation 27–29
plant-level compartmentation 27
root-level ion exclusion 27
natural variability, in rice germplasm 29
seed germination 105
seedling-stage salinity tolerance 32
semi-aquatic annual grass 2
sensing molecules 29
shoot K^+ concentration on chromosome 1 (SKC1) 32

shoot Na^+ concentration on chromosome 7 (qSNC7) 32
short interfering RNA (siRNA) 261
simple sequence repeats (SSR) 8
single nucleotide polymorphism (SNP) 8, 32, 171
  genotyping 56, 86
  mutation 146
single-nucleotide polymorphisms (SNPs) 39
single-seed descent method (SSD) 84
site-specific nucleases (SSNs) 66
Snorkel 1 (SK1) 81
Snorkel 2 (SK2) 81
SNP-SEEK 9
soil degradation 26
soil salinity 26
spikelet fertility (SF) 77
standardized reference emission scenarios (SRES) 6
STAY-GREEN (SGR) protein 205
substantial equivalence, golden rice 296–297
symplastic Fe transport 316

t
targeted environment (TE) 54
targeting induced local lesions in genomes (TILLING) 261–262
temperature gradient chambers (TGC) 100
thermal imaging 39
tissue culture 32
tocopherols/tocotrienols composition, antioxidant 365–366, 370–371
trait-based selection, for drought resistance 55
trait-specific SNPs 57
transcript-derived fragments (TDFs) 94
transcription activator-like effector nucleases (TALENs) 41, 66
transcription factors 29
transgenic approach, iron biofortification 323
transgenic biofortified crop development 323
transporter of mugineic acid (TOM1) 319
transporters, for Fe transport 318–319
trehalose-6-P phosphatase (TPP7) 207
tropospheric ozone 182–184
two-dimensional difference gel electrophoresis (2-D DIGE) 110
two-dimensional electrophoresis (2-DE) 108

u
ubiquitin-mediated 26S proteasome degradation 147

v
vacuolar H^+-ATPase subunit c1 (SaVHAc1) 36
Vacuolar Phosphate Transporter 1 (VPT1) 216
validation population (VP) 170
virus-induced gene-silencing (VIGS) machinery 260–261
vitamin A deficiency (VAD) 287
VIT gene, Fe rice grain 326–328
volatile organic compounds (VOCs) 180

w
water scarcity 2, 9
water stagnation 75
water use efficiencies (WUE) 385
weed problems, aerobic rice 390–391
whole-genome resequencing (WGR) 109
whole-genome sequencing (WGS) 3, 39, 109
wide compatibility genes (WCGs) 98
wild relatives, grain zinc 307–308
World Health Organization (WHO) 163

y
yellow stem borer (YSB) 15
YSL genem, Fe rice grain 329

z
zinc-finger nucleases (ZFNs) 41, 66
Zn deficiency 304
Zrt-/Irt-like protein (ZIP) 142